PROCEEDINGS

OF THE

1991 INTERNATIONAL CONFERENCE

ON

PARALLEL PROCESSING

August 12-16, 1991

Vol. III Algorithms & Applications
Dr. Kimming So, Editor
IBM, Austin, TX

Sponsored by

THE PENNSYLVANIA STATE UNIVERSITY

CRC Press
Boca Raton Ann Arbor Boston London

Catalog record is available from the Library of Congress
ISSN 0190-3918
ISBN 0-8493-0190-4 (set)
ISBN 0-8493-0191-2 (vol. I)
ISBN 0-8493-0192-0 (vol. II)
ISBN 0-8493-0193-4 (vol. III)
IEEE Computer Society Order Number 2355

Additional copies may be obtained from:

IEEE Computer Society Press
Customer Service Center
10662 Los Vaqueros Circle
P.O. Box 3014
Los Alamitos, CA 90720-1264

IEEE Computer Society
13, Avenue de l'Aquilon
B-1200 Brussels
BELGIUM

IEEE Computer Society
Ooshima Building
2-19-1 Minami-Aoyama,
Minato-Ku
Tokyo 107, JAPAN

PREFACE

The 1991 version of the International Conference on Parallel Processing is the 20th in this long and distinguished series of conferences. Over the past twenty years, the conference has grown enormously, both in terms of numbers of technical papers presented and in terms of topics and areas covered. The banquet on the Monday night before the conference celebrates the continuation of ICPP. This banquet was arranged by Professor Ming T. Liu, of Ohio State University, and features an address by Professor Duncan Lawrie, of the University of Illinois at Urbana-Champaign.

A special event on the program is the panel entitled Toward Teraflops Computing. We want to thank the six panelists for taking the time to present their ideas on this important topic. The six panelists are Chris Hsuing, Cray Research, Dave Patterson, University of California Berkeley, Justin Ratner, Intel, Burton Smith, Tera Computer, Guy Steele, Thinking Machines, and Steve Wallach, Convex Computer. This promises to be an outstanding part of the technical program.

The three Co-Chairs for the Technical Program have worked hard, to try to assure the best possible program. For the third time in the history of the conference, the number of submitted papers has exceeded 500. These papers were submitted to one of three tracks: architecture, software or applications/algorithms. Each paper was handled by one of the three Co-Chairs and was reviewed by at least two external referees. The accepted papers were classified as either a regular paper, a concise paper or a poster session paper. The brief table below summarizes the papers handled in each of the three tracks.

Area	Submitted	Accepted Regular	Concise	Poster
Architecture	253	42 (16%)	66 (26%)	45 (17%)
Software	113	25 (22%)	14 (12%)	24 (21%)
Applications/Algorithms	140	18 (13%)	30 (21%)	27 (19%)
All Areas	506	85 (17%)	110 (22%)	96 (19%)

There were 126 papers submitted from 25 foreign countries. Thus, the word "International" in the title is well justified. This participation by foreign authors and speakers is an important aspect of the broad nature of the conference.

This kind of conference depends on many volunteers. In addition to the three Co-Chairs for the Technical Program, literally hundreds of individuals have served as referees. This is a thankless, but absolutely essential, job in the structuring of a technical conference. We are very grateful to the many referees who agreed to review from one paper, to in some cases five or six papers, within our fairly tight deadlines. The degree of response helped give the conference its high degree of technical quality.

Finally, we have to thank *all* of the authors: those whose papers were accepted and those whose papers were not accepted. The authors of accepted papers had to, in many ases, alter their papers, to fit within the page limitations of concise and poster papers. Again, tight deadlines were an additional problem.

Each of the Co-Chairs needs to thank several individuals. First, Dr. Tse-yun Feng, the founder and sustainer of this conference has contributed much in the areas of local arrangements, publicity, and the overall management of this large conference. Next, Chita R. Das, of Pennsylvania State University, handled the papers submitted by authors with affiliations which created conflicts of interest. And finally, in the areas of Algorithms and Applications, my management at IBM have provided me the opportunity to serve on the program committee this year; their support throughout the process is greatly appreciated. My colleagues Jin Su, Wen-hann Wang, Anita Jindal, and Shing-Chong Chang have contributed a significant amount of time in selecting the referees and the papers. Special thanks go to Georgette Hardin for providing invaluable help in handling the lists of authors and papers.

The Program Co-Chairs hope that each attendee finds at least one item of value in the technical program. If each of you are able to do that, it was all worthwhile.

Kimming So
1991 Program Co-Chair: Algorithms and Applications
International Business Machines Corporation
11400 Burnet Road
Austin, TX 78758

LIST OF REFEREES
VOLUME III - ALGORITHMS AND APPLICATIONS

Aggarwal, J. K.	U. of Texas, Austin
Akers, S. B.	U. of Massachusetts
Akl, S. G.	Queen's U., Canada
Ananth, G. Y.	U. of Minnesota
Annapareddy, N.	IBM, Almaden
Appel, A. W.	Princeton U.
Austin, L. M.	Texas Tech U.
Azadegan, S.	U. of Minnesota
Bailey, D. H.	NASA Ames Res. Ctr.
Bailey, M. L.	U. of Arizona, Tucson
Bestavros, A.	Harvard U.
Bestul, T.	U. of Maryland
Brady, M.	Penn. State U.
Brahmadathan, K.	U. of Wyoming
Bukhres, O.	Moorhead State U., MN
Chang, S.-C.	IBM, Yorktown
Chen, L.	Ohio State U.
Cho, T.-W.	U. of Kentucky
Christidis, Z. D.	IBM, Yorktown
Christopher, T. W.	IIT, Chicago
Chronopoulos, A.	U. of Minnesota
Chung, M. J.	Michigan State U.
Cidon, I.	IBM, Yorktown
Colbrook, A.	MIT
Cole, R.	New York U.
Conley, B.	U. of Wisconsin, Green Bay
Cook, S. A.	U. of Toronto, Canada
Cypher, R.	IBM, Almaden
Damianakis, S.	IBM, Yorktown
Das, C. R.	Penn. State U.
Das, S.	Georgia Inst. of Tech.
Das, S. K.	U. of North Texas, Denton
de la Torre, P.	U. of New Hampshire
Dehne, F.	Carleton U., Canada
Deshpande, S. R.	IBM, Austin
Edirisooriya, G.	U. of Iowa
Efe, K.	U. of SW Louisiana, Lafayette
ElGindy, H.	McGill U., Canada
Fotouhi, F.	Wayne State U.
Foulser, D. E.	Yale U.
Fujimoto, R. M.	Georgia Inst. of Tech.
Gao, G. R.	McGill U., Canada
Gerasch, T. E.	MITRE, Washington
Gerth, J.	IBM, Yorktown
Geyer, S.	MRJ, Inc., Oakton
Ghafoor, A.	Purdue U.
Ghosh, J.	U. of Texas, Austin
Ghosh, K.	Georgia Inst. of Tech.
Ghosh, S.	Brown U.
Greenberg, A. G.	AT&T Bell Labs, Murray Hill
Grunwald, D.	U. of Colorado, Boulder
Gupta, R.	U. of Pittsburgh
Han, Y.	U. of Kentucky
Hanson, F. B.	U. of Illinois, Chicago
Herbordt, M. C.	U. of Massachusetts, Amherst
Ho, C.-T.	IBM, Almaden
Holey, J. A.	U. of Minnesota
Hong, Y.-C.	U. of California, Riverside
Horng, M.-Y.	U. of California, Los Angeles
Hsieh, W.	MIT
Hwang, I.	U. of Florida, Gainesville
Hwang, J. N.	U. of Washington
Ibarra, O. H.	U. of California, Santa Barbara
Iyengar, S. S.	Louisiana State U.
JaJa, J.	U. of Maryland
Jenq, J.-F.	U. of Minnesota
Jiang, H.	Texas A&M U.
Jindal, A.	IBM, Austin
Joe, B.	U. of Alberta, Canada
Johnson, R. C.	Cornell U.
Johnson, T.	U. of Florida, Gainesville
Jong, J. W.	U. of Southern California
Juang. F. L.	U. of Illinos, Urbana
Jun, M. S.	New Mexico State U.
Keefe, T. F.	Penn. State U.
Kim, K.	U. of Southern California
Koc, C. K.	U. of Houston
Kreulen, J. T.	Penn. State U.
Krishnamurthy, B.	Tektronix Lab., Beaverton
Krishnamurty, S.	U. of Maryland
Kruskal, C. P.	U. of Maryland
Kuhl, J. G.	U. of Iowa
Kumar, D.	Case Western Reserve U.
Kumar, V.	U. of Minnesota
Kuo, S.-Y.	U. of Arizona, Tucson
Kuszmaul, B. C.	MIT
Lai, T.-H.	Ohio State U.
Lee, S.-Y.	Cornell U.
Lee, Y.-H.	U. of Florida, Gainesville
Li, H.	IBM, Almaden
Lin, H.	Texas A&M U.
Lin, J. M.	U. of Michigan
Lin, J. Y.-B.	Bellcore, NJ
Lin, R.	SUNY, Geneseo
Lin, W.	U. of Hawaii, Manoa
Liu, J. W.-H.	York U., Canada
Liu, Y.	Savannah State College
Lu, M.	Texas A&M U.
Luby, M.	U. of California, Berkeley
Malony, A. D.	U. of Illinois, Urbana
Miller, R	SUNY, Buffalo
Moitra, A.	GE, Schenectady
Morse, S.	MRJ, Inc., Oakton
Moser, L. E.	U. of California, Santa Barbara

Narayanan, P. J.   U. of Maryland
Narayanaswami, C.   Rensselaer Poly. Inst.
Natarajan, K. S.   IBM, Yorktown
Norton, A.   IBM, Yorktown
Olariu, S.   Old Dominion U., Norfolk
Ouksel, M.   U. of Illinos, Chicago
Paden, R. L.   Andrews U., Berrien Springs
Palem, K. V.   IBM, Yorktown
Pan, V.   Lehman College, CUNY
Panda, D. K.   U. of Southern California
Pandian, R.   IBM, Kingston
Pardalos, P. M.   Penn. State U.
Prasad, S. K.   Georgia State U., Altanta
Raab, L. J.   Dartmouth College
Ramarao, K. V. S.   SW Bell Tech. Resources, Inc.
Ranka, S.   Syracuse U.
Ray, K.   U. of Texas, Austin
Reinhold, M.   MIT
Rudolph, L.   IBM, Yorktown
Saha, A.   Lehigh U.
Sahni, S.   U. of Florida
Samet, H.   U. of Maryland
Sarkar, D.   U. of Miami, Coral Gables
Scherson, I. D.   U. of California, Irvine
Scheuermann, P.   Northwestern U.
Sethu, H.   Lehigh U.
Shaaban, M.   U. of Southern California
Shea, D. G.   IBM, Yorktown
Shetty, B.   Texas A&M U.
Shiran, Y.   Silvar-Lisco, Sunnyvale
Silberger, A.   Cleveland State U.
Solowiejczyk, Y.   Oryx Corporation, Paramus
Somani, A.   U. of Washington, Seattle
Sprague, A. P.   U. of Alabama, Birmingham
Sridhar, M. A.   U. of South Carolina, Columbia
Srimani, P. K.   Colorado State U.
Stout, Q. F.   U. of Michigan
Su, J   IBM, Austin
Sunday, D. M.   Johns Hopkins U.
Takefuji, Y.   Case Western Reserve U.
Tan, H.-Q.   U. of Akron
Thomasian, A.   IBM, Yorktown
Tripathi, A.   U. of Minnesota
Tsai, J. J.   Georgia Inst. of Tech.
Turek, J.   New York U.
Tzong, N. F.   U. of SW Louisiana
Varadarajan, V.   U. of Florida, Gainesville
Wagh, M. D.   Lehigh U.
Wan, F.   Indiana U.
Wang, C.   U. of California, Riverside
Wang, P.   MIT
Wang, W.-H.   IBM, Austin

Watson, L.   Virginia Poly. Inst. and State U.
Weisbecker, J. R.   Penn. State U., Media
Wendt, P.   IBM, Austin
Whitman, S.   Ohio State U.
Wong, C.-K.   IBM, Yorktown
Woo, T.-K.   Jacksonville U., FL
Wu, E.   IBM, Yorktown
Wu, J.   Florida Atlantic U.
Xe, H.   SUNY, Buffalo
Xu, H.   U. of Illinois, Chicago
Yang, G.-C.   U. of Illinois, Urbana
Yang, M. K.   Penn. State U.
Yenamandra, M.   U. of Minnesota
Young, H.   IBM, Almaden
Zenios, S. A.   U. of Penn.
Zhang, G.   Vantage Analysis Sys., Fremont
Zhao, W.   Texas A&M U.

AUTHOR INDEX -- FULL PROCEEDINGS

Volume I -- Architecture
Volume II -- Software
Volume III -- Algorithms and Applications

vi

TABLE OF CONTENTS
VOLUME III - ALGORITHMS AND APPLICATIONS

(R): Regular Papers
(C): Concise Papers
(P): Poster Papers

xiii

A Parallel Algorithm for Exact Solution of Linear Equations

Çetin Kaya Koç   and   *Rose Marie Piedra*

Department of Electrical Engineering

University of Houston

Houston, TX 77204

Abstract

We present a parallel algorithm for computing the exact solution of a system of linear equations via the congruence technique. The basic idea of the technique is to convert the original system of equations into a system of congruences modulo various primes, and combine the solutions by the application of the Chinese remainder theorem. The proposed parallel algorithm requires only local communication among the processors and is particularly suitable for implementation on distributed-memory multiprocessors and systolic computing systems. We have implemented the parallel congruence algorithm on an Intel cube and obtained up to 92 % efficiency.

Key Words: Congruence techniques, Chinese remainder theorem, Gaussian elimination, mixed-radix conversion algorithm, systolic schedule.

1 Introduction

We consider the solution of the system of linear equations

$$\mathbf{Ax} = \mathbf{b} \ , \tag{1}$$

where $\mathbf{A}$ is a $k \times k$ invertible matrix, and $\mathbf{x}$ and $\mathbf{b}$ are $k \times 1$ vectors. This problem has been studied intensively. There are several methods to solve (1), most notably the Gauss and the Gauss-Jordan elimination methods. There are situations, however, in which these methods are inadequate; for example, when one is interested in the *exact* solution of (1) or when $\mathbf{A}$ is ill-conditioned.

The exact solution of (1) can be found by either the direct computation method using multiple-precision integer arithmetic or the residue arithmetic techniques. The residue arithmetic techniques find the result by either using multiple moduli (the congruence technique) or single modulus (*p*-adic expansions). In general, these methods require that the entries of $\mathbf{A}$ and $\mathbf{b}$ be integers. However, this is not a serious restriction, since the rational number or the floating-point number entries of $\mathbf{A}$ and $\mathbf{b}$ can be converted to integers by scaling.

Algorithms using *p*-adic expansions are given by Dixon [7] and by Gregory and Krishnamurthy [10]. Parallel implementations of the *p*-adix expansion and the direct computation methods are given by Villard [20]. In this paper, we focus on parallel implementation of the congruence technique which solves (1) by first solving $n + 1$ such systems in

$\mathcal{Z}_{m_i}$ (the ring of integers modulo m_i) for $0 \leq i \leq n$. These results are then combined using the Chinese remainder theorem to find the solution in $\mathcal{Z}_M$ where $M = m_0 m_1 m_2 \cdots m_n$. The reader is referred to the books by Knuth [13], Lipson [17], Mackiw [18], and Young and Gregory [21] (and the references therein) for discussions on the congruence technique and various algorithms for computing exact solution of linear equations with integer and rational entries.

We would like to state the following assumptions for the timing analysis of the algorithm presented in this paper.

- We choose the moduli $m_i < W$ for $0 \leq i \leq n$. W is the wordsize of the computer, i.e., the largest integer which can be operated on by the arithmetic unit of the computer.

- An arithmetic operation ($\in \{+, -, \times\}$) on integers $< W$ is assumed to take 1 unit of time, defined as an arithmetic step.

- For operations on integers $> W$, we implement multi-precision arithmetic.

- The multiplicative inverse of $a < W$ in $\mathcal{Z}_b$ exists if $\gcd(a, b) = 1$. It is defined as the integer $x < W$ such that
$$ax = 1 \pmod{b} \ ,$$
and can be computed using the extended Euclid algorithm [13, 17]. We denote the operation to find the inverse of a in $\mathcal{Z}_b$ by INVERSE (a, b). If $a, b < W$, then Euclid's algorithm requires $O(\log W)$ arithmetic operations to compute INVERSE (a, b). Since W is independent of n, we assume that INVERSE operation on single-precision numbers takes only $O(1)$ arithmetic operations.

- Our computer is also capable of carrying out arithmetic operations ($\in \{+, -, \times, \div\}$) on floating-point numbers. A floating-point arithmetic operation is also assumed to take 1 step.

For the parallel implementation of the congruence technique, we assume that we have a distributed-memory multiprocessor with p processors. The processors are connected with communication links, forming a network topology. Examples of network topologies are linear arrays, rings, trees, 2-dimensional and multi-dimensional meshes, and hypercubes. The communication cost of the parallel algorithm is measured by counting the total number of parallel routing steps, where a routing step is defined as the time required to send an operand (i.e., a single-precision integer) from a processor to one of its neighboring processors. If two communicating processors are not adjacent then the data is assumed to be forwarded by the processors on a path between these two processors, and the routing cost is taken to be the length of the path.

We describe the congruence technique in §2. When the solutions in $\mathcal{Z}_{m_i}$ are combined using the Chinese remainder theorem to find the solution in $\mathcal{Z}_M$, we may pick either the single-radix conversion algorithm or the mixed-radix conversion algorithm. The mixed-radix conversion algorithm, described in §3, is chosen for parallel implementation of the congruence technique. In §4, we present the parallel congruence algorithm and time-optimal and spacetime-optimal systolic schedules for parallel implementation of the mixed-radix conversion algorithm. We have implemented the congruence technique and obtained up to 92 % efficiency on an first generation Intel cube which is a very slow machine in terms of interprocessor communication. In §5, we report the implementation results. Finally, in §6 we discuss several variations of the parallel congruence algorithm and point out some future topics for research.

2   Congruence Algorithm

Let $m_0, m_1, m_2, \ldots, m_n$ be a set of pairwise relatively prime numbers and $M = m_0 m_1 m_2 \cdots m_n$. We denote the determinant of $\mathbf{A}$ by $d = \det(\mathbf{A})$ and the adjoint of $\mathbf{A}$ by $\mathbf{A}^{\mathrm{adj}}$. Since we assume $d \neq 0$, $\mathbf{A}^{\mathrm{adj}}$ is also a nonsingular integral matrix satisfying

$$\mathbf{A}\mathbf{A}^{\mathrm{adj}} = \mathbf{A}^{\mathrm{adj}}\mathbf{A} = d\,\mathbf{I} \ ,$$

where $\mathbf{I}$ is the $k \times k$ unit matrix. For a matrix $\mathbf{B} = [b_{ij}]$, we also define $\mathrm{MAX}(\mathbf{B}) = \max_{i,j} |b_{ij}|$. The solution of (1) can be written as

$$\mathbf{x} = \frac{1}{d}\mathbf{A}^{\mathrm{adj}}\mathbf{b} = \frac{1}{d}\mathbf{z} \ ,$$

where $\mathbf{z} = \mathbf{A}^{\mathrm{adj}}\mathbf{b}$.

The Congruence Algorithm

Step 0. Choose the moduli set $m_0, m_1, \ldots, m_n$ such that $M > 2\max(|d|, |\mathrm{MAX}(\mathbf{A})|)$ and $\gcd(M, d) = 1$.

Step 1. Solve the $n + 1$ systems $\mathbf{A}\mathbf{y}_i = \mathbf{b}$ (mod m_i) for $0 \leq i \leq n$. Compute the determinant $d_i = \det(\mathbf{A})$ (mod m_i) for $0 \leq i \leq n$. Also compute $\mathbf{z}_i = d_i \mathbf{y}_i$ (mod m_i) for $0 \leq i \leq n$.

Step 2. Use the Chinese remainder theorem to solve the simultaneous congruences $\mathbf{z} = \mathbf{z}_i$ (mod m_i) for $0 \leq i \leq n$. Also, the simultaneous congruences $d = d_i$ (mod m_i) for $0 \leq i \leq n$ are solved by the application of the Chinese remainder theorem.

Step 3. The solution of (1) is found as $\mathbf{x} = \frac{1}{d}\mathbf{z}$.

We have the following observations:

- The choice of m_i such that the conditions in Step 0 hold may be a difficult and time consuming process. It is possible to estimate the determinant by the use of Hadamard inequality. We then have to choose m_i such that $\gcd(m_i, d) = 1$ for $0 \leq i \leq n$. Newman suggests the use of a predetermined prime moduli set, with the full knowledge that in certain instances the method will fail [19]. By choosing large primes, the probability of the occurrence $\gcd(m_i, d) > 1$ can be made very small [19, 21].

- In Step 1, the operations are performed using single-precision integer arithmetic. We choose $m_i < W$ for $0 \leq i \leq n$, and then implement the Gaussian elimination algorithm in single-precision integer arithmetic. As matrix A is triangularized, determinant $d_i = \det(\mathbf{A})$ (mod m_i) is also computed in single-precision.

- In Step 2, the Chinese remainder theorem is used to find the weighted-radix representation of the residue numbers d_i and $\mathbf{z}_i$ for $0 \leq i \leq n$. The methods for conversion of a residue number to a weighted number system are based on two different constructive proofs of the Chinese remainder theorem. In the first case, the number is converted to a *single-radix* weighted number system, whereas in the second case it is converted to a *mixed-radix* weighted number system. Since during the computation of d or $\mathbf{z}$, the intermediate and the resulting values can be larger than W, the single-radix conversion algorithm requires implementation of multi-precision integer arithmetic.

- In Step 3, floating-point arithmetic is used to compute the solution $\mathbf{x}$

$$\mathbf{x} = [x_0, x_1, \ldots, x_{k-1}]^T = \frac{1}{d}[z_0, z_1, \ldots, z_{k-1}]^T$$

Theorem 1 *The congruence algorithm finds the solution of (1) using* $O(nk^3 + n^2 k)$ *arithmetic steps.*

Proof The solution of the systems

$$\mathbf{A}\mathbf{y}_i = \mathbf{b} \quad (\mathrm{mod}\ m_i)$$

and the computation of the determinant

$$d_i = \det(\mathbf{A}) \quad (\mathrm{mod}\ m_i)$$

for $0 \leq i \leq n$ are achieved by the use of the Gaussian elimination algorithm. First the coefficient matrix is triangularized; then backward-substitution is performed to solve for $\mathbf{y}_i$. During these computations, multiplicative inverses of the nonzero elements of $\mathbf{A}$ are computed using the extended Euclid algorithm. It becomes evident that one such system is solved in $O(k^3)$ arithmetic steps (see also, e.g., [19, 17]). Thus, the solution $n + 1$ of these systems requires $O(nk^3)$ arithmetic steps. Since the computation of $\mathbf{z}_i = d_i \mathbf{y}_i$ (mod m_i) requires $O(nk)$ arithmetic steps for

$0 \leq i \leq n$, the number of arithmetic operations required in Step 2 is $O(nk^3)$.

In Step 2, we use either the single-radix conversion algorithm or the mixed-radix conversion algorithm to find the weighted-radix representation of the numbers z_i and d_i for $0 \leq i \leq n$. The mixed-radix conversion algorithm (as well as the single-radix conversion algorithm) requires $O(n^2)$ arithmetic steps to find the integer d using the residue numbers $d_i = d \pmod{m_i}$ for $0 \leq i \leq n$. Since the vector z_i has k components, we see that Step 3 of the congruence algorithm requires $O(n^2k)$ arithmetic operations.

In Step 3, we compute the solution by performing k floating-point division operations. According to our assumption, this takes $O(k)$ arithmetic steps.

Thus, the total number of arithmetic operations required is found to be $O(nk^3 + n^2k)$. □

3 Mixed-Radix Conversion

We now describe the mixed-radix conversion algorithm and give a theorem regarding the number of arithmetic operations required. The detailed proof of this theorem can be found in [16, 2, 17]. We assume that we are given the residues u_i of a weighted number u with respect to each modulus m_i, i.e.,

$$u = u_i \pmod{m_i} \text{ for } 0 \leq i \leq n .$$

The mixed-radix conversion algorithm computes the mixed-radix coefficients $(v_0, v_1, \ldots, v_n)$ of u. Once the mixed-radix coefficients have been obtained, u is written in terms of these coefficients and the moduli as

$$u = v_0 + v_1 m_0 + v_2 m_0 m_1 + v_3 m_0 m_1 m_2 + \cdots$$
$$\cdots + v_n m_0 m_1 \cdots m_{n-1} . \quad (2)$$

The Mixed-Radix Conversion Algorithm

Step 1. Compute the inverses c_{ij} for $0 \leq i < j \leq n$ where

$$c_{ij} = \text{INVERSE} \ (m_i, m_j) .$$

Step 2. Set $v_0 = u_0 \pmod{m_0}$, and for $k = 1, 2, \ldots, n$ compute

$$v_k = (\cdots((u_k - v_0)c_{0k} - v_1)c_{1k} - \cdots$$
$$\cdots - v_{k-1})c_{k-1,k} \pmod{m_k} .$$

Theorem 2 *Given the moduli $m_0, m_1, \ldots, m_n$ and the remainders $u_0, u_1, \ldots, u_n$ such that $m_i < W$ for $0 \leq i \leq n$, the mixed-radix number representation $(v_0, v_1, \ldots, v_n)$ of u can be computed in $O(n^2)$ arithmetic steps with the mixed-radix conversion algorithm.*

The mixed-radix representation can be converted to single-radix representation by applying Horner's algorithm to formula (2). If $v_i, m_i < W$ for $0 \leq i \leq n$, then the application of Horner's algorithm to compute single-radix representation

of u requires $O(n^2)$ arithmetic steps using multi-precision arithmetic [17].

4 Parallel Congruence Algorithm

A remarkable property of the congruence algorithm is its parallelism. Since the solution of equation

$$A y_i = b \pmod{m_i}$$

is independent for every $i = 0, 1, 2, \ldots, n$, the algorithm is very suitable for implementation on parallel computers. Once the solutions in $\mathcal{Z}_{m_i}$ are computed, we need to apply a Chinese remaindering algorithm to compute the solution in $\mathcal{Z}_M$. Let $n + 1 = qp$ where $n + 1$ is the number of moduli, p is the number of processors and $q \geq 1$ is an integer. For the time being we assume that $q = 1$. We partition the moduli set in such a way that processor i executes Step 1 of the algorithm modulo m_i, and thus, solves system $A y_i = b \pmod{m_i}$ and computes determinant $d_i = \det(A) \pmod{m_i}$. It then proceeds to compute $z_i = d_i y_i \pmod{m_i}$. This computation is performed by all processors for $i = 0, 1, \ldots n$.

Thus at the end of Step 1, we will have a $k \times 1$ vector z_i and an integer d_i in processor i for all $0 \leq i \leq n$. We now need to apply the Chinese remainder theorem to compute a $k \times 1$ vector z and an integer d. Notice that if the single-radix conversion algorithm is implemented, then the components of z and the determinant d are multi-precision integers. If the mixed-radix conversion algorithm is employed then we can avoid the implementation of the multiple-precision arithmetic, and compute the mixed-radix coefficients in single-precision. We can then use floating-point arithmetic to compute the solution vector x with these mixed-radix coefficients.

Let the $(k+1) \times 1$ vector u_i be

$$u_i = \begin{bmatrix} z_i \\ d_i \end{bmatrix} .$$

This vector is available in processor i. We now use the mixed-radix conversion algorithm to compute the $(k+1) \times 1$ vector u such that $u = u_i \pmod{m_i}$.

The distributed mixed-radix conversion algorithm picks the rth element of each vector u_i from processor i for $0 \leq i \leq n$ and computes the mixed-radix coefficients of the rth element of the vector u for all $r = 1, 2, 3, \ldots, k + 1$. Denote the rth component of vector u_i with u_{ir}. We need to compute v_{ir} for $0 \leq i \leq n$ and $1 \leq r \leq k + 1$ such that

$$u_{ir} = v_{0r} + v_{1r} m_0 + v_{2r} m_0 m_1 + \cdots$$
$$+ v_{nr} m_0 m_1 \cdots m_{n-1} \pmod{m_i} .$$

We define the $(k + 1) \times 1$ vector V_{ij} for $0 \leq i < j \leq n$ such that $V_{-1,i} = V_i = u_i$ for $0 \leq i \leq n$ and $V_{i-1,i} = v_i$ for $0 \leq i \leq n$. V_{ij} for $0 \leq i < j \leq n$ are the temporary values of v_j resulting from the operations in Step 2 of the mixed-radix conversion algorithm. We construct a triangular table of values with diagonal entries $v_i = V_{i-1,i}$ for $0 \leq i \leq n$. For $n = 3$, it is as follows:

$$\mathbf{V}_{03} = (\mathbf{V}_3 - \mathbf{V}_0)c_{03} \quad (\text{mod } m_3) \quad \mathbf{V}_{13} = (\mathbf{V}_{03} - \mathbf{V}_{01})c_{13} \quad (\text{mod } m_3) \quad \mathbf{V}_{23} = (\mathbf{V}_{13} - \mathbf{V}_{12})c_{23} \quad (\text{mod } m_3)$$

$$\mathbf{V}_{02} = (\mathbf{V}_2 - \mathbf{V}_0)c_{02} \quad (\text{mod } m_2) \quad \mathbf{V}_{12} = (\mathbf{V}_{02} - \mathbf{V}_{01})c_{12} \quad (\text{mod } m_2)$$

$$\mathbf{V}_{01} = (\mathbf{V}_1 - \mathbf{V}_0)c_{01} \quad (\text{mod } m_1)$$

The mixed-radix conversion algorithm computes $\mathbf{V}_{ij}$ for $0 \le i < j \le n$ by performing the following operations on single-precision integer operands:

$$c_{ij} = \text{INVERSE } (m_i, m_j) , \tag{3}$$
$$\mathbf{V}_{ij} = (\mathbf{V}_{i-1,j} - \mathbf{V}_{i-1,i})c_{ij} \quad (\text{mod } m_j) , \tag{4}$$

where (4) is performed for all $k + 1$ entries of vector $\mathbf{V}_{ij}$.

The data dependences among the entries in the above table lend themselves to systolic implementation. The first step in achieving a systolic implementation is to form the *process dependence graph* of the mixed-radix conversion algorithm. Coefficient c_{ij} is in column i and row j. The positions of terms $\mathbf{V}_{ij}$ are arranged as follows: First, a term of the form $\mathbf{V}_{i-1,i}$ is computed on the diagonal, then this term is used in every operation along the ith column. Based on these observations, Figure 1 presents the process dependence graph of the mixed-radix conversion algorithm for $n = 7$. The graph is drawn on the (i, j) coordinate system. The nodes of this directed acyclic graph (dag) represent the operations, and the arcs correspond to dependences between the variables used in the operations. The node at point (i, j) computes $\mathbf{V}_{ij}$ by performing the operations given in (3) and (4).

Figure 1. The process dependence graph of the mixed-radix conversion algorithm for $n = 7$.

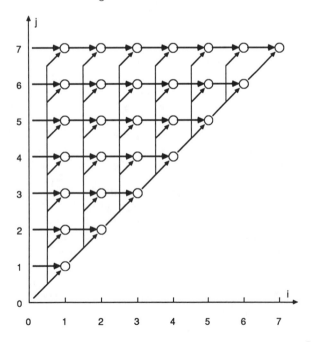

Several time-optimal and spacetime-optimal systolic arrays for the mixed-radix conversion algorithm are given by Koç and Cappello [14]. The arrays and their corresponding schedules in [14] assume that m_i and $\mathbf{u}_i$ are being fed to the array from the south-end. For our problem, however, this is not the case; m_i and $\mathbf{u}_i$ for $0 \le i \le n$ are already available in ith processor after Step 1 of the congruence algorithm. In the following, we introduce two new systolic arrays which use the data available in the processors.

As a first step, we modify the data dependence graph in Figure 1 in order to achieve local dependence, i.e., a node is dependent only on its neighboring nodes. Discussions and several examples on transformation of data dependence graphs to achieve local dependence can be found in [15]. The resulting dag representing the operations performed by the mixed-radix conversion algorithm is given in Figure 2. We then embed the localized process dependence graph of the mixed-radix conversion algorithm in spacetime to produce a time-optimal and a spacetime-optimal systolic array. The reader is referred to [15] (and the references therein) for spacetime embedding techniques.

Figure 2. The graph in Figure 1 with localized dependence.

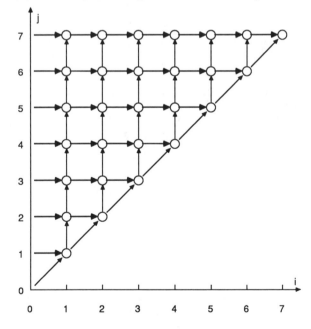

A Time-Optimal Systolic Schedule: We embed the process dependence graph for the mixed-radix conversion algorithm in spacetime. The abscissa is interpreted as time t; the ordinate as space s. The linear embedding E_1 is as follows:

Figure 3. A time-optimal systolic schedule for the mixed-radix conversion algorithm.

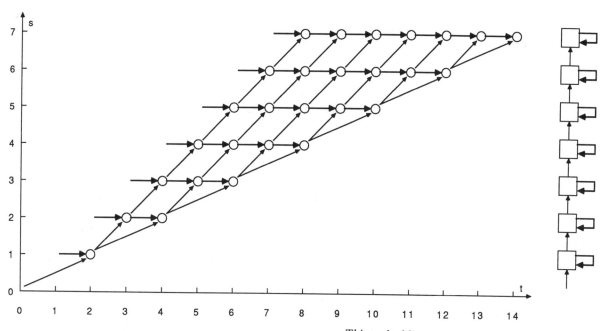

$$
\begin{aligned}
t &:= i + j \ ; \\
s &:= j \ .
\end{aligned}
$$

The result, depicted in Figure 3 for $n = 7$, is a time-optimal array. Data that flows west $\rightarrow$ east in Figure 2 flows in the direction of time (perpendicular to space) in this design: it is in the processors' memory. Data that flows south $\rightarrow$ north in Figure 2 flows up through the array. Data that flows south $\rightarrow$ east in Figure 2 also flows up through the array, but at half the speed of the south $\rightarrow$ north data.

Process (i, j) is executed at time $i + j$ in processor j. By inspection, we see that the array uses n processors, finishing the computation in $2n$ steps. The number of vertices (processes) in a longest directed path in any process dependence graph is a lower bound on the number of steps of any schedule for computing the processes. In our graph, the number of vertices in a longest path is $2n$. Thus, this array uses a spacetime embedding that is optimal with respect to the number of steps used. Such an embedding is referred to as *time-optimal*.

A Spacetime-Optimal Systolic Schedule: We now give another embedding that achieves spacetime-optimality, i.e., it is space-minimal among all designs that are time-optimal. We nonlinearly embed the process dependence graph as follows:

$$
\begin{aligned}
t &:= i + j \ ; \\
s &:= j \bmod \left\lfloor \frac{n}{2} \right\rfloor \ .
\end{aligned}
$$

This embedding E_2 is illustrated in Figure 4 for $n = 7$. Its data flow characteristics are identical to those of embedding E_1, except that the upper processor is attached to the lower processor forming a ring topology, and data movement wraps around the array.

This embedding results in a computation of the process dependence graph that uses $2n$ steps and a ring array of $\lceil \frac{n+1}{2} \rceil$ processors.

The embedding E_2 is space-minimal in addition to being time-optimal. In order to see this, consider time step 8 in which all 4 processors are used. To reduce the number of processors, it is necessary that the process depicted by node $(8, 3)$ be rescheduled onto another processor. However, node $(8, 3)$ is on a longest directed path in the process dependence graph. This means that it cannot be rescheduled for earlier completion without violating a dependence. Neither can it be scheduled for later completion without either violating a dependence, or extending the overall completion time, violating time-optimality. The number of processors therefore cannot be reduced: the design is spacetime-optimal.

Any spacetime embedding of this process dependence graph that completes in $2n$ cycles, must use at least $\lceil \frac{n+1}{2} \rceil$ processors [14].

Thus, the optimal value of q is equal to 2, i.e., we partition the moduli set such that each processor executes Step 1 of the congruence algorithm for 2 moduli.

Theorem 3 *The parallel congruence algorithm requires* $O(k^3 + nk)$ *arithmetic and* $2n(k+1)$ *routing steps on a ring array with* $p = \lceil \frac{n+1}{2} \rceil$ *processors.*

Proof Assuming that $n + 1$ is even, we allocate the moduli set and A and b such that processor i receives m_i and m_{i+p} for $i = 0, 1, 2, \ldots, p$ where $n + 1 = 2p$. Processor i computes $\mathbf{y}_i$ and $\mathbf{y}_{i+p}$ by solving the systems $\mathbf{A}\mathbf{y}_i = \mathbf{b} \pmod{m_i}$ and $\mathbf{A}\mathbf{y}_{i+p} = \mathbf{b} \pmod{m_{i+p}}$. Simultaneously, the determinants $d_i = \det(\mathbf{A}) \pmod{m_i}$ and $d_{i+p} = \det(\mathbf{A}) \pmod{m_{i+p}}$ are computed. Thus, the computation of $\mathbf{y}_i$ and d_i for all $0 \le i \le n + 1$ will take $O(k^3)$ arithmetic steps since all $p = \lceil \frac{n+1}{2} \rceil$ processors work simultaneously.

Figure 4. A spacetime-optimal systolic schedule for the mixed-radix conversion algorithm.

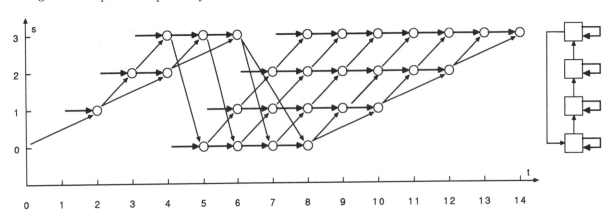

Similarly, the computation of $z_i = d_i y_i \pmod{m_i}$ $z_{i+p} = d_{i+p} y_{i+p} \pmod{m_{i+p}}$ takes $O(k)$ arithmetic steps.

We then start the parallel mixed-radix conversion algorithm which requires $2n$ steps, where at each step the operations given by (3) and (4) are performed, and a vector of dimension $k + 1$ may be sent. Since vectors $\mathbf{V}$ are of dimension $k + 1$, we perform $O(nk)$ arithmetic operations altogether. Furthermore, at most $2n(k+1)$ routing steps are needed by the parallel mixed-radix conversion algorithm.

After the execution of the mixed-radix conversion algorithm, processor $p - 1$ has vectors $\mathbf{V}_{i,i+1} = \mathbf{v}_i$ for $0 \le i \le n$. We compute $\mathbf{u}$ (i.e., $\mathbf{z}$ and d) in processor $p - 1$ using floating-point arithmetic with Horner's algorithm as

$$\mathbf{u} = \mathbf{v}_0 + \mathbf{v}_1 m_0 + \mathbf{v}_2 m_0 m_1 + \cdots + \mathbf{v}_n m_0 m_1 \cdots m_{n-1} .$$

This step is completely sequential and requires $O(nk)$ arithmetic operations.

Therefore, the parallel congruence algorithm requires a total of $O(k^3 + nk)$ arithmetic and $2n(k + 1)$ routing steps on a ring array of $\lceil \frac{n+1}{2} \rceil$ processors. □

5   Implementation

The parallel congruence algorithm described in this paper is suitable for implementation on distributed-memory multiprocessors and systolic computing systems. The processors of the parallel computer system must be powerful enough to solve a system of linear equations. Examples of commercially available distributed-memory multiprocessors are Intel iPSC series, NCUBE, and Ametek. Examples of software-oriented systolic/wavefront computing systems include an array of Transputers [12], the Warp [3], and the Matrix-1 [8].

We have implemented the parallel congruence algorithm on a first generation Intel hypercube with $p = 8$ processors (iPSC/d3). The ring topology required for the spacetime-optimal systolic implementation of the mixed-radix conversion algorithm is easily embedded in the hypercube by using the binary-reflected Gray code [6].

The timing results for the time-optimal and spacetime-optimal schedules are summarized in Figure 5. The values of $n + 1$ and k were limited to those given due to memory limitations (512 KBytes per node on this particular Intel cube).

The time for the sequential version of the algorithm is measured by executing the sequential congruence algorithm on one of the eight identical processors. In order to obtain a fair comparison between the sequential and the parallel algorithm, we measure the total execution time, but not the initial loading of the data and the final unloading of the results. The efficiency of the parallel congruence algorithm is plotted in Figures 6.

Theoretically, the parallel congruence algorithm should achieve linear speedup (constant efficiency). However, we observe a decline in the efficiency as n increases. This is due to the fact that $2n(k + 1)$ routing operations required by the parallel algorithm start taking a significant amount of time for large n. Furthermore, the slow communication (*store-and-forward*) mechanism of the first generation hypercubes constitutes an obstacle to high efficiency. We note that the newer message-passing multiprocessors have more advanced and thus much faster data communication mechanisms [4]. The efficiency becomes higher for larger k, since Step 1 (which requires no communication and $O(k^3)$ parallel arithmetic steps) starts dominating Step 2 (which requires $2n(k + 1)$ routing and $O(nk)$ parallel arithmetic steps).

6   Conclusions

Several extensions of the parallel congruence algorithm can be proposed. An important issue arises when the number of processors in the parallel computing system does not match the size of the problem to be solved. There are two cases:

Fewer processors ($2p < n + 1$): Let $n + 1 = qp$ where $q > 2$. We partition the moduli set into $2p$ groups where each group contains $\frac{q}{2}$ moduli. The allocation of the data is similar to the case $q = 2$, however, now processor i is assigned to perform computations using the moduli in groups i and $i + p$ for $i = 0, 1, \ldots, p - 1$.

More processors ($2p > n + 1$): In this case, the best approach is to exploit parallelism inherent in the Gaussian elimination step, and also in vector operations required by the congruence algorithm. As an example, assume that we have p processors where $2p = k(n + 1)$. We can allocate k processors for each of the linear systems of equations solved in Step 1 of the congruence algorithm. We can use these k processors to reduce the number of arithmetic operations required by the Gaussion elimination algorithm from $O(k^3)$ to $O(k^2)$. Furthermore, the parallel mixed-radix conversion algorithm will now require only $O(n)$ arithmetic operations instead of $O(nk)$. Parallel algorithms for Gaussian elimination and LU decomposition are well-known [6]. However, the parallel congruence algorithm requires Gaussian elimination over the ring of integers $\mathcal{Z}_{m_i}$ (or the Galois field $GF(m_i)$ when m_i is prime). Thus, pivoting or partial pivoting is necessary, since every m_ith element is zero. Systolic algorithms for Gaussian elimination without pivoting [9, 1] and with partial pivoting [5, 11] have been proposed. Thus, we see that, depending on the number of processors available in the parallel computer system, different levels of parallelism already inherent in the congruence algorithm can be exploited.

When the number of processors available is more than $\lceil \frac{n+1}{2} \rceil$, we can allocate more than one processor for each congruent linear system solved, and thus reduce the amount of memory required for each node. This is due to that fact each node receives fewer than k rows, allowing a larger system to be solved for a given memory space.

Before concluding, we note that the hypercube network is richer in connectivity than a linear (or ring) array. Thus, it seems worthwhile to investigate how the total number of routing operations required can be reduced by utilizing the additional connections. Other distributed-memory architecture topologies (e.g., two or three dimensional mesh, binary tree) can also be used for implementing the congruence algorithm.

References

[1] H. M. Ahmed, J. M. Delome, and M. Morf. Highly concurrent computing structures for matrix arithmetic and signal processing. *IEEE Computer*, 15:65–82, 1982.

[2] A. Aho, J. E. Hopcroft, and J. D. Ullman. *The Design and Analysis of Computer Algorithms*. Addison-Wesley Publishing Co., 1974.

[3] A. M. Annaratone, E. Arnould, T. Gross, H. T. Kung, M. Lam, O. Menzilcioglu, and J. Webb. The WARP computer: Architecture, implementation, and performance. *IEEE Transactions on Computers*, 36(12):1523–1538, December 1987.

[4] W. C. Athas and C. L. Seitz. Multicomputers: Message-passing concurrent computers. *IEEE Computer*, 21(8):9–25, August 1988.

[5] H. Barada and A. El-Amawy. Systolic architecture for matrix triangularisation with partial pivoting. *IEE Proceedings, Part E: Computers and Digital Techniques*, 135(4):208–213, July 1988.

[6] D. P. Bertsekas and J. N. Tsitsiklis. *Parallel and Distributed Computation, Numerical Methods*. Prentice-Hall, 1989.

[7] J. D. Dixon. Exact solution of linear equations using p-adic expansions. *Numerische Mathematik*, 40:137–141, 1982.

[8] D. E. Foulser and R. Schreiber. The Saxpy Matrix-1: A general-purpose systolic computer. *IEEE Computer*, 20(7):35–43, July 1987.

[9] W. M. Gentleman and H. T.Kung. Matrix triangularisation by systolic arrays. In *Proc. SPIE 298, Real-Time Signal Processing IV*, pages 19–26, San Diego, CA, 1981.

[10] R. T. Gregory and E. V. Krishnamurthy. *Methods and Applications of Error-Free Computation*. Springer-Verlag, 1984.

[11] B. Hochet, P. Quinton, and Y. Robert. Systolic Gaussian elimination over $GF(p)$ with partial pivoting. *IEEE Transactions on Computers*, 38(9):1321–1324, September 1989.

[12] INMOS Ltd., Almondsbury, Bristol, UK. *IMS T800 Transputer*, November 1986.

[13] D. E. Knuth. *The Art of Computer Programming, Volume 2: Seminumerical Algorithms*. Addison-Wesley Publishing Co., second edition, 1981.

[14] Ç. K. Koç and P. R. Cappello. Systolic arrays for integer Chinese remaindering. In M. D. Ercegovac and E. Swartzlander, editors, *Proceedings of the 9th Symposium on Computer Arithmetic*, pages 216–223, Santa Monica, California, September 6–8 1989. IEEE Computer Society Press.

[15] S. Y. Kung. *VLSI Array Processors*. Prentice-Hall, 1988.

[16] J. D. Lipson. Chinese remainder and interpolation algorithms. In *Proc. 2nd Symp. Symbolic and Algebraic Manipulation*, pages 372–391, 1971.

[17] J. D. Lipson. *Elements of Algebra and Algebraic Computing*. Addison-Wesley Publishing Co., 1981.

[18] G. Mackiw. *Applications of Abstract Algebra*. John Wiley and Sons, Inc., 1985.

[19] M. Newman. Solving equations exactly. *Journal of Research of the National Bureau of Standards*, 71B(4):171–179, October–December 1967.

[20] G. Villard. Exact parallel solution of linear systems. In J. Della Dora and J. Fitch, editors, *Computer Algebra and Parallelism*, pages 197–205. Academic Press, 1989.

[21] D. M. Young and R. T. Gregory. *A Survey of Numerical Mathematics*, volume 2. Dover Publications, Inc., 1988.

Figure 5. Timing results (in milliseconds) for the sequential (T_{seq}), the time-optimal (T_{to}), and the spacetime-optimal (T_{sto}) schedules.

$n+1$	$k = 10$			$k = 20$			$k = 30$		
	T_{seq}	T_{to}	T_{sto}	T_{seq}	T_{to}	T_{sto}	T_{seq}	T_{to}	T_{sto}
16	3205	615	600	16475	2450	2450	48440	6620	6550
32	8235	1700	1720	36450	5810	5700	102040	14560	14310
48	15135	3290	3365	59995	10100	9870	160920	23925	23315
64	23965	5450	5400	87210	15495	14870	225215	34890	33580
80	34765	8150	7800	118195	21920	20710	295075	47370	45225
96	47555	11390	10990	152980	29370	27675	370540	61370	58180
112	63050	15220	14875	192930	37955	35485	453635	77040	72275

Figure 6. Efficiency of the parallel congruence algorithm.

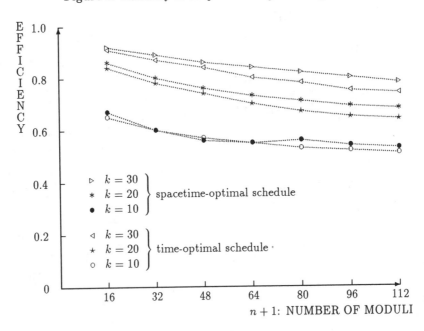

An Iterative Sparse Linear System Solver on Star Graphs

Kichul Kim and V. K. Prasanna Kumar

SAL-344

Department of Electrical Engineering-Systems

University of Southern California

Los Angeles, CA 90089-0781

Abstract

We present an efficient implementation of iterative sparse linear system solver on star graphs. Only non-zero entries of the iteration matrix are mapped onto a star graph. Each iteration of the computation can be done in $O(n^2)$ time for a star graph with $n!$ nodes. To solve the data transport problem arising in the mapping, new algorithms for star graphs are developed for routing, simultaneous broadcasting and simultaneous summation. The implementation can be easily modified to solve many problems based on directed graphs and matrix-vector multiplication.

1 Introduction

Star graphs were introduced by Akers and Krishnamurthy as an alternative to well known hypercubes [1, 2]. Being a special case of Cayley graphs, star graphs are very rich in symmetry and have very useful hierarchical structures [1]. Furthermore, star graphs have significant advantages over hypercubes in degree per node, diameter, average diameter, and fault-tolerance [1, 2].

In spite of their superiority over hypercubes in many graph theoretic properties, only few result on star graphs has been reported [3, 6, 7, 9]. Especially, parallel algorithms for star graphs have been developed only for basic problems including routing and sorting [2, 3, 6, 9]. Palis *et al.* developed an optimal randomized routing algorithm on star graphs [9]. Their algorithm can route data in $O(D)$ steps in the worst case with high probability, where D is the diameter of the star graph. Annexstein and Baumslag developed a deterministic routing algorithm for star graphs [3]. Their algorithm runs in $O(n)$ time on star graphs with $n!$ processing elements with powerful communication capability. Using a modified Shearsort, Menn and Somani succeed in developing a sorting algorithm on star graphs which is comparable in performance to the best known sorting algorithm for hypercubes [6].

In this paper, we present an efficient iterative solution to systems of linear equations with sparse coefficient matrices using star graphs. In our implementation, only non-zero entries of the iteration matrix are mapped onto a star graph with $N = m + e$ processing elements, where m is the number of equations and e is the number of non-zero entries in the iteration matrix. To solve the data transport problem arising in the mapping, new algorithms for star graphs are developed for routing, simultaneous broadcasting and simultaneous summation. These algorithms are based on multi-dimensional grids which can be emulated by star graphs without penalty in time complexity. Each iteration of the computation can be done in $O(n^2)$ time for a star graph with $n! = N = m + e$ nodes. Preprocessing takes $O(Nn^2)$ time. The implementation can be easily modified to perform many computations based on directed graphs and vector-matrix multiplication.

2 Preliminaries

This section introduces star graphs and iterative sparse linear system solvers which are based on matrix-vector multiplications.

2.1 Star Graphs

This subsection briefly introduces star graphs. More details on star graphs can be found in [1, 2].

Definition 1 *An n-star graph has n! nodes. The nodes are labelled by n! permutations on n different symbols. Two nodes u and v in a star graph are connected to each other if and only if the permutations of*

⁰This research was supported in part by the National Science Foundation under grant IRI-9145810 and in part by DARPA under contract F-33615-87-C-1436.

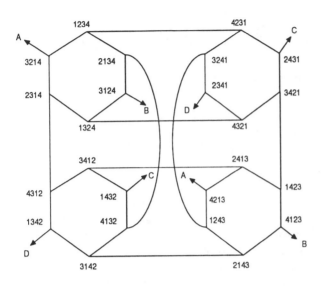

Figure 1: A 4-star graph

u and v differ in exactly two positions including the first position. A star graph with n! nodes is denoted by S_n.

Figure 1 shows an example of S_4. Each hexagon in Figure 1 is an example of S_3. For simplicity, we will use n-star to denote n-star graph. In the rest of this paper, we use numbers from 1 to n to denote the n different symbols and N represents n!.

A graph is vertex symmetric if, for any two nodes in the graph, there exists an automorphism of the graph that maps one node into the other. Edge symmetry is defined similarly. Star graphs are both vertex and edge symmetric as shown in the following lemma.

Lemma 1 *Star graphs are both vertex and edge symmetric. [2]*

Besides symmetry, star graphs also have a very useful hierarchical structure. Let $r_i, 1 \leq r, i \leq n$ denote the induced subgraph of a star graph consisting of all the permutations that contain the symbol r in the i-th position. For example, when n = 4, 4_4 is a subgraph of S_4 whose nodes are all the permutations that have the symbol 4 in the fourth position, i.e., {1234, 2134, 1324, 3124, 2314, 3214}. Note that 4_4 is a star graph by itself as shown in Figure 1. In general, for $2 \leq i \leq n$, the subgraph r_i is isomorphic to S_{n-1}. Since any of n symbols can be placed in the i-th position, we can partition an n-star into n (n-1)-stars, i.e., $1_i, 2_i, \ldots n_i$. Since n-1 positions, except the first position, can be used for fixing sym-

bols, there are n-1 ways to partition an n-star as summarized in the following lemma.

Lemma 2 *There are n-1 ways to partition an n-star into n (n-1)-stars. [2]*

We can also partition a star graph into $1_1, 2_1, \ldots, n_1$, i.e., by fixing symbols in the first position. The subgraphs are not isomorphic to S_{n-1}. It is easy to see that two nodes in r_1 are not connected to each other by an edge of a star graph. The following lemma shows an interesting property between r_1 and $r_i, 2 \leq i \leq n$.

Lemma 3 *In an n-star, each node in $r_i, 2 \leq i \leq n$ is connected to exactly one node in r_1. [2]*

2.2 Iterative Sparse Linear System Solver

A system of linear equations can be written in matrix form as:

$$Ax = b \tag{1}$$

where A represents the invertible coefficient matrix of order $m \times m$ and x is the solution vector of unknowns. x and b are both vectors of order $m \times 1$. The problem of solving these equations is one of determining a solution vector x for which the above equation holds. A sparse matrix is one that has enough zero entries to justify a special method of solution. Systems of linear equations with sparse coefficient matrices can be solved using direct methods including the well known Gaussian Elimination. They can also be solved with iterative methods which has the following merits [11].

- Iterative methods do not introduce fill-ins, the replacement of zero entries by non-zero entries. This is good for parallel implementations where one wants to map only the non-zero entries to the individual processing elements.

- Iterative methods are well suited for large sparse linear systems since they don't need very large storage.

As an example of iterative linear system solver, we explain Jacobi's method [8]. Assuming that the diagonal elements of A are all non-zero, equation (1) can be transformed to the equivalent linear one-point matrix iteration

$$\begin{aligned} x^{i+1} &= D^{-1}Bx^i + D^{-1}b \\ &= Px^i + q \end{aligned} \tag{2}$$

where D is the diagonal matrix made with the diagonal elements of A and $B = D - A$. The kth component of x^{i+1} can be represented as

$$x_k^{i+1} = \sum_{j=1}^{m} p_{kj} x_j^i + q_k \qquad (3)$$

where $\{p_{kj}\}$ represent the elements of the iteration matrix P. Equation (3), therefore, requires the calculation of m such inner products with a total number of multiplications equal to the number of non-zero entries in P. Our implementation can be applied to any iterative method which can be expressed with iterative matrix-vector multiplication as shown in equation (2).

3 Indexing Scheme and Mapping a Grid onto Star Graph

In this section, we briefly show an indexing scheme and a mapping of a grid onto a star graph used by Menn and Somani [6]. Interested readers can refer [6] for more details.

Definition 2 *Let* $f : V \to [0, |V| - 1]$ *be an isomorphism between a vertex set* V *and the range of integers* $[0, |V| - 1]$. *Assume that* V *is partitioned into* R *sets of equal size. These sets are called rows, and they are numbered from 1 to* R. *The map* f *is a row major indexing scheme if* $\forall r, 1 \leq r \leq (R-1)$, *a node* u *is in row* r, *and a node* v *is in row* $(r + 1)$ *implies* $f : u < f : v$. *[6]*

Let an n-star S_n be partitioned into n $(n-1)$-stars: $1_n, 2_n, \ldots, n_n$. The substars are of the same size and we can consider each of them as a row. If we map the nodes of substar $r_n, 1 \leq r \leq n$, to integers $[(r-1)(n-1)!, r(n-1)! - 1]$, this indexing scheme is a row major indexing scheme by definition 2. We can renumber the symbols $1, \ldots, r-1, r+1, \ldots, n$ of r_n to $1, 2, \ldots, n-1$ and apply the same indexing scheme. By applying the indexing scheme recursively to all substars, we have a completely defined row major indexing scheme. Menn and Somani called this scheme the Row Major Indexing Scheme (RMIS) [6]. Figure 2 shows the RMIS for a 4-star. It is noteworthy that getting permutation from an index is as easy as getting index from a permutation.

The following definitions show two mappings of $n \times (n-1)!$ grid onto n-star.

Definition 3 *RM (Row Mode) maps an* $n \times (n-1)!$ *grid onto an* n-star. *The rows of the grid are the* n

0	4321	1	3421	2	4231	3	2431	4	3241	5	2341
6	4312	7	3412	8	4132	9	1432	10	3142	11	1342
12	4213	13	2413	14	4123	15	1423	16	2143	17	1243
18	3214	19	2314	20	3124	21	1324	22	2134	23	1234

Figure 2: Row Major Indexing Scheme for 4-star

independent $(n-1)$-stars: $1_n, 2_n, \ldots, n_n$. *Two nodes in a pair of consecutive rows are in the same column if and only if their permutations differ exclusively in the last and any (one) other position. [6]*

Definition 4 *CM (Column Mode) also maps an* $n \times (n-1)!$ *grid onto an* n-star. *The rows of the grid are the* n *independent sets:* $1_1, 2_1, \ldots, n_1$. *Two nodes in a pair of consecutive rows are in the same column if and only if their permutations differ exclusively in the first and any (one) other position. By definition of the star graph, such two nodes are connected. Therefore, the columns in CM are connected in a linear array. [6]*

The mappings RM and CM are not unique since if we do any column permutation on a RM (CM) the resulting mapping will be also a RM (CM). As shown in lemma 3, each node in r_n is connected to exactly one node in r_1. Each node in the r-th row of RM is connected to exactly one node in the r-th row of CM. Therefore, transformation between any CM and any RM can be done in a single step. In other words, for all x and y, a datum in a processing element whose position is x-th row and y-th column in RM can be sent to a processing element whose position is x-th row and y-th column in CM in one parallel step. Suppose we map node u of S_n to a grid point whose row major index is $RMIS(u)$. The resulting mapping is shown in Figure 3 for a 4-star. It is easy to see that the mapping satisfy the definition of RM. Throughout this paper, this mapping will be called as RM for simplicity. If we exchange the first symbol and the last symbol of each permutation in RM, we obtain another mapping as shown in Figure 4. It is easy to see that the mapping satisfy the definition of CM and each column of the mapping is connected in a linear array. As same as RM, this mapping will be called as CM for simplicity. Readers can check that a node mapped to (x, y), x-th row and y-th column, in RM is connected to a node mapped to (x, y) in CM.

Since the transformation between RM and CM is possible in one parallel step and the columns of CM

4321	3421	4231	2431	3241	2341
4312	3412	4132	1432	3142	1342
4213	2413	4123	1423	2143	1243
3214	2314	3124	1324	2134	1234

Figure 3: Row Mode for a 4-star

1324	1423	1234	1432	1243	1342
2314	2413	2134	2431	2143	2341
3214	3412	3124	3421	3142	3241
4213	4312	4123	4321	4132	4231

Figure 4: Column Mode for a 4-star

$$P = \begin{bmatrix} x & 0 & 0 & 0 & x & 0 & 0 & 0 \\ 0 & 0 & 0 & x & 0 & 0 & x & 0 \\ 0 & 0 & x & 0 & 0 & 0 & 0 & 0 \\ x & 0 & 0 & 0 & 0 & 0 & 0 & 0 \\ 0 & x & x & 0 & 0 & x & 0 & x \\ x & 0 & 0 & 0 & 0 & x & 0 & x \\ x & 0 & 0 & 0 & x & 0 & 0 & 0 \\ 0 & 0 & x & 0 & 0 & 0 & 0 & 0 \end{bmatrix}$$

x1 q1	x2 q2	x3 q3	x4 q4	x5 q5	x6 q6
x7 q7	x8 q8	(1,1)	(4,1)	(6,1)	(7,1)
(5,2)	(3,3)	(5,3)	(8,3)	(2,4)	(1,5)
(7,5)	(5,6)	(6,6)	(2,7)	(5,8)	(6,8)

Figure 5: A sparse iteration matrix and the initial data mapping

are connected in linear arrays, RM can easily emulate $(n-1)!$ linear arrays of size n as follows,

1. Transform RM into CM.

2. Perform communications using linear arrays of size n in CM.

3. Transform CM back into RM

The only difference between "real" linear arrays and columns of RM is that we need two extra steps for transformation between RM and CM. Since each row of the RM is a star graph of smaller size, we can consider each row of RM as $(n-2)!$ linear arrays of size $(n-1)$. By applying this idea recursively, we can consider RM as an n-dimensional grid whose size is $n \times n - 1 \times n - 2 \times \ldots \times 1$. The difference between RM and a real n-dimensional grid is that we can not use two different kinds of link at the same time, e.g., some links in the 4-th dimension and some links in the 5-th dimension at the same time, and communication in each dimension takes two extra steps for transformations.

4 An Overview of the Algorithm

The elements of the initial solution vector x^0 are placed in the first m processing elements. The elements of the right hand side vector q are also placed in the first m processing elements. Non-zero entries of the iteration matrix are placed in the remaining processing elements in column major order. Thus the array has $N = n! = m + e$ processing elements. The processing element with the least index within a column is called the leader of the column. A 8×8 sparse iteration matrix and its initial data mapping is shown in Figure 5, where (i, j) represents $p_{i,j}$.

In each iteration step, the algorithm computes equation (3), which consists of the following steps.

Step 1: Elements of the solution vector x are routed to the leaders of appropriate columns of the iteration matrix P. This step is shown in Figure 6, for x_1 and the leader of the first column of P.

Step 2: Each column leader broadcasts the received value to the elements of the column. This step is illustrated in Figure 7.

Step 3: The products of p_{kj} and x_j are routed to processing elements such that the new distribution of the products forms a row major order. This step is shown in Figure 8, where p_{kj} and x_j in a processing element represents the product of the two items.

Step 4: The products in a row are summed to complete the inner product of equation (3).

Step 5: The sum S_k of the row k is routed to the processing element containing x_k and q_k where

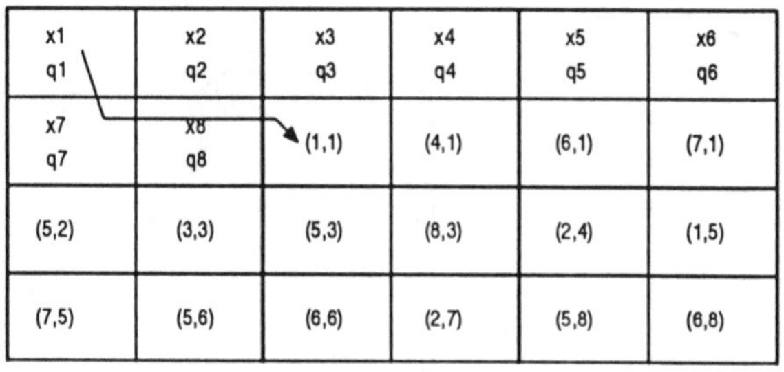

x1 q1	x2 q2	x3 q3	x4 q4	x5 q5	x6 q6
x7 q7	x8 q8	(1,1)	(4,1)	(6,1)	(7,1)
(5,2)	(3,3)	(5,3)	(8,3)	(2,4)	(1,5)
(7,5)	(5,6)	(6,6)	(2,7)	(5,8)	(6,8)

Figure 6: x_k is routed to the leader of column k

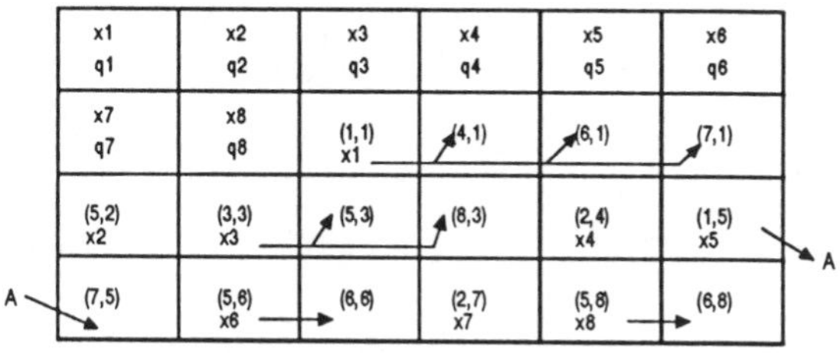

x1 q1	x2 q2	x3 q3	x4 q4	x5 q5	x8 q6
x7 q7	x8 q8	(1,1) x1	(4,1)	(6,1)	(7,1)
(5,2) x2	(3,3) x3	(5,3)	(8,3)	(2,4) x4	(1,5) x5
(7,5)	(5,6) x6	(6,6)	(2,7) x7	(5,8) x8	(6,8)

Figure 7: x_k is broadcast within column k of P

the sum is added to q_k to update x_k. This step is explained in Figure 9 for S_1 and x_1.

The number of iterations to be used depends on the particular iterative scheme being used and on the desired accuracy of the final result [8].

In our analysis, we use the standard assumption that a communication step as well as a computation step takes one unit of time. As shown in the next section, all data transport problem involved in the algorithm can be solved in $O(n^2)$ time. Hence, one iteration of the algorithm takes $O(n^2)$ time.

x1 q1	x2 q2	x3 q3	x4 q4	x5 q5	x6 q6
x7 q7	x8 q8	(1,1) x1	(1,5) x5	(2,4) x4	(2,7) x7
(3,3) x3	(4,1) x1	(5,2) x2	(5,3) x3	(5,6) x6	(5,8) x8
(6,1) x1	(6,6) x6	(6,8) x8	(7,1) x1	(7,5) x5	(8,3) x3

Figure 8: The distribution after the transformation to row major order

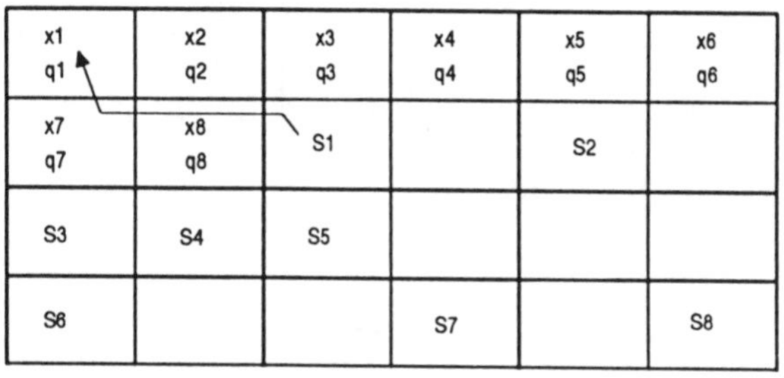

x1 q1	x2 q2	x3 q3	x4 q4	x5 q5	x6 q6
x7 q7	x8 q8	S1		S2	
S3	S4	S5			
S6			S7		S8

Figure 9: The sum of row k is routed to x_k

5 Data Transport

All the steps described in the previous section involve data transport among the processing elements in the array. Each of these data transport problems falls into one of the following three problems: data routing, simultaneous broadcasting within each group and simultaneous summation within each group. We show efficient solutions to these data transport problems in this section.

5.1 Data Routing

During each iteration, each element of the solution vector should be routed to the leader of an appropriate column of P in step 1. Then, in step 3, each product of an element of the solution vector and a member of a column should also be routed so that the resulting distribution forms a row major order. Finally, in step 5, the sum of row k should be routed to the processing element containing x_k. In the following, we show a routing algorithm which can route data in $O(n^2)$ time on star graphs with $n!$ processing elements. The algorithm is based on a routing algorithm for the well known three-stage Clos networks. A similar technique for two-dimensional mesh can be found in [10]. First we will briefly explain the routing algorithm three-stage Clos networks.

A three-stage Clos network, $C(2,2,4)$, of size 8 is shown in Figure 10 [4]. It consists of three stages. The first and last stages consist of four 2×2 switches which can realize either a parallel connection or a crossed connection. The middle stage consists of two 4×4 switches which can realize all 4! permutations. In general, a three-stage Clos network, $C(2,2,N/2)$, of size N has three stages. The first and last stages consist of $N/2$ 2×2 switches. The middle stage consists of two $N/2 \times N/2$ switches.

A matching M in a bipartite graph $G = (V_1, V_2, E)$ is a set of edges such that no two edges in M are

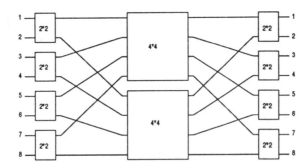

Figure 10: A Clos network of size 8, $C(2,2,4)$

incident on a same vertex. The size of a matching M is the number of edges in M. A complete matching is one whose size is $\min(|V_1|, |V_2|)$. An algorithm to determine the switch settings of a three-stage Clos network, $C(2,2,N/2)$, of size N is shown below [5].

1. Construct a bipartite graph $G = (V_1, V_2, E)$ such that $V_1 = V_2 = \{1, 2, \ldots, N/2\}$ and whenever an input of a switch i in the first stage should be connected to an output of a switch j in the last stage, an edge (i, j) is added to E. Notice that the degree of a node is two and there can be multiple edges between a pair of vertices. Hence, in general, G is a bipartite multigraph.

2. Find disjoint complete matchings M_1 and M_2 in G. The existence of such complete matchings is guaranteed by Hall's theorem.

3. Realize the connections in M_1 through the upper switch in the middle stage. Also realize the connections in M_2 through the lower switch in the middle stage. During this step, the switch setting for the first and last stages are obtained as well as the switch settings for the middle stage.

The complexity of the algorithm is dominated by finding complete matchings in G, which can be done in $O(N)$ time.

Suppose we know how to route data in $(n-1)$-star, S_{n-1}. Recall that we can regard columns of RM as linear arrays of size n and each row of RM is a S_{n-1}. This means that we can perform row-wise routing and column-wise routing in RM. Using these partial routings, we can perform routing in S_n as follows.

1. Construct a bipartite graph $G = (V_1, V_2, E)$ such that $V_1 = V_2 = \{1, 2, \ldots, (n-1)!\}$ and whenever a datum in column i should be routed to column j, an edge (i, j) is added to E. Hence, each node

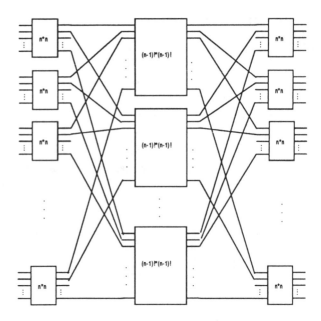

Figure 11: A Clos network of size $n!$, $C(n, n, (n-1)!)$

has n edges and G is a bipartite multigraph with $n!$ edges.

2. Find n disjoint complete matchings $M_1, \ldots, M_n$ in G.

3. Perform column-wise routing such that a datum in matching M_i is routed to the i-th row. This can be done easily by transforming RM to CM as explained in section 3.

4. Route data to their destination columns using row-wise routing on S_{n-1} of each row.

5. Route data to their final destination processing elements using column-wise routing.

It is easier to understand the above algorithm by visualizing the star graph (RM) as a three-stage Clos network shown in Figure 11. Each $n \times n$ switch in the first and last stage represents a linear array of size n and each $(n-1)! \times (n-1)!$ switch in the middle stage represents an $(n-1)$-star. There are $(n-1)!$ switches in the first and last stages and there are n switches in the middle stage.

We can get a complete routing algorithm for an n-star by recursively applying the above algorithm to smaller star graphs. Since finding complete matchings for an n-star takes $O(Nn)$ time, the whole preprocessing will take $O(Nn^2)$ time. When we use swapping operation between adjacent processing elements, routing in a linear array of size n takes $n-1$

time. Routing in columns of RM takes $n + 1$ time because of the transformation between RM and CM. Routing for n-star takes $R(n)$ time where,

$$R(n) = n + 1 + R(n-1) + n + 1 \qquad (4)$$
$$R(2) = 1 \qquad (5)$$

Hence the routing algorithm runs in $O(n^2)$ time. As mentioned in the introduction, two routing algorithms are known for star graphs. Palis's randomized routing algorithm can route data in $O(D)$ time for the worst input with high probability, where D is the diameter of the star graph [9]. Annexstein and Baumslag's routing algorithm can route data in $O(n)$ time using processing elements which can route n messages through all n ports in one unit of time [3]. If we allow routing through only two ports at a time, the routing algorithm results in $O(n^2)$ time complexity. The main difference between the routing algorithms in [3] and in this paper is that spoke structures are used for the basic routing in [3] and linear arrays are used for the basic routing in this paper.

5.2 Simultaneous Broadcasting and Summation

During each iteration, in step 2, the leader of a column of the iteration matrix P has to broadcast the value of an appropriate element of the solution vector to the members of the column. During each iteration, in step 4, the product terms in each row should be summed to complete the inner product of equation (3). Here, we show a simultaneous broadcasting algorithm and a simultaneous summation algorithm both with $O(n^2)$ time complexity.

In simultaneous broadcasting and summation, nodes in a star graph is divided into several groups of different size. The RMIS (Row Major Indexing Scheme) of members of a group are continuous. The processing element with the least index in a group is called the leader of the group. Each member of a group knows the size of the group and position of the group, i.e, the index of the leader. Figure 12 shows an example of grouping for simultaneous broadcasting and summation. The processing elements marked with 'X' are the leaders of each group. The problem of simultaneous broadcasting is how to send the datum in a group leader to all the members of the same group, for all groups simultaneously. The problem of simultaneous summation is how to sum the values in each group member into the group leader, for all groups simultaneously.

1	2	3	4	5	6	7	8	9	10	11	12	13	14	15	16	17	18	19	20	21	22	23	24
X					X						X	X					X						

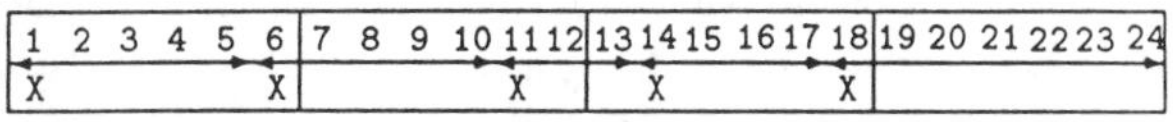

Figure 12: An example of grouping for simultaneous broadcasting and summation

Our approach to the simultaneous broadcasting is that a datum is broadcast to a larger substar until it is sent to all the group members. The actual broadcasting is performed by the linear arrays of CM. In this approach, a datum to be broadcast to a larger substar should be broadcast to the entire current substar before being broadcast to the larger substar. Note that each substar has only one datum to be broadcast to the larger substar since there is at most one leader in a substar which should broadcast to other substars. The broadcasting algorithm has $n-1$ stages. Broadcasting on i-star is performed in the $(i-1)$-th stage. Each datum from a leader has the information for the size and position of the group. The following outlines the $(i-1)$-th stage of the broadcasting.

1. Transform RM of i-star to CM.

2. Perform column broadcasting in each column of CM. To perform column broadcasting each processing element in a (column) linear array does the following $i-1$ times.

   (a) Get a datum from the upper processing element, i.e., an adjacent processing element with smaller index.

   (b) If the datum is from the right leader, take it.

   (c) If the datum is from the right leader and it should be sent to the next $(i-1)$-substar, send it to the lower processing element in the next iteration.

   (d) Get a datum from the lower processing, i.e., an adjacent processing element with larger index.

   (e) If the datum should be broadcast to the larger substar, i.e., $(i+1)$-substar, store it and send it to the upper processing element in the next iteration. Otherwise discard the datum.

3. Transform CM back to RM.

Assume that, at the start of $(i-1)$-th stage, every $(i-1)$-substar have right data to broadcast to other

$(i-1)$-substars in the same i-star. At the end of $(i-1)$-th stage of the broadcasting, we can see the following.

1. Each processing element has received a datum from the right leader if the leader is in the same i-substar.

2. Each processing element has a datum to be broadcast to the larger substar $((i+1)$-star). This datum is originated from the leader with the largest index among leaders in the i-star. Therefore, in the beginning of i-th stage, all the processing elements in a row (i-star) of $(i+1)$-star have the same datum to be broadcast.

At the beginning of the broadcasting, every 1-star have right data to broadcast to other 1-stars. Hence, the simultaneous broadcasting algorithm gives the right result.

The i-th stage of the simultaneous broadcasting takes $O(i)$ time. Hence the whole simultaneous broadcasting algorithm takes $O(n^2)$ time.

The simultaneous summation algorithm can be done in a similar way to the simultaneous broadcasting. Summation is done in smaller substars and partial sums of smaller substars are broadcast to larger substars until the leader of a group gets the right sum. The direction of data movement is opposite to that of simultaneous broadcasting, *i.e.*, partial sums are broadcast to leaders from processing elements with larger indices. At any stage, each substar has only one partial sum to be broadcast to the larger substar since there is at most one group whose leader is in another substar. Simultaneous summation can be performed in $O(n^2)$ time.

6 Conclusion

An efficient iterative sparse linear system solver for star graphs is shown in this paper. The number of processing elements used is $N = n! = m + e$, where m is the number of equations and e is the number of non-zero entries in the iteration matrix. Each iteration takes $O(n^2)$ time. The preprocessing takes $O(Nn^2)$ time on a serial computer.

New algorithms for routing, simultaneous broadcasting and simultaneous summation are developed for star graphs to solve data transport problems. These algorithms are based on multi-dimensional grids which can be emulated by star graphs without penalty in time complexity.

The method shown in this paper can be easily modified to solve various problems based on directed graphs and matrix-vector multiplication.

References

[1] S. Akers and B. Krishnamurthy, "A Group Theoretic Model for Symmetric Interconnection Networks," *International Conference on Parallel Processing*, pp. 216-223, 1986.

[2] S. B. Akers, D. Harel and B. Krishnamurthy, "The Star Graph: An Attractive Alternative to the n-Cube," *International Conference on Parallel Processing*, pp. 393-400, 1987.

[3] F. Annexstein and M. Baumslag, "A Unified Approach to Off-Line Permutation Routing on Parallel Networks," *ACM Symposium on Parallel Algorithms and Architectures*, pp. 398-406, 1990.

[4] V. E. Benes, "On Rearrangeable Three-Stage Connecting Networks," *Bell System Technical Journal*, Vol. 41, pp. 117-125, Sept. 1962.

[5] V. E. Benes, *Mathematical Theory of Connecting Networks and Telephone Traffic*, Academic Press, New York, 1965.

[6] A. Menn and A. Somani, "An Efficient Sorting Algorithm for the Star Graph Interconnection Network," *International Conference on Parallel Processing*, Vol III, pp. 1-8, 1990.

[7] M. Nigam, S. Sahni and B. Krishnamurthy "Embedding Hamiltonians and Hypercubes in Star Interconnection Graphs," *International Conference on Parallel Processing*, Vol III, pp. 340-343, 1990.

[8] J. M. Ortega, *Introduction to Parallel and Vector Solution of Linear Systems*, Plenum Press, New York, 1988.

[9] M. A. Palis, S. Rajasekaran and D. S. L. Wei, "General Routing Algorithms for Star Graphs," *International Parallel Processing Symposium*, pp. 597-611, 1990.

[10] C. S. Raghavendra and V. K. Prasanna Kumar, "Permutations on Illiac IV-Type Networks," *IEEE Transactions on Computers*, Vol. C-35, No. 7, pp 662-669, July 1986.

[11] P. S. Tseng, "Iterative Sparse Linear System Solvers on Wrap", *International Conference on Parallel Processing*, pp. 32-37, 1988.

Shared Memory Parallel Algorithms for Homotopy Curve Tracking

D. C. S. Allison, K. M. Irani, C. J. Ribbens and L. T. Watson

Department of Computer Science
Virginia Polytechnic Institute & State University
Blacksburg, VA 24061

Abstract.

Results are reported for a series of experiments involving numerical curve tracking on a shared memory parallel computer. HOMPACK is a mathematical software package implementing globally convergent homotopy algorithms for solving systems of nonlinear equations. The HOMPACK algorithms for sparse Jacobian matrices use a preconditioned conjugate gradient algorithm to compute the kernel of the Jacobian matrix, a required linear algebra step for homotopy curve tracking. A parallel version of the HOMPACK algorithms for sparse problems is implemented on a shared memory parallel computer with various levels and degrees of parallelism (e.g., linear algebra, function and Jacobian matrix evaluation). A detailed study is presented for each of these levels with respect to parallel speedup.

1. Introduction.

Homotopies are a traditional part of topology, and only recently have begun to be used for practical numerical computation in solving nonlinear systems of equations. The essence of homotopy methods is the construction of a special homotopy map, and the tracking of a smooth curve in the zero set of this map. The computational costs involved in curve tracking are often substantial, especially for large systems of equations. In this paper we describe shared-memory parallel implementations of such algorithms.

The homotopies considered here are called "artificial-parameter generic homotopies", in contrast to methods where the homotopy variable is a physically meaningful parameter. While natural-parameter homotopies are frequently of interest, the resulting homotopy zero curves often have properties that make curve tracking difficult (e.g., bifurcations or other singular or ill-conditioned behavior). The homotopy zero curves for artificial-parameter generic homotopies obey strict smoothness conditions, which means the curve-tracking algorithm can assume a well-behaved class of curves.

The theory and algorithms for functions $F(x)$ with small dense Jacobian matrices $DF(x)$ are well developed [1], [4]; parallel algorithms for these problems have also been studied [2],

[5]. In this paper we focus on shared-memory parallel algorithms for large sparse $DF(x)$. Solving large sparse nonlinear systems of equations via homotopy methods involves sparse rectangular linear systems and iterative methods for the solution of such systems. Preconditioning techniques are used to make the iterative methods more efficient.

In Section 2 we summarize the mathematics behind homotopy algorithms and survey the linear algebra algorithms required by such methods. Section 3 describes our numerical experiments, and results from various parallel implementations are presented and discussed in Section 4.

2. Homotopy algorithms and numerical linear algebra.

Let E^n denote n-dimensional real Euclidean space, and let $F : E^n \to E^n$ be a C^2 function. The fundamental problem is to solve the nonlinear system of equations $F(x) = 0$. The modern homotopy approach to solving the nonlinear system is to construct a C^2 map $\rho : E^m \times [0, 1) \times E^n \to E^n$, such that ρ and $\rho_a(\lambda, x) = \rho(a, \lambda, x)$ have the following properties: (1) the Jacobian matrix $D\rho$ has full rank on $\rho^{-1}(0)$, (2) $\rho_a(0, x) = 0$ has a unique solution $W \in E^n$, (3) $\rho_a(1, x) = F(x)$, and (4) $\rho_a^{-1}(0)$ is bounded. Given such a map, the supporting theory [6], [10] says that for almost all $a \in E^m$ there is a zero curve γ of $\rho_a(\lambda, x)$, along which the Jacobian matrix $D\rho_a(\lambda, x)$ has full rank, emanating from $(0, W)$ and reaching a zero $\bar{x}$ of F at $\lambda = 1$. Furthermore, γ has finite arc length if $DF(\bar{x})$ is nonsingular. The homotopy algorithm consists of following the zero curve γ of ρ_a emanating from $(0, W)$ until a zero $\bar{x}$ of $F(x)$ is reached (at $\lambda = 1$). This "globally convergent probability-one" homotopy algorithm has two important distinctions from classical continuation: (1) the homotopy parameter λ is not required to increase monotonically along γ, so turning points are permissible, and (2) the use of the random parameter vector a which guarantees the absence of bifurcations and singularities along γ with probability one.

The zero curve γ of the homotopy map $\rho_a(\lambda, x)$ can be tracked by many different techniques. The mathematical software package HOMPACK [12] provides three different algorithmic approaches to tracking γ: (1) an ODE-based algorithm, (2) a predictor-corrector algorithm whose iterates follow a trajectory normal to γ (a "normal flow" algorithm), and (3) a simple Newton algorithm on an augmented system (an "augmented Jacobian matrix" method). Each of these algorithms requires the solution of linear systems of equations involving $DF(x)$. Along with function and Jacobian matrix evaluation, these linear system solves dominate the computational costs.

The parallel experiments reported here were based on the normal flow codes in HOMPACK. The main computational linear algebra step in this algorithm is the solution of a nonsquare linear systems of equations for the normal flow iteration calculations. These nonsquare systems are converted to equivalent square linear systems of the form

$$Ay = \begin{pmatrix} B & f \\ c^t & d \end{pmatrix} y = b,$$

where the $n \times n$ matrix B is bordered by the vectors f and c to form a system of dimension $(n+1) \times (n+1)$. In the present context $B = D_x\rho_a(\lambda, x)$ is symmetric and sparse, but A is not necessarily symmetric.

Due to the size and sparsity of these linear systems, iterative methods are generally preferred. Unfortunately, most of the well-known iterative solvers depend on matrix properties that are clearly not present here (e.g., symmetry, positive-definiteness). One iterative method known to converge for general nonsymmetric problems is the conjugate gradient method [8] applied to the normal equations $A^t A y = A^t b$. A recent study [9] advocates the use of Craig's method [7], a variant of the conjugate gradient algorithm which solves the similar system $A A^t z = b$, where $y = A^t z$. Craig's method can be implemented in such a way that y is computed directly, with no reference to z and no explicit formation of $A A^t$. As is typical, a good preconditioner is crucial to the efficient performance of this algorithm. We use an incomplete LU (ILU) preconditioner in the experiments reported below.

The HOMPACK linear algebra algorithms seek to take advantage of the very special structure of the system $A y = b$ by splitting A into the sum of a symmetric matrix and a rank-one correction. Since the leading principal submatrix B is symmetric, this approach allows methods which exploit symmetry to be used. While Craig's method does not require symmetry, we do take advantage of this structure in terms of storage and in computing the ILU preconditioner. See [11] for a detailed discussion of the numerical linear algebra aspects of these methods.

3. Numerical experiments.

To understand the levels and degrees of parallelism, we first describe briefly the sequential HOMPACK code used as the basis for the parallel implementation. First, we need to compute the function values and the Jacobian matrix for the coefficient matrix A. Then to track the homotopy curve, we need to solve nonsquare linear systems of equations for the tangent vector and the normal flow iteration calculations. Subroutines PCGDS and PCGNS solve these rectangular systems by first converting them to equivalent square linear systems. The combination of PCGDS and PCGNS involves the solution of four independent linear systems. It is to these symmetric linear systems that Craig's method is applied to obtain the solution. The preconditioning matrix Q needs to be factored at each step of the curve-tracking. This is done only once for each pair of PCGDS and PCGNS calls, since both systems have the same coefficient matrix A.

Our study was carried out on a Sequent Symmetry S81 with ten processors using the system call m_fork and the compiler directive DOACROSS. Results are reported for two test problems: a shallow arch problem and a turning point problem. These problems are described briefly below and in more detail in [9].

Shallow arch problem. This is a relatively small but very difficult problem from structural mechanics. It is derived from equilibrium equations for a discretization of a shallow arch under an externally applied load. The results here are for an arch load just below the limit point. The Jacobian matrix $D_x \rho_a$ for the shallow arch problem has bandwidth five. The arch parameters and a complete derivation of the governing equations for a shallow arch are given in [9].

Turning point problem. This is an artificial problem defined by $F(x) = \big(F_1(x), \ldots, F_n(x)\big)^t = 0$, where $F_i(\mathbf{x}) = \tan^{-1}\big(\sin[x_i(i \bmod 100)]\big) - (x_{i-1} + x_i + x_{i+1})/20$, and $x_0 = x_{n+1} = 0$. The zero curve γ tracked from $\lambda = 0$ to $\lambda = 1$ corresponds to $\rho_a(x, \lambda) = (1 - .8\lambda)(x - a) + .8\lambda F(x)$, where a was chosen artificially to produce turning points in γ.

There are five different levels that we considered for the parallel implementation:

1. **Function and Jacobian matrix computations.** The DO loops that evaluate the components of the function $F(x)$ and Jacobian matrix $DF(x)$ are parallelized using the DOACROSS directive. The granularity is by columns; i.e., evaluating the components of a single column of $DF(x)$ is one parallel task. The shallow arch problem has very complex function evaluation computations (about 70% of the sequential execution time). Extensive efforts were made to parallelize these computations at a different level of granularity, but no improvement was seen.

2. **Low level linear algebra.** At this level, the lower level functions and linear algebra are implemented in parallel along with LINPACK functions and subroutines. These include copying, scaling, vector norms, inner products and matrix vector products.

3. **Computations with the preconditioner.** There are two subroutines which are candidates for parallelization at this level. The first one computes the ILU preconditioner. The second one computes $Q^{-1}f$ by applying forward and backward substitution to solve $Qx = f$. We have not shown the execution timings for this level in the tables because there was no speedup over the serial execution time. A brief explanation of this is given in Section 4.

4. **Linear solves within PCGDS and PCGNS.** At this level, PCGDS and PCGNS are executed serially one after the other. Within each, as explained earlier, two linear systems of equations need to be solved and these are done in parallel since they are independent of one another.

5. **PCGDS and PCGNS.** This is one level higher parallelism than the previous level. Here the subroutines PCGDS and PCGNS are executed in parallel. Note that this means that the two solves within each are still executed serially.

Levels 2–5 described above can be embedded within each other giving rise to varying degrees of parallelism. For example, if we combine level 4 and level 5, then we are executing PCGDS and PCGNS in parallel as well as the two linear solver algorithms within each of the subroutines in parallel. So actually, all four linear solves are being executed in parallel. This gives a higher degree of parallelism than simply implementing level 4 or 5 individually. For the experiments we wanted to include all possible degrees of parallelism arising from the levels of parallelism, starting from combining levels 2 and 3 and eventually implementing levels 2, 3, 4 and 5 together. Combining levels, in order to obtain the degrees of parallelism, involves implementing a DOACROSS/m_fork within a DOACROSS/m_fork. For example, combining levels 2 and 3 together involves implementing a DOACROSS within a DOACROSS. Unfortunately, all these degrees of parallelism could not be implemented because of the limitation of the Sequent parallel programming directives that within a m_fork or a DOACROSS, we cannot insert another DOACROSS or m_fork. So, due to these constraints, and in an effort to look at the most interesting combinations, we report experiments using the following combinations: (1) levels 4 and 5 together, i.e., all four solves in parallel; (2) levels 1, 4 and 5 together; and (3) levels 1 and 2 together.

In the tables below the following are used to identify the various levels of parallelism:

M1– Function and Jacobian matrix evaluations in parallel, with the Jacobian matrix done by columns.
M2– Lower level linear algebra in parallel.
M3– PCGDS and PCGNS in parallel.
M4– Within PCGDS and PCGNS, the two linear solves in parallel.
M5– M3 and M4 (all four linear solves in parallel).
M6– M1 and M5 combined.
M7– M1 and M2 combined.

4. Discussion and conclusions.

As can be observed from the tables, we have not included the timings for the third level, i.e., for the preconditioning computations in parallel. We performed the experiments for this level but did not get any speedup with either four or eight processors. The coefficient matrix for both test problems is sparse, which means that there are only a few nonzero entries in each row or column of the matrix. These matrices are stored in the packed skyline format. Hence, for all DO loop computations involving the coefficient matrix, the number of computations to be performed per iteration is quite small. This results in each processor not getting enough work to do to overcome the overhead cost of executing a loop in parallel.

Tables 1–4 show execution time in seconds and parallel efficiency for the two test problems with eight processors for all cases and four processors for the largest case. For the linear solver code only, the most efficient algorithm was M5 amongst algorithms M2, M3, M4 and M5 for both test problems. This is what one would expect since M5 has the highest degree of parallelism, being a combination of M3 and M4. Note also that the difference in timings for M3 and M4 is very small, since there are only a few computations to be done within PCGDS and PCGNS before executing the code for the two linear solves. If there were more code before the two linear solvers' code within each of PCGDS and PCGNS, one would expect the timings for algorithm M3 to be smaller than those for algorithm M4.

Note also the efficiency we obtained with algorithm M1 for the shallow arch problem as compared to the turning point problem. This is because about 83% of the serial execution time for the arch problem is spent computing the function values and the Jacobian matrix, whereas for the turning point problem the same number is less than 2%.

Overall, for both test problems, Algorithm M6 is the best algorithm in terms of timings and the speedup obtained by the parallel implementations. This is because M6 combines the most efficient parallel algorithm for the function values and the Jacobian matrix evaluations with that for the linear solver

TABLE 1
Execution time for shallow arch problem with p processors.

p	n	Serial	M1	M2	M3	M4	M5	M6	M7
8	29	440	86	435	432	433	425	69	82
8	47	5733	939	5590	5509	5558	5489	750	925
4	47	5733	1496	5623	5509	5558	5489	1467	1495

TABLE 2
Efficiency with p processors for shallow arch problem.

p	n	M1	M2	M3	M4	M5	M6	M7
8	29	0.640	0.126	0.127	0.127	0.129	0.797	0.671
8	47	0.763	0.128	0.130	0.129	0.131	0.955	0.775
4	47	0.958	0.255	0.260	0.258	0.261	0.977	0.958

TABLE 3
Execution time for turning point problem with p processors.

p	n	Serial	M1	M2	M3	M4	M5	M6	M7
8	20	5	5	5	3	3	2	2	5
8	125	50	46	34	28	29	20	15	29
8	250	87	79	56	49	50	34	26	48
8	500	168	153	105	94	97	65	50	91
8	1000	392	355	238	219	224	151	101	205
4	1000	392	364	265	219	224	151	121	234

TABLE 4
Efficiency with p processors for turning point problem.

p	n	M1	M2	M3	M4	M5	M6	M7
8	20	0.125	0.125	0.208	0.208	0.313	0.313	0.125
8	125	0.136	0.183	0.223	0.216	0.313	0.417	0.216
8	250	0.138	0.194	0.222	0.218	0.320	0.418	0.227
8	500	0.137	0.200	0.223	0.216	0.323	0.420	0.231
8	1000	0.138	0.206	0.224	0.219	0.325	0.485	0.239
4	1000	0.269	0.370	0.447	0.438	0.649	0.810	0.419

code. The tables also show the results for the same experiments with four processors, for the largest dimension n for each of the test problems. In terms of the most efficient algorithm, the same discussion holds as for eight processors. Comparing the efficiencies obtained with four processors to those obtained with eight processors, some very interesting observations can be made. First, for the turning point problem, the efficiency obtained with four processors is almost twice as good as that with eight processors. The same holds true for the shallow arch problem, for algorithms M2, M3, M4 and M5, i.e., the linear solver parallel algorithms. However, this is not true for algorithms M1, M6, and M7, i.e., all the algorithms involving the parallel function and Jacobian matrix evaluations for the shallow arch problem. The reason for this can be attributed to the fact that about 83% of the sequential execution time is spent executing the function and Jacobian matrix evaluation code. Hence the eight processors can be kept busy most of the time.

Amdahl's law provides a useful way of comparing the actual speedup attained by a parallel implementation to the maximum possible speedup, taking into consideration the fraction of the total execution time that is spent on sequential code. Comparing our best parallel implementation (Algorithm M6) with the theoretical bounds, we find that with four processors the actual speedup obtained is quite close to the theoretical speedup. For example, for $n = 47$ on the shallow arch problem, our speedup is 3.91 compared with a theoretical upper bound of 3.94; for the turning point problem ($n = 1000$) the corresponding numbers are 3.24 and 3.40. This explains why the overall speedup for the turning point problem is poor; for this problem, the fraction of time spent in serial execution is high. With eight processors, the actual speedup obtained on the turning point problem (3.88) is not close to the theoretical speedup (5.52) because algorithm M6 is a combination of M1 and M5, and M5 uses only four processors. This explains why the timings for algorithm M5 (as well as M3 and M4) are the same for eight processors and four processors, and why there is a significant gap between the theoretical and the actual speedup for eight processors. For the arch problem, the scenario is completely different (actual speedup 7.64 and theoretical bound 7.76) since compared with the turning point problem, a smaller proportion of the total time is spent in the linear solver code.

In another paper [3], we discuss in detail several other issues regarding the parallelization of the algorithms under consideration here. These issues include extra storage requirements, and the programming effort required for various levels of parallelization. Most of the conclusions drawn there and in the present paper reaffirm existing parallel computing theory and are very general. Regarding specific conclusions to be drawn for parallel HOMPACK, some levels of parallelization are simply not worth the time and effort required for the implementation, considering the speedup obtained. For example, Algorithm M2 (lower level linear algebra in parallel) was not worth the effort. It took 20 man-hours to obtain a maximum speedup of 1.65 using eight processors. Similarly, attempting to implement computations relating to the preconditioner in parallel is not worthwhile since we get no speedup at all. Regarding the other levels/degrees, the problem usually determines what level of parallelism is appropriate. For the turning point problem, the implementation of the function values and the Jacobian matrices was not worth the effort whereas the degrees/levels relating to the linear solver code did give a good speedup relative to the effort expended. For the arch problem it was exactly the opposite, although it could be debated that spending 100 hours to obtain a speedup of 6.1 with eight processors is not worth the effort. Regarding extra memory allocation for the parallel implementation, the parallel algorithms required only just a few extra $(n+1)$-vectors and hence memory is not an important issue for parallel HOMPACK. Hence a general purpose parallel HOMPACK, applicable to any problem, should implement the four linear solves in parallel. The parallelization of the function and Jacobian matrix evaluation subroutines will depend on the problem being solved.

Acknowledgement. The work of Irani, Ribbens and Watson was supported in part by Department of Energy grant DE-FG05-88ER25068.

References.

[1] E. L. Allgower and K. Georg, *Introduction to Numerical Continuation Methods*, Springer Verlag, Berlin, 1990.

[2] D. C. S. Allison, K. M. Irani, C. J. Ribbens and L. T. Watson, High dimensional homotopy curve tracking on a shared memory multiprocessor, *J. Supercomputing*, submitted.

[3] D.C.S. Allison, A. Chakraborty, and L. T. Watson, Granularity issues for solving polynomial systems via globally convergent algorithms on a Hypercube, *Proc. Third Conference on Hypercube Concurrent Computers and Applications*, Pasadena, CA (1988) 1463–1472.

[4] S.C. Billups, An augmented Jacobian matrix algorithm for tracking homotopy zero curves, M.S. Thesis, Dept. of Computer Sci., VPI & SU, Blacksburg, VA, 1985.

[5] A. Chakraborty, D. C. S. Allison, C. J. Ribbens, and L. T. Watson, Parallel orthogonal decompositions of rectangular matrices for curve tracking on a hypercube, *Proc. Fourth Conf. on Hypercube Concurrent Computers and Applications*, J. Gustafson (ed.), ACM, Monterey, CA, 1989.

[6] S. N. Chow, J. Mallet-Paret, and J. A. Yorke, Finding zeros of maps: Homotopy methods that are constructive with probability one, *Math. Comput.* **32** (1978) 887–899.

[7] E. J. Craig, *Iteration procedures for simultaneous equations*, Ph.D. thesis, MIT, Cambridge, 1954.

[8] M. R. Hestenes and E. Stiefel, Methods of conjugate gradients for solving linear systems, *J. Res. National Bureau of Standards* **49** (1952) 409–435.

[9] K. M. Irani, M. P. Kamat, C. J. Ribbens, H. F. Walker, and L. T. Watson, Experiments with conjugate gradient algorithms for homotopy curve tracking, *SIAM J. Optim.*, to appear.

[10] L.T. Watson, A globally convergent algorithm for computing fixed points of C^2 maps, *Appl. Math. Comput.* **5** (1979) 297–311.

[11] L.T. Watson, Numerical linear algebra aspects of globally convergent homotopy methods, *SIAM Rev.* **28** (1986) 529–545.

[12] L.T. Watson, S.C. Billups and A.P. Morgan, Algorithm 652: HOMPACK: A suite of codes for globally convergent homotopy algorithms, *ACM Trans. Math. Software* **13** (1987) 281–310.

Efficient Systolic Array for Matrix Multiplication

Fabian Klass †   *Uri Weiser* ‡

† Department of Electrical Engineering
Delft University of Technology
The Netherlands

‡ Intel - Israel

Abstract— Several approaches have been proposed as systolic solutions for the multiplication of matrices. They have used mesh-connected and hex-connected arrays. Solutions using mesh-connected arrays and fully dynamic computations have all incurred in contraflow schemes, with the consequence of a low processor utilization. In this paper, a previously reported contraflow network is transformed into a more efficient design. A two-stage transformation method is applied. In the first stage, unidirectional data flow is incorporated. In the second stage, data retiming is performed combined with further modifications in the clocking scheme. The resultant arrays provide an efficiency two times higher than the previous design.

I. INTRODUCTION

Matrix multiplication belongs to a special class of computation-bound algorithms where the total number of computations is greater than the total number of I/O operations. Systolic arrays, which combine parallelism and pipelining, are especially suitable to this class of algorithms, taking advantage of their regular and localized data flow. Basic "inner product" processing elements can be locally connected to perform matrix multiplication and other related operations. Based on different approaches for mapping algorithms into systolic arrays, several solutions have been proposed for the multiplication of matrices. They have included mesh-connected and hex-connected arrays (one-dimensional arrays are not considered here). In the case of band matrices, systolic solutions can be characterized as highly efficient [12]. However, in the case of square densely-populated matrices, or matrices with large bandwidth, the array utilization is significantly degraded. For instance, previous solutions using mesh-arrays have achieved an average efficiency of only 25%. Even those solutions have been produced by systematic design techniques, not always the techniques are tuned for maximal performance. In particular, mesh-connected arrays involving fully dynamic computations have all incurred in contraflow schemes. The term contraflow means data items moving in opposite directions along one or more dimensions of the array while computations are evaluated. As a direct consequence of this feature, the throughput rate of the system is usually slowed down, or equivalently, the processor utilization is degraded.

In this paper, a previously reported systolic array for matrix multiplication is transformed into a more efficient structure. To overcome performance degradation, the cause of the low performance, the data contraflow, is eliminated by performing a transformation which yields a unidirectional data flow design. By performing a time scaling, throughput rate and array utilization are significantly improved. To carry out such transformation, a two-stage method introduced in [4] is applied. The transformation produces a unidirectional data flow array with a processor utilization two times higher than the previous design. In terms of performance, not only is processor utilization

improved. Having unidirectional data flow has further advantages in terms of multilevel of pipelining and fault tolerance schemes [6]. Further applications of the two-stage method used here have been reported for band matrix multiplication [4], and a linear discriminant classifier [5]. An example of how an original architecture may be modified by incorporating unidirectional data flow can be found in [7], but no systematic tool for such transformation has been provided.

In the following, a formal presentation of the matrix multiplication problem is presented and previous solutions are discussed.

II. PREVIOUS SOLUTIONS

The problem of multiplying two matrices is considered. Suppose two rectangular matrices $A = (a_{ij})$ and $B = (b_{ij})$ whose dimensions are $N \times M$ and $M \times P$ respectively. The product matrix $C = (c_{ij})$ is a $N \times P$ matrix whose elements are computed as (See Fig. 1):

$$c_{ij} = \sum_{k=1}^{M} a_{ik} b_{kj} \quad , \quad 1 \leq i \leq N \text{ and } 1 \leq j \leq P \qquad (1)$$

$$\begin{bmatrix} a_{11} & a_{12} & a_{13} \cdots & a_{1M} \\ a_{21} & a_{22} & & \vdots \\ a_{31} & & & \vdots \\ \vdots & & & \\ a_{N1} & \cdots\cdots & & a_{NM} \end{bmatrix} \times \begin{bmatrix} b_{11} & b_{12} & b_{13} \cdots b_{1P} \\ b_{21} & b_{22} & & \\ b_{31} & & & \vdots \\ \vdots & & & \\ b_{M1} & \cdots\cdots & & b_{MP} \end{bmatrix} = \begin{bmatrix} c_{11} & c_{12} & c_{13} \cdots c_{1P} \\ c_{21} & c_{22} & & \vdots \\ c_{31} & & & \vdots \\ \vdots & & & \\ c_{N1} & \cdots\cdots & & c_{NP} \end{bmatrix}$$

Fig. 1. Multiplication of two matrices.

To implement the algorithm described by (1), a 100% efficient array which holds the elements of one input matrix static on the array has been proposed in [11]. Updating matrix coefficients in this type of design is difficult, extra I/O time and extra logic is needed. A fully dynamic solution using a mesh-connected array has been presented in [3], based on the approach given in [8]. A general scheme of that solution is shown in Fig. 2 for the case when $N = M = P = 3$. Data elements of matrices A and B are entered into the array from the top and bottom side respectively, flowing through the network in opposite directions, whereas elements of resulting matrix C move to the left. In terms of array efficiency, only one of two consecutive processors on the mid output path is active at any given time. This utility decreases linearly toward the input boundaries of the array. When computing large dimension matrices, an average efficiency of only 25% is achieved.

Based on another approach for mapping algorithms into systolic arrays, 'all' systolic solutions have been produced for the matrix multiplication problem [2]. Excluding those with static data allocation, all solutions using mesh-connected arrays have incurred in contraflow schemes, with a consequent average efficiency of only 25%. By incorporating unidirectional data flow, the number of possible systolic solutions for matrix multiplication is increased. Furthermore, a better array utilization can be achieved.

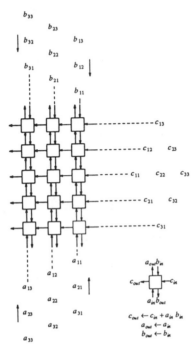

Fig. 2. Contraflow array for matrix multiplication.

III. Unidirectional Approach

The unidirectional approach proposed in this paper achieves an array efficiency two times higher than a previously reported design. To get this solution, a two-stage transformation method is applied. A more detailed discussion of this method is reported in [4]. The two stages are described below.

1) *Flow Inversion:* The contraflow network is converted into an analogous network which involves only unidirectional data streams. The resultant design performs the same computations as the original version, at the same throughput, except that all data elements flow in the same direction.

2) *Time Scaling:* A retiming of input data is performed on the design obtained in the previous step. This is combined with further modifications in the array clock scheme. As a result, throughput is increased.

The first stage, the flow inversion, is a delay transfer based methodology. Not only are delays transferred through the network, but also 'predictors'. This allows in turn the reversion of a data path in a contraflow network, as will be discussed later. Conditions which allow data flow inversion in a contraflow network are the associativity of the computations and the data-dependence exclusivity. The second condition means that no mutual data-dependence between contraflow data is allowed.

A. Wavefront Notation.

To represent more precisely a systolic network, a wavefront notation is adopted [12],[13]. A wavefront is an ordered set of data whose elements move uniformly in time or in space. Wavefront X is denoted as $\overline{X} = \{x(1, \underline{m}), x(2, \underline{m}), ..., x(N-1, \underline{m}), x(N, \underline{m})\}$. Time indexes are underlined, spatial indexes are not underlined.

Using this notation, a network is viewed as a particular transformation of sets of data - wavefronts - to produce a new result. These transformations are defined using a special operator D, called the delay operator [1]. To define D formally, suppose a sequence $a(\underline{1})$, $a(\underline{2})$, ..., $a(\underline{n})$ is given, where $a(\underline{i})$ precedes the arrival of $a(\underline{i+1})$ by n time steps, for every i such that $1 \le i \le n-1$. The delay operator is defined as $D^n[a(\underline{i})] = a(\underline{i-1})$ A dual operator is the predictor operator defined as $D^{-n}[a(\underline{i})] = a(\underline{i+1})$. The delay operator D^n, applied to a data element, gives the data element that was at the same point in the network n time steps before, whereas the predictor operator D^{-n} gives the data element that will be n time steps later. The delay operator can be implemented as a latch.

A complete set of transformation rules to formally represent networks is found in [13]. Assuming a wavefront $\overline{X}$ defined as $\overline{X} = \{x(1), x(2), ..., x(N-1), x(N)\}$ (time scripts are omitted), a subset of such transformations is given below.

1) Delayed Wavefront:

$$D[\overline{X}] = \{D[x(1)], D[x(2)], ..., D[x(N-1)], D[x(N)]\}$$

2) Positive Rotation:

$$R_+[\overline{X}] = \{x(1), D[x(2)], ..., D^{N-2}[x(N-1)], D^{N-1}[x(N)]\}$$

3) Negative Rotation:

$$R_-[\overline{X}] = \{D^{N-1}[x(1)], D^{N-2}[x(2)], ..., D[x(N-1)], x(N)\}$$

Properties concerning the composition of the above operators can be found in [4]. In particular, composition of functions R_+ and R_- results in a delay function:

$$R_+[R_-[\overline{X}]] = D^{N-1}[\overline{X}] \qquad (2)$$

B. Matrix multiplication using wavefront tools.

Using wavefront notation, components of the product matrix C are obtained as a particular transformation between wavefront $\overline{A}$, defined as a row vector in matrix A, and wavefront $\overline{B}$, defined as a column vector in matrix B. A function KM is used for merging the two wavefronts to produce a single result [12]. Any element in the main diagonal of matrix C is formulated as

$$c(\underline{m}, \underline{m}) = KM\{\overline{A}(\underline{m}), \overline{B}(\underline{m})\} \qquad (3)$$

where wavefronts $\overline{A}$ and $\overline{B}$ are defined as

$$\overline{A}(\underline{m}) = \{a(\underline{m}, N), a(\underline{m}, N-1), \cdots, a(\underline{m}, 2), a(\underline{m}, 1)\}$$

$$\overline{B}(\underline{m}) = \{b(N, \underline{m}), b(N-1, \underline{m}), \cdots, b(2, \underline{m}), b(1, \underline{m})\}$$

and KM is the inner product step function defined as

$$KM(a_{in}, b_{in}, c_{in}) = c_{in} + a_{in}\, b_{in}$$

where a_{in} belongs to $\bar{A}$, b_{in} belongs to $\bar{B}$, and c_{in} is a partial result from the previous computation.

An element not in the main diagonal of C, for instance $c(\underline{m}, \underline{m}+1)$, is formulated as

$$c(\underline{m}, \underline{m}+1) = KM\{\bar{A}(\underline{m}), \bar{B}(\underline{m}+1)\} \qquad (4)$$

Fig. 3 shows the network of Fig. 2 using wavefront notation (for $N = 3$). Each processing element, denoted with a circle, performs its computations with zero delay. Edges labeled with a capital letter D (or D^2, D^{-1}, etc.) represent the delay/predictor operator.

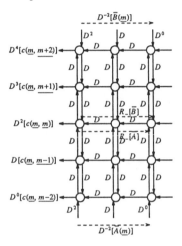

Fig. 3. Matrix multiplication using wavefront notation.

Progressions of input wavefronts on the $c(\underline{m}, \underline{m})$-path are $R_-[\bar{A}(\underline{m})]$ and $R_-[\bar{B}(\underline{m})]$. Thus, element $c(\underline{m}, \underline{m})$ is given by

$$D^2[c(\underline{m}, \underline{m})] = R_+[KM\{R_-[\bar{A}(\underline{m})], R_-[\bar{B}(\underline{m})]\}]$$

The above equation can be transformed into (3) if operator R_- is extracted out of the KM argument and (2) is applied.

A relevant feature of the design results from evaluating element $c(\underline{m}, \underline{m}+1)$. This term is obtained by interacting a delayed wavefront $R_-[\bar{A}(\underline{m})]$ and a predicted wavefront $R_-[\bar{B}(\underline{m})]$ (see Fig. 3). After a single mathematical transformation, element $c(\underline{m}, \underline{m}+1)$ is given by

$$c(\underline{m}, \underline{m}+1) = KM\{\bar{A}(\underline{m}), D^{-2}[\bar{B}(\underline{m})]\} \qquad (5)$$

Comparing (5) and (4), it follows that $D^{-2}[\bar{B}(\underline{m})] = \bar{B}(\underline{m}+1)$. This means that successive data items belonging to B have to be separated by two time steps, according to the predictor definition (the same holds for matrix A). This results in a throughput rate of one output every two cycles and a low average efficiency of 25% for the whole array.

C. Flow inversion.

The contraflow network shown in Fig. 3 is transformed into a more efficient array. In this step, unidirectional data flow is incorporated into the design. In turn, this allows achieving a better array utilization by performing a time scaling of the circuit. Data flow inversion is carried out in two substeps:

1) *Rotation:* Every contraflow path in the array is transformed in order to generate prediction elements in one of the paths.

2) *Path reversion:* Edges labeled with predictors are reverted and prediction elements are replaced by delays.

To implement the above transformations, a graphical representation is preferable to avoid the emerging complexity of a mathematical approach. Since the network is symmetric, reverting the a- or the b-path is equivalent. Reversion of the b-path is considered. In the first substep, every input/output in a node is embedded with an equal number of delay/prediction elements, preserving the equivalence of the design in such a way that reducing operators in the node yields a prediction element D^{-1} on the b-path edge. For symmetry, the operation begins at the central node and is propagated to the ends. The result after completion of this operation is shown in Fig. 4(a).

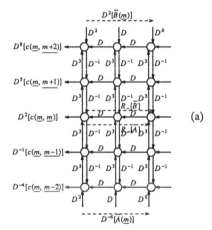

(a)

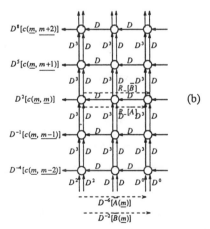

(b)

Fig. 4. Flow inversion: (a) generation of prediction elements, (b) path reversion.

In the second substep, edges labeled with prediction elements are reverted, whereas prediction elements are changed to delays. Input to the reverse path is modified according to the rule given by $D^{-i}[a(\underline{m})] = D^{N-i}[D^{-N}[a(\underline{m})]]$ [13]. Fig. 4(b) shows the network after path reversion.

D. Time Scaling.

The reversion of the data flow in the circuit does not modify its throughput. The unidirectional network of Fig. 4(b) remains as slow as the contraflow network of Fig. 3. A time scaling is required to improve the throughput of the array. This can be done by scaling up operator D in the b-path or scaling down operator D^3 in the a-path. Scaling down is adopted to reduce the number of delays and latency in the final stage of the design. If a variable j is introduced as a superscript of D^3 in the a-path (D^j instead of D^3) in the network of Fig. 4(b), element $c(\underline{m}, m+1)$ is given by

$$c(\underline{m}, \, \underline{m+1}) = KM\{\overline{A}(\underline{m}), \, D^{1-j}[\overline{B}(\underline{m})]\}$$

According to (4), the relationship $D^{1-j}[\overline{B}(\underline{m})] = \overline{B}(m+1)$ holds. By tuning input rates such that data items are preceded by $|i-j|$ time steps, different throughput rates can be achieved. In particular, a maximum rate is obtained when $1-j = -1$, resulting $j = 2$. Therefore, operator D^3 has to be scaled down to D^2 in the a-path as shown in Fig. 5.

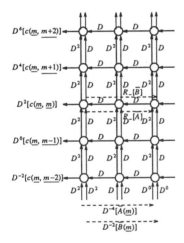

Fig. 5. Time scaling, $D^3 \rightarrow D^2$.

The resulting design is depicted in Fig. 6 using a more standard graphical convention. Elements of matrix A and B are both entered from the bottom side of the array. They move up through the network in the same direction, whereas elements of resultant matrix C move to the left. Comparing with the contraflow array of Fig. 2, data elements are not only differently arranged, but also input/output rate is increased. When many problem instances are processed by the same array, the average efficiency of this design is 50%, two times higher than the contraflow scheme.

IV. PERFORMANCE COMPARISON

There are many factors to take into account when evaluating the performance of a systolic array. Usually, the choice of the optimal design depends on the application. For matrix multipli-

cation, several architectures have been proposed. In this section, three designs are compared and evaluated.

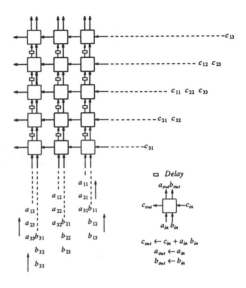

Fig. 6. Unidirectional flow array for matrix multiplication.

Fig. 7 shows schematically the three networks in the case that two successive problem instances are processed. They are: (a) a mesh-connected array using contraflow (see Fig. 2), (b) a mesh-connected array using unidirectional flow (see Fig. 6), and (c) a hex-connected array using unidirectional flow [2]. The following performance parameters are considered:

- *Computation Time* (C): time interval between the first and last computation of the problem instance.

- *Block Pipeline Time* (P): time interval between the initiation of two successive problem instances.

- *Throughput* (T): reciprocal of the time interval between the computation of two successive output results.

- *Array Size* (S): number of processing elements in the array.

- *Array Utilization* (U): average relation between the time processors are busy and the total processing time.

The above parameters are listed in Table I for each systolic solution shown in Fig. 7, assuming for simplicity $N \times N$ input matrices. Comparing mesh-arrays only, the unidirectional scheme is twice as efficient as the contraflow array, if many problem instances are pipelined and processed. However, the computation time is faster in the contraflow design. The unidirectional scheme offers the best performance when large matrix problems are partitioned into smaller subproblems [9],[10]. In this case, the new problem instance exhibits indefinite input data and a solution achieving maximum throughput is desired. The unidirectional data flow scheme achieves maximum throughput by data retiming. Comparing unidirectional data flow structures, mesh and hex-array are equivalent in terms of throughput and block pipeline time, although the mesh-array saves about half the number of processing elements which represents a better hardware cost effectiveness. By replacing processing elements which do not participate in the computations by delays (hatched areas in Fig.

7(c)), the efficiency of the hex-array can be increased up to 33%, still less than the 50% achieved by the mesh-array. The computation time is faster for the hex-array and equal to the time of the contraflow network.

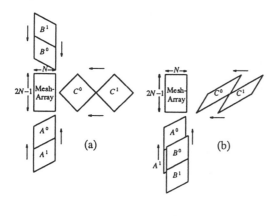

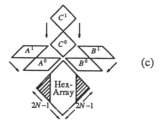

Fig. 7. Different approaches for matrix multiplication: (a) Mesh-array with contraflow, (b) Mesh-array with uni-directional flow (c) Hex-array with unidirectional flow

Table I. Comparison between different systolic solutions for matrix multiplication.

Network	C	P	T	S	U
Mesh-array with contraflow	$3N-1$	$2N$	$1/2$	$(2N-1)N$	$\dfrac{N}{2(2N-1)}$
Mesh-array with unidirectional flow	$4N-2$	N	1	$(2N-1)N$	$\dfrac{N}{(2N-1)}$
Hex-array with unidirectional flow	$3N-1$	N	1	$(2N-1)^2$	$\dfrac{N^2}{(2N-1)^2}$

V. CONCLUDING REMARKS

An efficient systolic array is proposed for matrix multiplication. To get this solution, a two-stage transformation method is applied to a given contraflow array. In the first stage, data flow inversion is performed to get a unidirectional flow array. In the second stage, time scaling is carried out in combination with data retiming. The resulting design offers computational throughput and efficiency two times higher than the previous design, although the number of delays and the computation time are slightly increased. Having unidirectional data flow has further advantages in terms of multilevel pipelining and fault tolerance schemes.

Comparing this approach with the classical ways of improving processor utilization, such as pipeline interleaving or processor sharing, this solution offers advantages. To get a higher throughput, there is no need of interleaving independent data or problems of different dimensions. To get a better processor utilization, there is no need for incorporating additional hardware, as in the case of processor sharing, to handle the new resource scheduling.

REFERENCES

[1] D. Cohen, "Mathematical Approach to Iterative Computation Networks," *Proceedings of the Fourth Symposium on Computer Arithmetic*, pp. 226-238, October, 1978.

[2] N. Faroughi and M. A. Shanblatt, "Systematic Generation and Enumeration of Systolic Arrays from Algorithms", *Proceedings of the 1987 International Conference on Parallel Processing*, University Park, P.A., USA, 17-21, August 1987, pp. 844-7.

[3] S. Horiike, S. Nishida, T. Sakaguchi, "A Design Method of Systolic Arrays under the Constraint of the Number of the Processors", *Proc. ICASSP April 1987, pp. 764-7, Vol. 2.*

[4] F. Klass, "Efficient Transformations for Systolic Arrays," M.Sc.Thesis, Department of Electrical Engineering, Technion, Israel, Oct. 1989.

[5] F. Klass, "Unidirectional-flow Systolic Array for Linear Discriminant Function Classifier," *Electronic Letters*, Vol. 26, No. 20, pp. 1702-1703, Sep. 1990.

[6] H. T. Kung and M. Lam, "Fault Tolerance and Two-level of Pipelining in VLSI Systolic Arrays," *Journal of Parallel and Distributed Computing 1*, pp. 32-63, 1984.

[7] J. V. McCanny, R. A. Evans, and J. G. McWhirter, "Use of Unidirectional Dataflow in Bit Level Systolic Arrays Chips," *Electron. Letts.*, Vol. 22, No. 10, pp. 540-541, 1986.

[8] D. I. Moldovan: "On the Design of Algorithms for VLSI Systolic Arrays", Proc. IEEE, 71, 1, pp. 113-120, 1983.

[9] D. I. Moldovan and J. A. B. Fortes, "Partitioning and Mapping Algorithms Into Fixed-Size Systolic Arrays", *IEEE Tral, Jan. 1986, pp.1-12.*

[10] J. J. Navarro, J. M. Llaberia, and M. Valero, "Partitioning: An Essential Step in Mapping Algorithms into Systolic Arrays Processors", *Computer*, Vol. 20, No. 7, July 1987, pp. 77-89.

[11] R. B. Urquhart and D. Wood, "Systolic Matrix and Vector Multiplication Methods for Signal Processing", *IEEE Proc. F, Commun., Radar & Signal Process.*, 1984, 131, pp. 623-631.

[12] U. Weiser and A. Davis "A Wavefront Notation for VLSI Array Design," in *VLSI Systems and Computations*, Computer Science Department, Carnegie-Mellon University, Computer Science Press, Rockville, Md., Oct. 1981, pp. 226-234.

[13] U. Weiser, "Mathematical and Graphical Tools for the Creation of Computational Arrays," Ph.D. Thesis, Department of Computer Science, University of Utah, July 1981.

Selection on the Reconfigurable Mesh

Hossam ElGindy   Paulina Węgrowicz

School of Computer Science
McGill University

Abstract

This paper presents a new parallel algorithm for finding the kth smallest element in a set of n or fewer items, which runs in $O(\log^2 n)$ time on the reconfigurable mesh architecture with n processors.

Keywords: selection, splitter, parallel algorithm, reconfigurable mesh.

1 Introduction

The desire to further reduce the running time for solving problems, beyond what can be achieved by sequential algorithms running on single processor architectures, has generated great interest in parallel algorithms which can exploit the advantages of multi-processor architectures. The difficulty in developing parallel algorithms is to efficiently perform computation in parallel while successfully avoiding the problem of resource contention. Two general models of parallel computation are usually considered when designing parallel algorithms, one in which the processors share a common memory and one in which the memory is distributed among the processors [Prep79].

The concurrent read exclusive write parallel random access machine (CREW PRAM) is an example of the shared memory model. All processors can simultaneously access the memory as long as no two attempt to write to the same memory location simultaneously. This provides for an essentially unconstrained exchange of data between processors. In contrast, the reconfigurable mesh is an example of the distributed memory model. The processors are interconnected in a network, with each processor having a relatively small (constant size) local memory. Not all processors can communicate directly thus constraining the flow of data and increasing the running time of many algorithms. It should be noted that the reconfigurable mesh is a practical example that is well suited to current technology, whereas the CREW PRAM is an idealized example which remains technologically infeasible to implement [Prep79].

This paper proposes a new parallel algorithm which finds the kth smallest element of a totally ordered set S of n elements, where the order is not known, in $O(\log^2 n)$ time. When $k = \lceil \frac{n}{2} \rceil$, the kth element is called the median. This problem, known as *selection*, has received considerable attention and a number of parallel solutions are available for the mesh and the CREW PRAM [Thom77, Stou83, Pras87, Plax89, Cole88, Węgr91]. It is also shown that a splitter can be obtained on the reconfigurable mesh after only one iteration of the selection algorithm, that is in $O(\log n)$ time.

One way of finding the kth smallest element is to compute the ranks of all elements and pick the kth element. This can be accomplished in $\Theta(n^{1/2})$ time on a mesh with no broadcasting buses with a sorting algorithm such as the Odd-Even Merge Sort described in [Thom77]. When only one processor is available, sorting requires $\Theta(n \log n)$ time, but the selection problem can be solved in $\Theta(n)$ time with an algorithm by Blum et al [Blum72]. An algorithm, based on [Blum72], exists for the selection problem which runs in $\Theta\left((n \log n)^{1/3}\right)$ time on a mesh with a single broadcast bus

[Stou83]. This is reduced even further on a mesh with multiple broadcasting, where an algorithm is available which runs in $O\left(n^{1/6}(\log n)^{2/3}\right)$ time [Pras87]. For the reconfigurable mesh, an algorithm can be implemented, based on the method in [Blum72], which runs in $O(n^\epsilon)$ time, where ϵ is a chosen constant, such that $0 < \epsilon \le \frac{1}{2}$. The running time of this algorithm is bounded below by the number of recursive calls made throughout its execution. Since two consecutive recursive calls are required at each level of recursion, this algorithm would not result in a polylogarithmic running time. Węgrowicz [Węgr91] gave the first algorithm to run in polylogarithmic time on an architecture based on a mesh of processors. The algorithm is based on a sequential technique due to Munro and Patterson [Munr80] and runs in $O(\log^3 n)$ time on the reconfigurable mesh of size n. An algorithm which achieves optimal efficiency [Akl85] was designed by Cole [Cole88] for an exclusive read exclusive write PRAM model. The algorithm runs in $O(n/p + \log p \log^* p)$ time, and achieves optimal efficiency for $n = \Omega(p \log p \log^* p)$, where p is the number of processors.

A recent result due to Plaxton [Plax89] gives a lower bound on the selection problem of $\Omega((n/p) \log \log p + \log p)$, where n is the number of elements to select from and p is the number of processors in the network. The result applies to a number of common network models including the reconfigurable mesh and the CREW PRAM.

This paper is organized into three sections following this introduction. Section 2 presents the reconfigurable mesh architecture. Section 3 presents the parallel selection algorithm on the reconfigurable mesh. Section 4 concludes this paper.

2   The reconfigurable mesh architecture

The reconfigurable mesh architecture [Mill88a], also called the mesh with a reconfigurable bus, combines the advantages of the mesh with the power and flexibility of a dy-
namically reconfigurable bus structure. A reconfigurable mesh of size N consists of a two-dimensional array of N processors arranged on a $N^{1/2} \times N^{1/2}$ rectilinear grid, overlaid by a reconfigurable broadcast bus of the same shape (Figure 1). Each processor has a fixed number of registers of $O(\log N)$ size on which it can perform standard arithmetic and logic operations. Each processor $P_{i,j}$ has stored in its registers its row and column indexes i and j, with $i, j \in \left[0, \ldots, N^{1/2} - 1\right]$, where for simplicity it is assumed that $N = 4^k$, for some positive integer k. Each processor is connected by local links to its neighbours. $P_{i,j}$ is connected to $P_{i\pm1,j\pm1}$, if they exist, with $i, j \in \left[0, \ldots, N^{1/2} - 1\right]$, and can send and receive data through these links.

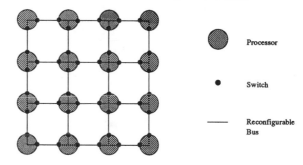

Figure 1: The reconfigurable mesh architecture.

In addition to being indexed by row and column numbers, processors can also be indexed by a chosen ordering scheme which represents a one-to-one mapping from $\{0, 1, \ldots, N^{1/2} - 1\} \times \{0, 1, \ldots, N^{1/2} - 1\}$ onto $\{0, 1, \ldots, N - 1\}$ [Thom77, Mill88c]. Some common ordering schemes are illustrated in Figure 2. The *row-major* ordering is obtained by numbering processors in each row left to right beginning with row 0 and ending with row $N^{1/2} - 1$. This is equivalent to the mapping $k = j + iN^{1/2}$, where i is the row number and j is the column number of a given processor. The *shuffled row-major* ordering is obtained by shuffling the binary representation of the row-major index, that is "abcdefgh" becomes "aebfcgdh". This ordering has the property that the first $N/4$ processors form the first quadrant, the second $N/4$ processors form the second quadrant and so on, with this property holding recursively in each quadrant. The *snake-like* ordering is a variation of the row major ordering obtained by reversing the ordering in the odd rows. This gives the property that processors with consecutive indices are adjacent on the mesh. The *proximity* ordering combines the properties of the shuffled row-major and the snake-like orderings. The proximity index of a processor can be computed in $O(\log n)$ time by that processor based on its row and column indices.

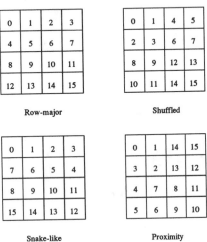

Figure 2: Indexing schemes for the processors of a mesh [Mil88c].

The proximity ordering will be used throughout this paper. Each processor can easily compute its proximity ordering index from its row and column indices and vice versa, but it is convenient to have both stored and available. Therefore, each processor of the mesh will contain a register initialized to represent its index in proximity ordering.

Each processor is also connected to the broadcast bus through four locally controllable switches (three for boundary and two for corner processors), as shown in Figure 1. Each processor can dynamically set its switches. By controlling the switches, the bus can be subdivided into independent connected components called subbuses. All processors connected to a subbus, or the whole bus, can simultaneously read a data value from it, but only one processor can write to a subbus at a time [Mill88a]. This is more restrictive than the model in [Mill88c] but is consistent with [Mill88a] where it is also shown that this model of

the reconfigurable mesh with exclusive write can simulate, without loss in time, a bus system where multiple identical values may be broadcast simultaneously on the bus. This is accomplished through a bus-splitting technique, which is given in [Mill88a].

The processors of the reconfigurable mesh operate synchronously in single instruction multiple data (SIMD) mode. That is, at each time unit all processors perform the same instruction, but each takes as operands the particular data stored in its registers. Each PE can perform a number of different primitive operations in unit time:

- carry-out arithmetic and logic operations on the contents of its registers,
- send or receive data from its neighbours through local communication links,
- set any of its four switches,
- send or receive data from the bus.

It is assumed, as in [Mill88a], that under the unit time delay model the data put on the reconfigurable bus reaches all processors in constant time.

The distinguishing characteristic of the reconfigurable mesh is the ability to dynamically obtain substructures consisting of groups of processors connected to an independent subbus. Each such substructure can function independently and has the same characteristics as the reconfigurable mesh (except possibly for its size and shape). For example, all the switches can be connected so that one global bus exists with all processors connected to it. Then any processor can broadcast a value to all others in one step. By connecting all column switches and disconnecting all row switches another configuration may be obtained with $O(N^{1/2})$ column buses. Such buses can be used similarly to static column buses, but have the advantage in that they can be subdivided (also recursively) to give for example $N^{1/2}$, $N^{1/4} \times N^{1/4}$ size meshes with column buses. This, of course, cannot be done with the mesh with row and column buses architecture as only a fixed $N^{1/2}$ buses exist there. Other dynamic configurations can be obtained, a few of which are illustrated in Figure 3.

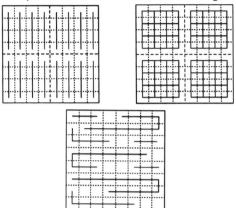

Figure 3: A variety of bus configurations.

By subdividing the bus, a large number of subbuses of some intermediate length or diameter can be created. For example, in creating $N^{1/2}$ groups of $N^{1/4} \times N^{1/4}$ processors with column buses, as in Figure 3, $N^{3/4} = N^{1/2} \times N^{1/4}$ column buses were created, each of length $N^{1/4}$. Since each bus can broadcast one piece of data in unit time, as many as $N^{3/4}$ values can move simultaneously over a distance of $N^{1/4}$ each. In general, as much data can be moved as there exist distinct subbuses, but the more subbuses that exist, the shorter they are.

The above observation was essential to developing the selection algorithm. It also leads to an understanding of the limitations of the reconfigurable mesh architecture in solving problems which require extensive data movement, such as sorting which still requires $\Omega(n^{1/2})$ time [Węgr91]. Thus, for some problems, especially those which require extensive data movement, the same asymptotic running time is required on the reconfigurable mesh as on a mesh with no buses. The selection algorithm will demonstrate, that better worst-case running times may be achieved for "easier" problems on the reconfigurable mesh.

2.1 Data movement operations on the reconfigurable mesh

Abstract data movement operations are commonly used by many parallel algorithms and will be used extensively in the parallel selection algorithm. In [Mill88a], an algo-

rithm is presented, which computes the maximum of a set of n or fewer values on a reconfigurable mesh of size n in $O(\log \log n)$ time. The algorithm is based on a technique called *bus-splitting*, which can be used to compute the maximum of $n^{1/2}$ or fewer values, using a reconfigurable mesh of size n in constant time. Another important and useful operation which will be employed in the selection algorithm is parallel prefix. It can be used to sum values, broadcast data or count and number active processors. Miller et al [Mill88a] describe in detail how to compute this operation on the reconfigurable mesh with processors in row-major ordering in optimal $O(\log n)$ time. The random access read (RAR) and random access write (RAW) operations, also used in the parallel selection algorithm are described in [Mill88a].

3 The selection algorithm

A sequential method due to Munro and Paterson [Munr80] will be shown to lead to a parallel selection algorithm for the reconfigurable mesh, with $O(\log^2 n)$ running time. Their algorithm, designed to select from a file stored on a read-only tape with limited amount of internal storage Q available for computation, runs in $O\left(n\left(\frac{\log n}{\log Q} + 1\right)\right)$ time. Two values a_u and a_v are chosen from $S = \{a_i\}$ to form a *filter* — an interval which is known to contain the kth element. Initially a_u and a_v are the minimum and maximum elements of S respectively. On each pass, elements within this interval are used to form a *sample* from which new values for a "narrower" filter, containing fewer elements, will be chosen. A sample is constructed recursively from a *population*—the remaining active values (those within the filters). For a fixed s, an s-sample at level i is a sorted set of s elements chosen from a population of $s2^i$ elements. At level 0 (the bottom) it is just the whole population ($s2^0$ elements) in sorted order. An s-sample at level $i + 1$ is formed by taking two samples at level i, each from half of the population $s2^{i+1}$. The level i samples are thinned by removing every second element of each sample with the remaining elements merged to form the $i + 1$ level sample.

At any iteration of the algorithm, a sample at level r is taken with the relationship $n' \leq s2^r$ so that all remaining n' elements are in the population. From this sample a new filter is chosen with a_u being the $\lceil \frac{k}{2^r} \rceil - r$ smallest element in the sample and a_v being the $\lceil \frac{k}{2^r} \rceil$ smallest one. It is shown in [Munr80] that at most $(2r-1)2^r$ candidates remain between the new filters.

3.1 The selection algorithm on the reconfigurable mesh

Given a set S of n or fewer elements, the objective is to find the kth smallest one. The elements are distributed in no particular order, one per processor of a reconfigurable mesh of size n. Call the processors holding an element "active" and the empty ones "inactive". As elements are eliminated as possible candidates for the kth element, more processors become inactive.

The algorithm will proceed, executing the following steps, until fewer than $\log^2 n$ candidates remain and the problem can be solved through sorting:

Procedure Select(n, k, S)

Step 1 Number the active processors by performing parallel prefix operation as addition with active processors holding a 1 and inactive ones a 0. Let m be the number of active processors, that is the number of elements remaining in S. If $m \bmod (4\log n)^2 \neq 0$ compute $\max\{S\}$ and place its value as data in sufficient number of processors, at most $(4\log n)^2$, to ensure that $m \bmod (4\log n)^2 \equiv 0$.

Step 2 If m is less than $\log^2 n$, arrange the remaining active values one per row in the first column of the mesh, then sort all values, pick the kth one and exit.

Step 3 Compute the sample by calling recursive procedure Sample($4\log n$, n, S).

Step 4 Choose the new filter values and broadcast them to all processors. Perform parallel prefix again with values less than a_u holding a 1 to compute l, the number of elements smaller than a_u which will no longer be active. Mark as inactive all elements outside the new filters. Perform parallel prefix again to compute n', the number of remaining active elements comprising S'.

Step 5 Call Select(n', $k - l$, S').

End Select

Let A, the data in the active processors, be indexed $a_1, \ldots, a_t$, as computed by the parallel prefix in step 1. Let s be the number of elements in the sample and t the number of elements in the population. Then, step 3 is performed by the following procedure for computing samples. The procedure efficiently sorts and merges samples by consolidating active elements in $s \times s$ size groups of processors.

Procedure Sample(s, t, A)

Step 1 Recursively subdivide the mesh according to the following procedure:

> **Procedure** RSubdivide(s,t,A)
>
> If $t > s^2$, alternating dimensions, divide the mesh in half to give $A_1 = \{a_1, \ldots, a_{t_1}\}$ and $A_2 = \{a_1, \ldots, a_{t_2}\}$, with $|A_1| = t_1$, $|A_2| = t_2$ and $t_1 + t_2 = t$. If $t_1 \neq 0$ call RSubdivide(s, t_1, A_1) and if $t_2 \neq 0$ call RSubdivide(s, t_2, A_2), in parallel.

Step 2 Let $A = \{a_1, \ldots, a_t\}$ be divided into groups of s^2 consecutive elements, in proximity ordering. Let group leaders be elements indexed i such that $i \bmod s^2 \equiv 1$.

For all odd groups in parallel, move all elements in a group to the submesh in which the group leader resides, by applying RAR and RAW operations, using subbuses formed along the paths specified by proximity ordering between processors in each pair of groups. Repeat this for all even groups.

At this point the mesh contains m/s^2 distinct groups of $s \times s$ size with exactly s^2 active elements in each group.

Step 3 For all groups in parallel, sort the elements in each group in ascending order, using Odd-Even Merge Sort [Thom77] and pick s evenly spaced elements to form a sample of size s.

While group size is less than $n^{1/2} \times n^{1/2}$, for all pairs of groups in parallel, thin samples by discarding every second element and merge each odd numbered sample with the following even numbered sample.

End Sample

Procedure RSubdivide, which constitutes the first step of computing a sample, decomposes the mesh, in $O(\log n)$ time, into groups of at least $s \times s$ processors with at most s^2 active elements in each group. The active elements are consolidated into some (possibly all) of these groups of processors so that each group contains exactly s^2 or 0 values. Given any two consecutive groups of processors of size $p \times p$ and $p' \times p'$, resulting from applying procedure RSubdivide, the properties of proximity ordering guarantee that $\min\{p, p'\}$ subbusses can be formed between the two groups. Furthermore, distinct subbusses can be formed for all pairs of groups, at the same time. These paths allow to move $O(s^2)$ values in $O(s)$ time, as illustrated in Figure 4.

The s-sample obtained from a group of s^2 elements by sorting and choosing s evenly spaced values, satisfies the

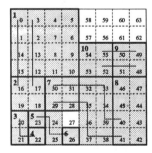 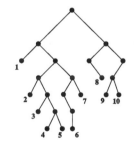

Figure 4: Groups of s^2 elements consolidated into submeshes.

properties of an $i = \log \log \frac{n}{s}$ level sample in [Munr80]. The sorting takes $O\left((s^2)^{1/2}\right) = O(\log n)$ time [Thom77]. The sample size $s = 4 \log n$ was chosen as to sort groups of data that are as small as possible, yet be able to show that

the size of the problem is diminishing after each iteration. Then the samples are merged in $O(\log n)$ stages. During each stage, for all pairs of samples, the samples are thinned by removing every second element, then the elements of one sample in the pair are ranked against the elements of the other sample and vice versa, which allows to merge the samples in $O(1)$ time. Although the elements of a sample which have been transported on subbuses may arrive in descending order, it is easy to verify, that in $O(1)$ time the order can be reversed by broadcasting using row and column subbuses.

The total time to compute the sample at level r, $r \geq \log n/s$, can be expressed as

$$\begin{cases} t(r) = t(r-1) + O(1) \\ t(0) = O(s) \end{cases}$$

which gives $t(r) = O(s) = O(\log n)$.

Going back to the algorithm Select, step 1 requires $O(\log n)$ time and step 2, which is executed once only, requires $O(\log n)$ time. For step 4, consider the jth largest element in a sample at level i. Let L_{ij} and M_{ij} respectively be the least number and most number of elements, from the corresponding population, which can be greater than the jth largest element in the sample. Lemma 2 in [Munr80] states that

$$L_{ij} = j2^i - 1 \quad \text{and} \quad M_{ij} = (i + j - 1)2^i.$$

In choosing the new filter, it must be ensured that the kth element is one of the filter values or lies between them, that is

$$k - 1 \geq M_{ru} = (r + u - 1)2^r$$

and

$$k - 1 \leq L_{rv} = v2^r - 1$$

The choice for the new filter will therefore be $u = \lceil \frac{k}{2^r} \rceil - r$ and $v = \lceil \frac{k}{2^r} \rceil$. Broadcasting these two values over the whole mesh and comparing with the data in the active processors, will allow the elimination of all values lying outside the new filters as candidates for the kth element, thus completing step 4 in constant time.

The remaining number of elements is at most

$$
\begin{aligned}
M_{rv} - L_{ru} - 1 &= (2r-1)2^r \\
&= \left(2\log\frac{n}{s} - 1\right)\frac{n}{s} \\
&= [2\left(\log n - \log(4\log n)\right) - 1]\frac{n}{4\log n} \\
&= \frac{n}{2} + \frac{n\log\log n}{2\log n} - \frac{5n}{4\log n} \\
&= n\left(\frac{1}{2} + \frac{\log\log n}{2\log n} - \frac{5}{4\log n}\right)
\end{aligned}
$$

and since $\frac{1}{4} \geq \frac{2\log\log n - 5}{4\log n}$, or equivalently $\log n \geq 2\log\log n - 5$ for all n, at most $\frac{3}{4}n$ elements remain.

The running time of algorithm Select is

$$
\begin{cases}
t(n) = t(\frac{3}{4}n) + O(\log n) \\
t((4\log n)^2) = O(\log n)
\end{cases}
$$

which gives $t(n) = (O\log^2 n)$.

Note that no stack is required. To keep track of the recursion only two registers per processor are needed. One register stores the current level of recursion, which is the same for all processors and one which stores the level at which the processor may become active again.

3.2 Finding a splitter on the reconfigurable mesh

In many instances it is not required that the data set S be divided in an exact manner. It suffices to find an element p of S, not of exact rank, but rather one for which it is known that at least a constant proportion α, $0 < \alpha < 1/2$, of the elements of S are greater than p and at least αn are smaller than p. Such an element will be called an α-splitter. Węgrowicz [Węgr91] has shown that a splitter can be found among the elements of the r level s-sample for any chosen $\alpha < \frac{3}{8}$. Hence, it can be obtained in $O(\log N)$ time after one call to procedure Sample.

4 Conclusions

The parallel selection algorithm presented here has achieved an improvement in running time by a factor of $\log n$ over the algorithm given in [Węgr91], which demonstrated that selection can be accomplished in polylogarithmic time on the reconfigurable mesh architecture. The parallel selection algorithm can also run in $O(\log^2 n)$ time on the 2-dimensional polymorphic-torus network [Li87], which is similar to the reconfigurable mesharchitecture [Mill88a]. It remains to be determined whether an algorithm can be obtained which matches the lower bound of $\Omega((n/p)\log\log p + \log p)$ [Plax89], or one which achieves optimal efficiency.

An immediate consequence of the results presented in this paper is the improvement in running time for the parallel linear programming algorithm given in [Węgr91] to $O(\log^2 n)$ for two dimensions and to $O(n^{1/3}\log^2 n)$ for three dimensions. The linear programming problem can be used in solving various image processing problems, namely linear separability, circular separability and the digital disk problem.

References

[Akl85]   S.G. Akl, *Parallel sorting algorithms,* Academic Press, Orlando, Florida (1985).

[Blum72]   M. Blum, R.W. Floyd, V.R. Pratt, R.L. Rivest, R.E. Tarjan, "Time bounds for selection," *Journal of Computer and System Sciences,* **7**(4), 448-461 (1972).

[Cole88]   R. Cole, "An optimally efficient selection algorithm," *Information Processing Letters,* **26**, 295–299 (1987/88)

[Fred83]   G.N. Frederickson, "Tradeoffs for selection in distributed networks," *Proceedings of the 15th Annual ACM Symposium on Theory of Computing,* 154–160 (1983).

[Li87]   H. Li and M. Maresca, "Polymorphic-torus network," *Proceedings of the International Conference on Parallel Processing,* 411–414 (1987).

[Mill88a]   R. Miller, V.K. Prasanna-Kumar, D.I. Reisis and Q.F. Stout, "Meshes with reconfigurable buses," *MIT Conference on Advanced Research in VLSI,* 163–178 (1988).

[Mill88b] R. Miller, V.K. Prasanna-Kumar, D.I. Reisis and Q.F. Stout, "Data movement operations and applications on reconfigurable VLSI arrays," *Proceedings of the International Conference on Parallel Processing, 1,* 205–208 (1988).

[Mill88c] R. Miller and Q.F. Stout, "Efficient parallel convex hull algorithms," *IEEE Transactions on Computers,* **37**(12), 1605–1618, (1988).

[Munr80] J.I. Munro and M.S. Paterson, "Selection and sorting with limited storage," *Theoretical Computer Science,* **12**, 315–323 (1980).

[Nash91] G. Nash, D. Shu and M. Eshaghian, "Finding connected components in a gated connected VLSI network," *Journal of VLSI Signal Processing,* to be published.

[Plax89] C.G. Plaxton, "On the network complexity of selection," *Proceedings of the 30th Annual IEEE Symposium on Foundations of Computer Science,* 396–401 (1989).

[Pras87] V.K. Prasanna Kumar and C.S. Raghavendra, "Array processors with multiple broadcasting," *Journal of Parallel and Distributed Computing,* **4**, 173–190 (1987).

[Prep79] F.P. Preparata and J. Vuillemin, "The cube-connected cycles," *Proceedings of the 20th Annual IEEE Symposium on Foundations of Computer Science,* 140–147 (1979).

[Stou83] Q.F. Stout, "Mesh-connected computers with broadcasting" *IEEE Transactions on Computers,* **9**, 826–830 (1983).

[Thom77] C.D. Thompson and H.T. Kung, "Sorting on a mesh connected parallel computer," *Communications of the ACM,* **20**(4), 263–271 (1977).

[Vali75] L.G. Valiant, "Parallelism in comparison problems," *SIAM Journal on Computing,* **4**(3), 348–355 (1975).

[Węgr91] P. Węgrowicz, "Linear programming on the reconfigurable mesh and the CREW PRAM," Masters Thesis, McGill University (1991).

Reconfigurable Mesh Algorithms For The Hough Transform*

Jing-Fu Jenq
University of Minnesota *and*

Sartaj Sahni
University of Florida

Abstract

We develop parallel algorithms to compute the Hough transform on a reconfigurable mesh with buses (RMESH) multiprocessor. The p angle Hough transform of an $N \times N$ image can be computed in $O(p\log(N/p))$ time by an $N \times N$ RMESH, in $O((p/N)\log N)$ time by an $N \times N^2$ RMESH with N copies of the image pretiled, in $O((p/\sqrt{N})\log N)$ time by an $N^{1.5} \times N^{1.5}$ RMESH, and in $O((p/N)\log N)$ time by an $N^2 \times N^2$ RMESH.

Keywords and Phrases

Hough transform, reconfigurable mesh with buses, parallel algorithms, complexity.

1 Introduction

Let (x_i, y_i) be any point on a straight line L in two dimensional space (Figure 1). The normal to L is a straight line that orignates at the origin $(0,0)$, terminates at a point on L, and is perpendicular to L. Let θ be the angle between the normal and the x-axis and let r be the length of the normal. From Figure 1, we see that regardless of the position of (x_i, y_i) on L, the following equality holds:

$$x_i\cos\theta + y_i\sin\theta = r \qquad (1)$$

This equality may be used to detect straight lines or edges in images. In the Hough transform method this is done by trying out a set $\{\theta_j \mid 0 \le j < p\}$ of p possible angles for the normal. An angle θ_j and point (x_i, y_i) uniquely define the line L and its normal. For any angle θ_j all image points that have the same normal length r as computed by Equation (1) lie on the same line L. If for a given θ_j we know how many image points have the same normal length r, we can determine the probability that these points actually define an edge of the image. Lines with many points on them have a higher likelihood of defining an edge in the image than those with fewer points.

Let $I[0..N-1, 0..N-1]$ be an $N \times N$ image such that $I[i,j] = 1$ iff the point (i,j) is a candidate for an edge point of the image. $I[i,j] = 0$ otherwise. The p angle Hough transform of I is an array H such that:

$$H[r,j] = \left| \{(x,y) \mid r = \left\lfloor x\cos\theta_j + y\sin\theta_j \right\rfloor \right.$$

$$\theta_j = \frac{\pi}{p}(j+1), \text{ and } I[x,y] = 1\} \right|$$

The second coordinate of H, j, corresponds to the p angles and is in the range 0 through $p-1$. Since $\theta_j = \frac{\pi}{p}(j+1)$, and $0 \le j < p$, $0 < \theta_j \le \pi$. Furthermore, since the image point coordinates x and y are in the range $0 \le x, y < N$. $r = \left\lfloor x\cos\theta_j + y\sin\theta_j \right\rfloor$ is in the range $-\sqrt{2}N$ through $\sqrt{2}N$.

* This research was supported in part by the National Science Foundation under grants DCR-84-20935 and MIP 86-17374

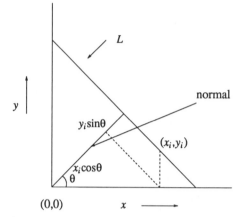

Figure 1 A line L and its normal

Hence H is at most a $2\sqrt{2}N \times p$ array. Pairs (r,j) for which $H[r,j]$ is greater than some threshold value define likely edges. The pair (r,j) defines the unique line L whose normal has length r and angle θ_j.

On a single processor computer, the Hough transform is easily computed in $O(N^2p)$ time. Parallel algorithms for mesh connected computers have been proposed by Rosenfeld et al. [ROSE88], Cypher et al. [CYPH87], Guerra and Hambrush [GUER87], and Silberberg [SILB85]. The algorithm of Cypher et al. uses a pipelined technique and has complexity $O(N+p)$ on an $N \times N$ mesh. Fisher and Highnam [FISH87] consider a scan line array. Their algorithm has time complexity $O(N^2p)$ and is suitable for VLSI implementation. The processor array size is $O(N)$ and each PE requires $O(p)$ space. Ranka and Sahni [RANK90] develop two $O(p+\log N)$ SIMD hypercube algorithms to compute H. Both of these use an N^2 processor SIMD hypercube. One uses $O(1)$ memory per processor while the other uses $O(\log N)$ memory per processor. They also develop algorithms for an MIMD hypercube and present experimental results on an NCUBE hypercube. The computation of the Hough transform on an SIMD tree machine is considered by Ibrahim et al. [IBRA86]. Rather than deal in (r,θ) space, their work uses the (m,c) space where m is the slope and c is the y-axis intercept of the line (i.e., the straight line equation $y = mx + c$ is used). A Hough transform algorithm for a polymorphic torus is developed in [LI89, MARE88, and MARE89] and a fast Hough transorm algorithm is given in [LI86].

In this paper we consider a variant of the mesh connected computer. This variant called "reconfigurable mesh with buses" (RMESH) was introduced by Miller, Prasanna

Kumar, Resis, and Stout [MILL88abc]. We develop algorithms to compute the p angle Hough transform of an $N \times N$ image on different size RMESHs. Our algorithm for an $N \times N$ RMESH has complexity $O(p\log(N/p))$ which is a significant improvement over the $O(p + N)$ complexity for an $N \times N$ mesh when $p \ll N$. On an $N \times N^2$ RMESH we can compute the Hough transform in $O((p/N)\log N)$ time with N copies of the image pretiled over the RMESH, and in times $O(p/\sqrt{N}\log N)$ and $O((p/N)\log N)$ on $N^{1.5} \times N^{1.5}$ and $N^2 \times N^2$ RMESHs, respectively.

2 The RMESH Model

The important features of an RMESH are [MILL88abc]:

1 An $N \times M$ RMESH is a 2-dimensional mesh connected array of processing elements (PEs). Each PE in the RMESH is connected to a broadcast bus which is itself constructed as an $N \times M$ grid. The PEs are connected to the bus at the intersections of the grid. A 4×4 RMESH is shown in Figure 2. Each processor has up to four bus switches that are software controlled and that can be used to reconfigure the bus into subbuses. The ID of each PE is a pair (i,j) where i is the row index and j is the column index. The ID of the upper left corner PE is $(0,0)$ and that of the lower right one is $(N-1, M-1)$.

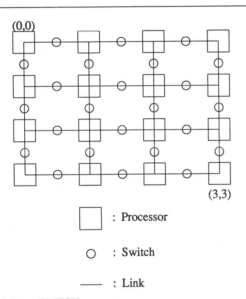

(0,0)

(3,3)

☐ : Processor

○ : Switch

—— : Link

Figure 2 4×4 RMESH

2 The up to four switches associated with a PE are labeled E (east), W (west), S (south) and N (north). Notice that the east (west, north, south) switch of a PE is also the west (east, south, north) switch of the PE (if any) on its right (left, top, bottom). Two PEs can simultaneously set (connect, close) or unset (disconnect, open) a particular switch as long as the settings do not conflict. The broadcast bus can be subdivided into subbuses by opening (disconnecting) some of the switches.

3 Only one processor can put data onto a given sub bus at any time

4 In unit time, data put on a subbus can be read by every PE connected to it. If a PE is to broadcast a value in register I to all of the PEs on its subbus, then it uses the command broadcast(I).

5 To read the content of the broadcast bus into a register R the statement R := content(bus) is used.

6 Row buses are formed if each processor disconnects (opens) its S switch and connects (closes) its E switch. Column buses are formed by disconnecting the E switches and connecting the S switches.

7 Diagonalize a row (column) of elements is a command to move the specific row (column) elements to the diagonal position of a specified window which contains that row (column). This is illustrated in Figure 3.

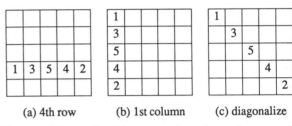

(a) 4th row   (b) 1st column   (c) diagonalize

Figure 3 Diagonalize 4th row or 1st column elements of a 5×5 window

3 N^2 Processor RMESH

As is the case in the algorithms of [CYPH87], [GUER87], and [RANK90], our N^2 processor RMESH algorithm divides the p angles into four classes C1-C4 as below:

$C1 = \{\theta_j \mid 0 < \theta_j \leq \pi/4\}$

$C2 = \{\theta_j \mid \pi/4 < \theta_j \leq \pi/2\}$

$C3 = \{\theta_j \mid \pi/2 < \theta_j \leq 3\pi/4\}$

$C4 = \{\theta_j \mid 3\pi/4 < \theta_j \leq \pi\}$

The algorithms for each of these classes are quite similar. So, we provide the details for just one of these, i.e., C3. The number of angles in each class is $q = p/4$ (for simplicity, we assume that 4 divides p). For any θ_j we may define a matrix V of normal vector lengths for lines that go through points (a,b), $0 \leq a, b < N$ and whose normal angle is θ_j. This matrix is defined as below:

$$V[a,b] = \left\lfloor a\cos\theta_j + b\sin\theta_j \right\rfloor, 0 \leq a,b < N$$

For any $N \times N$ image I, the j'th column, $H[*,j]$, of the Hough transform matrix can be computed using the equality

$$H[r,j] = |\{(a,b) \mid V[a,b] = r \text{ and } I[a,b] = 1\}|$$

We first consider some properties of V for the case $\pi/2 < \theta_j \leq 3\pi/4$. Figure 4 shows a line L whose normal angle is in this range. The following properties are easily established [RANK90, CYPH87, and GUER87].

P1: If $V[a,b] = V[a,b+z]$ for any $z > 0$, then $z = 1$.

P2: If $V[a,b] = V[a+1,c]$, then $c = b$ or $c = b+1$.

P3: If $V[a,b] = V[a,b+1] = V[a+1,c]$, then $c = b+1$.

P4: If $V[a,b] \neq V[x,b+1]$ for $x > a$, then $V[a,b] \neq V[x,y]$ for $y > b$.

Suppose we consider computing $H[r,j]$ for a fixed r and j such that $\theta_j \in C3$ by sending a token through every point (a,b) such that $V[a,b] = r$. This token begins with the value 0 and is incremented by 1 each time it visits a point (a,b) with $I[a,b] = 1$. Since $I[a,b] \in \{0,1\}$ we may simply increment the token by $I[a,b]$ each time it reaches a point (a,b) with $V[a,b] = r$. Since for every r the corresponding line L must cross the left or bottom boundaries of the image and since $|\cos\theta|$ and $|\sin\theta|$ are in the range $[0,1]$, $V[0..N-1,0]$ and $V[0,0..N-1]$ cover the range of possible r values in $V[0..N-1,0..N-1]$.

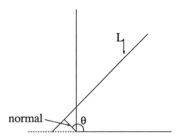

Figure 4 A line L with normal angle in the range $[\pi/2, 3\pi/4]$

Hence we may start our token for $H[r,j]$ at a left or bottom point (a,b) such that $V[a,b] = r$. From properties P1 - P4 it follows that the token needs to move according to the rule given in Figure 5.

Let (a,b) be the current position of the token.
Let $r = \left\lfloor a\cos\theta_j + b\sin\theta_j \right\rfloor$.
if $r = \left\lfloor a\cos\theta_j + (b+1)\sin\theta_j \right\rfloor$
then move the token to $(a,b+1)$
else if $r = \left\lfloor (a+1)\cos\theta_j + b\sin\theta_j \right\rfloor$
    then move the token to $(a+1,b)$
    else move the token to $(a+1,b+1)$

Figure 5 Rule to move a token

The token is moved until it falls off the image. This will happen after the token reaches either the top or right boundary of the image. At this time the token's value is $H[r,j]$.

Rather than send a single token through the array, we can simultaneously send several. The discipline we adopt is that at any time all active tokens in a column correspond to the same angle θ_j. A token becomes inactive when a move according to the rule of Figure 5 would cause it to fall off the image. For convenience, we assume that the processors in the $N \times N$ RMESH are indexed such that PE (i,j) corresponds to the image point (i,j) (see Figure 6).

Let us consider the token movement strategy for the case of a single angle θ_j, $\pi/2 < \theta_j \le 3\pi/4$. This is described in Figure 7. We assume one processor per pixel. The tokens for θ_j begin

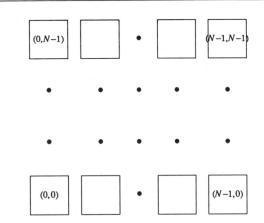

Figure 6 PE indexing scheme

in column 0. Each token corresponds to a distinct r value. From P1 we know that two image points in the same column can have the same r value only if they are adjacent. In step 1 of Figure 7 one token for each unique r in column 0 is created. These newly created tokens have the value 0. In step 2 the tokens are moved through all points (a,b) with the same $V[a,b]$ value. After incrementing the token values to account for the image values at their current locations in column k, the tokens that also correspond to the image point one up in the same column are moved one up and incremented. In case this requires the token in row $N-1$ to move, this token is deactivated as the move would cause it to fall off the image. Because of property P1, the tokens in column k are not to be moved to any other positions in the column.

The tokens now need to be moved to column $k+1$ (unless $k = N-1$). A token in PE (k,a) moves to either $(k+1,a)$ or $(k+1,a+1)$. This is resolved by computing the r value for position $(k+1,a)$. At column $k+1$ it is possible that position $(k+1,0)$ corresponds to a new line. This is true if and only if PE $(k+1,0)$ does not receive a token from column k. In this case this PE initializes a new token with value 0.

Following step 2 we have at most one deactivated token in each of the PEs in row $N-1$ (i.e., top row of PEs) and at most one active token in each of the PEs in column $N-1$ (i.e., right most column of PEs). These tokens correspond to distinct r values and define the column of the Hough transform matrix that corresponds to θ_j.

The computation for all angles θ_j, $\pi/2 < \theta_j \le 3\pi/4$, can be done in a pipelined fashion. Following the movement of the tokens for θ_j from column 0 to column 1, column 0 can initiate the tokens for the next angle θ_j. The scheme of Figure 7 is easily modified so that the PEs in a column know θ_j (or $\cos\theta_j$ and $\sin\theta_j$) for the tokens they currently hold. With this pipelining the Hough transform may be computed in $O(N+p)$ time. This is essentially the strategy of [CYPH87] and [GUER87].

The performance can be improved by employing the above strategy on $N \times (p/4)$ sub RMESH's only. Recall that we have assumed that the number of angles in each of C1, C2, C3, and C4 is $p/4$. We consider the $N \times N$ image as $4N/p$ independent $N \times (p/4)$ subimages and compute the Hough transform for each independently. Then the $4N/p$ Hough

Step 1

[Initialize column 0 tokens]

PE $(0,i)$ creates a token with value 0 if
$\lfloor i\sin\theta_j \rfloor \neq \lfloor (i-1)\sin\theta_j \rfloor$, $1 \leq i < N$

PE $(0,0)$ creates a token with value 0

Step 2

[Update and move tokens]

for $k := 0$ **to** $N-1$ **do**

begin

{tokens are in column k }

the PEs in column k that have a token, add their I value to it;

{ move some tokens up the column by 1 }

PE (k,i) determines if $\lfloor k\cos\theta_j + i\sin\theta_j \rfloor = \lfloor k\cos\theta_j + (i+1)\sin\theta_j \rfloor$.

This is done by all PEs in column k that have a token.

If the equality holds, the PE sends its token to PE $(k,i+1)$ unless $i+1 = N$.

In this latter case PE (k,i) saves the token as a deactivated token.

{ Increment moved tokens }

Each PE (k,i) in column k that received a token adds its I value to it;

{ token updating for column k has been completed }

{ advance tokens to next column }

if $k \neq N-1$ **then**

begin

every PE (k,i) that has an active token determines if $\lfloor k\cos\theta_j + i\sin\theta_j \rfloor = \lfloor (k+1)\cos\theta_j + i\sin\theta_j \rfloor$.

If so, it sends its token to PE $(k+1,i)$.

Otherwise it sends it to PE $(k+1,i+1)$ except when $i+1 = N$.

In this latter case the token is saved as a deactivated token by PE (k,i).

If PE $(k+1,0)$ does not receive a token, it creates one with value 0;

end;

end;

Figure 7 Token movement and updating for angle θ_j

---

transforms are combined to get the Hough transform for the original $N{\times}N$ image (actually this will only get us the transform for angles in C3; similar algorithms need to be run to get the Hough transform for the remaining angles).

Assume that following the application of Figure 7, in a pipelined fashion, for all θ_j in C3, the Hough transform matrix is stored in columns 0 through $p/4-1$ of each $N{\times}(p/4)$ sub RMESH. For this, when the tokens for the j'th angle reach column $p/4-1$ of the sub RMESH, they are broadcast along row buses to the processors in column j of the sub RMESH. Also, when tokens get deactivated in row $N-1$ they are transmitted to processors in columns dedicated for their angles (column j processors of each sub RMESH are dedicated to the j'th angle in C3). Note that the tokens that get simultaneously deactivated in row $N-1$ correspond to different angles in C3 as at most one token deactivates in each row $N-1$ processor at any time and each column corresponds to a different angle. The deactivated token (if any) in processor $(k,N-1)$ of the sub RMESH is routed to processor (j,k) of the sub RMESH where

θ_j is the angle corresponding to the token. This is accomplished in O(1) time as in Figure 8.

Step 1

Set up column buses

Step 2

Each PE $(k,N-1)$ of the sub RMESH broadcasts its deactivated token together with the corresponding r and j values;

Step 3

PEs (i,i) of the sub RMESH, $0 \leq i < p/4$ read their bus and are now the only PEs in the sub RMESH with deactivated tokens;

Step 4

Set up row buses local to each sub RMESH;

Step 5

The PEs with deactivated tokens broadcast tokens and the corresponding r and j values;

Step 6

All PEs read their bus. However, a PE stores the deactivated token value and r value read only if the PE is in column j of the sub RMESH (j is the third value read from the bus);

Figure 8 Redistributing deactivated tokens

Once the computation for all $p/4$ angles in C3 has been completed, each PE in column j of each $N{\times}(p/4)$ sub RMESH contains at most two token values. One received from the rightmost column in the sub RMESH and one from row $N-1$. The first is an active token and the latter a deactivated token. All tokens in column j of the sub RMESH correspond to the j'th angle in C3. The time needed to accomplish this is $O(p)$.

Now we need to combine together the partial Hough transform values computed in each $N{\times}(p/4)$ sub RMESH. Column j of each sub RMESH contains partial Hough transform values for the j'th angle in C3, $0 \leq j < p/4$. Across these columns, we need to add together values that correspond to the same r. Each processor has at most two tokens: active and deactivated. There are two quantities associated with each token. One is the r value and the other is a count of the number of pixels that have contributed to this token (this count has so far been referred to as the token value). Let us call these quantities $token.r$ and $token.count$, respectively. The sum of the $token.count$'s for the same angle and $token.r$ values can be obtained in $O(p\log(N/p))$ time by computing these sums for one angle at a time. This corresponds to considering all columns j with $j \bmod (p/4) = k$ for a fixed k in $(0,p/4-1)$ at the same time and adding together the $token.count$s in these columns for tokens that have the same r values. This summation is done in $O(\log(N/p))$ time by first summing up pairs of columns in adjacent blocks; then pairs of these results are summed, etc. Figure 9 shows the strategy for the case of 8 blocks each of size $N{\times}(p/4)$. The column j tokens of each block are to be summed. The leaves of the summation tree are labeled by the block number they represent. Blocks of the RMESH are numbered left to right 0 through $4N/p-1$. The input for each summation consists of two columns of tokens. The columns are initially $p/4$ processors apart; then, at the next level, they are $p/2$ processsors apart, then p; then $2p$; and so on. One of the two token columns is to the left of the other. This is called the L column and the other column is called the R column. On input, each processor contains at most two tokens; one active and one deactivated. The output of the summation operation is left in the input processor column corresponding to L. Again, each processor in this column will have at most two tokens; one active and the other deactivated. Furthermore the r values corresponding to the deactivated tokens (active tokens) decrease as we go down a column and the deactivated tokens have a larger r value than do the active tokens. When the columns being merged are p/4 apart (i.e., leaves of Figure 9), the sets of deactivated and active tokens are as defined earlier. For a pair of columns L and R, these sets at the parent node (cf. Figure 9) are given by the equalities:

deactivated $(L \cup R)$ = deactivated $(L) \cup$ deactivated(R)
active $(L \cup R)$ = active (R)

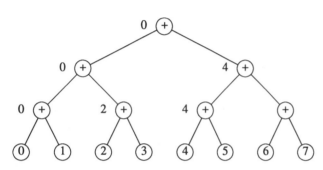

Figure 9 Summing the j'th column of 8 blocks

Let Z be a summation node (i.e., internal node) of Figure 9. Let the distance between the two columns L and R being summed at Z be s, $s \in \{p/4, p/2, \cdots, N/2\}$ It is easy to see that $|deactivated(Z)| \leq 2s \leq N$ and $|active(Z)| \leq N$. We assume that each token has two values: *count* and r associated with it. *Count* is the number of pixels that has contributed to it and r is the length of the normal to the line represented by this token. We observe that the *count* value of the deactivated tokens of L does not change as a result of the summation. In fact the *count* values can change only for those tokens that are deactivated or active tokens of R. To get the new count values for the deactivated tokens of R, we use the processor columns L and R together with the s−1 columns between them. Thus an $N \times (s+1)$ sub RMESH is used. The code for such a sub RMESH is given in Figure 10. Its complexity is O(1).

The deactivated tokens of R can next be compacted to lie in consecutive rows of L immediately following the last row of L that contains a deactivated token. This requires us to rank the deactivated tokens of R and then route these to row w of L where w is the token rank plus the number of deactivated tokens already in L. We assume that the deactivated tokens of L lie in consecutive rows of L beginning at row 0. This is not true for the leaf nodes of Figure 9. However, the deactivated tokens in these nodes may be compacted in O(1) time using the ideas used to route the deactivated tokens of R to L. The ranking and routing of the deactivated tokens of R can be done in O(1) time using the $N \times (s+1)$ processor sub RMESH

Step 1
Use column and row buses in the sub RMESH to obtain the data configuration:
PE (i,j) of the $N \times (s+1)$ sub RMESH contains the deactivated token (if any) from the row i processor of column R and the active token (if any) from the row j processor of column L, $0 \leq i < s+1, 0 \leq j < N$.
Step 2
if PE (i,j) of the sub RMESH has two tokens and both have the same r value
then update the count of the deactivated tokens to be the sum of the two token counts and destroy the active token of L by sending a signal down row bus j, $0 \leq i < s+1, 0 \leq j < N$.
Step 3
if PE (i,j) has a deactivated token with updated count
then the updated count is reported back to the PE in row i of column R.
Figure 10 Updating the count for deactivated tokens of R

Step 1
if PE (R,k) has a token
then it broadcasts it and the token's rank on its row bus, $0 \leq k < s$ PE (i,j) reads its bus, $0 \leq i < (s+1), 0 \leq j < s$.
Step 2
PE (i,j) retains the token (if any) read in Step 1 only if i equals the token rank, $0 \leq i < s+1, 0 \leq j < s$.
Step 3
The PEs that have tokens, broadcast these along column buses.
Step 4
Let t be the number of deactivated tokens in column L. PE $(j,t+j)$, $0 \leq j < q$ where q is the number of deactivated tokens in R, reads its bus.
Step 5
PE $(j,t+j)$, $0 \leq j < q$ broadcasts the token read on its row bus.
Step 6
PE $(L,t+j)$, $0 \leq j < q$ reads its bus.
Figure 11 Routing deactivated tokens of R to L

comprised of columns L and R and the s−1 processor columns between them. The ranking takes O(1) time (Appendix A) as we are ranking at most s tokens, one token to a row, in an $s \times (s+1)$ sub RMESH (the deactivated tokens lie in the first s rows of R). The routing scheme is described in Figure 11.

To update the count of the active tokens in R a different strategy is used. First note that the set of active tokens in R includes the remaining active tokens of L (i.e., those not destroyed in step 2 of Figure 10). Also, note that as tokens progress through the image (see algorithm of Figure 5), they can move up by at most two rows for each column they move right. Hence, if a token of L has the same r value as a token in R, then the two tokens must reside in rows of L and R that are at most 2s apart (recall that L and R are s columns apart). To update the active tokens of R, we employ the PEs in the two blocks which L and R belong to by moving L to the left most column in its block and R to its right most column in its block.

So, a total of $2s$ processor columns (i.e., an $N \times (2s)$ sub RMESH) are used. For the update, we need to bring active token pairs from L and R that have the same r value together. This is done using the strategy of Figure 12. We assume that L and R are the 2 extreme columns of the combined block. The broadcasting of step 2 is done in two stages. First, all PEs $[L,i]$ with i odd do this and then those with i even do it. Figure 13 (a) gives the bus structure used for the first stage and

Step 1
PE (R,i) broadcasts its token's (if any) count and r values to all processors on row i using a row bus, $0 \le i < N$

Step 2
PE (L,i) broadcast its token's (if any) count and r values to PEs $(L+q,i+q)$, $0 \le q \le \min\{N-1-i,2s\}$ using the bus structures of Figure 13, $0 \le i < N$

Step 3
Set up row buses.

Step 4
Each PE (i,j) in the $N \times (2s)$ sub RMESH that has two tokens with the same r value adds their count and disconnects its W switch and then broadcasts the new count and r value to PE (R,j).

Step 5
PE (R,j) broadcast its r and *count* values, $0 \le j < N$.

Step 6
PE (L,j) reads r and *count* from the row bus, $0 \le j < N$

Figure 12 Updating counts for active tokens of R

Figure 13 (b) gives the structure used in the second stage.

The complexity of the algorithm described to compute the Hough transform for the angles in C3 is $O(p\log(N/p))$. The time needed for the angles in each of the remaining sets C1, C2, and C4 is the same. So, the over all complexity of our N^2 processor RMESH Hough transform algorithm is $O(p\log(N/p))$.

4 N^3 Processor RMESH

If the RMESH is configured as an $N \times N^2$ array with one copy of the image in each $N \times N$ sub RMESH, then the Hough transform can be computed in $O((p/N)\log N)$ time by having each $N \times N$ sub MESH compute the transform for p/N angles using the algorithm of Section 2. In case the RMESH is configured as a $N^{1.5} \times N^{1.5}$ array with one pixel in each $\sqrt{N} \times \sqrt{N}$ sub MESH, the Hough transform can be computed in $O((p/\sqrt{N})\log N)$ time. For this, $p/\sqrt{N}$ passes are made. In each, the Hough transform for $\sqrt{N}$ angles is computed. The Hough transform for $\sqrt{N}$ angles can be computed in $O(\log N)$ time by having PE (i,i) of each $\sqrt{N} \times \sqrt{N}$ sub RMESH initiate a token with value equal to that of the image value at the corresponding point. The r value for the token in PE (i,i) is obtained by using the i'th angle. Now we need to add up the token values of all tokens with the same r and θ values. This can be done by combining tokens in pairs of sub RMESHs at a time. The strategy employed is similar to that used in [JENQ90]. Two kinds of combinations are needed: horizontal and vertical. In horizontal combining two sub RMESHs each containing $k \times k$ $\sqrt{N} \times \sqrt{N}$ sub RMESHs are combined to get a single sub RMESH which is a $k \times (2k)$ configuration of $\sqrt{N} \times \sqrt{N}$ sub

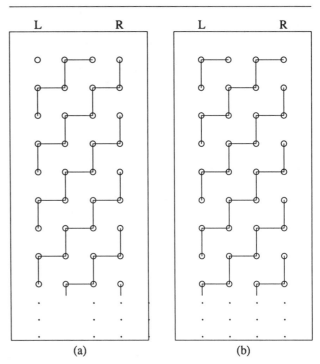

(a)   (b)

Figure 13 Bus structures used in step 2 of Figure 12

RMESHs. In a vertical combining, two $k \times (2k)$ arrays of $\sqrt{N} \times \sqrt{N}$ sub RMESHs are combined to get a single $2k \times 2k$ sub RMESH.

The initial configuration for a horizontal combine has the tokens in the PEs on the antidiagonals of the two $k \times k$ $\sqrt{N} \times \sqrt{N}$ sub RMESHs (Figure 14). As in the discussion of Section 2, assume that θ_j is in the range $\pi/2 < \theta_j \le 3\pi/4$. So, r is in the range $[-\sqrt{2}N/2, N]$. However for any fixed θ_j in the above range at most $\sqrt{2}N$ distinct integer r values are possible. The maximum number of tokens in each PE is therefore $\sqrt{2}k\sqrt{N}$ (at most $\sqrt{2}k$ per angle and $\sqrt{N}$ angles). The number of PEs on each antidiagonal is $k\sqrt{N}$. So, using k PEs per angle we need to store at most $\sqrt{2} < 2$ tokens in each PE. The final configuaration has at most $2\sqrt{2}k\sqrt{N}$ tokens stored at most 4 per PE in the antidiagonal PEs of the left sub RMESH A.

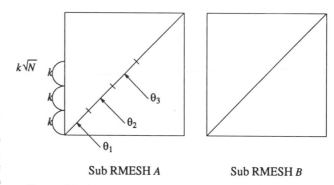

Sub RMESH A Sub RMESH B

Figure 14 Initial configuration for horizontal combining

For a vertical combine, the initial configuration is the final configuation of a horizontal combine and the final configuration is the initial configuation of a horizontal combine. We shall describe how to go from the initial configuration of a horizontal combine to its final configuration in $O(1)$ time. The method for a vertical combine is similar. Since $O(logN)$ parallel combine steps suffice to combine all tokens, the complexity is $O(logN)$.

A horizontal combine for any angle θ_j, uses the PEs on the k rows dedicated to this angle. So, a $k \times 2k\sqrt{N}$ sub RMESH is available. Since $\pi/2 < \theta_j \leq 3\pi/4$, the r values of the tokens in the k rows of the left sub RMESH A (call this sub RMESH KA) are $\geq$ the r values of the tokens in the same k rows of the right sub RMESH B (call this sub RMESH KB). Using column and row buses confined to the $k \times 2k\sqrt{N}$ sub RMESH formed by A and B we can obtain, in $O(1)$ time, a configuration in which PE (i,j) of a $k \times k$ sub RMESH of A contains the at most $2\sqrt{2}$ tokens in the i'th row of KA and j'th row of KB. In an additional $O(1)$ time, this PE can check if any of the tokens it has from KA have the same r value as any of the tokens it has from KB. If so, the KA token is deleted and the value of the KB token incremented by the value of the deleted KA token. The KA and KB tokens can be ranked in $O(1)$ time using the available processors and packed into the k antidiagonal processors of KA packing no more than four tokens per processor in an additional $O(1)$ time. Hence the horizontal combine can be completed in $O(1)$ time.

5 N^4 Processors

When N^4 processors are available, there is a rather straightforward $O((p/N)logN)$ algorithm to compute the Hough transform. We assume that the RMESH is configured as an $N^2 \times N^2$ array with the N^2 pixel values initially in row 0. Since there are only $2\sqrt{2}N$ different r values possible, we compute the Hough transform for $N/(2\sqrt{2})$ angles simultaneously. Thus, $2\sqrt{2}p/N$ iterations are needed. Each row of the $N^2 \times N^2$ RMESH is assigned the task of obtaining the value of $H(r,j)$ for one pair (r,j). Since we are working with $N/(2\sqrt{2})$ angles simultaneously, the number of (r,j) pairs is N^2 which equals the numbers of rows. The image can be broadcast to each row using column buses. The processors in a row use the assigned angle to obtain the r for their pixel. If this equals the r value assigned to that row and if the pixel value is 1, the processor sets its *count* variable to 1; otherwise it sets it to 0. Adding the *count* variables in a row gives the Hough transform value for the (r,j) pair assigned to the row.

6 References

[CYPH87]   R. E. Cypher, J. L. C. Sanz, and I. Snyder, "The Hough transform has $O(N)$ complexity on SIMD $N \times N$ mesh array architectures", Proceedings of IEEE 1987 Workshop on Computer Architecture for Pattern Analysis and Machine Intelligence, pp 115-121, 1987.

[FISH87]   A. Fisher and P. Highnam, "Computing the Hough transform on a scan line array processor", Proceedings of IEEE 1987 Workshop on Computer Architecture for Pattern Analysis and Machine Intelligence, pp 83-87, 1987.

[GUER87]   C. Guerra and S. Hambrush, "Parallel algorithms for line detection on a mesh", Proceedings of IEEE 1987 Workshop on Computer Architecture for Pattern Analysis and Machine Intelligence, pp 99-106, 1987.

[IBRA86]   H. A. Ibrahim, J. B. Kender, and D. E. Shaw, "On the application of massively parallel SIMD tree machine to certain intermediate-level vision tasks", Computer Vision, Graphics, and Image Processing, 36, 1986, pp 53-75.

[JENQ90]   J. Jenq and S. Sahni, "Reconfigurable mesh algorithms for the area and perimeter of image components and histogramming", submitted.

[LI86]   Li, Lavin, and LeMaster, "Fast Hough transform: A hierarchical approach", Computer Vision, Graphics, and Image Processing, 36, 3, Dec. 1986.

[LI89]   Li, and Maresca, "Polymorphic-torus architecture for computer vision", IEEE Trans. on PAMI, 11, 3, March 1989.

[MARE88]   Maresca, Lavin, and Li, "Parallel Hough transform algorithms on polymorphic torus", in High Level Vision in Multicomputers, Levialdi, (ed), Academic Press, 1988.

[MARE89]   Maresca, Li, and Sheng, "Parallel computer vision on polymorphic torus architecture", Intl. Jr. on Computer Vision and Applications, 2, 4, Fall 1989.

[MILL88a]   R. Miller, V. K. Prasanna Kumar, D. Resis and Q. Stout, "Data movement operations and applications on reconfigurable VLSI arrays", Proceedings of the 1988 International Conference on Parallel Processing, The Pennsylvania State University Press, 1988, pp 205-208.

[MILL88b]   R. Miller, V. K. Prasanna Kumar, D. Resis and Q. Stout, "Meshes with reconfigurable buses", Proceedings 5th MIT Conference On Advanced Research IN VLSI, 1988, pp 163-178.

[MILL88c]   R. Miller, V. K. Prasanna Kumar, D. Resis and Q. Stout, "Image computations on reconfigurable VLSI arrays", Proceedings IEEE Conference On Computer Vision And Pattern Recognition, 1988, pp 925-930.

[RANK90]   S. Ranka and S. Sahni, *Hypercube algorithms with applications to image processing and pattern recognition*, Springer-Verlag, New York, 1990, pp 145-166.

[ROSE88]   A. Rosenfeld, J. Ornelas, and Y. Hung, "Hough transform algorithms for mesh-connected SIMD parallel processors", Computer Vision, Graphics, and Image Processing, 41, 1988, pp 293-305.

[SILB85]   T. M. Silberberg, "The Hough transform in the geometric arithmetic parallel processor", Proceedings IEEE Workshop on Computer

Architecture and Image Database Management, 1985, pp 387-391.

Appendix A: O(1) Ranking

Consider an $N \times N$ RMESH in which each PE has a Boolean variable *selected*. If *selected* (i,j) is true then *rank* (i,j) is the number of PEs with *selected* (i,j) true that precede it in the defined linear ordering. If *selected* (i,j) is false, then *rank* (i,j) is undefined. Suppose that all the PEs with *selected* (i,j) = true are on row 0 (i.e. *selected* (i,j) = false, $i > 0$). Hence, at most, N elements are to be ranked. *rank* $(0,j)$, $0 \le j < N$, can be computed in $O(1)$ time using the steps of Figure A1.

Step 1   [rank even columns]
Compute $r(0,j)$ for j even where r is defined as
$r(0,j) = |\{q|q$ is even and *selected* $(0,q)$ and $q \le j\}|$

Step 2   [rank odd columns]
Compute $r(0,j)$ for j odd where r is defined as
$r(0,j) = |\{q|q$ is odd and *selected* $(0,q)$ and $q \le j\}|$

Step 3   [combine]
$$rank(0,j) = \begin{cases} r(0,0) - 1 & j = 0 \\ r(0,j) + r(0,j-1) - 1 & j > 0 \end{cases}$$

Figure A1 Steps in O(1) ranking

The algorithms for steps 1 and 2 are similar. So we describe the algorithm only for step 1. To compute $r(0,j)$, for even j, we set the bus switches as in Figure A2 (a) in case *selected* $(0,j)$ is true and as in Figure A2 (b) in case it is not. The switch settings are similar to those used to compute the exclusive or of 1's in [MILL88]. In this figure e denotes an even index and o an odd index. So, (e,j) denotes all PEs (i,j) with even i. Note that since j is even (e,j) is equivalent to (e,e) and $(e,j+1)$ to (e,o). Solid lines indicate connected (closed) switches and blanks indicate disconnected (open) switches.

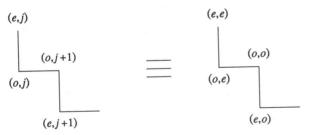

(a) Settings for *selected* $(0,j)$ = true

(b) Setting for *selected* $(0,j)$ = false

Figure A2 Switch settings to compute $r(0,j)$ for j even

The algorithm to implement this strategy is given in Figure A3. Its complexity is readily seen to be $O(1)$. As mentioned earlier, the algorithm for step 2 is similar. Step 3 simply requires a rightward shift of 1 which can be easily done in $O(1)$ time. Hence the entire ranking can be done in $O(1)$ time.

{ Compute $r(0,j)$ for j even }

Step 1 $t(0,j) := $ *selected* $(0,j)$, $0 \le j < N$;

Step 2   Set up column buses;

Step 3   Broadcast $t(0,j)$ on column bus j, $0 \le j < N$;

Step 4 $t(i,j) := $ content(bus); $0 \le i,j < N$;

Step 5   { send $t(i,j)$ for j even to $t(i,j)$ for j odd }
All *PEs* (i,j) with j even disconnect their N, S, W switches and connect their E switch;
All *PEs* (i,j) with j even broadcast $t(i,j)$;
All *PEs* (i,j) with j odd set $t(i,j)$ to their bus content;

Step 6   { set switches as in Figure A2 }
if $t(i,j)$ **then case** (i,j) **of**
(odd,odd),(even,even): PE (i,j) disconnects its E switch and connects its S switch;
else PE (i,j) connects its E switch and disconnects its S switch;
endcase
else case i **of**
odd : PE (i,j) disconnects its E and S switches;
else : PE (i,j) connects its E switch and disconnects its S switch;
endcase;

Step 7   PE (0,0) broadcasts a special value on its bus;

Step 8   All PEs (i,j) with i and j even read their bus;
If the special value is read, then they set their S value to true and r value to $i/2 + 1$;

Step 9   Set up column buses;

Step 10   PE (i,j) puts its r value on its bus if $S(i,j)$ is true;

Step 11 $r(0,j) = $ content(bus), j even;

Figure A3 RMESH algorithm to compute $r(0,j)$ for j even

CONFIGURATIONAL COMPUTATION: A NEW COMPUTATION METHOD ON PROCESSOR ARRAYS WITH RECONFIGURABLE BUS SYSTEMS

Biing-Feng Wang[+], Gen-Huey Chen[+], and Hungwen Li[++]

+Department of Computer Science and Information Engineering,
National Taiwan University, Taipei, Taiwan, Republic of China
++IBM Research Division, Almaden Research Center,
650 Harry Road, San Jose, CA 95120-6099, U.S.A.

Abstract -- In this paper, a new computation method on processor arrays with reconfigurable bus systems, called configurational computation, is introduced. The processor arrays with reconfigurable bus systems refer to a general class of computation models that each consist of a processor array and a reconfigurable bus system. Configurational computation solves problems mainly using the great flexibility in the configurations of reconfigurable bus systems. A sufficient condition for a problem to be solved in $O(1)$ time by configurational computation on a two-dimensional processor array with a reconfigurable bus system is also suggested. According to this sufficient condition, $O(1)$ time algorithms for solving many problems can be developed. These problems include computing the logical exclusive-OR of n bits, adding two n-bit binary integers, summing n bits, summing n m-bit binary integers and multiplying two n-bit binary integers.

1. Introduction

Many networks of processors have been supplemented with buses to decrease their diameters in order to enhance the system performance [1], [9], [12]. Buses give networks of processors greater communication capabilities and they allow broadcasting and long-distance communication to be completed in negligible time. Recently, some parallel computers have been further equipped with reconfigurable bus systems to solve problems more efficiently [6], [7], [10], [13], [16]. A *reconfigurable bus system* is a bus system whose configurations can be dynamically changed. Bus automaton [2], [13], polymorphic-torus network [7], [8], reconfigurable mesh [10] and mesh-connected array processors with bypass capability [6] are four examples.

The PARBS's have a very powerful computation capability and many problems have been solved efficiently on them. For example, on a 2-d PARBS, Miller *et al.* [10] have designed an $O(1)$ time maximum finding algorithm, and Wang and Chen [16] have designed an $O(1)$ time algorithm for solving the transitive closure problem. On a 3-d PARBS, Champion and Rothstein [2] have designed an $O(1)$ time algorithm for solving the longest common subsequence problem, and Wang *et al.* [18] have designed an $O(1)$ time sorting algorithm.

In this paper, a new computation method on the PARBS's, called *configurational computation*, is introduced. Far different from conventional computation, configurational computation solves problems mainly using the great flexibility in the configurations of reconfigurable bus systems. A sufficient condition for a problem to be solved in $O(1)$ time by configurational computation on a 2-d PARBS is also suggested. According to this sufficient condition, $O(1)$ time algorithms for solving many problems can be easily developed. These problems include computing the logical exclusive-OR of n bits, adding two n-bit binary integers, summing n bits, summing n m-bit binary integers and multiplying two n-bit binary integers.

The rest of this paper is organized as follows. In the next section, the PARBS's are described. In Section 3, configurational computation, which is a new computation method on the PARBS's, is introduced. In Section 4, a sufficient condition for a problem to be solved in $O(1)$ time by configurational computation on a 2-d PARBS is proposed. Then, in Section 5, we show that $O(1)$ time algorithms for some applications can be easily derived from the sufficient condition. In Section 6, we present two approaches that are useful in reducing the size of PARBS's, when we design algorithms according to the sufficient condition. Finally, in Section 7, we conclude this paper with some final remarks.

2. Processor arrays with reconfigurable bus systems

A *2-d $N_1 \times N_2$ PARBS* consists of an $N_1 \times N_2$ array of processors which are connected to a grid-shaped reconfigurable bus system. Each processor is identified by a unique index (i, j), $0 \leq i \leq N_1 - 1$, $0 \leq j \leq N_2 - 1$. The processor with index (i, j) is denoted by $P_{i,j}$. Within each processor, four ports, denoted by U, D, L and R (standing for up, down, left and right, respectively), are provided; ports L, R are built in i-direction, and ports U, D are built in j-direction. Through these ports, processors are connected to the reconfigurable bus system.

In Figure 1, a 6×4 PARBS is shown. (Note that for ease of description, all PARBS's shown in this paper have i-direction represent the horizontal direction and j-direction represent the vertical direction.) The configurations of the reconfigurable bus system are dynamically changeable by adjusting the local connections among ports within each processor. For example, by connecting port L to port R within each processor, horizontally straight buses are established to connect the processors of the same row together (see Figure 2(a)). The horizontally straight buses are further split into sub-buses, if some processors disconnect their established local connections (see Figure 2(b)). Besides, other configurations of the reconfigurable bus system are allowed, as long as they can be formed by properly adjusting the local connections among ports within each processor. When no local connection among ports is set within each processor, a square PARBS is functionally equivalent to a mesh-connected computer.

When a bus configuration is established, processors that are attached to a same bus can communicate with one another by broadcasting on the common bus. We assume that each broadcasting takes one time unit. When more than one processor attempts to broadcast values on a same bus simultaneously, a collision occurs and the final value received is unexpected.

In this paper, we use $\{g_1\}$, $\{g_2\}$, ..., $\{g_t\}$ to represent the local connection within a processor, where each g_i, $1 \leq i \leq t$, denotes a group of ports that are connected together within the processor. For example, in Figure 2(a), the local connection within each processor can be represented by $\{L, R\}$. If the local connection within a processor is represented by $\{U, L\}$, $\{D, R\}$, it means that two connections exist within the processor; one connects ports U and L together, and the other connects ports D and R together.

Although only the 2-d PARBS is introduced in this section, extension to higher dimensions is rather straightforward.

3. Configurational computation

In this section, based on the reconfigurability of the reconfigurable bus systems, a new computation method on the PARBS's, called *configurational computation*, is introduced.

Let us consider a PARBS with p processors, and denote the number of possible local connections within each processor by l. For example, $l=2$ for a linear PARBS, since each processor can either connect port L and port R together or not. The configurations of the reconfigurable bus system are determined by the underlying network topology and the local connections established within each processor. Thus, totally l^p different configurations can be formed. Let C denote the set of all the l^p configurations.

Suppose there is a problem to be solved on the PARBS, and the problem can be expressed as a function

$$F: X \longrightarrow Y,$$

where X is the set of possible input instances and Y is the set of corresponding answers (i.e. $Y=F(X)$). Assume $Y=\{y_0, y_1, y_2, ..., y_{d-1}\}$ and $d \leq |C|$. To solve the problem by configurational computation, we first define a *construction rule*, by which a configuration $c_x \in C$ can be established for each input instance x, and $c_x \neq c_{x'}$ if $F(x) \neq F(x')$, where x, $x' \in X$. Then, according to the construction rule, the set C is conceptually partitioned into $d+1$ subsets, denoted by C_{-1}, C_0, C_1, ..., C_{d-1}, as follows:

(1) For $0 \leq i \leq d-1$, $C_i = \{c_x \mid x \in X \text{ and } F(x)=y_i\}$;

(2) $C_{-1} = C - (C_0 \cup C_1 \cup ... \cup C_{d-1})$.

Then, the problem can be solved according to the following procedure. (Initially, an input instance x is stored in the PARBS.)

Phase 1: Establish c_x according to the construction rule.

Phase 2: Determine the subset, assume C_v, $0 \leq v \leq d-1$, which c_x belongs to.

Phase 3: Set y_v as the desired answer.

As will be seen later in this paper, the execution time of the above procedure is mainly dependent on the construction rule.

In the next section, a sufficient condition for a problem to be solved in $O(1)$ time by configurational computation on a 2-d PARBS is suggested.

4. A sufficient condition

Consider a 2-d $N_1 \times N_2$ PARBS. Conveniently, we denote the four ports U, D, L and R within each processor $P_{i,j}$, $0 \leq i \leq N_1-1$, $0 \leq j \leq N_2-1$, by $U_{i,j}$, $D_{i,j}$, $L_{i,j}$ and $R_{i,j}$ respectively. Each pair of ports $U_{i,j}$ and $D_{i,j+1}$, $0 \leq i \leq N_1-1$, $0 \leq j < N_2-2$ (or $R_{i,j}$ and $L_{i+1,j}$, $0 \leq i < N_1-2$, $0 \leq j \leq N_2-1$), are said

to be *externally connected*. On the other hand, two ports within the same processor are said to be *internally connected* with respect to a bus configuration, if they are connected together in the bus configuration. Any two externally connected ports are always connected together in the PARBS, while two ports are internally connected for some particular bus configurations. For example, ports $L_{0,0}$ and $R_{0,0}$ are internally connected in the bus configuration of Figure 2(b). And ports $L_{0,0}$ and $U_{0,0}$ are not.

Clearly, when a bus configuration is established, two ports, say O_0 and O_r, are connected to the same bus if and only if there exists a sequence of different ports ($O_0, O_1, O_2, ..., O_r$) such that O_k and O_{k+1}, $0 \leq k \leq r-1$, are either externally connected or internally connected. Besides, ($O_0, O_1, O_2, ..., O_r$) is called a *connected-path* from O_0 to O_r. For example, in Figure 2(b), ($L_{0,0}, R_{0,0}, L_{1,0}, R_{1,0}, L_{2,0}, R_{2,0}$) is a connected-path from $L_{0,0}$ to $R_{2,0}$ and thus $L_{0,0}$ and $R_{2,0}$ are connected to the same bus.

Definition 1: Let f be a function that maps a set of integers to another set of integers, and S be a subset of the domain of f. A *function-mapping configuration* of a 2-d $w \times h$ PARBS with respect to f and S, which is denoted by $FMC(f, S, w \times h)$, is a bus configuration established on a 2-d $w \times h$ PARBS that has the following four properties:

(1) $S \subseteq \{0, 1, ..., h-1\}$,

(2) $f(S) \subseteq \{0, 1, ..., h-1\}$, where $f(S)=\{f(j) \mid j \in S\}$,

(3) for each $j \in S$, ports $L_{0,j}$ and $R_{w-1,f(j)}$ are connected to the same bus, and

(4) no two ports $R_{w-1,j}$ and $R_{w-1,j'}$ are connected to the same bus, where $0 \leq j \leq h-1$, $0 \leq j' \leq h-1$ and $j \neq j'$.

As two illustrative examples, an $FMC(f(j)=2*j+1, \{0, 1, 2\}, 3 \times 6)$ and an $FMC(f(j)=j \text{ div } 2, \{0, 1, 2, 3, 4, 5\}, 3 \times 6)$ are depicted in Figure 3.

Note that an $FMC(f, S, w \times h)$ is also an $FMC(f, S', w \times h)$ for any set $S' \subseteq S$. Also, by definition, if processor $P_{0,j}$, $j \in S$, broadcasts a value on the bus to which port $L_{0,j}$ is connected, only one processor ($P_{w-1,f(j)}$) on column $w-1$ can receive the value from its port R ($R_{w-1,f(j)}$). Thus, an $FMC(f, S, w \times h)$ is very useful in computing the value $f(j)$ for each $j \in S$. For example, we can determine each value $f(j)$, $j \in S$, in $O(1)$ time by letting processor $P_{0,j}$ broadcast a signal on the bus to which port $L_{0,j}$ is connected. Then, $f(j)$ is determined as v if processor $P_{w-1,v}$ on column $w-1$ received the signal from port $R_{w-1,v}$.

We say that an $FMC(f_0, S_0, w_0 \times h)$ and an $FMC(f_1, S_1, w_1 \times h)$ are *composable* if $f_0(S_0) \subseteq S_1$. For example, the two function-mapping configurations depicted in Figure 3 are composable. Similarly, r function-mapping configurations $FMC(f_k, S_k, w_k \times h)$, $k=0, 1, ..., r-1$, are composable if $f_k f_{k-1} ... f_0(S_0) \subseteq S_{k+1}$ for $k=0, 1, ..., r-2$. (Throughout this paper we use $f_k f_{k-1} ... f_0(S_0)$ to denote $f_k(f_{k-1}(... (f_0(S_0)) ...))$.)

Let c_k denote an $FMC(f_k, S_k, w_k \times h)$, $0 \leq k \leq r-1$, and suppose $c_0, c_1, ..., c_{r-1}$ are composable. The *composition* of $c_0, c_1, ..., c_{r-1}$ is a bus configuration on a 2-d $w \times h$ PARBS, where $w=w_0+w_1+ ... +w_{r-1}$, which is established as follows:

(1) conceptually, divide the 2-d $w \times h$ PARBS into r 2-d PARBS's: $SUB\text{-}PARBS_0$, $SUB\text{-}PARBS_1$, ..., $SUB\text{-}PARBS_{r-1}$, where each $SUB\text{-}PARBS_k$, $0 \le k \le r-1$, is of dimension $w_k \times h$ and contains processors ranging from column $w_0+w_1+ ... +w_{k-1}$ to column $(w_0+w_1+ ... +w_k)-1$; (2) establish the bus configuration of each $SUB\text{-}PARBS_k$, $0 \le k \le r-1$, as c_k.

As an illustrative example, Figure 4 shows the composition of the two function-mapping configurations depicted in Figure 3. It is not difficult to see that the composition is indeed an $FMC(f(j)=(2*j+1)$ **div** 2, $\{0, 1, 2\}$, $6 \times 6)$. In fact, for any two composable function-mapping configurations, we have the following lemma.

Lemma 1: Let c_0, c_1 denote an $FMC(f_0, S_0, w_0 \times h)$, an $FMC(f_1, S_1, w_1 \times h)$, respectively, and suppose they are composable. The composition of c_0 and c_1 is an $FMC(f_1 f_0, S_0, w \times h)$, where $w=w_0+w_1$.

Proof: Conceptually, the 2-d $w \times h$ PARBS (the composition of c_0 and c_1) is partitioned into a 2-d $w_0 \times h$ PARBS, denoted by $SUB\text{-}PARBS_0$, and a 2-d $w_1 \times h$ PARBS, denoted by $SUB\text{-}PARBS_1$. The $SUB\text{-}PARBS_0$ ($SUB\text{-}PARBS_1$) consists of processors $P_{i,j}$'s, where $i=0, 1, ..., w_0-1$ ($i=w_0$, $w_0+1, ..., w-1$) and $j=0, 1, ..., h-1$. A connected-path $(O_0, O_1, O_2, ..., O_r)$ is said to be in $SUB\text{-}PARBS_0$ ($SUB\text{-}PARBS_1$) if each port O_k, $0 \le k \le r-1$, is within a processor of $SUB\text{-}PARBS_0$ ($SUB\text{-}PARBS_1$). Clearly, the composition of c_0 and c_1 satisfies the first two properties of an $FMC(f_1 f_0, S_0, w \times h)$.

Since c_0 is an $FMC(f_0, S_0, w_0 \times h)$, for each $j \in S_0$, there exists a connected-path $(L_{0,j}, O_1, O_2, ..., O_r, R_{w_0-1,f_0(j)})$ from port $L_{0,j}$ to port $R_{w_0-1,f_0(j)}$ in $SUB\text{-}PARBS_0$. Also, since c_1 is an $FMC(f_1, S_1, w_1 \times h)$ and $f_0(j) \in S_1$, there exists a connected-path $(L_{w_0,f_0(j)}, O'_1, O'_2, ..., O'_q, R_{w-1,f_1 f_0(j)})$ from port $L_{w_0,f_0(j)}$ to port $R_{w-1,f_1 f_0(j)}$ in $SUB\text{-}PARBS_1$. Consequently, $(L_{0,j}, O_1, ..., O_r, R_{w_0-1,f_0(j)}, L_{w_0,f_0(j)}, O'_1, ..., O'_q, R_{w-1,f_1 f_0(j)})$ is a connected-path from port $L_{0,j}$ to port $R_{w-1,f_1 f_0(j)}$ in the 2-d $w \times h$ PARBS. In other words, for each $j \in S_0$, port $L_{0,j}$ and port $R_{w-1,f_1 f_0(j)}$ are connected to the same bus in the 2-d $w \times h$ PARBS.

To complete the proof, we only need to show that no two ports on column $w-1$ are connected to the same bus. By contradiction, this can be proved as follows. Suppose there are two ports $R_{w-1,j}$ and $R_{w-1,j'}$ on column $w-1$ that are connected to the same bus, where $0 \le j \le h-1$, $0 \le j' \le h-1$ and $j \ne j'$. Equivalently, there is a connected-path $(R_{w-1,j}, O_1, O_2, ..., O_r, R_{w-1,j'})$ from port $R_{w-1,j}$ to port $R_{w-1,j'}$ in the 2-d $w \times h$ PARBS. Since c_1 is an $FMC(f_1, S_1, w_1 \times h)$, there is no connected-path from port $R_{w-1,j}$ to port $R_{w-1,j'}$ in $SUB\text{-}PARBS_1$. Thus, there must exist $2t$ ($t \ge 1$) integers k_0, $k_1, ..., k_{2t-1}$ such that $(R_{w-1,j}, O_1, ..., O_{k_0})$ is a connected-path in $SUB\text{-}PARBS_1$, $(O_{k_{l+1}}, ..., O_{k_{(l+1)}})$, $0 \le l < 2t-1$, is a connected-path in $SUB\text{-}PARBS_0$ ($SUB\text{-}PARBS_1$) if l is even (odd), and $(O_{k_{2t-1}+1}, ..., R_{w-1,j'})$ is a connected-path in $SUB\text{-}PARBS_1$.

Clearly, (O_{k_0}, O_{k_0+1}) must be $(L_{w_0,j_0}, R_{w_0-1,j_0})$ for some j_0 and (O_{k_1}, O_{k_1+1}) must be $(R_{w_0-1,j_1}, L_{w_0,j_1})$ for some j_1, where $0 \le j_0 \le h-1$, $0 \le j_1 \le h-1$ and $j_0 \ne j_1$. Consequently, $(O_{k_0+1}, ..., O_{k_1})$ is a connected-path from port R_{w_0-1,j_0} to port R_{w_0-1,j_1} in $SUB\text{-}PARBS_0$, which is impossible since c_0 is an $FMC(f_0, S_0, w_0 \times h)$. This completes the proof.   Q.E.D.

Let us consider Figure 4 again. According to *Lemma 1*, we know that if processor $P_{0,j}$, $j \in \{0, 1, 2\}$, broadcasts a signal on the bus to which port $L_{0,j}$ is connected, only one processor ($P_{5,(2j+1)}$ **div** 2) on column 5 can receive the signal from port R ($R_{5,(2j+1)}$ **div** 2). However, processor $P_{2,2j+1}$ is now no longer the unique one on column 2 that can receive the signal from port R, although the bus configuration established on the sub-PARBS that contains processors ranging from column 0 to column 2 is an $FMC(f(j)=2*j+1$, $\{0, 1, 2\}$, $3 \times 6)$. For example, both processors $P_{2,0}$ and $P_{2,1}$ can receive the signal from port $R_{2,0}$ and port $R_{2,1}$, respectively, if processor $P_{0,0}$ broadcasts a signal on the bus to which port $L_{0,0}$ is connected. Note that this situation occurs because the function $f(j)=j$ **div** 2 is not one-to-one.

By applying *Lemma 1* repeatedly, we have the following lemma.

Lemma 2: Let c_k denote an $FMC(f_k, S_k, w_k \times h)$, $0 \le k \le r-1$, and suppose c_0, c_1, ..., c_{r-1} are composable. The composition of c_0, c_1, ..., c_{r-1} is an $FMC(f_{r-1} ... f_1 f_0, S_0, w \times h)$, where $w= w_0+w_1+ ... +w_{r-1}$.

Theorem 1: Suppose there exists a problem that can be expressed as a function $F: X \longrightarrow Y$, where X is the set of all possible input instances and $Y=\{0, 1, 2, ..., h-1\}$ is the set of their corresponding answers. The problem can be solved by configurational computation in $O(1)$ time on a 2-d $w \times h$ PARBS if for each $x \in X$,

(1) there exist an integer j_x and r functions $f_{x,0}, f_{x,1}, ..., f_{x,r-1}$ such that $F(x)=f_{x,r-1} ... f_{x,1} f_{x,0}(j_x)$, where $0 \le j_x \le h-1$ and $r \ge 1$,
(2) there exist r composable function-mapping configurations $c_{x,0}$, $c_{x,1}$, ..., $c_{x,r-1}$, where $c_{x,k}$, $0 \le k \le r-1$, is an $FMC(f_{x,k}, S_{x,k}, w_{x,k} \times h)$, $w=w_{x,0}+w_{x,1}+ ... +w_{x,r-1}$ and $j_x \in S_{x,0}$, and
(3) in $O(1)$ time the value j_x can be known to processor P_{0,j_x} and the composition of $c_{x,0}$, $c_{x,1}$, ..., $c_{x,r-1}$ can be established on the 2-d $w \times h$ PARBS.

Proof: Based on configurational computation, the problem can be solved as follows: define the construction rule as the composition of $c_{x,0}$, $c_{x,1}$, ..., $c_{x,r-1}$ on the 2-d $w \times h$ PARBS for each $x \in X$, and then solve the problem by the following three-phase procedure. (Initially, an input instance x is stored in the PARBS.)

Phase 1: Establish a configuration of the reconfigurable bus system according to the construction rule. By *Lemma 2*, the established configuration is an $FMC(f_{x,r-1} ... f_{x,1} f_{x,0}, S_{x,0}, w \times h)$.
Phase 2: Determine the value j_x and let processor P_{0,j_x} broadcast a signal on the bus to which port L_{0,j_x} is

connected. (Assume processor $P_{w-1,v}$ on column w-1 receives the signal from port $R_{w-1,v}$.)

Phase 3: Each processor $P_{w-1,j}$, $0 \le j \le h$-1, establishes local connection $\{U, D\}$ to form a straight bus along j-direction, which connects all processors on column w-1 together. Then, processor $P_{w-1,v}$ transmits the value v to processor $P_{w-1,0}$ through the established bus.

After executing the above procedure, the answer $v=F(x)$, can be found in processor $P_{w-1,0}$. Clearly, the overall time complexity of the procedure is $O(1)$. The correctness of the procedure is assured by *Lemma 2*.   Q.E.D.

Indeed, *Theorem 1* suggests a sufficient condition for a problem to be solved in $O(1)$ time on a 2-d PARBS. As will be seen in the next section, the sufficient condition makes it easy to derive $O(1)$ time algorithms on 2-d PARBS's for many problems. These problems include computing the logical exclusive-OR of n bits, adding two n-bit binary integers, summing n bits, summing n m-bit binary integers and multiplying two n-bit binary integers. More applications are still to be found.

5. Applications

5.1 Computing the logical exclusive-OR of n bits

This problem can be expressed as a function $F: X \longrightarrow Y$, where $X=\{(b_0, b_1, ..., b_{n-1}) \mid b_k=0,1$ for $k=0,...,n-1\}$, $Y=\{0, 1\}$, and $F(b_0, b_1, ..., b_{n-1})=b_0 \oplus b_1 \oplus ... \oplus b_{n-1}$. Here, $\oplus$ denotes the logical exclusive-OR operation.

Define two functions XOR_0, XOR_1 as follows: $XOR_0=\{(0, 0), (1, 1)\}$ and $XOR_1=\{(0, 1), (1, 0)\}$ (i.e., for each $b \in \{0, 1\}$, $XOR_0(b)=b$ and $XOR_1(b)=1-b$). Then, $F(x)=XOR_{b_{n-1}}...XOR_{b_1}XOR_{b_0}(0)$ for each $x=(b_0, b_1, ..., b_{n-1}) \in X$ (i.e., $j_x=0$, $r=n$, and $f_{x,k}=XOR_{b_k}$ for $k=0,...,n-1$). A function-mapping configuration $c_{x,k}$, which is an $FMC(XOR_{b_k}, \{0, 1\}, 2 \times 3)$, can be established on a 2-d 2×3 PARBS as follows. (See Figure 5.)

Step 0: Initially, b_k is stored in processor $P_{0,0}$.

Step 1: // The value b_k is broadcast to all processors. //
Processor $P_{0,0}$ sends a copy of b_k to processor $P_{1,0}$. Then, all processors establish local connection $\{U, D\}$ to form straight buses along j-direction. And then, processors $P_{0,0}$ and $P_{1,0}$ broadcast b_k on the established buses.
After Step 1, each processor owns a copy of b_k.

Step 2: // Establish the bus configuration as an $FMC(XOR_{b_k}, \{0, 1\}, 2 \times 3)$. //
Each processor $P_{0,j}$, $0 \le j \le 2$, establishes local connection $\{L, R\}$ if $j \le 1$ and $b_k=0$, $\{L, U\}$ if $j=0$ and $b_k=1$, $\{L, R\}$, $\{U, D\}$ if $j=1$ and $b_k=1$, and $\{D, R\}$ if $j=2$ and $b_k=1$. And, each processor $P_{1,j}$, $0 \le j \le 2$, establishes local connection $\{L, R\}$ if $j \le 1$ and $b_k=0$, $\{U, R\}$ if $j=0$ and $b_k=1$, $\{L, D\}$, $\{U, R\}$ if $j=1$ and $b_k=1$, and $\{L, D\}$ if $j=2$ and $b_k=1$.

Since $XOR_{b_k}... XOR_{b_1}XOR_{b_0}(\{0, 1\}) \subseteq \{0, 1\}$ for $k=0,1,...,n$-2, $c_{x,0}, c_{x,1}, ..., c_{x,n-1}$ are composable and thus their composition is an $FMC(XOR_{b_{n-1}} ...$

$XOR_{b_1}XOR_{b_0}, \{0, 1\}, 2n \times 3)$. For example, the composition of $c_{x,0}, c_{x,1}, c_{x,2}, c_{x,3}$ for $x=(b_0, b_1, b_2, b_3)=(1, 0, 1, 1)$ is depicted in Figure 6. It is not difficult to check that in $O(1)$ time the value $j_x(=0)$ can be known and the composition of $c_{x,0}, c_{x,1}, ..., c_{x,n-1}$ can be established on a 2-d $2n \times 3$ PARBS if b_k, $0 \le k \le n$-1, is initially stored in processor $P_{2k,0}$. Thus, by *Theorem 1*, the problem of computing the logical exclusive-OR of n bits can be solved in $O(1)$ time on a 2-d $2n \times 3$ PARBS.

Note that since the functions XOR_0, XOR_1 are one-to-one, all the values $b_0 \oplus b_1 \oplus ... \oplus b_i$, $i=0, 1, ..., n$-2, are also computed simultaneously.

5.2 Adding two n-bit binary integers

Let $A=a_{n-1}a_{n-2} ... a_0$, $B=b_{n-1}b_{n-2} ... b_0$ be two n-bit integers, and $Q=q_{n-1}q_{n-2} ... q_0$ be their sum (i.e., $Q=(A+B)$ **mod 2^n**). It is known [5] that $q_t=a_t \oplus b_t \oplus e_t$, where $e_0=0$ and $e_t=(a_{t-1} \wedge b_{t-1}) \vee ((a_{t-1} \oplus b_{t-1}) \wedge e_{t-1})$ for $1 \le t \le n$-1. The boolean value e_t is called the *carry* to bit position t. The integer Q can be computed in $O(1)$ time if all the carries e_t, $t=1,...,n$-1, can be determined in $O(1)$ time.

The problem of computing each carry e_t, $1 \le t \le n$-1, can be expressed as a function $F: X \longrightarrow Y$, where $X=\{(a_{n-1}a_{n-2} ... a_0, b_{n-1}b_{n-2} ... b_0) \mid a_k=0,1$ and $b_k=0,1$ for $k=0,...,n$-1$\}$, $Y=\{0, 1\}$, and $F(a_{n-1}a_{n-2} ... a_0, b_{n-1}b_{n-2} ... b_0)=e_t$.

Define three functions qf_0, qf_1, qf_2 as follows: $qf_0=\{(0, 0), (1, 0)\}$, $qf_1=\{(0, 0), (1, 1)\}$, and $qf_2=\{(0, 1), (1, 1)\}$. Then, $F(x)=qf_{a_{t-1}+b_{t-1}} \cdots qf_{a_1+b_1}qf_{a_0+b_0}(0)$ for each $x=(a_{n-1}a_{n-2} ... a_0, b_{n-1}b_{n-2} ... b_0) \in X$ (i.e., $j_x=0$, $r=t$, and $f_{x,k}=qf_{a_k+b_k}$ for $k=0,...,t$-1). Note that only the rightmost t bits of A and B are required to determine e_t. It is not difficult to check that a function-mapping configuration $c_{x,k}$, which is an $FMC(qf_{a_k+b_k}, \{0, 1\}, 1 \times 2)$, can be easily established in $O(1)$ time on a 2-d 1×2 PARBS if a_k and b_k are initially stored in processor $P_{0,0}$ (see Figure 7).

Clearly, $c_{x,0}, c_{x,1}, ..., c_{x,t-1}$ ($1 \le t \le n$-1) are composable and their composition, which is an $FMC(qf_{a_{t-1}+b_{t-1}} \cdots qf_{a_1+b_1}qf_{a_0+b_0}, \{0, 1\}, t \times 2)$, can be established on a 2-d $t \times 2$ PARBS in $O(1)$ time if a_k and b_k, $0 \le k \le t$-1, are initially stored in processor $P_{k,0}$. Thus, by *Theorem 1*, the carry e_t can be computed in $O(1)$ time on a 2-d $t \times 2$ PARBS. Using this result, the problem of adding two n-bit binary integers can be solved in $O(1)$ time on a 2-d $n \times 2(n$-1$)$ PARBS if a_k and b_k, $0 \le k \le n$-1, are initially stored in processor $P_{k,0}$ and proper data routing is performed. Note that computing all carries $e_1, e_2, ..., e_{n-1}$ can not be done simultaneously on an $n \times 2$ PARBS since not all fuctions $qf_{a_k+b_k}$ are one-to-one.

5.3 Summing n bits

This problem can be expressed as a function $F: X \longrightarrow Y$, where $X=\{(b_0, b_1, ..., b_{n-1}) \mid b_k=0,1$ for $k=0,...,n$-1$\}$, $Y=\{0, 1, ..., n\}$, and $F(b_0, b_1, ..., b_{n-1})=b_0+b_1+ ... +b_{n-1}$.

Define two integer functions $INCR_0, INCR_1$ as follows: for each integer j, $INCR_0(j)=j$ and $INCR_1(j)=j+1$. Then, $F(x)=INCR_{b_{n-1}} ... INCR_{b_1}INCR_{b_0}(0)$ for each $x=(b_0, b_1, ..., b_{n-1}) \in X$, (i.e., $j_x=0$, $r=n$, and $f_{x,k}=INCR_{b_k}$ for $k=0,1,$

...,n-1). Clearly, a function-mapping configuration $c_{x,k}$, which is an $FMC(INCR_{b_k}, \{0, 1, ..., k\}, 1 \times (n+1))$, can be easily established in $O(1)$ time on a 2-d $1 \times (n+1)$ PARBS if b_k, $0 \leq k \leq n$-1, is initially stored in processor $P_{0,0}$. Figure 8 shows an example of $n=5$.

Since $INCR_{b_k} ... INCR_{b_1}INCR_{b_0}(\{0\}) \subseteq \{0, 1, ..., k+1\}$ for $k=0,1,...,n$-2, $c_{x,0}, c_{x,1}, ..., c_{x,n-1}$ are composable. It is not difficult to see that in $O(1)$ time the value $j_x(=0)$ can be determined and the composition of $c_{x,0}$, $c_{x,1}, ..., c_{x,n-1}$ can be established on a 2-d $n \times (n+1)$ PARBS if b_k, $0 \leq k \leq n$-1, is initially stored in processor $P_{k,0}$. Thus, by *Theorem 1*, the problem of summing n bits can be solved in $O(1)$ time on a 2-d $n \times (n+1)$ PARBS. Note that the derived $O(1)$ time algorithm can be easily adapted for the problem of computing the prefix sums of n bits since the functions $INCR_0$, $INCR_1$ are one-to-one.

5.4 Summing n m-bit binary integers and multiplying two n-bit binary integers

The multiplication of two n-bit binary integers $A=a_{n-1}a_{n-2} ... a_0$ and $B=b_{n-1}b_{n-2} ... b_0$ can be computed by summing n $2n$-bit integers $M_0, M_1, ..., M_{n-1}$, where $M_k=(A*b_k)*2^k$, $0 \leq k \leq n$-1. Therefore, we only show how to solve the problem of summing n m-bit integers.

Let $A_0=a_{0,m-1}a_{0,m-2} ... a_{0,0}$, $A_1=a_{1,m-1}a_{1,m-2} ... a_{1,0}$, ... , $A_{n-1}=a_{n-1,m-1}a_{n-1,m-2} ... a_{n-1,0}$ be n m-bit integers, and $S=s_{m-1}s_{m-2} ... s_0$ be their sum (i.e., $S=(A_0+A_1+ ... +A_{n-1})$ **mod** 2^m). Define $u_{-1}=0$ and $u_t=(a_{0,t}+a_{1,t}+ ... +a_{n-1,t})+(u_{t-1}$ **div** 2) for $t=0,1,...,m$-1, where u_{t-1} **div** 2 is the carry to bit position t. Then, $s_t=u_t$ **mod** 2, for $t=0,1,...,m$-1. Since the carry to bit position t is not greater that n, we have $u_t \leq 2n$-1. The problem of summing n m-bit integers can be solved in $O(1)$ time if all u_t's can be determined in $O(1)$ time.

The problem of computing each u_t, $0 \leq t \leq m$-1, can be expressed as a function $F: X \longrightarrow Y$, where $X=\{ (a_{0,m-1}a_{0,m-2} ... a_{0,0}, a_{1,m-1}a_{1,m-2} ... a_{1,0}, ... , a_{n-1,m-1}a_{n-1,m-2} ... a_{n-1,0}) \mid a_{k,l}=0,1$ for $k=0,...,n$-1 and $l=0,...,m$-1$\}$, $Y=\{0, 1, ..., 2n$-1$\}$, and $F(a_{0,m-1}a_{0,m-2} ... a_{0,0}, a_{1,m-1}a_{1,m-2} ... a_{1,0}, ... , a_{n-1,m-1}a_{n-1,m-2} ... a_{n-1,0})=u_t$.

Define $n+1$ functions $sf_0, sf_1, ..., sf_n$ as follows: for each $j \in \{0, 1, ..., 2n$-1$\}$, $sf_k(j)=k+(j$ **div** 2$)$, $k=0,1,...,n$. Then, $F(x)=sf_{a_{0,t}+a_{1,t}+...+a_{n-1,t}} ... sf_{a_{0,1}+a_{1,1}+...+a_{n-1,1}} sf_{a_{0,0}+a_{1,0}+...+a_{n-1,0}}(0)$ for each $x=(a_{0,m-1}a_{0,m-2} ... a_{0,0}, a_{1,m-1}a_{1,m-2} ... a_{1,0}, ... , a_{n-1,m-1}a_{n-1,m-2} ... a_{n-1,0}) \in X$ (i.e., $j_x=0$, $r=t+1$, and $f_{x,k}=sf_{a_{0,k}+a_{1,k}+...+a_{n-1,k}}$ for $k=0,1,...,t$. A function-mapping configuration $c_{x,k}$, which is an $FMC(sf_{a_{0,k}+a_{1,k}+...+a_{n-1,k}}, \{0, 1, ..., 2n-1\}, 2n \times 2n)$, can be established on a 2-d $2n \times 2n$ PARBS as follows. (See Figure 9, where an example with $n=3$, $a_{0,k}=1$, $a_{1,k}=0$ and $a_{2,k}=1$ is shown.)

Step 0: Initially, each $a_{i,k}$ is stored in processor $P_{i+n,0}$, $0 \leq i \leq n$-1.

Step 1: // The value $a_{i,k}$ is broadcast to each processor $P_{i+n,j}$, $0 \leq i \leq n$-1, $0 \leq j \leq 2n$-1. //
Each processor $P_{i+n,j}$, $0 \leq i \leq n$-1, $0 \leq j \leq 2n$-1, establishes local connection $\{U, D\}$ to form straight

buses along j-direction. Then, each processor $P_{i+n,0}$, $0 \leq i \leq n$-1, broadcasts $a_{i,k}$ on the established bus to which it is connected.
After Step 1, each processor $P_{i+n,j}$, $0 \leq i \leq n$-1, $0 \leq j \leq 2n$-1, owns a copy of $a_{i,k}$.

Step 2: // Establish the bus configuration as an $FMC(sf_{a_{0,k}+a_{1,k}+...+a_{n-1,k}}, \{0,1, ...,2n-1\}, 2n \times 2n)$.//
Each processor $P_{i,j}$, $0 \leq i \leq n$-1, $0 \leq j \leq 2n$-1, establishes local connection $\{L, D\}$, $\{U, R\}$ if $i<j$, $\{U, L, R\}$ if $i=j$, and $\{L, R\}$ if $i>j$. And, each processor $P_{i+n,j}$, $0 \leq i \leq n$-1, $0 \leq j \leq 2n$-1, establishes local connection $\{L, R\}$ if $a_{i,k}=0$, and $\{L, U\}$, $\{D, R\}$ if $a_{i,k}=1$.

Note that $c_{x,k}$ is really the composition of $n+1$ composable function-mapping configurations; the first one is an $FMC(f(j)=j$ **div** 2, $\{0, 1, ..., 2n$-1$\}, n \times 2n)$, and the l-th one is an $FMC(INCR_{a_{l-2,k}}, \{0, 1, ..., n+l-3\}, 1 \times 2n)$ for $l=2, 3, ..., n+1$. Therefore, $c_{x,k}$ is an $FMC(sf_{a_{0,k}+a_{1,k}+...+a_{n-1,k}}, \{0, 1, ..., 2n-1\}, 2n \times 2n)$.

Clearly, $c_{x,0}, c_{x,1}, ..., c_{x,t}$, $0 \leq t \leq m$-1, are composable. And, in $O(1)$ time the value $j_x(=0)$ can be determined and the composition of $c_{x,0}, c_{x,1}, ..., c_{x,t}$ can be established on a 2-d $2(t+1)n \times 2n$ PARBS if each $a_{i,k}$, $0 \leq i \leq n$-1, $0 \leq k \leq t$, is initially stored in processor $P_{(2k+1)n+i,0}$. Thus, by *Theorem 1*, the value u_t can be computed in $O(1)$ time on a 2-d $2(t+1)n \times 2n$ PARBS. Using this result, the problem of summing n m-bit integers can be solved in $O(1)$ time on a 2-d $2mn \times 2mn$ PARBS if each $a_{i,k}$, $0 \leq i \leq n$-1, $0 \leq k \leq m$-1, is initially stored in processor $P_{(2k+1)n+i,0}$ and proper data routing is performed.

Based on the above discussion, we know that the problem of multiplying two n-bit binary integers can be solved in $O(1)$ time on a 2-d $4n^2 \times 4n^2$ PARBS (the case of $m=2n$), if proper data routing is performed to create the necessary n $2n$-bit binary integers. We omit the routing procedure here. An interested reader can work out the details without much difficulty.

6. Further improvement

As we know from the previous discussion, it is possible for a given problem to derive more than one $O(1)$ time algorithm from *Theorem 1*. Different algorithms may require PARBS's of different sizes. For the purpose of cost-effectiveness, we wish to minimize the size of the PARBS (i.e, the number of used processors). So, in this section, we suggest two approaches that are useful in reducing the size of the PARBS, when we design algorithms according to *Theorem 1*.

The first approach is to decompose F, which expresses the input problem, into composable functions $f_{x,k}$, more carefully. It is clear that the decomposition of F is not unique, and different decompositions of F may result in PARBS's of different sizes. If we choose $f_{x,k}$'s carefully, it is very possible to obtain a better PARBS. For example, let us consider again the problem of computing the logical exclusive-OR of n bits. As shown in the previous section, if we choose $f_{x,k}=XOR_{b_k}$, $k=0, 1, ..., n$-1, the resulting PARBS is of size $2n \times 3$. If, instead, we let (assume n is even) $f_{x,k}=\{(0, 1), (1, 2)\}$ if k is even and $b_k=0$, $f_{x,k}=\{(0, 2), (1, 1)\}$ if k is even and $b_k=1$, $f_{x,k}=\{(1, 0), (2, 1)\}$ if k is odd and $b_k=0$, and $f_{x,k}=\{(1, 1), (2, 0)\}$ if k is odd and $b_k=1$, then the resulting PARBS is of size $n \times 3$, since each of $FMC(f_{x,2i}, \{0, 1\}, 1 \times 3)$ and $FMC(f_{x,2i+1}, \{1, 2\}, 1 \times 3)$ can be established in $O(1)$ time on a 2-d 1×3 PARBS.

The second approach is to transform the input problem into another problem, and then design algorithms, according to *Theorem 1*, for the new problem. The solution of the original problem can be obtained by solving the new problem. Besides, more importantly, solving the new problem requires fewer processors than solving the original one, according to *Theorem 1*. For example, let us consider again the problem of summing n bits, which we have solved in the previous section in $O(1)$ time on a 2-d $n \times (n+1)$ PARBS. In the following, we show that this problem can be solved with fewer processors.

Let $p_0, p_1, ..., p_{t-1}$ be mutually prime positive integers. According to the Chinese remainder theorem [4], a positive integer i smaller than the product $p_0 p_1 ... p_{t-1}$ can be uniquely determined if the values $i \bmod p_0$, $i \bmod p_1$, ..., $i \bmod p_{t-1}$ are known. Thus, choosing $t=2$, $p_0 = \lceil n^{1/2} \rceil$ and $p_1 = \lceil n^{1/2} \rceil + 1$, the sum of n bits $b_0, b_1, ..., b_{n-1}$ can be determined if the values $(b_0 + b_1 + ... + b_{n-1}) \bmod p_0$ and $(b_0 + b_1 + ... + b_{n-1}) \bmod p_1$ are computed. According to *Theorem 1*, these two values can be easily computed in $O(1)$ time on 2-d PARBS's of sizes $2n \times (p_0 + 1)$ and $2n \times (p_1 + 1)$, respectively. Thus, the problem of summing n bits can be solved in $O(1)$ time on a 2-d $2n \times (\lceil n^{1/2} \rceil + 2)$ PARBS. Indeed, by properly choosing $t, p_0, p_1, ..., p_{t-1}$, it is possible to compute the sum of n bits in $O(1)$ time on a 2-d PARBS of size $O(n \times n^\varepsilon)$ for any fixed $\varepsilon > 0$ (the interested readers may consult any book, e.g. [11], about the *theory of numbers* for useful properties of integers).

One more example is the problem of summing n m-bit integers $A_i = a_{i,m-1} a_{i,m-2} ... a_{i,0}$, $i=0, 1, ..., n-1$, which we have solved in the previous section in $O(1)$ time on a 2-d $2mn \times 2mn$ PARBS. Let $l = \lceil \log_2(n+1) \rceil$ and $X_j = x_{j,l-1} x_{j,l-2} ... x_{j,0}$ be the sum of $a_{n-1,j}, a_{n-2,j}, ..., a_{0,j}$ for $j=0, 1, ..., m-1$. We have $A_0 + A_1 + ... + A_{n-1} = X_0 * 2^0 + X_1 * 2^1 + ... + X_{m-1} * 2^{m-1}$. Also, let Y_i denote an integer value whose binary representation is obtained by packing $X_i, X_{i+l}, X_{i+2l}, ..., X_{i+\lfloor (m-1)/l \rfloor * l}$ into an $(m+l)$-bit integer word (i.e., $Y_i = X_i * 2^i + X_{i+l} * 2^{i+l} + X_{i+2l} * 2^{i+2l} + ... + X_{i+\lfloor (m-1)/l \rfloor * l} * 2^{i+\lfloor (m-1)/l \rfloor * l}$) for $i=0, 1, ..., l-1$. Then, we have $A_0 + A_1 + ... + A_{n-1} = Y_0 + Y_1 + ... + Y_{l-1}$, which implies that the problem of summing n m-bit integers can be transformed into the problem of summing l $(m+l)$-bit integers. The transformation requires the computation of all X_j's, which can be performed in $O(1)$ time on a 2-d PARBS of size $O(n \times mn^\varepsilon)$, according to the discussion of the previous paragraph. Then, Y_i's can be obtained by proper data routing (note that each Y_i is distributed over $l+m$ processors, each holding one bit). The routing procedure is rather lengthy, and is omitted in this paper. The interested readers are encouraged to work out the details. By applying the result of the previous section, the sum of $Y_0, Y_1, ..., Y_{l-1}$ can be computed in $O(1)$ time on a 2-d $2l(m+l) \times 2l(m+l)$ PARBS. Thus, the multiplication of two n-bit integers can be performed in $O(1)$ time on a 2-d PARBS of size $O(n \log n \times n^{1+\varepsilon})$.

7. Concluding remarks

In this paper, configurational computation, which is a new computation method on the PARBS, was introduced.

Based on this strategy, $O(1)$ time algorithms for computing the logical exclusive-OR of n bits, adding two n-bit binary integers, summing n bits, summing n m-bit binary integers and multiplying two n-bit binary integers, all on 2-d PARBS's, were derived. Many other problems, e.g., computing the logical AND of n bits, computing the logical OR of n bits, comparing two n-bit binary integers and determining the 2's complement of an n-bit binary integer, can also be solved similarly by configurational computation. Indeed, the $O(1)$ time algorithms proposed in [2], [16], [18] for solving the sorting problem, the longest common subsequence problem and the transitive closure problem can all be obtained by configurational computation.

Note that it is still possible to reduce the number of processors used in the $O(1)$ time algorithms that are derived from *Theorem 1*. For example, the proposed $O(1)$ time algorithm for adding two n-bit integers can be adapted to a linear PARBS of size n [17]. However, this improvement is problem dependent and there is no general way to do so.

It is worth mentioning that at least two VLSI implementations have be implemented to demonstrate the feasibility and benefits of the 2-d PARBS; one is the YUPPIE (Yorktown Ultra-Parallel Polymorphic Image Engine) chip [7], [8] and the other is the GCN (Gated-Connection Network) chip [15]. These implementations suggested that the broadcast delay, although not constant, is very small. For example, only 16 machine cycles are required to broadcast on a one million-processor YUPPIE. The GCN has further shorten the delay by adopting pre-charged circuits. This makes the algorithms based on configurational computation very efficient. Such a performance can be very useful for applications (such as digital signal processing) that required fast special-purpose implementation of arithmetic functions representable by *FMC*. Ideally, it has been shown in [14] that the $O(1)$ time claim may be made true if the reconfigurable bus system is implemented using Optical Fibers [3] as the underlying global bus system and Electrically Controlled Coupler Switches (ECS) [3] for connecting or disconnecting two fibers.

Finally, some future research topics are suggested as follows.

- Find more applications that can be solved efficiently by configurational computation.

- Find more rules, like *Theorem 1*, that are useful in designing efficient algorithms by configurational computation.

- Find rules that can be used to systematically (automatically) generate an *FMC* construction, to detect the existence of *FMC*, and to obtain an optimal *FMC* with respect to both time and number of processors.

- Study some theoretical properties about configurational computation, e.g., what is the time lower bound for a given problem to be solved by configurational computation? and if $O(1)$ time is possible, what is the processors lower bound?

Acknowledgement

This research is partially supported by Advanced Technology Center, Computer & Communication Research Laboratories, Industrial Technology Research Institute.

Reference

[1] A. Aggarwal, "Optimal bounds for finding maximum on array of processors with *k* global buses," *IEEE Transactions on Computers*, vol. C-35, no. 1, pp. 62-64, 1986.

[2] D. M. Champion and J. Rothstein, "Immediate parallel solution of the longest common subsequence problem," in *Proceedings of the 1987 International Conference on Parallel Processing*, 1987, pp. 70-77.

[3] D. G. Feitelson, *Optical Computing*, MIT press,1988.

[4] R. L. Graham, D. E. Knuth, and O. Patashnik, *Concrete Mathematics*, New York: Addison- Wesley, 1989.

[5] K. Hwang, *Computer Arithmetic*, New York: Wiley, 1979.

[6] D. Kim and K. Hwang, "Mesh-connected array processors with bypass capability for signal/image processing," in *Proceedings of the Hawaii Conference on System Science*, 1988.

[7] H. Li and M. Maresca, "Polymorphic-torus network," *IEEE Transactions on Computers*, vol. C-38, no. 9, pp. 1345-1351, 1989.

[8] M. Maresca and H. Li, "Connection autonomy in SIMD computers: a VLSI implementation," *Journal of Parallel and Distributed Computing*, vol. 7, no. 2, pp. 302-320, 1989.

[9] P. Mckinley, "Multicast routing in spanning bus hypercubes," in *Proceedings of the 1988 International Conference on Parallel Processing*, vol. 2, 1988, pp. 204-211.

[10] R. Miller, V. K. Prasanna Kumar, D. Reisis, and Q. F. Stout, "Data movement operations and applications on reconfigurable VLSI arrays," in *Proceedings of the International Conference on Parallel Processing*, vol. 1, 1988, pp. 205-208.

[11] I. Niven and H. S. Zuckerman, *An Introduction to the Theory of Numbers*, 3rd ed., New York: John Wiley and Sons, Inc, 1972.

[12] V. K. Prasanna Kumar and C. S. Raghavendra, "Array processor with multiple broadcasting," *Journal of Parallel and Distributed Computing*, vol. 4, no. 2, pp. 173-190, 1987.

[13] J. Rothstein, "Bus automata, brains, and mental models," *IEEE Transactions on Systems,Man, and Cybernetics*, vol. SMC-18, vol. 4, pp. 522-531, 1988.

[14] A. Schuster and Y. Ben-Asher, "Algorithms and optic implementation for reconfigurable networks," in *Proceedings of the 5th Jerusalem Conference on Information Technology*, October, 1990.

[15] D. B. Shu, L. W. Chow, and J. G. Nash, "A content addressable, bit serial associate processor," in *Proceedings of the IEEE Workshop on VLSI Signal Processing*, Montery CA, Nov. 1988.

[16] B. F. Wang and G. H. Chen, "Constant time algorithms for the transitive closure problem and some related graph problems on processor arrays with reconfigurable bus systems," *IEEE Transactions on Parallel and Distributed Systems*, vol. 1, no. 4, pp. 500-507.

[17] B. F. Wang, G. H. Chen, and H. Li, "Fast algorithms for some arithmetic and logic operations," Tech. Rep., Department of Computer Science and Information Engineering, National Taiwan University, 1990.

[18] B. F. Wang, G. H. Chen, and F. C. Lin, "Constant time sorting on a processor array with a reconfigurable bus system," *Information Processing Letters*, vol. 34, no. 4, pp. 187-192, 1990.

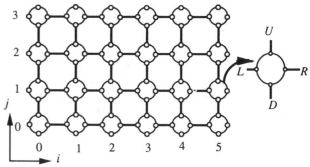

Figure 1. A 2-d 6x4 PARBS.

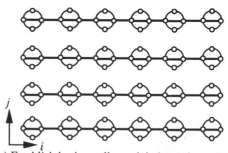

(a) Establish horizontally straight buses by connecting port *L* to port *R* within each processor.

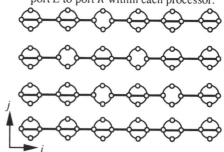

(b) Split the buses established in (a) into sub-buses by disconnecting some local connections between port *L* and port *R*.

Figure 2. Two configurations of the reconfigurable bus system shown in Figure 1.

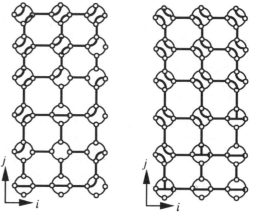

(a) An *FMC*(*f*(*j*)=2*j+1, {0, 1, 2}, 3x6).   (b) An *FMC*(*f*(*j*)=*j* **div** 2, {0, 1, ..., 5}, 3x6).

Figure 3. Two function-mapping configurations.

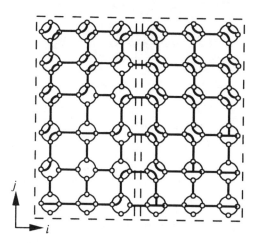

Figure 4. The composition of the two function-mapping configurations depicted in Figure 3.

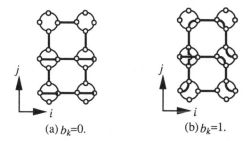

(a) $b_k=0$.   (b) $b_k=1$.

Figure 5. An $FMC(XOR_{b_k}, \{0, 1\}, 2\times3)$, $0 \le k \le n-1$.

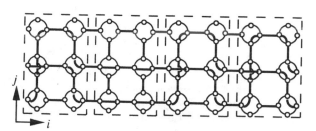

Figure 6. The composition of an $FMC(XOR_1, \{0, 1\}, 2\times3)$, an $FMC(XOR_0, \{0, 1\}, 2\times3)$, an $FMC(XOR_1, \{0, 1\}, 2\times3)$, and an $FMC(XOR_1, \{0, 1\}, 2\times3)$.

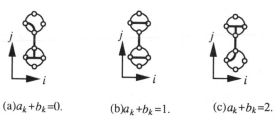

(a)$a_k+b_k=0$.   (b)$a_k+b_k=1$.   (c)$a_k+b_k=2$.

Figure 7. An $FMC(qf_{a_k+b_k}, \{0, 1\}, 2\times3)$, $0 \le k \le n-2$.

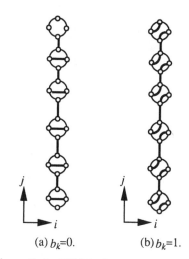

(a) $b_k=0$.   (b) $b_k=1$.

Figure 8. An $FMC(INCR_{b_k}, \{0, 1, ..., k\}, 1\times6)$, $0 \le k \le 4$.

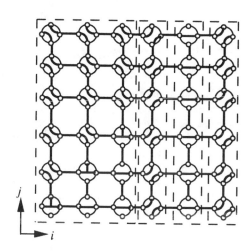

Figure 9. An $FMC(sf_{1+0+1}, \{0, 1, ..., 5\}, 6\times6)$.

Breadth-First Traversal of Trees and Integer Sorting in Parallel *

Calvin C.-Y. Chen and Sajal K. Das

Center for Research in Parallel and Distributed Computing

Department of Computer Sciences

University of North Texas

Denton, Texas 76203-3886

Abstract

We present a cost-optimal parallel algorithm for traversing a general tree in breadth-first fashion. For a tree with n nodes, this algorithm requires $O(n/p + \log n)$ time employing p processors on the EREW PRAM model. To the best of our knowledge, it is the first parallel algorithm for breadth-first tree-traversal, which has $O(\log n)$ time complexity and yet being cost-optimal for $p = O(n/\log n)$. Our approach is based on a novel idea which converts the breadth-first traversal problem into a parentheses matching problem. As an application of this technique, we design a simple but elegant parallel algorithm for sorting a special class of integers in which any two successive integers of the input sequence differ at most by unity. The proposed sorting algorithm is stable and may be applied to many potential areas. It achieves the same performance as that of our breadth-first traversal algorithm, and has merits toward an $O(\log n)$-time, optimal (deterministic) parallel algorithm for bucket-sorting n integers in the range $[1 .. n]$ on the EREW PRAM model.

Key Words: Tree traversal, breadth-first search, parallel algorithm, integer sorting, parentheses matching.

1 Introduction

The tree is a widely used data structure in computer science [AHU83, B86, K73], and traversal of trees is a fundamental operation in manipulating them. Tree traversals are generally classified into two broad types — depth-order and level-order traversals. The *depth-order* includes three commonly-used traversal techniques, namely pre-order, in-order, and post-order; and the *level-order* traversal includes the breadth-first and breadth-depth. There are optimal sequential algorithms for implementing them in $O(n)$ time for an n-node tree [AHU83].

In comparison with the parallel depth-order tree traversals [CD89, KB85, KRS86, RG78, TV85, W79], the parallel breadth-first traversal of trees has not gained that

much attention. Ghosh and Bhattacharjee [GB84] presented an $O(\lambda \log n)$ time algorithm for the latter problem using $O(n^{1+1/\lambda})$ processors, where $\lambda > 0$, on the concurrent-read and exclu-sive-write (CREW), parallel random access machine (PRAM) model. This algorithm is far from being cost-optimal. Recently Chen and Das [CD90] proposed an $O((n/p) \log n / \log(n/p))$ time algorithm using p processors on the exclusive-read and exclusive-write (EREW) PRAM model. Though it yields optimal speedup for $p \leq n^{1-\epsilon}$, where $0 < \epsilon \leq 1$, this algorithm does not achieve $O(\log n)$ time complexity employing $O(n/\log n)$ processors.

Because of important applications of integer sorting, numerous parallel algorithms have been developed for this problem. Here we are interested in sorting a special class of integers (defined later). Since a related concept is *bucket sorting* in which the integers are limited within a specified range, we give an overview of existing parallel algorithms for bucket sorting n integers. Kruskal *et al* [KRS86] and Wagner and Han [WH86] proposed algorithms for sorting integers in the range $[1 .. m]$, which require $O(\frac{(n/p) \log m}{\log(n/p)})$ and $O([\log m / \log(n/p + \log p)](n/p + \log p))$ time, respectively, on the p-processor EREW PRAM model. Cole and Vishkin [CV86] designed another EREW algorithm which runs in $O(\log n)$ time using $O(n/\log n)$ processors, but the integers belong to the range $[1 .. \log n]$. Hagerup's [H87] algorithm has $O(\log n)$ time complexity using $O(\frac{n \log \log n}{\log n})$ processors; and Reif [R85] developed the first cost-optimal (randomized) algorithm which runs in $O(\log n)$ time using $O(\frac{n}{\log n})$ processors. The algorithms in [H87, R85] sort n integers in the range $[1 .. n]$ on the CRCW PRAM. In summary, there is no known EREW-algorithm for bucket sorting n integers in the range $[1 .. n]$, requiring $O(\log n)$ time and $O(n/\log n)$ processors.

In this paper, we design a novel algorithm for breadth-first tree traversal by observing that this problem consists of three subproblems — parentheses matching, linked list ranking, and prefix-sum. For a general tree with n nodes, the proposed algorithm requires $O(n/p + \log n)$ time using p processors on the EREW PRAM model. To the best of our knowledge, this is the first parallel algorithm for breadth-first tree-traversal, achieving $O(\log n)$ time complexity and yet being cost-optimal. This result also unifies

*This work was partially supported by a Junior Faculty Summer Research Fellowship from the University of North Texas.

the performance (time and processor complexities) of both level-order and depth-order traversals of general trees. It can further be shown that the performance of these cost-optimal algorithms are independent of input data structures among the commonly-used ones, namely "leftmost child and right sibling", "parent-of relation with implicit ordering of children", and "parent-of with explicit ordering of children" [AHU83, CD89].

Applying the similar technique as that for obtaining the breadth-first tree-traversal, we present a novel parallel algorithm for sorting a special class of integers in which the input sequence is such that any two successive integers in it differ at most by unity[1]. This integer-sorting algorithm is stable and may be applied to many potential areas such as computer graphics and computational geometry. It also has merits toward a cost-optimal, deterministic parallel algorithm for the general case of sorting n integers in the range $[1 .. n]$ on the EREW PRAM model.

2 Breadth-First Tree-Traversal

Given a tree T, an edge (cn, fn) is assumed to define a pair of arcs (or directed edges) in opposite directions, where node cn is a child of its parent, fn. Specifically, we define a forward arc $Z = < fn, cn >$ from fn to cn, and a backward arc $\bar{Z} = < cn, fn >$ from the child-node, cn, to the parent-node, fn. For example, corresponding to an input tree in Fig. 1, the arcs associated with every edge are depicted in Fig. 2.

Our breadth-first algorithm uses an *Euler tour* of the input tree, which starts from a node, traverses each edge exactly once in both directions and finally returns to the starting node [TV85]. It is known that an input tree can be represented by any one of two relations — "leftmost child and right sibling" and "parent-of with explicit ordering of children" [AHU83, CD89]. The former relation is usually implemented by a data structure which specifies the leftmost child and right sibling of each tree-node. On the other hand, the latter relation is implemented by a data structure which specifies the parent and the rank (among siblings) of each node. The Euler tour implementation of an input tree using either of these representations has the same parallel time complexity as detailed in [CD89, CDA91]. We simply cite the result here.

Lemma 1 *The Euler tour of a tree with n nodes can be obtained in $O(n/p + \log n)$ time employing p processors on the EREW PRAM model.*

Proof. See [CD89, TV85]. □

The arcs traversed in the Euler tour form a sequence if we pick the root of the tree as the starting node and follow the rule that subtrees be visited recursively in a left-to-right manner[2]. We define this unique sequence as the *arc-sequence* of the input tree. The arc-sequence for the tree in Fig. 2 is recorded in Table 1. If Z and $\bar{Z}$ are arcs corresponding to an edge (cn, fn), then the *level* of Z or $\bar{Z}$, denoted as $l(Z)$ or $l(\bar{Z})$, is defined as the level of the child-node, cn, in the tree where the root is assumed to be at level 0. With respect to Fig. 2, $l(E) = level(f) = 1$, and $l(\bar{C}) = level(d) = 2$. A level (of the tree) with only one edge is designated as *trivial*. Otherwise, it is a non-trivial level.

Most of the existing algorithms solve a tree-traversal problem by first building a successor function (denoted by NEXT) which characterizes the successor of a node in the required order of traversal, and then applying a linked list ranking algorithm [CD89, CDA91]. Our approach is similar, but it utilizes a novel idea which converts a breadth-first tree-traversal problem into a parentheses matching problem. The conversion is performed by associating each forward and each backward arc (excluding the leftmost and the rightmost arcs at each level of the tree) of the arc-sequence with a right and a left parenthesis, respectively. A forward (backward) arc is called a *right (left)-parenthesis-associated arc*. In Theorem 1, it will be shown that this conversion produces a well-formed sequence of parentheses, called the *Euler sequence*, corresponding to an input tree. The Euler sequence for the tree in Fig. 1 is shown in Table 1. The leftmost and the rightmost arcs at each level of the tree are also identified. Fig. 3 shows the parentheses assigned to the arcs corresponding to the Euler sequence.

Computing the match of a left parenthesis with a right parenthesis in the Euler sequence enables *left-parenthesis-associated* arc, say $\bar{V}$, to be paired with *right-parenthesis-associated* arc, say U, in the following sense. The arc U is the closest (towards right in the arc-sequence) to arc $\bar{V}$ among all arcs of the tree which are at the same level as U and $\bar{V}$. In other words, starting from 1, if we number the arcs at each level from left to right and if $\bar{V}$ is the ith arc at a level, then U is the $(i+1)$th arc at the same level. With respect to such numbering on arcs, we define the *successor* of the kth arc at a level as the $(k+1)$th arc, if the kth arc is not the rightmost arc at that level. Otherwise the successor is the leftmost arc at the next higher level. Note that the pairing of left- and right-parenthesis-associated arcs gives rise to the value of NEXT function for the nodes (excluding the rightmost one) at a level, while that for the rightmost node at a level is obtained by considering the successor of the rightmost arc at that level. The *NEXT* value of the root-node is always given by the leftmost node at the first level, where the root is at level 0.

---

[1]It will be seen in Section 4 that this is a special case of bucket sorting, since the n integers in this class are limited within $[min .. min + n]$, where min is the smallest integer in the sequence.

[2]The subtree rooted at a node which has a rank of 1 among siblings is considered as the leftmost subtree.

Table 1

Euler tour (Arc-sequence)	A	B	$\bar{B}$	C	$\bar{C}$	$\bar{A}$	D	$\bar{D}$	E	F	$\bar{F}$	$\bar{E}$
Level of arcs	1	2	2	2	2	1	1	1	1	2	2	1
Leftmost arcs	×	×										
Rightmost arcs											×	×
Euler sequence			(	)	(	(	)	(	)	)		

The levels of arcs can be computed as follows. Let $(a_1, a_2, ..., a_{2(n-1)})$ be the arc-sequence of an input tree having n nodes. Assign a weight of $+1$ and -1 to each forward and backward arc, respectively, in the arc-sequence. Then the prefix sum for an arc a_i, denoted as $prefixsum(a_i)$, yields the level of the tree-node to which this arc leads. And the level of a_i, for $1 \leq i \leq 2(n-1)$, is given as

$$l(a_i) = \begin{cases} prefixsum(a_i) & \text{if } a_i \text{ is a forward arc} \\ prefixsum(a_i) + 1 & \text{if } a_i \text{ is a backward arc.} \end{cases}$$

For example, corresponding to the arc-sequence in Table 1, $l(\bar{C}) = prefixsum(\bar{C}) + 1 = (1+1-1+1-1)+1 = 2$, and $l(E) = prefixsum(E) = (1+1-1+1-1-1+1-1+1) = 1$.

Our aim now is to identify the leftmost arc at each level. For an arc a_i in the arc-sequence, let $L(a_i) = \max\{l(a_j) \mid 1 \leq j \leq i\}$, where $l(a_i)$ is the level of an arc a_i. Clearly, $L(a_i)$ is the largest level among arcs in the subsequence $(a_1, a_2, ..., a_i)$. Then a_i is the leftmost arc at level $l(a_i)$, if $[L(a_i) - L(a_{i-1})] > 0$, for $2 \leq i \leq 2(n-1)$. Note that a_1 is always the leftmost arc at level $l(a_1)$. The rightmost arc at each level can be similarly identified by reversing the arc-sequence and applying the procedure just described. The computation of $L(a_i)$, for $1 \leq i \leq 2(n-1)$, is essentially a parallel prefix-sum problem, since the "maximum" operation is associative.

The preceding ideas are formally expressed in the following algorithm for computing the breadth-first traversal of a tree.

Algorithm BREADTH-FIRST

Step 1. Compute the arc sequence corresponding to the Euler tour of an input tree.

Step 2. Compute levels of these arcs.

Step 3. Identify the leftmost and rightmost arcs at each level and delete them from the arc sequence.

Step 4. Assign a left (right) parenthesis to each backward (forward) arc in the sequence obtained in Step 3, and designate the parentheses sequence to be an Euler sequence.

Step 5. Solve the parentheses matching problem corresponding to the Euler sequence to obtain the successor of each backward arc except the rightmost one at each level. The successor of the rightmost arc a_i at level $l(a_i)$, excluding the highest level, is the leftmost arc at level $l(a_i) + 1$.

Step 6. Order the nodes by defining the NEXT function. Given a backward arc $\bar{X} = <x, y>$ and its successor $Y = <u, v>$, define $NEXT[x] = v$. Also define $NEXT[u] = v$ if $<u, v>$ is the first arc in the arc-sequence — this is for the root-node, u.

Step 7. Apply a linked list ranking algorithm to the NEXT function. The ranks of tree-nodes yield the breadth-first traversal, where the rank of a node is the number of nodes preceding it (including itself) in the linked list, NEXT.

Example 1.

Let us illustrate Steps 5 through 7 of the algorithm for the input tree in Fig. 1, with the help of Table 1. (Steps 1 through 4 have already been explained.) The output of these steps are recorded in Table 2. The NEXT function is based on pairs $(\bar{X}, Y)$, and as can be seen, the leftmost arc at level one is treated seperately.

Table 2

Arc and its successor	NEXT						
$-$	a	b	c	d	e	f	g
$(\bar{B}, C)$				d			
$(\bar{C}, F)$							g
$(\bar{A}, D)$					e		
$(\bar{D}, E)$						f	
$(\bar{E}, B)$							c
Arc A	b						

Node	a	b	c	d	e	f	g
NEXT	b	e	d	g	f	c	nil
Rank	1	2	5	6	3	4	7

After computing the ranks of tree-nodes from the linked list, NEXT, the breadth-first traversal of the input tree in Fig. 1 is given as a, b, e, f, c, d, g.

3 Correctness and Complexity

To guarantee the correctness of the BREADTH-FIRST algorithm, we need to prove Theorem 1. But let us first define the *nesting level* of a parenthesis in a well-formed sequence of n parentheses. Assigning $+1$ and -1 to each left and right parenthesis, respectively, the prefix sum of these integers for each left parenthesis is its nesting level. The nesting level of a right parenthesis is defined to be that of its matching left parenthesis. For example, for the following sequence of eight parentheses, nesting-level(p_2) = nesting-level(p_3) = nesting-level(p_4) = 2; and nesting-level(p_5) = nesting-level(p_6)= 3.

$$(\quad (\quad) \quad (\quad (\quad) \quad) \quad)$$
$$p_1 \quad p_2 \quad p_3 \quad p_4 \quad p_5 \quad p_6 \quad p_7 \quad p_8$$

Theorem 1 *For a given input tree, the following claims are true.*

1. *The Euler sequence, generated by the BREADTH-FIRST algorithm, is well-formed.*

2. *Each of the parentheses (in the Euler sequence) associated with arcs at the highest non-trivial level of the tree, has a nesting level equal to 1.*

3. *The arc associated with the mate of a left parenthesis is the closest (towards right) to the arc associated with the left parenthesis among all arcs in the arc-sequence, which are at the same level.*

Proof. We prove this theorem by an induction on the height, m, of an input tree. When $m = 1$, the tree is of the form as shown in Fig. 4 and the corresponding Euler sequence is ( )( )( ) ... ( )( )( ). Thus all claims in Theorem 1 are trivially true for $m = 1$. Assume the theorem to be true for $m = k$. Consider an input tree T (as shown in Fig. 5) of height $(k + 1)$. Let T' be the tree in Fig. 6, obtained by deleting the edges and nodes at level $(k + 1)$ in tree T. Without loss of generality, assume there are r tree-nodes at level k having (a) child(ren). Since the Euler tour of T traverses through the subtrees in a left-to-right fashion, the arcs of each subtree rooted at a node N_i (for $1 \leq i \leq r$) of level k are placed in the tour between a forward arc $Z = <fn, cn>$ and a backward arc $\bar{Z} = <cn, fn>$ of level k, where these two arcs share the same child-node, cn, and parent-node, fn, in opposite directions. The parentheses corresponding to these arcs in the subtree rooted at a node N_i form a subsequence of Euler sequence of T and it has one of the following three forms[3].

(α) [ ] ... [ ] [

(β)] [ ] ... [ ] [

(γ)] [...] [ ],

Forms (α) and (γ) are respectively chosen when node N_i is the leftmost and rightmost nodes at level k having children, and (β) is chosen if node N_i is neither the leftmost nor the rightmost node at level k but it has children. Since node N_i is a leaf-node in tree T', arcs Z and $\bar{Z}$ are in consecutive positions in the arc-sequence of T'. The parentheses associated with them are given as ") (", each of which has a nesting level of 1 in the Euler sequence of T'. Hence the Euler sequence of tree T can be obtained from that of T' by appropriately inserting the Euler subse-

---

[3]Here a parenthesis corresponding to an arc at level $(k + 1)$ is distinguished from others by a square bracket "[" or "]", which actually represents the parenthesis "(" or ")" accordingly.

quences of the subtrees rooted at nodes N_i, for $1 \leq i \leq r$. In fact, the Euler sequence of T can be written as

(w)(w) ... (w)[ ] [ ] [(w)(w) ... (w)] [ ] [ ] [(w) ... (w)] [ ] [ ] [(w) ... (w)] [ ] [ ](w) ... (w),

where the entity w stands for a well-formed subsequence of parentheses. Thus, by the above form of the Euler sequence of tree T and by our inductive hypothesis, the three claims of the theorem are true for height $m = (k + 1)$ of the tree. Hence the proof. □

Theorem 2 *The breadth-first traversal of a tree with n nodes can be obtained in $O(n/p + \log n)$ time, using p processors on the EREW PRAM model.*

Proof. Step 1 of algorithm BREADTH-FIRST requires $O(n/p + \log n)$ time by Lemma 1. The time complexity for each of Steps 2 and 3 is the same as that for a parallel prefix-sum algorithm. Each of Steps 4 and 6 requires $O(n/p)$ time using p processors on the EREW PRAM model. Steps 5 and 7 require $O(n/p + \log n)$ time on the same model due to the application of the parentheses matching [CD91, DP91, TLC89] and linked list ranking [AM88, CV86, CV88] algorithms, respectively. Therefore, the overall time complexity of the algorithm BREADTH-FIRST is $O(\frac{n}{p} + \log n)$ employing p processors on the EREW PRAM model. □

4 Integer Sorting

The technique presented in Section 2 can be generalized to sort a special class of integers (SCI), in which any two successive integers differ either by 0 or 1. The sequence (4, 5, 6, 5, 5, 5, 6, 5, 4, 4, 3, 2) is an instance of SCI. In general, given a sequence $(\alpha_1, \alpha_2, \cdots, \alpha_n)$ of n integers belonging to SCI, we consider it as a sequence of $n - 1$ pairs given by (α_i, α_{i+1}), for $1 \leq i \leq n-1$, occuring in consecutive places. Let us now consider a restricted subclass of integers (RSCI) of this special class such that

(i) the input sequence starts with a pair of integers $(0, 1)$ and terminates with a pair $(1, 0)$, and the integer 0 cannot occur anywhere else in the sequence,

(ii) no integer of the sequence repeats in two consecutive places, *i.e.* any two successive integers differ exactly by 1.

For example, $(0, 1, 2, 3, 4, 5, 4, 5, 4, 3, 2, 1, 2, 3, 2, 1, 0)$ gives an instance of such a sequence in RSCI. One interesting property of this subclass is stated in the following theorem.

Theorem 3 *There exists a tree such that the level numbers of nodes visited while traversing the tree according to its arc-sequence, correspond to a given sequence of integers in the subclass, RSCI.*

Proof. The proof is clear from the following simple (sequential) algorithm which constructs such a tree using a stack, where the tree is represented by "leftmost child and right sibling" information for each node. □

Algorithm TREE-CONSTRUCTION

Step 1 Read the first integer of the input sequence;
  Get a new node, nd;
  value[nd] ← 0;
  leftmost-child[nd] ← nil;
  rightmost-child[nd] ← nil;
  current ← nd; /* Current node */

Step 2 Read the next (input) integer into variable, nx;
  while $nx \neq 0$ **do**
   begin
   if $nx >$ value[current] **then**
    begin
    Get a new node, nd;
    leftmost-child[nd] ← nil;
    rightmost-child[nd] ← nil;
    value[nd] ← nx;
    if leftmost-child[current] = nil **then**
     begin
     rightmost-child[current] ← nd;
     leftmost-child[current] ← nd;
     end
    else right-sibling[rightmost-child[current]] ← nd;
    Push current into the stack;
    current ← nd;
    end
   else Pop top of the stack and assign it to current;
   Read the next integer of the sequence into nx;
   end.

Theorem 3 enables us to view a pair of consecutive integers, say (α_i, α_{i+1}) where $\alpha_{i+1} = \alpha_i \pm 1$, in a given sequence of RSCI as follows. They, respectively, denote the level numbers of nodes n_i and n_{i+1} corresponding to an arc from n_i to n_{i+1} in a conceptual tree. An arc is forward (or backward) if α_i is greater (or less) than α_{i+1}. Such an arc will be represented as $< l\alpha_i, l\alpha_{i+1} >$, where $l\alpha_i$ stands for the label[4] of the ith input integer, α_i. Note that these labels actually depend on the positions of integers in the input sequence rather than on their values. In other words, $l\alpha_i \neq l\alpha_j$ even if $\alpha_i = \alpha_j$ for $i \neq j$. Also, by our construction, different labels may represent the same node of the conceptual tree.

In Section 2, it has been shown that by associating a left (or right) parenthesis with a backward (or forward) arc and by solving the corresponding parentheses matching problem, we obtain the successor of each backward arc (if it is not the rightmost arc) at a level of the tree. Fur-

thermore, the successor of the rightmost (backward) arc at any level is the leftmost arc at the next higher level. Now the successor of each forward arc can be computed in a similar way, but by associating each forward and backward arcs (including the leftmost and rightmost ones) at each level with a left and right parentheses, respectively. The sequence of parentheses thus obtained is designated as a *complementary Euler (or C-Euler) sequence*.

The parent- and child-nodes of these successor arcs impose an ordering of labels for input integers. However, an ordering obtained only from parent-nodes (or child-nodes for that matter) is partial in the sense that different labels may represent the same node in the conceptual tree. This is illustrated in Example 2. A complete ordering is obtained from the following subalgorithm by computing $succ[l\alpha_i]$, the successor of each input label $l\alpha_i$ for $1 \leq i \leq n$, with the help of both the parent- and child-nodes of successor arcs.

Subalgorithm ORDERING

Step 1. Initialize entries of an array *succ* to "∗", which stands for a "nil" value.

Step 2. For each arc $< l\alpha_i, l\alpha_{i+1} >$ for $1 \leq i < n$, order the input-labels according to their successors as follows. Let the successor of arc $< x, y >$ be $< u, v >$ then

  1. $succ[x] = v$ if $< x, y >$ is a forward arc and $x \neq v$;

  2. $succ[x] = v$ if $< x, y >$ is a backward arc but not the rightmost arc at the highest level;

  3. $succ[y] = u$ if $< x, y >$ is the rightmost arc at level 1.

Example 2.

Let us illustrate the major steps of the preceding ideas for sorting a sequence of integers in the subclass RSCI. Figure 7 depicts a conceptual tree corresponding to the given input sequence. The Euler and C-Euler sequences are shown in Table 3. As a convention, the level of an arc $< l\alpha_i, l\alpha_{i+1} >$ is written under the input-label $l\alpha_{i+1}$. The contents of array *succ*, after the execution of Step 2 of subalgorithm ORDERING, is shown in Table 4. Note that after Substep (1), there are undefined successor values for some input-labels. The complete ordering is obtained after executing Substeps (2) and (3). Thus the sorted list of input-labels is $(a, o, b, h, j, n, c, e, g, i, k, m, d, f, l)$ and the corresponding (sorted) output sequence of integers is $(0, 0, 1, 1, 1, 1, 2, 2, 2, 2, 2, 2, 3, 3, 3)$.

The correctness of the proposed algorithm for (stable) sorting our subclass of integers depends on the following two lemmas.

Lemma 2 *The C-Euler sequence corresponding to a tree constitutes a well-formed sequence of parentheses.*

---

[4]We introduce the concept of "label" for input integers in order to avoid confusion with other integer values, namely levels of arcs or ranks of tree-nodes.

Proof. This can be easily proved by induction on the height of the given tree. □

Lemma 3 *The mate of a left parenthesis in the C-Euler sequence is associated with a backward arc which is the successor of the left-parenthesis-associated arc.*

Proof. From the definitions of nesting level of a parenthesis and the arc-sequence of the given tree. □

The preceding integer-sorting algorithm can be modified to relax restriction (ii) mentioned at the beginning of this section, thus including the case in which an integer may repeat in consecutive places of an input sequence. In fact, we build a linked list consisting of these repeated integers and use a *representative* for this list. By this mechanism, we convert the original input sequence into one belonging to the subclass RSCI. After sorting, we replace the representative by the linked list in order to obtain the desired sorting of input integers.

Restriction (i) can also be relaxed by finding the smallest element of the sequence, denoted by min, and subtracting $(min - 1)$ from each element. This mechanism does not destroy the relative ordering among input integers. Let α_f and α_l be, respectively, the first and the last elements of the sequence after the subtraction operation. Now, if necessary, we append integers $(0, 1, \ldots, \alpha_f - 1)$ and $(\alpha_l - 1, \alpha_l - 2, \ldots, 1, 0)$, respectively, to the beginning and the end of the sequence. Then the sorting algorithm for the RSCI is applied to this new sequence, and the appended elements are deleted in order to obtain the desired output. Finding the smallest element and deleting the appended integers are parallel prefix-sum problems of size $O(n)$. Therefore, we have the following.

Theorem 4 *The special class of integers (SCI) of size n can be sorted in $O(n/p + \log n)$ time using p processors on the EREW PRAM model, where n is the total number of integers.*

5   Conclusions

We have presented adaptive parallel algorithms for the breadth-first tree traversal and a special class of integer-sorting problems on the EREW PRAM model. Each of the two algorithms requires $O(n/p + \log n)$ time using $O(p)$ processors, thus achieving optimal speedup for $p \le n/\log n$. To the best of our knowledge, the proposed breadth-first traversal algorithm is the first such parallel algorithm which achieves $O(\log n)$ time complexity employing $O(n/\log n)$ processors. Using this result, we have shown that the parallel time complexity of tree-traversal − both level-order and depth-order − algorithms remain the same, no matter which one of the three commonly-used input representations, namely "leftmost child and right sibling", "parent-of with implicit ordering of children", and "parent-of with explicit ordering of children" is used [AHU83, CD89].

Our novel approach is further generalized to sort a special class of integers, where any two successive integers differ at most by unity. This sorting algorithm is stable and achieves cost-optimality using $p \le n/\log n$ processors, where n is the number of integers. Although the sorting algorithm applies only to a special class of integers, it has potential applications in such areas as computer graphics and computational geometry. It also has merits toward an $O(\log n)$-time and cost-optimal (deterministic) parallel algorithm for the general case of bucket sorting n integers in the range $[1 .. n]$. (It is easy to see that the range of integers considered in our algorithm is no wider than $[min .. min + n]$.)

The breadth-first tree-traversal problem is identified as a subproblem for sorting integers belonging to the identified special class, and parentheses matching is in turn considered as a subproblem for our breadth-first traversal solution. Currently, we are investigating other class of problems that can be efficiently solved in parallel, using parentheses matching and/or breadth-first traversal as subalgorithms. We believe that the underlying approach might lead to a general paradigm for designing cost-optimal parallel algorithms for various problems related to trees.

References

[**AHU83**] A. V. Aho, J. E. Hopcroft and J. D. Ullman, *Data Structures and Algorithms* (Addison Wesley, Reading, Mass. 1983).

[**AM88**] R. J. Anderson and G. L. Miller, "Deterministic Parallel List Ranking", *Lecture notes in Computer Science*, Vol. 319, VLSI Algorithms and Architecture (Ed. J. Reif), 3rd Aegean Workshop on Computing (June-July, 1988) pp. 81-90.

[**B86**] A. Berztiss, "A Taxonomy of Binary Tree Traversals", *BIT*, Vol. 20, 1986, pp. 266-276.

[**CD89**] C. C-Y. Chen and S. K. Das, *Traversing Trees in Parallel*, Tech. Rep. #N-89-006, Dept. Computer Science, Univ. North Texas, Denton, TX, Sep. 1989. Also, to appear in *Advances in Parallel Computing* (Ed. D. J. Evans), JAI Press, London.

[**CD90**] C. C-Y. Chen and S. K. Das, "Parallel Breadth-First and Breadth-Depth Traversals of General Trees", *Proc. Int. Conf. Computing and Information* May 1990, Niagara Falls, Canada, pp. 383-387. Also, *Lecture Notes in Computer Science*, Vol. 468 (Eds. S. G. Akl, et al), pp. 395-404, 1990, Springer-Verlag.

[**CD91**] C. C-Y. Chen and S. K. Das, "A Cost-Optimal Parallel Algorithm for the Parentheses Matching Problem on an EREW PRAM", *Proc. Int. Parallel Process. Symp.*, Anaheim, California, Apr. 30 - May 2, 1991, to appear.

[CDA91] C. C-Y. Chen, S. K. Das, and S. G. Akl "A Unified Approach to Parallel Depth-First Traversals of General Trees", *Inform. Process. Letters*, to appear.

[CV86] R. Cole and U. Vishkin, "Deterministic Coin Tossing with Applications to Optimal Parallel List Ranking", *Inform. Control*, Vol. 70, pp. 32-53, 1986.

[CV88] R. Cole and U. Vishkin, "Approximate Parallel Scheduling. Part I: The Basic Technique with Application to Optimal Parallel List Ranking in Logarithmic Time", *SIAM J. Comput.*, Vol. 17, No. 1, Feb. 1988, pp. 128-142.

[DP91] N. Deo and S. Prasad, "Two EREW Algorithms for Parentheses Matching", *Proc. Int. Parallel Process. Symp.*, Anaheim, California, Apr. 30 - May 2, 1991, to appear.

[GB84] R. K. Ghosh and G. P. Bhattacharjee, "Parallel Breadth-First Search Algorithms for Trees and Graphs", *Int. J. Comput. Math.*, Vol. 15, 1984, pp. 255-268.

[H87] T. Hagerup, "Towards Optimal Parallel Bucket Sorting", *Inform. and Computat.*, Vol. 75, 1987, pp. 39-51.

[KB85] N. C. Kalra and P. C. P. Bhatt, "Parallel Algorithms for Tree Traversals", *Parallel Computing*, Vol. 2, 1985, pp. 163-171.

[K73] D. E. Knuth, *The Art of Computer Programming. Vol. 1, Fundamental Algorithms* (Addison Wesley, Reading, Mass. 1973).

[KRS86] C. P. Kruskal, L. Rudolph, and M. Snir, "Efficient Parallel Algorithms for Graph Problems", *Proc. Int. Conf. Parallel Process.*, 1986, pp. 869-876.

[R85] J. H. Reif, "An Optimal Parallel Algorithm for Integer Sorting", *Proc. 26th Annual Symp. Foundations of Computer Science*, 1989, pp. 308-313.

[RG78] E. Reghbati and D. G. Corneil, "Parallel Computations in Graph Theory", *SIAM J. Comput.*, Vol. 7, No. 2, May 1978, pp. 308-313.

[TLC89] W. W. Tsang, T. W. Lam, and F. Y. L. Chin, "An Optimal EREW Parallel Algorithm for Parentheses Matching", *Proc. Int. Conf. Parallel Process.*, Vol. 3 (Eds. F. Rigs and P. M. Kogge), Aug. 1989, pp. 185-192.

[TV85] R. E. Tarjan and U. Vishkin, "An Efficient Parallel Biconnectivity Algorithm", *SIAM J. Comput.*, Vol. 14, Nov. 1985, pp. 862-874.

[WH86] R. A. Wagner and Y. Han, "Parallel Algorithms for Bucket Sorting and the Data Dependent Prefix Problem", *Proc. Int. Conf. on Parallel Process.*, 1986, pp. 924-930.

[W79] J. Wyllie, *The Complexity of Parallel Computations*, Ph.D. Thesis, Cornell Univ., Ithaca, NY, 1979.

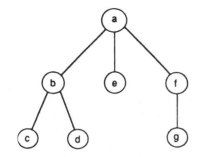

FIG. 1: Input Tree

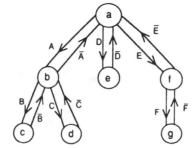

Fig. 2: Arc Sequence

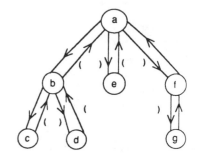

Fig. 3: Euler Sequence

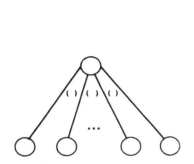

FIG. 4: A Tree of Height One

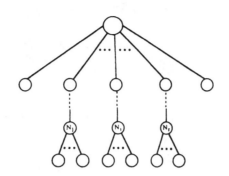

FIG. 5: A Tree, T

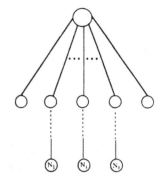

FIG. 6: Tree T', a Subtree of T

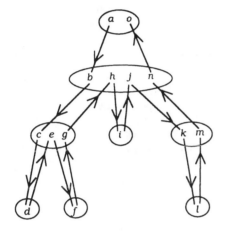

Fig. 7: A Conceptual Tree

Table 3															
Input sequence	0	1	2	3	2	3	2	1	2	1	2	3	2	1	0
Label	a	b	c	d	e	f	g	h	i	j	k	l	m	n	o
Level of arcs	-	1	2	3	3	3	3	2	2	2	2	3	3	2	1
Leftmost arcs	-	×	×	×											
Rightmost arcs	-												×	×	×
Euler sequence	-			(	)	(	(	)	(	)	)				
C-Euler sequence	-	(	(	(	)	(	)	)	(	)	(	(	)	)	)

Table 4															
After Substep (1) of Step 2															
input label	a	b	c	d	e	f	g	h	i	j	k	l	m	n	o
succ	o	h	e	*	g	*	*	j	*	n	m	*	*	*	*

After Substeps (2) and (3) of Step 2															
input label	a	b	c	d	e	f	g	h	i	j	k	l	m	n	o
succ	o	h	e	f	g	l	i	j	k	n	m	*	d	c	b
rank	1	3	7	13	8	14	9	4	10	5	11	15	12	6	2

Bitonic Sort with an Arbitrary Number of Keys

Biing-Feng Wang, Gen-Huey Chen, and Cheng-Chung Hsu
Department of Computer Science and Information Engineering,
National Taiwan University,
Taipei, Taiwan, Republic of China

Abstract -- The previous implementation of bitonic sort on the hypercube requires the number of input keys to be a power of 2. For sorting an arbitrary number of keys, one commmon way is to add dummy keys with maximal values to the end of the input to make the total number of keys a power of 2. By doing so, m keys can be sorted on a hypercube with $2^{\lceil \log m \rceil}$ processors in $\lceil \log m \rceil (\lceil \log m \rceil + 1)(2T_t + T_{ce})/2$ time, where T_t is the time for a processor to transmit a key to one of its neighbors and T_{ce} is the time for a processor to compare two local keys and exchange them if necessary. In this paper, it is shown that bitonic sort can be executed with any m input keys on an incomplete hypercube with exactly m processors. The total execution time is also $\lceil \log m \rceil (\lceil \log m \rceil + 1)(2T_t + T_{ce})/2$.

1. Introduction

Bitonic sort [2] is an efficient parallel sorting scheme, which was first proposed by Batcher to construct a sorting network. Later, bitonic sort has been adapted to a variety of parallel computers such as hypercube, perfect shuffle [12], mesh [9], [13], mesh-of-tree [10], cube-connected-cycles [11], etc. All these implementations of bitonic sort requires the number of input keys to be a power of 2. For sorting an arbitrary number of keys, one common way is to add some dummy keys with maximal values to the end of the input to make the total number of keys a power of 2.

The hypercube is a high-connectivity and regular multiprocessor computer system. An n-dimensional hypercube consists of 2^n identical processors, denoted by $P_0, P_1, ..., P_{2^n-1}$, and there exists a communication link between two processors P_i and P_j if and only if the binary representations of i and j differ in exactly one bit position. These links facilitate the implementation of bitonic sort on the hypercube. It is easy to show (see [1], [6]) that m keys can be sorted on a $\lceil \log m \rceil$-dimensional hypercube in $\lceil \log m \rceil (\lceil \log m \rceil + 1)(2T_t + T_{ce})/2$ time (throughout this paper, the logarithm has a base 2), where T_t is the time for a processor to transmit a key to one of its neighbors and T_{ce} is the time for a processor to compare two local keys and exchange them if necessary. Note that in case m is not a power of 2, more processors than the input keys are needed.

A problem with the hypercube is its inflexibility in the number of processors used. A hypercube multiprocessor system must be built with 2^k processors, where k is a positive integer. This restriction has severely limited its use in practical applications. A way to overcome this restriction is the *incomplete hypercube* [7], which is essentially a hypercube except for some missing processors. Unlike the hypercube, an n-dimensional incomplete hypercube may contain any m processors, denoted by $P_0, P_1, ..., P_{m-1}$, where $2^{n-1} < m \leq 2^n$. Compared with an n-dimensional hypercube, the $2^n - m$ missing processors in an n-dimensional incomplete hypercube have the first $2^n - m$ greatest indices.

In this paper, we first show that an m-key descending-ascending sequence can be sorted non-decreasingly in $\lceil \log m \rceil (2T_t + T_{ce})$ time on an incomplete hypercube with m processors. Using this result, we further show that any m keys can be sorted in $\lceil \log m \rceil (\lceil \log m \rceil + 1)(2T_t + T_{ce})/2$ time on an incomplete hypercube with m processors.

The rest of this paper is organized as follows. In Section 2, we introduce some notations and definitions that are used throughout this paper. In Section 3, the previous implementation of bitonic sort on the hypercube is briefly reviewed. In Section 4, a $\lceil \log m \rceil (\lceil \log m \rceil + 1)(2T_t + T_{ce})/2$ time algorithm that sorts m keys on an incomplete hypercube with m processors is presented. Finally, in Section 4, concluding remarks are given.

2. Notations and definitions

An *ascending sequence*, represented as /-*sequence*, is a non-decreasing sequence of keys and a *descending sequence*, represented as \-*sequence*, is a non-increasing sequence of keys. An *ascending-descending sequence*, represented as ∧-*sequence*, is a concatenation of an /-sequence and a \-sequence, and a *descending-ascending sequence*, represented as ∨-*sequence*, is a concatenation of a \-sequence and an /-sequence. A sequence of keys is a *bitonic sequence* if and only if it can be obtained from an ∧-sequence by cyclic shifts. Note that both an ∧-sequence and a ∨-sequence are instances of bitonic sequences.

Throughout this paper, we use k_i $(i \geq 0)$ to denote the $(i+1)$-th bit from the right in the binary representation of a nonnegative integer k (i.e., the q-bits binary representation of k is $k_{q-1}k_{q-2} ... k_0$, where $q \geq \lceil \log(k+1) \rceil$), and use $k^{(j)}$, $j \geq 0$, to denote the integer that is obtained by reversing the bit k_j of the integer k. A sequence $S = (s_0, s_1, ..., s_{m-1})$ is *well-stored* in a hypercube (or an incomplete hypercube) with m processors if and only if each s_k is stored in processor P_k, $0 \leq k \leq m-1$. By T_t we denote the time for a processor of a hypercube (or an incomplete hypercube) to transmit a key to one of its neighbors, and by T_{ce} we denote the time for a processor to compare two local keys and exchange them if necessary.

3. Review of bitonic sort on hypercubes

Let $A = (a_0, a_1, ..., a_{2r-1})$ be a bitonic sequence and let $A_L = (\min\{a_0, a_r\}, \min\{a_1, a_{r+1}\}, ..., \min\{a_{r-1}, a_{2r-1}\})$ and $A_H = (\max\{a_0, a_r\}, \max\{a_1, a_{r+1}\}, ..., \max\{a_{r-1}, a_{2r-1}\})$. Batcher [2] has shown that both A_L and A_H are bitonic sequences and the maximal key in A_L is smaller than or equal to the minimal key in A_H. This fact implies that a bitonic sequence of length 2^n can be transformed into a concatenation of two bitonic subsequences each of length 2^{n-1} in one compare-exchange step, in which 2^{n-1} compare-exchange operations (each compares two keys and exchanges them if necessary) are performed simultaneously. Thus, by repeatedly applying the compare-exchange step to the obtained bitonic subsequences, the original bitonic sequence can be sorted. A sorting network based on this scheme has been built and for convenience, we refer to such a sorting network as a *bitonic-sorter⁺* (*bitonic-sorter⁻*), if it can sort a bitonic sequence non-decreasingly (non-increasingly). Note that the number of input keys to a bitonic-sorter⁺ (or bitonic-sorter⁻) must be a power of 2.

Figure 1 shows an 8-key bitonic-sorter⁺, which can sort an 8-key bitonic sequence into non-decreasing order in three compare-exchange steps. In this diagram, horizontal bold lines indicate data paths through which keys flow from left to right, and vertical arrows indicate comparators. The head and the tail of each arrow indicate the pair of keys to be compared and after comparison, the larger one is kept at the head and the smaller one is kept at the tail. A key is at position k, $0 \le k \le 7$, if it appears on the $(k+1)$-th data path from the top. Note that if the arrows of Figure 1 are reversed, then an 8-key bitonic-sorter⁻ results.

It is not difficult to see that an n-dimensional hypercube can simulate a 2^n-key bitonic-sorter⁺ and a 2^n-key bitonic-sorter⁻. In the simulation algorithm, the input bitonic sequence is initially well-stored in the n-dimensional hypercube and then each processor P_k, $0 \le k \le 2^n-1$, holds the key at position k as the simulation proceeds. Before presenting the simulation algoritrhm, we first define a procedure, *Cube-Compare-Exchange(k,j,mask)*, as follows.

Procedure *Cube-Compare-Exchange(k, j, mask)*;
/* Let key_k and $key_{k(j)}$ be the two keys held by processors P_k and $P_{k(j)}$ respectively. In this procedure, P_k cooperates with $P_{k(j)}$ in performing a compare-exchange operation on key_k and $key_{k(j)}$. */
  if $k_j=1$ **then** send key_k to $P_{k(j)}$;
  if $k_j=0$ **then**
  begin
    compare key_k and $key_{k(j)}$;
    if *mask*=0 **then** keep the smaller one and send the
        other one to $P_{k(j)}$
        else keep the larger one and send the other
        one to $P_{k(j)}$
  end

The simulation algorithm consists of n iterations; each simulates a compare-exchange step. The following algorithm is executed by an n-dimensional hypercube to simulate a 2^n-key bitonic-sorter⁺.

Algorithm 1: /* Simulation of a 2^n-key bitonic-sorter⁺ by an n-dimensional hypercube. The input is a bitonic sequence, which is initially well-stored in the hypercube.*/
  for $j:=n-1$ **downto** 0 **do**
    for all P_k, $k=0,1,...,2^n-1$, **do in parallel**
      Cube-Compare-Exchange(k, j, 0)

If the parameter 0 is changed to 1, then a 2^n-key bitonic-sorter⁻ is simulated. Since each execution of *Cube-Compare-Exchange(k, j, 0)* takes $2T_t+T_{ce}$ time, the total execution time of the simulation algorithm is $n(2T_t+T_{ce})$.

Lemma 1: On an n-dimensional hypercube, a 2^n-key bitonic sequence can be sorted into non-decreasing (or non-increasing) order in $n(2T_t+T_{ce})$ time. Initially, the input sequence is well-stored in the hypercube.

Clearly, two sorted sequences that are in opposite order can be merged by sorting their concatenated sequence, which is an $\wedge$-sequence or a $\vee$-sequence. Therefore, using *Algorithm 1* as a merge procedure, we can obtain a parallel

merge sort, the *bitonic sort*, on the hypercube, as follows.

Algorithm 2:/*Bitonic sort is executed on an n-dimensional hypercube. The input is an arbitrary sequence of length 2^n, which is initially well-stored in the hypercube. The sorted sequence is in nondecreasing order. */
  for $i:=1$ **to** n **do**
  begin
    for all P_k, $k=0,1,...,2^n-1$, **do in parallel**
      if $k_i=0$ **then** *mask*:=0 **else** *mask*:=1;
    for $j:=i-1$ **downto** 0 **do**
      for all P_k, $k=0,1,...,2^n-1$, **do in parallel**
        Cube-Compare-Exchange(k, j, mask)
  end

The total execution time is $(1+2+ ... +n)(2T_t+T_{ce})= n(n+1)(2T_t+T_{ce})/2$. Note that the overhead to set *mask* is not included in the time analysis, since it is negligible as compared with the execution time of *Cube-Compare-Exchange(k, j, mask)*.

4. Bitonic sort on incomplete hypercubes

4.1 Sort an m-key $\vee$-sequence on an incomplete hypercube with m processors

It has been shown in [8] that a $\vee$-sequence $S=(s_0, s_1, ..., s_{m-1})$ can be sorted into non-decreasing order by first sorting the two $\vee$-subsequences $(s_0, s_2, s_4, ..., s_{2*\lfloor (m-1)/2 \rfloor})$ and $(s_1, s_3, s_5, ..., s_{2*\lceil (m-1)/2 \rceil-1})$ independently, and then performing compare-exchange operations on $s_0{:}s_1$, $s_2{:}s_3$, ..., $s_{2*\lceil (m-1)/2 \rceil-2}{:}s_{2*\lceil (m-1)/2 \rceil-1}$. Based on this scheme, a network for sorting a 6-key $\vee$-sequence into non-decreasing order is depicted in Figure 2.

Recently, a simple approach to construct a network for sorting an $\wedge$-sequence of arbitrary length into non-decreasing order has been suggested by Batcher [3]. Suppose Q is an existing network that can sort an $(r+1)$-key $\wedge$-sequence into non-decreasing order. Also, when two keys are compared in Q by a comparator, the smaller one is always moved to a higher position and the larger one to a lower position (here, "higher" means "closer to the top" and "lower" means "closer to the bottom"). Batcher has shown that a network for sorting an r-key $\wedge$-sequence can be obtained from Q by removing the first key and all comparators whose one input key is at position 0. This is due to the following observation: an r-key $\wedge$-sequence $(s_0, s_1, ..., s_{r-1})$ can be sorted into non-decreasing order by using Q to sort the $(r+1)$-key $\wedge$-sequence $(-\infty, s_0, s_1, ..., s_{r-1})$. Since $-\infty$ will be kept at position 0 throughout the execution of the sorting network, all comparators incident on it can be removed. By a similar argument, we have the following lemma.

Lemma 2: Let Q denote a sorting network that can sort an $(r+1)$-key $\vee$-sequence into non-decreasing order. If each comparator in Q always moves the smaller key to a higher position and the larger key to a lower position, then a network that sorts an r-key $\vee$-sequence into non-decreasing order can be obtained from Q by removing the last key and all comparators whose one input key is at position r.

By applying *Lemma 2* repeatedly, a network that sorts an m-key $\vee$-sequence into non-decreasing order can be obtained from a $2^{\lceil \log m \rceil}$-key bitonic-sorter⁺ by removing the last $2^{\lceil \log m \rceil}-m$ keys and all comparators whose one input key is among the last $2^{\lceil \log m \rceil} - m$ positions. This network is referred to as m-key $\vee$-sorter$^+$ in the following discussion. In fact, although with a different interpretation, the sorting network depicted in Figure 2 is a 6-key $\vee$-sorter$^+$, which can be obtained from an 8-key bitonic-sorter$^+$ (as depicted in Figure 1) by removing the last two keys and all comparators whose one input key is among the last two positions.

Since *Algorithm 1* implements a 2^n-key bitonic-sorter$^+$ on an n-dimensional hypercube, by *Lemma 2* a modification of *Algorithm 1* can implement an m-key $\vee$-sorter$^+$ on an incomplete hypercube with m processors. The necessary modification is to eliminate the execution of *Cube-Compare-Exchange*$(k, j, 0)$ for $k=m,...,2^{\lceil \log m \rceil}-1$, since they simulate the execution of the comparators whose one input key is among the last $2^{\lceil \log m \rceil}-m$ positions. Clearly, after the modification, the processors P_k, $k=m,...,2^{\lceil \log m \rceil}-1$, become idle throughout the execution of *Algorithm 1*. Hence, these processors can be removed from the $\lceil \log m \rceil$-dimensional hypercube, which results in an incomplete hypercube with m processors. The modified algorithm is shown below.

Algorithm 3: /* Sort an m-key $\vee$-sequence into non-decreasing order on an incomplete hypercube with m processors. The input sequence is initially well-stored in the incomplete hypercube. */

for $j:=\lceil \log m \rceil-1$ **downto** 0 **do**
    for all $P_k, k=0,1,...,m-1$ and $k^{(j)} \leq m-1$, **do in parallel**
        Cube-Compare-Exchange$(k, j, 0)$

Lemma 3: On an incomplete hypercube with m processors, an m-key $\vee$-sequence can be sorted non-decreasingly in $\lceil \log m \rceil(2T_t+T_{ce})$ time. Initially, the input sequence is well-stored in the incomplete hypercube.

4.2 Sort m keys on an incomplete hypercube with m processors

For convenience, we define $L(S, i)$ and $S_{i,j}$ for a sequence $S=(s_0, s_1, ..., s_{m-1})$, where $0 \leq i \leq \lceil \log m \rceil$ and $0 \leq j \leq L(S, i)-1$, as follows:

$L(S, i)$ $= \lceil |S|/2^i \rceil = \lceil m/2^i \rceil$ and

$S_{i,j}$ denotes the subsequence $(s_{j*2^i}, s_{j*2^i+1}, ..., s_{(j+1)*2^i-1})$ when $0 \leq j \leq L(S, i)-2$, and the subsequence $(s_{j*2^i}, s_{j*2^i+1}, ..., s_{m-1})$ when $j=L(S, i)-1$.
(S is divided into $L(S, i)$ subsequences each of length 2^i, except for the rightmost subsequence whose length may be smaller than 2^i.)

Also, S is 2^i-ready, $0 \leq i \leq \lceil \log m \rceil$, in an incomplete hypercube with m processors if and only if (1)S is well-stored in the incomplete hypercube, and (2) Each $S_{i,j}$, $0 \leq j \leq L(S, i)-1$, is an /-sequence if $L(S, i)-j$ is odd, and a \-sequence otherwise (see Figure 3).

Since a 1-key sequence is both an /-sequence and a \-sequence, any sequence well-stored on an incomplete hypercube is 2^0-ready. On the other hand, by definition, an m-key sequence is $2^{\lceil \log m \rceil}$-ready only if it is an /-sequence.

Property 1: Suppose $S=(s_0, s_1, ..., s_{m-1})$ is a 2^{i-1}-ready sequence, $0<i \leq \lceil \log m \rceil$, in an incomplete hypercube with m processors. Then, each $S_{i,j}$, $0 \leq j \leq L(S, i)-1$, is a bitonic sequence. Also, $S_{i,L(S,i)-1}$ is a $\vee$-sequence.

Proof: We consider two cases: $L(S, i-1)$ is even and $L(S,$

---

i-1) is odd.
Case 1. $L(S, i-1)$ is even.

By definition, each $S_{i-1,2*j}$ is a \-sequence and each $S_{i-1,2*j+1}$ is an /-sequence, $0 \leq j \leq L(S, i-1)/2-1$. Thus, each $S_{i,j}$, $0 \leq j \leq L(S, i)-1$, which is the concatenation of $S_{i-1,2*j}$ and $S_{i-1,2*j+1}$, is a $\vee$-sequence.
Case 2. $L(S, i-1)$ is odd.

In this case, each $S_{i-1,2*j}$, $0 \leq j \leq (L(S, i-1)-1)/2$, is an /-sequence and each $S_{i-1,2*j+1}$, $0 \leq j \leq (L(S, i-1)-1)/2-1$, is a \-sequence. Thus, each $S_{i,j}$, $0 \leq j \leq L(S, i)-2$, is an $\wedge$-sequence and $S_{i,L(S,i)-1}$ is an /-sequence (an /-sequence can be regarded as a $\vee$-sequence, in which the first key is a 1-key \-sequence).   Q.E.D.

Property 2: Suppose $S=(s_0, s_1, ..., s_{m-1})$ is a 2^i-ready sequence, $0 \leq i \leq \lceil \log m \rceil$, in an incomplete hypercube with m processors. Then, for each s_k, $0 \leq k \leq m-1$, the subsequence (assume $S_{i,r}$, $0 \leq r \leq L(S, i)-1$) that contains s_k is an /-sequence if and only if $k_i=(m-1)_i$. (Recall that $(m-1)_i$ denotes the $(i+1)$-th bit from the right in the binary representation of $m-1$.)

Proof: Since s_k belongs to the subsequence $S_{i,r}$, we know $r*2^i \leq k \leq (r+1)*2^i-1$, from which

$$r_0=k_i, \qquad\qquad (1)$$

is derived. From (1), we have

$$(L(S, i)-1)_0=(m-1)_i, \qquad\qquad (2)$$

since s_{m-1} belongs to $S_{i,L(S, i)-1}$.

Moreover, since S is 2^i-ready, we know that $S_{i,r}$ is an /-sequence if and only if

$$L(S, i)-r \text{ is odd.} \qquad\qquad (3)$$

Clearly, (3) is true if and only if $r_0=(L(S, i)-1)_0$, which by (1) and (2), is true if and only if $k_i=(m-1)_i$.   Q.E.D.

Lemma 4: If $S=(s_0, s_1, ..., s_{m-1})$ is a 2^{i-1}-ready sequence, $0<i \leq \lceil \log m \rceil$, in an incomplete hypercube with m processors, then in $i(2T_t+T_{ce})$ time, S can be rearranged into a 2^i-ready sequence.

Proof: By definition, if each $S_{i,j}$, $0 \leq j \leq L(S, i)-1$, is sorted properly, then the resulting sequence can be 2^i-ready. This can be done as follows: sort $S_{i,j}$ into an /-sequence if $L(S, i)-j$ is odd, and a \-sequence otherwise. Since, each $S_{i,j}$, $0 \leq j \leq L(S, i)-2$, is a bitonic sequence (from *Property 1*) and is well-stored in an i-dimensional (sub-)hypercube, by *Lemma 1* it can be sorted either non-decreasingly or non-increasingly in $i(2T_t+T_{ce})$ time. Moreover, since $S_{i,L(S,i)-1}$ is a $\vee$-sequence (from *Property 1*) and is well-stored in an incomplete (sub-)hypercube, by *Lemma 3* it can be sorted non-decreasingly in at most $i(2T_t+T_{ce})$ time. Thus, the main difficulty lies in determining for each $S_{i,j}$, $0 \leq j \leq L(S, i)-1$, whether it should be sorted non-decreasingly or non-increasingly. In other words, each processor P_k, $0 \leq k \leq m-1$, must know whether the subsequence containing s_k is to be sorted non-decreasingly or non-increasingly. Fortunately, this can be known from *Property 2*.   Q.E.D.

Using *Lemma 4*, a sequence of length m can be sorted non-decreasingly on an incomplete hypercube with m processors in $\lceil \log m \rceil$ phases as follows. Initially, the input sequence is 2^0-ready and in the i-th phase a 2^i-ready sequence is obtained by rearranging the 2^{i-1}-ready sequence that is obtained in the $(i-1)$-th phase, $1 \leq i \leq \lceil \log m \rceil$. The algorithm is shown below.

Algorithm 4: /* Sort an arbitrary m-key sequence non-decreasingly on an incomplete hypercube with m processors. The input sequence is initially well-stored in the incomplete hypercube. */

for $i:=1$ **to** $\lceil \log m \rceil$ **do**
begin
    for all P_k, $k=0,1,...,m$-1, **do in parallel**
      if $k_i=(m$-$1)_i$ **then** *mask*:=0 **else** *mask*:=1;
    for $j:=i$-1 **downto** 0 **do**
    for all P_k, $k=0,1,...,m$-1 and $k^{(j)} \leq m$-1, **do in**
    parallel
      Cube-Compare-Exchange(k, j, mask)
end

It is not difficult to check that the total execution time of the algorithm is $\lceil \log m \rceil (\lceil \log m \rceil + 1)(2T_t+T_{ce})/2$.

It was seen that *Algorithm 4* was designed to sort an arbitrary m-key sequence into non-decreasing order. In fact, it can sort the input sequence into non-increasing order as well, if *mask* is set 1 as $k_i=(m$-$1)_i$ and 0 otherwise.

Theorem 1: On an incomplete hypercube with m processors, an arbitrary m-key sequence can be sorted non-decreasingly or non-increasingly in $\lceil \log m \rceil (\lceil \log m \rceil+1)(2T_t+T_{ce})/2$ time. Initially, the input sequence is well-stored in the incomplete hypercube.

5. Concluding Remarks

In this paper, due to the observation that an m-key $\vee$-sequence can be sorted using exactly m processors, we have successfully developed a bitonic sort algorithm that sorts arbitrary m keys using exactly m processors. This algorithm can be executed on an incomplete hypercube with m processors or on an incomplete part of a hypercube that contains m processors. Besides, the algorithm can be easily adapted to other parallel computation models such as the EREW PRAM and the magnetic bubble memory systems [4]. Recently, Lin [5] has further shown that this algorithm can also be executed on mesh-connected computers and the STAR machine.

Refferences

[1] S. G. Akl, *Parallel Sorting Algorithms*, Academic, Orlando, Fl., 1985.

[2] K. E. Batcher, "Sorting networks and their applications," in *Proceedings of the AFIPS Spring Joint Computer Conference*, 1986, vol. 32, pp. 307-314.

[3] K. E. Batcher, "On bitonic sorting network," in *Proceedings of the 1990 International Conference on Parallel Processing*, 1990, vol. 1, pp. 376-379.

[4] K. M. Chung and F. Luccio, "On the complexity of sorting in magnetic bubble memory systems," *IEEE Transactions on Computers*, vol. C-29, no. 7, pp. 553-562, 1980.

[5] C. S. Lin, "Bitonic sort on a mesh-connected computer of an arbitrary size," Master thesis, National Taiwan University, 1991.

[6] S. L. Johnsson, "Combining parallel and sequential sorting on a boolean n-cube," in *Proceedings of the 1984 International Conference on Parallel Processing*, 1984, pp. 444-448.

[7] H. P. Katseff, "Incomplete hypercubes," *IEEE Transactions on Computers*, vol. C-37, no. 5, pp. 604-608, 1988.

[8] D. E. Knuth, *The Art of Computer Programming, vol. 3: Sorting and Searching*, pp. 232-233, Addision-Wesley, Reading, MA, 1973.

[9] D. Nassimi and S. Sahni, "Bitonic sort on a mesh-connected parallel computer," *IEEE Transactions on Computers*, vol. C-27, no. 1, pp. 2-7, 1979.

[10] D. Nath, S. N. Maheshwari, and P. C. P. Bhatt, "Efficient VLSI networks for parallel processing based on orthogonal trees," *IEEE Transactions on Computers*, vol. C-32, no. 6, pp. 569-581, 1983.

[11] F. Preparata and J. Vuillemin, "The cube-connected cycles: a versatile network for parallel computation," *Communications of the ACM*, vol. 24, no. 5, pp. 300-309, 1981.

[12] H. S. Stone, "Parallel processing with the perfect shuffle," *IEEE Transactions on Computers*, vol. C-20, no. 2, pp. 153-161, 1971.

[13] C. D. Thompson and H. T. Kung, "Sorting on mesh-connected computers," *Communications of the ACM*, vol. 20, no. 4, pp. 263-271, 1977.

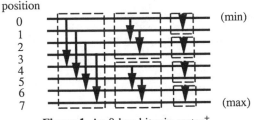

Figure 1. An 8-key bitonic-sorter$^+$.

Figure 2. A network for sorting a 6-key $\vee$-sequence into non-decreasing order.

$S_{i,j}$	$S_{i,0}$	$S_{i,1}$	$S_{i,2}$	$S_{i,3}$	$S_{i,4}$	• • •	$S_{i,L(S,i)-4}$	$S_{i,L(S,i)-3}$	$S_{i,L(S,i)-2}$	$S_{i,L(S,i)-1}$
/-sequence (+) or \-sequence (-)	-	+	-	+	-	• • •	-	+	-	+

(a) $L(S, i)$ is even.

$S_{i,j}$	$S_{i,0}$	$S_{i,1}$	$S_{i,2}$	$S_{i,3}$	$S_{i,4}$	• • •	$S_{i,L(S,i)-4}$	$S_{i,L(S,i)-3}$	$S_{i,L(S,i)-2}$	$S_{i,L(S,i)-1}$
/-sequence (+) or \-sequence (-)	+	-	+	-	+	• • •	-	+	-	+

(b) $L(S, i)$ is odd.

Figure 4. A 2^i-ready sequence S.

Experiments in Parallel Heuristic Search

(Extended Abstract)

Lélio de P. Sá Freitas
Valmir C. Barbosa

Programa de Engenharia de Sistemas e Computação, COPPE
Universidade Federal do Rio de Janeiro
Caixa Postal 68511
21945 Rio de Janeiro - RJ, Brazil

Summary

We introduce four parallel variants of the A^* and BS^* heuristic search algorithms, two based on A^* and two based on BS^*. In BS^* the search is bidirectional, thus hinting at a possibly higher amenability to parallel processing than in the case of A^*. Two of the variants we introduce employ a centralized-list approach, and the other two employ a distributed-list approach. Experiments with randomly generated instances of the Route Planner Problem on an eight-processor, T-800 Transputer hypercube have indicated a generally superior performance of the distributed-list approaches, and between these a very clear superiority of the variant based on the A^* search.

Introduction

Consider the problem whose solution space is the directed graph $D = (N, E)$, where each member of N stands for a partial solution, or, equivalently, a sub-problem, and the edges in E are directed in order to indicate increasing progress in the description of a complete solution, or, equivalently, decreasing complexity of the sub-problems. A node $n \in N$ may have predecessors ($n' \in N$ such that $(n' \to n) \in E$) and successors ($n' \in N$ such that $(n \to n') \in E$).

To solve the problem whose solution space is D is to find a directed path from the starting node s to one of the nodes in the set of goal nodes G, in such a way as to satisfy a certain optimization criterion, usually the minimization of this s-to-G path's "cost." Often D is a very large graph, and is not entirely available when the search for an optimal s-to-G path is started. Rather, its nodes are generated as needed.

We will in this paper restrict ourselves to the cases in which the "cost" of a directed path is given by the sum of the individual "costs" of its directed edges. With each edge $(n \to n') \in E$ we then associate a cost $c(n, n')$, and with each node $m \in N$ we associate a cost $f^*(m)$, which is the sum of $c(n, n')$ for all edges $(n \to n')$ on the optimal s-to-G path that includes m. Of course the value of f^* is the same for all nodes on a same optimal s-to-G path.

A very successful "best-first" strategy to determine an s-to-G path that minimizes the value of f^* is the A^* algorithm [1,2], which maintains a tree rooted at s. This tree grows outward from s by the generation of all successors of a selected leaf. Whenever a node $n \in N$ is generated a pointer $p(n)$ is set to the predecessor of n from which it was generated.

When during the A^* search a leaf is selected which is found to be one of the nodes of G, the search terminates and that node is proclaimed the optimal solution to the problem. Depending on the criterion for selecting leaves for the generation of successors, that node in G is indeed optimal in the sense that an s-to-G path where f^* is minimum has been found (and is, in fact, given by an s-to-G path in the tree that the search maintained).

When G is a singleton whose only member, say t, is known beforehand, two concurrent A^* searches may be advantageous, one starting at s and one at t, thereby concurrently generating two trees that somehow meet "halfway" between s and t in the solution space. The potential gain with this dual approach comes from the realization that if a tree of height h has a number of nodes which is $O(H)$, then a tree of height $h/2$ has a number of nodes which is $O(\sqrt{H})$, so at least in principle many fewer nodes are generated with this bidirectional search.

The most successful bidirectional search procedure in the literature (among those which do not employ the so-called "wave-shaping" techniques) is the BS^* algorithm [3], which is a sequential search algorithm, but whose potential for parallelism is obvious. BS^* is essentially composed of two instances of the A^* search, one that generates a tree from s toward t, and another generating a tree from t toward s. In this paper we describe an experimental evaluation of two parallel versions of A^* and two parallel versions of BS^* on the Route Planner Problem (RPP, to be described later), using a T-800 Transputer hypercube [4] programmed in Occam2 [5].

Although BS^* has in principle a much greater potential for parallelism than A^* does, it is not clear at all which one will perform better in a real distributed parallel imple-

---

This work was supported by the Brazilian agencies FINEP, CAPES, and CNPq.

mentation. The main reason why one should not trust a hasty *a priori* judgement concerning the two methods' relative performance is that BS^*, although capable of generating many fewer nodes due to the existence of the two search trees, has to somehow control the parallel expansion of the two trees so that they are led to meet in the "middle" of the solution space, thereby imposing an additional cost which may affect performance.

Both the A^* and the BS^* search methods maintain their trees' leaf and nonleaf nodes in lists, which apparently constitutes an inherent "sequential bottleneck" of the methods. As in previous parallel implementations of branch-and-bound techniques [6–9], where similar lists are also kept, a choice has to be made concerning the use of a centralized- or a distributed-list approach. These are then the four parallel heuristic search methods we evaluate: A^* with centralized list and distributed list, and BS^* with centralized lists and distributed lists. As we shall see, an appropriate partitioning of the node set N has led to our conclusion that for RPP the A^* method with distributed list is by far the best of the four approaches.

The paper has for lack of space been considerably shortened; a complete version with details of the sequential and parallel algorithms, as well as a more thorough experimental evaluation, can be found in [10].

In what follows, the tree nodes whose successors have already been generated are called closed nodes; the others are referred to as open nodes.

The Parallel Versions of A^*

Our parallel algorithms employ a set of τ concurrent tasks $T = \{t_1, \ldots, t_\tau\}$, each of which responsible for a subset $N_i \subseteq N$ of nodes in the solution space of the problem, $1 \le i \le \tau$. These subsets N_i exhaust the space N and are usually pairwise disjoint, although the latter is not a requirement. The idea behind this partition of N into τ sets is that task t_i will only generate nodes in N_i.

Parallel A^* with Centralized List (PACL)

In a centralized-list, parallel implementation of A^* an additional concurrent task, t_M, the master task, is utilized. What this task does is to select an open node and send it to each of the tasks in T for successor generation. Those tasks in turn send back to t_M the successors generated with their corresponding attributes. After the distributed parallel A^* search is terminated, t_M may then send the solution node $n \in G$ it found (if any) to any $t_i \in T$ for retrieval of the entire optimal s-to-G path. PACL can be implemented on a hypercube of 2^δ processors by letting the number τ of tasks in T be $\tau = 2^\delta - 1$. The master task t_M is then allocated to one of the processors, and each of the $t_i \in T$ is allocated to one of the remaining processors.

Parallel A^* with Distributed List (PADL)

In contrast with PACL, in a distributed-list, parallel implementation of A^* only the τ identical tasks in T are used, and employ a distributed extreme-finding procedure to determine the open node to have its successors generated.

When the algorithm terminates at all tasks $t_i \in T$, the optimal s-to-G path it found (if any) is readily available at all tasks. On a hypercube with 2^δ processors an implementation of PADL employs $\tau = 2^\delta$ tasks, each one running on each of the processors.

The Parallel Versions of BS^*

For the parallel versions of BS^* two sets of concurrent tasks will be employed, each with τ tasks. These are $T_s = \{t_{s,1}, \ldots, t_{s,\tau}\}$, for the search that starts at s, and $T_t = \{t_{t,1}, \ldots, t_{t,\tau}\}$, for the search that progresses in the opposite direction from t. As in the case of A^*, a subset N_i of N is associated with $t_{s,i}$ and $t_{t,i}$, which will then both generate nodes in N_i only.

A task $t_{s,i} \in T_s$ will often have to communicate with its counterpart $t_{t,i} \in T_t$. In our parallel algorithms this will happen whenever a new list of successors is generated by $t_{s,i}$ and $t_{t,i}$, so the two trees will be expanded in a relatively synchronized fashion, which is also convenient from the standpoint of the intuitive expectation that for efficiency the two trees should meet approximately "halfway" between s and t.

Parallel BS^* with Centralized Lists (PBSCL)

For a parallel version of BS^* with centralized lists two master tasks, $t_{s,M}$ and $t_{t,M}$, are used. As in the case of A^*, these tasks interact with those in T_s and T_t for node generation. When the algorithm terminates, the optimal s-to-t path it found (if any) can be traced back from the meeting node of the two tree expansions by any pair $t_{s,i}$, $t_{t,i}$ of tasks. In a 2^δ-processor implementation of PBSCL we let $\tau = 2^{\delta-1} - 1$. The tasks can then be allocated as follows. $t_{s,M}$ runs on processor 0 and $t_{t,M}$ on processor $2^{\delta-1}$. Allocate each of the $t_{s,i}$ (in increasing order of i) to each of the processors $1, \ldots, 2^{\delta-1} - 1$, and each of the $t_{t,i}$ (in increasing order of i) to each of the processors $2^{\delta-1} + 1, \ldots, 2^\delta - 1$. Then for any i there exists a communication channel connecting the processor on which $t_{s,i}$ runs to that where $t_{t,i}$ runs.

Parallel BS^* with Distributed Lists (PBSDL)

As in the case of the parallel algorithm for A^* with distributed lists, here too the tasks in T_s and T_t execute a distributed procedure to determine the open node to have its successors generated on the trees rooted at s and t, respectively. Like PBSCL, here too pairs of tasks $t_{s,i}$ and $t_{t,i}$ communicate. When the algorithm terminates, the optimal s-to-t path it found (if any) can be traced back from the meeting node of the two tree expansions by any pair $t_{s,i}$, $t_{t,i}$ of tasks, as in the case of PBSCL. A hypercube implementation with 2^δ processors employs $\tau = 2^{\delta-1}$ tasks for each tree expansion. Each task $t_{s,i}$ (in increasing order of i) can be allocated to each of the processors $0, \ldots, 2^{\delta-1} - 1$, and each $t_{t,i}$ (in increasing order of i) to each of the processors $2^{\delta-1}, \ldots, 2^\delta - 1$. As in the case of PBSCL, a communication channel exists between the processor at which $t_{s,i}$ runs and that at which $t_{t,i}$ runs.

Experimental Evaluation on RPP

Given a set of points Z and two distinguished points $z, z' \in Z$, RPP asks for the shortest route between z and z' with respect to a set of positive distances $dist(a, b)$ between each two points $a, b \in Z$.

The set of distances $dist(a, b)$ implies an undirected graph whose vertices are the elements of Z and whose edges correspond to those pairs of points $a, b \in Z$ such that $dist(a, b) < \infty$. In terms of our heuristic search terminology, this undirected graph is the underlying structure of the solution space D, where $N = Z$ and $E = \{(a \to b), (b \to a) | dist(a, b) < \infty\}$.

In this paper we take the points in Z to be points on a plane, and the distance $dist(a, b)$ to be the Euclidean distance between a and b if $dist(a, b) < \infty$. In terms of our previous notation, we then have $s = z$, $t = z'$, $c(n, n') = dist(n, n')$ for all $(n \to n') \in E$, and the function f^* shortest distances.

The experiments we report on here have been carried out on an eight-processor, T-800 Transputer hypercube, with the four parallel algorithms programmed in Occam2. Our experiments have been performed on random undirected graphs with points in Z generated randomly inside a 100×100 square, and pairs of points (independently) interconnected with probability P_c. We report on experiments for $|Z| \in \{50, 60, 70, 80, 90, 100\}$, $P_c \in \{0.1, 0.3, 0.5\}$, and $2^\delta = 8$.

Once the number of tasks has been determined, the specification of each N_i constitutes another crucial step toward a real implementation. As each N_i contains the set of nodes to be generated by a task, its determination is in fact a load balancing procedure that should attempt to get as close as possible to the ideal situation in which all tasks are required to do approximately the same amount of node-generation work.

For our experiments on RPP we have chosen to proceed as follows. As the points in Z are randomly generated, they are assigned circularly to the tasks (the N_i's), i.e., the first point is assigned to the first task, the second point to the second task, and so on, until all tasks have one point, then the first task receives another point, etc. Points are then interconnected pairwise with probability P_c, so the average number of points interconnected to a point $a \in Z$ and allocated to a certain task t_i is the same for all tasks. This means that the successors of a node $n \in N$ are on average distributed evenly among the tasks, whose processors are then on average loaded equally.

Each run of an algorithm in our experiments involves taking s and t to be all the possible pairs of nodes, thereby avoiding a possible biased choice for s and t in the randomly generated set of points. Partial results are shown in Figures 1(a) through 1(c), where the speedup of the algorithms with respect to the sequential A^* search is shown on eight processors organized as a hypercube.

A general trend in the performance of the four parallel methods is that it tends to increase with the problem's density (i.e., the value of P_c) and with problem size. This is in a way expected, since as either P_c or the problem's size in-

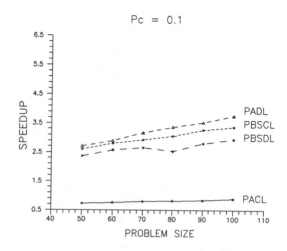

Figure 1(a). Performance of the parallel algorithms on eight processors with $P_c = 0.1$

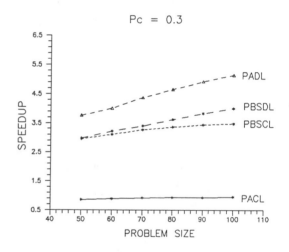

Figure 1(b). Performance of the parallel algorithms on eight processors with $P_c = 0.3$

creases the methods become more "computation-intensive" with the generation of more nodes, and consequently the speedup degradation caused by interprocessor communication is less effective.

Another general trend that can be inferred is the superiority of the distributed-list approaches over their centralized-list counterparts. This too is expected, in view of the obvious communication bottlenecks with the master tasks in the centralized-list approaches. An exception to this general observation has been the case of instances with $P_c = 0.1$ on eight processors (Figure 1(a)), where PBSCL exhibits a performance slightly better than PBSDL. This may be an indication (confirmed by the results for

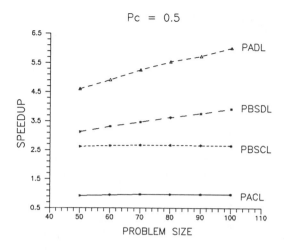

Pc = 0.5

Figure 1(c). Performance of the parallel algorithms on eight processors with $P_c = 0.5$

other network sizes) that for low densities the communication between tasks operating on different "wave fronts" of BS^* is the predominant source of communication overhead, so the centralized- and distributed-list approaches tend to perform equivalently to each other.

If we restrict ourselves to observing the centralized-list methods, then clearly the BS^* implementation is superior to that of A^*. We believe the reason for this to be twofold. First of all, each tree expansion in PBSCL spans half of the processors used by PACL, so the communication bottleneck with the master task is much less prominent in the former case. Secondly, we might also think of attributing a fraction of this superiority of PBSCL over PACL to the very nature of the BS^* search, with the expected superior performance of two concurrent tree expansions. However, if now we restrict ourselves to observing the performance of the distributed-list methods alone, then we are led to conclude that perhaps this second reason is in fact of minor importance after all, as PADL is in general far superior to all others.

Concluding Remarks

We have in this paper investigated four parallel algorithms for A^* and BS^* heuristic searches. Two of these algorithms follow a centralized-list approach (PACL for A^*, PBSCL for BS^*), and two employ distributed lists in conjunction with a distributed extreme-finding procedure (PADL for A^*, PBSDL for BS^*).

Experiments have been described for instances of the Route Planner Problem (RPP) on an eight-processor, T-800 Transputer hypercube. These experiments have indicated a tendency for increased performance of all four algorithms as either the problem's density or problem size increases. They also reveal a general superiority of the distributed-list approaches over the centralized-list approaches, although apparently this tends to be the other

way around for BS^* when dealing with low-density instances of RPP. Within the framework of the centralized-list methods, BS^* seems to be superior to A^*. When restricted to the distributed-list approaches, however, A^* is considerably superior to BS^*.

Although the performance of heuristic search methods is very dependent upon the particular problem being treated and the particular instances of that problem as well, and although (potentially serious) limitations have been imposed by our hardware's relatively small number of processors and memory size per processor, the overall superiority of PADL seems to be well established.

References

1. N. J. Nilsson, *Principles of Artificial Intelligence*, Tioga Publishing Company, Palo Alto, CA, 1980.

2. J. Pearl, *Heuristics*, Addison-Wesley Publishing Company, Reading, MA, 1984.

3. J. B. H. Kwa, "BS^*: an admissible bidirectional staged heuristic search algorithm," *Artificial Intelligence*, **38**, 95–109 (1989).

4. Núcleo de Computação Paralela, COPPE/UFRJ, *NCP-I — Project Documentation*, Rio de Janeiro, Brazil, 1989–90.

5. A. Burns, *Programming in Occam2*, Addison-Wesley Publishing Company, Wokingham, England, 1988.

6. T. S. Abdelraham and T. N. Mudge, "Parallel branch-and-bound algorithms on hypercube multiprocessors," in *Proc. Third Conference on Hypercube Concurrent Computers and Applications*, vol. *3*, ACM Press, New York, NY, 1988, pp. 1492–1499.

7. E. W. Felten, "Best-first branch-and-bound on a hypercube," in *Proc. Third Conference on Hypercube Concurrent Computers and Applications*, vol. *3*, ACM Press, New York, NY, 1988, pp. 1500–1504.

8. R. P. Ma, F. S. Tsung and M. H. Ma, "A dynamic load balancer for a parallel branch-and-bound algorithm," in *Proc. Third Conference on Hypercube Concurrent Computers and Applications*, vol. *3*, ACM Press, New York, NY, 1988, pp. 1505–1513.

9. R. P. Pargas and D. E. Wooster, "Branch-and-bound algorithms on a hypercube," in *Proc. Third Conference on Hypercube Concurrent Computers and Applications*, vol. *3*, ACM Press, New York, NY, 1988, pp. 1514–1519.

10. L. de P. S. Freitas and V. C. Barbosa, "Experiments in parallel heuristic search," *Technical Report ES-231/90*, Programa de Engenharia de Sistemas e Computação, COPPE/UFRJ, Rio de Janeiro, Brazil, 1990. Submitted for journal publication.

Efficient Parallel Sorting and Merging Algorithms for Two-Dimensional Mesh-Connected Processor Arrays

Sy-Yen Kuo
Department of Electrical Engineering
National Taiwan University
Taipei, Taiwan
R.O.C.

Sheng-Chiech Liang
Department of Electrical
and Computer Engineering
University of Arizona
Tucson, Arizona

Abstract- The row-column sort algorithms on mesh-connected processor arrays have the properties of very simple control hardware and ease of implementation. However, they are based on the odd-even transposition sort such that half of the processors are idle during each basic comparison-interchange step. Therefore, a modified odd-even merge procedure is presented to sort two sets of data inputs concurrently by utilizing the idle processors and then merge them together. This procedure is further generalized to merge m sorted input sets ($m>2$). A speedup of $O(\log_2 m)$ over the previous merge-split method is achieved.

I. INTRODUCTION

Parallel sorting algorithms for two-dimensional mesh-connected processor arrays have been intensively studied in [1,2] and more recently, in [3,4]. Efficient implementations of these sorting algorithms in two-dimensional VLSI models are presented in [4,5]. These early efforts modify inherently parallel algorithms, such as the odd-even merge sort and the bitonic sort [6], and apply them on the mesh-connected array in an efficient manner such that the time complexity does not exceed $O(n)$. However, these implementations spend most of the time in routing data to appropriate processors, and the complicated data movements during each iteration are very difficult to control. In general, complex control structures are needed and thus, offset the advantage of simple interconnections.

Recently, Sado and Igarashi [7], and Scherson and Sen [8] presented two parallel sorting algorithms, the *parallel bubble sort* and the *shear sort*, respectively, for the two-dimensional SIMD model. These sorting algorithms basically apply the *bubble sort* technique [9] repetitively on the rows and columns of the array to be sorted. Two similar approaches, one for VLSI implementation and the other for constructing a sorting networks using the k-sorter, were presented in [10] and [11], respectively. An optimal version of the *shear–sort* for VLSI networks with complexity $O(n\sqrt{n})$ was presented in [12]. The optimal iterative $O(n)$ algorithm on a mesh-connected computer based on the *shear–sort* was shown in [8]. Furthermore, the authors recently proposed a new two-dimensional sorting algorithm, the *trapezoid sort* [13]. It is similar to the *shear–sort* and preserves the properties of simple control hardware and ease of implementation. However, the time complexity is improved approximately by a factor of 2.

In this paper, we will refer any of the above algorithms as the *row–column sort* algorithm since they all contain two basic operations: the row-sort and the column-sort. However, the row-column sort algorithm is designed to sort $N = n{\times}n$ inputs only, where N is the number of processors in the mesh array. If the number of elements to be sorted is larger than N, the row-column sort algorithm can not be applied directly. To overcome this, the method in [1,8] uses the merge-split operation to replace the compare-interchange (or compare-and-swap in this paper) operation. Although that method is simple, it is not efficient. A novel *modified odd–even merge* method is proposed here which can merge m sorted sets with an $O(\log_2 m)$ improvement in time complexity.

II. MODEL OF COMPUTATION

Let $Q=[Q_{ij}]$ be an $n \times n$ array of identical processors onto which we have mapped an input sequence S. The interconnections between processors are mesh-connected without wrap-around connections on the boundaries and similar to that of the ILLIAC IV [14]. Sorting the sequence S is then equivalent to sorting the elements of Q in some

predetermined indexing scheme. We use the *snake–like row major* (*SLRM*) indexing scheme as shown in Fig. 1 to order the elements of Q.

In the following sections, we will use the *trapezoid sort* [13] to illustrate the merging algorithm. However, we can also use the *parallel bubble sort* or the *shear sort*, but the order of speedup will be the same. The *trapezoid sort* is presented in Fig. 2 with *PASCAL*-like notations. In this procedure, the ith row and the jth column of the array Q are denoted by $Q[i, 1 \cdots n]$ and $Q[1 \cdots n, j]$, respectively. A row vector is sorted in *nondecreasing order* from left to right by the procedure *row–sort* and a column vector is sorted in *nondecreasing order* from top to bottom by the *column–sort* procedure. Both the *row–sort* and the *column–sort* procedures are implemented based on the odd-even transposition sort [9]. The procedure *row–sort* sorts a row vector of Q in an opposite order to the *row–sort* procedure, i.e., nonincreasingly from left to right. The parameter $l(\approx\sqrt{n})$ which was derived in [13] is used to control the number of iterations in step 3 of the algorithm to obtain a *snake–like row major* ordered output array.

III. SORTING MORE THAN *N* INPUTS

The row-column sort algorithm can only handle N input elements which is equal to the number of processors. If the number of elements to be sorted is larger than N, the row-column sort algorithm can not be applied directly. To overcome this, the method in [1,8] distributes the elements evenly among the processors and apply the merge-split operation instead of the compare-interchange (or compare-and-swap in this paper) operation. The "merge-split" operation is described as follows. First, processor P_1 sends its largest element to P_2 and P_2 sends its smallest element to P_1. Then this process repeats until the largest element in P_1 is not greater than the smallest element in P_2 [15]. For example, if there are only two elements in each processor, this process can be implemented by the following substeps: (1) sort the elements in each processor, (2) route the largest element in P_1 to its neighbor processor P_2, and P_2 routes its smallest element to P_1, (3) sort the elements in each processor, (4) route the largest element in P_1 to P_2 and P_2 routes its smallest element to P_1.

1	2	3	4	5
10	9	8	7	6
11	12	13	14	15
20	19	18	17	16
21	22	23	24	25

Fig. 1. An example of the snake-like row major indexing scheme.

A compare-and-swap operation on two data elements in adjacent processors can be implemented by the following sequence: route left, compare, and route right. The time for a compare-and-swap is $t_{cs}=2t_r+t_c$ where t_r is the time to route and t_c is the time to compare. Therefore, $2t_{cs}$ is required to execute a merge-split operation if there are two elements in each processor. If there are m elements in each processor, $O(m\log_2 m)$ compare-and-swap steps are required if an optimal sequential sorting algorithm is used in substeps (1) and (3) of the "merge-split" operation.

In addition, during each compare-swap operation in the row-column sort algorithm half of the processors are idle. This inefficiency can be improved based on the fact that all the elements move in the same direction at a time and processors are idle in an alternating manner. With some modification on the substeps of a compare-swap operation, the idle

Procedure *Trapezoid Sort* (Q, l);
begin
    /* step 1 */
    for all $t := 1$ to n do in parallel
        row−sort $Q[t, 1 \cdots n]$;
    /* step 2 */
    for all $t := 1$ to n do in parallel
        cyclic shift right $Q[t, 1 \cdots n]$ by $(t-1)$ positions;
    /* step 3 */
    for $i := 1$ to $\lceil \log_2 l \rceil + 1$ do   /* column-row sort module */
        begin
            for all $t := 1$ to n do in parallel
                column−sort $Q[1 \cdots n, t]$;
            for all $t := 1$ to n do in parallel
            /* two adjacent rows are sorted in opposite directions */
            if odd(t)
                then *row−sort* $Q[t, 1 \cdots n]$
                else $\overline{row−sort}$ $Q[t, 1 \cdots n]$;
        end;
end.

Fig. 2. The trapezoid sort algorithm.

processors can sort another set of input data at the same time. That is, assuming that there are two registers, R_A and R_B, in each processor and two sets of input data, N_1 and N_2, are preloaded in the array, then instead of routing right and left in the row-sort (or up and down in the column-sort), an exchange (or swap) operation is performed between two neighbor processors and this is referred as the *modified compare−and−swap* operation. As shown in Fig. 3, in every data routing substep of a modified compare-and-swap step, the content in R_B of the upper (left) processor is exchanged with the content in R_A of the lower (right) processor in the column (row) sort and all processors execute the same instruction at the same time. In a row-sort (column-sort) operation, if processors in the odd-numbered rows (columns) are processing N_1, processors in the even-numbered rows (columns) are processing N_2 at the same time. Therefore, at the time when the first input data set N_1 is being sorted and stored in R_A registers of the processors in snake-like row major order, the second input data set N_2 is also being sorted into snake-like row major order but stored in R_B registers of the processors.

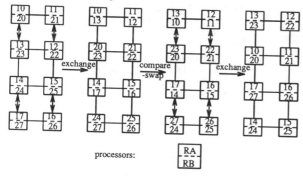

Fig. 3. A modified compare-and-swap step in the column-sort.

A. Sorting 2N inputs

Let the processor in the (i, j) position of the array be represented by $PE_{i,j}$, $1 \leq i$, $j \leq n$. In the first step, the "interleaving" of the odd-even merge is executed as shown in Fig. 4(a). If i is odd, the content in R_B of $PE_{i,j}$ is swapped with the content in R_A of $PE_{i,j+1}$, the content in R_B of $PE_{i,j+2}$ is swapped with the content in R_A of $PE_{i,j+3}$, and there is no swapping between $PE_{i,j+1}$ and $PE_{i,j+2}$, for all odd j. If i is even, the content in R_A of $PE_{i,j}$ is swapped with the content in R_B of $PE_{i,j+1}$, the content in R_A of $PE_{i,j+2}$ is swapped with the content in R_B of $PE_{i,j+3}$, and there is no swapping between $PE_{i,j+1}$ and $PE_{i,j+2}$, for all odd j. The configuration

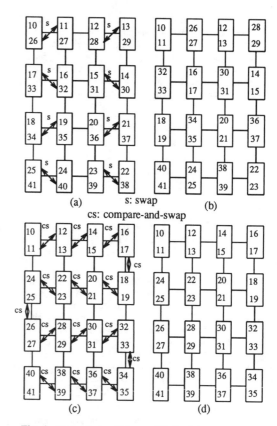

s: swap
cs: compare-and-swap

Fig. 4. An example of the modified odd-even merge.

after the interleaving process is shown in Fig. 4(b), where the two interleaved sequences are stored in R_A registers and R_B registers of the processors, respectively. Let these two interleaved sequences be represented by L_A and L_B.

In the second step, "sorting the two interleaved sequences" is performed. This scenario can be viewed as that the two random input sequences, L_A and L_B, are preloaded in the array. However, these two sequences L_A and L_B are not random input sequences, since they have already been sorted in some order. For example, for the sequence L_A, if i is odd, the content in R_A of $PE_{i,j}$ is no greater than the content in R_A of $PE_{i,j+2}$, and if i is even, the content in R_A of $PE_{i,j}$ is not less than the content in R_A of $PE_{i,j+2}$. Furthermore, L_A (or L_B) itself is generated by two shuffled sorted sequences. Let the two sorted sequences which form L_A be represented as M_1 and M_2. Using the serial bubble sort, the worst case scenario in sorting L_A is that the largest element in M_1 is less than the smallest element in M_2, since M_1 and M_2 are shuffled to form L_A and this case has the maximum distance $(n/2)$ to move an element to its final position. This is also the worst case in the row-column sort which is implemented based on the odd-even transposition sort. The detail reasoning based on the {0-1} principle [9] is in the following.

If the largest element in M_1 is less than the smallest element in M_2, then M_1 can be assumed to have all zeros and the largest element of M_1 will be in $PE_{n,n-1}$ and M_2 can be assumed to have all ones and the smallest element of M_2 will be in $PE_{1,2}$. In this situation, instead of using $\lceil \log_2 l \rceil + 1$ iterations of row-column sort followed by a row-sort, a single column-sort followed by a single row-sort is sufficient to sort the sequence. We designate the operation of a single column-sort followed by a row-sort as a *single−column−row sort*. An example worst case of an 7×7 L_A is shown in Fig. 5. Any change of an element from 1 to 0 in making L_A such that it is no longer a worst case should be performed at $PE_{1,2}$ first. Similarly, any change of an element from 0 to 1 should be done at $PE_{n,n-1}$ first, since M_1 and M_2 are two sorted sequences. As shown in Fig. 5, any change from 0 to 1 or 1 to 0 will not increase the number of steps required to sort L_A, that is, $2n$ steps of the odd-even transposition sort are sufficient to sort the sequence L_A. These $2n$ steps include n steps of compare-and-swap in the column-sort and n steps of compare-and-swap in the row-sort. However, based on the fact that M_1 and M_2 are interleaved to form L_A, $n/2$ steps are sufficient to clean a column in the column-sort since the maximum distance required for any 0 or 1 to move to its final destination is $n/2 - 1$. Similarly, $n/2$ steps in the following row-sort can clean every row after the column sort. We call this

0	1	0	1	0
1	0	1	0	1
0	1	0	1	0
1	0	1	0	1
0	1	0	1	0

Fig. 5. An example 5×5 L_A.

operation the *reduced single−column−row sort* to represent the fact that instead of using n steps of the odd-even transposition sort for each of the column-sort and the following row-sort, only $n/2$ steps are sufficient for each.

In the third step, "merging of the two sorted sequences" is executed. This process is implemented by two compare-and-swap steps as shown in Fig. 4(c). Since concurrent data movements are allowed in the same direction only, the compare-and-swap step that compares two neighbor processors in the same column can not be executed until the two neighbor processors in the same row finish sorting. It should be noted that in the original odd-even merge method, another interleaving step is required between step 2 and step 3. However, for the current application, this step can be combined with the merging operation and implemented directly by comparing R_B of $PE_{i,j}$ with R_A of $PE_{i,j+1}$ (Fig. 4(c)).

Therefore, with one step to perform the interleaving operation, $n/2+n/2$ steps to sort the interleaved sequence, and two more steps to complete the merging, the two sorted sequences N_1 and N_2 can be merged as a sorted 2N-output sequence. Thus, instead of using $2 \times T_N$ (T_N is the time complexity required to sort N inputs) compare-and-swap steps to sort the 2N-input sequence by the *merge−split* operation, only T_N+n+3 steps are required by the *modified odd−even merge sort*.

B. Sorting mN inputs

In order to simplify the analysis, we will assume that (1) instead of only two registers R_A and R_B as in subsection A, there are m registers, R_1, R_2, ..., R_m, in each processor (or each processor has a local memory that can store m elements) and m is a power of two, (2) each processor can access its registers (or memory) with the same speed, and (3) the mN inputs are equally distributed among processors.

Let the input sequence S with mN elements be represented as m N-element sequences. The first sequence stored in R_1 registers of the processors is represented as Q_1, the second sequence stored in R_2 registers is represented as Q_2, ..., and the mth sequence stored in R_m registers is represented as Q_m. From subsection A, we know that two sets of inputs can be sorted concurrently and then merged together to form a sorted 2N-element sequence. Therefore, there will be $m/2$ sorted 2N-element sequences generated after the first merge. These $m/2$ sequences can be merged again based on the modified odd-even merge to form $m/4$ sorted 4N-element sequences, and so on until the sequence S is sorted. That is, the *modified odd−even merge* is applied recursively to merge the m sorted sequences two at a time until all of them are merged.

The generalized *modified odd−even merge* procedure is described in Fig. 6. Let $|S|$ represent the number of elements in the input sequence S. If there are only 2N inputs, as described in subsection A, the procedure will completely sort these 2N inputs. If $|S|>2N$, the sequence S will be equally divided into two subsequences S_1 and S_2. The subsequence S_1 which includes Q_1 to $Q_{m/2}$ will be sorted first by recursively calling the procedure and then followed by sorting the subsequence S_2 which includes $Q_{m/2+1}$ to Q_m. After the two subsequences are sorted, the *modified odd−even merge* procedure will interleave (or shuffle) the two sorted sequences to be merged. When implemented in a mesh-connected

array, this step means that the contents in $R_{m/4+1}$ to $R_{m/2}$ are exchanged with the contents in $R_{m/2+1}$ to $R_{3m/4}$. Thus, the subsequence formed by Q_1 to $Q_{m/2}$ can be further decomposed into two sorted subsequences, Q_1 to $Q_{m/4}$ and $Q_{m/4+1}$ to $Q_{m/2}$. Therefore, the *modified odd−even merge* can be executed again to sort the two interleaved sequences in Q_1 to $Q_{m/2}$ and $Q_{m/2+1}$ to Q_m, separately. The function *Reduced Single−Column−Row Sort* in Fig. 6, as described in subsection A, is implemented by a column-sort with only $n/2$ steps followed by a row-sort with the same number of steps.

An example of sorting 4N inputs is shown in Fig. 7. The random input sequence with 4N inputs are sorted two subsequences at a time. At the beginning, the two subsequences in R_1 registers and R_2 registers of all processors, respectively, are sorted concurrently and merged into a sorted 2N-output sequence. Then the next two subsequences in R_3 registers and R_4 registers are processed. Two sorted sequences, S_1 and S_2, with 2N elements each are stored in the processor array as shown in Fig. 7(a). The sequence S_1 consisting of 10, 11, 12, ..., 40, 41 is stored in R_1 and R_2 registers of all processors and ordered in *snake−like row major ordering*. In the same way, the sequence S_2 with 42, 43, ..., 73 is stored in R_3 and R_4 registers.

In the second step, "interleaving" (or shuffling) of the two sequences is performed as shown in Fig. 7(b). By exchanging the contents in each pair of R_2 and R_3, the two sequences in R_1 registers and R_2 registers can be viewed as two sorted sequences and the combination of these two sequences is an interleaved sequence. An example of this interleaved sequence is shown in Fig. 9(b) as 10, 42, 12, 44, ..., 38, 70, 40, 72.

In the third step, the *modified odd−even merge* procedure merges 2N elements. The contents of R_1 registers are interleaved again with those of R_2 registers as the process in Fig. 4(a). Then the contents in R_1 registers and R_2 registers are sorted by the modified column-row sort concurrently and rearranged in snake-like row major order as the process in Fig. 4(c). Let these two sorted sequences be represented as L_1 and L_2. L_1 and L_2 are then merged to form a sorted 2N-output sequence. The ordered 2N-output sequence, 10, 12, 14, ..., 70, 72, is stored in the R_1 and R_2 registers of the processors. After L_1 and L_2 have been merged, the

---

Procedure *Modified Odd−Even Merge* (S);
  begin
    /* divide S into two subsequences, S_1 and S_2, of equal sizes */
    S_1 = contents in R_1 registers through $R_{m/2}$ registers;
    S_2 = contents in $R_{m/2+1}$ registers through R_m registers;
    if $|S|>2N$ then   /* $m = |S|/N$ */
      begin
        interleave S_1 and S_2;
        Modified Odd-Even Merge (S_1);
        Modified Odd-Even Merge (S_2);
        merge the two sorted sequence S_1 and S_2 into S;
      end
    else
      begin
        interleave the two sorted sequence S_1 and S_2;
        do in parallel
          begin
            Reduced Single-Column-Row Sort (S_1);
            Reduced Single-Column-Row Sort (S_2);
          end
        merge the two sorted sequence S_1 and S_2 into S;
      end
  end

Fig. 6. Modified odd-even merge procedure.

same sorting and merging process can be repeated on contents in R_3 and R_4 registers. Therefore, two sorted sequences, 10, 12, ..., 72 and 11, 13, ..., 73 are stored in the array as shown in Fig. 7(c). Finally, merging of these two sorted sequences is done by comparing (and exchange if necessary) the contents in each pair of R_2 and R_3 with no interleaving required before merging.

IV. ANALYSIS AND DISCUSSION

Let T_{mN} represent the number of modified compare-and-swap steps required to sort mN inputs, $i.e.$, the time complexity of the procedure $merge\ sort$, and C_{mN} represent the time complexity of the procedure $modified\ odd-even\ merge$. Then, from Fig. 6 we have

$$T_{mN}=2T_{\frac{m}{2}N}+C_{mN} \tag{1}$$

and

$$C_{mN}=\frac{m}{4}+2C_{\frac{m}{2}N}+\frac{m}{4} \tag{2}$$

which implies

$$T_{mN}=2(T_{\frac{m}{2}N}+C_{\frac{m}{2}N})+\frac{m}{2}. \tag{3}$$

The first $m/4$ in (2) represents the number of exchange steps (it should be noted that the amount of time required by an exchange step is less than that of a compare-and-swap step) required to exchange the contents in $R_{m/4+1}$ through $R_{m/2}$ with the contents in $R_{m/2+1}$ through $R_{3m/4}$, respectively. An example is shown in Fig. 7(b). Since $m=4$, one step is required to exchange the content in R_2 with that in R_3. The second $m/4$ in (2) represents the number of compare-and-swap steps required in the last merging step (as shown in Fig. 7(d)) to compare and exchange the contents in $R_{m/4+1}$ through $R_{m/2}$ with the contents in $R_{m/2+1}$ through $R_{3m/4}$, respectively. If there are only $2N$ inputs, from III.A we know that two sets of data can be processed concurrently by a mesh-connected processor array and therefore,

$$T_N=2(\lceil \log_2 l \rceil +1)n+n \text{ and } C_N=\frac{n}{2}+\frac{n}{2},$$
$$T_{2N}=T_N+C_{2N} \text{ and } C_{2N}=C_N+3.$$

This implies that

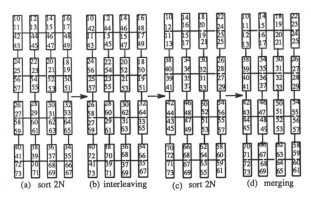

(a) sort 2N   (b) interleaving   (c) sort 2N   (d) merging

Fig. 7. An example of sorting 4N inputs.

$$T_{mN}\approx2(T_{\frac{m}{2}N}+C_{\frac{m}{2}N})+\frac{m}{2}\approx\frac{m}{2}\times[T_{2N}+(\log_2 m-1)\cdot C_{2N}]+\frac{m}{2}+m\cdot(\log_2 m-2).$$

When n is large and $m\ll n$, we have
$$T_N\gg\log_2 m\cdot C_N\gg m\cdot\log_2 m.$$
Also we know that $T_{2N}\approx T_N+C_N$ and therefore,

$$T_{mN}\approx2(T_{\frac{m}{2}N}+C_{\frac{m}{2}N})\approx\frac{m}{2}\times[T_{2N}+(\log_2 m-1)\cdot C_{2N}]\approx\frac{m}{2}\times[T_N+(\log_2 m-1)\cdot C_N].$$

That is, instead of an $O(m\log_2 m)\cdot T_N$ time complexity to sort mN data inputs with the "merge-split" operation, only $O(m/2)\cdot(T_N+n\cdot\log_2 m)$ $(T_N\gg n\cdot\log_2 m)$ time complexity is sufficient to sort mN data inputs by the $modified\ odd-even\ merge$ procedure. Therefore, we have achieved an $O(\log_2 m)$ order improvement.

V. CONCLUSIONS

A novel $merge\ sort$ method for the mesh-connected processor arrays is presented in this paper. Instead of an $O(m\log_2 m)T_N$ time complexity to sort mN input elements by the previous merge-split operation, only $O(m/2)T_N$ time complexity is sufficient to sort mN random inputs where T_N is the time complexity of the row-column sort algorithm. Therefore, we have achieved an $O(\log_2 m)$ order of improvement in time complexity to sort mN inputs. Other advantages of the proposed method include the simplicity of the architecture and efficient data movements with only near-neighbor communications. Therefore, it is very suitable for sorting more than two sets of input elements in mesh-connected processor arrays.

REFERENCES

[1] C. D. Thompson and H. T. Kung, ''Sorting on a mesh-connected parallel computer,'' $Commun.\ Ass.\ Comput.$, vol. 20, pp. 263-271, Apr. 1977.

[2] D. Nassimi and S. Sahni, ''Bitonic sort on a mesh-connected parallel computer,'' $IEEE\ Trans.\ Comput.$, vol. c-27, pp. 2-7, Jan 1979.

[3] M. Kumar and D. S. Hirschberg, ''An efficient implementation of Batcher's odd-even merge algorithm and its application in parallel sorting schemes,'' $IEEE\ Trans.\ Comput.$, vol. c-32, pp. 254-264, Mar. 1983.

[4] H. -W. Lang, M. Schimmler, H. Schemeck, and H. Schroder, ''Systolic sorting on a mesh-connected network,'' $IEEE\ Trans.\ Comput.$, vol. c-34, pp. 652-658, July 1985.

[5] H. Schmeck, H. Schroder, and C. Strake, ''Systolic s^2-way merge sort is optimal,'' $IEEE\ Trans.\ Comput.$, vol. c-38, pp. 1052-1056, July 1989.

[6] K. E. Batcher, ''Sorting networks and their applications,'' $Proc.\ AFIPS\ Conf.$, vol. 32, pp. 307-314, 1968.

[7] K. Sado and Y. Igarashi, ''Some parallel sorts on a mesh-connected processor array and their time efficiency,'' $Journal\ of\ Parallel\ and\ Distribut.\ Comput.$, vol. 3, pp. 398-410, 1986.

[8] I. D. Scherson and S. Sen, ''Parallel sorting in two-dimension VLSI models of computation,'' $IEEE\ Trans.\ Comput.$, vol. c-38, pp. 238-249, Feb. 1989.

[9] D. E. Knuth, $The\ art\ of\ computer\ programming\ -\ searching\ and\ sorting$. Addison-Wesley, 1973.

[10] T. Leighton, ''Tight bounds on the complexity of parallel sorting ,'' $IEEE\ Trans.\ Comput.$, vol. c-34, pp. 344-354, Apr. 1985.

[11] B. Parker and I. Parberry, ''Constructing sorting networks from k-sorters ,'' $Information\ Processing\ Letters$, vol. 33, pp. 157-162, Nov. 1989.

[12] I. D. Scherson, S. Sen, and A. Shamir, ''Shear sort: a true two-dimension sorting technique for VLSI networks,'' $International\ Conference\ on\ Parallel\ Processing$, pp. 903-908, 1986.

[13] S.-C. Liang, $High\ yield\ and\ reliable\ sorting\ networks\ for\ VLSI/WSI\ implementations$. University of Arizona, Tucson, Arizona: Ph.D. Dissertation, Department of Electrical and Computer Engineering, 1991.

[14] G. H. Barnes and et al. , ''The ILLIAC IV computer,'' $IEEE\ Trans.\ Comput.$, vol. c-17, pp. 746-757, 1968.

[15] U. Manber, $Introduction\ to\ algorithms$. Addison-Wesley, 1989.

Optimal Parallel External Merging
under Hardware Constraints*

Jin-Yuan Fu and Ferng-Ching Lin

Department of Computer Science and Information Engineering
National Taiwan University
Taipei, Taiwan, R.O.C.

Abstract

External multiway merging is a critical part of external mergesort. It also finds applications in database management systems. The three factors most affecting the process of external merging—merge order, block size, and buffering level—are considered and tuned to get shorter running time in the environment of parallel processing with practical hardware constraints. Under the limitedness of available resources of processing elements and memory, the best achievable running time is analysed. A parallel algorithm for loosely coupled multiprocessors is designed to achieve the best performance, which keeps both the CPU time and the I/O time to the minimal and garantees perfect read/write/compute overlapping with minimal number of buffers. Insights for the usage of parallel computers to solve external merging and other I/O-intensive problems emerge naturally during the analysis and algorithm design.

1 Introduction

The problem of merging has been studied for a long time. It arises in several applications like database management systems. It is also a critical step of mergesort, which is a standard approach for external sorting [3]. Mergeseort consists of two phases: sort phase and merge phase. If the file is large, the most cost lies in the merge phase [7].

Many parallel algorithms for merging have appeared in the literature (see [2,10] and the references therein). Some discussed the parallelization of 2-way merging. Others addressed the problem of multiway merging. Many of them approached the problem under the assumption of huge number of processing elements (PEs) or on an architecture of shared-memory model. Merging in distributed systems was discussed by some authors, for examples, in [8,9].

In a system with external (hierarchical) memory, the I/O complexity becomes a crucial issue. Because of the characteristics of random-access disks, the number of required block references is the most dominating factor for performance measurement. Disk models with capability of parallel block transfers were studied, and tight bounds of I/O complexity were derived for external sorting [1,12].

Three parameters have the most influences on the performance of exteranl merging. They are the order of merge, the block size, and the buffering level [7]. By tuning these parameters to balance the subsystems involved, better performance can be obtained. The algorithm adopted to implement the merging operation is also influencing. It determines the degree of read/write/compute overlapping for various patterns of input.

In this paper, we take a hard look at the problem of external merging with the considerations of I/O bottleneck, limited number of PEs, limited local memory, and the characteristics of random-access disks. A procedure to tune the merge order and the block size is derived to attain to the best possible performance under these hardware constraints. Then we present a (generic) parallel algorithm to achieve the optimal running time, which guarntees perfect read/write/compute overlapping with minimal number of buffers. The contribution of the paper lies not only in a more practical environmental settings, but also in the insights shed for the parallel processing of I/O-intensive problems.

The rest of the paper is organized as follows. Section 2 introduces the problem of parallel external multiway merging, along with our model, notations and assumptions. Section 3 derives the tuning procedure of finding optimal setting of merge order and block size. Then in Section 4, a generic algorithm which is adjustable to any setting is designed. Combining the procedure and the algorithm, we get an algorithm being tuned up to the best running time. Some concluding remarks are given in Section 5.

2 Notations and the Problem

Exteranl merging is needed if the collection of records to be merged is too large to fit in the memory available within the system. We focus on the architectures of distributed-memory model, where the system's internal memory is evenly distributed among the PEs. There is also a secondary mass storage, assumed to be disks, where the sorted runs of records to be merged are stored. The final single sorted run generated by the merging algorithm will be put back to it also.

Let N denote the whole size of the file and n the number of sorted runs in the very beginning. Records are transferred (read and written) between disks and the system in blocks. We denote the block size by β. Further, μ specifies how many runs are merged in a single merging process, $\mu \geq 2$. Such a μ-way merging is repeatedly applied to the sorted runs until the whole file becomes a single sorted run.

Here we define the parallel model of the hardware configurations.

Definition A *system configuration* is a 4-tuple $< P, M, \sigma_1, \sigma_2 >$, where

P is the number of PEs,

M is the number of records fit in local memory of a PE,

σ_1 is the intra-PE 1-selection average computing time function,

σ_2 is the inter-PE 1-selection average computing time function.

$\sigma_1(\mu)$ denotes the time reguired in average to select the smallest out of μ records within a PE, while $\sigma_2(\mu)$ that of selecting the smallest out of μ records, each in a different PE. Throughout the paper, it is assumed that $\sigma_1(\mu) = C_1 \lg \mu$.[1] Depending on the connection topology of the system, $\sigma_2(\mu)$ could take any form. Although we make the assumption that $\sigma_2(\mu) = C_2 \lg \mu$, where $C_2 > C_1$, any increasing function $\sigma_2(\mu)$ can be used in the analysis later.

Definition An *I/O configuration* is a 4-tuple $< I, r, s, t >$, where

I is the number of blocks can be transferred in each I/O operation,

r is the average rotational latency,

s is the average seek time,

t is the bulk transfer time per record.

*This research was partially supported by National Science Council of R.O.C. under contract NSC79-0408-E002-19.

---

[1] $\lg x = \log_2 x$.

Sorted runs are placed along contiguous locations in disk(s), such that, transferring I blocks, each with β records, needs $(r+s)+\beta t$ time. The external I/O is presumed to allow independent reading and writing by I channels. The model is similar to the one defined in [1].

When a system configuration and an I/O configuration are given, none of the hardware factors is changeable. But we can adjust some *software factors* to achieve better performance. The merge order and the block size are two of the most critical software factors for external merging [7].

Definition A merging process is called a (μ, β)-*merge* if it contains μ-way merges, and in each I/O operation each channel can transfer β records.

Buffering level is also important. Buffers are allocated for each merging process. To facilitate overlapping of CPU and I/O operations, at least two additional buffers, one for input and one for output, are needed. A merging algorithm should manage the buffers carefully in order to get more degree of overlapping. A μ-way merging process with block size β requires at least a memory allocation of size $(\mu + 2)\beta$. Now we can state the problem in our terminology as follows.

Problem. *Given a system configuration, an I/O configuration, and a file of size N which is divided into n sorted runs, find (μ, β) with $\mu \geq 2$, $\beta \geq 1$ and $(\mu + 2)\beta \leq PM$, such that (μ, β)-merge leads to minimal running time, and design an algorithm to realize the optimal (μ, β)-merge with only 2 extra buffers in each merging process.*

3 Tuning merge order and block size

If the system performs a (μ, β)-merge and $(\mu + 2)\beta < PM$, there can more than one independent *merging processes* (MPs) running concurrently. Each MP does a μ-way merging without correlation with one another. There is thus at most $\frac{PM}{(\mu+2)\beta}$ MPs.[2] Notice that if $(\mu + 2)\beta < M$, there are $\frac{M}{(\mu+2)\beta}$ MPs in one PE. The merging operations for each MP are sequentially performed within one PE. MPs in different PEs, however, can operate in parallel.

3.1 CPU time and I/O time

A standard technique to do multiway merging is replacement selection [3]. We assume a heap is used. If $\mu \leq M$, it takes $C_1 \lg \mu$ CPU time to output the smallest element and readjust the heap. If $\mu > M$, it takes $C_1 \lg M + C_2 \lg(\mu/M)$ CPU time. If there are more than one MPs residing in one PE, they are performed sequentially. Therefore, there are three cases for the CPU time to process one block for *every* MP:

$$\frac{M}{\mu+2} C_1 \lg \mu \qquad \text{if } (\mu+2)\beta \leq M,$$
$$\beta C_1 \lg \mu \qquad \text{if } (\mu+2)\beta > M, \ \mu \leq M,$$
$$\beta C_1 \lg M + C_2 \lg(\mu/M) \quad \text{if } \mu > M.$$

Such parallel block operations will be performed $\frac{N}{PM/(\mu+2)}$ times for the whole file to be processed once. The file needs to be processed in passes to get the final one sorted run. The number of passes is $\log_\mu n = \lg n / \lg \mu$. The total CPU time can be obtained from these quantities. Let $(\mu + 2)\beta = \alpha M$. We denote the required CPU time by $t_c(\mu, \beta)$.

$$t_c(\mu, \beta) = \begin{cases} \frac{N \lg n}{P} C_1 & \text{if } (\mu+2)\beta \leq M, \\ \frac{N \lg n}{P} \alpha C_1 & \text{if } (\mu+2)\beta > M, \ \mu \leq M, \\ \frac{N \lg n}{P} \alpha \frac{C_1 \lg M + C_2 \lg(\mu/M)}{\lg M + \lg(\mu/M)} & \text{if } \mu > M. \end{cases} \tag{1}$$

Note that if α is fixed and $C_2 > C_1$, then the second case always outperforms the third case, because

$$\alpha \frac{C_1 \lg M + C_2 \lg(\mu/M)}{\lg M + \lg(\mu/M)} = \alpha C_1 + \alpha \frac{(C_2 - C_1) \lg(\mu/M)}{\lg M + \lg(\mu/M)} > \alpha C_1. \tag{2}$$

---

[2]Throughout the paper, discrete quantities are treated as continuous ones for presentational simplicity.

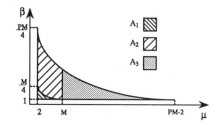

Figure 1: Possible settings for (μ, β).

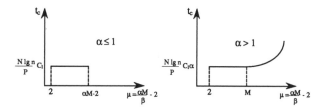

Figure 2: The CPU time when α is fixed.

The three cases divide the possible settings into areas, as shown in Fig. 1. We name these areas as $A = \{(\mu, \beta) : \mu \geq 2, \ \beta \geq 1, (\mu+2)\beta \leq PM\}$, $A_1 = \{(\mu, \beta) \in A : (\mu+2)\beta \leq M\}$, $A_2 = \{(\mu, \beta) \in A : (\mu+2)\beta > M, \ \mu \leq M\}$, and $A_3 = \{(\mu, \beta) \in A : \mu > M\}$.

There is no difference with I/O time for these three cases. To read/write one block from/to disks takes $(r + s) + \beta t$ of I/O time. There are $\frac{PM}{(\mu+2)\beta}$ MPs and I parallel I/O channels. One half of the channels are assigned to perform simultaneous reading, while the other for writing. The I/O time to read/write one block for every MP becomes

$$\frac{PM}{(\mu+2)\beta}[(r+s) + \beta t]\frac{1}{I/2}.$$

The total required I/O time can be obtained. Denote it by $t_{io}(\mu, \beta)$.

$$t_{io}(\mu, \beta) = \frac{N \lg n}{\lg \mu} \left[\frac{r+s}{\beta} + t \right] \frac{1}{I/2}. \tag{3}$$

3.2 Optimal running time

From t_c and t_{io}, we can get the *lower bound* of total running time. The best running time, $t(\mu, \beta)$, which could be achieved is the maximum of t_c and t_{io}, i.e.,

$$t(\mu, \beta) = \max\{t_c(\mu, \beta), \ t_{io}(\mu, \beta)\}.$$

This represented a complicated surface very hard to analyse by brute force. It is not trivial to find μ^* and β^* such that $t(\mu^*, \beta^*) \leq t(\mu, \beta)$ for all $(\mu, \beta) \in A$. But it is easier to figure out the optimal $t(\mu, \beta)$, when $(\mu+2)\beta = \alpha M$ for some fixed α, $0 < \alpha < P$. Let A^α denote the set $\{(\mu, \beta) \in A : (\mu+2)\beta = \alpha M\}$, $0 < \alpha \leq P$, and let $A_i^\alpha = A^\alpha \cap A_i$. Given a fixed α, by (1), $t_c(\mu, \beta)$ is constant over $(\mu, \beta) \in A_1^\alpha \cup A_2^\alpha$. It is also implied by (2) that $t_c(\mu', \beta') > t_c(\mu, \beta)$ if $(\mu, \beta) \in A_2^\alpha$ and $(\mu', \beta') \in A_3^\alpha$. Viewed along the curve $(\mu+2)\beta = \alpha M$, t_c has one out of the two forms shown in Fig. 2.

Let us see the behavior of $t_{io}(\mu, \beta)$ for $(\mu, \beta) \in A^\alpha$ by differentiating $t_{io}\left(\mu, \frac{\alpha M}{\mu+2}\right)$ and letting $\frac{d}{d\mu} t_{io}\left(\mu, \frac{\alpha M}{\mu+2}\right) = 0$. After simplification, we have the following equation:

$$\nu \ln \nu = k, \tag{4}$$

where $\nu = \mu/e$ and $k = (\frac{\alpha M t}{r+s} + 2)/e$. There is no closed-form solution for the above equation up to the authors' knowledge. But it is not difficult to write a numerical algorithm to estimate the solution. Denote the solution of (4) by $\hat{\mu}_\alpha$, and let $\hat{\beta}_\alpha = \alpha M/(\hat{\mu}_\alpha + 2)$. Substituting α in (4) with $M/(\mu+2)\beta$, we get

$$\ln \mu = \left[1 + \frac{2}{\mu} \right]\left[1 + \frac{\beta t}{r+s} \right], \tag{5}$$

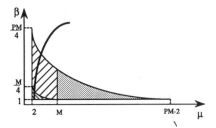

Figure 3: Minimum I/O curve.

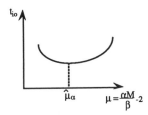

Figure 4: The I/O time when α is fixed.

which defines a curve in the μ-β plane. We shall call it the *minimum I/O curve*. Fig. 3 shows the curve. Now let us view t_{io} along the curve $(\mu + 2)\beta = \alpha M$. It has the form shown in Fig. 4. Revealed by Fig. 2 and Fig. 4, the following lemma is obvious.

Lemma 1 *If $\hat{\mu}_\alpha \leq M$, then $t(\hat{\mu}_\alpha, \hat{\beta}_\alpha) \leq t(\mu, \beta)$ for all $(\mu, \beta) \in A^\alpha$.*

From (4) we can derive an equivalent condition for the inequality $\hat{\mu}_\alpha \leq M$ to hold.

Lemma 2 $\hat{\mu}_\alpha \leq M$ *if and only if $\alpha \leq \frac{r+s}{t}(\ln M - \frac{2}{M} - 1)$.*

To get an impression of the quantity $\frac{r+s}{t}(\ln M - \frac{2}{M} - 1)$, we give a typical example. Consider the case where the size of a record is 50 bytes, $(r + s) \approx 2 \times 10^{-2}$sec, $t \approx 50 \times \frac{1}{3 \times 10^6} = (50/3) \times 10^{-6}$sec, $M = \frac{1}{50} \times 2 \times 10^6 = 4 \times 10^4$ records (with transfer rate 3Mbytes/sec, and 2 Mbyte local memory per PE). Then $\frac{r+s}{t}(\ln M - \frac{2}{M} - 1)$ is in the order of 10^4. Because $\alpha \leq P$, Lemma 2 tells that the optimal $\hat{\mu}$ for t_{io} is possible to be greater than M only when the number of PEs is substantially large, or more precisely, $P > \frac{r+s}{t}(\ln M - \frac{2}{M} - 1)$. In fact, such a great number of PEs with 2 Mbyte memory may even make external merging unnecessary.

As mentioned before, $t(\mu, \beta)$ is too complicated. The following lemma is the key to simplify the analysis of the optimality of $t(\mu, \beta)$. It guarantees that if the number of PEs is not too large, the search space of optimal setting can be narrowed down to within the minimum I/O curve.

Lemma 3 *If $P \leq \frac{r+s}{t}(\ln M - \frac{2}{M} - 1)$, then there is an α, $0 < \alpha \leq P$, such that $t(\hat{\mu}_\alpha, \hat{\beta}_\alpha) \leq t(\mu, \beta)$ for all $(\mu, \beta) \in A$.*

Proof. By contradiction. Assume that there exists $(\tilde{\mu}, \tilde{\beta}) \in A$ such that $t(\tilde{\mu}, \tilde{\beta}) < t(\hat{\mu}_\alpha, \hat{\beta}_\alpha)$ for all α, $0 < \alpha \leq P$. Let $(\tilde{\mu} + 2)\tilde{\beta} = \tilde{\alpha}M$, where $0 < \tilde{\alpha} \leq P$. By Lemma 2, the condition $\tilde{\alpha} \leq P \leq \frac{r+s}{t}(\ln M - \frac{2}{M} - 1)$ implies $\hat{\mu}_{\tilde{\alpha}} \leq M$. By Lemma 1, $t(\hat{\mu}_{\tilde{\alpha}}, \hat{\beta}_{\tilde{\alpha}}) \leq t(\mu, \beta)$ for all $(\mu, \beta) \in A^{\tilde{\alpha}}$. Since $(\tilde{\mu}, \tilde{\beta}) \in A^{\tilde{\alpha}}$, $t(\hat{\mu}_{\tilde{\alpha}}, \hat{\beta}_{\tilde{\alpha}}) \leq t(\tilde{\mu}, \tilde{\beta})$, a contradiction to the assumption. ∎

On the minimum I/O curve, both $\hat{\mu}_\alpha$ and $\hat{\beta}_\alpha$ increase at the same time when α increases. Formula (3) shows that t_{io} is a strictly decreasing function of both μ and β. So $t_{io}(\hat{\mu}_\alpha, \hat{\beta}_\alpha)$ turns out to be a strictly decreasing function of α when $\hat{\mu}_\alpha \leq M$. On the other hand, (1) shows that $t_c(\hat{\mu}_\alpha, \hat{\beta}_\alpha)$ is an increasing function of α when $\hat{\mu}_\alpha \leq M$. Fig. 5 illustrates a typical case of $t_c(\hat{\mu}_\alpha, \hat{\beta}_\alpha)$ and $t_{io}(\hat{\mu}_\alpha, \hat{\beta}_\alpha)$. The optimal setting, $\mu = \hat{\mu}_{\alpha^*}$ and $\beta = \hat{\beta}_{\alpha^*}$ occurs when $t_c = t_{io}$, or in other words, the computation is *balanced*.[3]

---
[3] A computation is balanced if its CPU time equals its I/O time.

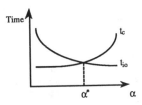

Figure 5: CPU and I/O time viewed along the minimum I/O curve.

3.3 The tuning procedure

There are cases that the computation will never be balanced even tuning the software factors μ and β. If $\frac{C_1}{P} > \frac{(r+s)+t}{I/2}$, then the CPU time is always larger than the I/O time, and any value of $(\mu, \beta) \in A_1$ is a possible setting to have the optimal running time. The following lemma depicts the reverse unbalanced case where the I/O time dominates.

Lemma 4 *If $P \leq \frac{r+s}{t}(\ln M - \frac{2}{M} - 1)$ and $C_1 < \frac{1}{\lg e}\left[\frac{r+s}{\hat{\beta}_P}\right]\frac{1}{I/2}$, then $t(\hat{\mu}_P, \hat{\beta}_P) \leq t(\mu, \beta)$ for all $(\mu, \beta) \in A$.*

Two lemmas are given below to describe the conditions for balanced computation to exist and where an optimal setting can be found in A_1 or A_2, respectively.

Lemma 5 *If $\frac{(r+s)+t}{I/2} \geq \frac{C_1}{P} \geq \frac{1}{\lg e}\left[\frac{r+s}{\hat{\beta}_1}\right]\frac{1}{I/2}$, then the equation (1)=(3) for $(\mu, \beta) \in A_1$*

$$\frac{N \lg n}{P}C_1 = N \ln n \left[\frac{r+s}{\hat{\beta}_\alpha}\right]\frac{1}{I/2}$$

has an unique solution α^*, $0 < \alpha^* \leq 1$.

In fact, any element from the set $B = \{(\mu, \beta) \in A : \frac{N \lg n}{P}C_1 \geq \frac{N \lg n}{\lg \mu}\left[\frac{r+s}{\beta} + t\right]\frac{1}{I/2}\}$ is a candidate for the optimal t. Notice that $(\hat{\mu}_1, \hat{\beta}_1)$ is always belonging to the set.

Lemma 6 *If $\frac{C_1}{P} \leq \frac{1}{\lg e}\left[\frac{r+s}{\hat{\beta}_1}\right]\frac{1}{I/2}$ and the solution of the equation (1)=(3) for $(\mu, \beta) \in A_2$*

$$\frac{N \lg n}{P}C_1\alpha = N \ln n \left[\frac{r+s}{\hat{\beta}_\alpha}\right]\frac{1}{I/2}$$

is $\alpha^* \leq \frac{r+s}{t}(\ln M - \frac{2}{M} - 1)$, then $t(\hat{\mu}_{\alpha^*}, \hat{\beta}_{\alpha^*}) \leq t(\mu, \beta)$ for all $(\mu, \beta) \in A$.

The inequality $\alpha^* \leq \frac{r+s}{t}(\ln M - \frac{2}{M} - 1)$ implies $\frac{C_1}{P}\bar{\alpha} \geq \frac{1}{\lg e}\left[\frac{r+s}{\hat{\beta}_\alpha}\right]\frac{1}{I/2}$, where $\bar{\alpha} = \frac{r+s}{t}(\ln M - \frac{2}{M} - 1)$. We are now ready to present the procedure of finding optimal (μ, β)-merge.

1. If $\frac{C_1}{P} > \frac{(r+s)+t}{I/2}$, then any element of A_1 is a good choice.

2. If $\frac{(r+s)+t}{I/2} \geq \frac{C_1}{P} \geq \frac{1}{\lg e}\left[\frac{r+s}{\hat{\beta}_1}\right]\frac{1}{I/2}$, then any element of the set B is a possible candidate. The setting $(\hat{\mu}_1, \hat{\beta}_1)$ always works.

3. If $\frac{C_1}{P} \leq \frac{1}{\lg e}\left[\frac{r+s}{\hat{\beta}_1}\right]\frac{1}{I/2}$ and $\frac{C_1}{P}\bar{\alpha} \geq \frac{1}{\lg e}\left[\frac{r+s}{\hat{\beta}_\alpha}\right]\frac{1}{I/2}$, $\bar{\alpha} = \frac{r+s}{t}(\ln M - \frac{2}{M} - 1)$, then choose $(\hat{\mu}_{\alpha^*}, \hat{\beta}_{\alpha^*})$ as the optimal setting, where α^* is the solution of the equation (1)=(3).

4. If $P \leq \frac{r+s}{t}(\ln M - 1)$ and $C_1 < \frac{1}{\lg e}\left[\frac{r+s}{\hat{\beta}_P}\right]\frac{1}{I/2}$, then $t(\hat{\mu}_P, \hat{\beta}_P)$ is the choice.

5. Otherwise, the optimal setting lies in A_3 and the procedure has no idea about its exact location.

The integer points near the chosen optimal setting should be searched to get a suitable setting for implementation. Treating the I/O time as continuous may cause some problems, especially when some of the characterizing factors considered in the paper are relatively small. A more thorough search is recommended.

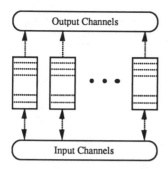

Figure 6: A system with multiple linear arrays of PEs.

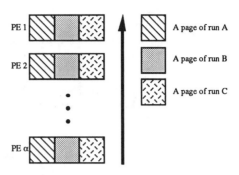

Figure 7: Initial allocation of blocks of 3 runs.

4   Parallel Perfect Overlapping Merge

We can design a merging algorithm to achieve the best running time each time when specific system configuration and I/O configuration are given. Another approach, which we take here, is to design a generic one which is able to conform to any reasonable setting of (μ, β). Such a generic algorithm must satisfy some conditions to assure the optimal performance. Denote the CPU time, I/O time, and total running time of an algorithm by T_c, T_{io}, and T, respectively.

Definition An algorithm is called *overlapping* if one of the CPU time and the I/O time is (almost) totally overlapped by the other.

Definition A generic merge algorithm is called *perfect overlapping* if for any fixed (μ, β) the algorithm can be tuned to be overlapping and $T_c(\mu, \beta) = t_c(\mu, \beta)$, $T_{io}(\mu, \beta) = t_{io}(\mu, \beta)$ (approximately).

The terms "almost" and "approximately" are inserted in the definitions, because there are some inevitable but small overheads and non-overlappabilities in the algorithm design. In this section a parallel perfect overlapping algorithm for external merging is presented. Combined with the procedure of finding optimal setting, the algorithm can be tuned to achieve the best running time.

4.1   2-level forecasting merging

When $P = 1$, $2\mu + 2$ buffers for an MP with the technique of *forcasting* are sufficient to make replacement selection to be perfect overlapping [3]. The amount of internal memory for buffers can be reduced to $(\mu + 2)\beta$ if the buffers for sorted runs are of varying size, with small overheads incurred by an additional link field and dynamic buffer management [11].

Now we extend the technique of forecasting to the environment of parallel processing. Consider the μ sorted runs of an MP in some pass. The runs are divided into *pages* of equal size (the size will be determined later). Let the key field of the first record of a page be called the *critical element*.[4] Let $(\mu + 2)\beta = \alpha M$. When $\alpha \leq 1$, there are $\frac{1}{\alpha}$ MPs in one PE, and the page size is taken the same as the block size β. Blocks are read into internal memory by comparing their critical elements for each MP. The P PEs function independently in this case.

When $\alpha > 1$, the PEs are organized into multiple linear arrays, as illustrated in Fig. 6. In each linear array, only the first PE does merging operations, the others bypass pages to their preceding PEs. The page size now equals β/α. Initially, the first blocks of all runs are loaded. They are divided into pages and located in the PEs' memories. Fig. 7 shows a three-run example. Not shown in the figure are the input and output buffers, which serve to overlap read/write/compute and also have page size. Besides the first records of leading blocks of current runs are stored in the memory of PE α for the purpose of forecasting the next fetched block.

The algorithm we propose here turns out to be a *2-level forecasting*, one in the page level and the other in the block level. PE 1 maintains the heap of leading elements of all runs. The memory allocation

---
[4]Traditionally, the critical element is the key field of the *last* record.

needed for the heap is relatively small because μ is small often. So our analysis is not nullified by this space overhead. PE α issues command to read the next preselected block, which is the one with the smallest key field of the first record among all blocks of runs not completely read in yet. At the same time, PE i, $i = 2, \ldots, \alpha$, outputs the page with the smallest critical element among the pages within the PE's memory. Two buffers of page size are used in each PE to facilitate data reading and writing. After a page is output to the preceding PE, its memory space is used as the next input buffer of the PE. With the same buffer management strategy, PE α uses two buffers of page size to read in records of block size. The reason why it works is the following: Because the inter-PE communication is often faster than external I/O, the communication between PEs is assumed to be totally overlapped by external I/O. While reading the second page of a block, the page in PE α with the smallest critical element is sending to PE $\alpha - 1$. After having read the second page, the space for the output page becomes free to be reallocated for the third page, if any, of the block. We specify the parallel merging more formally in the following algorithm. It does not consider end of runs just for presentational simplicity. Besides, the management of buffers is not detailed.

Algorithm for PE 1
    Initial data loading
    Initialize the heap of the first elements
    Repeat
        Concurrently do I/O and CPU operations
            {I/O part} Read from PE 2 a page
            Repeat $\frac{\beta}{\alpha}$ times {CPU part}
                Output the smallest from the heap
                Adjust the heap

Algorithm for PEs $2, \ldots, \alpha - 1$
    Initial data loading
    Repeat
        Concurrently do input and output
            {input} Read from PE $i + 1$ a page
            {output}
                Compare the critical elements of pages in memory
                Send to PE $i - 1$ the page of smallest crt elmt

Algorithm for PE α
    Initial data loading
    Repeat
        Compare the last elements of blocks
        Issue read to fetch a block from the preselelcted run
        Repeat α times
            Concurrently do input and output
                {input} Get $\frac{\beta}{\alpha}$ data and take it as a page
                {output}
                    Compare the critical elements of pages in memory
                    Send to PE $\alpha - 1$ the page of the smallest crt elmt

4.2 Correctness of the algorithm

We give a numbering of the pages. The pages in PE 1 in the initial phase have no associated numbers. A page is numbered i if it is *supposed* to be the i-th page read into PE 1 after initialization. We also call the time period to read/write/compute one block of records as one block step. Further, one page step equals one α-th of a block step, numbered from 0. That the fetching order is *kept* is equivalent to that page i has been completely read into PE 1 at the beginning of page step i for all i.

Lemma 7 *The parallel replacement selection with 2-level forecasting does merging correctly.*

Proof. We proceed to prove that page i has been completely read into PE 1 at the beginning of page step i for all i. First, consider $i < \alpha$. It is easy to see that page i sites in one of the PEs $2, \ldots, i + 1$ at the beginning of page step 0. After a page step, page 1 enters PE 1 and page i, $2 \le i < \alpha$, is contained within PEs $2, \ldots, i$. Similarly, page i arrives at PE 1 at the beginning of page step i.

Now consider page i, $\alpha \le i < 2\alpha$. If the page has been in the system at the beginning, there is no problem for it to arrive PE 1 at the proper time. If the page has not been in the system but contained in the next preselected block, there will be no problem either. In fact, if such a page i is not in the system at the beginning, it must be in the preselected one because of the following reason. The key field of the first record of a block is also the critical element for the first page of that block. Consider any two of the first pages of next blocks to be read. Because there are at least $2\alpha - 2$ pages preceding them, at most one of them is associated with a number less than 2α. This implies that page i, $\alpha \le i < 2\alpha$, could only appear in the preselected block, if it is not in the system yet.

We have proved the lemma for page i, $1 \le i < 2\alpha$. By similar reason, the lemma is also true for all i. Page i will be completely sent into PE 1 at the beginning of page step i. ∎

Because the overhead of heap maintenance and buffer management is large only in extreme cases, we can say that $T_c(\mu, \beta) \approx t_c(\mu, \beta)$ and $T_{io}(\mu, \beta) \approx t_{io}(\mu, \beta)$. The algorithm is also an overlapping one for (μ, β)-merge. In other words, it is a perfect overlapping algorithm for external merging and leads to the optimal running time. Following is the main result of the paper.

Theorem 1 *Given a system configuration and a disk configuration with $P \le \frac{r+s}{t}(\ln M - \frac{2}{M} - 1)$, by applying the procedure of finding optimal (μ, β)-merge and tuning the parallel replacement selection with 2-level forecasting, an algorithm can be obtained to have the shortest running time over all algorithms implementing (μ, β)-merge.*

5 Concluding remarks

At the first glance, it is strange to have a parallel algorithm in which most PEs only serve for data routing. This can be explained by the notions of I/O complxity and balanced computation. Recall again the tight bound for the number of I/O operations for sorting (and similar for merging):

$$\Theta\left(\frac{N}{I\beta} \frac{\log(N/\beta)}{\log(PM/\beta)}\right).$$

Because P is usually much smaller than M, increasing the number of PEs does not help much in reducing the I/O complexity. Block size β and parallel I/O capacity I are more relevant factors for such I/O-intensive problems. From the viewpoint of balanced compuation, exponential growth of memory size is needed to balance an increasing of CPU power [4]. Multiple PEs serve more as a large memory rather than multiple CPUs. So it is very natural for the data only to be stored and bypassed in and through PEs' memories. A single PE with large memory is more suitable than multiple PEs with the same amount of total memory.

Using external merging in external mergesort, balance between sort phase and merge phase should be considered. The most important factor is $\rho = N/n$, the length of the sorted runs just before the merge phase. The notion of balanced computation and the approach of analysis we adopt here could be helpful to find optimal (ρ, μ, β)-mergesort.

Lemma 2 tells us that if $p \le \frac{r+s}{t}(\ln M - \frac{2}{M} - 1)$, then the optimal (μ, β) lies in $A_1 \cup A_2$. Under the assumption that inter-PE communication is faster than external I/O, the cost of page movement between PEs is totally overlapped by the I/O time. The simple structure of linear arrays of PEs is sufficient to do (μ, β)-merge. All of these tell that, for the problem of external merging, network topology (communication pattern) is not of much importance. This differs with the common opinions. We expect that other kinds of I/O-intensive problems have the same property.

Another problem about the I/O time model is on the evaluation of parameters C_1, r, s, and t. It could be done by simulation on the target machine and I/O peripherals. Because they are random variables, their effects on the optimal setting (μ, β) might be very complex and impossible to be fully catched only by their mean values [5]. A more realistic model for t_{io} is demanded. But the optimal settings can be reasonably assumed to site on the new minimum I/O curve again.

There is a problem which has not been touched at all in this paper: presortedness. It is an interesting and open problem whether the concept of balanced computation helps to derive optimal algorithms with the considerations of presortedness.

References

[1] Aggarwal, A., and Vitter, J. S., "The Input/Output Complexity of Sorting and Related Problems," *Comm. of ACM*, **31**(9), pp. 1116–1127 (1988).

[2] Dekel, E., and Ozsvath, I., "Parallel External Merging," *J. Parallel and Distributed Computing*, 6, pp. 623–635 (1989).

[3] Knuth, D. E., *The Art of Computer Programming*, Volume 3, *Sorting and Searching*. Addison-Wesley, Reading, MA (1973).

[4] Kung, H. T., "Memory Requirements for Balanced Computation," *J. Complexity*, 1, pp. 147–157 (1985).

[5] Kwan, S. C., and Baer, J. L., "The I/O Performance of Multiway Mergesort and Tag Sort," *IEEE Trans. Computers*, **C34**(4), pp. 383–387 (1985).

[6] Lin, F. C., and Shieh, J. C., "Space and Time Complexities of Balanced Sorting on Processor Arrays," *J. Complexity* 6, pp. 365–378 (1990).

[7] Lorin, H., *Sorting and Sort Systems*. Addison-Wesley, Reading, MA (1975).

[8] Luk, W. S., and Ling, F., "An Analytic/Empirical Study of Distributed Sorting on a Local Area Network", *IEEE Trans. Soft. Eng.*, **15**(5), pp. 575–586 (1989).

[9] Rotem, D., Santoro, N., and Sideny, J. B., "Distributed Sorting," *IEEE Trans. Computers*, **C34**(4), pp. 372–375 (1985).

[10] Varman, P.J., Iyer, B. R., and Scheufler, S. D., "A Multiprocessor Algorithm for Merging Multiple Sorted Lists," *Proc. 1990 Intl. Conf. on Parallel Processing*, pp. III-22–26, (1990).

[11] Verkamo, A. I., "Performance Comparison of Distributive and Mergesort as External sorting Algorithms," *J. Systems and Software*, **10**, pp. 187–200 (1989).

[12] Vitter, J. S., and Shriver, E. A. M., "Optimal Disk I/O with Parallel Block Transfer," *Proc. 22nd Annual IEEE Symposium on Theory of Computing*, Baltimore, MD (1990).

ALGORITHMS FOR DETERMINING OPTIMAL PARTITIONS
IN PARALLEL DIVIDE-AND-CONQUER COMPUTATIONS

Arindam Saha and Meghanad D. Wagh

Department of Computer Science and Electrical Engineering

Lehigh University, Bethlehem, PA 18015

Abstract. Parallel divide-and-conquer computations, encompassing a wide variety of applications, can be modeled to include the significant parallel computing overheads, as

$$T(n) = \begin{cases} t_0, & \text{for } n < n_0, \\ \min_{0 < r < n}\{\max\{T(n-r), T(r) + kr^s\} + \lambda\}, & \text{otherwise.} \end{cases}$$

The optimal partition size (solution r of the above equation) is nontrivial and is very different from the adhoc $n/2$ value conventionally used. Using the optimal partitions at every stage of the recursion, the performance is greatly enhanced. In this paper we develop a generalized algorithm to compute the optimal partition size given any problem size. We also study two special cases that result in simpler implementations.

I. Introduction

Many parallel processing algorithms are based on the divide-and-conquer paradigm [1]. Divide-and-conquer allows partitioning of a problem in (typically) two smaller instances of itself, and combining the partial results to get the final solution. Since these smaller instances can themselves be recursively partitioned into even smaller sizes, the procedure reduces to solving minimal sized problems and recombining their results. The divide-and-conquer technique is ideal for parallel processing because the smaller problems are mutually independent and can be executed concurrently in different processors.

The performance of any parallel divide-and-conquer algorithm is affected by the architectural and algorithmic overheads inevitable in any realistic parallel computing environment [2]. These overheads, among others, include the cost of interprocessor communication to distribute data and collect results, the cost of combining the partial results and the costs associated with converting one or both subproblems to conform to the exact form of the original problem. Conventionally, these overheads are neglected in the design of algorithms [3-6]. This leads to adhoc equal partitioning and nonoptimal performance of the divide-and-conquer technique on real world parallel machines.

We have recently [7] shown that by proper choice of partitions, one may extract optimal performance from parallel divide-and-conquer algorithms in the presence of overheads. The complexity of such optimal algorithms shows a considerable improvement over algorithms using equal partitions [7]. Unfortunately, the choice of optimal partition sizes is greatly influenced by the nature of the overheads. It is therefore imperative to correctly model the architectural and algorithmic overheads and then efficiently determine the optimal partitions given any problem size . This paper develops procedures to obtain such optimal partitions.

Section II of this paper describes the model for parallel divide-and-conquer algorithm incorporating both the symmetric and nonsymmetric overheads inherent in any parallel computing environment. The algorithm to compute optimal partitions in a general case is derived in Section III. This algorithm separates problem sizes into classes of problems that may use the same optimal partition size. The time to obtain the optimal partition for any given problem size n is shown to be $O(n^{1/(1+s)})$, where s is a positive integer characteristic of the nonsymmetric overheads. Determination of the optimal partitions when the overheads are constant or are simply related, is presented in Section IV. Finally, Section V provides a discussion of the results obtained.

II. Parallel Divide-and-Conquer Model

The complexity of a parallel divide-and-conquer algorithm may be described by the following equation.

$$T(n)=\begin{cases} t_0, & \text{for } n < n_0, \\ \min_{0<r<n}\{\max\{T(n-r),T(r)+kr^s\}+\lambda\}, & \text{otherwise.} \end{cases} \quad (1)$$

where, $T(n)$ is the complexity of a size n problem, n_0 is some small problem size below which the recursion (1) is not applied and all problems have the same complexity t_0, and the kr^s and λ are the partition and recombination overheads respectively.

Recursion (1) is graphically depicted in Fig. 1. From this figure, one can easily see that since the partitioned problems of size r and n−r are executed concurrently in different sets of processors, the total time $T(n)$ is based only upon the larger of the two. The minimization in (1) ensures that the best possible partitions are chosen to optimize $T(n)$. The asymmetric partition overhead, kr^s, which is always nonzero, is generally related to the costs incurred during interprocessor communication as well as any extra computations one of the partitions may require. On the other hand, the symmetric recombination overhead, λ, often characterizes the costs associated with the recombination of the partial results.

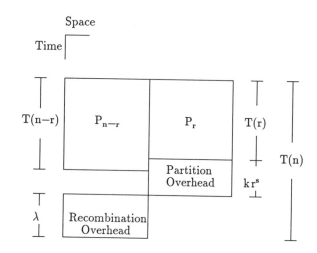

Fig. 1. Computational model of parallel Divide-and-Conquer solving size n problem P_n.

As an example, consider the problem P_n of finding the three maximum elements in a collection of n elements. P_n assigned to processor A can be partitioned so that it only solves P_{n-r} and some other processor B solves P_r. In this case, the partition overhead is the cost of communicating three top elements from processor B to processor A and can be modeled as some constant k. The recombination cost is the time required to choose the three winners from the six intermediate winners and can be characterized by a constant λ. Note that the nature of the problem enables one to further recursively partition the subproblems till size 3 or less is achieved. Thus this computation can be correctly modeled by (1) with s=0, $n_0=3$ and $t_0=0$. Further, if the preloading of the processors with data is not assumed, then the partition overhead will be dominated by the cost of communicating the r data elements required by processor B. Such a situation can be modeled by using s=1. In general, proper choice of k, λ, s, n_0 and t_0 enables a variety of applications to be accurately modeled by (1) [8-13].

The aim of this paper is to enable the determination of optimal r values in the solution of (1)

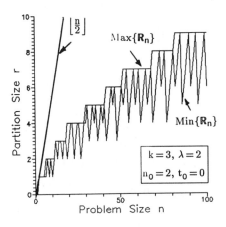

Fig. 2. A plot of optimal partition versus problem sizes.

for all problem sizes of interest. Such r values will be referred to as the optimal partitions. Fig. 2 shows the a typical plot of optimal partitions for different problem sizes. One can see from this figure that the optimal partitions are very different from the adhoc equal partition n/2.

III. General Partition Algorithm

For ease of reference we list all the relevant symbols used in this paper and the corresponding definitions in Table 1.

This section deals with the case of nonzero s. When s is zero, the partition sizes may be computed rather easily as shown in Section IV. We begin the derivation of the partition algorithm by noting that for any problem size $n \in S_m$, the set of optimal partition sizes is given by $\mathbb{R}_n = \{p \mid (n - \eta_{m-\lambda}) \leq p \leq r_m\}$, where r_m is given by [7]

$$r_m = \eta_m - \eta_{m-\lambda}. \tag{2}$$

Clearly, this r_m is the optimal partition for all elements in the complexity class S_m. Henceforth, in this paper, we will ignore other elements of $\mathbb{R}_n$ and concentrate only on the calculation of r_m, which is the unique partition size at $n = \eta_m$. The values of r_m change slowly with the

complexity. In fact, it can be shown [7] that for $s \neq 0$,

$$r_m - r_{m-g} = \begin{cases} 1 & \text{if } m = \mu_{r_m}, \\ 0 & \text{otherwise,} \end{cases} \tag{3}$$

where, S_{m-g} and S_m are two consecutive complexity classes, and μ_{r_m} is the smallest complexity m such that r_m is an optimal partition for all elements of S_m.

Table 1. Symbols and their definitions

Symbol	Definition
g	$\gcd(k, \lambda)$.
η_m	Largest n such that complexity $T(n)=m$.
S_m	Set of problem sizes of same complexity m, ($\{n \in \mathbb{Z} \mid T(n)=m\}$).
$\mathbb{R}_n$	Set of optimal partition sizes for n.
r_m	Unique optimal partition size for $n=\eta_m$.
μ_p	Smallest m such that p is an optimal partition for all elements of S_m.
$\mathbb{M}_p$	Set of m's such that p is an optimal partition for all elements of S_m.
J_m	$\eta_m - \eta_{m-g}$ for some complexity value m.
$\eta_{max}(p)$	Largest problem size for which p is an optimal partition size.

As (3) shows, r_m may not change for every change of complexity m. We therefore define $\eta_{max}(p)$ to be the maximum problem size that uses optimal partition p. Thus to obtain required partition sizes, it is sufficient to determine all η_{max}'s. We compute these by first obtaining the difference between consecutive η_{max}'s. This difference can be easily represented as a summation of jumps $J_m = \eta_m - \eta_{m-g}$. The computation of these jumps is crucial to our algorithm. J_m's have two interesting properties which simplify their computation. Firstly, (2) shows that

$$\begin{aligned} J_m - J_{m-\lambda} &= (\eta_m - \eta_{m-g}) - (\eta_{m-\lambda} - \eta_{m-\lambda-g}), \\ &= (\eta_m - \eta_{m-\lambda}) - (\eta_{m-g} - \eta_{m-\lambda-g}), \\ &= r_m - r_{m-g}. \end{aligned}$$

Application of (3) then gives

$$J_m = \begin{cases} J_{m-\lambda} + 1 & \text{if } m = \mu_p \text{ for some p,} \\ J_{m-\lambda} & \text{otherwise.} \end{cases} \tag{4}$$

Thus, the jumps in η_m's repeat cyclically with a periodicity λ in m. The only exception to this is if the corresponding complexity m is the smallest element of some class $\mathbb{M}_p$. Since m can change only by integral multiples of $\gcd(k,\lambda) = g$ (because every time a new complexity is obtained by merely adding λ or $\lambda + kr^s$ to some previous complexity value), and λ is also a multiple of g, one can keep track of all the jumps by remembering only (λ/g) jumps at complexities distinct modulo (λ/g). These jumps are represented by $jump(l)$, $l=0,1,\ldots,$ $(\lambda/g)-1$ in the algorithm and defined through $J_m = jump(m \bmod (\lambda/g))$. In order to get the correct jumps at all times, one must ensure (in accordance with (4)) that when m increases into a new class $\mathbb{M}_p$, the jump corresponding to $m = \mu_p$ must be incremented by 1.

The second interesting property of jumps relates to the (λ/g) consecutive complexities m, m+g, m+2g, ..., $m+(\lambda/g)-1$ that belong to some set $\mathbb{M}_p$. In this case, one has

$$\sum_{i=0}^{(\lambda/g)-1} J_{m+ig} = p. \tag{5}$$

The relation (5) is simple to obtain. Expanding every individual jump from its definition, the sum in (5) telescopes and gives $\eta_{m+\lambda} - \eta_m$. Application of (2) then directly gives (5).

We are now ready to compute the differences between consecutive η_{max}'s. From the definitions of η_{max}, $\mathbb{M}_p$ and J_m, one immediately has

$$\eta_{max}(p) - \eta_{max}(p-1) = \sum_{m \in \mathbb{M}_p} J_m. \tag{6}$$

Since $\mathbb{M}_p = \{ m < \mu_{p+1} \mid m = \mu_p + ig, \text{ for } i = 0,1,\ldots,|\mathbb{M}_p|-1 \}$ where $|\mathbb{M}_p| = (\mu_{p+1}-\mu_p)/g$, the sum in (6) can be split into two parts as follows:

$$\eta_{max}(p) - \eta_{max}(p-1) = \sum_{i=0}^{(|\mathbb{M}_p| \bmod (\lambda/g))-1} J_{\mu_p+ig}$$

$$+ \sum_{i=|\mathbb{M}_p| \bmod (\lambda/g)}^{|\mathbb{M}_p|-1} J_{\mu_p+ig}. \tag{7}$$

The number of consecutive jumps added in the second sum is an integral multiple $(=\lfloor |\mathbb{M}_p|/(\lambda/g) \rfloor)$ of (λ/g). Since all these jump indices belong to the same set $\mathbb{M}_p$, we see from (5) that the second sum equals $\lfloor |\mathbb{M}_p|/(\lambda/g) \rfloor * p$. The first sum in (7) can be evaluated from the (λ/g) values of jumps that are on record. Equation (7) thus transforms into:

$$\eta_{max}(p) - \eta_{max}(p-1) = \sum_{l=0}^{(|\mathbb{M}_p| \bmod (\lambda/g))-1} jump((\mu_p + lg) \bmod (\lambda/g))$$

$$+ p*\lfloor |\mathbb{M}_p|/(\lambda/g) \rfloor. \tag{8}$$

In order to use (8), one needs to know the values of μ_p for every partition size p of interest. In order to estimate the largest partition size one may need, one may use the following upper bound on p in terms of the problem size n [7] :

$$r < \left((\lambda/k)\left(\frac{n-n_0-1}{A}\right)^{s/(s+1)} + 2(\lambda/k) + 1 + (1/k) \right)^{1/s}, \tag{9}$$

where, $A = (1/s)[\lambda/(\lambda+k)]^{1/s}$.

Based on the discussion of this section, we now present a generalized algorithm, Algorithm-A, in Fig. 3 to compute all the relevant η_{max}'s till some partition size N. Besides the initialization step, this algorithm proceeds in two stages — namely the Precomputation and the Computation stages. The purpose of the Precomputation stage is to compute the quantities μ_p for $0 \leq p \leq N+1$, and the jumps (one or more of which may be zero) $jump(i)$, for $0 \leq i \leq (\lambda/g)-1$. However, as shown in [7], μ_p has the form

$$\mu_p = T(p) + kp^s + \lambda. \tag{10}$$

Thus the precomputation stage should also obtain the complexities $T(j)$ for $n_0 \leq p \leq N+1$.

For the sake of efficiency, we deal with only the nonempty complexity classes. The necessary and sufficient condition for this is [7]

$$|S_m| \neq 0 \text{ iff } |S_{m-\lambda}| \neq 0 \text{ or for some } p, \mu_p = m. \quad (11)$$

Algorithm-A.

Step 1. (Initialization)

$b \leftarrow 0; \; i \leftarrow 1;$

$m(0) \leftarrow k + \lambda; \; r(0) \leftarrow 1; \; eta(0) \leftarrow n_0;$

for $j=0$ to n_0-1 **do** $T(j) \leftarrow t_0; \quad T(n_0) \leftarrow t_0 + k + \lambda;$

for $j = 0$ to $(\lambda/g)-1$ **do** $jump(j) \leftarrow 0;$

$jump(m(0) \bmod (\lambda/g)) \leftarrow 1; \; mu \leftarrow \mu_2;$

Step 2. (Precomputation)

$nonzero\_complexity \leftarrow m(b) + \lambda;$

$m(i) = \min[\; nonzero\_complexity, \; mu\;];$

if $m(i) = mu$ **then**

$\qquad \{ eta\_max(r(i-1)) \leftarrow eta(i-1);$

$\qquad\quad$ **if** $eta\_max(r(i-1)) > N+1$ **then goto** Step 3;

$\qquad\quad r(i) \leftarrow r(i-1) + 1;$

$\qquad\quad jump(mu \bmod (\lambda/g))$ ++; $\}$

$eta(i) \leftarrow eta(i-1) + jump(m(i) \bmod (\lambda/g));$

for $j = eta(i-1)+1$ to $eta(i)$ **do** $T(j) \leftarrow m(i);$

if $r(i) > r(i-1)$ **then** $mu \leftarrow \mu_{1+r(i)};$

if $m(i) = nonzero\_complexity$ **then** b ++;

i ++;

goto Step 2;

Step 3. (Computation)

for $j = r(i-1)+1$ to N **do**

$\qquad \{ cardinality\_of\_M(j) \leftarrow (\mu_{j+1} - \mu_j)/g;$

$\qquad\quad start\_index(j) \leftarrow (\mu_j \bmod (\lambda/g));$

$\qquad\quad jump(start\_index(j))$ ++;

$\qquad\quad sum \leftarrow 0;$

$\qquad\quad$ **for** $l=0$ to $(cardinality\_of\_M(j) \bmod (\lambda/g))-1$ **do**

$\qquad\qquad sum \leftarrow sum + jump((start\_index(j) + l) \bmod(\lambda/g));$

$\qquad\quad eta\_max(j) \leftarrow eta\_max(j-1)$

$\qquad\qquad\quad + j * \lfloor cardinality\_of\_M(j)/(\lambda/g) \rfloor + sum; \}$

Fig. 3. Algorithm to compute the maximum problem size

for a given partition size.

The corresponding two possible complexities are represented by $nonzero\_complexity$ and mu in the algorithm. Amongst these, we choose the smaller one to be the next complexity of interest at step i, $m(i)$. Variable b in this implementation represents the index corresponding to an older step which provides a contender $m(b)+\lambda$ for $m(i)$. If the chosen complexity marks the beginning of a new class M_r, we appropriately update eta_max, r, and the corresponding $jump$. The rest of the Precomputation stage evaluates eta, T's, and mu, and then updates the indices b and i. This stage is exited when eta_max exceeds $N+1$. In the second stage, called the Computation stage, we directly compute the remaining η_{max}'s using (8). We also update the appropriate $jump$ in this stage.

The results from a sample execution of Algorithm-A are shown in Tables 2 and 3. In this example the largest partition size N of interest is assumed to be 20. The η_{max} values obtained from these tables allow one to choose the required partition sizes since they are the problem size boundaries of optimal partitions. For example, given a problem size n=200 which is bounded as $\eta_{max}(19) = 194 < 200 \leq 204 = \eta_{max}(20)$, the optimal partition size is 20.

Estimating the complexity of Algorithm-A is not difficult. During Step 2, the complexity m and therefore the quantity η_m necessarily increase at every increase of index i. We end Step 2 when the η_m exceeds N+1 and all computations corresponding to that last partition=p are completed. The number of i values involved in the class of partition p equals

$$|M_p| = (\mu_{p+1}-\mu_p)/g = (T(p+1)-T(p))/g + k((p+1)^s - p^s)/g.$$

Table 2. Results of the Precomputation step with $k=2$, $\lambda=3$, $n_0=2$, $t_0=0$, $s=1$. Note that the boxed eta's are $\eta_{max}(r)$'s and boxed complexities are μ_r's.

eta	r	complexity	jump(0)	jump(1)	jump(2)
2	1	$\boxed{5}$	0	0	1
3	1	8	0	0	1
$\boxed{4}$	1	11	0	0	1
5	2	$\boxed{12}$	1	0	1
6	2	14	1	0	1
$\boxed{7}$	2	15	1	0	1
9	3	$\boxed{17}$	1	0	2
10	3	18	1	0	2
12	3	20	1	0	2
$\boxed{13}$	3	21	1	0	2
14	4	$\boxed{22}$	1	1	2
16	4	23	1	1	2
$\boxed{17}$	4	24	1	1	2
19	5	$\boxed{25}$	1	2	2
21	5	26	1	2	2
22	5	27	1	2	2
$\boxed{24}$	5	28	1	2	2

Table 3. Results of Computation step with $k=2$, $\lambda=3$, $n_0=2$, $t_0=0$, $s=1$.

r	$|M_r|$	jumps(0,1,2)			sum	start_index	η_{max}
6	3	1	2	3	0	2	30
7	4	1	2	4	4	2	41
8	2	2	2	4	4	0	45
9	3	2	2	5	0	2	54
10	4	2	2	6	6	2	70
11	2	3	2	6	5	0	75
12	3	3	2	7	0	2	87
13	3	3	2	8	0	2	100
14	3	3	2	9	0	2	114
15	2	3	2	10	13	2	127
16	3	3	3	10	0	1	143
17	3	3	4	10	0	1	160
18	2	3	5	10	15	1	175
19	3	4	5	10	0	0	194
20	2	5	5	10	10	0	204

From [7] one knows that the difference in two consecutive complexities is bounded above by λ and p, the optimal partition for problem size $N+1$ is of $O(N^{1/(s+1)})$. Using this, one can conclude that $|M_p|$ is $O(N)$. Thus the number of iterations in Step 2, which equals the maximum index i is $O(N)$. Since in each of these iterations we do constant work, the complexity of Step 2 is $O(N)$. The number of iterations and hence the complexity of Step 3 is also clearly $O(N)$. Finally, since N is $O(n^{1/(1+s)})$, one can conclude that the complexity of Algorithm-A is $O(n^{1/(1+s)})$.

Once the table of $\eta_{max}(p)$'s is created, the time to locate a given problem size between two consecutive η_{max}'s and thereby predict the optimum partition, is proportional to the length N of this table and is therefore $O(n^{1/(1+s)})$.

IV. Two Special Cases

This section deals with two special cases corresponding to $\lambda|k$ and $s=0$. In both these cases, the partition algorithm given in Section III, is greatly simplified.

Note that when $\lambda|k$ one has $g=\lambda$. Hence for every nonempty complexity class, S_m, $\lambda|(m-t_0)$. But as noted earlier, the difference in the complexities of two successive problem sizes can be at most λ. Thus in this case, the successive nonempty complexity classes must have complexities differing exactly by λ. In other words if $\lambda|k$, then

$$T(1 + \eta_m) = m + \lambda. \qquad (12)$$

and for every m satisfying $g|(m-t_0)$, $|S_m| \neq 0$. This brings about two simplifications. Firstly, since $(\lambda/g)=1$, $|M_p| \bmod (\lambda/g)=0$ for all p. Thus the first summation in (7) drops out and (7) converts only to

$$\eta_{max}(p) = \eta_{max}(p-1) + |M_p|p. \qquad (13)$$

Secondly, because (12) says that the complexities of consecutive problems differ by 0 or λ,

$$(\mu_{p+1} - \mu_p) = (T(p+1) - T(p)) + ((p+1)^s - p^s).$$

The calculation of $|M_p|$ becomes rather direct from this and yields (since $g = \lambda$,):

$$
\begin{aligned}
|M_p| &= \mu_{p+1} - \mu_p \\
&= h(p) + 1 + (k/\lambda), \quad \text{if } p = n_0 - 1, \\
&= h(p) + 1, \quad \text{if p is some previous } \eta, \\
&= h(p), \quad \text{otherwise}, \quad (14)
\end{aligned}
$$

where, $h(p) = k((p+1)^s - p^s)/\lambda$.

Using the values of $|M_p|$ given by (14) and the fact that $\eta_{max}(0) = n_0 - 1$, recursion (13) can give all the η_{max} values of interest. However, the expression for $|M_p|$ as given in (14) requires one to know if p was some previous η. The Algorithm-B shown in Fig. 3 therefore maintains a Boolean array g(p) whose p-th component is true if p is an η.

It can be seen that although Algorithm-B has the same order of complexity as that of Algorithm-A, it is clearly superior because of its simplicity.

Algorithm-B.

Step 1. (Initialization)
> **for** $i := 2$ **to** N **do** $g(i) \leftarrow$ *false;*
> *eta_ max*(0) $\leftarrow$ $n_0 - 1$;

Step 2. (Computation)
> **for** *newr* $:= 1$ **to** N **do**
>   { $y \leftarrow (k/\lambda)((newr+1)^s - newr^s)$;
>   **if** *newr* $= n_0 - 1$ **then** $y \leftarrow y + (k/\lambda) + 1$
>   **else if** *g(newr)* $=$ *true* **then** $y \leftarrow y++$;
>   *eta_ max (newr)* $\leftarrow$ *eta_ max (newr−1)* $+ y * newr$;
>   **if** *eta_ max (newr)* $\leq$ N **then**
>   **for** $i := 0$ **to** $y - 1$ **do**
> $g (eta\_ max (newr) - newr * i) \leftarrow$ *true*; }

Fig. 4. An algorithm to calculate the $\eta_{max}(p)$ for a given p when $\lambda | k$.

We next consider the case of $s = 0$. In this event, the η's are related by the following simple recurrence [14-15]

$$\eta_m = \eta_{m-\lambda-k} + \eta_{m-\lambda}. \quad (14)$$

Relation (2) shows that $\eta_{m-\lambda-k}$ can now be used as the optimal partition size for any $n \in S_m$. Thus for this special case of constant overheads one can simply compute all the η's using recursion (14) in $O(\log n)$ time. Given any problem size n, one should determine two consecutive η's η_{m_1} and η_{m_2} such that $\eta_{m_1} \leq n < \eta_{m_2}$ and use $\eta_{m_2-\lambda-k}$ as the optimal partition size for that n.

V. Conclusions

Algorithms based upon the divide-and-conquer paradigm are ideal for execution on a parallel architecture because it allows concurrent execution of all the leaves of a divide-and-conquer tree. However, the performance of such algorithms is limited by the various algorithmic and architectural overheads of parallel computing. By proper choice of partitions, the effect of these overheads may be minimized. Even though optimal algorithms have been studied earlier [7,14-15], no efficient methods to compute the necessary optimal partitions are available in the literature.

This paper presents procedures to evaluate optimal partitions for parallel divide-and-conquer algorithms. To apply these procedures, one should first model the parallel computing environment as in Fig. 1. The model presented here allows one to incorporate a partition size (r) dependent nonsymmetric overhead, kr^s, that applies only to one partition as well as a constant symmetric overhead, λ, that applies to both the partitions. It also allows one to specify a problem size n_0, below which all problems have the same complexity t_0. This general model of divide-and-conquer execution is applicable in a wide variety of problems [8-13].

The optimal partitioning algorithms separate the problem sizes into classes that may use the same optimal partition size, and find the boundaries of these classes in $O(n^{1/(1+s)})$ time. Given any problem size one can then search the table containing these boundaries in $O(n^{1/(1+s)})$ time to obtain the corresponding optimal partition. When $\lambda | k$, one may use a relatively simple algorithm presented in Fig. 4. Even though this algorithm has the same *order* of complexity, it can be executed much faster than the general algorithm. For constant overheads the computation of the optimal partitions can be greatly simplified and accomplished in only $O(\log n)$ time.

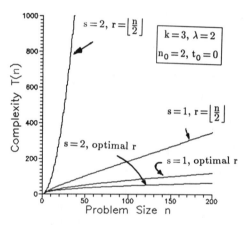

Fig. 5. Parallel divide-and-conquer performance using optimal and the adhoc choice of equal partitions.

Fig. 5 shows that the complexity of the parallel divide-and-conquer algorithm can be significantly reduced by using partitions obtained through the procedures given. Analysis presented in [7,14] shows that such optimal divide-and-conquer has a complexity of only $O(n^{s/(s+1)})$ when $s \neq 0$. This may be compared with the complexity $O(n^s)$ due to adhoc equal partitions.

References

[1] E. Horowitz and S. Sahni, *Fundamentals of Computer Algorithms*, Computer Science Press, Potomac, Maryland, 1978.

[2] D. P. Bertsekas and J. N. Tsitsiklis, *Parallel and Distributed Computing: Numerical Methods*, Prentice Hall, N.J., (1989).

[3] A. Zorat, *Divide-and-Conquer Computer*, Ph. D. Dissertation, University of Southern California, 1979.

[4] M. J. Atallah, R. Cole and M. T. Goodrich, *Cascading Divide-and-Conquer: A technique for designing parallel algorithms*, SIAM J. Comput. , 18 (1989), pp. 499-532.

[5] Q. F. Stout, *Supporting Divide-and-Conquer algorithms for image processing*, J. Parallel and Distributed Comput., 4 (1987), pp. 95-115.

[6] B. Abramson, *Divide and conquer under global constraints: A solution to the N-Queens problem*, J. Parallel and Distributed Comput., 6 (1989), pp. 649-662.

[7] A. Saha and M. D. Wagh, *Partitioning strategies for parallel recursive algorithms Part I: variable partition overhead*, in review.

[8] Z. Li and E. M. Reingold, *Solution of a Divide-and-Conquer maximin recurrence*, SIAM J. Comput., 18 (1989), pp. 1188-1200.

[9] J. M. Hammersley and G. R. Grimmett, *Maximal solutions of the generalized subadditive inequality*, Stochastic Geometry, eds. E. F. Harding and D. G. Kendall, John Wiley, London, 1974, pp. 270-284.

[10] C. J. K. Batty, M. J. Pelling and D. G. Rogers, *Some recurrence relations of recursive minimization*, SIAM J. Algebraic and Discrete Methods, 13 (1982), pp. 13-29.

[11] M. L. Fredman and D. E. Knuth, *Recurrence relations based on minimization*, J. Math. Analysis and Appl., 48 (1974), pp. 534-559.

[12] C. J. K. Batty and D. G. Rogers, *Some maximal solutions of the generalized subadditive inequality*, SIAM J. Algebraic and Discrete Methods, 13 (1982), pp. 369-378.

[13] B. S. Veroy, *Average complexity of Divide-and-Conquer algorithms*, Info. Processing Lettters, 29 (1988), pp. 319-326.

[14] M. D. Wagh and G. Bakdash, *Optimality in parallel Divide-and-Conquer algorithms in the presence of overheads*, in review.

[15] S. Kapoor and E. M. Reingold, *Optimum lopsided binary trees*, J. ACM, 36 (1989), pp. 573-590.

Efficient Parallel Computation of Hamilton Paths and Circuits in Interval Graphs

M. A. Sridhar and S. Goyal

Dept. of Computer Science

University of South Carolina

Columbia, SC 29208

sridhar@usceast.cs.scarolina.edu

Abstract. We develop an efficient parallel algorithm for computing hamilton paths in interval graphs. When given an n-vertex interval graph, the algorithm uses $O(\log n)$ parallel time and n^2 processors on an EREW PRAM. Sequential algorithms have been reported for this problem [11, 15], but no parallel algorithm has been reported. Our approach is based on a new characterization of interval graphs containing hamilton paths. We also outline an extension of our algorithm to compute hamilton circuits.

1 Introduction

Interval graphs are a family of graphs with a wide range of applications [7]. There has been significant research activity recently in the area of algorithmic problems relating to interval graphs, on both sequential and parallel models of computation [1, 17, 16, 12].

One of the interesting features of interval graphs is that many of the problems that are *NP*-complete for general graphs are solvable in polynomial time for interval graphs. Examples include computing maximal cliques, coloring and computing maximum independent sets [1, 12]. Most of these problems are solvable for interval graphs in linear time on a sequential machine. There are, however, some simple generalizations of interval graphs for which these problems become intractable [2].

One particular problem of significant interest is that of computing a *hamilton path* in a graph, i.e., a path on which every vertex of the graph occurs exactly once. It is well-known that this problem is *NP*-complete [5]. But there are many special cases, such as planar graphs

and interval graphs, for which this problem can be solved very efficiently [10, 11]. The problem of computing hamilton paths and circuits in interval graphs has been investigated in [11, 15]. Both these papers present linear-time sequential algorithms for this problem; their approaches, however, are different from each other and from ours. Moitra and Johnson [16] have shown a parallel algorithm for the hamilton path problem for proper interval graphs, but not for arbitrary interval graphs.

In this paper, we present a new solution to the problem of computing hamilton paths in interval graphs. We construct our solution by first obtaining a characterization of exactly those interval graphs which contain hamilton paths. This characterization is related to the decomposition of the interval graph into maximal cliques. Using this characterization, we construct a parallel algorithm for computing hamilton paths. When given an interval graph with n vertices, the algorithm runs in $O(\log n)$ time on an EREW PRAM with n^2 processors. Subsequently, we show how to extend these ideas to the problem of computing hamilton circuits in interval graphs.

2 Background

An *interval graph* $G = (V, E)$ is a graph defined by a collection of intervals on the real line, as follows [7]. Given a set $\mathcal{I} = \{[l_1, u_1], [l_2, u_2], \ldots, [l_n, u_n]\}$ of intervals, the corresponding interval graph has one vertex corresponding to each interval in $\mathcal{I}$, and two vertices are adjacent whenever the corresponding intervals have at least one point in common. The set $\mathcal{I}$ is called the *interval representation* of the graph G.

We may assume without loss of generality that no two intervals share a common end-point. Since there are n intervals with two end-points each, we can relabel the end-points of the intervals with the numbers $\{1, 2, \ldots, 2n\}$; thus it is clear that every interval graph with n vertices corresponds to a collection of intervals with end-points labeled by the first $2n$ positive integers. The related problem of constructing this interval representation from the adjacency list representation has been addressed elsewhere in the literature [3, 13, 18]. Henceforth, we will assume that the interval representation of the graph is given, with the end-points of the intervals labeled in this form. We will use the term 'vertex' synonymously with 'interval' unless otherwise indicated.

Given a vertex v corresponding to an interval $[l, u]$, we refer to l as $left(v)$ (the *left end* of v), and to u as $right(v)$ (the *right end* of v).

Consider an interval graph defined by the family $\mathcal{I}$ of intervals. Assume that the intervals are sorted based on their left ends. Suppose we scan the set of end-points from left to right, and keep track of the current number of intervals that are "active" at the current end-point of scan (i.e., the current point in the scan lies in all the active intervals). Clearly, at any given point during the scan, the collection of currently-active intervals forms a clique. So suppose we invent a function *active*, where $active(i)$ tells us the number of active intervals at end-point i. Then the maximal cliques must correspond to those points in the scan at which the number of active intervals attains a local maximum, i.e., such that $active(i) \geq \max(active(i-1), active(i+1))$. Any right endpoint i that satisfies this condition will be called a *clique endpoint*, because it represents a maximal clique.

As an example, consider the set of intervals and corresponding interval graph depicted in Figure 1. In this picture, the dotted vertical lines indicate the endpoints of the intervals, and the numbers underneath them indicate the *active* values at the endpoints; the positions of the clique endpoints are marked by arrows. Thus the maximal cliques in this graph are $C_1 = \{1, 2, 3, 4, 5\}$, $C_2 = \{3, 4, 5, 6\}$, $C_3 = \{3, 5, 6, 8\}$ and $C_4 = \{3, 6, 7, 8\}$.

Also, if we scan any given interval from its left end to its right, we will progress through all those maximal cliques in which x participates. This immediately yields

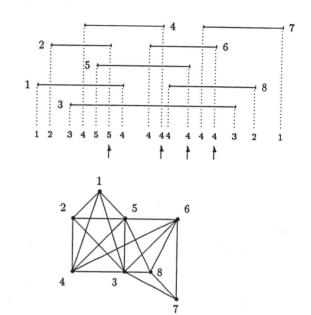

Figure 1: An example: interval family and corresponding graph

the Gilmore-Hoffman theorem [6]:

Theorem 1. [Gilmore-Hoffman theorem.] The maximal cliques of an interval graph can be linearly ordered so that for any given interval (vertex) x, the set of cliques in which x occurs appear consecutively in the linear order.

$\square$

(In fact, the converse is also true [7]: any graph whose maximal cliques can be so linearly ordered is an interval graph.) We will refer to this ordering of the maximal cliques of an interval graph as the *canonical clique decomposition* of the interval graph.

There are a few other results that are relevant to this paper, and for convenience we summarize them here. The first concerns *systems of distinct representatives* (SDR). Given a collection $\mathcal{F} = (F_1, F_2, \ldots, F_t)$ of subsets of a finite set X, a system of distinct representatives for $\mathcal{F}$ is a function $h : \mathcal{F} \to X$ that associates a representative element $x_i \in F_i$ with each subset F_i, such that all the representatives are distinct. A necessary and sufficient condition for the existence of an SDR [8] is that if we choose any subcollection $\mathcal{F}'$ of the given collection $\mathcal{F}$, there must be at least as many elements in the union of the subsets in $\mathcal{F}'$ as there are subsets in $\mathcal{F}'$. More formally:

Hall's theorem. A family $\mathcal{F} = (F_1, F_2, \ldots, F_t)$ of subsets of a finite set X admits a system of distinct representatives iff for every subfamily $\mathcal{F}' \subseteq \mathcal{F}$,

$$\left| \bigcup_{F \in \mathcal{F}'} F \right| \geq |\mathcal{F}'|.$$

A second result we need concerns *parallel prefix computation* [14]. It concerns the following problem: given an array $X[1 \ldots n]$ of elements and an associative binary operation $\circ$ on the elements of X, we need to compute and output an array $Y[1 \ldots n]$ where, for each i,

$$Y[i] = X[1] \circ X[2] \circ \cdots \circ X[i].$$

It is well known that this problem can be solved in $O(\log n)$ parallel time using $n/\log n$ processors on an EREW PRAM, assuming that each operation takes one unit of time to evaluate.

3 Finding hamilton paths

In this section, we will develop a characterization of those interval graphs that contain hamilton paths. We will then use this characterization for constructing a parallel algorithm for this problem.

First, consider the simple special case where the given interval graph is a clique. Here, the problem of finding a hamilton path through the clique that has two given vertices as its endpoints, is trivial; we can enumerate the vertices of the clique in arbitrary order. This suggests that we might be able to compute hamilton paths in interval graphs by first constructing the canonical ordering $C_1, C_2, \ldots, C_k$ of the maximal cliques of the given interval graph, choosing appropriate vertices $t_1, t_2, \ldots, t_{k-1}$, where t_i is some vertex common to C_i and C_{i+1}, and then constructing, for each i, an arbitrary path P_i that begins at t_{i-1}, ends at t_i, and has as its intermediate vertices all the elements of the clique $C_i - C_{i+1}$. (We may pick t_0 to be an arbitrary vertex in $C_1 - C_2$, and t_k to be an arbitrary vertex in $C_k - C_{k-1}$.)

With this approach, if we can guarantee that all the chosen t_i's are distinct, then the concatenation of the individual hamilton paths would be a hamilton path through the whole graph. This turns out to yield the characterization we seek.

Before we prove this claim, consider the way this would be done in the example of Figure 1. The linkage sets in this case are $C_1 \cap C_2 = \{3, 4, 5\}$, $C_2 \cap C_3 = \{3, 5, 6\}$ and $C_3 \cap C_4 = \{3, 6, 8\}$. If we choose $t_1 = 3$, $t_2 = 5$ and $t_3 = 8$, then we can construct $P_1 = (1, 2, 3)$, $P_2 = (3, 4, 5)$, $P_3 = (5, 8)$ and $P_3 = (8, 6, 7)$. Clearly, the concatenation of these paths is a hamilton path through the graph.

We will now formally define the notions used here. First, we define the family of *linkage sets* $L_i = C_i \cap C_{i+1}$, for $1 \leq i \leq k-1$; L_i is the set of vertices common to C_i and C_{i+1}. Let $\mathcal{L}$ denote the family of linkage sets, i.e.,

$$\mathcal{L} = (L_1, L_2, \ldots, L_{k-1})$$

and let L denote the union of the linkage sets, i.e.,

$$L = \bigcup_{1 \leq i \leq k-1} L_i.$$

Notice that a vertex v is in a linkage set $L_i = C_i \cap C_{i+1}$ iff v straddles the clique end-points of both C_i and C_{i+1}.

We are now ready to prove our main characterization theorem.

Theorem 2. Given an interval graph G with family of linkage sets $\mathcal{L} = (L_1, L_2, \ldots, L_{k-1})$, the graph contains a hamilton path iff the family $\mathcal{L}$ admits a system of distinct representatives (SDR), i.e., there is a one-one function $f : \mathcal{L} \to \bigcup_{1 \leq i \leq k-1} L_i$ such that $f(L_i) \in L_i$ for $1 \leq i \leq k-1$.

To establish this result, we will need the following lemma.

Lemma 3. For an arbitrary graph G, if G contains a hamilton path, then for every subset S of the vertices of G, the graph $G - S$ (obtained by removing all the vertices in S as well as edges incident on vertices in S) has at most $|S| + 1$ connected components.

Proof. Suppose G contains a hamilton path H, which is an edge subgraph of G. Then the removal of any given subset S of vertices must split H into at most $|S| + 1$ connected components; and each connected component in $H - S$ is (part of) a connected component in $G - S$.

Proof of Theorem 2. If $\mathcal{L}$ admits an SDR, then for each of the cliques $C_i - C_{i+1}$ $(1 \leq i \leq k-1)$, we construct hamilton paths P_i with endpoints t_{i-1} and t_i such that the intermediate vertices of P_i are all elements of $C_i - C_{i+1}$. This can be done arbitrarily, since $C_i - C_{i+1}$ is a clique. (We define t_0 to be an arbitrary element of $C_1 - C_2$, and t_k to be an arbitrary element of $C_k - C_{k-1}$.) We can then "patch together" a hamilton path for the whole graph by simply concatenating the P_i's in order.

Conversely, if $\mathcal{L}$ does not admit such an SDR, then by Hall's theorem, there exists a subfamily $\mathcal{L}' \subseteq \mathcal{L}$ such that

$$|L'| \leq |\mathcal{L}'| - 1, \quad \text{where} \quad L' = \bigcup_{L_i \in \mathcal{L}'} L_i.$$

Now if L_i is any linkage set in the family $\mathcal{L}'$, then the removal of the vertices in L_i from G would disconnect C_i from C_{i+1} as well as all subsequent cliques in the canonical ordering. Thus the removal of the subset L', containing $|L'|$ vertices, from G would produce at least $|\mathcal{L}'| + 1 \geq |L'| + 2$ distinct connected components, and so by lemma 2, G cannot have a hamilton path. $\qquad\square$

The problem of computing a system of distinct representatives is of course the same as that of bipartite matching; and there are not many good parallel algorithms known for this problem. Intuitively, however, it seems that we might not need the full power of bipartite matching, given that we are dealing with linkage sets arising from interval graphs; we should be able to apply some kind of greedy approach to this problem. This indeed turns out to be the case, as we will now proceed to show.

Given an interval graph G with linkage sets $L_1, L_2, \ldots, L_{k-1}$, define the *target* t_i of the linkage set L_i to be that vertex in L_i with least right endpoint which has not already been chosen as the target of some earlier linkage set.

More formally, given a set of intervals S, let us say that $minright(S)$ denotes that interval in S whose right end-point occurs earliest among all the intervals in S. Then the *target* t_i of the linkage set L_i (for $1 \leq i \leq k-1$) is defined inductively as

$$t_i = minright(L_i - \{t_1, t_2, \ldots, t_{i-1}\}).$$

Then we claim the following result. We omit its proof, as it is somewhat technical, and due to space constraints.

Theorem 4. Given an interval graph G with linkage sets $L_1, L_2, \ldots, L_{k-1}$, the graph has a hamilton path iff it is possible to find a target for each of the linkage sets. $\qquad\square$

Thus the sequential algorithm for finding a hamilton path in a given interval graph G is as follows.

(1) Find the canonical clique decomposition $C_1, C_2, \ldots, C_k$ of the maximal cliques of G. Compute the linkage sets $L_i = C_i \cap C_{i+1}$, for $1 \leq i \leq k-1$.

(2) Let v_0 be an arbitrary vertex in $C_1 - C_2$, and let v_k be an arbitrary vertex in $C_k - C_{k-1}$. Find, if possible, the target t_i of each linkage set L_i, for $1 \leq i \leq k-1$. If it is not possible to find a target for some linkage set, then the graph does not have a hamilton path.

(3) For each clique C_i, $1 \leq i \leq k-1$, compute (arbitrarily) a hamilton path between the end-points t_{i-1} and t_i, such that the intermediate vertices of the path are in $C_i - C_{i+1}$. Also compute a hamilton path between t_{k-1} and t_k whose intermediate vertices are in $C_k - C_{k-1}$.

(4) Output the composite hamilton path through G.

It is not hard to show that this algorithm can be implemented on a sequential machine in time $O(n+m)$, where m is the number of edges in the interval graph.

3.1 Parallelization

We now show how to implement this algorithm on an EREW PRAM using n^2 processors, in $O(\log n)$ parallel time.

The chief difficulty with the parallelization arises, as expected, in attempting to compute the targets; this is because our "greedy" definition requires that in order to compute the target of a given linkage set, we need the targets of all the previous linkage sets. We now show how to surmount this difficulty.

Suppose that we first compute the canonical clique decomposition and the linkage sets, and then we compute the union L of all of the linkage sets. Suppose further that we sort the vertices in L into ascending order based on their *right* ends; denote the resulting sequence by $(v_1, v_2, \ldots, v_p)$. Thus $right(v_1) < right(v_2) < \cdots < right(v_p)$.

Consider a pair of consecutive nodes v_{i-1}, v_i in this ordering such that v_i is the target of some linkage set L_j. Observe that since v_i is the target of L_j, the previous node v_{i-1} must not have been chosen as target of L_j because one of the following two cases is true:

(a) either $v_{i-1} \notin L_j$, or else

(b) v_{i-1} is already chosen as target of some earlier linkage set L_{j-l}.

Now we claim that in case (b), L_{j-l} must actually be L_{j-1}, i.e., we cannot have $l > 1$; for, if say $l = 2$, then we would have v_{i-1} as target of L_{j-2}, and v_i as target of L_j, which leaves L_{j-1} without a target, i.e., L_{j-1} is empty, which means of course that the graph does not have a hamilton path.

We thus have the following result:

Lemma 4. If v_i is the target of L_j, then either v_{i-1} is the target of L_{j-1}, or else $v_{i-1} \notin L_j$. $\qquad\square$

Applying this observation repeatedly, we see that if v_i is the target of L_j, then as long as case (b) holds, v_{i-1} must be the target of L_{j-1}, v_{i-2} must be the target of L_{j-2}, and so forth, until at some point case (a) prevails, which means that for some $s \geq 1$, we find that v_{i-s} is the target of L_{j-s} and $v_{i-s-1} \notin L_{j-s}$.

Our algorithm uses this observation in a "backwards" fashion, as follows. It begins by first marking all those pairs $\langle v_i, L_j \rangle$ such that $v_i \in L_j$ and $v_{i-1} \notin L_j$. Every such pair is potentially the beginning of a sequence $\langle v_i, L_j \rangle, \langle v_{i+1}, L_{j+1} \rangle, \ldots$ of pairs of target vertex and corresponding linkage set. Moreover, we claim that the converse is also true: if a vertex v_i must be the target of a linkage set L_j, then $\langle v_i, L_j \rangle$ must be in some such marked sequence, because either $v_i = minright(L_j)$ (in which case $v_{i-1} \notin L_j$, so that our claim is satisfied), or else some of the earlier elements of L_j (in the linear ordering) are targets of previous linkage sets, in which case we have the above case (b) prevailing, and our claim is still satisfied.

In the next step, the algorithm identifies, for each vertex v_i, the earliest linkage set L_j such that $\langle v_i, L_j \rangle$ is a marked pair. Finally, the algorithm does the inverse assignment of targets to linkage sets.

In summary, the algorithm consists of the following steps.

Step 1. Compute the maximal cliques of the given graph:

1a. Construct a function $IsLeft(i)$ defined for each of the $2n$ interval endpoints ($i \in \{1, 2, \ldots, 2n\}$), to be

$$IsLeft(i) = \begin{cases} 1 & \text{if } i \text{ is the left end of} \\ & \text{some interval} \\ -1 & \text{if } i \text{ is the right end} \\ & \text{of some interval.} \end{cases}$$

This function can be computed in constant time by n processors.

1b. For each left endpoint i, compute the value

$$active(i) = \sum_{1 \leq j \leq i} IsLeft(i).$$

Similarly, for each right endpoint i, compute the value

$$active(i) = \sum_{1 \leq j \leq i-1} IsLeft(i).$$

This value gives us the number of active intervals at the endpoint i. (The two cases are distinguished to make sure that each interval includes its two endpoints.) The function *active* can be computed in $O(\log n)$ parallel time by n processors, using prefix computation [14].

1c. For each right endpoint i, if $active(i)$ is larger than both of $active(i-1)$ and $active(i+1)$, then mark i as a clique endpoint.

1d. For each clique endpoint i, compute the number of clique endpoints among the endpoints $1, 2, \ldots, i$. This value tells us the rank of the clique represented by i in the canonical ordering. This computation can be done in $O(\log n)$ parallel steps by $n/\log n$ processors, using prefix computation. This step also yields the number k of maximal cliques in the graph.

Create an array CE of clique endpoints, where $CE[i]$ is the clique endpoint of the i-th clique in the canonical ordering.

Step 2. Compute the set $L = \bigcup_{1 \le i \le k-1} L_i$ of all vertices in all the linkage sets, as follows. For each vertex v and each clique endpoint $CE[i]$, determine whether the interval v contains both the clique endpoints $CE[i]$ and $CE[i+1]$; if so, mark v as being a part of L, since v is in the linkage set L_i. This step can obviously be achieved in constant time using n^2 processors; this can also be done in $O(\log n)$ time with $n^2/\log n$ processors, using the standard work-efficient simulation idea [4].

Step 3. Sort the set of vertices marked as being in L based on their right ends. Since there are at most n of these vertices, this can be achieved in $O(\log n)$ time by n processors on an EREW PRAM, using parallel bucket sort [9]. (This works because we are only sorting distinct values here.) Suppose the resulting ordered sequence of vertices in L is $L = (v_1, v_2, \ldots, v_p)$.

Step 4. Compute the targets of each linkage set:

4a. Construct a $p \times (k-1)$ matrix M (recall that $p = |L|$ is the number of vertices in the union of all the linkage sets, and $k-1$ is the number of linkage sets). The rows of this matrix are indexed by the vertices $v_1, \ldots v_p$ in order, and the columns by the linkage sets in order. This matrix is used to keep track of the marked pairs.

Set $M[i,j] \leftarrow 1$ if $v_i \in L_j$ and either $i = 1$ or $v_{i-1} \notin L_j$, and 0 otherwise. This can be done in constant time by n^2 processors.

4b. For each cell $M[i,j]$ such that $v_i \in L_j$, scan (in parallel) along the up-and-left diagonal consisting of the cells $M[i-1, j-1], M[i-2, j-2], \ldots$, to determine whether there is some $M[i-t, j-t]$ containing a 1. If so, then compute the boolean AND

$$(v_{i-t+1} \in L_{j-t+1}) \wedge (v_{i-t+2} \in L_{j-t+2})$$
$$\wedge \cdots \wedge (v_{i-1} \in L_{j-1})$$

and assign this value to $M[i,j]$.

The determination of whether there is such a cell $M[i-t, j-t]$ is merely an OR operation, and can be carried out in $O(\log n)$ steps with n^2 processors on an EREW PRAM, using standard parallel processing techniques. But the computation of the boolean AND's needs to be done using prefix computation.

4c. For each row i of M, compute the leftmost column j such that $M[i,j] = 1$, and set $Target(L_j) \leftarrow v_i$. This step can be done in $O(\log n)$ time with n^2 processors.

Step 5. If any of the linkage sets was not assigned a target in step 4c, then the graph does not have a hamilton path. Otherwise, compute the intermediate hamilton paths P_i for each $C_i - C_{i+1}$, using the above targets; output the composite hamilton path consisting of the concatenation of the P_i's in order. (Note that the computation of the intermediate hamilton path for $C_i - C_{i+1}$ involves listing its elements in arbitrary order, the only constraint being that the list should begin with $Target(L_{i-1})$.)

4 Computing hamilton circuits

Here we present a brief outline of how to extend the ideas of the previous section to computing hamilton circuits in interval graphs.

Suppose it is possible to choose, from each linkage set L_i, a pair of target elements $T_i = \{a_i, b_i\}$ with the conditions that (a) any two target pairs T_i and T_j have at most one element in common, and (b) each T_i contains at least one vertex not included in any other target pair. Then we can construct a hamilton circuit through the interval graph as follows.

First, note that if a vertex a is in T_i as well as in T_j, for $i < j$, then it must be an element of all intervening linkage sets $L_{i+1}, L_{i+2}, \ldots, L_{j-1}$ as well. Therefore, for each of those target pairs $T_{i+1}, T_{i+2}, \ldots, T_{j-1}$ in which a has not been chosen, we replace an arbitrary element of the target pair by a. This will not affect the fact that any two of the target pairs have at most one common element.

Next, in each target pair T_i, we identify a *forward target* $f_i \in T_i$ such that $f_i \notin T_{i+1}$. We refer to the vertex $b_i \in T_i$ that is not the forward target of T_i as its *backward target*.

For each $i \in \{2, 3, \ldots, k-1\}$, we construct an arbitrary path P_i between the forward target f_{i-1} in T_{i-1} and the forward target f_i in T_i. (Clearly $f_{i-1} \neq f_i$, since $f_{i-1} \notin T_i$.) The intermediate vertices of P_i are all of the vertices in $C_i - C_{i+1} - T_{i-1}$, in arbitrary order. We also compute P_1 to be a path between f_1 and b_1 whose intermediate vertices are all those in $C_1 - C_2$, in arbitrary order, and a path P_k between f_{k-1} and b_{k-1} whose intermediate vertices are all of the elements of $C_k - T_{k-1}$, in arbitrary order. This gives us a composite path $P_1, P_2, \ldots, P_k$ whose endpoints are b_1 and b_k. We can complete this path into a circuit by adding edges $[b_i, b_{i+1}]$ wherever $b_i \neq b_{i+1}$ (there is always such an edge, since both b_i and b_{i+1} are in C_{i+1}). It is not hard to see that this is a hamilton circuit in the interval graph.

To illustrate this method, consider once again the example of Figure 1. The maximal cliques here are $C_1 = \{1, 2, 3, 4, 5\}$, $C_2 = \{3, 4, 5, 6\}$, $C_3 = \{3, 5, 6, 8\}$ and $C_4 = \{3, 6, 7, 8\}$. The linkage sets are $L_1 = \{3, 4, 5\}$, $L_2 = \{3, 5, 6\}$ and $L_3 = \{3, 6, 8\}$. Suppose we choose $T_1 = \{3, 4\}$, $T_2 = \{3, 5\}$ and $T_3 = \{3, 8\}$. This would mean that $f_1 = 4$, $b_1 = 3$, $f_2 = 5$, $b_2 = 3$, $f_3 = 8$ and $b_2 = 3$. Then we can construct $P_1 = (3, 1, 2, 4)$, $P_2 = (4, 5)$ (there are no intermediate vertices in $C_2 - C_3 - T_1$), $P_3 = (5, 8)$ and $P_4 = (8, 6, 7, 3)$. In this example it turns out that $b_1 = b_{k-1}$, so that we do not have to add any edges; thus the hamilton circuit is $(3, 1, 2, 4, 5, 8, 7, 6, 3)$.

We have just shown that whenever it is possible to find a collection of target pairs from the linkage sets, we can construct a hamilton circuit in the interval graph. It is possible to prove the converse result, that such a collection of target pairs can be found in any interval graph containing a hamilton circuit; we omit the proof due to space constraints.

It is also possible to use the greedy approach to construct a parallel algorithm for this problem, along the same lines as the algorithm for section 2.

5   Conclusions

We have shown an efficient parallel algorithm for computing a hamilton path in an interval graph, if one exists. The result is based on a new characterization of interval graphs containing hamilton paths. We have also outlined an extension of our ideas to computing hamilton circuits in interval graphs. Our algorithm uses n^2 processors and $O(\log n)$ parallel time; however, it seems straightforward to modify it to use no more than $n^2/\log n$, by applying the work-efficient simulation idea due to Brent [4].

It is straightforward to extend the techniques of this paper to computing all hamilton paths, or all hamilton circuits, in an interval graph, as well as to computing longest paths. Related problems solvable using these methods include the computation of hamilton paths and circuits in circular-arc graphs.

Acknowledgement

We are very thankful to one of the referees for constructive criticisms and suggestions.

6   References

[1] A. Aboelfotoh and C. Colbourn, "Efficient algorithms for computing the reliability of permutation and interval graphs," *Networks* 20 (1990), pp. 883–898.

[2] A. Bertossi and Bonnucelli, "Hamiltonian circuits in interval graph generalizations," *Inf. Proc. Lett.* 23 (1986), pp. 195–200.

[3] K. S. Booth and G. S. Lueker, "Testing for the consecutive ones property, interval graphs and graph planarity using PQ-tree algorithms," *J. Comput. Syst. Sci.* 13 (1976), pp. 335–339.

[4] R. P. Brent, "The parallel evaluation of general arithmetic expressions," *J. ACM* 21:2 (1974), pp. 201–206.

[5] M. R. Garey and D. S. Johnson, *Computers and intractability: a guide to the theory of NP-completeness*, W.H. Freeman and Co., New York, 1979.

[6] P. C. Gilmore and A. J. Hoffman, "A characterization of comparability graphs and of interval graphs," *Canad. J. Math.* 16 (1964), pp. 539–548.

[7] M. C. Golumbic, *Algorithmic graph theory and perfect graphs*, Academic Press, New York, 1980.

[8] P. Hall, "On representations of subsets," *J. London Math. Soc.* 10 (1935), pp. 26–30.

[9] D. S. Hirschberg, "Fast parallel sorting algorithms," *Comm. ACM* 21 (1978), pp. 657–661.

[10] J.E. Hopcroft and J.K. Wong, "Linear time algorithm for isomorphism of planar graphs," *Proc. ACM Symp. Theory of Computing*, 1974, pp. 172 – 184.

[11] J. M. Keil, "Finding hamiltonian circuits in interval graphs," *Inf. Proc. Lett.* 20 (May 1985), pp. 201–206.

[12] P. N. Klein, "Efficient parallel algorithms for chordal graphs," *Proc. Foundations of Computer Science*, 1988, pp. 150–161.

[13] D. Kozen, U. V. Vazirani, and V. V. Vazirani, "NC algorithms for comparability graphs, interval graphs and unique perfect matchings," *Lecture Notes in Computer Science* 206 (1985), pp. 496–503.

[14] R. E. Ladner and M. J. Fischer, "Parallel prefix computation," *J. ACM* 27 (1980), pp. 831–838.

[15] G. A. Manacher, T. A. Mankus, and C. J. Smith, "An optimal $\Theta(n \log n)$ algorithm for finding a canonical hamiltonian path and a canonical hamiltonian circuit in a set of intervals," *Information Processing Letters* 35 (August 1990), pp. 205–211.

[16] A. Moitra and R. Johnson, "PT-optimal algorithms for interval graphs," *Proc. Allerton Conf. on Communication, Control and Computing*, Sept 1988.

[17] A. Moitra and R. Johnson, "A parallel algorithm for maximum matching on interval graphs," *Proc. Int. Conf. on Parallel Processing* III (1989), pp. 114–120.

[18] G. Ramalingam and C. Pandu Rangan, "New sequential and parallel algorithms for interval graph recognition," *Information Processing Letters* 34 (1990), pp. 215–219.

SOLVING A LOAD BALANCING PROBLEM USING BOLTZMANN MACHINES

Injae Hwang   **Ravi Varadarajan**

Computer and Information Sciences Department
University of Florida, Gainesville, Florida 32611
E-mail: ih@cis.ufl.edu   ravi@cis.ufl.edu

Abstract - We investigate the Boltzmann machine *parallel* networks for solving a load balancing problem that frequently arises in distributed systems. Our approach not only guarantees *feasible* solutions but can provide *close to optimal* solutions, as demonstrated by the experimental results. There is inherent parallelism in our approximation approach.

Introduction

The Boltzmann machine, introduced by Hinton and Sejnowski [2], has the same architecture as Hopfield and Tank network, but it uses the idea of simulated annealing [5] to escape from local minimal energy states. It was first demonstrated by Hopfield and Tank in 1985 [3] that neural networks can solve *hard combinatorial optimization* problems by mapping the well known Travelling Salesman problem onto the neural networks. Since then, attempts have been made to map many problems onto neural networks, including maximum independent set and graph coloring problems [6].

In this paper, we demonstrate how the Boltzmann machines can be used to solve a particular *load balancing problem* that arises in determining the resource migration activities of a distributed system. In this problem, there is a fixed number of resource units that are initially located in some nodes with no more than one resource per node. Due to the change in the resource access requests from different nodes, the resources need to migrate among the nodes. The load balancing problem is one of determining the new locations of the resources and the assignment of job requests to the resource nodes so as to achieve a desirable combination of the two criteria, namely resource migration costs and average job response time. File migration in a distributed file system or database is an example of this type of resource migration. The formulation of this problem was first given in [7] and its intractability had also been proved. We investigate the use of Boltzmann machines primarily due to its inherent parallelism.

This paper is organized as follows: First, we describe our load balancing problem and discuss how it is mapped onto Boltzmann machines. We evaluate our approximation approach using experiments. We also indicate how our approximation algorithm can be easily parallelized. We omit the proofs of our results due to lack of space. See [4] for these proofs.

Load balancing problem

In this section, we explain our load balancing problem in the context of a distributed database or a file system. First we need the following notations:

V — set of nodes ($P(V)$ is power set of V)

I', I — set of nodes containing file copies before and after file migration

$f : (I - I') \to I'$ — migration function specifying how the files migrate

$g : V \to I$ — specifies where file access requests from a node are processed

$h : I \to P(V)$ — inverse function of g

b_x — file access request rate from node x

q — processing capacity of a node

$D_{x,y}, E_{x,y}$ — cost for file access and file migration respectively from node x to node y

$\delta : I \to Z^+$ — processing delay for file access requests

$R(I, f)$ — file migration cost

$T(I, g)$ — average job response time

The load balancing problem involves determining I, f and g, for a given I' so as to achieve a desirable combination of $R(I, f)$ and $T(I, g)$ defined as follows:

$$R(I, f) = \max_{y \in I} E_{f(y), y}$$
$$T(I, g) = \max_{\{x \in V | b_x > 0\}} [D_{x, g(x)} + \delta(g(x))]$$

Here $\delta(y) = \frac{\sum_{x \in h(y)} b_x}{q}$. Without loss of generality, we assume q to be unity. We implemented a branch and bound algorithm to find an optimal solution to this problem. Even for reasonable size problems with 12 nodes and 3 file copy nodes, we found the branch and bound approach to be computationally expensive. Parallelizing the branch and bound algorithm will reduce the execution time but still there will be inefficiencies due to synchronization and communication overhead. For this reason, we need to explore efficient algorithms that find reasonable approximate solutions and yet can be efficiently parallelized. In this paper, we describe one such approach that uses Boltzmann machine models.

Boltzmann machine mapping

The Boltzmann machine is a network represented by an undirected graph $(\mathcal{U}, \mathcal{E})$, where $\mathcal{U} = (\mu_0, \ldots, \mu_{n-1})$ denotes the set of units and $\mathcal{E} \subseteq \mathcal{U} \times \mathcal{U}$ the set of connections between the units with connection strengths $S(\mu_i, \mu_j) \in R$. The state of a unit μ in a configuration h is denoted by $a_h(\mu)$, where $a_h(\mu) \in \{0, 1\}$. A connection (μ_i, μ_j) is said to be activated in a given configuration h if $a_h(\mu_i)a_h(\mu_j) = 1$. The consensus $C(h)$ of a configuration h denotes the overall desirability of all the activated connections in the given configuration and is defined as

$$C(h) = \sum_{(\mu_i, \mu_j) \in \mathcal{E}} S(\mu_i, \mu_j)a_h(\mu_i)a_h(\mu_j).$$

The consensus maximization is achieved by the process of simulated annealing. The simulated annealing algorithm is implemented on the Boltzmann machine by updating the states of units according to the probabilistic update rule. The probability of acceptance of a new state is a function of the consensus difference between the current and the new states and a temperature parameter. The temperature is lowered according to a cooling schedule that we will specify later.

The Boltzmann machine model we use, has two types of units which we call as "u-units" and "v-units" respectively. There are $M \cdot N$ u-units and N^2 v-units where M is the number of file copies and N is the number of nodes. Unit u_{ij} when activated indicates that the i-th file migrates from its present location to the node labelled j. Similarly, unit v_{ij} when activated indicates that the file access requests from the node labelled i are assigned to the node labelled j assuming that a file copy exists at that node. There are *eight* types of connections among these units which we denote as $C_b^u, C_b^v, C_1^u, C_2^u, C_1^v, C_2^v, C_1^{uv}$ and C_2^{uv}. C_b^u and C_b^v are the sets of bias connections of the u and v-units respectively. C_1^u and C_2^u are the sets of pairwise connections between the u-units in a row (i.e. between u_{ij} and u_{ik}, $j \neq k$) and in a column (i.e. between u_{ij} and u_{pj}, $i \neq p$) respectively. Similarly, C_1^v and C_2^v are the sets of pairwise connections between the v-units in a row and in a column respectively. C_1^{uv} is the set of pairwise connections between u and v-units whose column indices match (for example, between u_{ij} and v_{pj}) while C_2^{uv} is the set of pairwise connections between u and v-units whose column indices do not match.

The strengths of these various connections are determined by the following considerations: (1) All the local maxima of the consensus (energy) function of the Boltzmann machine correspond to only feasible solutions. (2) The quality of the solution as determined by the job response time and the file migration costs is closely related to the consensus function of the Boltzmann machine. The purpose of the two types of bias connections, namely C_b^u and C_b^v, is to help achieve a proper distribution of files and file access requests so as to minimize file and job migration costs; "job migration cost" refers to the cost of processing a file access request on a remote node. Thus their strengths should reflect the costs of migration between the corresponding nodes. The connections C_2^v are used to balance the file access request load among the different nodes that the files move to. The remaining five types of connections $C_1^u, C_2^u, C_1^v, C_1^{uv}, C_2^{uv}$ are used to enforce various constraints for obtaining feasible solutions to the load balancing problem.

We established the following relationships between the strengths of different types of connections.

1. The positive bias of a u-unit must be large enough to avoid incomplete solutions for file migration that are likely to occur due to inhibitory connections C_2^{uv}.

$$\forall (u_{ij}, u_{ij}) \in C_b^u, \quad S(u_{ij}, u_{ij}) > -\sum_{i=1}^{N} i\text{-th smallest}$$

connection strength in C_2^{uv} (1)

2. C_1^u and C_2^u connections enforce the constraints that a file copy is not located in more than one node and that not more than one file copy exists in any node.

$$\forall (u_{ip}, u_{iq}) \in C_1^u, \; S(u_{ip}, u_{iq}) < -\min\{S(u_{ip}, u_{ip})$$

$$+ \sum_{k=1}^{N} S(u_{ip}, v_{kp}), S(u_{iq}, u_{iq}) + \sum_{k=1}^{N} S(u_{iq}, v_{kq})\} \;\; (2)$$

$$\forall (u_{ip}, u_{jp}) \in C_2^u, \; S(u_{ip}, u_{jp}) < -\min\{S(u_{ip}, u_{ip}),$$

$$S(u_{jp}, u_{jp})\} - \sum_{k=1}^{N} S(u_{ip}, v_{kp}) \;\; (3)$$

3. The connection C_1^v enforces the constraint that file access requests from a node (job) are assigned to not more than one node.

$$\forall (v_{ip}, v_{iq}) \in C_1^v, \quad S(v_{ip}, v_{iq}) < -\min\{S(v_{ip}, v_{ip})$$

$$+ \max_{1 \leq k \leq M}(u_{kp}, v_{ip}), S(v_{iq}, v_{iq}) + \max_{1 \leq k \leq M} S(u_{kq}, v_{iq})\} \;\; (4)$$

4. The connections C_1^{uv} and C_2^{uv} are complementary as they cancel each other in a feasible configuration. These connections ensure that jobs get assigned to nodes with file copies. Note that for a given $1 \leq j, p \leq N$, all the connections $(u_{ip}, v_{jp}) \in C_1^{uv}$ must have the same strength and all the connections $(u_{iq}, v_{jp}) \in C_2^{uv}$ must have the same strength.

$$\forall (u_{ip}, v_{jp}) \in C_1^{uv} \text{ and } \forall (u_{mn}, v_{jp}) \in C_2^{uv},$$

$$S(u_{ip}, v_{jp}) = -(M-1)S(u_{mn}, v_{jp}) \;\; (5)$$

$$\forall (u_{ip}, v_{jq}) \in C_2^{uv}, S(u_{ip}, v_{jq}) < -S(v_{jq}, v_{jq})/M \;\; (6)$$

Now we have the following theorem.

Theorem 1 *All the local maxima of the consensus function correspond to feasible solutions in our Boltzmann machine model.* ∎

In the formulation of the problem, job migration cost $T(I, g)$ consists of two parts. One is proportional to the migration distance and it can be incorporated in the network by giving proper strengths to the v-unit bias connections. This part will be discussed later. The other is proportional to the file access request load of the node to which a job is sent. To minimize this cost, file access request load needs to be balanced among the different nodes that the files move to. Here, the inhibitory connections C_2^v, which connect v-units in the same column pairwise, are used to give proper penalty for heavily loaded nodes.

$$\forall (v_{ip}, v_{jp}) \in C_2^v, S(v_{ip}, v_{jp}) = -(b_i + b_j) \times F \;\; (7)$$

where b_x is the file access rate of the job submitted at node x. We call F the "load balance factor".

In the case that the number of jobs is larger than the number of files, more than one job is assigned to a node, and hence at least one connection of type C_2^v is activated. This may cause the network to settle in a configuration corresponding to an incomplete solution where fewer than N jobs are assigned to file copy nodes. Proper value should be given to the variable F to prevent incomplete solutions. Before we state our results regarding the upper bound for F to guarantee complete and hence feasible solutions, we introduce some additional notations. Let $k = \lfloor (N-1)/M \rfloor$ be the average number of jobs per file copy node. Without loss of generality, assume that $b_1 \geq b_2 \geq \ldots \geq b_N$ and let $w = \sum_{i=2}^{k+1} b_i + kb_1$. Also let F^* be equal to the minimal v-unit bias divided by w, when $k \neq 0$, and be equal to ∞ otherwise. Now we are ready to state the theorem for the upper bound of F.

Theorem 2 *When $F \leq F^*$, the configurations, in which fewer than N jobs are assigned, do not correspond to the local maxima of the consensus function.* ■

Now, we explain how to use the bias connection strengths to incorporate the migration costs in the network. If we need to minimize the sum of the migration costs of all the files, then we can use $(L - E_{i,j})$ to be the bias strength of unit u_{ij}, where L is some integer greater than $\max_{i,j} E_{i,j}$. In this case, maximizing the consensus amounts to minimizing the sum of the file migration costs. Since our objective is to minimize bottleneck migration costs, we assign bias unit strengths based on the ordering of the migration costs. In the remainder of this section, we explain how the u-unit bias connection strengths are determined. The v-unit bias connection strengths are determined in a similar manner. To determine the u-unit bias strengths, we use the sequence defined below.

Definition 1 *Sequence $Q_n(i)$ is defined as follows:*

$$Q_n(i) = i \quad (1 \leq i \leq n)$$
$$Q_n(i) = \sum_{j=i-n}^{i-1} Q_n(j) \quad (i > n)$$

The following theorem shows that it is more appropriate to use this sequence instead of the actual file migration cost values for u-unit bias, if the bottleneck criterion is of interest. In the future discussion, we assume that u-units are sorted in non-decreasing order of the file migration costs $E_{i,j}$ and indexed according to this sorted order.

Theorem 3 *Suppose the u-unit bias connection strengths are defined as follows :*

$$S(u_i, u_i) = L - Q_M(i), \quad 1 \leq i \leq NM$$

where $L \geq \max_{1 \leq i \leq NM} Q_M(i)$. Then, for any configuration, ignoring v-unit connections, the consensus is maximized if and only if the corresponding bottleneck file migration cost is minimized. ■

The sequence $\{Q_N(i), 1 \leq i \leq N^2\}$ can be used in a similar fashion for v-units for job migration. Then, maximization of the consensus is approximately equivalent to minimization of the sum of the bottleneck file migration and job migration costs.

Choice of key parameters

In mapping the load balancing problem onto the Boltzmann machines, we had identified two major parameters; "load balance factor" (denoted by F) and "trade-off factor" (denoted by α). In this section, we discuss the implications of these parameters and explain how we determine the proper values for these parameters.

In order to avoid incomplete job assignments, we had determined a theoretical upper bound, given by Theorem 2, for the strength of the connections in C_2^v. This upper bound is usually too low to be of any practical use. When the average file access request rate is high, these connection strengths must be large enough to permit proper load balancing of file access requests among the file copy nodes. On the other hand, when the average file access request rate is small compared with the job migration cost, load balancing

becomes less important and hence the connection strengths in C_2^v must be small. Hence we use the *balance factor F* for tuning the connection strengths to the average file access request rate. Specifically, we relate the ratio (A) as determined by the balance factor divided by the average v-unit bias strength to the ratio (B) as determined by the average file access request rate divided by the average file access cost (job migration cost). The proper value of the ratio A/B can be obtained experimentally for a given network.

In our load balancing problem, there are two objective criteria, namely job response time and file migration cost. The proper trade-off between these two criteria is determined by the relative importance of one over the other. In our Boltzmann machine model, we adjust the strength of u-unit bias connections (within certain bounds) so as to make file migration less or more desirable. For this purpose, we use the *trade-off factor α*; larger the value of α, less desirable to migrate the files. For a given desired response time, we can arrive at the proper value of α by using *interpolation search* between the lower and upper bounds. The lower bound is determined by the feasibility consideration and it was established by condition (1) stated in Section 3. The upper bound is the value below which the files start to migrate and it is found by experimentation, whereby α is increased little by little from the lower bound until the files stop migrating. It needs to be determined only once for a given network.

Approximation algorithm

Based on all the ideas we had mentioned before, we had developed an approximation algorithm for solving the load balancing problem that uses the Boltzmann machine simulator. We assume that a desired response time T^* is given. Let LB denote the lower bound for relative strength (α) of u-units which guarantees complete solution. Let UB denote the upper bound for relative strength of u-units above which files start to migrate.

Procedure LoadBalance;
// Returns an approximate solution consisting of file and job migration functions f and g //
   begin
      1. Simulator(UB);
         Let T_U be the job response time returned
         by the simulator.
      2. If $(T_U \leq T^*)$ then return the solution found
         by the simulator as optimal solution;
      3. Simulator(LB);
         Let T_L be the job response time returned
         by the simulator.
      4. If $(T_L > T^*)$ then declare "no feasible solution";
      5. If $(T_L = T^*)$ then return the solution found
         in step 3 as optimal solution;
      6. While $|UB - LB| \geq \epsilon$
      // ϵ is a small number > 0 //
        begin
          7. $M \leftarrow \left(\frac{T^* - T_L}{T_U - T_L}\right)(UB - LB) + LB;$
          8. Simulator(M);
             Let T be the response time
             returned by the simulator.
          9. If $(T < T^*)$
             then $[LB \leftarrow M; T_L \leftarrow T];$

else if $(T > T^*)$ then $[UB \leftarrow M; T_U \leftarrow T]$;
else return solution found in step 8;

   end;

  10. Return solution found in step 8.

end.

Procedure Simulator (α);
// Uses α as the trade off factor for Boltzmann machine mapping. Then, it uses the Boltzmann machine simulator to determine the configuration of maximal consensus. The file and job migrations, the job response time, and the bottleneck file migration cost associated with this configuration are also found. //

Experimental Results

Computer simulations of the Boltzmann machine model were performed to evaluate the effectiveness of this approach. In the simulator, each run uses a seed for generating the values of the random variable in the Boltzmann distribution. For each problem, we run the experiment four times and the solution with the smallest file migration cost (in case of a tie, the one with the best response time) was selected.

The annealing schedule used in the experiments is as follows :

1. initial temperature t_0: $2 \sum_{ij} | S(\mu_i, \mu_j) | / $(No. of units)
2. no. of cycles: 100
3. no. of trials per cycle: no. of units $\times$ no. of nodes
4. final temperature t_{100}: 0.01

After each cycle, the temperature decreases linearly. At the final temperature of 0.01, we run the simulator 10 more cycles at that temperature to make sure that the Boltzmann machine settles at a configuration corresponding to a local maximum of the consensus. This ensures that the simulator always produces feasible solutions. At very low temperature, since a hill climbing search is employed, a local maximum will be reached within a finite number of iterations. Note that in all our experiments, we obtained feasible solutions.

We use a branch and bound program to determine the optimal solutions to be compared with the solutions obtained by our approximation algorithm. We tested problem instances with 12 and 15 nodes. We fixed the number of file copies to be 3 and 4 for the problem instances containing 12 and 15 nodes respectively. The file migration cost of our approximate solution for 10 out of 12 instances with 12 nodes, was within 3 units of the optimal cost. For one instance, our algorithm gave a solution that had less cost than the optimal at the expense of increased response time. We had somewhat similar results for the 15 node case. On the other hand, the average error in response time was 11% for 12 nodes instances and 12% for 15 node instances. These results are quite encouraging.

Parallel implementation

The Boltzmann machine is a massively parallel network in which all the units can change their states simultaneously. But, in proving the feasibility property of the local maxima, it was assumed that only one unit changes its state at a time. Allowing more than one unit to change their

states simultaneously may cause erroneously calculated differences in consensus and may appear to violate the feasibility constraint. But, the asymptotic convergency property, mentioned in [6], alleviates this problem thus enabling the successful parallelization of simulated annealing in the Boltzmann machine. Parallelization of neural networks by mapping them onto the multicomputer architectures was studied by Ghosh and Hwang [1]. This parallelization will reduce the execution time significantly. With the advent of neural network hardware, there is a potential for achieving a million fold reduction in execution time for instances having 2000 units; see [6] for implications of neural network hardware.

Conclusions

In this paper, we had demonstrated how an efficient approximation algorithm based on Boltzmann machines can provide a good quality solution to the load balancing problem that frequently arises in distributed systems. Our preliminary experiments indicate that reasonable approximate solutions can be obtained by this method. We are planning to investigate this approach for solving other load balancing problems that frequently need to be solved in parallel computers.

References

[1] J. Ghosh and K. Hwang, "Mapping neural networks onto message-passing multicomputers," *Journal of Parallel and Distributed Computing*, Vol.6, No.2, April 1989, pp.291-330.

[2] G. E. Hinton and T. J. Sejnowski, "Learning and relearning in Boltzmann machines," in *Parallel and Distributed Processing : Explorations in the microstructure of cognition*, Vol.1, Rumelhart,D.E.,McClelland,J.L., and the PDP Research Group (Eds.), 1986.

[3] J. J. Hopfield and D. W. Tank, "Neural computations of decisions in optimization problems," *Biol. Cybernet.*, 52(1985), 141.

[4] I. Hwang and R. Varadarajan, "Solving a Load Balancing Problem using Boltzmann Machines," TR-91-09, The University of Florida, 1991.

[5] S. Kirkpatrick, C. D. Gelatt and M. P. Vecchi, "Optimization by simulated annealing," *Science*, 220(1983), 671.

[6] J. H. M. Korst and E. H. L. Aarts, "Combinatorial optimization on a Boltzmann machine," *Journal of Parallel and Distributed Computing*, Vol.6, No.2, April 1989, pp.331-357.

[7] R. Varadarajan and Y. E. Ma, "An Approximate Load Balancing Model with Resource Migration in Distributed Systems," *International Conference on Parallel Processing*, August 1988, pp.13-17.

Acknowledgement: This work is supported by a planning grant from the Florida High Technology and Industrial Council.

Study of an Inherently Parallel Heuristic Technique

Ira Pramanick and Jon G. Kuhl

Department of Electrical & Computer Engineering
University of Iowa
Iowa City, Iowa - 52242

ABSTRACT

A non-traditional application of parallel processing, known as *Parallel Dynamic Interaction (PDI)*, is discussed. PDI is a heuristic solution framework, applicable to certain types of exponentially hard problems. From a parallel processing standpoint, PDI is interesting for two reasons: it appears to be an inherently parallel technique; and, the use of parallel processing is not primarily aimed at achieving computational speedup, but rather is aimed at achieving a higher quality solution than could be obtained in comparable time using serial methods. This paper gives empirical results of the application of PDI to the *job-shop scheduling problem*, and focuses specifically on the parallel processing aspects of PDI. A study of 30 examples is reported, in which PDI produced solutions within an average of 5.1% of optimal for the job-shop problem. Another experiment supports the claim that PDI is an inherently parallel technique. In this study, solution quality degraded by about an average of 25% when the fully parallel PDI implementation was changed to a serial approximation.

I. Introduction

This paper discusses a non-traditional application of parallel processing, known as *Parallel Dynamic Interaction* (PDI). PDI is a heuristic solution framework, applicable to certain types of exponentially and superexponentially hard problems. The technique shows promise as a practical problem-solving method, based upon empirical study of several example application domains.

The PDI method directly attacks the exponent of the complexity figure of an NP-hard problem by dividing the problem into a number of lower-order subproblems, each defined over a subset of the constraints and/or input space of the original problem. These subproblems are then solved simultaneously, through the use of parallel processing. To account for the potentially numerous constraints or interrelationships among the subproblems, a limited *global* state information is maintained through which subproblems can identify potential points of conflict or inconsistency with other subproblems as they incrementally build their local solutions. When such potential conflicts are identified, the local subproblem solution processes may spawn *alternative* solution instances to account for (or hedge against) the potential conflict. In this way, each local subproblem builds a limited *local* solution tree containing several dynamically determined local solution alternatives. This process of constructing the local solution trees is called the *booking phase*. A subsequent *commitment phase* then simultaneously traverses these local solution trees to incrementally construct a consistent solution to the global problem. Since the constraint resolution process during booking has oriented the local solution trees towards global constraints, the local trees represent a highly constrained solution space from which a consistent, and hopefully high quality, global solution can quickly be extracted.

The use of parallel processing during the booking phase is important in order to provide a level of interaction among subproblems that fairly and naturally distributes constraint resolution (spawning of alternative local solution processes) among subproblems and uniformly orients local solution trees toward global constraints. Without this *dynamic interaction* — e.g., if subproblem solutions were pursued in some arbitrary serial order — the order of subproblem solution would impose monotonically increasing responsibility for constraint resolution upon subproblems according to their order of solution.

Parallel processing is also central to the commitment phase. By simultaneously traversing the local solution trees and incrementally committing a global solution, it is possible to dynamically redirect the trajectory of the commitment path through each local tree to account for the effects of incremental commitment progress by other subproblem instances. In this way, the overall commitment can maintain an orientation toward the desired global objective.

From the parallel processing standpoint, PDI is of importance for two reasons:

i) It appears to be an *inherently parallel* technique. That is to say, the PDI methodology is based upon the dynamic interplay among simultaneously executing components. As such, PDI does not appear to have a direct serial analog, other than one which explicitly or implicitly emulates the parallel interaction present in the parallel version.

ii) The use of parallel processing in PDI is not primarily aimed at achieving computational speedup, but rather is aimed at achieving a *higher quality* solution than could be obtained in comparable time using serial methods.

To date, PDI has been applied to the heuristic solution of three NP-hard problems: **flow-shop scheduling, job-shop scheduling** and **vertex cover**. On approximately 70 example instances of these problems, for which it has been possible to exhaustively determine the optimal solution for comparative purposes, the PDI heuristic has been able to consistently find high quality solutions, averaging within 5% of optimal. The time required by PDI to find these solutions was as much as 5 orders of magnitude faster than the time required for a parallel branch and bound algorithm running on the same hardware to find a first solution of comparable quality.

In addition to presenting the empirical results of the application of PDI to a particular problem area — classical job-shop scheduling [1 - 4] — this paper focuses specifically on the parallel processing aspects of PDI. It is difficult to quantify the parallel processing performance of PDI in traditional terms, since it does not have a serial analog against which to compare computational speedup. Nor is there a straightforward notion of a "best serial heuristic" against which to compare PDI, since different heuristics for a given problem tend to give significantly different solution qualities for a given problem instance, thus making direct computational efficiency comparisons unfair or meaningless.

It is possible, however, to empirically investigate the importance of parallel processing to the PDI technique, by examining the effect upon PDI performance (in terms of solution quality) resulting from limiting the level of parallel interaction that is allowed to occur. That is the approach taken herein. In particular, we focus on a detailed study of the application of PDI to job-shop scheduling. For this study, a set of 30 example job-shop scheduling problems were solved using the PDI method on a 14 processor shared memory multiprocessor system. Then a series of experiments were conducted in which the *granularity* at which subproblem instances were allowed to interact with one another was artificially limited at several levels, including a version in which booking and commitment of subproblem instances proceeds in a strictly serial order. The effect of various restrictions in interaction granularity upon solution quality was noted for each of the 30 problem instances.

The results of this study, reported in Section IV of this paper, clearly indicate that PDI solution quality depends directly upon sub-problem interaction during booking and commitment and that limiting this interaction, in general, substantially mitigates the effectiveness of this technique. These results support the claim that PDI is an "inherently parallel technique", at least for the particular application studied. That is to say, in order to construct a serial version for the technique capable of producing a comparable level of solution quality, the incremental progress, and interaction, of subproblems would have to be maintained at a sufficiently fine level of granularity so as to approximate the effect of parallel processing among the instances.

We cannot make any explicit claims regarding speedup of the parallel version of PDI versus a comparable version that emulates the parallel interactions (we have not attempted to construct such a serial version). The motivation for employing parallel processing in this problem was the natural framework it provided for supporting the dynamic interaction, rather than any overt speedup considerations. However, it is at least interesting to note that this application provides a rare perspective of a problem for which a serial implementation may have to incur specific overheads (i.e., those of emulating the parallel processing context switch among subproblem instances) *not present* in the parallel implementation.

II. Overview of PDI & Its Application to Job-shop Scheduling

This section provides a brief overview of PDI and describes its specific application to the job-shop scheduling problem. A more detailed description of PDI as a general solution framework, and additional example applications, can be found in [5 - 7].

The job-shop scheduling problem is an important problem from classical scheduling theory [2 - 4]. The elements of a job-shop scheduling problem are a set of machines and a collection of jobs to be scheduled, each job consisting of several tasks to be performed on these machines. In the worst case, for the job-shop problem, there could be a total of $(n!)^m$ schedules.

The *job-shop* scheduling problem can be defined as

Given:

$m \geq 1$ different processors or machines;
$n \geq 1$ jobs,
$l \geq 1$ the maximum number of tasks that a job can have;
t_{ij} ($1 \leq i \leq l$, $1 \leq j \leq n$), the processing time required for task i of job j;
p_{ij} ($1 \leq i \leq l$, $1 \leq j \leq n$), the processor required for task i of job j;
Each job has its own precedence constraint for its tasks, and it may be different from those of other jobs. A job can also visit a machine more than once, or not visit a machine at all.

Objective:

to form a schedule that minimizes a given performance measure.

The most commonly used performance measure employed is **make-span** or finish-time which is the time taken for the completion of all the tasks of all the jobs.

This problem is specified in terms of two matrices: the machine constraint matrix, the entries of which are given by p_{ij}s; and the time constraint matrix, the entries of which are given by t_{ij}s. Extensive research has been done for this problem, and several branch and bound formulations exist for it [2, 3]. No tightly bounded heuristic algorithm exists for the general case.

PDI consists of three phases: **the partitioning phase, the booking phase, and the commitment phase**. Each of these phases together with the specific details with respect to the job-shop problem is discussed below.

Partitioning Phase: The problem P must first be divided into n suitable *subproblems* or **instances** denoted by $P_1, P_2, \cdots, P_n$. In general, each P_i is obtained by limiting the original problem P to a smaller input space and a smaller set of constraints. That is, P_i is specified in terms of an input space I_i that is a subset of the input space I of the original problem P, and a constraint set C_i which is a subset of the global constraint set C of P. The constraints that must hold in P for the inputs of P_i should also be imposed on P_i. The constraints that are relaxed are with respect to the inputs of P that are not included in the input space of P_i.

The heuristic that will be used in this phase to decide the partitioning strategy is referred to as the **partitioning heuristic**. Some problems exhibit *natural* and obvious partitions. For example, in scheduling problems involving n jobs, each consisting of several tasks, it is natural to partition the original problem jobwise. For the multidimensional knapsack problem, each instance could correspond to a knapsack. For some other problems like the vertex cover problem, several *natural partitionings* (viz. dividing the graph into various subgraphs) exist, and it is not obvious what criterion should be used to select a particular partition. For yet others, like the traveling salesperson problem and various routing problems, such natural partitionings may not exist at all and the problem may not be a good candidate for PDI solution.

The partitioning heuristic used for the job-shop problem was to divide the problem into n instances, each corresponding to a job and consisting of at most l tasks. Here, an instance j ($1 \leq j \leq n$) is responsible for the tasks of job j, and has to satisfy only the constraints specified by the jth columns of the machine and time constraint matrices.

Booking Phase: Parallel execution begins at this point in the computation. After the instances have been defined, each instance proceeds to form several *tentative plans*. The planning of each instance is modeled as an asynchronous, incremental process. It involves moving step by step toward a solution to the subproblem, and it is tentative because certain assumptions are made during the execution of the processes regarding the constraints which may later have to be retracted.

Each instance begins as a single process so that, initially, the number of processes is equal to the number of instances. An instance and its children processes (that will be started later as a result of the competitive, dynamic interaction between instances) form the *alternatives* for that instance. The initial alternative corresponds to the leftmost branch of the local solution tree (LST) of the respective subproblem. This phase consists of building the "good" parts of the LST of each instance. Each instance dynamically interacts with other instances and, based on heuristics, begins execution of other alternatives to account for potential conflicts with other planning instances.

An alternative forms a tentative plan by *booking resources*. For the job-shop problem, a *<machine, time-interval>* constitutes a resource. A *resource* may be in one of three states, viz., *free, booked* or *committed*. The *free* state implies that this resource has not as yet been included in any of the plans of any instance. The *booked* state implies that this resource has been included in one or more plans belonging to one or more instances. And, the *committed* state indicates that the resource is part of the final global solution, and may lead to a potential conflict. Resources reach the committed state during the commitment phase (described below) which may partially overlap with the booking phase. Each resource also has a *bywhom* list, which stores the identifiers of all instances that have booked that resource. The information regarding the resources must be globally accessible since these resources are common to multiple instances. For job-shop, a *<machine, time-interval>* pair is *free* only if the machine in question has not yet been included in any plan of any instance for the entire time-interval. Booking or commitment of the machine for a part or whole of the time-interval makes the state of the resource *booked* or *committed* respectively.

An alternative extends its solution by booking a required resource. The state of the resource is checked. If it was *free*, the state of the resource is changed to *booked*. In addition, the identifier of the instance of this alternative is appended to the *bywhom* list of this resource. If the state of the resource was *booked*, then the instance identifier is appended to its *bywhom* list. This list is checked to see if this resource was booked by an alternative of another instance, indicating the possibility of a conflict. Whenever there is a possibility of a conflict, another alternative may be started (based on heuristics). If the state was *committed*, then the alternative in question can not book this resource and has to extend its solution by booking another resource. Hence, heuristics are employed here to *explore* the search space, but in a very limited fashion. The instances thus interact with each other dynamically, via changes made to the states of the various resources.

The heuristics that are used at potential conflict points serve two purposes: to decide whether an alternative should at all be started, and if so, to decide how the potential conflict should be resolved in the alternative plan. These heuristics will be referred to as *search-directing heuristics*. These heuristics are typically associated with an upper limit on the number of alternatives that an instance can have.

For the job-shop problem, when an alternative finds a resource state to be *committed*, it books the corresponding machine for the required interval after the conflict period. If it finds the state to be *booked* by another instance and if the number of alternatives already started for this instance is less than a user-defined maximum *maxalt*, then another alternative is started with an added delay corresponding to the conflict period. The original alternative proceeds to book this resource. For the example problems discussed later in this paper, a value of *maxalt* in the range of 10 to 20 was used.

Ideally, when a potential conflict is detected during booking, all instances involved in the conflict should be made aware of it and given the opportunity to use *search-directing heuristics* to start alternatives, if required. This would imply the need for an instance that creates a potential conflict by booking a resource to notify all instances that had previously booked that resource. However, this is not practical to implement, especially since the exact point of conflict in the local solution trees of the previous instances would have to be identified. On the other hand, failure to account for conflicts in the earlier instances could potentially result in an instance having too narrow a breadth in the upper levels of its local solution tree.

A pragmatic alternative is to use some form of *anticipation* during the booking phase. That is, at upper levels of its local solution tree, an instance anticipates (based on some heuristics) a potential conflict if one does not exist, and starts off alternatives if necessary. The heuristics used for starting alternatives in anticipation are referred to as *anticipating heuristics*. Anticipating heuristics are associated with the level of the local solution tree above which they should be used. These increase the breadth of the corresponding local solution tree at its upper levels, if necessary.

For the job-shop problem, if the alternative is at or below a user-defined level *priorl* in the tree and does not find any possibility of a conflict, it *anticipates* a conflict. It starts another alternative, subject to *maxalt*, with an added delay equal to the minimum of those entries in the time-constraint matrix that correspond to the use of the machine in question. The value of *priorl* was kept at about a third of *l*, the maximum number of tasks.

An alternative stops when it has booked all the required resources. For the job-shop problem, an alternative stops when it has booked the machine for the required time interval for the last task (and hence, all previous tasks) of its instance.

The local solution tree of an instance represents solutions to the corresponding subproblem, and only satisfies its local constraints. However, it is formed as a result of the potential conflicts with respect to the global constraints. Hence, the formation of the local solution trees is a function of the global constraints and the dynamic interaction between the instances, and is thus oriented towards the global constraints. For the job-shop problem, each local solution tree obeys only the constraints regarding the ordering of the various tasks of that instance, and their time intervals. The global constraint of using a machine only for one task at a time is not satisfied by the set of local trees generated. However, the formation of these tree has been in response to this constraint.

The local solution trees generated by the instances, in general, can not be obtained from the global search tree. In fact, since these local solution trees are generated under assumptions that may not hold for the global environment, these local solution trees may be *illegal* with respect to the global search tree. Additionally, these local solution trees are much smaller than the *nth* fraction of the global search tree. For the job-shop problem, a global search tree would contain on the order of $(n!)^m$ nodes, whereas each local tree contains of the order of $l \times maxalt$ nodes, giving a total of $n \times l \times maxalt$ nodes. For instance, for a job-shop problem with ten jobs, five machines, and five tasks for each machine, the global search tree would contain of the order of $(10!)^5 = 6.3 \times 10^{32}$ nodes. For a *maxalt* of 20 for PDI, the local solution trees generated by PDI would have a total of $10 \times 5 \times 20 = 1000$ nodes.

Empirical analysis shows that simple, local heuristics can work very well towards forming good tentative plans, which are later combined to yield a good global solution. The local application of these heuristics coupled with the competitive nature of the instances orient the solution towards the global constraints. That is, unlike the traditional use of globally applied heuristics which almost always affect the global solution obtained, these local heuristics that are used at the boundary of the subproblems may have less effect on the quality of the overall global solution. For the job-shop problem, the local heuristics employed were simple, and as the results show, did not give rise to significantly different global solutions.

Commitment Phase: The commitment phase for an instance begins after some or all of its plans have been formed, i.e. there may be some amount of overlap between the booking and commitment phases of an instance. Also, this phase may be started independently or simultaneously for all the instances. The commitment phase is also executed in parallel for the instances. It consists of *finalizing* a global solution by combining ''good'' local solutions (or *portions* thereof) to yield a coherent, global solution. Heuristics are used to determine when to start this phase and to choose the plan to *commit*. These heuristics will be referred to as *committing heuristics*. Each instance contributes toward the final solution by committing one (or more) of its local solutions completely (or partly).

At a given point during its commitment phase, an instance chooses its best (based on some heuristic) local solution, and picks the most necessary unit of this local solution for commitment. The commitment of a resource involves checking its state and changing it, if possible. If the state is *committed*, then it has to choose an alternative resource for commitment. If the state is not *committed*, then the instance changes the state to *committed*. This commitment of a basic unit of a plan has to be atomic. However, the entire commitment phase is *non-atomic*. That is, an instance does not commit the units of one or more of its plans all at once, and local plan commitment is not serialized among instances.

Because instances are committing their plans concurrently, the current best plan of an instance may become devalued due to the commitment of a unit (or units) of another instance(s). It may even become unacceptable or illegal because of a conflict with the committed unit(s) of the other instance(s). However, because alternative local plans have been generated during the *booking* phase to account for such conflicts, there should be branch points in the local solution space that represent legal (or better) solutions in the presence of the conflict. Any of these alternatives could now become the best candidate for the continuation of the commitment of this instance. Hence, the commitment process switches over to the alternative local solution or branch of the local solution tree which represents the best current solution in light of the state of partial commitment of this instance as well as those of other instances.

So, after committing a unit of the current best plan, an instance reevaluates its local solutions' worth (which may have changed because of the non-atomic nature of the commitment process and the resulting commitment of unit(s) of local solutions of other instances), and picks the new best plan. That is, the local solutions from which the units are drawn to form the global solution are determined dynamically, depending on interaction from other instances. Hence, non-atomicity during commitment can also help orient solutions toward a global optimum rather than simply adopting a given valid local subsolution. An instance is finished when it has contributed a complete local solution to the global solution.

For the job-shop problem, each instance could enter its commitment phase independent of other instances, or the commitment phase could be started simultaneously for all the instances. Both the strategies were tried. At the beginning of the commitment phase, the first resource required by the leftmost branch of the local solution tree is picked for commitment. Thereafter, the plan with the least delay in the valid part of the local solution tree is chosen and the resource at the first uncommitted level (representing the first uncommitted task of the job) is picked for commitment. An instance may need to form additional partial plans because all the resources at this uncommitted level have been rendered invalid by the commitment of parts of plans of other instances.

An additional heuristic that was applied for the job-shop problem was: assign a priority index to each task before the booking phase. This priority index is a function of the number of tasks that could be in conflict with this task and the time taken by the remaining tasks of this job. The priority index of a task is 1 (higher priority) if the total time of the remaining tasks of the corresponding job is greater than the average remaining time for all tasks that use the same machine as this task. Else, the priority index of the task is 0. During the commitment phase, if the task being committed has priority 0, then its commitment is deferred for a user-controlled time interval.

For the job-shop problem, an instance stops when it has committed the required time intervals for each of the machines required by its tasks, under the precedence constraint for that instance.

Since PDI is a non-deterministic procedure, both with respect to the booking and commitment phases, two executions of PDI on the same problem, using the same heuristics, may yield different solutions. In fact, it is desirable to run PDI several times with slight perturbations in starting conditions, in order to generate the best possible solution. In addition, the solutions may also differ depending on the variety of heuristics applied. However, as the empirical data for job shop illustrates, the method is relatively insensitive to starting conditions or choice of heuristics, in the sense that it tends to consistently generate high quality solutions.

III. Empirical Study of PDI for Job-Shop Scheduling

The PDI version of job-shop scheduling was implemented on a 14-processor shared-memory Encore Multimax multiprocessor system. While it is likely that some of the larger problems could have benefited from the availability of more parallelism, this machine was used since it was the largest shared-memory system available for the study. A total of 30 example problem instances were chosen for the experiment. The entries of the time and machine constraint matrices for these examples were randomly generated. For each example, PDI was run using various combinations of the search-directing, anticipating and committing heuristics described earlier. To assess the absolute quality of the PDI solutions, the optimal solutions was determined for each example. This was done by conducting an exhaustive branch and bound search for each problem, using the best available bounding functions [2, 3], and an initial seed provided by the best PDI solution. For some of the larger problems these searches required as much as four months of background computation, even with the problem split across several workstations.

It was also desired to compare the PDI heuristic to other reported heuristics for the job-shop problem. Six unbounded heuristics reported for this problem in [3] were used here for this purpose. No bounded heuristics are known for the general job-shop problem [8, 9]. Our intent is not to claim that PDI unequivocally represents the most superior heuristic for the job-shop problem. Rather, we present the comparison to illustrate the viability of PDI as compared to common heuristics used for the problem. In [8], the difficulty of finding heuristics with worst case bounds less than m, the number of machines is discussed.

The six unbounded heuristics used for the job-shop problem are [3]: *Shortest Processing Time* (SPT), *First-Come First-Served* (FCFS), *Most Work Remaining* (MWKR), *Most Operations Remaining* (MOPR), *Least Work Remaining* (LWKR), and *Random* (RNDM). Details of these heuristics will not be given here. The interested reader is referred to [3].

For the 30 example problems, the number of jobs, n, varied from four to twelve; the number of machines, m, varied from three to eight; and the number of tasks, l, varied from three to eight. The percentage deviation from optimal of the best PDI solution ranged from 0% (for the five cases where PDI found the optimal solution) to 14.5% over all 30 examples, averaging at within 5.1% of optimal. The time taken by a PDI run was minimal, in the range 0.3 sec. to 2.0 sec. A complete tabulation of the results can be found in [7].

In order to compare the execution efficiency of PDI with more conventional approaches capable of producing comparable solution quality, a parallel branch and bound version of job-shop scheduling, using the best available bounding functions, was implemented on the same 14 processor computer. This algorithm was used on 10 of the examples, with the stopping criterion being the discovery of the first solution equal to or better than the PDI solution. Its execution time for these examples ranged from 42 to 134,992 seconds, with an average execution time of 24,219 seconds.

In comparison with the classical job-shop heuristics listed above, PDI produced a higher quality solution than all of the other heuristics in 20 of the 30 examples. In three cases its solution tied for best with one or more of the other heuristics. The best of other six heuristics (MWKR) found the best solution in only three of the 30 cases.

As compared to the average percentage deviation from optimal of 5.1% for PDI, the percentage deviation averaged at 36.0% for SPT, 14.3% for FCFS, 15.2% for MWKR, 14.2% for MOPNR, 47.3% for LWKR, and 30.6% for RNDM. As compared to the maximum percentage deviation from optimal of 14.5% for PDI, the maximum percentage deviation from optimal was 59.3% for SPT, 37.2% for FCFS, 41.1% for MWKR, 39.7% for MOPNR, 87.2% for LWKR, and 63.8% for RNDM. Details of this comparison can be found in [7].

IV. Role of Parallelism in PDI

PDI is designed to be an inherently parallel technique. The dynamic interaction between instances is allowed to proceed asynchronously and non-deterministically. The local search trees are built depending upon this interaction, and so is the determination of the trajectory of a local solution through this tree. Hence, PDI appears to have no serial analog, except for the one in which this interaction is explicitly simulated. As a result PDI solution quality should be expected to deteriorate if the degree to which instances are allowed to freely interact is restricted.

Several methods of experimentally limiting parallel interaction were considered. One possibility considered was varying the number of processors to produce varying degrees of available parallelism. This approach was decided against since the example problems are of varying sizes and hence employ varying numbers of subproblem instances, they would be expected to respond differently to reductions in the number of processors employed. Also, reducing the number of processors induces additional factors, such as operating system induced time-slicing among instances, that are not directly controllable.

To provide a more fair and directly interpretable experiment, an alternative method of limiting parallel interaction was employed. The experiment consisted of deliberately constraining the ability of instances to interact. Each instance was required to proceed for a certain number (user-controlled) of steps before any other instance was allowed to execute, each step corresponding to the booking or commitment of a single resource. The number of steps was varied from one to the maximum number of tasks for an instance. The overall algorithm proceeded by time-slicing among instances at this granularity. In addition, a completely serial approximation to PDI was constructed, in which execution of instances was sequentially ordered (this order being random). Here later instances could utilize the results of earlier instances, but not vice versa.

In addition to directly assessing the importance of full dynamic interaction to PDI performance, this experimental methodology allows assessing of the level of interactive granularity which a true serial approximation of PDI would need to provide, in order to approximate the performance level of parallel PDI. That is, the level of interaction granularity translates, at least qualitatively, into the levels of overhead that the serial version would accrue in subproblem context switching to provide the necessary interaction at this granularity.

Experimental Results

The grain sizes were divided into small, medium and large sizes, depending on the number of steps (mentioned above) for which an instance was allowed to execute. The small size roughly corresponded to the lower third of this number, the medium size to the middle third and the large size to the upper third. For instance, for example 3 for which l, the maximum number of tasks is six, the small grain sizes corresponded to the number of steps being 1 and 2, the medium grain sizes corresponded to the number of steps being 3 and 4, and the large grain sizes corresponded to the number of steps being 5 and 6. PDI was run at each granularity level using the heuristic combinations described earlier.

The results indicate that, in general, PDI solution quality degrades substantially as the size of the granularity of interaction increases, and is the worst for the serial approximation of PDI. For the fully parallel PDI version, the percentage deviation of the best PDI solution from the optimal solution averaged 5.1%, the maximum deviation being 14.5%. For small grain sizes, the average percentage deviation was 17.0%, the maximum deviation being 35.7%; for medium grain sizes, the average percentage deviation was 21.3%, the maximum deviation being 37.3%; for large grain sizes, the average percentage deviation was 22.4%, the maximum deviation being 47.1%; and for the serial PDI approximation, the average percentage deviation was 30.5%, the maximum deviation being 52.9%. The minimum deviation for fully parallel PDI was 0% (cases where an optimal solution was found by it); for small grain sizes, it was 4.3%; for medium grain sizes, it was 6.3%; for large grain sizes, it was 8.6%; and for the serial approximation of PDI, the minimum deviation was 11.4%. A complete listing of the results of this study can be found in [7].

The results indicate that the dynamic interaction among subproblem instances provided by parallel processing is, in fact, central to the good performance of PDI. In this study, the solution quality of a strictly serial approximation of PDI (with no attempt to emulate parallel interaction) produced solutions averaging nearly 25% poorer than those produced by a fully parallel version of PDI. Although PDI can occasionally produce good solutions with restrictions on the level of parallel interaction, in general, a fully parallel implementation seems necessary to ensure generation of uniformly high quality solutions. Another observation is that the range of PDI solution quality across problem instances becomes wider, in general, as the grain size is increased, indicating that the confidence level for obtaining good solutions decreases with increasing grain sizes.

V. Discussion

From a parallel processing perspective, PDI is interesting because it appears to be inherently parallel in nature. The technique also represents a very nontraditional application of parallel processing, since the use of parallelism is not directly for purposes of com-

putational speedup. The empirical results have shown that PDI performance on an example application degrades substantially as the level of dynamic interaction is restricted. This supports the claim that PDI requires a parallel processing implementation to be fully effective (i.e. to produce highest quality results). Further empirical study of this issue is underway.

We believe that PDI shows considerable promise as a practical problem-solving technique. For the job-shop scheduling problem studied in this paper, it produced solutions which were on an average within 5.1% of the optimal solution. For 77% of the cases, it produced a solution equal to or better than that of any of the other comparative heuristics that were tried, while no other heuristic performed best in more than 23% of the cases. PDI should be applicable to any problem for which a reasonable partitioning into subproblem instances can be obtained.

PDI implementations have also been done for two other problems, namely flow-shop scheduling and vertex cover. For the flow-shop scheduling problem, PDI found the optimal solution for more than one-third of 40 example cases. The percentage deviation of its best solution from the optimal averaged 2.5%, with the maximum deviation being 10.9%. For the vertex cover problem, for 9 example graphs for which the optimal solution could be determined, PDI found the optimal solution for two-thirds of the cases. The deviation of its best solution from the optimal averaged 1.9%, with the maximum deviation being 7.7%. For 11 larger graph examples for which the optimal is unknown, PDI consistently produced better solutions than several heuristics, which included tightly and loosely bounded heuristics [6, 7]. Additionally, the computational cost of PDI for these problems was low. Complete details are provided in [5 - 7]. Several other exponentially hard problems exist for which such partitionings are possible. These include problems in areas such as VLSI placement and routing, logic minimization and testing.

References

[1] M. Garey and D. S. Johnson, *Computers and Intractability*, Freeman, San Francisco, 1979.

[2] S. French, *Sequencing and Scheduling: An Introduction to the Mathematics of the Job-Shop*, Ellis Horwood, Chichester, 1982.

[3] K. E. Baker, *Introduction to Sequencing and Scheduling*, Wiley, New York, 1974.

[4] J. Blazewicz et al, ''Scheduling Under Resource Constraints — Deterministic Models,'' *Annals of Operations Research*, Vol. 7 (1-4), 1986.

[5] I. Pramanick and J. G. Kuhl, ''An Inherently Parallel Method for Heuristic Problem Solving: Part I — General Framework,'' submitted to the *IEEE Transactions on Parallel and Distributed Systems*.

[6] I. Pramanick and J. G. Kuhl, ''An Inherently Parallel Method for Heuristic Problem Solving: Part II — Example Applications,'' submitted to the *IEEE Transactions on Parallel and Distributed Systems*.

[7] I. Pramanick, ''Ph.D. Thesis,'' *in preparation*, Dept. of Electrical and Computer Engg., The University of Iowa.

[8] T. Gonzalez and S. Sahni, ''Flowshop and Jobshop Schedules: Complexity and Approximation,'' *Operations Research*, Vol. 26(1), January-February, 1978, pp. 36 - 52.

[9] M. O'hEigeartaigh, J. K. Lenstra and A. H. G. Rinnooy Kan, eds., *Combinatorial Optimization: Annotated Bibliographies*, Wiley, New York, 1985.

The support of the University of Iowa Center for Advanced Studies is gratefully acknowledged.

A PARALLEL ALGORITHM FOR COMPUTING FOURIER TRANSFORMS ON THE STAR GRAPH

by

Paraskevi Fragopoulou and Selim G. Akl

Department of Computing & Information Science
Queen's University
Kingston, Ontario, Canada, K7L 3N6

Abstract

The n-Star graph, denoted by S_n, is one of the graph networks that have been recently proposed as attractive alternatives to the n-Cube topology for interconnecting processors in parallel computers. In this paper we present a parallel algorithm for the computation of the Fourier transform on the Star graph. The algorithm requires $O(n^2)$ multiply-add steps for an input sequence of $n!$ elements and to the best of our knowledge is the first algorithm for the computation of the Fourier transform on the Star graph.

Key Words and Phrases: Fast Fourier Transform, Parallel Algorithm, Interconnection Network, Star Graph, Cost Optimality.

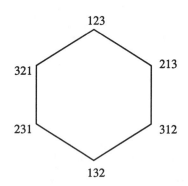

Figure 1: The 3-Star: A Hexagon.

1 Introduction

The Star graph was proposed in [Ake87] as an attractive alternative to the n-Cube (or hypercube) topology for interconnecting processors in parallel computers. It offers a network with fewer interconnection edges and smaller communication delays compared to the hypercube. Due to its rich symmetry the Star graph has various decomposition schemes.

In this paper we present a parallel algorithm for the computation of the Fourier transform on the Star graph. To the best of our knowledge this is the first such algorithm to be published. The algorithm is based on a method initially presented by Cooley and Tukey in 1964 [Coole]. We use a specific decomposition of the Star graph which combines two known decomposition schemes. These are the decomposition of the n-Star graph into n subgraphs S_{n-1}, and into $(n-2)!$ disjoint cycles of length $(n-1)n$ each. Applications of the Fourier transform, such as signal processing, image analysis, spectral analysis, speech processing, filter design etc., are described in [Brigh].

The paper is organized as follows. In section 2 we give a description of the Star graph as well as some basic properties. In section 3 we describe a scheme for decomposing the Star graph into disjoint cycles. In section 4 we present the Cooley-Tukey method. Our parallel algorithm appears in section 5, and in section 6 we give a brief description of

the Linear Array Fourier transform algorithm used in substeps of the main algorithm. Finally in the last section we analyze the algorithm and compare it with other parallel algorithms.

2 The Star graph

The n-Star graph, denoted by S_n, has $n!$ vertices corresponding to the $n!$ permutations of n distinct symbols. A vertex corresponding to permutation $a_1 a_2 \ldots a_{i-1} a_i a_{i+1} \ldots a_n$ is connected to those vertices corresponding to permutations $a_i a_2 \ldots a_{i-1} a_1 a_{i+1} \ldots a_n$ for $2 \leq i \leq n$, (i.e. those permutations which result from interchanging the first symbol in the permutation $a_1 a_2 \ldots a_{i-1} a_i a_{i+1} \ldots a_n$ with any of the remaining $n-1$ symbols). The edge connecting the original permutation to that permutation resulting from the interchange between the first and the ith symbols is called the ith dimension. In this way, every vertex is an endpoint of $n-1$ edges corresponding to the $n-1$ symbols that can be interchanged with the symbol in the first position of the associated permutation, as shown in Fig. 1 for the three symbols 1, 2 and 3 [Ake87].

As shown in [Ake87, Ake89], the n-Star graph enjoys a number of properties desirable in interconnection networks. These include regularity, vertex and edge symmetry, maximal fault tolerance, and strong resilience. Because of its rich symmetry, the graph is easily extensible, can be decomposed in various ways, and allows for simple routing algorithms. The graph was shown to be Hamiltonian in [Akl90, Nigam], and efficient sorting algorithms were developed for it in [Menn, Qiu90b].

---

[1]This research was supported by the Telecommunications Research Institute of Ontario and the Natural Sciences and Engineering Research Council of Canada.

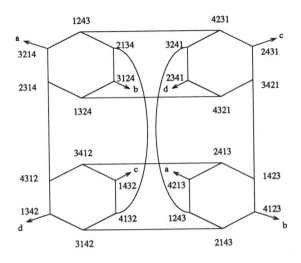

Figure 2: The 4-Star: 4 Interconnected 3-Stars.

In addition, the n-Star graph is superior to the n-Cube [Akl89] with respect to three key properties: total number of vertices, degree (number of edges at each vertex), and diameter (maximum distance between any two vertices). For any $n > 3$, the n-Star graph has more vertices than the n-Cube ($n!$ vertices versus 2^n), smaller degree (n-1 versus n), and smaller diameter ($\lfloor \frac{3(n-1)}{2} \rfloor$ versus n).

The n-Star graph can be decomposed into n subgraphs S_{n-1} by fixing each different symbol in one particular position 1 to n [Ake87]. If we fix a symbol in the last position we observe that there are $(n-1)!$ permutations that constitute an S_{n-1}. For example, if the symbol in the last position is held fixed with any symbol, then we get $(n-1)!$ permutations (i.e. an S_{n-1}) for every one of the n symbols. Thus the vertices of the S_n can be partitioned into n groups each containing $(n-1)!$ permutations and each being isomorphic to S_{n-1} as shown in Fig. 2.

3   A Decomposition scheme for the Star Graph

Recently, a new decomposition scheme was proposed in [Qiu90a], which partitions the n-Star graph into $(n-2)!$ disjoint cycles of length $(n-1)n$ each.

Each cycle can be defined by its initial permutation. We now show how, given the initial permutation of a cycle, the other permutations of the cycle can be generated in the right order. Assume that the initial permutation of a cycle in S_n is denoted by $1a_2 a_3...a_{n-1}n$. Then the other permutations can be generated by visiting dimensions a_2, a_3, ..., a_{n-1}, n repeatedly until we reach again the initial permutation of the cycle, $1a_2 a_3...a_{n-1}n$. For example, the cycle generated by the initial permutation 1324 in S_4 consists of the following permutations:

$$1324, 2314, 3214, 4213, 1243, 2143,$$
$$3142, 4132, 1432, 2431, 3421, 4321$$

We now give a recursive method that generates the starting permutations of the cycles in the Star graph. If we have an initial permutation used to generate a cycle in S_{n-1} denoted by $1a_1 a_2...a_{n-2}(n-1)$ then in S_n the permutations used to generate cycles can be produced by cyclically shifting to the right the previous permutation (without the initial unity) and adding the symbol n to the end of each resulting permutation. The resulting permutations are:

$$1a_2 a_3...a_{n-2}(n-1)n$$
$$1a_3...a_{n-2}(n-1)a_2 n$$
$$.........................$$
$$1a_i...a_{n-2}(n-1)a_2...a_{i-1}n$$
$$.........................$$
$$1(n-1)a_2 a_3...a_{n-2}n$$

The cycles resulting from the above construction can be easily shown to be disjoint and to consist of $(n-1)n$ permutations.

4   A Decomposition Scheme for the Fourier Transform

Given a sequence of numbers $\{x(1), x(2),...,x(N)\}$, its Fourier transform (FT) is the sequence resulting from evaluating the expression:

- $X(j) = \sum_{k=1}^{N} x(k)w^{(k-1)(j-1)}$ for $j = 1, 2, ..., N$,

where w is a primitive N^{th} root of unity, i.e. $w = e^{2\pi i/N}$ with $i = \sqrt{-1}$.

Suppose N, the number of input elements, is composite with only two factors $N = r_1 \times r_2$. If we arrange the N elements into an $r_1 \times r_2$ array using row major order, the Fourier transform can be written as:

- $X(j_1, j_0) = \sum_{k_0} \sum_{k_1} x(k_1, k_0)w^{j[(k_1-1)r_2+k_0]}$
  $= \sum_{k_0} \sum_{k_1} x(k_1, k_0)w^{j(k_1-1)r_2} w^{jk_0}$

where $x(k_1, k_0)$ refers to the element in row k_1 and column k_0,

$$j = (j_1 - 1)r_1 + j_0, \qquad j_0, k_1 = 1, 2, ..., r_1,$$
$$k = (k_1 - 1)r_2 + k_0, \quad \text{and} \quad j_1, k_0 = 1, 2, ..., r_2.$$

Since,

$$w^{(j_1-1)(k_1-1)r_1 r_2} = w^{(j_1-1)(k_1-1)N} = \left(e^{2\pi i}\right)^{(j_1-1)(k_1-1)} = 1$$

we have:

- $X(j_1, j_0) = \sum_{k_0} w^{j_1 k_0 r_1}[w^{j_0 k_0} \sum_{k_1} x(k_1, k_0)w^{j_0(k_1-1)r_2}]$

According to this decomposition method, the Fourier transform can be computed in three steps [Coole]:

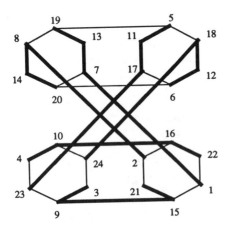

Figure 3: Star Graph FFT Algorithm. Step 1: After the elements have been arranged initially as shown, this step computes 4-length FT's among the elements connected by boldface lines.

1. $X'(j_0, k_0) = \sum_{k_1} x(k_1, k_0) w^{j_0(k_1-1)r_2}$

   (Fourier transform of each column of the array. Notice that there is a constant factor r_2 involved in the exponent of w.)

2. $X''(j_0, k_0) = w^{j_0 k_0} X'(j_0, k_0)$
   (Local multiplications.)

3. $X(j_1, j_0) = \sum_{k_0} X''(j_0, k_0) w^{(j_1-1)k_0 r_1}$

   (Fourier transform of each row of the array. Notice that there is a constant factor r_1 involved in the exponent of w.)

It is important to emphasize that at the end of the execution of the algorithm the resulting matrix X is transposed.

This approach to computing the FT is referred to as a Fast Fourier Transform (FFT) algorithm. It requires Nr_1 operations to obtain X'', and Nr_2 operations to calculate X from X'', for a total of $N(r_1 + r_2)$ operations. We can see that successive applications of this algorithm, when N can be decomposed into m factors $N = r_1 \times r_2 \times ... \times r_m$, give an m-step algorithm requiring $N(r_1 + r_2 + ... + r_m)$ operations. When N is a power of 2 then $r_i = 2$ for every i and we get the classical and more well known version of the FFT algorithm requiring $NlogN$ operations [Cochr]. Another example of particular interest in this paper is when $N = 1 \times 2 \times 3 \times ... \times n$, in which case the number of operations required is $O(Nn^2)$. For this value of N, an optimal parallel implementation would have a *processor × time* product of $O(Nn^2)$.

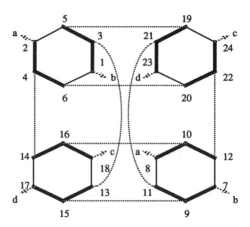

Figure 4: Star Graph FFT Algorithm. Step 2: After rearrangement over the 4-Star connections (dotted lines) at the end of the previous step, this step computes 3-length FT's among the elements connected by boldface lines.

5 The Fourier Transform on the Star Graph

We are given a sequence of numbers $\{x(1), x(2), ..., x(n!)\}$, and $n!$ processors interconnected in an n-Star topology with a fixed indexing of the processors P_i, $1 \le i \le n!$, such that each processor holds one element of the sequence. An algorithm is now presented which computes the Fourier transform of the sequence on the Star graph.

In what follows when we refer to the *rearrangement of elements over the ith dimension*, we mean that all the vertices connected through the ith dimension exchange their elements.

5.1 Mapping of the Fourier Transform Decomposition to the Star Graph

Assume that we want to compute the Fourier transform of a sequence of length N, where $N=n!$, on the Star graph S_n, using the decomposition method described above. From its definition, $n!$ can be written as:

$$N = n! = 1 \times 2 \times 3 \times 4 \times ... \times (n-1) \times n$$

In other words, N is a number composed of n factors. If we denote the ith factor by r_i and exclude the initial unity, then $n!$ can be written as:

$$N = n! = r_2 \times r_3 \times r_4 \times ... \times r_{(n-1)} \times r_n$$

The Fourier transform that we use is based on a decomposition of N, the number of elements, into $n-1$ factors [Coole]. So the $n!$-length Fourier transform can be computed as:

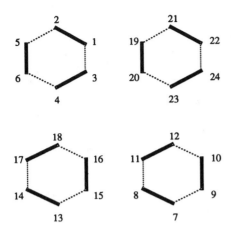

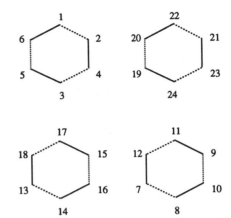

Figure 5: Star Graph FFT Algorithm. Step 3: After rearrangement over the 3-Star connections (dotted lines) at the end of the previous step, this step computes 2-length FFT's among the elements connected by boldface lines.

Figure 6: Star Graph FFT Algorithm. Final arrangement of the elements after the rearrangement over the 2-Star connections (dotted lines) at the end of step 3.

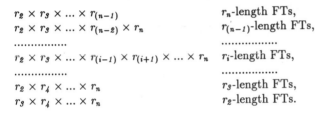

$$r_2 \times r_3 \times ... \times r_{(n-1)} \qquad r_n\text{-length FTs,}$$
$$r_2 \times r_3 \times ... \times r_{(n-2)} \times r_n \qquad r_{(n-1)}\text{-length FTs,}$$
$$.................... \qquad ....$$
$$r_2 \times r_3 \times ... \times r_{(i-1)} \times r_{(i+1)} \times ... \times r_n \qquad r_i\text{-length FTs,}$$
$$.................... \qquad ....$$
$$r_2 \times r_4 \times ... \times r_n \qquad r_3\text{-length FTs,}$$
$$r_3 \times r_4 \times ... \times r_n \qquad r_2\text{-length FTs.}$$

step	number of cycles	length of cycle
1	$(n-2)!$	$(n-1)n$
2	$(n-3)! \times n$	$(n-2) \times (n-1)$
3	$(n-4)! \times (n-1) \times n$	$(n-3) \times (n-2)$
⋮	⋮	⋮
i	$(n-i-1)! \times (n-i+2) \times$ $... \times n$	$(n-i) \times$ $(n-i+1)$
⋮	⋮	⋮
$n-2$	$4 \times 5 \times ... \times (n-1) \times n$	2×3
$n-1$	$3 \times 4 \times ... \times (n-1) \times n$	1×2

Table 1: The n-Star viewed as sets of disjoint cycles of different lengths.

According to the decomposition described in [Qiu90a] and the known decomposition of S_n into n subgraphs S_{n-1} [Ake87], we can easily see that S_n can be decomposed into $(n-2)!$ disjoined cycles of length $(n-1)n$, or into $(n-3)!n$ disjoined cycles of length $(n-2)(n-1)$, or into $(n-4)!(n-1)n$ disjoined cycles of length $(n-3)(n-2)$, etc.

If we adopt the previous notation we can see that in the Star graph there are:

number of disjoined cycles:	length of cycle:
$r_2 \times r_3 \times ... \times r_{(n-2)}$	$r_{(n-1)} \times r_n$
$r_2 \times r_3 \times ... \times r_{(n-3)} \times ... \times r_n$	$r_{(n-2)} \times r_{(n-1)}$
$.................$	$.................$
$r_2 \times r_3 \times ... \times r_{(i-2)} \times r_{(i+1)} \times ... \times r_n$	$r_{(i-1)} \times r_i$
$.................$	$.................$
$r_4 \times r_5 \times ... \times r_n$	$r_2 \times r_3$
$r_3 \times r_4 \times ... \times r_n$	r_2

5.2   Description of the Algorithm

The algorithm consists of $n-1$ steps as shown in Table 1. In step i, $(n-i+1)$-length Fourier transforms are computed among $(n-i+1)$ adjacent processors, on the disjoint cycles of the Star graph, starting in each cycle from the vertex represented by its initial permutation. Note that in step i, $(n-i)$ sets of $(n-i+1)$ adjacent processors form a cycle of length $(n-i)(n-i+1)$.

In order to always have the appropriate elements in adjacent positions in the cycles of the Star graph, a special initial arrangement of the elements is needed and every Fourier transform step is separated from the next by a rearrangement of the elements. Thus step i is followed by a transfer of the elements over the $(n-i+1)^{st}$ dimension of the Star graph. If we imagine that the elements are arranged into a $2 \times 3 \times ... \times n$, $(n-1)$-dimensional array and indexed in row major order, this rearrangement is needed in step i in order to bring together $(n-i+1)$ elements which are adjacent over dimension $(n-i+1)$. For example in the two dimensional case, this rearrangement corresponds to a change from columns to rows.

We are now ready to give the algorithm for computing the FT on the Star graph in a more compact form.

After the initial loading of the elements $\{x(1), ..., x(n!)\}$ into the processors of the Star graph in the right order, described in the next section, the algorithm essentially repeats three main steps $n-1$ times as follows:

III-103

1234	1243	1342	2341
2134	2143	3142	3241
1324	1423	1432	2431
3124	4123	4132	4231
2314	2413	3412	3421
3214	4213	4312	4321

Table 2: Ordering of the permutations given by the algorithm in section 5.3.

4123	(1)	3124	(7)	2134	(13)	1234	(19)
4213	(2)	3214	(8)	2314	(14)	1324	(20)
4132	(3)	3142	(9)	2143	(15)	1243	(21)
4312	(4)	3412	(10)	2413	(16)	1423	(22)
4231	(5)	3241	(11)	2341	(17)	1342	(23)
4321	(6)	3421	(12)	2431	(18)	1432	(24)

Table 3: The initial ordering of the vertices of the 4-Star graph for the Fourier transform algorithm.

For $i = 1, 2, ..., n - 1$ do

1. Perfom Fourier transforms of length $(n - i + 1)$ on groups of $(n - i + 1)$ processors that are adjacent on the $(n - i)(n - 1 + 1)$-length cycles starting from the initial node of each cycle, as defined in the decomposition scheme in [Qiu90a].

2. Multiply each element locally by an appropriate coefficient.

3. Rearrange the elements over the $(n - i + 1)^{st}$ dimension.

The steps of the algorithm for the 4-Star are illustrated in Figs. 3-6.

5.3   Ordering of the Processors

In this section we give an ordering for the vertices of the n-Star graph and a function which maps this ordering to the positive integers. This ordering allows input and output of the elements in an appropriate manner.

We start from the sorted permutation and exchange each of the symbols (starting with the one in the second position) with all previous elements (starting with the element immediate preceding it). Each time we exchange a symbol in position i with one of its predecessors, we start again with the exchange of the second symbol, and continue until we reach symbol i again.

If we denote by q_{ij} the operator that exchanges symbols a_i with a_j ($i > j$) in a permutation, then this ordering scheme can be described by the following recursive algorithm. If we start from the sorted permutation $1234...n$, and for the l-Star graph the sequence of operators that produce the permutations in the right order is denoted by A_l, we have:

$$A_2 = q_{21}$$
$$A_3 = A_2 q_{32} A_2 q_{31} A_2$$
$$\vdots$$
$$A_n = A_{n-1} q_{n(n-1)} A_{n-1} q_{n(n-2)} ... A_{n-1} q_{n1} A_{n-1}$$

where it is assumed that a new permutation is obtained by applying each operator q_{ij} to the permutation generated $(i - 1)!$ steps earlier in the current sequence.

Assume that the symbols used in the permutations are the first n positive integers, and that the permutations

are listed in the order obtained from the above algorithm. For example, for $n=4$ this ordering of the permutations is given in Table 2 (in column major order). We now describe a function which maps the permutations into the first $n!$ positive integers. This mapping is needed in order for each processor to be able to calculate from the permutation it represents, the index of the element it should read in the first phase of the algorithm. This initial arrangement of the elements, followed by rearrangement during the algorithm, allows the elements to be reported in the right order (i.e. the order in Table 2) when the algorithm terminates.

If $a_1 a_2 ... a_n$ is a permutation then for each symbol a_i of the permutation we define π_i to be:

$$\pi_i = \left| a_i - i - \sum_{j=i+1}^{n} [a_i > a_j] \right| (i - 1)! \quad \text{for } 2 \le i \le n$$

where $[a_i > a_j]$ equals 1 when $a_i > a_j$ and 0 otherwise.

Then the number of permutation $a_1 a_2 ... a_n$ is:

$$\Pi = 1 + \pi_2 + \pi_3 + ... + \pi_n$$

This mapping function resembles the one appearing in [Akl89] for the ordering of permutations in lexicographic order.

This ordering leads to an appropriate entry of the elements. Initially, we want the processor indexed by the permutation $a_1 a_2 a_3 ... a_n$ to have the element whose position results from the number of the permutation $a_2 a_3 ... a_n a_1$, obtained from the mapping function. So we have to shift once to the left the permutation representation of each vertex and then apply the mapping function. For example, for $n = 4$ Table 3 shows the initial ordering of the vertices (in column major order), along with the index of the element each processor contains (see Fig. 3). Subsequently, the elements are rearranged during the execution of the algorithm as illustrated in Figs. 4, 5 and 6. Therefore, when the algorithm terminates, each processor has the element indexed by the number of its permutation obtained from the mapping function.

The main observation is that if we input the elements in the Star graph according to the ordering described in this section, and if we assume that the $n!$ elements are arranged in a $2 \times 3 \times ... \times n$, $(n - 1)$-dimensional array indexed in row major order, then after every rearrangement of the elements over dimensions n, $n-1$,..., 3, 2 of

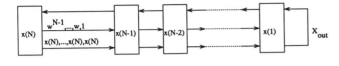

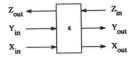

Figure 7: The Linear Systolic Array for the Fourier Transform.

the Star graph, elements that are adjacent in the $(n-1)$-dimensional array over dimensions n, n-1,..., 3, 2 appear in adjacent processors of the Star graph that form the disjoint cycles.

6 The Fourier Transform on the Linear Array

The algorithm we use in each of the substeps of the previous algorithm to compute the Fourier transforms of lengths $r_n, r_{(n-1)}, ..., r_2$ is the Linear Array algorithm which we briefly describe in this section.

The N-point Fourier transform of an input sequence $\{x(1), x(2), ..., x(N)\}$ can be viewed as the evaluation of the following polynomial:

- $x(N)a^{N-1} + x(N-1)a^{N-2} + ... + x(2)a + x(1)$

at $a = 1, w, w^2, ..., w^{N-1}$, which by Horner's rule can be written as:

$$(...((x(N)a + x(N-1))a + x(N-2))a + ...x(2))a + x(1)$$

This algorithm can be executed by a Linear Array of N multiply-add cells as shown in Fig. 7 [Zhang]. The basic cell of this array can be seen in the same figure and its function is:

1. $Z_{out} = Z_{in}$

2. $Y_{out} = Y_{in}$

3. $X_{out} = X_{in} Y_{in} + x$

The X_{out} that come out of the rightmost cell are the FT of the input sequence. All the X_{out}'s are computed in a pipeline fashion in $O(N)$ multiply-add steps.

This algorithm can be used for the substeps of the main algorithm presented in this paper. In each of the substeps the processors that must perform Fourier transforms are connected in a Linear Array. The only difference is that the FT's that are output by the rightmost cell must be fed back to the Linear Array.

7 Conclusions

Because the linear array algorithm used to compute the Fourier transform runs in $O(k)$ multiply-add steps for an input sequence of length k, the algorithm of section 5 runs in:

$$r_n + r_{n-1} + ... + r_3 + r_2$$

multiply-add steps, and since for each i, $r_i \leq n$ the algorithm runs in $O(n^2)$ multiply-add steps for an input sequence of length $n!$. Input sequences of length other than $n!$ can be clearly be padded with zeros to form sequences of $n!$ elements.

As pointed out in section 4 the running time of the sequential algorithm is $N(2 + 3 + ... + n) = O(Nn^2)$ when $N = 2 \times 3 \times ... \times n$. The cost of our parallel implementation on the Star graph, i.e. the *processor $\times$ time* product, is $O(Nn^2)$. This means that the parallel algorithm is *cost optimal* with respect to the sequential algorithm on which it is based.

Compared with other parallel algorithms for the Fourier transform, this algorithm is quite efficient because it uses the same number of processors as the number of input elements and its running time is $O(n^2)$ for an input sequence of length $n!$.

The best known parallel algorithms for the Fast Fourier transform are the algorithms on the hypercube [Prepa] and the Perfect Shuffle [Stone]. Both of these networks use the same number of processors as the number of input elements. Their running time is $O(logN)$ multiply-add steps for an input sequence of length N. When $N = n!$, these algorithms run in $O(logn!) = O(nlogn)$ multiply-add steps. However, these networks are useful only when the number of input elements is a power of 2.

The Fourier transform on networks like the Omega Network [Parke], the Indirect Binary n-Cube [Pease, Parke], the R network [Parke], the Butterfly network [Thomp] and the Cube-Connected-Cycles (CCC) [Parke, Prepa], need the same number of multiply-add steps, i.e. $O(nlogn)$ for $n!$ elements, but use a larger number of processors, namely $O(n!logn!)$. The same holds for the Mesh of Trees FT [Akl89] with $O((n!)^2)$ processors. These networks also require the number of input elements to be a power of 2.

Algorithms like the Linear Array Multiplication [Akl89], the Linear Array FT [Thomp], the Linear Systolic Array [Zhang] and the Cascade FT implementation [Wold] which use the same number of processors as the number of input elements, involve an $O(n!)$ factor in their running time.

References

[Ake87] S. B. Akers, D. Harel, B. Krishnamurthy. *The Star Graph: An Attractive alternative to the n-Cube*. Proceedings of the International Conference in Parallel Processing, pp. 393-400, 1987.

[Ake89] S. B. Akers, B. Krishnamurthy. *A Group-Theoretic Model for Symmetric Interconnection Networks,* IEEE Transactions on Computers, Vol. 38, No. 4, April 1989.

[Akl89] S. G. Akl. *The Design and Analysis of Parallel Algorithms,* Prentice Hall, Englewood Cliffs, New Jersey 1989.

[Akl90] S. G. Akl, J. Duprat, A. G. Ferreira. *Building Hamiltonian Circuits and Paths in Star Graphs,* Technical Report, Laboratoire de l' Informatique du Parallelisme, Ecole Normale Superieure de Lyon, Lyon, France, March 1990.

[Brigh] E. Oran Brigham. *The FFT and its Applications,* Prentice-Hall, 1988.

[Cochr] W. T. Cochran, J. W. Cooley, D. L. Favin, H. D. Helms, R. A. Kaenel, W. W. Lang, G. C. Maling, D. E. Nelson, C. M. Rader, P. D. Welch. *What is the Fast Fourier Transform?* IEEE Transactions on Audio and Electroacoustics, Vol. AU-15, No. 2, June 1967.

[Coole] J. W. Cooley, J. W. Tukey. *An Algorithm for the Machine Calculation of Complex Fourier Series,* Math. of Comput., Vol. 19, pp. 297-301, April 1965.

[Menn] A. Menn, A. K. Somani. *An Efficient Sorting Algorithm for the Star Graph Interconnection Network,* Proceedings of the International Conference on Parallel and Distributed Processing, St. Charles, Illinois, August 1990, Vol. III, pp. 1-8.

[Nigam] M. Nigam, S. Sahni, B. Krishnamurthy. *Embedding Hamiltonians and Hypercubes in Star Interconnection Graphs,* Proceedings of the International Conference on Parallel and Distributed Processing, St. Charles, Illinois, August 1990, Vol. III, pp. 340-343.

[Parke] D. S. Parker. *Notes on Shuffle/Exchange-Type Switching Networks,* IEEE Transactions on Computers, Vol. C-29, No. 3, pp. 213-223, March 1980.

[Pease] M. C. Pease. *The Indirect Binary n-Cube Microprocessor Array,* IEEE Transactions on Computers, Vol. C-26, No. 5, pp. 458-473, May 1977.

[Prepa] F. P. Preparata, J. Vuillemin. *The Cube-Connected Cycles,* Communications of the ACM, Vol. 24, No. 5, pp. 300-309, May 1981.

[Qiu90a] K. Qiu, H. Meijer, S. G. Akl. *Decomposing a Star Graph into Disjoint Cycles,* Proceedings of the Second Canadian Conference on Computational Geometry, Ottawa, August 1990, pp. 70-73. TR.90-278, Department of Computing and Information Science, Queen's University at Kingston.

[Qiu90b] K, Qiu, H. Meijer, S. G. Akl. *A Parallel Sorting Algorithm on the Star Graph,* Technical Report No. 90-286, Department of Computing and Information Science, Queen's University, Kingston, Ontario, Canada, September 1990.

[Stone] H. S. Stone. *Parallel Processing with the Perfect Shuffle,* IEEE Transactions on Computers, Vol. C-20, No. 2, pp. 153-161, February 1971.

[Thomp] C. D. Thompson. *Fourier Transforms in VLSI,* IEEE Transactions on Computers, Vol. C-32, No. 11, pp. 1047-1057, November 1983.

[Wold] E. H. Wold, A. Despain. *Pipeline and Parallel-Pipeline FFT Processors for VLSI Implementations,* IEEE Transactions on Computers, Vol. C-33, No. 5, pp. 414-426, May 1984.

[Zhang] C. Nian Zhang, D. Y. Y. Yun. *Multi-Dimensional Systolic Networks for Discrete Fourier Transform,* The 11th Annual Int. Symp. on Comp. Arch., June 5-7, 1984, Ann Arbor, Michingan, pp. 215-222.

A Parallel Algorithm for the PROFIT/COST Problem

Yijie Han

Department of Computer Science
University of Kentucky
Lexington, KY 40506

Abstract

We present a parallel algorithm for the PROFIT/COST problem with time complexity $O(\log n)$ using a linear number of processors on an EREW PRAM. The design of this algorithm employs both the derandomization technique and the pipeline technique. The algorithm can be used to partition the vertices of a graph into two sets such that the number of edges incident with vertices in both sets is at least half of the total number of edges in the graph. Parallel algorithms for the PROFIT/COST problem have known applications in the design of parallel algorithms for several graph problems.

Keywords: Design of algorithms, parallel algorithms, graph algorithms, data structures, derandomization.

1 Introduction

The *PROFIT/COST* problems as formulated by Luby[6] can be described as follows.

Let $\vec{x} = < x_i \in \{0,1\} : i = 0, ..., n-1 >$. Each point $\vec{x}$ out of the 2^n points is assigned probability $1/2^n$. Given function $B(\vec{x}) = \sum_{i,j} f_{i,j}(x_i, x_j)$, where $f_{i,j}$ is defined as a function $\{0,1\}^2 \to \mathcal{R}$. The PROFIT/COST problem is to find a good point $\vec{y}$ such that $B(\vec{y}) \geq E[B(\vec{x})]$. B is called the BENEFIT function and $f_{i,j}$'s are called the PROFIT/COST functions.

The size m of the problem is the number of PROFIT/COST functions present in the input. The input is dense if $m = \theta(n^2)$ and is sparse if $m = o(n^2)$.

The vertex partition problem is a basic problem which can be modeled by the PROFIT/COST problem[6]. The vertex partition problem is to partition the vertices of a graph into two sets such that the number of edges incident with vertices in both sets is at least half of the number of edges in the graph. Let $G = (V, E)$ be the input graph. $|V|$ 0/1-valued uniformly distributed mutually independent random variables are used, one for each vertex. The problem of partitioning vertices into two sets is now represented by the 0/1 labeling of the vertices. Let x_i be the random variable associated with vertex i. For each edge $(i,j) \in E$ a function $f(x_i, x_j) = x_i \oplus x_j$ is defined, where $\oplus$ is the exclusive-or operation. $f(x_i, x_j)$ is 1 iff edge (i,j) is incident with vertices in both sets. The expectation of f is $E[f(x_i, x_j)] = (f(0,0) + f(0,1) + f(1,0) + f(1,1))/4 = 1/2$. Thus the BENEFIT function $B(x_0, x_1, ..., x_{|V|-1}) = \sum_{(i,j) \in E} f(x_i, x_j)$ has expectation $E[B] = \sum_{(i,j) \in E} E[f(x_i, x_j)] = |E|/2$. If we find a good point p in the sample space such that $B(p) \geq E[B] = |E|/2$, this point p determines the partition of vertices such that the number of edges incident with vertices in both sets is at least $|E|/2$.

The PROFIT/COST problem is a basic problem in the study of derandomization, *i.e.*, converting a randomized algorithm to a deterministic algorithm. The importance of the PROFIT/COST problem lies in the fact that it can be used as a key building block for the derandomization of more complicated randomized algorithms[2][7]. Recently we have succeeded in applying PROFIT/COST algorithms in the derandomization of Luby's randomized algorithms for the $\Delta + 1$ vertex coloring problem, the maximal independent set problem and the maximal matching problem and obtained more efficient deterministic algorithms for the three problems[2]. These results are summarized in Table 1.

Luby[6] gave a parallel algorithm for the PROFIT/COST problem with time complexity $O(\log^2 n)$ using a linear number of processors on the EREW PRAM model[9]. He used a sample space with $O(n)$ sample points and designed $O(n)$

Problem	Time Complexity	Processor Complexity	Model	Reference
$\Delta +1$ vertex coloring	$O(\log^3 n \, \log\log n)$	$O(m+n)$	CREW	6,7
	$O(\log^3 n)$	$O(m+n)$	CREW	3
	$O(\log^2 n \, \log\log n)$	$O((m+n)/\log\log n)$	CREW	2
	$O(\log^2 n)$	$O(mn^\varepsilon)$	CREW	2
Maximal independent set	$O(\log^3 n)$	$O((m+n)/\log n)$	EREW	1
	$O(\log^2 n)$	$O(mn^2)$	EREW	5
	$O(\log^2 n)$	$O(n^{2.376})$	CREW	2
	$O(\log^{2.5} n)$	$O((m+n)/\log^{0.5} n)$	EREW	2
	$O(\log^{2.5} n)$	$O((m+n)/\log^{1.5} n)$	CREW	2
Maximal matching	$O(\log^3 n)$	$O(m+n)$	CRCW	4
	$O(\log^{2.5} n)$	$O((m+n)/\log^{0.5} n)$	EREW	2
	$O(\log^2 n)$	$O(n^{2.376})$	CREW	2

Table 1.

uniformly distributed pairwise independent random variables on the sample space. His algorithm was obtained through a derandomization process in which a good sample point is found by a binary search of the sample space.

A set of n 0/1-valued uniformly distributed pairwise independent random variables can be designed on a sample space with $O(n)$ points[6]. Let $k = \lceil \log n \rceil$. The sample space is $\Omega = \{0, 1\}^{k+1}$. For each $a = a_0 a_1 ... a_k \in \Omega$, $Pr(a) = 2^{-(k+1)}$. The value of random variables x_i, $0 \le i < n$, on point a is $x_i(a) = (\sum_{j=0}^{k-1}(i_j \cdot a_j) + a_k) \bmod 2$, where i_j is the j-th bit of i starting with the least significant bit. It is not difficult to verify that x_i's are the desired random variables. Because $B(x_0, x_1, ..., x_{n-1}) = \sum_{i,j} f_{i,j}(x_i, x_j)$, where $f_{i,j}$ depends on two random variables, pairwise independent random variables can be used in place of the mutual independent random variables. A good point can be found by searching the sample space. Luby's scheme[6] uses binary search which fixes one bit of a at a time (therefore partitioning the sample space into two subspaces) and evaluates the conditional expectations on the subspaces. His algorithm[6] is shown below.

Algorithm Convert:
for $l := 0$ **to** k

begin

$B_0 := E[B(x_0, x_1, ..., x_{n-1}) \mid$
$\qquad a_0 = r_0, ..., a_{l-1} = r_{l-1}, a_l = 0];$
$B_1 := E[B(x_0, x_1, ..., x_{n-1}) \mid$
$\qquad a_0 = r_0, ..., a_{l-1} = r_{l-1}, a_l = 1];$
if $B_0 \ge B_1$ **then** $a_l := 0$ **else** $a_l := 1;$
/*The value for a_l decided above is denoted by r_l. */
end
output$(a_0, a_1, ..., a_k);$

Since each time the sample space is partitioned into two subspaces the subspace with larger expectation is preserved while the other subspace is discarded, the sample point $(a_0, a_1, ..., a_k)$ found must be a good point, i.e., the value of B evaluated at $(a_0, a_1, ..., a_k)$ is $\ge E[B]$.

By the linearity of expectation, the conditional expectation evaluated in the above algorithm can be written as $E[B(x_0, x_1, ..., x_{n-1}) \mid a_0 = r_0, ..., a_l = r_l] = \sum_{i,j} E[f_{i,j}(x_i, x_j) \mid a_0 = r_0, ..., a_l = r_l]$. It is assumed[6] that constant operations(instructions) are required for a single processor to evaluate $E[f_{i,j}(x_i, x_j) \mid a_0 = r_0, ... a_l = r_l]$. Algorithm Convert uses a linear number of processors and $O(\log^2 n)$ time on the EREW PRAM model.

III-108

Recently Han and Igarashi gave a CREW PRAM algorithm for the PROFIT/COST problem with time complexity $O(\log n)$ using a linear number of processors[3]. They used a sample space of $O(2^n)$ points. The problem is also solved by locating a good point in the sample space. They obtained time complexity $O(\log n)$ by exploiting the redundancy of a shrinking sample space and the mutual independence of random variables. In the next section we shall give a brief review of their algorithm.

In this paper we give an EREW parallel algorithm for the PROFIT/COST problem with time complexity $O(\log n)$ using a linear number of processors. The EREW algorithm is obtained by using a pipeline to remove the concurrent read feature from Han and Igarashi's algorithm[3] and by building row and column trees to help disseminating the bit setting information. The pipeline used has inevitably made our EREW algorithm rather complicated. Since EREW PRAM is weaker than the CREW PRAM, the introduced complication is worthwhile for removing the concurrent read feature from Han and Igarashi's algorithm[3]. In section 3.1 we will first present an EREW algorithm for the dense case. In section 3.2 the algorithm is then converted for the general case by using row and column trees.

2 Preliminaries

In [3] Han and Igarashi formulated the PROFIT/COST problem as a tree contraction problem[8]. Assuming without loss of generality n is a power of 2. n 0/1-valued uniformly distributed mutually independent random variables r_i, $0 \leq i < n$, are used. A *Random variable tree* T is built for $\vec{x}$. T is a complete binary tree with n leaves plus a node which is the parent of the root of the complete binary tree (thus there are n interior nodes in T and the root of T has only one child). The n variables x_i, $0 \leq i < n$, are associated with n leaves of T and the n random variables are associated with the interior nodes of T. The n leaves of T are numbered from 0 to $n-1$. Variable x_i is associated with leaf i.

Variables x_i, $0 \leq i < n$, are randomized as follows. Let $\vec{r} = < r_i : i = 0, ..., n-1 >$ and let $r_{i_0}, r_{i_1}, ..., r_{i_{\log n}}$ be the random variables on the path from leaf i to the root of T. Random variable x_i is defined to be $x_i(\vec{r}) = (\sum_{j=0}^{\log n - 1} i_j \cdot r_{i_j} + r_{i_{\log n}}) \bmod 2$, where i_j is the j-th bit of i starting with the least significant bit. We now show that random variables x_i are mutually independent random variables.

Lemma 1: Random variables x_i, $0 \leq i < n$, are uniformly distributed mutually independent random variables.
Proof: By flipping the random bit at the root of the random variable tree we see that each x_i is uniformly dis-

tributed in $\{0, 1\}$. To show the mutual independence we note that the mapping $m(\vec{r}) \mapsto \vec{x}$, where $x_i = (\sum_{j=0}^{\log n - 1} i_j \cdot r_{i_j} + r_{i_{\log n}}) \bmod 2$, is a one to one mapping. Thus $Pr(x_{i_1} = a_1, x_{i_2} = a_2, ..., x_{i_k} = a_k) = 2^{n-k}/2^n = 1/2^k = Pr(x_{i_1} = a_1) Pr(x_{i_2} = a_2) \cdots Pr(x_{i_k} = a_k)$. $\square$

Due to the linearity of expectation we have $E[B(\vec{x})] = \sum_{i,j} E[f_{i,j}(x_i, x_j)] = \sum_{i,j} E[f_{i,j}(x_i(\vec{r}), x_j(\vec{r}))] = E[B(\vec{x}(\vec{r}))]$. The problem now is to find a sample point $\vec{r}$ such that $B|_{\vec{r}} \geq E[B] = \frac{1}{4} \sum_{i,j} (f_{i,j}(0,0) + f_{i,j}(0,1) + f_{i,j}(1,0) + f_{i,j}(1,1))$.

Han and Igarashi's algorithm[3] fixes random variables r_i (setting their values to 0's and 1's) one level in a step starting from the level next to the leaves (level 0) and going upward on the tree T until level $\log n$. Since there are $\log n + 1$ interior levels in T all random variables will be fixed in $\log n + 1$ steps.

Let random variable r_i at level 0 be the parent of the random variables x_i and $x_{i\#0}$ in the random variable tree, where $i\#j$ is a number obtained by complementing the j-th bit of i. r_i will be fixed as follows. Compute $f_0 = f_{j,j\#0}(0,0) + f_{j,j\#0}(1,1) + f_{j\#0,j}(0,0) + f_{j\#0,j}(1,1)$ and $f_1 = f_{j,j\#0}(0,1) + f_{j,j\#0}(1,0) + f_{j\#0,j}(0,1) + f_{j\#0,j}(1,0)$. Note that here $f_k = 2E[f_{j,j\#0}(x_j, x_{j\#0}) + f_{j\#0,j}(x_{j\#0}, x_j)|r_i = k]$, $k = 0, 1$. If $f_0 \geq f_1$ then set r_i to 0 else set r_i to 1. Thus the subspace with larger expectation remains while the subspace with smaller expectation is discarded. All random variables at level 0 will be fixed in parallel in constant time using n processors. The fixing results in a smaller space with higher expectation for B. Therefore this smaller space contains a good point.

If r_i is set to 0 then $x_i = x_{i\#0}$, if r_i is set to 1 then $x_i = 1 - x_{i\#0}$. Therefore after r_i is set, x_i and $x_{i\#0}$ can be combined. The n random variables x_i, $0 \leq i < n$, can be reduced to $n/2$ random variables. PROFIT/COST functions $f_{i,j}$, $f_{i\#0,j}$, $f_{i,j\#0}$, and $f_{i\#0,j\#0}$ can also be combined into one function. It can be checked that the combining can be done in constant time using a linear number of processors.

During the combining process variables x_i and $x_{i\#0}$ are combined into a new variable $x_{\lfloor i/2 \rfloor}^{(1)}$, functions $f_{i,j}$, $f_{i\#0,j}$, $f_{i,j\#0}$, and $f_{i\#0,j\#0}$ are combined into a new function $f_{\lfloor i/2 \rfloor, \lfloor j/2 \rfloor}^{(1)}$. After combining a new function $B^{(1)}$ is formed which has the same form of B but has only $n/2$ variables. As we stated above, $E[B^{(1)}] \geq E[B]$.

What we have explained is the first step of the algorithm in [3]. This step takes constant time using a linear number of processors. After k steps the random variables at levels 0 to $k-1$ in the random variable tree are fixed, the n random variables $\{x_0, x_1, ..., x_{n-1}\}$ are reduced to $n/2^k$ random variables $\{x_0^{(k)}, x_1^{(k)}, ..., x_{n/2^k-1}^{(k)}\}$, functions

$f_{i,j}$, $i, j \in \{0, 1, ..., n-1\}$, have been combined into $f_{i,j}^{(k)}$, $i, j \in \{0, 1, ..., n/2^k - 1\}$.

After $\log n$ steps $B^{(\log n)} = f_{0,0}^{(\log n)}(x_0^{(\log n)}, x_0^{(\log n)})$. The bit at the root of the random variable tree is now set to 0 if $f_{0,0}^{(\log n)}(0, 0) \geq f_{0,0}^{(\log n)}(1, 1)$, and 1 otherwise. Thus Han and Igarashi's algorithm[3] solves the PROFIT/COST problem in $O(\log n)$ time with a linear number of processors.

Let $n = 2^k$ and A be an $n \times n$ array. Elements $A[i, j]$, $A[i, j\#0]$, $A[i\#0, j]$, $A[i\#0, j\#0]$ form a *gang* which is denoted by $g_A[\lfloor i/2 \rfloor, \lfloor j/2 \rfloor]$. All gangs in A form array g_A.

When visualized on a two dimensional array A (as shown in Fig. 1), a stage of Han and Igarashi's algorithm can be interpreted as follows. Let $f_{i,j}$ be stored in $A[i, j]$. Setting the random variables at level 0 of the random variable tree is done by examining the PROFIT/COST functions in the diagonal gang of A. Function $f_{i,j}$ then gets the bit setting information from $g_A[\lfloor i/2 \rfloor, \lfloor i/2 \rfloor]$ and $g_A[\lfloor j/2 \rfloor, \lfloor j/2 \rfloor]$ to determine how it is to be combined with other functions in $g_A[\lfloor i/2 \rfloor, \lfloor j/2 \rfloor]$. Han and Igarashi's algorithm[3] uses concurrent read because all PROFIT/COST functions in row i (column j) have to get information from $g_A[\lfloor i/2 \rfloor, \lfloor i/2 \rfloor]$ ($g_A[\lfloor j/2 \rfloor, \lfloor j/2 \rfloor]$) in constant time.

A *derandomization tree* D can be built which reflects the way the PROFIT/COST functions are combined. D is of the following form. The input PROFIT/COST functions are stored at the leaves, $f_{i,j}$ is stored in $A_0[i, j]$. A node $A_l[i, j]$ at level $l > 0$ is defined if there exist input functions in the range $A_0[u, v]$, $i * 2^l \leq u < (i+1) * 2^l$, $j * 2^l \leq v < (j+1) * 2^l$. A derandomization tree is shown in Fig. 2. As can be seen, the derandomization tree is contracted bottom-up as functions stored in the derandomization tree are combined. A node at level l represents a function $f_{i,j}^{(l)}$. The total number of nodes in the derandomization tree is $O(m \log n)$. It can be reduced to $O(m)$ by eliminating nodes which has only one child[3].

3   An EREW Algorithm

We show how to obtain an EREW PRAM algorithm for the PROFIT/COST problem with time complexity $O(\log n)$ using a linear number of processors.

3.1   The Dense Case

To remove the concurrent read feature in Han and Igarashi's algorithm[3] we design a pipeline for disseminating the bit setting information.

We assume all n^2 PROFIT/COST functions $f_{i,j}$ are present in the input. We store these functions in an $n \times n$ array A_0. Function $f_{i,j}$ is stored in $A_0[i, j]$. The leaves of the the derandomization tree D are the elements of array A_0. The l-th level of D is an $\frac{n}{2^l} \times \frac{n}{2^l}$ array A_l. The children of $A_l[i, j]$ is $A_{l-1}[2i, 2j]$, $A_{l-1}[2i+1, 2j]$, $A_{l-1}[2i, 2j+1]$ and $A_{l-1}[2i+1, 2j+1]$. Assign one processor to each element in each array. The total number of processors used is $\sum_i \frac{n^2}{4^i} = O(n^2)$.

The problem in obtaining an EREW algorithm is how to combine functions without using concurrent read. Consider the processors working on array A_0. The random bits at level 0 of the random variable tree are set by the processors in the diagonal gang of A_0. There are a total of $O(n)$ processors in the diagonal gang. These processors know the bit setting information when they set the random bits. However, in order for functions in A_0 to be combined into functions in A_1 all processors in A_0 need to know the bit setting information. There are $O(n^2)$ processors in A_0. Therefore it takes $O(\log n)$ time for the processors in the diagonal gang to disseminate the bit setting information to all processors in A_0 if concurrent read is not allowed. If functions in A_0 are combined into functions in A_1 after all processors in A_0 get the bit setting information then no function in A_1 will be defined until processors in A_0 spend $O(\log n)$ steps for acquir-

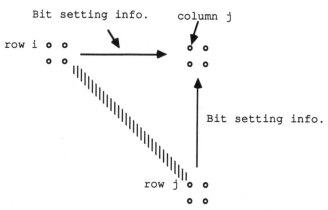

Fig. 1.

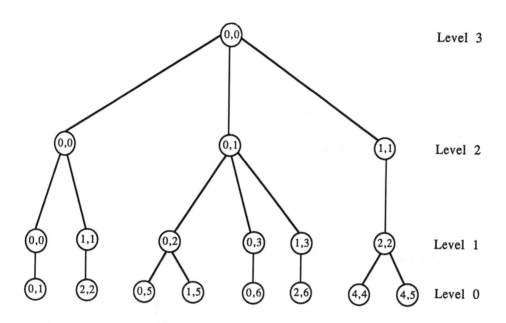

Fig. 2. A derandomization tree. Pairs in the circles
are the subscripts of PROFIT/COST functions.

ing the bit setting information. It will lead to an $O(\log^2 n)$ time algorithm because level l requires $O(\log(n/2^l))$ steps to disseminate bit setting information.

We observe that functions in A_0 which are close to the diagonal need to be combined sooner than functions which are far away from the diagonal (those which are close to the upper right corner and lower left corner). This is because functions in the diagonal gang of A_1 are obtained from combined functions close to the diagonal in A_0, while functions which are far away from the diagonal will not be combined into a function in a diagonal gang until at a higher level in the derandomization tree. This observation enables us to use a pipeline to disseminate the bit setting information.

Consider the action of processors at level l (those working on A_l). We define $n/2^{l+1}$ groups for elements in A_l (this defines groups for processors at level l as well). $A_l[i,j]$ is in group $|\lfloor i/2 \rfloor - \lfloor j/2 \rfloor|$. Refer to Fig. 3. Define step 0 for level l as the step immediately after $A_l[i,i]$, $0 \le i < n/2^l$, are defined (they come from the combination of functions at level $l-1$). The algorithm for processors at level l is shown below.

Step 0.
(* At the beginning of this step elements (functions) in $A_l[i,i]$, $0 \le i < n/2^l$, are defined. *)
No action is taken in this step.

Step 1.
(* At the beginning of this step elements in $A_l[i,j]$, $|i-j| < 2$, are defined. Therefore elements in group 0 are defined. *)
Processors in group 0 determine how the random variables at l-th level of the random variable tree are set and send the bit setting information (for levels 0 to l) to elements in group 1. They also combine functions in group 0 and send combined functions and the bit setting information to level $l+1$.
(* This action defines elements in $A_{l+1}[i,i]$, $0 \le i < n/2^{l+1}$. Note also that the bit setting information need not be sent to level $l+1$ in the dense case we are dealing with here, but it needs to be sent to level $l+1$ in the general case. *)

Step t (t > 1).

(* At the beginning of this step elements in $A_l[i,j]$, $|i-j| < 2^t$, are defined. Therefore elements in groups i, $0 \le i < 2^{t-1}$, are defined. Elements in groups i, $0 \le i < 2^{t-1}$, also have acquired the bit setting information. Besides, elements in groups i, $0 \le i < 2^{t-2}$, have already sent combined functions to level $l+1$. *)
Processors in groups i, $0 \le i < 2^{t-1}$, send the bit setting information to elements in groups i', $2^{t-1} \le i' < 2^t$. Processors in groups i, $2^{t-2} \le i < 2^{t-1}$, combine functions in these groups and send combined functions to level $l+1$.

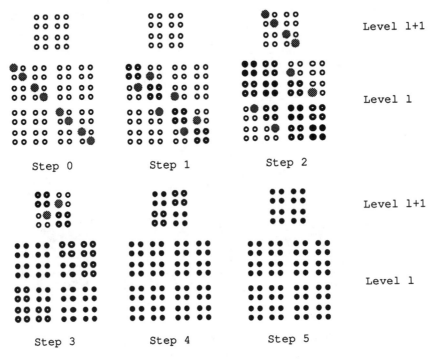

o   Undefined node.

◉   Defined node.

◦   Node received bit setting information.

●   Node which has been combined.

Fig. 3.

(* This action defines elements in $A_{l+1}[i,j], 2^{t-2} \leq |i-j| < 2^{t-1}$. *)

An example of the execution of the above algorithm is shown in Fig. 3.

Theorem 1: The PROFIT/COST problem can be solved in $O(\log n)$ time and $O(n^2)$ space with $O(n^2)$ processors on the EREW PRAM.

Proof: As can be seen in the algorithm, step t at level l is step $t - 2c$ at level $l + c$ for $c \geq 0$. Thus step $2 \log n + 1$ at level 0 is step 1 at level $\log n$. Since there is only one function left at level $\log n$ it takes one step to finish the computation at that level. Therefore the time complexity of the algorithm is $O(\log n)$.

To prove the correctness of the algorithm we need to show, at step t of level 0, that the elements in groups i, $0 \leq i < 2^{t-2l-1}$, at level l are defined and have acquired the bit setting information, that processors in groups $i, 0 \leq i < 2^{t-2l-2}$, have already sent combined functions to level $l+1$. These can be proved by induction as follows. For $t = 1$, the

elements in group 0 at level 0 are defined and no combined function has been sent to level 1. Assume that for t the hypothesis is true. At step $t + 1$, at each level the number of groups defined and the number of processors which have already sent the combined functions to a higher level will double according to the algorithm, therefore at level l the elements in groups i, $0 \leq i < 2^{t-2l}$, are defined and the processors in groups i, $0 \leq i < 2^{t-2l-1}$, have already sent combined functions to level $l+1$. When $t = 2l$ the elements in group 0 at level l will be defined in step $t + 1$.

Note that the number of groups which have received the bit setting information doubles in one step, therefore there is no need to use concurrent read to disseminate the bit setting information. □

3.2 The General Case

The algorithm for the dense case can not be used directly for the general case if we intend to use a linear number of processors and achieve $O(\log n)$ time. The reason is that many elements in A_l may not be present and therefore no processors are allocated to these elements, while the dissemination of the bit setting information in the algorithm for the dense case assumes that all these processors are available. Another problem is that the algorithm for the dense case uses $O(n^2)$ space which we want to avoid for the general case. As will be seen that $O(m \log n)$ space is sufficient for the general case. The saving in space is significant if m is much less than n^2.

In the general case if each node in the derandomization tree receives the bit setting information no later than it would receive the information in the algorithm for the dense case, then the pipeline we designed will still be valid. We shall call this condition the synchronization condition. Therefore our algorithm for the dense case can be used for the general case if we have a scheme for disseminating the bit setting information which satisfies the synchronization condition. In this subsection we show how to adapt our algorithm for the dense case to the general case by using *row and column trees* to disseminate the bit setting information.

We first build the derandomization tree D. This is done by sorting the input into file-major indexing and constructing the tree bottom up. The details are explained in [3]. We then build row and column trees for nodes at each level of the derandomization tree for disseminating the bit setting information.

The row and column trees at level l are built for nodes at level l of D. The column trees are built for disseminating the bit setting information from gang $g_{A_l}[j,j]$ to all the gangs in the same column ($g_{A_l}[k,j], 0 \le k < n/2^{l+1}$) while the row trees are for disseminating the information from gang $g_{A_l}[i,i]$ to all the gangs in the same row ($g_{A_l}[i,k], 0 \le k < n/2^{l+1}$). Without loss of generality we may assume that all

$A_l[i,i], 0 \le l \le \log n, 0 \le i < n/2^l$, are present in the input. If some of them are not present we use zero functions to represent them. Because we are adding no more than $O(n)$ nodes to the derandomization tree, the time complexity of our algorithm will not be affected. The construction and the function of row and column trees are similar, so we only discuss how to construct and use row trees. For nodes in A_l we only consider how to build row trees for gangs in $g_{A_l}[i,j], j \ge i$, and disseminate information on the tree. Again the tree for gangs $g_{A_l}[i,j], j \le i$, can be constructed and used similarly.

Assign index value $index(i,j) = i*n + i + sign(j - i)*brv(|j - i|, \log(n/2^{l+1}))$ to $g_{A_l}[i,j]$ if it is not empty (*i.e.*, at least one node in the derandomization tree D is in $g_{A_l}[i,j]$), where $sign$ is the sign function and $brv(i,j)$ is the bit reversal function which takes j least significant bits from i and reverses these bits to get the function value. Now sort elements in g_{A_l} by the $index$ value. $i*n$ in the index ensures that after sorting elements in the same row are consecutive in the sorted array. i and $sign(j - i)$ in the index ensures that elements in row i are arranged by the sorting so that $g_{A_l}[i,j]$ is before $g_{A_l}[i,i]$ if $j < i$ and $g_{A_l}[i,j]$ is after $g_{A_l}[i,i]$ if $j > i$.

For each row i a row tree R is built for $g_{A_l}[i,j], i \le j < n/2^{l+1}$. The tree can be built bottom-up. $g_{A_l}[i,j], i \le j < n/2^{l+1}$, are stored at level 0, *i.e.*, leaves, of R. A node v at level k of R is created if both ranges, $n*2^{k-1} \le index(i,j) < (n + 1)*2^{k-1}$ and $(n + 1)*2^{k-1} \le index(i,j) < n*2^k$, are not empty (*i.e.*, there are input functions whose index value fall into these two ranges). The construction of R is straightforward after $g_{A_l}[i,j]$'s are sorted by the index values. An example of such a tree is shown in Fig. 4.

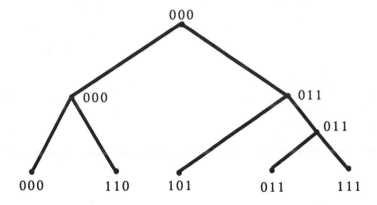

Fig. 4. A row tree. The numbers in binary are the |j-i| value. They are sorted by the bit reversal value.

Each node in R is labeled. Leaf $g_{A_l}[i,j]$ is labeled with (i,j). A parent is labeled with the label of one of its children which has the smaller $|j-i|$ value. Thus the root is labeled with (i,i) which represents $g_{A_l}[i,i]$. The bit setting information propagates from the root down to the leaves. $g_{A_l}[i,j]$ gets the information at the smallest depth of the tree where label (i,j) appears. Thus $g_{A_l}[i,i]$ gets the bit setting information at depth 0 and the children of the root of R get the bit setting information at depth 1. Again we define step 0 for level l of D as the step immediately after $A_l[i,i]$, $0 \le i < n/2^l$, are defined. In step t for level l of D the bit setting information is transmitted from nodes at depth $t-1$ to nodes at depth t on the row tree R. It is easy to see that if all $g_{A_l}[i,j]$, $i \le j < n/2^{l+1}$, are present, the bit setting information is disseminated on the row tree in exactly the same way as it is disseminated in the algorithm for the dense case.

Theorem 1′: The PROFIT/COST problem can be solved in $O(\log n)$ time and $O(m \log n)$ space with $O(m)$ processors on the EREW PRAM.

Proof: The time complexity for building the derandomization tree and row and column trees is $O(\log n)$ with a linear number of processors. The correctness of the algorithm is now proved by an induction showing at step t at level 0, that the elements in groups i, $0 \le i < 2^{t-2l-1}$, at level l are defined and have acquired the bit setting information, and that elements in groups i, $0 \le i < 2^{t-2l-2}$, have already sent combined functions to level $l+1$. We note that the only difference between the dense case and the general case is that the bit setting information is transmitted on the row and column trees in the general case while there is no need to explicitly construct the row and column trees in the dense case. In the algorithm for the general case a function to be combined with other functions will be defined and receive the bit setting information at a step that is no later than the step it would be defined and receive the bit setting information in the algorithm for the dense case we described in the last subsection. Thus the action taken by a processor at a node in D is the same as that in the algorithm for the dense case except the transmission of the bit setting information is now carried on the row and column trees. □

4 Conclusions

The PROFIT/COST problem is an important problem for which best parallel algorithm ought to be sought. The algorithm presented in this paper achieves time $O(\log n)$ using a linear number of processors. It is not clear whether time $O(\log n)$ can be achieved on the EREW PRAM using an optimal number $O(n/\log n)$ processors. The problem may be linked to the problem of whether it is possible to perform integer sorting with the same complexity on the EREW PRAM. Since we have used a pipeline and row and column trees the structure of the algorithm obtained is rather complicated. It is desirable to obtain an EREW algorithm which has a simpler structure.

Several applications of the PROFIT/COST problem are outlined in [2][7]. We expect that many more important applications of the PROFIT/COST problem will be found.

References

[1] M. Goldberg, T. Spencer. Constructing a maximal independent set in parallel. SIAM J. Dis. Math., Vol 2, No. 3, 322-328(Aug. 1989). Science Dept., Technion, Haifa, Israel, 1984.

[2] Y. Han. A fast derandomization scheme and its applications, TR No. 180-90, Dept. Computer Sci., Univ. of Kentucky.

[3] Y. Han and Y. Igarashi. Derandomization by exploiting redundancy and mutual independence. Proc. SIGAL 1990, LNCS 450, 328-337.

[4] A. Israeli, Y. Shiloach. An improved parallel algorithm for maximal matching. Information Processing Letters 22(1986), 57-60.

[5] M. Luby. A simple parallel algorithm for the maximal independent set problem. SIAM J. Comput. 15:4, Nov. 1986, 1036-1053.

[6] M. Luby. Removing randomness in parallel computation without a processor penalty. Proc. 1988 IEEE FOCS, 162-173.

[7] M. Luby. Removing randomness in parallel computation without a processor penalty. TR-89-044, Int. Comp. Sci. Institute, Berkeley, California.

[8] G. L. Miller and J. H. Reif. Parallel tree contraction and its application. Proc. 1985 IEEE FOCS, 478-489.

[9] S. Fortune and J. Wyllie. Parallelism in random access machines. Proc. 10th Annual ACM Symp. on Theory of Computing, San Diego, California, 1978, 114-118.

PARALLEL ALGORITHMS ON OUTERPLANAR GRAPHS

Hyung Ah Choi

Department of Computer Science
George Washing University
Washington DC 20052

Moon Jung Chung

Department of Computer Science
Michigan State University
East Lansing, MI 48824

ABSTRACT

This paper presents parallel algorithms for some outerplanar graph problems. Recently, it has been shown that the Hamiltonian cycle can be found efficiently for the class of biconnected outerplanar graphs. Based on this algorithm, a parallel algorithm to construct a maximal outerplanar graph is presented. Using these algorithms, we give parallel algorithms for node-coloring, one-to-all shortest paths, breadth-first-search, depth-first-search, and recognition of outerplanar graphs. All of our algorithms can be implemented in $O(\log n)$ time with $O(n)$ processors on a CRCW PRAM. The same algorithms can be implemented in max $\{f(n), O(\log n)\}$ time with $O(n)$ processors on an EREW PRAM, where $f(n)$ is the parallel time complexity for finding biconnected components and a spanning tree of an outerplanar graph; currently, $f(n)$ is known to be $O(\log^2 n)$.

1. Introduction

A planar graph G is called *outerplanar* if there exists a planar embedding of G in such a way that all nodes lie on the exterior face. An outerplanar graph is called *maximal outerplanar* if no edges between non-adjacent nodes can be added without losing outerplanarity.

It is well-known that a large number of NP-complete graph problems can be solved sequentially in linear time if an input graph is restricted to the class of outerplanar graphs. Researchers have developed sequential algorithms for finding the Hamiltonian cycle of an outerplanar graph [2], for recognizing outerplanar graphs, and for node-coloring outerplanar graphs [10], among others.

This paper studies parallel algorithms for some outerplanar graph problems on a Parallel Random Access Machine (PRAM) model. The shared memory PRAM models can be classified according to the ways they resolve read/write conflicts. They are Exclusive-Read Exclusive-Write (EREW) model, Concurrent-Read Exclusive-Write (CREW) model and Concurrent-Read Concurrent-Write (CRCW) model. The EREW and the CRCW models, respectively, are the weakest and the strongest ones among the above models. (See [8] for a recent survey paper for parallel algorithms on PRAM models.)

The class of biconnected outerplanar graphs is a subclass of (undirected) *series-parallel* graphs, and each biconnected component can be represented by a directed acyclic *two-terminal* series-parallel (TTSP) graph, in many different ways, by assigning the directions of the edges. He and Yesha [5,6] studied parallel algorithms for directed acyclic TTSP graphs. In their algorithms, they first construct a binary decomposition tree of the input graph (which is given as a directed acyclic TTSP) in $O(\log^2 n)$ parallel time. It is noted that if a biconnected outerplanar graph is given as a directed acyclic TTSP, one can easily

find a spanning tree and DFS in a constant time by choosing one incoming edge at each node. However, if the graph is given as an undirected graph, the best known parallel algorithms for finding a depth-first-search tree and a breadth-first-search tree both take $O(\log n)$ time using $O(n^3)$ processors on a CRCW PRAM [4]. In our parallel algorithms, we do not require the input to be directed acyclic specifying two terminal nodes. Our algorithms work for any outerplanar graph when the graph is given as an undirected graph.

It is known that every biconnected outerplanar graph has a unique Hamiltonian cycle, HC [2]. Recently, Chung and Choi [3] presented a parallel algorithm for finding the Hamiltonian cycle of a biconnected outerplanar graph G in $O(\log n)$ time with $O(n)$ processors on an EREW PRAM when a spanning tree of G is given.

Theorem 1.1. Given a biconnected outerplanar graph, the unique Hamiltonain cycle can be found in $O(\log n)$ time with $O(n)$ processors on a CRCW PRAM or in $O(\log^2 n)$ time using $O(n)$ processors on an EREW PRAM. However, if a spanning tree of the input graph is given, the Hamiltonian cycle can be found in $O(\log n)$ time with $O(n)$ processors on an EREW PRAM.

Based on the algorithm for finding the Hamiltonian cycle, a parallel algorithm to construct a maximal outerplanar graph is given. Using these algorithms, we present parallel algorithms for node-coloring, one-to-all shortest paths, breadth-first-search, depth-first-search, and recognition of outerplanar graphs. All of our algorithms can be implemented in max $\{f(n), O(\log n)\}$ time with $O(n)$ processors on an EREW PRAM, where $f(n)$ is the parallel time complexity for finding biconnected components and a spanning tree of an outerplanar graph; currently, $f(n)$ is known to be $O(\log^2 n)$ [10]. The same algorithms can also be implemented in $O(\log n)$ time with $O(n)$ processors on a CRCW PRAM.

2. Maximal Outerplanar Graph

In this section, we discuss how to construct maximal outerplanar graphs. In what follows, we first present an algorithm when the input graph is biconnected outerplanar, and then using this algorithm, we give an algorithm for arbitrary outerplanar graphs.

The algorithm $MAX-B-OUTER(G)$, which takes a biconnected outerplanar graph G as an input and produces a maximal outerplanar graph G_{max} as an output, is in two steps. In the first step, all the faces are recognized using the Eulerian technique [11]. In the second step, a maximal outerplanar graph is constructed by triangulating each face by adding edges between a designated node and every other node.

Let G be a biconnected outerplanar graph such that the nodes in G are numbered as $1, 2, \cdots, n$ according to the order in the Hamiltonain cycle of G. Such a numbering can be done using the algorithm HC in Theorem 1.1. We next describe how to construct a directed graph G' from G such that G' consists of edge-disjoint directed cycles (*i.e.*, each node in G' has indegree 1 and outdegree 1), and each cycle in G' corresponds to each face of G. (Note that the Hamiltonian cycle corresponds to the exterior face.) The node set of G' is defined such that for each node $v \in V(G)$, there is a corresponding block B_i, consisting of t nodes in $V(G')$, where t is the degree of v in G. That is, if node i is adjacent to nodes $j_1, j_2, \cdots, j_t$ in G such that $j_1 < j_2 < \cdots < j_t$, then in G', there is a block of t nodes $B_i = \{b^i_{j_1}, b^i_{j_2}, \cdots, b^i_{j_t}\}$. For each block $B_i \subseteq V(G')$, the edge set of G', whose tails are all in B_i is defined as follows : for each node $b^i_{j_l}$ $(1 \le l \le t)$, there is an incoming edge from node $b^{j_{l+1}}_i$ in block $B_{j_{l+1}}$, where we assume that $j_{t+1} = j_1$. Figure 1 illustrates the construction of G'.

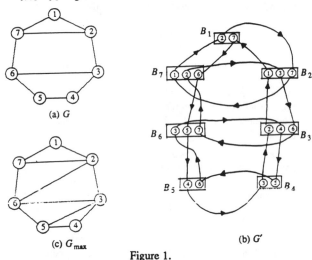

(a) G

(c) G_{max}

(b) G'

Figure 1.

It is observed that there is a cycle $C_0 = \{b^1_n, b^n_{n-1}, b^{n-2}_{n-2}, \cdots, b^2_1, b^1_n\}$ in G', which corresponds to the Hamiltonian cycle of G. From the above discussion, we establish the following lemma.

Lemma 2.1. There is a one-to-one correspondence between the cycles of G' and the faces of G.

Let $f = \{i_1, i_2, \cdots, i_k, i_1\}$ be a face in G such that $i_1 < i_2 < \cdots < i_k$. Then, face f can be identified by two nodes i_1 and i_k having the smallest and the largest numbers. Thus, we denote face f by $<i_1, i_k>$.

Algorithm $MAX-B-OUTER(G)$

Input : A biconnected outerplanar graph $G(V,E)$ such that the nodes of G are numbered as $1, 2, \cdots, n$ according to the order in the HC

Output : A maximal outer planar graph G_{max}.

(Note that the edges of G_{max} will be stored in the processors, such that if there is an edge (i,j) in G_{max}, then there is a processor whose content is (i,j).)

1. Find all the faces of G as follows.

   1.1 Construct G' from G.

   1.2 Allocate one processor to each edge of G'.

1.3. Using the Eulerian tour technique, find all the cycles of G'.

2. Triangulate each face by adding edges to G as follows.

   2.1 From each cycle of G', find the two nodes which have the smallest and the largest block numbers, say i_1 and i_k, respectively. Let each edge in cycle i keep the information $f_i = <i_1, i_k>$.

   2.2 For each edge (b^q_s, b^p_r) which belongs to face f_i (*i.e.*, cycle i), add edge (q, i_1) to G, if $(q, i_1) \notin E(G)$.

3. Let the resulting graph be G_{max}.

The above procedure is depicted in Figure 1. For example, a graph G in Figure 1 (a) has three faces $f_1 = (1, 2, 7, 1) = <1, 7>$, $f_2 = (2, 3, 6, 7, 2) = <2, 7>$, $f_3 = (3, 4, 5, 6, 3) = <3, 6>$. After we triangulate the graph G, G_{max} has five faces as shown in Figure 1 (c).

To analyze the complexity, note that we can find the cycles of G' using list linking in $O(\log n)$ time with $O(n/\log n)$ processors in an EREW PRAM. Thus, we can find all the faces with the same complexity. Adding new edges after obtaining the faces can be done in $O(1)$ time with $O(n)$ processors on an EREW PRAM. Hence, the algorithm $MAX-B-OUTER(G)$ can be implemented in $O(\log n)$ time with $O(n)$ processors, if a spanning tree of G is given.

Now, suppose G is an arbitrary outerplanar graph. Then, a maximal outerplanar graph G_{max} can be constructed by the following algorithm using $MAX-B-OUTER(G)$ as a subroutine.

Algorithm $MAX-OUTER(G)$

Input : an outerplanar graph $G(V,E)$

Output : a maximal outerplanar graph G_{max}

1. Find all the biconnected components of G.

2. For each biconnected component B of G, where $|V(B)| \ge 3$, construct B_{max} by calling the algorithm $MAX-B-OUTER(B)$.

3. For each articulation point $a \in V(G)$, do the following :

   3.1 Let $B^a_1, B^a_2, \cdots, B^a_k$ be the biconnected components of G which all include articulation point a.

   3.2 For $1 \le i \le k$, do the following : Let $h^a_{i,1}$ and $h^a_{i,2}$ denote the two nodes which are incident with two Hamiltonian edges $(h^a_{i,1}, a)$ and $(a, h^a_{i,2})$ of B^a_i, if $|V(B^a_i)| \ge 3$. If $|V(B^a_i)| = 2$, let $h^a_{i,1}$ and $h^a_{i,2}$ denote the same node other than a in B^a_i.

   3.3 Let $E^a = \{(h^a_{i,2}, h^a_{(i+1),1}) | 1 \le i \le k-1\}$.

4. Let G_{max} be a graph defined such that $V(G_{max}) = V(G)$ and $E(G_{max}) = \bigcup_B E(B_{max}) \cup_a E^a$, where B and a are a biconnected component and an articulation point of G, respectively.

Figure 2 shows the construction of a maximal outerplanar graph G_{max} of an outerplanar graph G by following the above procedure. To analyze the complexity of the algorithm, Step 1 can be done using the algorithm by Tarjan and Vishkin [11] in $O(\log^2 n)$ time with $O(n)$ processors on an EREW PRAM or in $O(\log n)$ time with $O(n)$ processors on a CRCW PRAM. Steps 3 and 4 can be done in $O(1)$ time with $O(n)$ processors on an

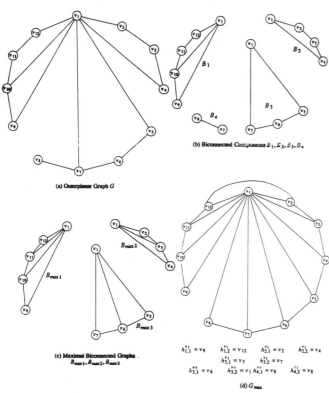

(a) Outerplanar Graph G

(b) Biconnected Components B_1, B_2, B_3, B_4

(c) Maximal Biconnected Graphs
$B_{max\,1}, B_{max\,2}, B_{max\,3}$

$h_{1,1}^{v_1} = v_9$ $h_{1,2}^{v_1} = v_{12}$ $h_{2,1}^{v_1} = v_2$ $h_{2,2}^{v_1} = v_4$

$h_{3,1}^{v_1} = v_5$ $h_{3,2}^{v_1} = v_7$

$h_{3,1}^{v_7} = v_6$ $h_{3,2}^{v_7} = v_1$ $h_{4,1}^{v_7} = v_8$ $h_{4,2}^{v_7} = v_8$

(d) G_{max}

Figure 2. Construction of a maximal outerplanar graph G_{max}.

EREW PRAM. Therefore, we establish Theorem 2.1.

Theorem 2.1. Given an outerplanar graph, a maximal outer-planar graph can be constructed in $O(\log^2 n)$ time with $O(n)$ processors on an EREW PRAM or in $O(\log n)$ time with $O(n)$ processors on a CRCW PRAM. However, if a spanning tree of the input graph is given, a maximal outerplanar graph can be constructed in $O(\log n)$ time with $O(n)$ processors on an EREW PRAM.

3. Node-Coloring

The problem of node-coloring of a graph is an assignment of colors to the nodes of the graph such that no two adjacent nodes receive the same color. The minimum number of colors for the node-coloring of a graph is called the *chromatic number*. Computing the chromatic number is NP-hard even for very restricted classes of graphs. However, if the input graph is restricted to the class of outerplanar graphs, the chromatic number is known to be 3 and an $O(|V|)$ sequential algorithm was developed by Proskurowski and Syslo using the dual graph [9]. In this section, we present a parallel algorithm for the node-coloring of outerplanar graphs without using the dual graph.

Algorithm $NC(G)$

Input :   An outerplanar graph $G(V,E)$

Output :   A function $f : V \rightarrow \{0, 1, 2\}$ such that $f(u) \neq f(v)$ for each edge $(u,v) \in E$

1. Construct a maximal outerplanar graph G_{max} of G.

2. Compute a set $S = \{<z,w> | z, w \in V(G_{max})$ and there exist two nodes $u, v \in V(G_{max})$ such that edges $(w,u), (w,v), (z,u), (z,v) (u,v) \in E(G_{max})\}$.

3. Construct a graph G^* such that $V(G^*) = V(G)$ and $E(G^*) = \{(z,w) | <z,w> \in S\}$.

4. Find the three connected components G_0^*, G_1^*, and G_2^* of G^*.

5. Let $f(v) = 0, 1,$ or 2, respectively, if $v \in V(G_0^*), V(G_1^*),$ or $V(G_2^*)$.

It is noted that a maximal outerplanar graph G_{max} can be constructed as described in Section 3 after finding the Hamiltonian cycle using the algorithm described in Section 2. Both processes can be done in $O(\log n)$ time with $O(n)$ processors on an EREW PRAM when a spanning tree of G is given.

In order to explain the idea behind Step 2, let u and v be nodes in a maximal outerplanar graph G_{max} such that edge $(u,v) \in E(G_{max})$. Suppose there are two other nodes z and w in G_{max} such that each of $u-z-v$ and $u-w-v$ forms a triangle. Since the chromatic number of every outerplanar graph is at most 3 and nodes $u, z,$ and v (and also $u, w,$ and v) must receive three different colors, nodes z and w must receive the same color. All those pairs of nodes are collected in S in parallel. Now, let us explain how we can do Step 2 in parallel in a constant time. For each non-Hamiltonian edge of G_{max}, there are exactly two faces which include this edge. Note that each face has only three edges since the graph is a maximal outerplanar. Thus, if we assign one processor to each face and to each edge, Step 2 can be done in a constant time as explained below.

Let $f = (a, b, c, a)$ be a face such that $a < b < c$ (We assume that nodes are numbered according to the order in which they appear in the HC). Then, the processor which is assigned to the face f sends data as follows: first, sends c (the node number c) to the processor which is assigned to the edge (a, b); next, sends a to the processor assigned to the edge (b, c); finally, sends b to the processor assigned to the edge (a, c). In this way, each non Hamiltonian edge (u, v) receives two nodes z and w such that (u, v, z) and (u, v, w) are faces without any read/write conflict. For example, consider a non-Hamiltonian edge (a, b). Let $f = (a, b, c, a)$ and $f' = (a, d, b, a)$ be two faces with the edge (a, b) such that $a < b < c$. Note that we have $a < d < b$ because of the numbering of nodes and the outerplanar property of G. Let $PE(f), PE(f'), PE(a, b), PE(b, c),$ and $PE(a, c)$ are processors assigned to the face f, face f', the edge (a, b) and edge (a, c), respectively. First, $PE(f)$ sends the data c to $PE(a, b)$ and $PE(f')$ sends data b to $PE(a, d)$. Next, $PE(f)$ sends a to $PE(b, c)$ and $PE(f')$ sends a to $PE(b, d)$. Finally, $PE(f)$ sends b to $PE(a, c)$ and $PE(f')$ sends d to $PE(a, b)$. Thus, $PE(a, b)$ receives data c and d at different time and there is no write-conflict.

To analyze Steps 3 and 4, it is noted that any pair of adjacent nodes in G^* constructed in Step 3 must receive the same color. Such a coloring can be done by finding all the connected components. In fact, graph G^* is always a forest having three trees. Hence, Steps 3 and 4 can be done in $O(\log n)$ time with $O(n)$ processors on an EREW PRAM using the Eulerian tour technique [11].

The above discussion leads to Theorem 3.1. In order to explain the above procedure, consider the outerplanar graph G shown in Figure 2 (a). A maximal subgraph G_{max} of G constructed in Step 1 is shown in Figure 2 (d), and the set S computed in Step 2 is $S = \{$ $<v_9, v_{11}>$, $<v_{12}, v_{10}>$, $<v_{11}, v_2>$, $<v_{12}, v_3>$, $<v_2, v_4>$, $<v_3, v_5>$, $<v_4, v_6>$, $<v_5, v_7>$, $<v_6, v_8>$ $\}$. The graph G^* defined in Step 3 is shown in Figure 3 (a), and Figure 3 (b) shows the color $f(v)$ of each node $v \in V(G)$ obtained by the above procedure.

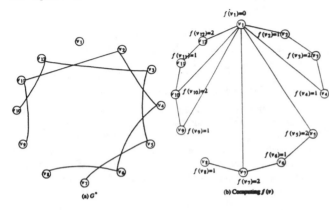

Figure 3. Node coloring.

Theorem 3.1 Given an outerplanar graph G, there exists a parallel algorithm for the node-coloring of G using 3 colors, which runs in $O(\log^2 n)$ time with $O(n)$ processors on an EREW PRAM or in $O(\log n)$ time with $O(n)$ processors on a CRCW PRAM. However, if the biconnected components and a spanning tree of G are given, the node-coloring can be done in $O(\log n)$ time with $O(n)$ processors on an EREW PRAM.

4. Other Applications

In this section, we state other applications of our results obtained in the previous sections.

Theorem 4.1. One-to-all shortest paths of an edge weighted outerplanar graph can be found in $O(\log^2 n)$ time with $O(n)$ processors on an EREW PRAM or $O(\log n)$ time with $O(n)$ processors on a CRCW PRAM. However, if the biconnected components and a spanning tree of the input graph are given, one-to-all shortest paths can be found in $O(\log n)$ time with $O(n)$ processors on an EREW PRAM.

Proof. Let G be an outerplanar graph such that a positive number $w(e)$ is associated with with each edge $e \in E(G)$. Construct a maximal outerplanar graph G_{max} of G using the algorithm $MAX-OUTER(G)$ in Section 3, and assign the weight to each edge in G_{max} such that the weight of edge e is equal to $w(e)$ if $e \in E(G)$, and ∞ if $e \notin E(G)$.

Suppose the nodes in G_{max} are numbered as $1, 2, \cdots, n$ according to the order in which they appear in the Hamiltonian cycle of G_{max}. Consider a face $f = \{i_1, i_2, ..., i_k, i_1\}$ of G_{max} such that $i_1 < i_2 < \cdots < i_k$. As explained in Section 3, we denote face f as $f = <i_1, i_k>$. We say that two different faces of G_{max} are adjacent if they have a common edge. Consider two adjacent faces $f = <s, t>$ and $f' = <s', t'>$. Then either (i) $s \le s'$ and $t \ge t'$ or (ii)

$s \ge s'$ and $t \le t'$. Moreover if (i) is true then the edge (s', t') is in the face f, and if (ii) is true then the edge (s, t) is in f'. We say that, for two different faces $f = <s, t>$ and $f' = <s', t'>$, $f < f'$ if $s \le s'$ and $t \ge t'$. Note that the relation "$<$" between faces is a partial order relation.

We next proceed to show how to compute the distance (i.e., the length of a shortest path) of each node from node 1. Let $d(i)$ denote the distance between node 1 and node i (i.e., the length of a shortest path between node 1 and node i), and $d(i,j)$ denote the distance between node i and node j (i.e., the length of a shortest path between node i and node j.) Consider three faces $f_1 = <s_1, t_1>$, $f_2 = <s_2, t_2>$ and $f_3 = <s_3, t_3>$, such that $f_1 < f_2 < f_3$. Note that the nodes appear in the order of $s_1, s_2, s_3, t_3, t_2, t_1$ in the Hamiltonian cycle. Suppose that $d(s_1)$ and $d(t_1)$ are known. Then we find the distances $d(s_2)$, $d(s_3)$, $d(t_2)$ and $d(t_3)$ as follows.

$$d(s_2) = \min (d(s_1) + d(s_1, s_2), d(t_1) + d(t_1, s_2)) \text{ and}$$
$$d(t_2) = \min (d(s_1) + d(s_1, t_2), d(t_1) + d(t_1, t_2)).$$

In general, for any three faces $f_1 = <s_1, t_2>$, $f_2 = <s_2, t_2>$, and $f_3 = <s_3, t_3>$, if $f_1 < f_2 < f_3$, then

$$d(s_3) = \min (d(s_1) + d(s_1, s_3), d(t_1) + d(t_1, s_3)) \text{ and}$$
$$d(t_3) = \min (d(s_1) + d(s_1, t_3), d(t_1) + d(t_1, t_3)),$$

where$d(s_1, s_3) = \min (d(s_1, s_2) + d(s_2, s_3), d(s_1, t_2) + d(t_2, s_3))$
$$d(s_1, t_3) = \min (d(s_1, s_2) + d(s_2, t_3), d(s_1, t_2) + d(t_2, t_3))$$
$$d(t_1, s_3) = \min (d(t_1, s_2) + d(s_2, s_3), d(t_1, t_2) + d(t_2, s_3))$$
$$d(t_1, t_3) = \min (d(t_1, s_2) + d(s_2, t_3), d(t_1, t_2) + d(t_2, t_3))$$

Now, let us describe how we can compute these values using the parallel tree contraction. We construct a rooted tree $T = (V', E', r)$ as follows. The set V' is the set of pairs of two adjacent faces (f, f') such that $f < f'$. There is an arc from a node (f, f') of T to a node (f'', f''') if and only if $f' = f''$. Formally,

$$V' = \{(f, f') \mid f \text{ and } f' \text{ are adjacent such that } f < f'\} \text{ and}$$
$$E' = \{((f, f'), (f', f'')) \mid (f, f') \text{ and } (f', f'') \text{ are in } V'\}.$$

Note that the root r of T is the face which has edge $(1, 2)$. For each node $x = (f, f')$ of T, where $f = <s, t>$ and $f' = <s', t'>$, we assign a label $l(x)$ which is a quadruple $l(x) = (w_1, w_2, w_3, w_4)$ defined as follows:

$$w_1 = d(s, s'), \quad w_2 = d(s, t'), \quad w_3 = d(t, s') \text{ and } w_4 = d(t, t').$$

For any two labels $l = (w_1, w_2, w_3, w_4)$ and $l' = (w'_1, w'_2, w'_3, w'_4)$, we define an operation $\bigcirc$ as follows:

$$l \bigcirc l' = (w_1'', w_2'', w_3'', w_4''), \text{ where}$$
$$w_1'' = \min (w_1 + w'_1, w_2 + w'_3), \quad w_2'' = \min (w_1 + w'_2, w_2 + w'_4),$$
$$w_3'' = \min (w_3 + w'_1, w_4 + w'_3), \quad w_4'' = \min (w_3 + w'_2, w_4 + w'_4).$$

Note that for $f = <s, t>$, $f' = <s', t'>$, and $f'' = <s'', t''>$ such that $(f, f'), (f', f'') \in V'$, we have the following relation.

$$l((f, f')) \bigcirc l((f', f'')) = (d(s, s''), d(s, t''), d(t, s''), d(t, t'')).$$

Now, consider a node $x = (f, f')$ of T, where $f' = <s', t'>$. Let $P = y_1, ..., y_k$, where $y_1 = r$ and $y_k = x$, be the path from the root r of T to a node x (using tree edges of T) and $L(x) = l(y_1) \bigcirc l(y_2) \bigcirc \cdots l(y_k) = (w_1, w_2, w_3, w_4)$. We then have $d(1, s') = w_1$ and $d(1, t') = w_2$.

Let us discuss how we can compute $L(x)$ in parallel. For each node $x = (f, f')$ of T, where $f = <s, t>$ and $f' = <s', t'>$, nodes

s' and t' are also in the cycle $f = <s,t>$. Thus, we can compute the label $l(x)$ of each x in parallel $O(\log n)$ time using $O(n/\log n)$ processors. Then, $L(x)$ can be computed for all nodes of T in $O(\log n)$ time using $O(n/\log n)$ processors on an EREW PRAM using the tree contraction.

After we have computed $d(i)$ for all $i \in V$, we can easily find the edges which have contributed in the computation of $d(i)$. This completes the proof of the theorem. $\square$

As a special case of Theorem 5.1, when all the edge-weights are equal to 1, we can establish the following theorem.

Theorem 4.2. A breadth-first-search tree of an outerplanar graph can be found in $O(\log^2 n)$ time with $O(n)$ processors on an EREW PRAM or $O(\log n)$ time with $O(n)$ processors on a CRCW PRAM. However, if the biconnected components and a spanning tree of the input graph are given, a breadth-first-search tree can be found in $O(\log n)$ time with $O(n)$ processors on an EREW PRAM. $\square$

As the third application of our previous results, we consider how to find a depth-first-search tree of an outerplanar graph. Before we describe our result, it is noted that Hagerup [4] developed an algorithm for finding a depth-first-search tree of a planar graph, which runs in $O(\log n)$ time with $O(n^3)$ processors on a PRIORITY CRCW PRAM.

Theorem 4.3. A depth-first-search tree of an outerplanar graph can be found in $O(\log^2 n)$ time with $O(n)$ processors on an EREW PRAM or in $O(\log n)$ time with $O(n)$ processors on a CRCW PRAM. However, if the biconnected components and a spanning tree of the input graph are given, a depth-first-search tree can be found in $O(\log n)$ time with $O(n)$ processors on an EREW PRAM.

Proof : Let G be an outerplanar graph. Find the biconnected components $B_1, B_2, \cdots, B_k$ of G. Construct a directed tree T defined as follows. The node set is is defined to $V(T) = A \cup C$, where $A = \{a \mid a$ is an articulation point of $G\}$ and $C = \{b_i \mid b_i$ corresponds to biconnected component B_i of $G\}$. The edge set of T is initially defined as the set of undirected edges $E_0 = \{(b_i, a) \mid$ articulation point a is included in biconnected component $B_i\}$. Choose an arbitrary node in C, say b_1, as the root of T and then assign the directions of edges in E_0 so that T becomes a directed rooted tree rooted at b_1. It is noted that each biconnected component B_i of G may have more than one articulation points, and in T, each non-root node b_i has exactly one incoming edge $e = (a, b_i)$ for some $a \in V(T)$. For each non-root node $b_i \in C$ in T ($i.e.$, $i = 1, 2, \cdots, k$), let a_i denote the parent of b_i. Let a_1 denote an arbitrary node in A such that $(b_1, a_1) \in E(T)$. For each i, $1 \le i \le k$, find the Hamiltonian path P_i of B_i, which starts at the articulation point corresponding to a_i. Now, the collection of all those P_i's gives a depth-first-search tree of G. $\square$

Finally, we present the following theorem.

Theorem 4.4. Outerplanar graphs can be recognized in $O(\log n)$ time with $O(n)$ processors on a CRCW PRAM or in $O(\log^2 n)$ time with $O(n)$ processors on an EREW PRAM. However, if the input graph is known as biconnected and a spanning tree is given, determining whether the graph is outerplanar can be done in $O(\log n)$ time with $O(n)$ processors on an EREW PRAM.

Proof : Note that a graph is outerplanar if and only if every biconnected component of the graph is outerplanar. Therefore, in order to determine whether a given graph is outerplanar, we only need to check whether each of its biconnected components of G is outerplanar.

Let $B(V,E)$ be a biconnected graph. Note that if B is outerplanar, then we can find the Hamiltonian cycle using the algorithm HC in Section 2. However, B may have a Hamiltonian cycle even if B is not outerplanar. Thus, if the output of algorithm HC is not a Hamiltonian cycle, then we can say that B is not outerplanar. Otherwise, $i.e.$, the output is a Hamiltonian cycle, we still need to check whether B is outerplanar or not. This can be accomplished by the following observation :

> Let HC be a Hamiltonian cycle of B such that $HC = (1, 2, \cdots, n, 1)$. Then, B is outerplanar if and only if there exist a pair of edges $e_1 = (i_1, j_1)$ and $e_2 = (i_2, j_2)$ such that either $i'_2 < j'_1 < j'_2$ or $j'_2 < j'_1 < i'_2$, where $x' = (x - i_1) \bmod n$ for $x \in \{i_2, j_1, j_2\}$.

This completes the proof of the theorem. $\square$

5. Conclusion

In this paper, we have shown that some outerplanar graph problems can be solved efficiently in parallel. The algorithms use the Hamiltonian cycle of biconnected outerplanar graph. The time complexity of all our algorithms is dominated by that of finding a spanning tree and biconnected components of the input graph. Currently, all algorithms have parallel time complexity $O(\log^2 n)$ using $O(n)$ processors on EREW model and time complexity $O(\log n)$ using $O(n)$ processors on CRCW model.

References

[1] K. Abrahamson, N. Dadoug and D. K. Kirkpatrick, "A Simple Parallel Tree Contraction Algorithm," *J. Algorithm*, 10 (1989), pp. 287-302.

[2] T. Beyer, W. Jones, and S. L. Mitchell, "Linear Algorithms for isomorphism of maximal outerplanar graphs," *J. ACM* 26 (1979), pp. 603-610.

[3] M. J. Chung and H.-A. Choi, "Parallel Algorithms for the Hamiltonian Cycle on Outerplanar Graphs," Technical Report, Department of Computer Science, Michigan State University, 1990.

[4] T. Hagerup, "Planar Depth-First Search in $O(\log n)$ Parallel Time," *SIAM J. Comput.*, August 1990, Vol. 19, No. 4, pp. 678-704.

[5] X. He and Y Yesha, "Binary Tree Algebraic Computation and Parallel Algorithms for Simple Graphs," *J. Algorithms*, 1986.

[6] X. He and Y Yesha, "Parallel Recognition and Decomposition of Two Terminal Series-Parallel Graphs," *Information and Computation*, 75 (1987), pp. 15-38.

[7] R. Karp and V. Ramachandran, "A Survey of Parallel Algorithms for Shared-Memory Machines," Technical Report.

[8] G. L. Miller and J. Reif, "Parallel Tree Contraction and Its Application," *26th IEEE Symp. on Foundations of Computer Science*, 1985, pp. 852-865.

[9] A. Proskurowski and M. M. Syslo, "Efficient Vertex- and Edge-Coloring of Outerplanar Graphs," *SIAM J. Alg. Desc. Meth.*, Vol. 7, No. 1, January 1986, pp. 131-136.

[10] Y. Shiloach and U. Vishkin, "An $O(\log n)$ Parallel Connectivity Algorithm," *J. Algorithm*, 3, 1982, pp. 57-67.

[11] R. E. Tarjan and U. Vishkin, "Finding Biconnected Components and Computing Tree Functions in Logarithmic Time," *Proc. 25th IEEE Found. of Computer Sci.*, 1984, pp. 12-20.

Graph-partitioning on shared-memory multiprocessor systems

K. S. Natarajan

IBM Research
T. J. Watson Research Center
Yorktown Heights, NY 10598

Abstract

In this paper we consider the problem of partitioning the nodes of a graph into two subsets of given sizes such that the total number of edges with end vertices in different subsets is minimized. We develop parallel, randomized, iterative-improvement based algorithms for the graph partitioning problem. We present the measured performance of the parallel algorithm that was implemented on a shared-memory multiprocessor system. Experiments were run on graphs with upto 200 nodes. The quality of results produced by a 8-processor parallel algorithm on large, dense graphs were comparable to those of the serial algorithm. In contrast, the parallel algorithm did not perform as well on small, sparse graphs.

1. Introduction

The problem of partitioning of graphs has numerous applications in electronic design automation [HU85]. In physical design automation, typically, the design task includes partitioning the logic into blocks of components, placing the components and interconnecting them according to given wiring requirements such that the number of interconnections crossing partition boundaries are minimized. This gives rise to a problem where the set of vertices of a graph have to be partitioned in such a way that the number of edges whose end nodes are in different subsets of the partition is minimized. In a more general formulation of the graph-partitioning problem, one or more of the following additional variations are often required:

- Weights are associated with the vertices and upper limits are placed on the sum of the weights associated with the vertices in each subset

- Arbitrary nonnegative weights are associated with the edges of the graph, and the objective is to find a partition that minimizes the sum of the weights of edges whose end nodes are in different subsets,

- Certain vertices are constrained to be in specific partitions.

Applications that include the more general kinds of constraints arise in parallel processing systems [BOK87]. In such systems the efficient mapping of an application onto a multiprocessor system requires the assignment of modules of the parallel program to processors such that the total inter-processor communication is minimized.

We consider how the parallel processing capability of a multiprocessor system can be used for solving graph partitioning problems. Specifically, we consider graphs whose vertices have unit weight and edges have nonnegative integer weights. We seek parallel algorithms that can partition the vertex set into two equal subsets (or, nearly equal if the number of vertices is odd) in such a way that the sum of the weights of edges that have end vertices in both subsets is a minimum.

We focus on parallel search algorithms that use the ideas of iterative improvement and randomization to achieve near-optimal solutions. The basic idea of the parallel algorithm is to *simultaneously* generate and evaluate several potential perturbations (or moves) of a partition on different processors. In this paper, we consider a move to be an interchange of a vertex from one subset with a vertex from the other subset of the partition. When several moves are considered simultaneously, it is likely that some of the moves interact with each other. If M_1, M_2 are two interacting moves, then the benefit (i.e., incremental change in the cost of a partition) of making M_1 can depend on whether move M_2 is made or not. Parallel algorithms that allow interacting moves to be evaluated and *accepted* simultaneously are likely to yield solutions that are worse than those produced by a 1-processor version of the same algorithm. Moreover, performance of the parallel partitioning algorithm depends on the density of graphs being partitioned. For dense graphs, our experiments indicate that simultaneous acceptance of interacting moves has less impact on quality of partition produced by parallel algorithm.

The rest of the paper is organized as follows. In Section 2 we state the version of the graph-partitioning problem considered in this paper. In Section 3, we outline the proposed parallel algorithm and describe the key aspects of a shared-memory multiprocessor implementation of the algorithm. In Section 4, we present detailed experimental results on the performance of the parallel partitioning algorithm. In Section 5, we present our conclusions.

2. Model

Let $G = (V,E)$ be an undirected graph with V, a set of vertices, and E a set of edges. Let s be a function defined on the set of vertices such that $s(v_i)$ is the size of vertex v_i. Let c be a function defined on the set of edges such that $c(v_i, v_j)$ is a cost associated with edge (v_i, v_j). If $W \subseteq V$, then the size of W is defined as $|W| = \sum_{v_i \in W} s(v_i)$.

Given a positive integer M, a 2-way partition of the graph is a collection $\{L, R\}$ of subsets of V such that $L \subseteq V$, $R \subseteq V$, $L \cup R = V$, and $L \cap R = \phi$, $|L| \leq M$ and $|R| \leq M$ where M is the maximum size of a partition. The cost associated with the partition is defined as $COST(L,R) = \sum_{v_i \in L} \sum_{v_j \in R} c(v_i, v_j)$

Consider two vertices $v_i \in L$ and $v_j \in R$. The external costs of v_i and v_j are defined as:

$$EC(v_i) = \sum_{v_k \in R} c(v_i, v_k)$$

$$EC(v_j) = \sum_{v_k \in L} c(v_j, v_k)$$

The internal costs of vertices v_i and v_j are:

$$IC(v_i) = \sum_{v_k \in L} c(v_i, v_k)$$

$$IC(v_j) = \sum_{v_k \in R} c(v_k, v_j)$$

The cost of the partition, denoted $COST(L,R)$ is:

$$COST(L,R) = \sum_{v_i \in L} EC(v_i) = \sum_{v_j \in R} EC(v_j)$$

A partition of G is optimal if it has the minimum cost among all partitions of G. The problem of partitioning a graph is NP-complete [GAR79, p.209]. Hence, efforts to solve graph partitioning problem have been primarily devoted to development of heuristic algorithms that produce good quality solutions. A classical sequential algorithm approach that has been used for graph partitioning is based on the idea of iterative improvement [KER70]. One begins with an initial feasible partition (L,R) and repeatedly attempts to improve upon it by directed perturbations (such as selective exchanges of pairs of vertices or displacement of vertices from one subset to the other) of the current partition.

Suppose vertices v_i and v_j are interchanged so that (L', R') is a modified feasible partition, where $L' = L - \{v_i\} + \{v_j\}$ and $R' = R - \{v_j\} + \{v_i\}$. The cost of the modified partition, $COST(L', R')$ is:

$$COST(L', R') = COST(L,R) + IC(v_i) + IC(v_j)$$

$$- EC(v_i) - EC(v_j) + 2 c(v_i, v_j)$$

The incremental change in costs, ΔC is:

$$\Delta C = IC(v_1) + IC(v_j) - EC(v_i) - EC(v_j) + 2 c(v_i, v_j)$$

The parallel partitioning algorithm that we develop in this paper is based on the method of optimization by simulated annealing [KIR83]. A detailed experimental study of the application of simulated annealing to the graph partitioning problem is reported in [JOH89].

3. Parallel Partitioning Algorithm

The main idea of the parallel algorithm we develop is to let multiple processors perturb and modify a partition at the same time. Next we consider what can happen when multiple processors attempt to improve a partition by simultaneously exchanging different pairs of vertices. Define the following terminology. Let the set of vertices adjacent to vertex v_i be denoted as $N(v_i)$. Let $M_1(v_1, v_2)$ be a move that interchanges vertices v_1 and v_2 from an existing partition. Let $Dom(M)$ be the set of vertices affected by move M. Then, $Dom(M_1) = N(v_1) \cup N(v_2)$. Let $M_2(w_1, w_2)$ be a move that interchanges vertices w_1 and w_2 from a partition. Then, $Dom(M_2) = N(w_1) \cup N(w_2)$. The two moves, M_1 and M_2, are independent if $Dom(M_1) \cap Dom(M_2) = \phi$. Otherwise, the two moves are *interacting* (or interfering).

Assume P_1 and P_2 are two processors that attempt to modify a partition concurrently. If processor P_i attempts move M_i, $i = 1, 2$, and the two moves are independent, the net effect on the cost of a partition due to M_1 is the same regardless of whether Move M_2 actually takes place or not. However, for some pair of dependent moves M_1, M_2 the following can be true:

$$\Delta C(M_1 \mid M_2 \text{ takes place}) \neq$$
$$\Delta C(M_1 \mid M_2 \text{ does not take place})$$

We next outline the main steps of a parallel algorithm for solving the graph partitioning problem stated in Section 2. If the algorithm is run with just one processor, the algorithm is the same as a sequential simulated annealing algorithm for the problem.

The state of a partition is captured by the values of the internal costs and external costs of the nodes in the graph. In the inner loop, a move (i.e., a perturbation of an initial partition) is proposed and its effect on the cost of the perturbed partition is evaluated. For the partition problem considered in this paper, the move set consisted of pairwise interchange of two randomly chosen vertices from an existing partition. Evaluation of a move consists of incremental computation of the changes in the internal and external costs of vertices affected as a result of a proposed move. If the move causes the cost function to decrease, then it is accepted.

Acceptance of a proposed move consists of *modifying* the internal and external costs of vertices that are affected by the move as well as cost of the configuration. If the cost function does not decrease, then the proposed move is accepted with a certain probability determined by the magnitude of the change in cost and the temperature T. The cycle of Propose-Evaluate-Accept-or-Reject move is repeated until an inner loop criterion specified by the user is satisfied.

```
repeat {
  repeat_in_parallel {
    Propose_in_parallel a move;
    Evaluate ΔC;
/* increase in cost due to the proposed move */
    if ΔC ≤ 0
    then Accept_in_parallel the proposed move;
else Accept_in_parallel with prob. e^(-ΔC/kT);
        /* T is temperature , k is a constant */
  } until Inner loop criterion is satisfied;
    Update_in_serial T and global state;
} until Outer loop condition is satisfied.
```

Outline of Parallel Partitioning Algorithm

The state of a partition is captured by *globally-shared* values of:

- Internal costs, i.e., IC_i, $i = 1, \dots, N_V$, and
- External costs, i.e., EC_i, $i = 1, \dots, N_V$.

If $k = \infty$, then only downhill moves are accepted. Computation in the inner loop is shared among multiple processors as follows. Each processor proposes a move (a pair of vertices for interchange) in parallel with other processors. Since multiple processors attempt to exchange vertex pairs, we must ensure that no more than one processor attempts to move a vertex at any instant of time. We associate a lock with each vertex. The lock is set whenever a vertex is being considered for a potential interchange with a vertex from the other set. Before a processor proposes to move a vertex, it first checks to see if the lock associated with the vertex is free. If so, it sets the lock and proceeds further. Otherwise, the processor tries at random to find another vertex whose lock is free. The integrity of the partition is thus maintained by allowing only exclusive access to the vertices in a partition.

The effect of a move proposed by a processor is evaluated by it under the "independence" assumption that no other vertex pairs are being interchanged at the same time by any other processor. If a move is accepted by a processor, it executes the **Accept_in_parallel** step. This step consists of *modifying* the globally shared internal and external cost values of the vertices that are affected by the move. In a multiprocessor implementation, several processors may attempt to modify the shared internal and external cost values at the same time. In order to ensure correct implementation

of the parallel algorithm, we use locks to obtain exclusive access to the internal and external cost values of vertices, update them and finally release their locks. The cycle of Propose-in-parallel-Evaluate-Accept-in-parallel move is repeated until the inner loop criterion specified by the user is satisfied. When the inner loop criterion is satisfied, then a designated processor executes the **Update_in_serial** step. The processor updates in serial the temperature T, evaluates the current state of the partition and its associated cost. If the outer loop criterion is satisfied, then execution stops. Otherwise, the partition resulting from the current iteration is used as a starting point and parallel execution of the next iteration of the inner loop is begun.

4. Experimental Results

The parallel partitioning algorithm was implemented on ACE, a shared-memory multiprocessor workstation developed at the IBM T. J. Watson Research Center at Hawthorne, NY [GAR89]. The system is composed of eight ROMP based processor nodes, each with 8 Mbytes of local memory and floating point hardware support. In addition, a processor can access the local memory of other processors as well as shared-memory cards via a custom, 32-bit, 80 Mbyte/sec inter-processor communication bus. The ACE-system on which the experiments were run has an enhanced version of the MACH UNIX operating system with extensions to support the ACE architecture. For further details on the ACE-system, the reader is referred to [GAR89].

The parallel program was written in PREFACE FORTRAN, an extension of FORTRAN that allows parallel constructs. The user marks up a sequential FORTRAN program indicating in essence the following: a) the data to be shared, b) the work that can be executed in parallel, and c) appropriate synchronization requirements. Parallel programs written in this extended-FORTRAN language are recognized and translated by PREFACE, a FORTRAN preprocessor for parallel workstation systems [BER88]. In order to assess the performance of the parallel partitioning algorithm, we considered the following instances of the partitioning problem. The problems are described below:

- Problem $GRID(m,n)$ is a grid graph with mn vertices arranged along m rows and n columns.
- Problem $RANDOMGRAPH(N,p)$ is a random graph with N vertices for which the probability of an edge between any pair of vertices is p.

4.1 Measured results on Problem GRID(m,n)

In all our experiments, we used a geometric annealing schedule with the following annealing parameters: $N_1, N_2, N_3,$ and α. For a specific value of the control parameter T, the program attempted N_1 moves per vertex; If N_2 moves per vertex were accepted at a given value of T, then the value of

T is reduced to $T' = \alpha T$, where $0 < \alpha < 1.0$. If there is no improvement in the score at N_3 consecutive values of T or the value of T is less than $T_{threshold}$, then the algorithm is deemed to have converged. Following values were used: $N1 = 50$, $N2 = 5$, $N3 = 10$, $\alpha = 0.90$. Each problem was solved 50 times using different starting seeds but with the same initial partition. The average cost of partition with 50 independent runs of the serial algorithm was 10.2 (cost of optimal partition = 10). The average amount of CPU time per serial run was 19.7 seconds.

In order to assess the performance of the parallel algorithm, we kept all annealing parameters the same as in the sequential execution and executed 2-processor parallel runs on Problem GRID(10,10). Measurements were again conducted for 50 independent runs of the parallel algorithm. The average cost of partition with 50 independent 2-processor runs was 10.31 and the standard deviation was 0.70. The total amount of CPU time (summed over two processors) per problem was 28.4 seconds. Experiments were repeated with 50 independent 4-processor runs on Problem GRID(10,10). The solutions had an average cost = 12.04 and a standard deviation = 3.74. The total amount of CPU time (summed over the 4 processors) per run was 27.5 seconds. The measurements, summarized in Table 1, revealed an increase in both the mean and the standard deviation of the final score achieved with the parallel algorithm. Since solution quality became worse with increasing parallelism, no runs were attempted with more than four processors in a run.

4.2 Measured results on Problem RANDOMGRAPH(N,p)

We conducted experiments with random graphs with N vertices for which the probability of an edge between a vertex pair is p. The performance of the algorithm was studied for different graph sizes (N values), graph density (p values) and number of processors.

Graph with 50 vertices: A random graph with 50 vertices was generated with $p = 0.2$. Five independent runs each were conducted with different number of processors (varied from 1 to 8) per run. The results are shown in Table 2.

Graphs with 100 vertices: Random graphs with 100 vertices were generated. Different values of $p(= 0.1, 0.2, 0.5)$ were used to generate graphs with different density. For each graph, five independent runs were conducted for a given number of processors per run. The results corresponding to the three values of p are shown in Tables 3, 4 and 5.

We observed that the quality of solutions produced by the parallel algorithm was not very sensitive to the graph density. For a graph with high density, say, $p = 0.5$, parallel moves are extremely likely to interfere with one another. One might

expect that accepting such parallel interfering moves should significantly degrade the quality of partitions produced by a parallel run using many processors. In our experiments, we did not observe such partition quality degradation. An intuitive explanation for this is the following. In dense graphs, there are possibly several partitions that have comparably good quality. The parallel algorithm appears to succeed in finding one of the many good partitions even though it may not be the same one found by the sequential algorithm. The effects of accepting interacting moves in dense graphs are not as severe as one might have expected.

Graphs with 200 vertices: A random graph with 200 vertices was generated with $p = 0.03$. Five independent runs each were conducted with different number of processors (varied from 1 to 6) per run. The results are shown in Table 6.

5. Conclusions

We developed and implemented a parallel, heuristic algorithm for the problem of partitioning the nodes of a graph into subsets of given sizes such that the total number of edges with end vertices in different subsets is minimized. The algorithm uses the ideas of iterative improvement and controlled, probabilistic hill-climbing. When run with 1 processor the algorithm reduces to a sequential simulated annealing algorithm. We presented the measured performance of the parallel algorithm that was implemented on a shared-memory multiprocessor system. Graphs of different sizes and density were used in the experiments. The quality of results produced by the parallel algorithm showed dependence on both size and density of graphs. Experiments using varying number of processors (1 to 8) were run on graphs with upto 200 nodes. The quality of results produced by a 8-processor parallel algorithm on large, dense graphs were comparable to those of the serial algorithm. In contrast, the parallel algorithm did not perform as well on small, sparse graphs.

References

[BER88] D. Bernstein and K. So, "PREFACE - A FORTRAN preprocessor for parallel workstation systems," *IBM Research Report,* RC 13600, March 1988.

[BOK87] S. Bokhari, "Assignment problems in parallel and distributed computing," *Kluwer Academic Publishers,* Norwell, Mass., 1987.

[GAR89] A. Garcia, R. Freitas and D. J. Foster, "The ACE multiprocessor workstation," *IBM Research Report RC14491,* 1989.

[GAR79] M. R. Garey and D. S. Johnson, "Computers and Intractability: A Guide to the Theory of NP-Completeness ," W. H. Freeman and Co., San Francisco, CA, 1979.

[HU85] T. C. Hu and E. S. Kuh, "VLSI Circuit Layout," IEEE Press, New York.

[JOH89] D. S. Johnson, C. A. Aragon, L. A. McGeoch and C. Schevon, "Optimization by simulated annealing: An experimental evaluation, Part I (Graph partitioning)," *Operations Research,* Dec. 1989.

[KER70] B. W. Kernighan and S. Lin, "An efficient heuristic procedure for partitioning graphs," *Bell Systems Technical Journal*, vol. 49, no. 2, pp.291-307, 1970.

[KIR83] S. Kirkpatrick, C. D. Gelatt, Jr., and M. P. Vecchi, "Optimization by simulated annealing," *Science*, 220, pp.671-680, 1983.

Table 1

GRID (10,10) (Ave. degree = 3.6)
$N1 = 50$ (Moves attempted per vertex)
$N2 = 5$, $N3 = 10$, $\alpha = 0.9$
Number of trials = 50

	Number of processors		
	1	2	4
Ave. Cost	10.2	10.3	12.0
Std. Devn.	0.3	0.7	3.74
CPU Secs	19.7	284	27.5

Table 2

50 Vertices, p = 0.2 (Ave. degree = 10)

	1	2	4	8
1	75	77	79	75
2	79	80	75	75
3	75	78	78	78
4	77	80	77	81
5	75	75	76	77
Ave. Cost	71.2	78	77	77.2
CPU Secs	10.4	14.5	17.4	49.5

Table 3

100 Vertices, p = 0.1 (Ave. degree = 10)

	1	2	4	8
1	143	144	150	147
2	141	149	145	152
3	143	142	143	143
4	142	142	147	151
5	141	143	146	143
Ave. Cost	142	145	146.2	147.2
CPU Secs	25.4	35.6	42.7	124

Table 4

100 Vertices, p = 0.2 (Ave. degree = 20)

	1	2	4	8
1	351	353	349	356
2	351	349	350	359
3	353	350	351	359
4	349	349	350	356
5	351	350	359	352
Ave. Cost	351	350.2	351.8	356.2
CPU Secs	21.4	30.1	36.1	86.2

Table 5

100 Vertices, p = 0.5 (Ave. degree = 50)

	1	2	4	8
1	1056	1052	1055	1058
2	1058	1055	1055	1057
3	1056	1053	1056	1070
4	1055	1059	1052	1058
5	1058	1055	1056	1066
Ave. Cost	1056.6	1054.8	1054.8	1061.8
CPU Secs	19.0	27.7	32.9	91.8

Table 6

200 Vertices, p = 0.03 (Ave. degree = 6)

	1	2	4	6
1	44	44	45	48
2	39	40	42	46
3	44	51	45	44
4	40	40	39	46
5	39	43	46	45
Ave. Cost	41.2	43.6	43.4	45.8
CPU Secs	119	133	170	150

Parallel Join Algorithms for SIMD Models

Shiva Azadegan and Anand Tripathi
Department of Computer Science
University of Minnesota
Minneapolis, MN 55455

Abstract: This paper presents two SIMD join algorithms for the Connection Machine (CM-2). These algorithms are hash based, i.e. the tuples in a relation are divided into different buckets based on the hash value of the join attribute. We present here results of the experimental evaluations of these algorithms for different values of design parameters and work-load. In the first algorithm the buckets are maintained in a centralized fashion in the sense that the tuples in a bucket are maintained at one CM processor. In the second algorithm the buckets are maintained in a distributed fashion, i.e., the tuples in a bucket are stored in an array of processors. Two different approaches to the creation of buckets are presented. We also evaluate the impact of different CM programming primitives in implementing these algorithms. The experimental results indicate that in general the distributed bucket table scheme performs better than the centralized bucket table scheme by an order of magnitude. Also, in general, having a hash function with a larger range, i.e. having a larger number of buckets, gives better performance.

Keywords: Join operator, relational database, hashing, parallel algorithms, SIMD models, Connection Machine

1. Introduction

The *join* operation is one of the most important and costly operations in relational database models. It combines two relations over their common attribute(s), and it is commonly used in the data retrieval queries which reference more than one relation .

During the last decade, significant amount of work has been done in developing efficient *join* algorithms. A number of parallel join algorithms [10, 8, 13] have been also developed for multiprocessor systems or computer networks based on the MIMD models [3] of parallel computing. Due to the recent advances in computer technology, large scale massively parallel computers are becoming commercially available. In such environments, where a very large number of relatively simple and cheap processors are available, we can have a fine grained distribution of data as compared to the approaches which distribute relations across disk clusters.

The motivation behind our work is to design efficient SIMD [3] join algorithms for a massively parallel machine. In this paper, we present the design, implementation, and evaluation of two parallel join algorithms for the Connection machine®[6]. Conceptually, these parallel algorithms are based on the techniques used in many of the sequential join

Connection Machine usage for this work was supported by the Army High Performance Computing Research Center, University of Minnesota.

® Connection Machine is a registered trademark of Thinking Machines Corporation.

algorithms. These algorithms are experimentally evaluated for different design parameters, implementation choices in terms of the CM programming primitives, and work-load.

The remainder of this paper is organized as follows. In section 2 a brief overview of the existing join algorithms is presented. The computational model of the CM and the description of the primitive operations that were frequently used in our algorithms are presented in section 3. Section 4 contains the notation and assumptions made in our system. In section 5 we discuss two SIMD join algorithms. The empirical results and comparisons of these algorithms are presented in section 6, and the conclusions of our study are presented in section 7.

2. Overview of Join Algorithms

In a uniprocessor environment, the following three types of *join* algorithms[9] have been proposed.

Nested Loops: In this scheme, each tuple in one relation is sequentially matched against every tuple in the other relation, and a matching pair with the same values of join attribute(s) is combined to produce the output tuple.

Sorted-Merge: In this scheme, first both relations are sorted based on one of the join attributes, then the sorted relations are merged and the tuples with matching join attribute(s) are combined to produce the output tuple.

Hashing: In this scheme, one relation is stored in a hash table by hashing on the join attribute(s). Then the hash values of the tuples in the other relation are used to probe the hash table. The tuples with matching join attribute(s) are combined to produce the output tuple.

Hash-based join algorithms have been shown [7] to be more efficient under most conditions; moreover, they also have the property of being easy to parallelize. Several variations of sequential hash-based join algorithms have been proposed and studied. Among which are *simple−hash/classic−hash* [11], *Grace−hash* [7], and *hybrid−hash* [7]. Some parallel versions of the above algorithms for the MIMD models have also been presented in the literature [4, 12]

3. Connection Machine Execution Model

The Connection Machine Model CM-2 [6, 1] is a SIMD machine. At the heart of the CM is the parallel processing unit (PPU), which is configured to have as many as 4 sections and each section may be composed of 8K or 16K processors (PEs). Each PE has 64K or 256K bits of bit-addressable memory. A processor may be set into an active or inactive state based on a test across all processors. The PPU is designed to operate under the programmed control of a front-end computer which issues instructions to the PPU.

The Connection Machine provides the virtual processor facility, whereby each PE is used to simulate some number of virtual processors. Every processor (physical/virtual) can be identified by a unique *send address*. We refer to each virtual processor as a *node*.

There are two types of interprocessor communication facilities available in the CM. The general communication is performed by a special-purpose hardware called *router*, which uses the hypercube interconnection. All processors can simultaneously send data to the local memory of other processors, or fetch data from the local memories of other processors into their own; message routing takes place in parallel. The NEWS communication organizes the processors into a multidimensional rectangular grid and supports message passing between the neighbors on the grid.

Following is a brief description of some of the primitives in the parallel instruction set (Paris)[2] of the CM that were extensively used in implementing the parallel join algorithms. In the description of the algorithms we show only the following parameters for the send and get instructions: 1) *dest−field*: is the identifier of the field where the content of the message is to be deposited; 2) *dest−addr*: is the address of destination or source processor in the *send* and *get* instructions, respectively; 3) *src−field*: is the identifier of the field from which the information is obtained.
SEND: This primitive is used for remote write operations. Every selected processor sends a message to a specified destination processor, and messages are all delivered to the same memory address within each receiving processor; a processor may receive messages even if it is not active in the current cycle. If a processor receives more than one message, then the message data received by the processor will not be predictable.
GET: This primitive is used for remote read operations. Every selected processor sends a message to a specified source processor, and messages are all delivered to the same memory address within each receiving processor; a processor may supply messages even if it is not active in the current cycle. It is possible that more than one processor request data from the same source processor in which case the same data is sent to each of the requesting processors.
SCAN: In general the *scan* operation [5] takes a binary operator @ with identity i, and an ordered set $[a_0, ..., a_{n-1}]$ and it returns the ordered set $[i, a_0, (a_0 @ a_1), ..., (a_0 @ a_1 @ \cdots @ a_{n-2})]$. In our implementation we used *segmented scan* operations [5]. In these operations we break the linear ordering of the nodes into segments, and a *scan* operation starts again at the beginning of each segment.
ENUMERATE: The CM-enumerate operation returns the number of active processors below or above every selected processor in some ordering of the processors. In the description of our algorithms, we show only one parameter for the enumerate instruction. The *dest−field* that is the identifier of the field where the enumeration value is to be deposited.
RANK: The CM-rank operation provides a built-in sorting primitive. The operation takes a corresponding value from each processor and returns the position of that value in a sorted ordering of the entire set of values. In the description of our algorithms, we show only two parameters for the rank instruction. The *dest−field* that is the identifier of the field on which the rank operation is performed, and the *rank−field* that is the identifier of the field where the rank value is to be deposited

In reference[5] it is argued that certain scan operations, also known as prefix computations, are as efficient as parallel memory references in the PRAM model. Therefore one can assume that the cost of such operations is the same as the unit cost assumed for parallel memory access in the PRAM model.

4. Notations and Assumptions

Let **R** and **S** be the two relations to be joined. The number of tuples in these two relations is denoted by |R| and |S|, respectively. We assume that **S** is the larger relation, i.e. $|R| \leq |S|$. Also, we assume that the join is performed on only one attribute. Our algorithms can be easily extended to multiple attributes.

The notation [P]:Q in the description of the algorithms means that all the nodes which satisfy condition P execute the statement Q. Let N denote the maximum number of nodes available in our machine. We assume that $|S| \leq N$. Considering that we are dealing with virtual processors and are not limited by the number of physical processors, this is a reasonable assumption. For instance, if we allot 1k bytes of memory to each node, on a fully configured CM with 64k processors we can represent relations of size over two million tuples.

5. SIMD Join Algorithms

This section describes two hash-based SIMD join algorithms. The performance comparisons of these algorithms is presented in the next section. Similar to the other hash-based join algorithms [11], we partition the relations **R** and **S** into disjoint subsets called *buckets*. To partition these relations, first we have to choose a hash function h. Let the set $\{H_0, ..., H_{n-1}\}$ represent the range of the function h. Using this hash function we partition the relations **R** and **S** into corresponding *buckets* $R_0, ..., R_{n-1}$, and $S_0, ..., S_{n-1}$, respectively, such that a tuple r of **R** is in R_i if the hashed value of its join attribute is equal to H_i. Clearly, since the same partitioning scheme is used for both relations, tuples in bucket R_i only have to join with the tuples in S_i.

Figure 1 shows the general structure of our algorithms. Steps 1 and 5 are the same for all the algorithms. Therefore, for comparative evaluation of different algorithms, we choose to exclude the time required by these steps when evaluating the algorithms. In step 1 the tuples in each relation are loaded into the machine, such that each tuple from a relation is stored on a different *node*. That is, *node$_i$* contains either both tuples r_i and s_i for $i \leq |R|$, or only tuple s_i for $i > |R|$. The output tuples are also maintained by each node, separately. In the case where the size of output tuples is greater than the available space, the generated output is written periodically to the parallel disk, called DataVault[1].

```
BEGIN
    step1: Input-data();
    step2: Apply-hash-fnc();
    step3: Construct-buckets();
    step4: Perform-join();
    step5: Output-result();
END
```

Figure 1: Outline of SIMD Join Algorithms

Step 2 is also the same in both algorithms. In this step, shown in Figure 2, we apply the hash function h to the join attribute. Due to the data parallelism inherent in SIMD models, the time required by this step is independent of the size of the relations. It is proportional to the complexity of the hash function. Moreover, in both algorithms we choose the hash function such that $H_0, ..., H_{n-1}$ are either valid node addresses, or they can be uniquely translated to node addresses. In the following sections we discuss steps 3 and 4 for each algorithm.

5.1. Centralized Bucket Table Scheme

In this scheme the bucket for each hash function value is maintained as an array in one node's memory. We refer to these arrays as bucket tables. This section describes the construct-bucket and perform-join procedures for the centralized bucket table join algorithm.

```
Apply-hash-fnc()
BEGIN
    [All the Nodes Containing R Tuples]:
        R-hashed-attr= h(R-join-attr);

    [All the Nodes Containing S Tuples]:
        S-hashed-attr= h(S-join-attr);
END
```

Figure 2: Hashing Step

To construct the buckets in step 3, all the nodes that contain the smaller relation, **R**, participate. The bucket table for each R_i is constructed at the node whose address is equal to H_i. Each node whose $R-hashed-attr$ value is equal to H_i writes to a different element of the bucket table at node H_i. In order to allow concurrent writes to the bucket tables, we first compute the index of bucket table element to which a particular node is to write. We have experimentally evaluated two approaches for computing these indices:

[1] **Unsorted Bucket Table:** The indices are generated in such a way that the order of the tuples in the bucket table is the same as the order of the address of the nodes on which the tuples reside. For the comparative performance analysis of the CM primitives, we implement the index computation in two ways. In the first scheme, sequentially for each hash value H_i, we enumerate the nodes belonging to that bucket. The enumeration value for each node becomes its tuple's index into the bucket table. We refer to this scheme, shown in Figure 3, as sequential bucket construction (SBC).

In the second scheme, referred to as parallel bucket construction (PBC) and shown in Figure 4, we use the CM *rank* operation to rank the tuples based on hashed value of the join

```
Construct-bucket()
BEGIN
    [All the Nodes Containing R Tuples]:
        for (i=0; i<n; i++)
        {
            [R-hashed-attr = $H_i$]:
                CM_enumerate(My-index);
        }
        CM_send(Dest-addr: R-hashed-attr,
            Dest-field:Bucket-table[My-index],Src-field:R-tuple)
END
```

Figure 3: Sequential Bucket Construction for
Unsorted Centralized Bucket Table Scheme

attribute. Therefore, the tuples that belong to the same bucket have consecutive rank values.

Next step is to compute each tuple's index within its bucket. This is accomplished by identifying the starting (minimum) rank value in each bucket. For this purpose, we use as communication channels the nodes whose addresses are equal to the hash function values. Each node first writes, using the $CM-send-with-min$ instruction, its rank value to the node whose address is equal to the hashed value of its join attribute. It then obtains the starting rank by reading from the same node the result of the preceding concurrent write operation. To compute the bucket table index, each node subtracts the obtained starting rank value from its own rank.

[2] **Sorted Bucket Table:** The indices are generated in such a way that the tuples in the bucket table are sorted based on the join attribute. Shown in Figure 5 is the PBC procedure for generating the indices in the sorted approach. It is similar to the PBC procedure in the unsorted approach. The only

```
Construct-bucket()
BEGIN
    [All the Nodes Containing R Tuples]:
        CM-rank(R-hashed-attr, My-rank);
        CM-send-with-min(Dest-addr:R-hashed-attr;Dest-field:Offset,
            Src-field: My-rank);
        CM-get(Dest-addr:R-hashed-attr;Dest-field:Offset,
            Src-field: offset);
        My-index = My-rank - offset;
        CM_send(Dest-addr: R-hashed-attr,
            Dest-field:Bucket-table[My-index],Src-field:R-tuple)
END
```

Figure 4: Parallel Bucket Construction for
Unsorted Centralized Bucket Table Scheme

difference is the field that is used in the rank operation. The field *pair−attrs* in each node contains the concatenation of the hashed value of the join attribute starting at the most significant bit and the join attribute at the lower significance bits. We rank the tuples based on the pair-attrs field. Therefore, the tuples that belong to the same buckets are sorted based on their join attribute.

The SBC scheme in this approach is similar to that of unsorted approach. We iteratively rank the tuples that have the same hashed value of the join attribute. Since we are performing an expensive operation(i.e. rank operation) sequentially, this method becomes a very inefficient way to construct the buckets. Thus, we do not consider the SBC method for the sorted approach as a candidate method. However, to give our readers an idea about its execution time, we included the experimental results in section 6.

After the indices are computed all nodes write in parallel their tuple to the corresponding bucket table element.

```
Construct-bucket()
BEGIN
    [All the Nodes Containing R Tuples]:
        Pair-attrs = <R-hashed-attr, R-join-attr>;
        CM-rank(Pair-attrs, My-rank);
        CM-send-with-min(Dest-addr: R-hashed-attr;
            Dest-field: Offset, Src-field: My-rank);
        CM-get(Dest-addr: R-hashed-attr;
            Dest-field: Offset, Src-field: offset);
        My-index = My-rank - offset;
        CM_send(Dest-addr:R-hashed-attr,
            Dest-field:Bucket-table[My-index],Src-field:R-tuple)
END
```

Figure 5: Bucket Construction for
Sorted Centralized Bucket Table Scheme
(Parallel Bucket Construction)

Figure 6 presents the join algorithm for the centralized bucket table scheme. In this figure, any of the above *construct−bucket* procedures can be used in step 3. In our current implementation, we store the complete tuple into the bucket table. However, in the case where the tuple size is very large, we can store only the address of the node which contains the tuple as the tuple ID. Upon a match the tuple can be retrieved from the given address.

To perform the join operation in step 4, all the nodes that contain the **S** relations participate. Each participating node retrieves the elements of the bucket table stored in the node whose address is equal to the $s−hashed−attr$, and it compares its own join attribute with that of the retrieved tuple. The output tuple is generated if a match is found. In the case where the entries in the bucket table are unsorted (shown in Figure 6), this procedure is repeated as many times as the maximum size of the bucket tables. However,

Centralized_Bucket-Table()
BEGIN

Step1: Load-data();

Step2: Apply-hash-fnc();

Step3: /* *Any of the construct bucket methods can be used here.* */
Construct-bucket();

Step4: [All the Nodes Containing S Tuples]:
    for (i=0; i<Max-BT-size; i++)
    {
        CM-get(Dest-addr: S-hashed-attr,
            Dest-field: Bucket-table[i], Src-field: Temp-tuple)
        If (Temp-tuple.join-attr = S-join-attr)
        {
            Step 5:Result-table[j] = Join(Temp-tuple, S-tuple)
                j++;
                If (j > Max-space)
                    Write Result-table to Disk;
        }
    }
END

Figure 6: Join Algorithm with the
Centralized Bucket Table

in the sorted case, a node stops when it retrieves a tuple whose join attribute value is greater than its own tuple's. (We assume the entries are in non-decreasing order.) The output tuple is inserted into the result table, which is maintained at each node. The result tables that are full are periodically written to the disk.

Note that the maximum size of the bucket table can be determined dynamically after the bucket table indices are computed. We use $scan-with-max$ primitive in order to identify the maximum size required by the bucket tables. The centralized bucket table algorithm poses a problem if the bucket size exceeds the maximum allocated space for the bucket table at a node. Because the goal of our experiments was to perform comparative evaluation of various algorithms for the average cases where the tuples in a relation nearly evenly distribute over different buckets, during our experiments we could afford to ignore this problem temporarily. It is also important to note here that the distributed bucket table algorithm, presented in the next section, does not pose this kind of space allocation problem. The performance evaluation experiments, presented in section 6, clearly show that the distributed bucket table algorithm is far superior to the centralized one. Hence, the distributed scheme is the algorithm of choice among the two. Addition of any dynamic space allocation mechanisms to the centralized bucket table algorithm will only degrade its performance.

5.2. Distributed Bucket Table Scheme

In this scheme the buckets are maintained distributedly; they are spread over several consecutive nodes. That is, if R_0 contains 10 tuples and R_1 contains 15 tuples, nodes 0 to 9 contain the tuples for R_0 and nodes 10 to 24 contain the tuples for R_1.

To construct the buckets in step 3, first we rank the nodes based on their hashed value of the join attribute using the CM $rank$ operation. Next, we rearrange the tuples using the node's rank value in order to place the tuples with the same hashed value of the join attribute on consecutive nodes. Figure 7 depicts an example of tuple configuration before and after the buckets are constructed. In this scheme

	node0	node1	node2	node3	node4	node5
R-tuple	r(a,b)	r(a,c)	r(v,w)	r(z,x)	r(m,n)	r(p,q)
R-hashed-attr	5	5	2	10	2	6
Rank	2	3	0	5	1	4
S-tuple	s(a,x)	s(m,d)	s(a,p)	s(a,w)	s(n,p)	s(i,j)
S-hashed-attr	5	15	5	5	2	1
Rank	2	5	3	4	1	0

(a) Tuple Configuration before the Bucket Construction Step

	node0	node1	node2	node3	node4	node5
R-tuple	r(v,w)	r(m,n)	r(a,b)	r(a,c)	r(p,q)	r(z,x)
R-hashed-attr	2	2	5	5	6	10
Rank	0	1	2	3	4	5
S-tuple	s(i,j)	s(n,p)	s(a,x)	s(a,p)	s(a,w)	s(m,d)
S-hashed-attr	1	2	5	5	5	15
Rank	0	1	2	3	4	5

(b) Tuple Configuration After the Bucket Construction Step

Figure 7

the buckets are constructed for both relations.

Before we perform the join operation, we have to delimit the buckets' boundaries, and inform the nodes in each bucket of those boundaries. Then, the nodes belonging to the same bucket can proceed and perform the join operation.

Construct_buckets()
BEGIN
[All the Nodes Containing R Tuples]:
    construct-bucket-step1(R-hashed-attr,R-rank,R-tuple,
            R-starting-flag,R-ending-flag);
[All the Nodes Containing S Tuples]:
    construct-bucket-step1(S-hashed-attr,S-rank,S-tuple,
            S-starting-flag,S-ending-flag);
    /* *Mark & communicate the bucket Boundaries.* */
[R-Starting-flag = True]:
    CM-send(Dest-addr:R-hashed-attr,Dest-field:Starting-rank,
            Src-field:R-Starting-flag);
[R-Ending-flag = True]:
    CM-send(Dest-addr:R-hashed-attr,Dest-field:Ending-rank,
            Src-field:R-Ending-flag);
[All the Nodes Containing S Tuples]:
    CM-get(Dest-addr:S-hashed-attr,Dest-field:Starting-rank,
            Src-field:Starting-rank);
    CM-get(Dest-addr:S-hashed-attr,Dest-field:Ending-rank,
            Src-field:Ending-rank);
END
construct-bucket-step1(Hashed-attr,Rank,Tuple,
                Starting-flag,Ending-flag)
BEGIN
    CM-rank(Hashed-attr,Rank);
                    /* *Construct Buckets.* */
    CM-send(Dest-addr:Rank,Dest-field:Tuple,
            Src-field:Tuple);
                    /* *Delimit the Bucket Boundaries.* */
    CM-NEWS-get(Direction:Left,Dest-field:Temp-hashed-attr,
                Src-field:Hashed-attr);
    If (Hashed-attr <> Temp-hashed-attr)
        Starting-flag = True;
    CM-NEWS-get(Direction:Right,Dest-field:Temp-hashed-attr,
                Src-field:Hashed-attr);
    If (Hashed-attr <> Temp-hashed-attr)
        Ending-flag = True;
END

Figure 8: Bucket Construction Algorithm for
Distributed Bucket Table Scheme

To delimit the buckets, we mark the *starting* and *ending* nodes in each bucket. To identify the *starting* nodes for relation **R** (**S**) in each bucket, a node obtains the $R-hashed-attr$ ($S-hashed-attr$) from its left neighbor. If the obtained value is different from its own $R-hashed-attr$ ($S-hashed-attr$), then the $R-starting-flag$ ($S-starting-flag$) is set to *True*.

To identify the *ending* nodes for relation **R** (**S**) in each bucket, the hashed value of the join attribute is obtained from the right neighbor. The same procedure as above is followed to set the $R-ending-flag$ ($S-ending-flag$) to *True*.

The last step in the construction of the buckets is to communicate the boundaries of the buckets for relation **R** to relation **S**. This step is performed by using as communication channels the nodes whose addresses are equal to H_0, ..., H_{n-1}. That is, *starting* and *ending* nodes for bucket R_i write their rank values to the node whose address is H_i. Subsequently, all the nodes for bucket S_i obtain these values from node H_i. The reason we require that all the nodes obtain the *starting* and *ending* rank values is for termination purposes. I.e., during the execution of the join algorithm (shown in Figure 9) a node disables itself after it has performed join with the tuple stored in the *ending* node. Figure 8 depicts the algorithm for constructing the buckets.

To perform the join operation in step 4, shown in Figure 9, the *starting* node in S_i iteratively obtains a copy of each tuple in R_i, and then broadcasts it to the other nodes in S_i. We used the CM *segmented scan* primitive operations to broadcast the tuples. As stated previously, the scan operations are very efficient in the sense that they have similar execution time as memory accesses in PRAM models. A node, upon receipt of a new tuple, compares its own join attribute to that of the new tuple. An output tuple is generated and inserted into the result table if a match is found.

```
Distributed-bt-join()
BEGIN
REPEAT

[All the Nodes Containing S Tuples]:
        If (Starting-rank > Ending-rank)
                Disable;

/* Only the starting nodes in each bucket participate in this phase. */
[S-Starting-flag = True]:
        /* Compute the send-address of the node with the given rank. */
        addr = CM-send-address(Starting-rank);
        CM-get(Dest-addr: addr, Dest-field: Temp-tuple; Src-field: R-tuple);
        /*Broadcast the retrieved tuple to the rest of nodes in the bucket. */
        CM-scan-copy(Temp-tuple);

[All the Nodes Containing S Tuples AND Not disabled]:
        If (Temp-tuple.joining-attr = S-joining-attr)
        {
                Result-table[j] = Join(Temp-tuple, S-tuple);
                j++;
                If (j > Max-space)
                        Write Result-table to Disk;
        }
        /* Compute the rank for the next node in the matching bucket. */
        Starting-rank++;
UNTIL (All the Nodes In S_i s Are Disabled)
END
```

Figure 9: Join Algorithm for the Distributed Bucket Table Scheme

As shown in Figure 9, the time complexity for the join step is proportional to the size of the largest R_i. Therefore, to enhance the efficiency of our algorithm, we can pick the smaller relation as the matching relation in each bucket, instead of the default relation **R**. (The matching relation is the one whose tuples are bucketized and then are retrieved by the other relation)

It has been pointed out by DeWitt [8] that one of the problems with hash-based join algorithms is to guarantee that a chosen partitioning of hash values results in buckets that fit in the memory. In the distributed bucket table scheme we do not encounter this problem. This is due to the fact that the nodes for the buckets are allocated dynamically and each bucket is allotted the exact number of nodes it requires. Reiterate that the buckets are maintained over the same array of nodes that is initially used to store the relations, hence, there is always sufficient number of nodes for the buckets.

6. Empirical Results

In conducting the experiments described in this section our primary goal was to investigate the variation in execution times as (1) the algorithm parameters and the problem size were altered, (2) different CM primitive operations were used. In our experiments, we assume a *uniform* data distribution for the relation **R**; that is, tuples from this relation are evenly distributed among the buckets.

We use the C/Paris interface in our implementation. Paris is a low-level instruction set for programming the Connection Machine. The Paris user interface consists of a set of macros, functions, and variables to be called from user code. One of such interfaces is C/Paris in which the program is written in the C language augmented with Paris calls as needed to express parallel operations.

We have conducted three sets of experiments. In the first set we study the effect of the hash function range on the execution time of our algorithms. In the second set, we investigate the effect of the size of the matching relation on the execution time of our algorithms. In the third set, we examine the effect of different CM primitives on the bucket construction time. In Tables 1 and 2, the blank entries are for the cases where the hash function range is of no particular interest for that relation size. Each table entry contains the bucket construction time and the join operation time, obtained by averaging results from 100 runs. The hashing step is almost the same for all the relation sizes and it takes less than 1 msec. The given time is the CM time and is measured in milliseconds. As stated before, the execution times for steps 1 and 5 are the same in both algorithms and are not included in our comparative studies. In our experiments we used only one section of the CM which contains 8K processors.

6.1. Effect of Different Hash Function Range

Tables 1 and 2 show the effect of hash function range on the performance of these parallel join algorithms. Obviously, a larger hash function range results in a smaller bucket size (assuming uniform data distribution); which in turn reduces the join operation time. Reiterate that the join operation time is a linear function of the bucket size. This correlation is clearly shown in our empirical results for both schemes. As the hash function range is increased by a factor of 2, the join operation time is decreased by the same factor.

It is also worth noting that independent of the relation size, the join operation time is almost the same for the buckets of the same size. That is, the join operation time of relation size 1024 and hash function range 64 is almost the same

HF Range → Rel Size ↓	16	32	64	256	512
128	12.52 91.10	19.74 41.79	34.76 24.46		
1024	41.30 1396.64	33.40 631.84	45.58 281.33	138.25 60.19	283.3 29.15
4096	138.78 6906.45	81.89 3246.79	67.33 1504.01	149.61 291.50	282.35 125.84
8192	269.29 14273.61	147.91 7173.50	107.79 3388.88	156.65 743.30	283.31 326.42

(Time unit = millisec; the top number in each box is the Bucket Construction time & the bottom number is the Join Operation time. Both **R** & **S** have the same size.)

Table 1: Effect of Hash Function Range for
Unsorted Centralized Bucket Table Scheme
(Sequential Bucket Construction)

as that of relation size 4096 and hash function range 256. Note that on average the buckets for both cases contain 16 entries. The discussion on the bucket construction time is incorporated with the analysis of the experiments on the effect of CM primitives (section 6.3).

HF Range → Rel Size ↓	16	32	64	256	512
128	84.39 22.98	87.97 13.28	97.67 8.11		
1024	85.00 174.99	88.70 89.14	94.36 46.67	110.13 17.40	93.96 10.93
4096	86.05 716.43	97.57 367.84	87.54 175.24	90.46 50.31	94.74 28.05
8192	89.78 1401.80	90.06 700.77	91.69 356.57	95.86 94.85	97.45 50.51

(Time unit = millisec; The top number in each box is the Bucket Construction time & the bottom number is the Join Operation time. Both **R** & **S** have the same size.)

Table 2: Effect of Hash Function Range for
Distributed Bucket Table Scheme

6.1.1. Centralized Bucket Table vs. Distributed Bucket Table

As we expected, the distributed bucket table scheme outperforms the centralized bucket table scheme (in most cases by a factor of 10). The underlying reasons for the speed-up are as follows. First, the communication traffic in this scheme is lower than the centralized one, since only one node in each bucket, as opposed to all the nodes in the centralized bucket table scheme, retrieves the tuples from the matching relation. This reduces the communication load by

a factor equal to the average bucket size. Moreover, the way we allocate the buckets in distributed scheme scatters the *starting* nodes more evenly in the system. Consequently, it reduces the degree of congestion in the communication subsystem. Note that the extra operation that we need to perform to propagate the retrieved tuples to the other nodes is the *scan* operation. As pointed out before, this operation is very efficient and can be regarded as a unit time operation in parallel models.

6.2. Effect of Different Relation Size

In this section we investigate the possible effect of the matching relation size on the execution time. In Table 3, we compare the performance of the join operation in the unsorted centralized bucket table scheme for two cases. In the first case the smaller relation is selected as the matching relation and it is bucketized. In the second case the larger relation is designated as the matching relation and it is bucketized. (To read the table, note that the **R** relation is always the one which is selected as the matching relation and is bucketized.) For example, consider the joining of two relations of sizes 8192 and 128. In the first case where the smaller relation is the matching one (lower-most left corner entry) the join operation takes much less time than the second case where the larger relation is selected as the matching relation (top-most right corner entry). One may expect a higher speed-up for the first case as the bucket size is reduced from 128 tuples to 2 tuples. In the ideal situation where the communication cost is assumed to be independent of the communication load, one should get a speed-up of 64. However, this assumption is far from real in existing systems and as shown in this experiment we only get tenfold speed-up. Note that for the smaller hash function range, more messages are sent to the node containing the bucket table.

R Size → S Size ↓	128	1024	8192
128	34.76 24.46	41.41 112.10	102.76 811.54
1024	35.86 48.90	45.58 281.33	102.01 1899.18
8192	34.07 73.12	42.15 414.12	107.79 3388.88

(Time unit = millisec; The top number in each box is the Bucket Construction time and the bottom number is the Join Operation time. The hash function range size in these experiments is 64.)

Table 3: Effect of Relation Size for
Unsorted Centralized Bucket Table Scheme
(Sequential Bucket Construction)

In Table 4, we show the same comparison for the distributed bucket table scheme. In this scheme, for the above example we get much higher speed-up, almost 40 times. This is due to the generally low communication load of this scheme. Thus, in both schemes it is always better to designate the smaller relation as the matching relation.

R Size → / S Size ↓	128	1024	8192
128	97.67 8.11	89.61 53.44	89.85 414.89
1024	89.30 9.65	94.36 46.67	88.46 379.1
8192	91.10 9.63	89.95 45.54	91.69 356.57

(Time unit = millisec; The top number in each box is the Bucket Construction time and the bottom number is the Join Operation time. The hash function range size in these experiments is 64.)

Table 4: Effect of Relation Size for
Distributed Bucket Table Scheme

6.3. Effect of Different CM Primitives

In this section we study the effect of the CM primitives on the bucket construction time for each scheme. Reiterate that to construct the buckets we need to compute the bucket table indices for each participating node. In the centralized bucket table scheme the bucket tables are maintained within the nodes; whereas, in the distributed bucket table the bucket tables are scattered over an array of nodes.

Unsorted Bucket Tables (Centralized): Tables 1 and 5 show the experimental results for the unsorted centralized bucket table with sequential bucket construction (SBC) and

HF Range → / Rel Size ↓	16	32	64	256	512
128	56.54 91.04	59.57 41.40	63.13 24.73		
1024	101.07 1398.12	91.42 632.23	77.64 278.76	64.22 58.51	64.7 24.51
4096	211.22 6935.74	140.73 3288.41	113.64 1522.62	72.08 291.07	67.48 125.82
8192	332.22 13985.84	209.00 7101.44	150.77 3372.50	82.90 739.68	75.13 319.73

(Time unit = millisec; The top number in each box is the Bucket Construction time & the bottom number is the Join Operation time. Both **R** & **S** have the same size.)

Table 5: Effect of Hash Function Range for
Unsorted Centralized Bucket Table Scheme
(Parallel Bucket Construction)

parallel bucket construction (PBC), respectively. For smaller hash function ranges, the SBC scheme performs better than the PBC scheme. This is due to the heavy communication load in the PBC scheme, which dominates the bucket construction time for the smaller hash function ranges. Therefore, the PBC method should be used only for the larger hash function ranges. Note that the rank operation time in PBC time is the same for each row in the table. The remaining portion of the PBC time is due to the

communication load. It can be clearly observed that the communication time becomes the dominating factor for the smaller hash function ranges.

HF Range → / Rel Size ↓	16	32	64	256	512
128	683.49 74.56	1463.61 37.67	3465.49 22.83		
1024	821.15 1018.85	1543.33 464.24	3673.99 207.67	14053.26 50.66	23318.04 24.59
4096	986.86 5348.82	1886.16 2512.40	4009.20 1185.64	12377.73 219.24	22731.34 99.47
8192	1285.27 12138.79	2132.39 5698.73	3808.56 2604.40	12251.69 516.26	23813.39 234.28

(Time unit = millisec; The top number in each box is the Bucket Construction time & the bottom number is the Join Operation time. Both **R** & **S** have the same size.)

Table 6: Effect of Hash Function Range for
Sorted Centralized Bucket Table Scheme
(Sequential Bucket Construction)

Sorted Bucket Tables (Centralized): Tables 6 and 7 show the experimental results for the sorted centralized bucket table with SBC and PBC, respectively. In the sorted centralized bucket table scheme, the PBC method is far superior to

HF Range → / Rel Size ↓	16	32	64	256	512
128	97.15 74.74	108.51 41.22	101.04 22.57		
1024	137.13 1008.55	129.03 473.12	123.32 199.77	113.53 49.94	107.86 24.81
4096	249.42 5355.25	179.62 2464.28	151.69 1128.60	123.77 214.94	115.87 103.15
8192	376.59 11647.13	246.59 5544.02	195.91 2607.21	150.17 522.25	136.02 235.97

(Time unit = millisec; The top number in each box is the Bucket Construction time & the bottom number is the Join Operation time. Both **R** & **S** have the same size.)

Table 7: Effect of Hash Function Range for
Sorted Centralized Bucket Table Scheme
(Parallel Bucket Construction)

the SBC method. Obviously, because we are performing an expensive operation (i.e. rank operation) sequentially.

Sorted vs. Unsorted Bucket Tables (Centralized): As one intuitively expects in the centralized bucket table scheme, the sorted bucket table performs better only for larger bucket sizes. That is, for bucket sizes of 32 or more we get 20% improvement on the total time (bucketization+join). For the smaller bucket sizes the performance gain on the join operation is nullified with the extra time needed to construct the buckets.

Sorted Bucket Tables (Distributed): In the sorted distributed bucket table scheme, the tuples within each bucket are also ranked based on their key attribute. The tuples of relation **R** are sorted in non-decreasing order and those of the **S** relation are sorted in non-increasing order. Hence, if the starting node in a bucket retrieves a tuple whose key is larger than its own key, it does not broadcast it to the other nodes; moreover, it signals all the nodes in the bucket to disable themselves.

Sorted vs. Unsorted Bucket Tables (Distributed): Tables 2 and 8 show the experimental results for the unsorted and sorted bucket tables in distributed bucket table scheme. The bucket construction time in this scheme contains the bucketization time for both relations. The difference in the bucket construction time of tables 2 and 8 is due to the ranking time required for the key attribute. As shown in both tables, the bucket construction time is nearly the same for all the cases. One of the reasons for this uniformity is that the communication primitive used here is a one-to-one CM-send operation which has a smaller and more uniform execution time than the many-to-one CM-send-with-min operation, which was used in the centralized bucket table scheme.

Unlike the centralized bucket table scheme, we do not benefit from the sorted bucket tables in this scheme. In some cases, even, a slight increase is observed in the join time of sorted version as compared to that of unsorted version. The extra comparisons of the key attributes necessary for early termination detection accounts for the increased time. The main reason for the absence of performance gain in sorted version is that the communication load in this scheme is relatively small, and the communicating nodes are scattered all over the system. Therefore, the early termination detection

HF Range → Rel Size ↓	16	32	64	256	512
128	175.49 29.06	168.50 13.48	168.64 7.40		
1024	189.29 189.09	188.27 95.81	203.18 65.92	203.47 20.71	206.11 13.39
4096	192.36 747.79	194.52 401.86	200.87 208.67	203.56 54.66	204.33 34.34
8192	193.63 1552.67	187.53 733.07	174.67 358.83	176.50 90.59	181.91 50.44

(Time unit = millisec; The top number in each box is the Bucket Construction time & the bottom number is the Join Operation time. Both **R** & **S** have the same size.)

Table 8: Effect of Hash Function Range for Sorted Distributed Bucket Table Scheme

does not make any significant impact on the communication time. The only performance gain will be due to the scan operations, which are already very fast operations.

In the majority of cases the increase in the bucket construction time by far outweighs the possible gain in the join operation. Unless the range of the join attribute of the relations are considerably different so that it results in early termination of all the buckets, the unsorted version outperforms the sorted version.

7. Joining Large Relations

In the previous sections we presented the experimental results for the relations of the size 8k or less. In this section, to give our readers a better understanding about the effect of virtual processor degree on the performance of our algorithm, we conducted a set of experiments for very large relations - as large as half a million tuples per relation. We conducted these experiments only for the distributed bucket table join scheme. We have also included the load data time and output result time. The data is read from and written to the DataVault. The preprocessing time required to convert the files from sequential format to the parallel format and vice versa is not included in the load data time and output result time.

As shown in Table 9, the bucket construction time and the join operation time, which constitute a large portion of the total response time, grows at a lower rate than the size of relations, particularly, when the virtual processor degree is less than or equal to 4. One reason for better performance is due to a reduction in the communication load. When the virtual processor degree is greater than one, some of the previously remote communications remain within the physical processor.

In Table 10, we show that the determining factor for the bucket construction and join operation time is the virtual processor degree. In this experiment we used two sections of the machine with a total of 16k processors. The bucket construction and join operation time of relations of size 16k (32k) in this experiment are nearly the same as those of relations of size 8k (16k) in the previous experiment. Thus one can expect that joining relations of size 4 million tuples on a fully configured CM, which contains 64k processors, takes approximately the same time as joining relations of size

Rel Size → (VPR)	8K (1)	16K (2)	32K (4)	64K (8)	128K (16)	256K (32)	512K (64)
Load Data	124.95	95.83	107.10	215.82	285.92	522.00	948.09
Hashing Step	2.17	4.19	8.2	16.2	32.38	64.65	129.09
Bucket Constr	166.65	222.30	366.69	798.76	1624.20	3640.32	6413.58
Join Operation	405.92	530.02	805.47	1404.52	2549.67	4969.37	9829.91
Output Result	135.57	268.27	533.37	1066.16	2131.27	4257.28	8515.20

(Time unit = millisec; No of processors = 8k; Both **R** & **S** have the same size. Average number of elements per bucket = 128)

Table 9: Join Operation Results for Large Relations (Distributed Bucket Table Scheme)

512k on a 8k processor machine.

8. Conclusions

We have presented here two schemes for SIMD join algorithms for the Connection Machine. The first scheme is based on centralized bucket tables, whereas the second scheme maintains the buckets in a distributed fashion over an array of processors. For all these algorithms, we have

Rel Size → (VPR)	16K (1)	32K (2)
Load Data	91.46	116.71
Hashing Step	2.16	4.17
Bucket Constr	175.3	233.2
Join	409.31	582.37
Output Result	319.82	660.92

(Time unit = millisec; No of processors = 16k; Both **R** & **S** have the same size. Average number of elements per bucket = 128)

Table 10: Effect of Virtual Processor Degree

evaluated performance of two important steps, which are different in these algorithms; these are *bucketization* and *join*. In our experimental studies we have evaluated the algorithms for different work-load and different values of the hash function range. We evaluated performance of these algorithms for average cases by selecting the work-load characteristics such that the tuples in a relation nearly uniformly distributed over all the buckets. From this study we have made the following observations:

• The distributed bucket table scheme performs better than the centralized scheme by at least an order of magnitude.

• For the bucket sizes of 32 or more, the centralized bucket table scheme with sorted bucket tables performs better than that with unsorted bucket tables.

• In distributed bucket table scheme, we do not benefit from sorted buckets. This is mainly due to the fact that the communication load in this scheme is relatively small; hence, early termination of some buckets has none or minuscule impact on the execution time.

• In both schemes the larger hash function ranges, which result in smaller bucket sizes, perform considerably better than the smaller hash function ranges.

• In both schemes it is better to select the smaller relation as the matching relation.

• Virtual processor degree greater than one may lessen the execution time growth, as some of the previously remote communication now becomes local within a physical processor.

• The execution time is a function of virtual processor degree. Different relation sizes on different machine sizes with the same virtual processor degree have nearly the same execution time.

• The communication component of both algorithms, particularly the degree of message congestion, has a significant impact on the performance of the algorithms. Hence, in designing the parallel algorithms one should give this factor thorough considerations.

We also have extended the distributed bucket table scheme to perform join operation on multiple sets of independent joining relations. The only step in our algorithm that requires some modification is the bucket construction step. In this step we have to ensure that the buckets for different sets are maintained correctly and independently. Due to lack of space we did not include the details of the modified algorithms in this paper. This feature is very desirable when one is working with relatively small size relations.

References

1. *The Connection Machine Systems: Release Notes,* Version 5.2 edition, Thinking Machine Corporation, Cambridge, MA, June 1989.

2. *The Connection Machine: Parallel Instruction Set,* Version 5.2 edition, Thinking Machine Corporation, Cambridge, MA, June 1989.

3. AKL, SELIM G., *The Design and Analysis of Parallel Algorithms,* Prentice Hall, 1989.

4. BITTON, DINA, HARAN BORAL, DAVID J. DEWITT, AND KEVIN WILKINSON, "Parallel Algorithms for the Execution of Relational Database Operation," *ACM Transaction on Database Systems*, vol. 8, No 3, pp. 324-353, Sep. 1983.

5. BLELLOCH, GUY E., "Scans as Primitive Parallel Operations," *IEEE Transactions on Computers*, vol. 38, pp. 1526-1538, Nov. 1989.

6. D., HILLIS, W., *The Connection Machine,* MIT Press, Cambridge, Mass., 1985.

7. DEWITT, DAVID J., RANDY H. KATZ, FRANK OLKEN, LEONARD D. SHAPIRO, MICHAEL STONEBRAKER, AND DAVID WOOD, "Implementation Techniques for Main Memory Database Systems," *Proceeding of SIGMOD*, pp. 1-8, ACM, Boston, 1984.

8. DEWITT, DAVID J. AND ROBERT GERBER, "Multiprocessor Hash-Based Join Algorithms," *Proceedings of Very Large Databases*, pp. 151-164, Stockholm, 1985.

9. HURSON, A.R., L.L. MILLER, AND S.H. PAKZAD, *Parallel Architectures For Database Systems,* IEEE Computer Society Press, 1989.

10. OMIECINSKI, EDWARD R. AND EILEEN TIEN LIN, "Hash-Based and Index-Based Join Algorithms for Cube and Ring Connected Multicomputers," *IEEE Transactions on Knowledge and Data Engineering*, vol. 1, No 3, pp. 329-343, Sep. 1989.

11. SHAPIRO, LEONARD D., "Join Processing in Database Systems with Large Main Memories," *ACM Transactions on Database Systems*, vol. 11, No 3, pp. 239-264, Sep. 1986.

12. VALDURIEZ, PATRICK AND GEORGES GARDARIN, "Join and Semijoin Algorithms for a Multiprocessor Database Machine," *ACM Transaction on Database Systems*, vol. 9, No 1, pp. 133-161, Mar. 1984.

13. WANG, XIAO AND W. S. LUSK, "Parallel Join Algorithms on a Network of Workstations," *International Symposium on Databases In Parallel & Distributes Systems*, pp. 87-95, Austin, Texas, Dec. 1988.

Massively Parallel Algorithms for Network Partition Functions

Albert G. Greenberg (1), and Isi Mitrani (2)

(1) AT&T Bell Laboratories, Murray Hill, New Jersey 07974, USA
(2) University of Newcastle, Newcastle upon Tyne, NE1 7RU, UK

ABSTRACT

Fast parallel algorithms are presented for the solution of closed product-form queueing networks. The algorithms are parallel counterparts of serial convolution algorithms for solving the networks via computation of the network partition function (table of normalization constants). The algorithms are simple, efficient, and well-suited for massively parallel SIMD machines. In particular, a load dependent, N server, R chain network with at most M jobs per chain, can be solved in time $O(R \log M \log N)$ using $O(M^R N/\log N)$ processors.

1. INTRODUCTION

We are concerned with the numerical solution of large, separable closed queueing network models. In general, these may involve multiple job chains and state-dependent center service rates. The theoretical solution of such models is of course known: If the network has N centers and R job chains, with chain k containing M_k jobs ($k = 1, 2, \ldots, R$), then its equilibrium distribution is of the form

$$P(S_1, \ldots, S_N) = G_N(M_1, \ldots, M_R)^{-1} \prod_{i=1}^{N} f_i(S_i). \quad (1.1)$$

Here, S_i describes the state of center i and $f_i(\cdot)$ is a function of that state. The normalizing constant $G_N(M_1, M_2, \ldots, M_R)$ is known as the "partition function" of the network; its value is determined by the requirement that the right-hand side of (1.1) should sum up to 1. Most performance measures that are of interest in connection with a closed queueing network are easily expressed and computed in terms of its partition function.

The computation of the partition function for a network with many centers, many job chains and large job populations, is a non-trivial task. The algorithm most widely used for this purpose, known as the *convolution algorithm*, relies on multidimensional recurrences with respect to the numbers of centers and the numbers of jobs in each chain (Buzen [1], Reiser and Kobayashi [2]). On a single processor, its time complexity is order $N(M_1 M_2 \ldots M_R)^2$ if the naive algorithm is used to compute convolutions and is order $NM_1 M_2 \ldots M_R \sum_{1 \le i \le R} \log M_i$ if the Fast Fourier Transform (FFT) [3, 4, 5] is used. RECAL, another recursive algorithm, has a complexity of order $\binom{M_1 + M_2 + \cdots + M_R}{N}$.

Our object here is to propose an algorithm that imple-

ments the convolution recurrences and allows a large number of processors to be used in parallel with near-optimal efficiency. More precisely, we show that the partition function can be computed in time $O[log(M_1 M_2 \ldots M_R)logN]$, using $O(M_1 M_2 \ldots M_R N/logN)$ processors. Of course, if fewer processors are available, then the task takes correspondingly longer. By "computing the partition function" we mean that the values of $G_i(j_1, j_2, \ldots, j_R)$ are determined for all $i = 1, 2, \ldots, N$ and all $j_k = 1, 2, \ldots, M_k$, $k = 1, 2, \ldots, R$.

To our knowledge, such a parallelization has not been achieved or attempted before. A parallel implementation of the *RECAL* algorithm was carried out by Greenberg and McKenna [6]. However, because of the different nature of the recurrences, the lowest time complexity that could be realized with that implementation was on the order of $O(M_1 + M_2 + \cdots + M_R)$.

Our approach is based on two ideas:

(a) One-step convolution, or polynomial multiplication, can be performed efficiently in parallel;

(b) Multi-step convolution, like multiple partial products, is a special case of the *prefix operation*, which can be performed efficiently in parallel.

The application of these ideas is illustrated first in the case of single chain networks with state-dependent service rates (Section 2). Scaling the computation for numerical stability [7] can be dovetailed into the parallel computation, though this is not described here. In the full paper, we discuss how to derive efficiently in parallel the usual performance measures of practical interest from the partition functions, such as the marginal distributions at particular centers, and the moments associated with these distributions. In Section 3, we show how the algorithm generalizes to networks with multiple closed chains.

2. SINGLE CHAIN NETWORKS

Let us consider a closed product-form queueing network [8, 9] having

- N service centers, numbered $i = 1, 2, \ldots, N$;

- M jobs, numbered $j = 1, 2, \ldots, M$;

- a single routing chain, whose invariant vector is $(e_1, e_2, \ldots, e_N)$ where e_i denotes the proportion of visits a job makes to center i;

- a speed vector $(\mu_i(1), \mu_i(2), \ldots, \mu_i(M))$ for each service center i, where $\mu_i(j)$ is the instantaneous rate of service at center i when j jobs are present.

The probability that m_i jobs reside at center i, $i = 1, 2, \ldots, N$, in equilibrium is given by [8, 9]: for each $m_i \geq 0$, $\sum m_i = M$,

$$P(m_1, m_2, \ldots, m_N) = \frac{1}{G_N(M)} \prod_{i=1}^{N} \alpha_{i,m_i}$$

where

$$\alpha_{i,j} = e_i^{\,j} / \prod_{k=1}^{j} \mu_i(k) \;\; ; \;\; 1 \leq i \leq N, 1 \leq j \leq M$$
$$\alpha_{i,0} = 1 \;\; ; \;\; 1 \leq i \leq N$$

and $G_N(M)$ is the normalization constant forcing the probabilities to sum to one:

$$G_N(M) = \sum_{(m_1, m_2, \ldots, m_N)} \prod_{i=1}^{N} \alpha_{i,m_i}$$

where the sum is over the $\binom{N + M - 1}{M - 1}$ possible partitions of the M jobs among the N centers. This expression explains the term "partition function", by which the normalization constant is also known.

The convolution algorithm computes $G_i(j)$ by solving a two-dimensional recurrence relation with respect to the number of centers, i, and the number of jobs, j. That relation can be written as, for $2 \leq i \leq N$, $0 \leq j \leq M$,

$$G_i(j) = \sum_{k=0}^{j} G_{i-1}(k) \alpha_{i,j-k} \qquad (2.1)$$
$$G_1(j) = \alpha_{1,j}.$$

The serial solution of the above recurrences is achieved in time $O(NM^2)$ using the naive convolution algorithm, or in time $O(NM \log M)$ using the FFT.

To develop a parallel counterpart, let us define the generating functions

$$g_i(x) = \sum_j G_i(j) x^j \;\; ; \;\; 1 \leq i \leq N$$
$$a_i(x) = \sum_j \alpha_{i,j} x^j \;\; ; \;\; 1 \leq i \leq N. \qquad (2.2)$$

so that we can rewrite (2.1) as

$$g_i(x) = g_{i-1}(x) a_i(x) \;\; ; \;\; 1 \leq i \leq N$$

and telescope to obtain

$$g_i(x) = a_1(x) a_2(x) \ldots a_i(x) \;\; ; \;\; 1 \leq i \leq N \qquad (2.3)$$

We wish to use (2.3) to compute the quantities (2.1), i.e., the coefficients of x^j in $g_i(x)$, for $i = 1, 2, \ldots, N$; $j = 1, 2, \ldots, M$. In that computation, it is clearly sufficient to truncate the infinite series (2.2) to polynomials of degree M. More-

over, when multiplying two such polynomials, all powers of x higher than M can be ignored, so that the result is also a polynomial of degree M. This interpretation of (2.3) will be adopted from now on. In particular, whenever polynomial multiplication is mentioned below, the truncated version of that operation will be implied.

Our parallel algorithm is simple:

1. In parallel, compute $a_1, a_2, \ldots, a_N$.

2. In parallel, compute $g_1 = a_1$, $g_2 = a_1 \cdot a_2$, ..., $g_N = a_1 \cdot a_1 \cdot \ldots \cdot a_N$.

The second step of the algorithm is the most interesting and the most difficult computationally, so let us first describe how to implement this step efficiently. The key is that computing the partial products $a_1 \cdot a_2 \cdot \ldots \cdot a_i$, for $i = 1, \ldots, N$, is a *parallel prefix* problem [10], involving polynomial multiplication (equivalently, vector convolution). The general parallel prefix problem is to compute the N partial products $x_1, x_1 \circ x_2, \ldots, x_1 \circ x_2 \circ \ldots \circ x_N$ of inputs $x_1, x_2, \ldots, x_N$, where $\circ$ is an arbitrary associative operator. Using P processors, this problem can be solved [10, 11] in time proportional to

$$C_\circ(N/P + \log P)$$

where $C_\circ$ is the time to compute the product $a \circ b$, for any a and b in the problem domain. Thus, with $P = O(N/\log N)$ processors, the time is $O(C_\circ \log N)$.

Here, the inputs are the polynomials $a_i(x)$, and the associative operator is polynomial multiplication. Using Q processors, the product of two degree M polynomials can be computed in time either

$$C_\circ = O(M \log M / Q + \log M)$$

using the parallel FFT [3, 4, 5], or

$$C_\circ = O(M^2 / Q + \log M)$$

using the parallel counterpart of the naive convolution algorithm.

Let us assume that the FFT is used. Then, with PQ processors, we can complete the second step of the algorithm in time

$$O((M \log M / Q + \log M)(N/P + \log P)).$$

Letting $Q = O(M)$, $P = O(N/\log N)$, the time becomes $O(\log M \log N)$ on $O(NM/\log N)$ processors.

To initialize the computation, we must compute the a_i's. To this end, note that, for $1 \leq i \leq N$, $1 \leq j \leq M$,

$$\alpha_{i,j} = (e_i/\mu_i(1)) \cdot (e_i/\mu_i(2)) \cdot \ldots \cdot (e_i/\mu_i(j))$$
$$\alpha_{i,0} = 1..$$

Thus, for fixed i, computing $\alpha_{i,1}, \alpha_{i,2}, \ldots, \alpha_{i,M}$ is another parallel prefix problem, solvable with P processors in time $O(M/P + \log M)$. We have N such problems, one for each

$i = 1, \ldots, N$. With P processors, all N can be solved in time $O(NM/P + \log M)$, which is lower order than the time used for the second step of the calculation.

To summarize, the total time to solve a single chain, closed, product-form queueing network, consisting of N load-dependent service centers and M jobs, is

$$O((\log M + M \log M/Q)(\log P + N/P)).$$

using PQ processors, and so is

$$O(\log M \log N)$$

using $O(MN/\log N)$ processors.

3. MULTIPLE CHAIN NETWORKS

We now generalize the treatment of single chain networks given in Section 2 to multiple chain networks. To distinguish between vectors and scalars, vectors will be underscored. For vectors $\underline{x} = (x_1, x_2, \ldots, x_k)$ and $\underline{y} = (y_1, y_2, \ldots, y_k)$ of identical dimension k, let

$$\underline{x} \leq \underline{y}$$

denote $x_i \leq y_i$ for every $i = 1, 2, \ldots, k$.

Let us consider a closed product-form [9] network with parameters:

- N service centers, numbered $i = 1, 2, \ldots, N$;

- R routing chains, numbered $r = 1, 2, \ldots, R$;

- a population vector $\underline{M} = (M_1, M_2, \ldots, M_R)$ where M_r denotes the number of jobs in chain r, so

$$M = \sum_{r=1}^{R} M_r$$

  is the total number of jobs in the network;

- for each routing chain r, an associated invariant vector $(e_{1,r}, e_{2,r}, \ldots, e_{N,r})$, where $e_{i,r}$ denotes the proportion of visits of a chain r job to service center i;

- for each service center, i, where the scheduling strategy is symmetric in the sense of Kelly [13] (e.g., Processor Sharing, Last-In-First-Out Preemptive-Resume or Infinite Server), and for each routing chain r, an average required service time $s_{i,r}$; at FIFO centers, $s_{i,r} = s_i$ for all r;

- for each service center i, a speed vector $(\mu_i(1), \mu_i(2), \ldots, \mu_i(M))$, where $\mu_i(j)$ denotes the instantaneous rate of service at center i when a total of j jobs are present. Let

$$P(\underline{m}_1, \underline{m}_2, \ldots, \underline{m}_R)$$

denote the probability that at equilibrium, for each $i = 1, 2, \ldots, N$, the population of jobs at center i is given by $\underline{m}_i = (m_{i,1}, m_{i,2}, \ldots, m_{i,R})$, where $m_{i,r}$ denote the number of jobs belonging to chain r present at the center. Then

$$P(\underline{m}_1, \underline{m}_2, \ldots, \underline{m}_R) = \frac{1}{G_N(\underline{M})} \prod_{i=1}^{N} \alpha_{i,\underline{m}_i}$$

where, for $\underline{j} = (j_1, j_2, \ldots, j_R)$, $j = j_1 + \ldots + j_R$, $1 \leq i \leq N$, $\underline{0} \leq \underline{j} \leq \underline{M}$,

$$\alpha_{i,\underline{j}} = \frac{j!}{\prod_{k=1}^{j} \mu_i(k)} \prod_{r=1}^{R} \frac{(e_{i,r} s_{i,r})^{j_r}}{j_r!} \tag{3.1}$$

$$\alpha_{i,\underline{0}} = 1,$$

and $G_N(\underline{M})$ is determined by the condition that the probabilities sum to one.

The multiple chain counterpart of (2.1) is, for $2 \leq i \leq N$, $\underline{0} \leq \underline{j} \leq \underline{M}$

$$G_i(\underline{j}) = \sum_{\underline{k}=\underline{0}}^{\underline{j}} G_{i-1}(\underline{k}) \alpha_{i,\underline{j}-\underline{k}} \tag{3.2}$$

$$G_1(\underline{j}) = \alpha_{1,\underline{j}}.$$

Let us rewrite (3.2) as

$$g_i(\underline{x}) = g_{i-1}(\underline{x}) a_i(\underline{x})$$

$$= a_1(\underline{x}) a_2(\underline{x}) \ldots a_i(\underline{x}) \quad ; \quad 1 \leq i \leq N \tag{3.3}$$

where

$$g_i(\underline{x}) = \sum_{\underline{j}} G_i(\underline{j}) \underline{x}^{\underline{j}} \quad ; \quad 1 \leq i \leq N \tag{3.4}$$

$$a_i(\underline{x}) = \sum_{\underline{j}} \alpha_{i,\underline{j}} \underline{x}^{\underline{j}} \quad ; \quad 1 \leq i \leq N. \tag{3.5}$$

A serial solution of the recurrences (3.3) yields the partition function in time $O(N(M_1 M_2 \ldots M_R)^2)$ if the naive convolution algorithm is used, or in time $O(N M_1 M_2 \ldots M_R \sum_{1 \leq i \leq R} \log M_i)$ if the FFT is used. The network performance measures of greatest interest can then be determined easily [12], as described in the full paper.

Again, we seek the parallel counterpart. We will use (3.3) to compute the coefficients of the $g_i(\underline{x})$. It suffices to truncate the infinite series (3.5) to keep only terms $\underline{x}^{\underline{j}}$ where $\underline{0} \leq \underline{j} \leq \underline{M}$. When multiplying two such polynomials, all powers $\underline{j} > \underline{M}$ can be ignored, so the result only retains powers $\underline{j} \leq \underline{M}$. This interpretation of (3.5) and of polynomial multiplication will be adopted from now on.

The parallel multiple chain algorithm is identical in form to the single chain algorithm:

1. In parallel, compute $a_1, a_2, \ldots, a_N$.

2. In parallel, compute $g_1 = a_1$, $g_2 = a_1 \cdot a_2$, $\ldots$, $g_N = a_1 \cdot a_2 \cdot \ldots \cdot a_N$.

The first step is now more interesting. Let us assume that $1 \leq i \leq N$, temporarily drop the subscript i from all variables, and write out $\underline{j}$ as $j_1, j_2, \ldots, j_R$. It is not hard to check that the following recurrence satisfies (3.1):

$$\alpha_{j_1, j_2, \ldots, j_R} = \frac{e_R s_R}{\mu(j_1 + \ldots + j_R)} \frac{j_1 + \ldots + j_R}{j_R} \alpha_{j_1, j_2, \ldots, j_R - 1}$$

which leads to

$$\alpha_{j_1,0,\ldots,0} = \frac{e_1 s_1}{\mu(1)} \cdot \frac{e_1 s_1}{\mu(2)} \cdot \ldots \cdot \frac{e_1 s_1}{\mu(j_1)} \qquad (3.6)$$

$$\alpha_{j_1,j_2,0,\ldots,0} = \alpha_{j_1,0,\ldots,0} \cdot \frac{e_2 s_2}{\mu(j_1+1)} \frac{j_1+1}{1}$$
$$\cdot \ldots \cdot \frac{e_2 s_2}{\mu(j_1+j_2)} \frac{j_1+j_2}{j_2} \qquad (3.7)$$

$$\vdots$$

$$\alpha_{j_1,j_2,\ldots,j_R} = \alpha_{j_1,j_2,\ldots,j_{R-1},0}$$
$$\cdot \frac{e_R s_R}{\mu(j_1+\ldots+j_{R-1}+1)} \frac{j_1+\ldots+j_{R-1}+1}{1}$$
$$\vdots$$
$$\cdot \frac{e_R s_R}{\mu(j_1+\ldots+j_{R-1}+j_R)} \frac{j_1+\ldots+j_{R-1}+j_R}{j_R} \qquad (3.8)$$

To compute $\alpha_{j_1,j_2,\ldots,j_R}$, for every $\underline{0} \leq \underline{j} \leq \underline{M}$, first compute $\alpha_{j_1,0,\ldots,0}$ for all $0 \leq j_1 \leq M_1$. By (3.6), this is a parallel prefix computation, which can be completed in time $O(M_1/P + \log M_1)$ using P processors. Second, using the values $\alpha_{j_1,0,\ldots,0}$, compute $\alpha_{j_1,j_2,0,\ldots,0}$ for all $0 \leq j_1 \leq M_1$ and all $0 \leq j_2 \leq M_2$. By (3.7), this is yet another parallel prefix computation, which can be completed in time $O(M_1 M_2/P + \log M_2)$ using P processors. Similarly, using the values $\alpha_{j_1,j_2,0,\ldots,0}$, compute $\alpha_{j_1,j_2,j_3,0,\ldots,0}$, and so forth to produce the final results $\alpha_{j_1,j_2,\ldots,j_R}$. This task is performed for each center in the network, $i = 1, 2, \ldots, N$. Using P processors the total time is order

$$N \sum_{i=1}^{R} \frac{M_i \cdot \ldots \cdot M_1}{P} + \sum_{i=1}^{R} \log M_i.$$

Now, consider the second step of the algorithm: computing the partial products of the a_i's. This is a parallel prefix problem where the associative operator is multivariate polynomial multiplication (R-dimensional convolution.) With Q processors, the product of any polynomials in $\underline{x}$ in which all terms $\underline{x}^{\underline{j}}$ satisfy $\underline{0} \leq \underline{j} \leq \underline{M}$ can be computed in time either

$$O\left(\left(\frac{M_1 \cdot \ldots \cdot M_R}{Q} + 1\right) \sum_{i=1}^{R} \log M_i\right)$$

using the FFT, or

$$O\left(\frac{(M_1 \cdot \ldots \cdot M_R)^2}{Q} + \sum_{i=1}^{R} \log M_i\right)$$

using the parallel counterpart of the naive convolution algorithm. Let us assume the FFT is used. Reasoning as in Section 2, step 2 can be completed with PQ processors in time order

$$\left(\frac{M_1 \cdot \ldots \cdot M_R \sum_{i=1}^{R} \log M_i}{Q}\right)(\log P + N/P). \qquad (3.9)$$

This dominates the time needed for step 1.

To summarize, a load dependent, R chain product-form queueing network with M_i jobs in chain i, $1 \leq i \leq R$, can be solved in time given by (3.9) using PQ processors. In particular, if M_{max} is the largest M_i and $P \cdot Q = O(M_{max}^R \cdot N/\log N)$ processors are used, then the network can be solved in time

$$O(R \log M_{max} \log N).$$

Alternatively, if P is large, some of the partial products in the right-hand sides of (3.6), (3.7), and (3.8) can be carried out in parallel, rather than in sequence. We do not pursue this avenue, because the whole algorithm is dominated by the computation of the g_i's and not by that of the a_i's.

References

[1] J.P. Buzen. Computational algorithms for closed queueing networks with exponential servers. *Communications of the ACM*, 16:527–531, 1973.

[2] M. Reiser and H. Kobayashi. Queueing networks with multiple closed chains: Computational algorithms. *IBM Journal of Research and Development*, 19(3):283–294, May 1975.

[3] S.G. Akl. *The Design and Analysis of Parallel Algorithms*. Prentice Hall, Englewood Cliffs, 1989.

[4] A. Aho, J. Hopcroft, and J. Ullman. *The Design and Analysis of Computer Algorithms*. Addison Wesley, New York, 1974.

[5] T. Leighton. An introduction to the theory of networks, parallel computation and VLSI design, 1989. draft.

[6] A.G. Greenberg and J. McKenna. Solution of closed, product form, queueing networks via the recal and tree-recal methods on a shared memory multiprocessor. In *1989 ACM Sigmetrics Performance Evaluation Review and Performance '89*, volume 17(1), pages 127–135, Berkeley, CA, May 1989. ACM Press.

[7] S.S. Lam. Dynamic scaling and growth behavior of queueing network normalization constants. *Journal of the Association for Computing Machinery*, 29:492–513, 1982.

[8] W.J. Gordon and G.F. Newell. Closed queueing systems with exponential servers. *Operations Research*, 15(2):254–265, April 1967.

[9] F. Baskett, K.M. Chandy, R.R. Muntz, and F. Pallacios. Open, closed, and mixed networks of queues with different classes of customers. *Journal of the Association for Computing Machinery*, 22(2):248–260, April 1975.

[10] R.E. Ladner and M.J. Fischer. Parallel prefix computation. *Journal of the ACM*, 27:831–838, 1980.

[11] C. P. Kruskal, L. Rudolph, and M. Snir. The power of parallel prefix. *IEEE Transactions on Computers*, C-34(10), October 1985.

[12] S.C Bruell and G. Balbo. *Computational Algorithms for Closed Queueing Networks*. North-Holland, 1980.

[13] F.P. Kelly. *Reversibility and Stochastic Networks*. Wiley, 1979. Reprinted 1987.

An Algorithm for Concurrent Search Trees

Adrian Colbrook[*] Eric A. Brewer[†] Chrysanthos N. Dellarocas [‡] William E. Weihl [§¶]

MIT Laboratory for Computer Science
545 Technology Square, Cambridge, MA 02139
email: colbrook@lcs.mit.edu

Abstract

In this paper we describe a new algorithm for maintaining a balanced search tree on a distributed-memory MIMD architecture. A $2^{B-2} - 2^B$ search tree is introduced, where a linear array of $\mathcal{O}(\log n)$ processors stores n entries. Update operations use a bottom-up node-splitting scheme, which we show leads to significant improvements in both throughput and response time for dictionary operations when compared to top-down search tree algorithms.

1   Introduction

In this paper we describe a new algorithm for maintaining a balanced search tree on a distributed-memory MIMD architecture. The algorithm is based on a linear array of processors, each with a large local memory; such arrays are easily emulated on most MIMD architectures. The algorithm performs updates using a bottom-up node-splitting scheme and gives significant improvements in both throughput and response time when compared to similar top-down algorithms [1, 2].

The search tree algorithm described here offers several advantages over similar schemes:

- improvements in both throughput and response time are achieved;

- simple variations in the branching factor allow the performance to be optimized for a given architecture;

- a hysteresis behavior is introduced for successive insertion and deletion operations, which is not present in some other balanced tree algorithms (notably B-trees [3] and 2-3-4 trees [1]).

- only $\mathcal{O}(\log n)$ processors are required to store a search tree of n entries as opposed to the $\mathcal{O}(n)$ processors used in some other schemes [4]. Fisher [5]

has noted that the $\mathcal{O}(n)$ schemes are not always the best route to high-performance systems.

Search trees are widely used for fast implementations of dictionary abstract data types. A dictionary is a partial mapping from keys to data that supports three operations: *insert*, *delete* and *search*. For simplicity we will assume that the dictionary stores no data with the keys and so may be viewed as a set of keys.

We introduce the $2^{B-2} - 2^B$ search tree [2], a variation of the B-tree. A $2^{B-2} - 2^B$ search tree (for $B \geq 3$) is a tree in which every node which is not the root or a leaf has between 2^{B-2} and 2^B sons, every path from the root to a leaf is of the same length, and the root has between 2 and 2^B sons. A search tree of n entries is implemented on an array of up to $[\log_2 n/(B-2)] + 1$ processors. Each processor holds a level of the tree structure in local memory, and the last processor stores the actual keys. Therefore, the memory required to store the search tree increases by a factor of 2^B between adjacent processors down the linear array.

A leaf node stores between 2^{B-2} and 2^B keys, while a non-leaf node contains redundant index information used to find the appropriate leaf. Each leaf node stores key values within a contiguous range; the ranges of all leaf nodes partition the set of possible key values and are in ascending order from the leftmost leaf to the rightmost. The index information stored at a non-leaf node is a value lower than or equal to the lowest key value associated with its sons.

In our new algorithm a dictionary is represented by a modified $2^{B-2} - 2^B$ search tree in which links are added from each node to its left and right neighbors in the same level. This is similar to the algorithm of Lehman and Yao [6] where right links between neighboring nodes are added to a B-tree. In addition, each non-root node contains a pointer back to its parent.

Carey and Thompson [1] implemented a 2-3-4 search tree using a linear array of $\mathcal{O}(\log n)$ processors. A 2-3-4 tree is a balanced tree in which each non-leaf node has two, three or four sons. That structure allows update operations to be performed using the top-down node-splitting scheme presented by Guibas and Sedgewick [7]. We first implemented a $2^{B-2} - 2^B$ search tree using

---

[*]Supported by a Science and Engineering Research Council Postdoctoral Fellowship.

[†]Supported by an Office of Naval Research Fellowship.

[‡]Supported by a Starr Foundation Fellowship.

[§]Supported by the National Science Foundation under grant CCR-8716884 and by the Defense Advanced Research Projects Agency (DARPA) under Contract N00014-89-J-1988.

[¶]The work was also supported by an equipment grant from Digital Equipment Corporation.

a top-down node-splitting scheme similar to that used in [1]. We then implemented the search tree using the bottom-up algorithm. Both of these algorithms have been implemented using *Proteus*, a multiprocessor simulator [8, 9] developed at MIT. The throughput and response times for varying processor array lengths, tree branching factors and query mixes were determined. These measurements allow the relative performance of the algorithms to be compared.

The remainder of this paper is structured as follows. Sections 2 and 3 describe the top-down and bottom-up algorithms. Section 4 describes the *Proteus* simulator. Section 5 presents the performance evaluation of both algorithms. Finally, conclusions are drawn in Section 6.

2 The Top-Down Algorithm

The top-down algorithm allows insertions, deletions and exact-match searches on a $2^{B-2} - 2^B$ search tree. The *search* operation is a simple version of the normal B-tree search operation [3]. The *insert* and *delete* operations are based upon a top-down node-splitting scheme [7], in which transformations are applied during a single traversal of the tree for an update operation. The tree is traversed from the root downward and transformations are applied between adjacent levels.

The *insert* operation performs node splitting on encountering a 2^B-branch node, other than the root, forming two 2^{B-1}-branch nodes. This transformation ensures that any future node-splitting does not cause upward propagation in the tree structure, which allows the transformation to be applied in a top-down fashion.

When a *delete* operation encounters a 2^{B-2}-branch node, other than the root, one of two general deletion transformations is applied. If a neighboring node has less than or equal to 2^{B-1} branches, the node and its neighbor are merged to form a single node. Otherwise the branches of the node and its neighbor are redistributed evenly between the two nodes. The neighbor relationship used in the deletion algorithm relates a node to its right sibling in the sub-tree or in the case of the rightmost node, to its left sibling. These transformations ensure that any future merging of nodes will not cause upward propagation of transformations.

Transformations applied to a root node cause the tree to grow (for *insert*) and shrink (for *delete*). A stabilizing (hysteresis) effect occurs in the $2^{B-2} - 2^B$ tree, as the insertion and deletion transformations are not inverses of one another. This property is described in [2].

The top-down algorithm is a *lock-coupling* algorithm. A processor is locked after sending an update message to the processor below it in the array. A reply message causes the lock to be released and is mandatory even if no actual transformation occurs. For n operations the top-down algorithm requires approximately $n \log_B n$ downward messages and approximately $n \log_B n$ upward messages.

3 The Bottom-Up Algorithm

The bottom-up algorithm is based upon a similar algorithm developed by Lehman and Yao [6] for B-trees. In their algorithm right links are added to adjacent nodes in a B-tree. In our new algorithm both left and right links are added between adjacent siblings in a $2^{B-2} - 2^B$ search tree. These links provide an additional method of reaching a node. The intent of this scheme is to make all nodes in a single level reachable from any other node at that level. If the wrong node is selected during a traversal then the correct node can be found by traversing the links. This allows changes to the tree structure to be propagated to higher levels in a lazy manner.

The algorithm again allows insertions, deletions and exact-match searches on a $2^{B-2} - 2^B$ search tree. Each of these operations begins by calling a *find* operation, which traverses the tree from the root until it reaches the leaf node that may store the specified key. For a *search* operation the keys stored at the leaf node are then searched for the specified key and the result is sent to the inquiring process.

An *insert* operation attempts to add the specified key to the keys stored at the leaf. If the leaf already contains 2^B keys and the specified key is not one of these, an insertion transformation is applied, which splits the leaf node into two 2^{B-1}-key nodes. The new node becomes the right sibling of the original leaf node. The specified key is then added to the keys stored at the appropriate node and the *insert* operation returns. The work required to propagate the split to higher levels is carried out as a background task. This task adds a pointer to the newly created node at the appropriate node in the next level. This in turn may cause an insertion transformation and this split is propagated until a level is reached where no split occurs. Should the root of the tree be split by such a propagation, a new root is created and the height of the tree increased by one.

A *delete* operation attempts to remove the specified key from the keys stored at the leaf. If the leaf contains 2^{B-2} keys, one of which is the specified key, then one of two transformations is applied to the node and its right neighbor in the tree (the left neighbor for the rightmost leaf). These transformations are identical to those used in the top-down algorithm. The *delete* operation then returns and the propagation of the transformation to higher levels is again carried out as a background task.

Since the changes to non-leaf levels caused by a transformation are carried out as a background task, the *find* operation must guarantee that the correct leaf node is reached, even if the tree traversal is made between a transformation at a leaf node and the completion of the subsequent transformations at the non-leaf levels. To achieve this we introduce a notion of reachability. A non-leaf node has associated with it two values, the minimum key value i that may be stored in the subtree of which the node is the root, and the minimum key value j that may be stored in the subtree of which the node's right neighbor is the root. A key k is said to be reachable from a node x if and only if x's index labels indicate that the leaf that may store k is contained in the subtree of which x is the root. That is, $reachable(k, x) \Rightarrow x.i \le k < x.j$. When a *find* operation encounters a non-leaf node that does not reach the specified key then the level is traversed from the node to either the left (if $x.i > k$) or to the right (if $x.j \le k$) until the node that reaches k is found.

In the case of a deletion transformation that causes two nodes to be merged, the neighbor is not removed immediately; it is flagged as deleted, and the links pointing to it from other nodes at the same level are updated. This allows references to a deleted node to be made by *find* until the result of the transformation has propagated to the higher level. When a deleted node is encountered by the *find* operation the traversal immediately begins at the left neighbor of the deleted node.

For n operations the bottom-up algorithm requires approximately $n \log_B n$ downward messages. However, upward messages only occur following a transformation.

4   The *Proteus* Simulator

The *Proteus* simulator simulates the events that take place in a parallel machine at the level of individual machine instructions. The user writes a parallel program using a simple superset of the C programming language and a set of supported simulator calls. The parts of the user program executed locally on each processor are written in standard C and translated by the C compiler into machine code. All nonlocal interactions are performed by the supporting simulator calls, which correspond to the machine code instructions that perform nonlocal interactions in real parallel machines.

We assume a reduced-instruction-set processor, in particular, each instruction requires one cycle. The network is packet switched and uses wormhole routing. We assume that a message fits in one packet. Synchronous messages are used for the communication between processors. [1]

We simulate a multiprocessor configured as a bidirectional ring. Each node consists of a processor, local memory, and a network chip that routes messages without using processor cycles.

5   Relative Performance

In this section we compare the relative performance of the two algorithms using the results of simulations. We present the results of two sets of simulations for each of the algorithms using a range of values for the branching constant B in each case. B was varied between 2 and 8 (a 64-256 tree). [2] The query throughputs and response times were measured. The throughput was measured in

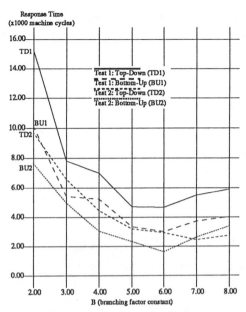

Figure 1: The response times for the search trees

terms of the average number of tree operations that complete for every 100,000 machine cycles and the response time was measured in terms of the average number of machine cycles between sending a query and receiving the corresponding reply.

The first set of simulations, *Test 1*, measured the performance of the algorithms when 50,000 *insert* operations using random key values were applied to an initially empty tree. For the second set of simulations, *Test 2*, 1,000 *insert* operations using random keys were first applied to an empty search tree followed by 10,000 randomly selected operations. The percentages of *insert*, *delete* and *search* operations in this random selection were 50%, 30% and 20%, respectively. Each time a *delete* operation was applied the key value closest to the selected key value was deleted.

The response times and throughputs are shown in Figures 1 and 2, respectively. The response times reach a minimum value as B increases. The value of B governs the number of processors used; as B increases the number of processors decreases since a greater number of key values are stored in each node. For increasing values of B below this minimum the improvement in response time is caused by the reduction in the number of inter-processor hops required for a single query. For increasing values of B greater than the minimum the degradation in response time occurs due to an increase in the processing time at a single node caused by transformations. Therefore there is a trade-off between the computation at a processor and the amount of communication to achieve the optimal response time.

In all cases the throughput improves for increasing values of B up to 4 (or 5 for *Test 2* applied to the bottom-up algorithm). This peaked response is caused by a trade-off between the number of transformations and the process-

---

[1] Using asyschronous messages results in an additional improvement in performance for the bottom-up algorithm.

[2] $B=2$ is not strictly a $2^{B-2} - 2^B$ search tree and actually represented the 2-3-4 tree used by Carey and Thompson [1].

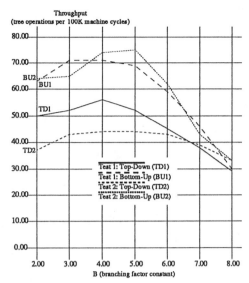

Figure 2: The throughputs for the search trees

ing time for searching the key values stored at a node. For low values of B insert and delete operations cause more transformations to the tree structure and lead to an increase in the average processing time at a single node. For higher values of B the time required to search and update the keys stored at a node increases. Both these effects lead to a reduction in throughput. Therefore an optimal value for B exists where the sum of these two effects is minimum.

When the two algorithms are compared, the bottom-up algorithm performs better in both throughput and response time for all cases. These improvements arise because nodes need not wait for replies, and upward messages in the bottom-up algorithm occur only when a transformation is required. This results in significantly fewer upward messages for the bottom-up algorithm, as demonstrated by the measurements in Table 1.

B	Top Down	Bottom Up	B	Top Down	Bottom Up
2	65213	2556	6	26143	71
3	48805	884	7	19654	36
4	37329	345	8	19422	15
5	27568	150			

Table 1: Upward messages required during *Test 2*

6 Conclusions

A $2^{B-2} - 2^B$ search tree of n entries is implemented on a linear array of up to $[\log_2 n/(B-2)]+1$ processors, where each processor stores a level of the tree structure. Such a linear array may be physically mapped onto processors in two or three dimensions on most available architectures and has been shown to be efficient when compared to the $\mathcal{O}(n)$ processor schemes.

The top-down node-splitting algorithm applies transformations during a single traversal of the tree structure.

However, processors are required to wait for replies, most of which merely verify that no transformation occurred.

The introduction of side links between adjacent nodes at the same level within the tree permits the use of a bottom-up algorithm. This algorithm allows the transformations resulting from changes to the tree structure to be performed asynchronously from the leaf nodes upwards, while guaranteeing the correctness of other operations concurrently executing on the data structure.

In a series of simulations conducted for both algorithms, the bottom-up approach gives significantly better query throughput and response time than the top-down algorithm. The number of upward messages (and hence the contention) between adjacent processors in the linear array is significantly less for the bottom-up algorithm. In systems where an asynchronous message-passing paradigm is available further improvements in the throughput and response time of the bottom-up algorithm can be achieved. Furthermore, the bottom-up algorithm shows increasingly superior performance as the one-way message delay between adjacent processors increases. These results are described in a longer version of this paper available from the authors.

The bottom-up algorithm for the $2^{B-2} - 2^B$ search tree has been shown to provide a highly efficient and flexible implementation of dictionary abstract data types on distributed-memory MIMD architectures. For a given cost ratio of computation to communication the value of B may be varied to maximize performance.

We thank Anant Agarwal and Wilson Hsieh for their comments and suggestions on this work.

References

[1] M.J. Carey and C.D. Thompson. An efficient implementation of search trees on [lgn+1] processors. *IEEE Transactions on Computers*, 33(11):1038–1041, 1984.

[2] A. Colbrook and C. Smythe. Efficient implementation of search trees on parallel distributed-memory architectures. *IEE Proceedings Part E*, 137(5):394–400, 1990.

[3] D. Comer. The ubiquitous B-tree. *Computer Surveys*, 11(2):121–137, 1979.

[4] T.A. Ottman, A.L. Rosenberg, and L.J. Stockmeyer. A dictionary machine for VLSI. *IEEE Transactions on Computers*, 33(9):892–897, 1984.

[5] A.L. Fisher. Dictionary machines with a small number of processors. In *Proceedings of the 11th Annual International Symposium on Computer Architectures*, pages 151–156, 1984.

[6] P.L. Lehman and S.B. Yao. Efficient locking for concurrent operations on B-trees. *ACM Transactions on Database Systems*, 6(4):650–670, 1981.

[7] L.J. Guibas and R. Sedgewick. A dichromatic framework for balancing trees. In *Proceedings of the 19th Annual IEEE Computer Society Symposium on the Foundations of Computer Science*, pages 8–21, 1978.

[8] C.N. Dellarocas. A high-performance retargetable simulator for parallel architectures. Master's thesis, MIT Laboratory for Computer Science, 1991.

[9] E.A. Brewer. Aspects of a high-performance parallel-architecture simulator. Master's thesis, MIT Laboratory for Computer Science, 1991.

OPTIMAL DATA PARALLEL METHODS FOR STOCHASTIC DYNAMICAL PROGRAMMING

H. H. Xu, F. B. Hanson and S.-L. Chung
Laboratory for Advanced Computing
University of Illinois at Chicago
P. O. Box 4348; M/C 249
Chicago, IL 60680

Abstract. The general optimal feedback control problems for nonlinear, continuous time dynamical systems have been further studied in the stochastic environment. The stochastic component includes both Gaussian noise for modeling background fluctuations and Poisson noise for modeling the more severe random fluctuations. In this paper, the optimal data parallel algorithms and data structures have been developed to numerically solve the larger scale stochastic optimal control problem. Two techniques called broadcasting and operation decomposition are introduced in terms of the special characteristics of the stochastic control problems. These techniques have resulted in significant improvements for both storage and time performances. Furthermore, some comparisons between CM-2 and Cray 2S systems demonstrate that data parallel computations are suitable in solving large scale control problems.

Keywords: Massively data parallel, stochastic dynamic programming, optimal control.

I. Introduction

The optimal feedback control problems for nonlinear, continuous time dynamical systems, perturbed by Poisson as well as Gaussian random white noise are applicable to many problems, such as aerospace dynamics, chemistry, economics, physics, robotics and transportation. The numerical approaches with modified Crank Nicholson methods have been studied for solving the optimal control problems [1]. It is well known that the numerical solutions suffer from computational difficulties in the multi-state and multi-control decision process, the so-called *Bellman's curse of dimensionality*. To alleviate the computational difficulties, some advanced supercomputing techniques called vectorization and parallelization have been implemented for dynamic programming [2].

The massive parallel computation has been investigated in this paper in order to improve the parallel computational abilities. The data parallel computer Connection Machine CM-2, which we have used through out this paper, is perhaps the most massively parallel architecture among all the parallel systems which have been recently developed [3, 10]. Many applications with the CM-2 indicate that data parallel methods can raise the performance necessary to solve computationally intense problems. McBryan [4] discusses an early implementation of fine grain decomposition, data parallel algorithms for using SOR, multigrid conjugate gradient methods to solve partial differential equations on the pre-floating point Connection Machine CM-1. Mathur and Johnsson [5] present a data parallel algorithm for the finite element methods on the Connection Machine CM-2 using virtual processors to map nodes unto physical processors.

For large scale stochastic optimal control problems, computations using the CM-2 have interesting aspects. As we will discuss, stochastic optimal control problems have a special data structure and properties in data parallel computations. By taking advantage of these features, a parallel computational procedure for the stochastic optimal control problems is successfully developed. We also introduce broadcasting techniques in this paper, in order to provide methods of greatly reducing the computational requirements but retaining general applicability to problems. Indeed, it is the goal of

This work was supported by the National Science Foundation Computational Mathematics Program under grant DMS-88-06099 at the University of Illinois at Chicago, by the Argonne National Laboratory Advanced Computing Research Facility, by the Los Alamos National Laboratory Advanced Computing Laboratory, by the University of Illinois at Urbana National Center for Supercomputing Applications, by the UIC Software Technologies Research Center, and by the UIC Workshop Program on Scientific Supercomputing.

this paper to show that the data parallel computation has significant potential for solving larger control problems.

II. Stochastic Dynamical Programming for Optimal Control

Our interest here is the Markov dynamical system, which is governed by the stochastic differential equation (SDE):

$$dy(s) = \mathbf{F}(\mathbf{y}, s, \mathbf{u})ds + G(\mathbf{y}, s)d\mathbf{W}(s) + H(\mathbf{y}, s)d\mathbf{P}(s) \,, \quad (2.1)$$

with initial values $\mathbf{y}(t) = x$, $0 < t < s < t_f$, $\mathbf{y}(s) \in \mathcal{D}_y$, $\mathbf{u} \in \mathcal{D}_u$, where $\mathbf{y}(s)$ is the $m \times 1$ state vector at time s starting at time t, $\mathbf{u} = \mathbf{u}(\mathbf{y}, s)$ is the $n \times 1$ feedback control vector, $\mathbf{F}$ is the $m \times 1$ deterministic nonlinearity vector, $\mathbf{W}$ is the r-dimensional normalized Gaussian white noise vector, $\mathbf{P}$ is the independent q-dimensional Poisson white noise vector with jump rate vector $[\lambda_i]_{q \times 1}$, G is an $m \times r$ diffusion coefficient array, and H is an $m \times q$ Poisson amplitude coefficient array.

The control criterion is the optimal expected cost performance,

$$V^*(\mathbf{x}, t) = \min_{\mathbf{u}} \left[\underset{\{\mathbf{P}, \mathbf{W}\}}{\mathrm{MEAN}}[V[\mathbf{y}, s, \mathbf{u}, \mathbf{P}, \mathbf{W}]|\mathbf{y}(t) = \mathbf{x}] \right] \quad (2.2)$$

over some specified optimal control set $\mathcal{D}_u$, where the total cost is

$$V[\mathbf{y}, t, \mathbf{u}, \mathbf{P}, \mathbf{W}] = \int_t^{t_f} ds \, C(\mathbf{y}(s), s, \mathbf{u}(\mathbf{y}(s), s)) \,, \quad (2.3)$$

on the time horizon (t, t_f). The instantaneous cost function $C = C(\mathbf{x}, t, \mathbf{u})$ is assumed to be a quadratic function of the control,

$$C(\mathbf{x}, t, \mathbf{u}) = C_0(\mathbf{x}, t) + \mathbf{C}_1^T(\mathbf{x}, t)\mathbf{u} + \frac{1}{2}\mathbf{u}^T C_2(\mathbf{x}, t)\mathbf{u}. \quad (2.4)$$

The unit cost of the control increases with $\mathbf{u}$ assuming C_2 is positive definite. For example, the cost criterion could be minimal fuel consumption, minimum distance to target, minimum time to target, or minus the maximum return on financial instruments. No final salvage value is assumed at final time for simplicity, so V is zero at $t = t_f$. In addition, the deterministic, nonlinear dynamics in (2.1) are assumed to be linear in the controls,

$$\mathbf{F}(\mathbf{x}, t, \mathbf{u}) = \mathbf{F}_0(\mathbf{x}, t) + F_1(\mathbf{x}, t)\mathbf{u} \,, \quad (2.5)$$

but nonlinear in the multibody state variable $\mathbf{x}$.

For numerical purposes, it is more convenient to convert equations (2.1)-(2.5) to an effectively deterministic partial differential equation using Bellman's principle of optimality. The Bellman functional PDE of stochastic dynamic programming,

$$
\begin{aligned}
0 &= \frac{\partial V^*}{\partial t} + L[V^*] \\
&\equiv \frac{\partial V^*}{\partial t} + \mathbf{F}_0^T \nabla V^* + \frac{1}{2}GG^T(\mathbf{x}, t) : \nabla \nabla^T V^* \\
&\quad + \sum_{l=1}^q \lambda_l \cdot [\, V^*(\mathbf{x} + \mathbf{H}_l(\mathbf{x}, t), t) - V^*(\mathbf{x}, t) \,] \\
&\quad + C_0 + (\frac{1}{2}\mathbf{U}^* - \mathbf{U}_R)^T C_2 \mathbf{U}^* \,,
\end{aligned}
\quad (2.6)
$$

follows from the generalized *Itô* chain rule for Markov SDEs as in [6], where $\mathbf{U}^*$ is the optimal feedback control computed by constraining the unconstrained or regular control,

$$\mathbf{U}_R(\mathbf{x},t) = -C_2^{-1}(\mathbf{C}_1 + F_1^{\ T}\nabla V^*) , \qquad (2.7)$$

to the control set $\mathcal{D}_u$. In general, the Bellman equation (2.7) is nonlinear with discontinuous coefficients due to the last term, $(\frac{1}{2}\mathbf{U}^* - \mathbf{U}_R)^T C_2 \mathbf{U}^*$, and due to the compact relationship between the constrained, optimal control and the unconstrained, regular control,

$$U_i^*(\mathbf{x},t) = \min[U_{max,i}, \min[U_{min,i}, U_{R,i}(\mathbf{x},t)]], \qquad (2.8)$$

for $i = 1$ to n controls under hypercube type constraints for $\mathcal{D}_u$, where $\mathbf{U}_{min}$ is the minimum control constraint vector and $\mathbf{U}_{max}$ is the maximum. As the constraints are attained, the optimal control $\mathbf{U}^*$, changes from the regular control, $\mathbf{U}_R$, to the single bang control values, $\mathbf{U}_{min}$ or $\mathbf{U}_{max}$, which in general are functions of state and time. In (2.7), the symbol (:) denotes the trace of a matrix product $A : B = \sum_{i=1}^{m}\sum_{j=1}^{m} A_{ij}B_{ij}$. It is important to note that the principal equation, the Bellman equation (2.7), is an exact equation for the optimal expected value V^* and does not involve any sampling approximations such as the use of random number generators in simulations.

Since there is no final salvage value and since (2.7) is a backward equation (unlike the usual diffusion equation, which is a forward equation), the final condition is that $V^*(\mathbf{x},t_f) = 0$ using (2.2) and (2.3). The boundary conditions for the PDE of stochastic dynamic programming (2.7) can not be stated in a simple form. This is because the boundary conditions vary significantly with the form the deterministic linearity function $\mathbf{F}$, the Gaussian noise W, and the Poisson noise P. Thus for treatment of general boundary conditions, it is most practical to directly integrate (2.7) for the boundary values of $\mathbf{x}$, or to use the objective functional directly as defined in (2.2) and (2.3).

III. Numerical Methods and Vectorized Computations

The integration of the Bellman equation (2.7) is backward in time, because V^* is specified finally at the final time $t = t_f$, rather than at the initial time. A summary of the discretization in states and backward time, with a compact vector subscript notation for the state variables to reduce the number of displayed subscripts, is given below :

$$\mathbf{x} \ \rightarrow \ \mathbf{X_j} = [X_{i1} + (j_i - 1)\cdot DX_i]_{m\times 1},$$
$$\mathbf{j} \ = \ [j_i]_{m\times 1}, \text{where } j_i = 1 \text{ to } M_i,$$
$$\text{for } i = 1 \text{ to } m;$$
$$s \ \rightarrow \ T_k = t_f - (k-1)\cdot DT,$$
$$\text{for } k = 1 \text{ to } K ;$$
$$V^*(\mathbf{X_j},T_k) \ \rightarrow \ V_{\mathbf{j},k};$$
$$V_t^*(\mathbf{X_j},T_k) \ \rightarrow \ \frac{V_{\mathbf{j},k+1/2+1/2} - V_{\mathbf{j},k+1/2-1/2}}{-2\cdot DT};$$
$$\nabla_x V^*(\mathbf{X_j},T_k) \ \rightarrow \ \mathbf{DV_{jk}};$$
$$\nabla_x \nabla_x^T V^*(\mathbf{X_j},T_k) \ \rightarrow \ \mathbf{DDV_{jk}};$$
$$V^*(\mathbf{X_j} + \mathbf{H_{ij}},T_k)_{q\times 1} \ \rightarrow \ \mathbf{HV_{jk}};$$
$$U_R(\mathbf{X_j},T_k) \ \rightarrow \ \mathbf{UR_{jk}};$$
$$U^*(\mathbf{X_j},T_k) \ \rightarrow \ \mathbf{U_{jk}},$$

where DX_i is the mesh size for state i and DT is the step size in backward time.

The numerical algorithm is a modification of the predictor corrector, Crank Nicholson methods for nonlinear parabolic PDEs in [7]. Modifications are made for the switch term and delay term calculations. Derivatives are approximated with an accuracy that is second order in the local truncation error, $O^2(\Delta x)$, at all interior and boundary points. The Poisson induced functional or delay term, $V^*(\mathbf{x} + \mathbf{H}_l, t)$, changes the local attribute of the usual PDE to a global attribute, such that the value at a node $[\mathbf{X} + \mathbf{H}_l]_{\mathbf{j}}$ will in general not be a node. Linear interpolation of second order for the delay term maintains the numerical integrity that is compatible with the numerical accuracy of the derivative approximations. Even though the Bellman equation (2.7) is a single PDE, the process of solving it not only produces the optimal expected value V^*, but a second result, also the optimal expected control law U^*. This is because the Bellman equation is a functional PDE, in which the computation of the regular control is fed back into the optimal value and the optimal value feeds back into regular control through its gradient.

Prior to calculating the values, $V_{\mathbf{j},k+1}$, at the new $(k+1)^{st}$ time step for $k = 1$ to $K-1$, the old values, $V_{\mathbf{j},k}$ and $V_{\mathbf{j},k-1}$, are assumed to be known, starting with $V_{\mathbf{j}0} \equiv V_{\mathbf{j}1}$. The algorithm begins with an convergence accelerating *extrapolator (x) start*:

$$V_{\mathbf{j},k+\frac{1}{2}}^{(\mathbf{x})} = \frac{1}{2}(3\cdot V_{\mathbf{j},k} - V_{\mathbf{j},k-1}) , \qquad (3.1)$$

which are then used to compute updated values of the gradient DV, the second order derivatives DDV, the Poisson functional terms (V^* at $(\mathbf{x} + \mathbf{H})$), the regular controls UR, the optimal controls U^*, and finally the new value of the Bellman equation spatial functional $L_{\mathbf{j},k+0.5}$. These evaluations are used in the *extrapolated predictor (xp) step*:

$$V_{\mathbf{j},k+1}^{(\mathbf{xp})} = V_{\mathbf{j},k} + DT\cdot L_{\mathbf{j},k+\frac{1}{2}}^{(\mathbf{x})} . \qquad (3.2)$$

which are then used in the *predictor evaluation (xpe) step*:

$$V_{\mathbf{j},k+\frac{1}{2}}^{(\mathbf{xpe})} = \frac{1}{2}(V_{\mathbf{j},k+1}^{(\mathbf{xp})} + V_{\mathbf{j},k}) , \qquad (3.3)$$

an approximation which preserves numerical accuracy and is used to evaluate all terms comprising $L_{\mathbf{j},k+0.5}$. The evaluated predictions are used in the *corrector (xpec) step*:

$$V_{\mathbf{j},k+1}^{(\mathbf{xpec},\gamma + 1)} = V_{\mathbf{j},k} + DT\ \cdot\ L_{\mathbf{j},k+\frac{1}{2}}^{(\mathbf{xpe},\gamma)} \qquad (3.4)$$

for $\gamma = 0$ to γ_{max} until the stopping criterion is met, with *corrector evaluation (xpece) step*:

$$V_{\mathbf{j},k+\frac{1}{2}}^{(\mathbf{xpece},\gamma + 1)} = \frac{1}{2}(V_{\mathbf{j},k+1}^{(\mathbf{xpec},\gamma + 1)} + V_{\mathbf{j},k}) . \qquad (3.5)$$

The predicted value is taken as the zero-th correction. The stopping criterion for the corrections is formally derived from a comparison to a predictor corrector convergence criterion for a linearized, constant coefficient PDE [8]. A robust mesh selection method is used to determine the stopping criterion, so that only a couple of corrections are needed, except at the first time step. The proper selection of the mesh ratio guarantees that the corrections for the comparison equation converge, whether the Bellman equation (2.7) is parabolic-like (with Gaussian noise) or hyperbolic-like (without Gaussian noise), according to whether or not an explicit second derivative is in the equation.

The implementation of above algorithm has been done by methods of parallelization and vectorization on NCSA Cray X-MP/48, Cray Y-MP/4-64 and Cray 2s/4-128. For more general multi-state and multi-control applications, a flexible general vectorized data structure with global vector state index jv is presented in [1] for the problems arrays, F, G and H, as well as for the solution arrays, V along with its derivatives DV and DDV as well as the optimal control law U. Namely, the data structures are either of the form

$$FV(is, jv), \qquad (3.6)$$

for the nonlinearity function, for example, or as

$$V(jv) \qquad (3.7)$$

for the solution V^*, rather than the hypercube-type data structure such as

$$FV(is, js(1), js(2), js(3), \cdots, js(m)) \qquad (3.8)$$

and

$$V(js(1), js(2), js(3), \cdots, js(m)) \qquad (3.9)$$

where $is = 1$ to m for each state equation and $js(is)$ is the current node index for state is. If it is assumed that there are a common number $M = M_1 = M_2 = \cdots = M_m$ of nodes per state, then $js(is) = 1$ to M points for $is = 1$ to m. The mapping from (3.6-3.7) to (3.8-3.9) is computed by an index transformation

$$jv = \sum_{i=1}^{m} (js(i) - 1) \times M^{i-1} + 1 \qquad (3.10)$$

where $jv = 1$ to M^m is the global state node index over all state nodes. Chung et al. [2] implement a method based on the above data structure, which optimally partitions the scalar computation part and the vectorization-parallelism computation part. This can be illustrated in the program that follows with the work-load of a typical loop running at full vector-concurrency:

```
       do 3 jv = 1, M**m    ! vector-concur.
CVD$L  NONVECTOR
       do 3 i = 1, m        ! scalar loop
3          FV(i,jv) = .....
```

It must be pointed out that the advantages of the algorithm and the implementation are to permit the treatment of general continuous time Markov noise or deterministic problems without noise in the same code, and to produce very vectorizable and parallelizable code of best performance.

IV. Data Parallel Methods

The famous *Bellman's curse of dimensionality* is a major property in this optimal control problem since its vector states with mesh points for each of the state component gives multi-states optimal controls. [1] and [2] have used supercomputing methods. However, the storage and execution time requirements of the supercomputing computer systems are relatively high and programs becoming serious for the case that number of states is greater than 4 or and mesh size is greater than 32. The requirements of large CM storage arise from evaluations for the state space function in high dimension cases. The functions are of form $\mathcal{F}(ms, nx)$ for DV, DDV, UR, and U^* and of form $\mathcal{V}(ms)$ for V^*, with $ms = M^m$ and $nx = m$. Therefore, the computation memory requirements are at least of order $m \cdot M^m$.

The initial implementation of the problem with data parallel computation is to locate all the arrays of form $\mathcal{F}(ms, nx)$ and form $\mathcal{V}(ms)$ on the Connection Machine memory. The nonconstant coefficient vectors, such as F_0, C_0 and DX, etc., are also located in the memory of Connection Machine with CM Fortran directives *LAYOUT* and *ALIGN*. The objective for this memory reallocation is to reduce both chip-wise *NEWS* and inter-chip *router* communications.

The storage locations of short vectors are important in this application, because they may result in large communication costs due to special handling required for odd length vectors. Let us look at an example from the discretization equations (3.1), where $V_{\mathbf{j},k}$ represents the values of optimal objective function V^* over all M^m grid points for fixed time k. The central difference approximation of the ith component the gradient is represented away from the boundaries by $DV_{i,\mathbf{j},k} = (V_{\mathbf{j}+\mathbf{e}_i,k} - V_{\mathbf{j}-\mathbf{e}_i,k})/(2 \cdot DX_i)$, evaluated over all M^m grid points. The mesh vector $DX = (DX_1, DX_2, \cdots, DX_m)$ is a relatively short vector. The operation "/" for DV acts on the two different size operands V and DX. In general, these two operands are arrays of quite different sizes, one is order $m \cdot M^m$ and another is order m. The operation is thus not efficient for the CM-2 in contrast to binary operations on arrays of the same size and causes much system communication overhead. Since similar unbalanced operations occur not only in the predictor step, but also in the corrector step of the modified Crank Nicholson method, the cumulative performance is seriously affected in terms of both in timing and storage.

The possibilities of saving CM-2 storage with more highly accurate numerical approximations that require fewer nodes, such as multigrid and finite element methods, are being studied in [9]. However, as far as the implementation techniques to this optimal control problem with data parallel computation methods, we really expect to save the storage as much as possible and also attain the best execution timing performance. In order to achieve the task, we introduce a communication technique called broadcasting. We illustrate this technique by following code fragment example, in which a very large array is multiplied by a very short vector.

```
       real a(nx)
       real u(ms,nx)
CMF$   ALIGN a(I) WITH u(1,I)
       u = u*spread(a,dim=2,ncopies=ms)
       sum(u,dim=1)
```

where nx and ms are defined as before. The *spread* function causes the vector a to be spread out or expanded so that both u and a have same size. The *sum* function adds the elements of u by column. This is a typical method used in a vector and matrix multiplication on the CM-2 [11]. The different sizes of arrays lead to a very unsuitable CM communication patterns and also cost a considerable amount of CM memory. The short vectors only occupy several CM processors, but the large size arrays occupy a large set of processors. Therefore, the *spread* function must do a lot of communications in the case that ms is large.

A modification to this implementation is to use another broadcasting technique, which locates only the large size array u in CM memory, and puts the short vector a in the slower front-end machine memory. The operation between them can be done by broadcasting the short vector element by element to the CM-2. We describe this by the sample fragment of code:

```
       a(nx)
       real u(ms),uu(ms)
       uu = 0.0
       do 3 L = 1,nx
          uu = u*a(L) + uu
3      continue
```

This implementation reduces storage requirements to $O(M^m)$. This also reduces the communication cost, by simply broadcasting scalar values to the CM-2 form the front-end. When the size of short vector is 3, 4 or 5, which are very practical numbers, the broadcasting technique becomes more useful.

To apply these broadcasting techniques to the stochastic optimal control problem, we have to decompose the operators of our original Bellman equation (2.7) problem into subproblems, so that the number of subproblems are equal to the number of states m. More formally, we have the gradient in equation (2.7),

$$\nabla V^* = \left(\frac{\partial V^*}{\partial x_1}, \frac{\partial V^*}{\partial x_2}, \cdots, \frac{\partial V^*}{\partial x_m} \right), \qquad (4.1)$$

where x_i is the ith component of the state vector $\mathbf{x}$, and each component of $\frac{\partial V^*}{\partial x_i}$ for $i = 1$ to m is a function of data structure $\mathcal{V}(ms)$. We can further decompose the nonlinearity coefficient $\mathbf{F}_0^T$ vector into m components, namely $F_{0,1}, F_{0,2}, \cdots, F_{0,m}$. The computation of the drift product $\mathbf{F}_0^T \nabla V^*$ then can be done as

$$\sum_{i=1}^{m} F_{0,i} \frac{\partial V^*}{\partial x_i} \qquad (4.2)$$

Similarly by (2.7) and (2.8), we can decompose the regular control term by

$$U_{R,i} = -C_{2,i}^{-1}(C_1 + F_1^T \nabla V^*) \qquad (4.3)$$

and the optimal control becomes

$$U_i^*(\mathbf{x}, t) = \min[U_{max,i}, \max[U_{min,i}, U_{R,i}(\mathbf{x}, t)]], \qquad (4.4)$$

for $i = 1$ to n controls. It is assumed, as usual, that the Gaussian and Poisson noise components are independent random processes. Further, $d\mathbf{W(t)}$ is the differential of zero mean, normalized Weiner process. Therefore,

$$\text{MEAN}[d\mathbf{W(t)}] = 0 \ \& \ \text{COVAR}[d\mathbf{W(t)}, d\mathbf{W(t)}] = I_r dt \qquad (4.5)$$

where I_r is the identity of order r. With this property, the term $\frac{1}{2}GG^T : \nabla\nabla^T V^*$ is decomposed into

$$\sum_{i=1}^{m} \frac{1}{2} G_i^T G_i DDV_i, \qquad (4.6)$$

where DDV_i denotes $\frac{\partial^2 V^*}{\partial z_i \partial z_i}$ and G_i is the ith row of matrix G.

By using the *operator decomposition*, the Bellman equation (2.7) can be rewritten as

$$0 = V_t + \sum_i [L_i(\mathbf{x}, \mathbf{u}, t)]. \qquad (4.7)$$

Further, by defining the partial sum

$$LS_i(\mathbf{x}, \mathbf{u}, t) = \sum_{k=1}^{i} L_k(\mathbf{x}, \mathbf{u}, t),$$

we have the recursive computation

$$0 = V_t + LS_{m-1}(\mathbf{x}, \mathbf{u}, t) + L_m(\mathbf{x}, \mathbf{u}, t). \qquad (4.8)$$

It must be noted that in [2] the solution V^* needs CM storage of order M^m and DV, DDV, U and UR need CM storage of $m \cdot M^m$. They are not balanced in size for the locations in CM memory. However, by the decomposition method, all the terms of equation (4.8) can be computed in $O(M^m)$ storage. The decomposition and broadcasting techniques result in the following outline of code implementation:

```
do 3 k = 1, N  !time loop
      :
   {compute v, vm = 1/2(v+vo)}  !predictor
   set LS = 0.0
do 111 L = 1, nx
   {compute dv, ddv, vh along dim. L}
   {compute ur, u, is along dimen. L}
   LS = LS + {LS along dim. L}
111 continue
   do 4 lc = 1, nc   !corrector
      {compute v, vm}
      set LS = 0.0
      do 222 L = 1, nx
         {compute dv, ddv, vh along dim. L}
         {compute ur, u, is along dim. L}
         LS = LS + {LS along dim. L}
222   continue
4   continue
   {similarly, compute LS for next loop}
   set LS = 0.0
   do 333 L = 1,nx
         :
      LS = LS + {LS along dim. L}
333 continue
3   continue
```

V. Performance Analysis and Results Discussion

Table 1 shows preliminary results from implementing the problem on the massively data parallel Connection Machine CM-2, without using decomposition and broadcasting techniques. This CM-2 is at NCSA and has 32K or 32,768 bit processors and one single precision floating point processor for every 32 bit processors. Double precision is accomplished by two single precision segments.

Upon using the above mentioned operator decomposition and broadcast techniques, the results in Table 2 show that the CPU timings slowly increase as the number of states increase. In addition, these measurements exhibit better performance for the larger size problems. For instance, the case $m = 4$ states and $M = 16$ nodes takes about half the time with decomposition and broadcasting, than without these techniques.

Table 3 from [2] summarizes the CPU time for implementing the problem on Cray 2S/4-128 at NCSA. It must be noted that the timings grow exponentially as either the state dimension m or logarithm of the number of nodes per state M increase. This exponentially growth in the state dimension m corresponds to the order of magnitude of the state space data structure,

$$m \cdot M^m = m \cdot e^{m \ln(M)}, \qquad (5.1)$$

which symbolizes the exponential growth of *Bellman's curse of dimensionality*. The Cray 2S performs very well on the smaller size problems, but the computational costs increase drastically as the problem size grows. This is because the Cray is a shared memory type vector supercomputer. It has very fast computation units, communication bus and channels, but for large problems more time is required for computing due to its limited number of arithmetic operation units, either scalar or vector. However, the NCSA Cray 2S has a very large memory (128 million 64 bit words), which permits the computation of a larger problem, $m = 4$ states and $M = 32$ nodes, than on the NCSA CM-2. It should be mentioned that that the Cray 2S timings are for 64 bit (Cray single) precision, while the CM-2 results are for the 64 bit precision by segmentation of the NCSA CM-2's 32 bit floating point units.

The results indicate, by comparing Table 2 with Table 3, that when the problem size increases, the Connection Machine exhibits a much smaller exponential growth, if not linear growth, in the execution time, than the Cray 2S. In fact, extrapolation for the missing data point at $m = 4$ and $M = 32$ in the CM-2 case with both techniques, indicates that it will be closer to the $M = 16$ Cray 2S value than that of the $M = 32$ value. The missing data points will be measured soon when a larger CM-2 becomes available.

Table 1: CPU time T (seconds) for different state dimensions and different mesh sizes on NCSA Connection Machine 32K CM-2 (CM Time) *without* decomposition and broadcasting techniques *in segmented CM double precision.*

m, State Variables	M, Mesh Points		
	8	16	32
2	9.54	20.45	42.20
3	14.04	29.01	221.24
4	21.40	——	——

VI. Conclusions

Stochastic dynamical programming is a very important method for stochastic optimal control problems. This paper indicates that massively data parallel computation methods are a powerful tool for computing the large size control problems. It is potentially more powerful than the vector supercomputer systems due to the slower growth of computational requirements with the growth in problem size. The techniques explored can be extended to the application to other control fields. In order to handle a large number of state variables in stochastic optimal control problems, a large number of parallel processors would be desirable.

Table 2: CPU time T (seconds) for different state dimensions and different mesh sizes on NCSA Connection Machine 32K CM-2 (CM Time) *with* decomposition and broadcasting techniques *in segmented CM double precision.*

m, State Variables	M, Mesh Points		
	8	16	32
2	15.26	31.98	67.04
3	32.53	71.82	148.10
4	100.40	411.91	——

Table 3: CPU time T (seconds) for different state dimensions and different mesh sizes on the NCSA Cray 2S/4-128 with Multitasking from [2] *in Cray single precision.*

m, State Variables	M, Mesh Points		
	8	16	32
2	0.033	0.130	0.685
3	0.104	1.169	24.626
4	2.527	60.151	2338.290

REFERENCES

[1] F. B. Hanson, *Computational Dynamic Programming for Stochastic Optimal Control on a Vector Multiprocessor,* **Argonne National Laboratory, Mathematics and Computer Science Division, Technical Memorandum No. 113**, Argonne, Ill, 1988.

[2] S.L. Chung, F.B. Hanson and H.H. Xu, *Supercomputing Optimizations for Stochastic Optimal Control Applications,* in **Proc. 4th Workshop on Computational Control of Flexible Aerospace Systems**, NASA, Williamsburg, July 1990.

[3] L.W. Tucker and G.G. Robertson, *Architecture and Applications of the Connection Machine,* **IEEE Computer, vol. 21, No. 8**, 1988, pp. 26-38.

[4] O.A. McBryan, *The Connection Machine: PDE solution on 65536 Processors,* **Parallel Computing, vol. 9**, 1988/89, pp. 1-24.

[5] K.K. Mathur and S.L. Johnsson, *The Finite Element Method on a Data Parallel Computing System,* **Int. J. High Speed Computing, vol. 1, No. 1**, 1989, pp. 29-44.

[6] I.I. Gihman and A.V. Skorohod, **Controlled Stochastic Processes.** New York: Springer-Verlag, 1979.

[7] J. Douglas, Jr., and T. DuPont, *Galerkin methods for parabolic equations,* **SIAM J. Num. Anal., vol. 7**, 1970, pp. 575-626.

[8] K. Naimipour and F.B. Hanson, *Convergence of a numerical method for the Bellman equation of stochastic optimal control with quadratic costs and constrained control,* In Preparation, 1991.

[9] Xu H.H., *On optimal control and supercomputing,* Ph.D Thesis in preparation, Univ. Illinois at Chicago.

[10] G.E. Blelloch, **Vector Models for Data-Parallel Computing.** Cambridge, MA: MIT Press, 1990.

[11] **CM Fortran Release Notes, Version 0.7-f**, Cambridge, MA: Thinking Machines Corporation, July 1990.

Triangulation, Voronoi Diagram, and Convex Hull in k-Space on Mesh-Connected Arrays and Hypercubes*

J. Andrew Holey
Department of Computer Science
University of Minnesota
Minneapolis, Minnesota   55455
Tel.: 612-866-8087
holey@umn-cs.cs.umn.edu

Oscar H. Ibarra
Department of Computer Science
University of California
Santa Barbara, California   93106
Tel.: 805-893-4171
ibarra@cs.ucsb.edu

Abstract:

We present new algorithms for the Delaunay triangulation of points in k-space. These include algorithms for two-way linear arrays, multi-dimensional mesh-connected arrays, and hypercubes. These algorithms can be used to calculate the convex hull and the Voronoi diagram from a Delaunay triangulation using a constant factor of additional time. We know of no other parallel algorithms for these problems in dimension higher than 3. Our algorithms obtain nearly optimal speedup over the best known sequential algorithms, with the hypercube algorithm using $O(\log^2 n)$ expected time and $O(\log^3 n)$ time in the worst case.

Keywords:

Computational geometry, convex hull, Delaunay triangulation, hypercube, mesh-connected arrays, parallel algorithms, Voronoi diagram, geometry in k-space.

1.   Introduction

One of the oldest problems in the field of computational geometry is the triangulation of an arbitrary set of points in k-space. The first important statement of this problem was made for 3-space by Boris N. Delaunay in 1934 [3]. Since then, the Delaunay triangulation in 2-space has been studied extensively. The triangulation of points in higher dimension spaces is a much more difficult problem, and there are relatively few papers on this subject, in part because of the unavoidably high computational costs in the worst case instances. Triangulations have important applications in robotics, computer vision, and computer-aided design.

Delaunay himself observed that the Delaunay triangulation is the dual of the Voronoi diagram which was first proposed by George Voronoï in 1908 [3, 12]. Avis and Bhattacharya give algorithms for Voronoi diagram and Delaunay triangulation in k-space [1], and Avis give an algorithm for computing arbitrary triangulations of simplicial point sets in k-space [2]. There is also a relation between triangulations and convex hull, and there are a few papers devoted to computing the convex hull in k-space [10, 11]. In another paper, we give a dynamic sequential algorithm for Delaunay triangulation in k-space which can also be used to compute the Voronoi diagram and convex hull [7]. We have found no papers which present parallel algorithms for any of these problems in higher dimensional spaces.

In an earlier paper, we used an iterative method to compute the triangulation and Voronoi diagram of points in the plane and the convex hull of points in 3-space on mesh-connected arrays [8], and that paper builds on an earlier paper which presents algorithms

for the convex hull of points in a plane [6]. This paper extends the results of our previous triangulation papers to give parallel algorithms for triangulation, Voronoi diagram, and convex hull in k-space. We will highlight new techniques needed to extend our algorithms to k-space: a data representation scheme for triangulations in arbitrary dimensions, the methods needed to generate new connections among points, and the way that these structures and methods can be handled on mesh-connected arrays. The portions of our algorithms which are simple extensions of the 2-space algorithms will be described only briefly.

We believe that these are the first parallel algorithms which solve these problems in dimension higher than 3. They are further distinguished by the fact that they make efficient use of resources. As will be shown, the combinatorial complexity of computing and representing these problems in k-space can vary greatly. Our algorithms use only as much computing resources as are necessary for a given instance of the problem. They perform well in both expected and worst-case situations.

Section 2 describes the mesh-connected array architectures used in this paper. In Section 3 gives formal definitions along with a discussion of data structures and their operations. The algorithms for triangulation are presented in Section 4. The adaptations to these algorithms needed to compute the Voronoi diagram and convex hull are sketched in Section 5. Section 6 gives some concluding remarks. The details of the algorithms will be given in the full paper.

2.   Mesh-Connected Arrays

Mesh-connected arrays are a family of parallel computer architectures which are characterized by a large number of fairly simple processors, each with a fixed number of registers, connected in an orthogonal lattice of one or more dimensions.

A two-way cellular array (CA) is a linear array of processors with two-way communications between processors. Input is to all processors in parallel, and the output is from all processing nodes in parallel. A d-CA is a d-dimensional two-way cellular array. Like the CA, it has parallel input to all processors along with parallel output. For our purposes, we constrain the number of processors in a d-CA so that there is an equal number of processors per row in each dimension, and thus the number of processors must be a perfect dth power of some integer.

A hypercube is a special case of a d-CA. In a hypercube the number of processors is always exactly 2^d, but the dimension of the hypercube will vary with the problem size. Any d-CA algorithm which runs on a mesh of arbitrary dimension may be run on a hypercube by letting the dimension of the hypercube be $\lceil \log p \rceil$ where p is the number of processors required for the given problem size.

* This research was supported in part by NSF Grants CCR89-18409 and DCR90-96221.

3. Preliminaries

The three problems discussed in this paper are closely related to each other. In order to make these relationships clear, we must define a number of geometric terms as well as the problems themselves. With these definitions, we can then discuss how these problems are interrelated and how they can be represented in a computer.

3.1. Definitions

The *convex hull* of a finite set of points **P** in *k*-space is the smallest convex set which contains **P**. We will denote the convex hull of a set as conv(**P**). The convex hull of **P** is a convex polytope, and the *extreme points* of **P** are the vertices of that polytope. If **P** is an affinely independent set, then the convex hull of **P** is a *simplex*. In particular, if $|\mathbf{P}| = i + 1$, then the convex hull of **P** is an *i-simplex*. We will represent a simplex as a list of the points, sorted lexicographically, which are the vertices of the simplex, and for notational purposes, we enclose the list in angle brackets.

Let **P** be a set of points in *k*-space, and let the affine hull of **P** be an *i*-flat, $i \leq k$. A set of *i*-simplices **T** is a *triangulation* of **P** if

a) for every simplex s in **T**, if **p** is a vertex of s then $p \in \mathbf{P}$,

b) for every s and t in **T**, $s \cap t = \emptyset$ or $s \cap t$ is a common *j*-face of s and t, and

c) for every s in **T** and for every (*j*-1)-face **f** of s, either there is a t in **T** such that $s \cap t = f$ or all the vertices of **f** are extreme points of conv(**P**).

T is a *Delaunay triangulation* of **P** if **T** is a triangulation of **P** and for every s in **T**, no point of **P** is in the interior of of s's circumsphere.

In 2-space, if $|\mathbf{P}| = n$, then the number of simplices in a triangulation **T** of **P** will be linear in *n*, since the edges of the triangulation form a planar graph. There is no known general formula for the number of simplices in a triangulation of a set of points in *k*-space. In the worst case, the number of simplices in a triangulation will be $O(n^{\lceil k/2 \rceil})$ if *k* is fixed [9]. However, for points which are uniformly distributed in a spherical region of *k*-space, it has been shown that the expected number of simplices is $O(n)$ with a constant of proportionality which is exponential in *k* [4].

A set of *k*-polyhedra **V** is called the *first-order Euclidean Voronoi diagram* of a set of points **P** in *k*-space if there is a one-to-one correspondence *v* from **P** to **V** such that for every **p** in **P**, the interior of *v*(**P**) is the set of all points closer to **p** that to any other point in **P**. We note that some of the polyhedra in **V** will be unbounded, and that these polyhedra are exactly those associated with points lying on conv(**P**). The intersection of any two distinct *k*-polyhedra in **V** will be either the empty set or a *j*-polyhedron, where $j < k$. The first-order Euclidean Voronoi diagram is also called a *Dirichlet tessellation*, but we will use the term *Voronoi diagram* in this paper. We will denote the Voronoi diagram of a set **P** as vor(**P**), and by extension, vor(**p**) will denote the Voronoi region associated with the point **p**.

3.2. Data Structures

One representation scheme for planar triangulations, first proposed by P. D. Gilbert in 1979 [5] stores the edges incident to each point in a data structure which is called a *star*. For each point in the triangulation, there is an associated star which stores the points

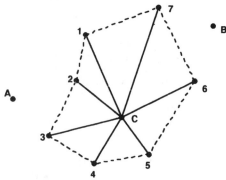

Figure 3.2.1.   A Star in 2-Space.

connected to that point in the triangulation in a circular, counter-clockwise list. Thus, each point in the list together with the central point determines one edge in the triangulation. Figure 3.2.1 shows a graphical representation of a star centered on Point **C**.

We would like to generalize the notion of a star to *k*-space. There are two difficulties with this approach. First, it is not sufficient simply to list for each point all the other points in the set to which it is connected by an edge in the triangulation; we need information about how the edges together form triangles, how the triangles together form tetragons, and in fact how the *k*-simplices are formed. Second, in dimensions higher than the plane, there is no meaningful way to order the points. The solution to these problems is to define a star in terms of (*k*-1)-simplices and (*k*-2)-simplices.

Let **P** be a set of points in *k*-space and let **T** be a triangulation of **P**. For each (*k*-2)-simplex **c** in **T**, we define the *k-star* s of **c** to be a circular list of all the points in **P** which together with **c** determine a (*k*-1)-simplex of **T**. The order of points in the circular list of s is determined by projecting those points onto a plane orthogonal to the (*k*--2)-flat spanned by **c** and intersecting **c**. The points are stored in counter-clockwise order around the point where **c** and the (*k*-2)-flat intersect. For convenience, we superimpose a rectangular coordinate system on the projection plane with the intersection point at the origin and consider the point on or immediately counter-clockwise from the positive *y*-axis to be the first point in the list. We will refer to the (*k*-2)-simplex **c** of s as the *core* of s. We will use the terms *star* for *k*-star whenever *k* is understood. We order stars lexicographically by their cores.

In order to obtain an acceptable speedup on a mesh-connected array, the representation of a triangulation must be distributed among the processors. One strategy suggests itself immediately: we can distribute one star to each processor. This approach has one significant problem. The size of a star's list in a triangulation can vary from $O(1)$ to $O(n)$ points, so that some stars are very small and some are very large. Thus, the space required for each processor can vary greatly.

In our parallel algorithms, we use an approach which allows us to make tradeoffs between the amount of space used per processor and the number of processors used. Furthermore, the amount of space will be roughly the same for all processors no matter how we choose to make the tradeoffs. We will allow stars to be broken into segments called *wedges*. Each wedge consists of the core of the star and a contiguous portion of its list. Using this representation, a large star may be distributed over several processors while several small stars may be stored in a single processor. With a suitable load balancing scheme, we can then insure that space required is about the same for all processors. We will consider the size of a wedge to be the number of list points it contains, and when we divide a wedge or combine wedges with a common core, we will ignore the need to eliminate or duplicate copies of the core since this matter does not affect the asymptotic space complexity.

The major difficulty with this data representation scheme is that the list of points in a star is supposed to be circular and all its points are supposed to be contiguous, but when the wedges of a particular star are distributed over several neighboring processors, some adjacent points in the list will be stored in different processors. In particular, the first point in the list may be several processors away from the last point. We overcome this problem by having each adjacent pair of wedges keep one list point in common. Thus, if the 2-star in Figure 3.2.1 were divided into three wedges, the wedges might be <c-1, 2, 3>, <c-3, 4, 5>, and <c-5, 6, 7, 1>.

For any star, it is necessary for a triangulation algorithm to determine what points should be included in the list. Observe that in Figure 3.2.1, Points A and B are not included in the star centered on C. They are omitted because they lie outside the circumcircle of the core point and the two adjacent points in the star's list. Thus, in a Delaunay triangulation, it is fairly simple to determine if a point should be added to a star's list.

Essentially, we determine if a point should be inserted in a star by checking the Delaunay criterion for the point and the star's core against those points already in the list. If we determine that the point should be inserted in the star, then we may need to remove some other points from the list. This process may result in the creation of new stars. We may also determine that the current star itself is no longer useful and delete if from the list.

4. The Triangulation Algorithms

Our parallel algorithms divide the stars of the triangulation into wedges which are small enough to fit within the space limit of each processor. The fundamental operation on collections of wedges is the same in all the algorithms. Each processor stores a set of wedges. When it receives another set of wedges from a neighbor, it merges the two sets of wedges, updating those wedges and possibly generating new wedges. It then splits the new set of wedges, keeping some of them and passing others to a neighboring processor. The processors repeat this operation in parallel until the triangulation is complete. Each merge operation brings the current state of the computation closer to the desired triangulation by eliminating simplices which do not belong in the final triangulation and generating new simplices which are more likely to be part of the final triangulation.

4.1. The CA Algorithm

Our CA algorithm forms the basis for our d-CA algorithm in the next section. At the beginning of the algorithm, one point is loaded into each of the n leftmost processors, and connections are made between a processor's initial point and the points of its neighbors. In each cycle, if a processor does not have an excess of wedges, it merges half of its wedges with half of the those of each of its neighbors and then merges

the results, sorting as it goes. After $O(p(n))$ cycles, all the stars of the final triangulation have been generated as wedges, although some list points in wedges with a common core may occur more than once and some unwanted list points and wedges may still be present. Another $O(p(n))$ cycles are performed during which all unwanted wedges are eliminated, duplicate list points are consolidated, and the sorting is completed. This algorithm is similar to the planar CA algorithm, but it allows for dynamic allocation of additional processors or additional storage space in each processor.

Theorem 4.1.1. A CA can compute the triangulation of a set of n points in k-space in $O(n)$ expected time using $O(n)$ processors. The worst case running time will be $O(n^{\lceil k/2 \rceil})$ using $O(n)$ to $O(n^{\lceil k/2 \rceil})$ processors with $O(n^{\lceil k/2 \rceil - 1})$ to $O(1)$ space per processor.

4.2. The Mesh Array Algorithm

Our d-CA algorithm is exactly like the CA algorithm except that each processor exchanges data with up to $2d$ neighbors, and the dynamic allocation scheme requires a few more details to implement. We embed a Hamiltonian path in the mesh in snake-like row major order. Each node knows for all its neighbors which are prior and which are posterior on the Hamiltonian path, and in each dimension, if a node has two neighbors, one will be prior and one posterior on the path. A dimension counter is added which is incremented mod d at each iteration of the algorithm. Communication during a given iteration is in the dimension of this counter. Analysis of the algorithm gives us

Theorem 4.2.1. A d-CA can compute the triangulation of a set of n points in k-space in $O(n^{1/d})$ expected time using $O(n)$ processors with $O(1)$ memory words per processor. In the worst case, the algorithm will require $O(n^{\lceil k/2 \rceil - 1 + 1/d})$ time if $O(n)$ processors are used or $O(n^{\lceil k/2d \rceil})$ if $O(n^{\lceil k/2 \rceil})$ processors are used.

4.3. The Hypercube Algorithm

The d-CA algorithm can be used on the hypercube with a few simple modifications. Initially we allocate a hypercube of dimension $\lceil \log n \rceil$ and load the points into the n lowest processors on the Hamiltonian path. Whenever a request for additional processors is issued, the global control doubles the dimension of the hypercube which has the effect of squaring the number of processors. We can simply substitute appropriate values into the formulas for the d-CA algorithm and add an extra factor of d for an extra loop that eliminates duplicate wedges to get

Theorem 4.3.1. A hypercube of dimension $\lceil \log n \rceil$ can compute the triangulation of a set of n points in k-space in $O(\log^2 n)$ expected time with $O(1)$ memory words per processor and in $O((\log^3 n)n^{\lceil k/2 \rceil - 1})$ time in the worst case. If additional processors can be allocated dynamically, the worst case time is reduced to $O(\log^3 n)$ on a hypercube of dimension $\lceil k/2 (\log n) \rceil$.

5. Voronoi Diagram and Convex Hull

All of the triangulation algorithms presented here can be used to compute the Voronoi diagram and the convex hull of a set of points in k-space. The additional time and space required will be no more than a constant factor times the time and space needed to compute the triangulation alone. Both the Voronoi diagram and the convex hull can be computed in a straightforward manner, but in the case of the convex hull, we can improve on the simplest strategy by exploiting a relationship between triangulation in k-space and convex hull in $(k+1)$-space.

5.1. Voronoi Diagram

As we have already noted, the Voronoi diagram is the straight line dual of the Delaunay triangulation. In the plane, each Voronoi region not associated with a point on the boundary of the convex hull is a convex polygon and can be represented as a circular list of points. In k-space, a bounded Voronoi region is a convex k-polytope, and representing such a polytope is not a trivial task. However, with a small amount of additional computation, we can use our triangulation data structures to describe the Voronoi regions of the point set, and we obtain the following corollaries of our previous theorems:

Corollary 5.1.1. A CA can compute the Voronoi diagram of a set of n points in k-space in $O(n)$ expected time using $O(n)$ processors. The worst case running time will be $O(n^{\lceil k/2 \rceil})$ using $O(n)$ to $O(n^{\lceil k/2 \rceil})$ processors.

Corollary 5.1.2. A d-CA can compute the Voronoi diagram of a set of n points in k-space in $O(n^{1/d})$ expected time using $O(n)$ processors with $O(1)$ memory words per processor. In the worst case, the algorithm will require $O(n^{(\lceil k/2 \rceil - 1 + 1/d)})$ time if $O(n)$ processors are used or $O(n^{(\lceil k/2 \rceil/d)})$ if $O(n^{\lceil k/2 \rceil})$ processors are used.

Corollary 5.1.3. A hypercube of dimension $\lceil \log n \rceil$ can compute the Voronoi diagram of a set of n points in k-space in $O(\log^2 n)$ expected time with $O(1)$ memory words per processor and in $O((\log^3 n)n^{\lceil k/2 \rceil - 1})$ time in the worst case. If additional processors can be allocated dynamically, the worst case time is reduced to $O(\log^3 n)$ on a hypercube of dimension $\lceil k/2 (\log n) \rceil$.

5.2. Convex Hull

To compute the convex hull, we make use of the fact that the convex hull of a set of points in k-space is combinatorially equivalent to the triangulation of a set of points in $(k-1)$-space. Briefly, the method is to project the points onto the hyperplane $x_k = 0$ and triangulate the projection using a criterion which eliminates points not in the convex hull in place of the Delaunay criterion. This method produces

Corollary 5.2.1. A CA can compute the convex hull of a set of n points in k-space in $O(n)$ expected time using $O(n)$ processors. The worst case running time will be $O(n^{\lfloor k/2 \rfloor})$ using $O(n)$ to $O(n^{\lfloor k/2 \rfloor})$ processors.

Corollary 5.2.2. A d-CA can compute the convex hull of a set of n points in k-space in $O(n^{1/})$ expected time using $O(n)$ processors with $O(1)$ memory words per processor. In the worst case, the algorithm will require $O(n^{(\lfloor k/2 \rfloor - 1 + 1/d)})$ time if $O(n)$ processors are used or $O(n^{\lfloor k/2 \rfloor/d})$ if $O(n^{\lfloor k/2 \rfloor})$ processors are used.

Corollary 5.2.3. A hypercube of dimension $\lfloor \log n \rfloor$ can compute the convex hull of a set of n points in k-space in $O(\log^2 n)$ expected time with $O(1)$ memory words per processor and in $O((\log^3 n)n^{\lfloor k/2 \rfloor - 1})$ time in the worst case. If additional processors can be allo-

cated dynamically, the worst case time is reduced to $O(\log^3 n)$ on a hypercube of dimension $\lfloor k/2 (\log n) \rfloor$.

6. Conclusions

We have shown new algorithms for the triangulation of point sets in k-space on CA's, d-CA's, and hypercubes. To our knowledge, these are the first parallel algorithms for these problems. These algorithms can also be used to compute the Voronoi diagram and convex hull.

An important feature of these algorithms is that their performance is nearly optimal for both time and space complexity in the expected case. Even in the worst case, there is minimal space overhead. For the linear array algorithm, the worst case running time is only a factor of n from asymptotic optimality. With higher dimensional meshes, the performance is even better. No problem instance can cause the algorithms' performance to degrade beyond these limits.

The method of parallel algorithm design used in this and two earlier papers [6, 8] has proved very useful in obtaining a variety of useful algorithms. We hope to use this method to design algorithms for other problems in the area of computational geometry. We also expect to implement some of these algorithms on currently available machines.

References

1. Avis, D. and B. K. Bhattacharya. "Algorithms for Computing d-Dimensional Voronoi Diagrams and Their Duals." *Advances in Computing Research*. 1: 159–180, 1983.

2. Avis, D. and H. ElGindy. "Triangulating Point Sets in Space." *Discrete & Comput. Geom.* 2(2): 99-111, 1987.

3. Delaunay, B. N. "Sur la Sphère Vide." *Izvest. Akad. Nauk SSSR., Ser. 7, Otd. Mat. i Est. Nauk.* (6): 793-800, 1934.

4. Dwyer. "Higher-Dimensional Voronoi Diagrams in Linear Expected Time." *5th Annual Symposium on Computational Geometry*. 1989 326-333.

5. Gilbert, P. D. Master of Science Thesis. *New Results on Planar Triangulations*. University of Illinois at Urbana-Champaign. 1979.

6. Holey, J. A. and O. H. Ibarra. "Iterative Algorithms for Planar Convex Hull on Mesh-Connected Arrays." *International Conference on Parallel Processing*. III: 102-109, 1990.

7. Holey, J. A. and O. H. Ibarra. "A Dynamic Algorithm for Triangulation, Voronoi Diagram, and Convex Hull in k-Space." *Manuscript*.

8. Holey, J. A. and O. H. Ibarra. "Triangulation in a Plane and 3-D Convex Hull on Mesh-Connected Arrays and Hypercubes." *Fifth International Parallel Processing Symposium*. 1991.

9. Klee, V. "On the Complexity of d-Dimensional Voronoi Diagrams." *Arch. Math.* 34: 75-80, 1980.

10. Rosen, J. B. and G.-L. Xue. "A General Convex Hull Algorithm with Linear Expected Time." University of Minnesota *Tech. Report* TR 89-33. May 1989.

11. Rosen, J. B., G.-L. Xue and A. T. Phillips. "Efficient Computation of Convex Hull in R^d." University of Minnesota *Tech. Report* TR 89-51. August 1989.

12. Voronoï, G. "Nouvelles Applications des Paramètres Continus à la Théorie des Formes Quadratiques." *J. reine angew. Math.* 134: 198-287, 1908.

PARALLELIZING SPICE2 ON SHARED–MEMORY MULTIPROCESSORS

Gung–Chung Yang
Center for Supercomputing Research and Development
University of Illinois at Urbana–Champaign
Urbana, IL 61801

Abstract. This paper presents a general method to parallelizing SPICE2, the *de facto* standard circuit simulator in industry, on shared–memory multiprocessors. The method extracts parallel tasks at the algorithmic level for each CPU–intensive module and is designed for a wide range of high–performance computer systems. The implementation of the method in SPICE2 resulted in a portable parallel direct circuit simulator, PARASPICE. The superior performance of PARASPICE is demonstrated on an eight–CE Alliant FX/80 using a number of benchmark circuits.

1. Introduction

Standard circuit simulators like SPICE2 [Nage75], ASTAP [WJMM73], and ADVICE [Nage80] are widely used in industry and have proven to be reliable and accurate. Unfortunately, these simulators are considered too inefficient to analyze VLSI circuits on conventional computers. Thus, there exists a strong incentive to increase the speed of circuit simulation using high–performance computers such as Cray X–MP, Alliant FX, and Convex C–2.

A number of approaches have been developed to improve the speed of circuit simulation by exploiting the vector and concurrent processing capabilities of high–performance computer systems for the standard circuit simulators ([Vlad82][YaTa85][JaNP86][SaVi87]) or resorting to relaxation methods [NeSa83]. In this paper, we describe a general method to parallelizing standard circuit simulators on shared–memory multiprocessors. Specifically, we use the SPICE2 program as the baseline serial code and an Alliant FX/80 as the test–bed multiprocessor.

There are three goals in this investigation. The first goal is the study of potential parallelism in the direct method. The computation times of a standard circuit simulator are usually dominated by three modules: device model evaluation (LOAD), direct solution of sparse linear systems (SOLVE), and local truncation error estimation (TRUNC), which account for at least 95 percent of the total job time. The degree of parallelism inherent in these three modules provides a theoretical limit on the speedup achievable by using parallel processing. The second goal is the investigation of parallel algorithms that (collectively) will form the basis of the parallel direct method in circuit simulation. These algorithms should exploit the parallelism at the problem level. Various schemes or models representing trade–offs between parallelism and other factors should be studied thoroughly. The third goal is the development of a portable parallel direct circuit simulator based on the aforementioned algorithms. The software should maintain the general applicability of SPICE2 and be readily ported to a wide range of shared–memory multiprocessors.

This paper is divided into seven sections. In Section 2, timing profiles of SPICE2 on the Alliant FX/80 are presented to highlight the computational structures of a standard circuit simulator. Sections 3 through 5 describe the design of a parallel direct circuit simulator called PARASPICE and its implementation on the Alliant FX/80. The overall performance of PARASPICE as compared with SPICE2 on a number of benchmark circuits is presented in Section 6. Section 7 concludes this paper with a discussion of the major results from this investigation and several directions for future research.

2. Timing Profile of SPICE2

Five circuits (most of which come from industrial designs) are used as benchmark circuits throughout this paper. Table 2.1 shows the timing profile of the largest benchmark circuit MOS330 (CMOS circuit with 330 MOSFET transistors) on the Alliant FX/80. In terms of computational structure, the majority of the time can be separated into: (1) "LOAD," 43.8% ; (2) "SOLVE," 45.4% ; and (3) "TRUNC," 5.5% of total job time. Experiments performed on the rest of the benchmark suite resulted in similar distributions and, hence, are not presented here.

These results basically characterize the sequential nature of the SPICE2 program, even though the Alliant FX/80 are high–performance shared–memory multiprocessors. To benefit from the computing power of these computer systems, algorithm restructuring and new methods are needed and the methods should be directed at a higher level of computational structure; that is, at the LOAD, SOLVE, and TRUNC modules.

3. The Parallel SOLVE Module

Transient analysis in circuit simulation requires the solution of a sequence of structurally identical sparse linear systems. The time spent in the SOLVE module often comprises a major portion of the total job time. In fact, there are two types of problems that need to be solved in the SOLVE module: (1) the ONE–OFF problem (solving the very first linear system); and (2) the 2–OFF problem (solving the remaining of the linear systems). Because the ONE–OFF problem does not provide *a priori* knowledge about the structure and numerical values of the linear system, dynamic pivoting and data structure are needed for the sparse solver to be effective. While solving 2–OFF

Table 2.1 Timing Profile of SPICE2
on Alliant FX/80

Phase	MOS330	
	Time (s)	Percent
(1) Readin & Errchk	9.50	2.3
(2) Setup	4.95	1.2
(3) Analysis	402.12	96.4
3.1 LOAD	182.78	43.8
3.2 SOLVE	189.27	45.4
3.3 TRUNC	22.99	5.5
(4) Output	0.59	0.1
Total Job Time	417.19	100.

problems can take advantage of the pivotal order and the associated LU map from the solution of the **previous** ONE–OFF problem, a fast solution procedure can be applied.

In PARASPICE, a unified computational model [Yang90a] has been employed and the solution is carried out in three phases: ONE–OFF, ANALYZE, and 2–OFF. In the first phase, the first sparse linear system is solved via parallel algorithms at the three basic stages: pivoting, reordering, and elimination. The primary issue in this phase is the pivoting scheme for stability, fill–in, and parallelism. In the second phase, a task graph is built through data dependence analysis and a task queue is subsequently constructed by a parallelism extraction algorithm. The ANALYZE phase is performed sequentially and represents a cost for the speed of the last phase. Finally, in the 2–OFF phase, the remaining sparse linear systems are solved by a fast parallel procedure based on the task graph.

Using this solution strategy, a class of new algorithms has been developed for PARASPICE and the parallel algorithms can be divided into three categories according to the task granularity:

(1) the Fine–grain model at the elemental operation level [HuWi79].

(2) the Medium–grain model at the row operation level.

(3) the Coarse–grain model at the pivot operation level.

The performance of the parallel SOLVE module depends upon the degree of parallelism extracted by the algorithm and the matching of the task granularity at the algorithm level to that of the underlying architecture.

3.1. Characteristics of Sparse Matrices

For a given circuit, choosing an appropriate algorithm requires a detailed characterization of the sparse matrix arising from circuit simulation. To characterize the sparse matrices, two metrics: Fill–ratio (the ratio of Fill–in to Nonzeroes before Factorization) and Density (number of nonzeroes per row or column after

Data Structure for Algorithm PEFGM

A	LU map from a ONE–OFF problem.
b	Right–hand side.
N	Order of the sparse matrix.
DEPA	Working storage, same size as A.
DEPb	Working storage, same size as b.
id	Task id.
TASK(id)	Task
DEPTH(id)	Depth of task id.
Dtask	Depth of the task graph.

Algorithm PEFGM /*Parallelism Extr in Fine–grain Model*/

```
clear DEPA; clear DEPb; id=0;
for k :=1 to N–1 do /* Gaussian Elimination */
   id=id+1 ;
   TASK(id)= {A_kk} ; /*store task id rep by op in { } */
   DEPA(k,k)=DEPA(k,k)+1 ;
   DEPTH(id)=DEPA(k,k) ;
   forall (i > k ∧ A_ik ≠ 0) do
      id=id+1 ;
      TASK(id)= {A_ik = A_ik / A_kk } ;
      DEPA(i,k)=Max{DEPA(i,k), DEPA(k,k)} + 1 ;
      DEPTH(id) = DEPA(i,k) ;
      id = id + 1 ;
      TASK(id) = {b_i = b_i − A_ik * b_k } ;
      DEPb(i) = Max{DEPb(i), DEPA(i,k), DEPb(k)} + 1 ;
      DEPTH(id) = DEPb(i) ;
      forall (j > k ∧ A_kj ≠ 0) do
         id=id+1 ;
         TASK(id) = {A_ij = A_ij − A_ik *A_kj} ;
         DEPA(i,j)=Max{DEPA(i,j), DEPA(i,k), DEPA(k,j)}+1 ;
         DEPTH(id) = DEPA(i,j) ;
      endforall
   endforall
endfor
for k:=N downto 1 do /*Back Substitution */
   id=id+1 ;
   TASK(id) = {b_k = b_k/A_kk} ;
   DEPb(k) = Max {DEPb(k), DEPA(k,k)} + 1 ;
   DEPTH(id) = DEPb(k) ;
   forall (i < k ∧ A_ik ≠ 0) do
      id = id + 1 ;
      TASK(id) = {b_i = b_i − A_ik *b_k} ;
      DEPb(i) = Max {DEPb(i), DEPA(i,k), DEPb(k)} + 1 ;
      DEPTH(id) = DEPb(i) ;
   endforall
endfor
Dtask = Max{DEPTH(1:id)} ;
```

Figure 3.1 Data Structure and PEFGM Algorithm
in the Fine–grain Model.

Factorization), play a crucial role. An experiment conducted on the five benchmark circuits shows that the Fill–ratio ranges from 0.16 to 0.53, which reflects the relative growth in the nonzero entries. The small values of Fill–ratio show the efficiency of the pivoting scheme employed in SPICE2. As a result, the factored

Table 3.1 Parallelism in the SOLVE Module
using Fine-grain Model

Circuit	N	Depth	Task	Tsk/Dep
MOS164	116	146	1826	13
BJT041	177	158	3040	19
MOS028	204	167	3364	20
MOS165	431	286	6232	22
MOS330	856	489	12437	25

matrices have a Density value of 6 despite the increasing order of the matrices. Such characteristics implies that the Fine-grain model is most suitable as long as the storage is not a limiting factor. To investigate the applicability of the Fine-grain model, the PEFGM Algorithm (described in Figure 3.1) is applied to the LU map of the factored matrix obtained as a result of solving the first ONE-OFF problem. The results conducted on the benchmark circuits are given in Table 3.1.

In the Fine-grain model, parallelism is exploited at the elemental operation level where task graphs are made of three basic operations: pivot check, divide, and update. The total number of tasks increases linearly with respect to the order of the matrix. Such results directly contradict the well-known assertion that it grows superlinearly. Because of this, the storage requirements are rather modest for the Fine-grain model. Column 3 of Table 3.1 gives the Depth of the task graph using PEFGM. All matrices have a Depth value well below their upper bound (i.e., 5N), where N is the order of the matrix. (Note that Depth is a function of the pivoting scheme used in the ONE-OFF phase and the problem of optimizing Depth is NP-complete.) As the size of matrix increases, Depth also grows, but at a slower rate, which implies that more parallelism is exploitable for large circuits as validated in column 5 where the average number of parallel tasks is listed. The number of tasks per depth level shown in column 5 is an upper bound for speedup. Because the Fine-grain model exploits maximum parallelism among three models, this also bounds the speedup achievable on any parallel computer system. The Medium-grain and Coarse-grain models will extract fewer parallel tasks and thus be limited by a smaller theoretical upper bound.

3.2. Design and Implementation

As discussed above, the Fine-grain model is apparently the best choice for our benchmark circuits. To employ the model, the strategy used in SPICE2 is no longer suitable. The new strategy adopted for PARASPICE can be summarized as follows.

(1) The symbolic preordering used in SPICE2 is very effective as a preprocessing for the pivoting scheme and therefore should be adopted in PARASPICE as well.

(2) The first Newton Raphson(NR) iteration at t=0.0 is solved as a ONE-OFF problem. This is followed by a procedure based on the PEFGM Algorithm. The remaining NR iterations are treated as 2-OFF problems. Because the parallel algorithm employed in the 2-OFF phase is unable to perform pivoting on the fly, a backup scheme is needed. The backup scheme used for the initial DC solution is to switch to a new ONE-OFF process. The goal of the scheme is to maintain the robustness of the solver.

(3) All NR iterations arising from the transient analysis are treated as 2-OFF problems. Again, a backup scheme is required if the algorithm fails during the execution. The backup scheme used for the transient analysis is to reject the current timepoint and restart with a reduced stepsize. The goal of the scheme is to maintain the speed.

The strategy works quite well for the benchmark circuits. In fact, no backup operations were observed during the execution.

3.3. Experimental Results

An experiment with the module on the benchmark circuits is conducted on an Alliant FX/80 using one CE, four CEs, and eight CEs, and all produced the same solution as SPICE2. Therefore, the speedup represents the genuine performance gains without sacrificing the accuracy.

The overall performance of the SOLVE module is given in Table 3.2. The speedup ranges from 2.63 to 3.09 using four CEs and from 3.47 to 4.54 using eight CEs. The speedup is primarily due to the parallel algorithms employed in the ONE-OFF and 2-OFF phases. For the SOLVE module, the speedup (slowdown) of SPICE2 versus PARASPICE using one CE is primarily the result of two competing effects: the price paid in executing the ANALYZE procedure and the overheads incurred from scheduling in the ONE-OFF and 2-OFF phases for PARASPICE, and the savings gained in address searching time during the 2-OFF phase of PARASPICE. Clearly, the second effect becomes increasingly dominant as the size of the circuit increases. Therefore, the overall speedup of the SOLVE module in PARASPICE versus SPICE2 is expected to be higher for large circuits. For example, MOS330 achieved a speedup of 4.51 using eight CEs, while it achieved a speedup of 5.18 compared to SPICE2.

4. The Parallel LOAD Module

At each NR iteration, the LOAD module is called to perform the model evaluation. The execution is carried out sequentially by evaluating one element at a time. For each element, the computations are often divided into the following five steps.

Table 3.2 Timing and Speedup of the SOLVE Module
on Alliant FX/80 using Fine-grain Model

Circuit	SPICE2	PARASPICE		
		1 CE	4 CEs	8 CEs
MOS164	14.72 (1.0)	14.75 (1.0)	5.20 (2.84)	3.87 (3.81)
BJT041	53.29 (0.81)	43.07 (1.0)	13.97 (3.08)	9.49 (4.54)
MOS028	16.99 (0.76)	12.84 (1.0)	4.88 (2.63)	3.70 (3.47)
MOS165	78.47 (0.91)	71.03 (1.0)	23.09 (3.08)	15.92 (4.46)
MOS330	189.27 (0.87)	164.54 (1.0)	53.17 (3.09)	36.51 (4.51)

(1) Model the element in terms of basic components.

(2) Replace the energy-storage components with resistive components and independent sources via stiffly-stable numerical integration formulae.

(3) Replace nonlinear resistive components with conductors and independent sources via the NR method.

(4) Compute the contributions to the Jacobian and the RHS via Modified Nodal Analysis (MNA).

(5) Update the sparse linear system.

The first four steps are usually dominated by scalar floating-point operations, whereas the last step only consists of a number of address fetch and fetch-and-add operations. For most semiconductor devices such as BJT and MOSFET, the majority of the time is spent in steps 1–4.

4.1. Design and Implementation

The computations described above can be lumped together naturally as a single task. Using this computational model, a task is identified with its associated element and all tasks can be executed in parallel if the dependences between them are properly synchronized. From an algorithm's point of view, the first four steps are strictly local to the associated element and thus require no synchronization. The only dependence occurs at the last step when the local contributions are added to the sparse linear system. Because the order of updates is immaterial, the dependency is basically a critical region relation.

However, the original SPICE2 program contains many spurious dependences in the LOAD module. For most devices, these spurious dependences can easily be broken by a careful analysis of the code. The most complicated, and therefore notable, example is the MOSFET, which uses the Common-block /MOSARG/ as working storage. Because this Common-block is used only for local information, its usage creates undesirable dependence between tasks. A solution for this dilemma is to replace the Common-block

/MOSARG/ with local variables and pass them as arguments. This technique has been implemented in PARASPICE. Note that on the Cray computer systems, the problem can easily be solved using TASK COMMON, a multitasking facility provided on the Cray machines.

Once the spurious dependences are removed, the update operation at step 5 becomes the only dependence between tasks. In PARASPICE, two schemes are employed to parallelize the LOAD module.

(1) Lockless-scheme.
(2) Lock-scheme.

Their implementations are discussed below.

The Lockless-scheme uses the task fission technique [Yang90b] to break and eventually eliminate the dependence. It exploits as many parallel tasks as possible at the algorithm level. In this scheme, the original task is divided into two subtasks and the process is carried out in two phases. The first subtask, which consists of steps 1–4, computes the local contributions and stores them in their private locations. In the first phase, all these subtasks are executed in parallel without the need for synchronization. The degree of parallelism is the number of elements in the circuit. The second subtask, which is associated with an entry in the sparse linear system, accumulates the local contributions from adjacent elements. In the second phase, all these subtasks are executed in parallel without the need for synchronization. The degree of parallelism is the number of entries in the sparse linear system. The key issue of implementing the Lockless-scheme is to expand the pointer system for the sparse linear system into the depth dimension. The original pointer system is a two-dimensional row and column data structure. The new pointer system creates an orthogonal dimension to accommodate the partial contributions in a linked-list.

The Lock-scheme uses the locking mechanism to ensure parallel processing at the device level. In this scheme, all five steps associated with an element are lumped together as a task and the last step is enclosed in a critical region. With such synchronizations, all tasks can now be executed in parallel. The degree of parallelism is the number of elements in the circuit. This scheme requires no changes to the data structure and is relatively easy to implement. In theory, it is feasible to design the critical region down to the individual entry level by employing multiple locks. However, this approach incurs excessive overhead in memory and execution time and thus may not be suitable for the Lock-scheme. In PARASPICE, a single lock is assigned to the entire sparse linear system and used by all elements in the circuit. The key issue of implementing the Lock-scheme is to reduce the sizes of the critical regions. Several techniques have been implemented in PARASPICE to achieve this goal.

To illustrate these techniques, consider the following MOSFET device

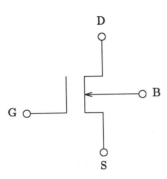

Then the corresponding stamp becomes

$$
\begin{array}{c c}
& \begin{array}{c c c c c c} D\ G\ S\ B\ D'\ S' \end{array} \quad \text{RHS} \\
\begin{array}{c} D \\ G \\ S \\ B \\ D' \\ S' \end{array}
&
\left[
\begin{array}{c c c c c c}
X & & & & X & \\
& X & & X & X & X \\
& & X & & & X \\
& X & & X & X & X \\
X & X & & & X & X \\
& X & X & X & X & X
\end{array}
\right]
\left[
\begin{array}{c}
\ \\
X \\
\ \\
X \\
X \\
X
\end{array}
\right]
\end{array}
$$

where D' and S' are the two internal nodes.

The stamp has 22 entries in the sparse matrix and four entries in the RHS. A simple approach, which requires one address fetch for each matrix entry, produces a critical region containing 48 statements. By removing the address fetches out of the critical region, the number is reduced to 26. Because no other elements in the circuit can connect to the internal nodes, there is no need to synchronize the corresponding entries. Accordingly, the number is further reduced to 8. Compared to the simple approach, a factor of 6 reduction has been achieved. These techniques can be applied to other devices like BJTs and JFETs and result in a similar reduction.

Both the Lockless–scheme and Lock–scheme are implemented in PARASPICE. The Lockless–scheme usually requires more memory (by a factor of 30 percent), but it exploits more parallel tasks. These two schemes provide a trade–off between memory and parallelism.

4.2. Experimental Results

An experiment with the module using the Lock–scheme is conducted on an Alliant FX/80 and the results are given in Table 4.1. The speedup ranges from 3.21 to 3.74 using four CEs and from 3.64 to 6.20 using eight CEs. The result clearly shows the effectiveness of the algorithm and, because no special hardware features are used in the implementation, should be applicable to other similar machines, such as the Cray Y–MP computer systems.

Table 4.1 Timing and Speedup of the LOAD Module on Alliant FX/80 using Lock–scheme

Circuit	SPICE2	PARASPICE		
		1 CE	4 CEs	8 CEs
MOS164	37.41 (1.16)	43.50 (1.0)	11.93 (3.65)	7.97 (5.46)
BJT041	18.00 (1.06)	19.05 (1.0)	5.86 (3.25)	4.49 (4.24)
MOS028	5.96 (1.23)	7.32 (1.0)	2.28 (3.21)	2.01 (3.64)
MOS165	80.59 (1.19)	96.05 (1.0)	25.68 (3.74)	15.62 (6.15)
MOS330	182.78 (1.18)	216.19 (1.0)	58.05 (3.72)	34.86 (6.20)

Similar experiments with the Lockless–scheme result in a speedup slowed by a factor of 30 percent. This is primarily due to the inefficient implementation of the assembly phase for the Lockless–scheme. However, with an improved version, the Lockless–scheme can be more suitable for massively parallel processing because of its large pool of parallel tasks that require no synchronization. In the current version of PARASPICE, the Lock–scheme is implemented as the default method and the Lockless–scheme is used as an option.

5. The Parallel TRUNC Module

At each timepoint of transient analysis, after the solution is obtained via the NR method, the TRUNC module is called to determine the new transient stepsize by a sequence of TERR operations. TERR estimates the LTE for a particular energy–storage component and then computes the appropriate local stepsize to be used. For each energy–storage component, the local stepsize is computed as

$$
h = \left| \text{TRTOL} * \frac{\text{TOL}}{\text{Max}\left\{ \epsilon'_a, \left| C_k \right| \left| DD_{k+1}(t_n) \right| \right\}} \right|^{\frac{1}{k}}
$$

where C_k is a constant depending on the order of the integration method and TRTOL is a user specified parameter. The computations associated with the TERR operation can be summarized as follows.

(1) Compute tolerance bound TOL.

(2) Determine Divided Differences (DD) recursively for the k–th order numerical integration formulae.

(3) Compute the appropriate local stepsize h.

(4) Update the new transient stepsize.

5.1. Design and Implementation

The computations described above can be lumped together naturally as a single task. According to this computational model, a task is identified with its associated energy-storage component and all tasks can be executed in parallel if the dependences are properly synchronized. As with the parallel LOAD module, the Lockless-scheme and Lock-scheme can be devised to parallelize the TRUNC module.

The Lockless-scheme uses the task fission technique to break the data dependence. In this scheme, the original task is divided into two subtasks and the process is carried out in two phases. The first subtask, which consists of steps 1 through 3, computes the local stepsize and stores the value in a private location. Consequently, in the first phase, all these subtasks are executed in parallel without the need for synchronization. The degree of parallelism is the number of energy-storage components in the circuit. The second subtask, which consists of step 4, updates the new transient stepsize with the local stepsize. Since the goal of these second subtasks is to find the smallest local stepsize, the subtasks can in fact be lumped together as a reduction operation like summation. If the local stepsizes obtained in the first phase are stored in consecutive locations, the reduction becomes an array operation that can usually achieve high performance on most parallel system. The key issue of implementing the Lockless-scheme is in the data structure's accommodating the local stepsizes.

The Lock-scheme uses the locking mechanism to ensure parallel processing at the energy-storage component level. In this scheme, the original task is treated as a unit and the last step is enclosed in a critical region. With such synchronization, all tasks can now be executed in parallel. The degree of parallelism is the same as in the Lockless-scheme. This scheme requires no local storage for the stepsizes and is quite easy to implement.

Both schemes are implemented in PARASPICE. The Lockless-scheme requires more storage, but no synchronization in parallel execution. The two schemes provide a trade-off between memory and parallelism.

5.2. Experimental Results

An experiment with the module on the benchmark circuits is conducted on an Alliant FX/80 using one CE, four CEs, and eight CEs. The timing results, as well as speedup, using the Lockless-scheme are given in Table 5.1. The speedup ranges from 3.70 to 3.89 using four CEs and from 6.71 to 7.37 using eight CEs. The results clearly show the effectiveness of the algorithm.

Table 5.1 Timing and Speedup of the TRUNC Module on Alliant FX/80 using Lockless-scheme

Circuit	SPICE2 Time (s)	PARASPICE		
		1 CE	4 CEs	8 CEs
MOS164	8.55 (1.69)	14.48 (1.0)	3.74 (3.87)	1.99 (7.28)
BJT041	3.17 (1.69)	5.37 (1.0)	1.44 (3.73)	0.8 (6.71)
MOS028	1.02 (1.67)	1.70 (1.0)	0.46 (3.70)	0.25 (6.80)
MOS165	10.07 (1.72)	17.35 (1.0)	4.47 (3.88)	2.39 (7.26)
MOS330	22.99 (1.66)	38.16 (1.0)	9.81 (3.89)	5.18 (7.37)

The high performance achieved by the parallel TRUNC module is primarily due to good load balancing, as exemplified by the fact that the highest speedup (3.89 for four CEs and 7.37 for eight CEs) is achieved with circuit MOS330 which has the largest number of parallel tasks. Because all parallel tasks are scheduled at the same time, the code is easily adapted to any number of processors and the speedup is expected to increase linearly to a saturation point, after which no additional processors will be able to contribute the performance. For example, circuit MOS330's speedup is leveled off with 990 processors. As a result, the small discrepancy from the ideal speedup, as observed in Table 5.1, can be partially attributed to the system, particularly the behavior of the cache on the Alliant systems.

Similar experiments with the Lock-scheme result in a speedup slowed by a factor of 50 percent. This is primarily due to the relatively large overhead incurred by the locking mechanism. Furthermore, because the tasks are quite uniform in size, the probability of collisions due to the serialization of the critical region increases dramatically as the number of processors increases. Therefore, the Lock-scheme may not be suitable for massively parallel processing. Nevertheless, the task coalescing technique can also be applied to partially offset both the scheduling and synchronization overheads. In the current version of PARASPICE, the Lockless-scheme is implemented as the default method and the Lock-scheme is used as an option.

6. Overall Performance Evaluation

This section is devoted to the analysis of the overall performance of PARASPICE. It is important to emphasize that the experiments with PARASPICE on the benchmark circuits result in the same solutions as SPICE2. This clearly shows that the speedup represents genuine performance gains without compromising the accuracy. In this section, timing results from PARASPICE are obtained with the following options.

Parallel SOLVE Module : Fine–grain Model
Parallel LOAD Module : Lock–scheme
Parallel TRUNC Module : Lockless–scheme

Table 6.1 shows the speedup of PARASPICE versus SPICE2 on an eight–CE Alliant FX/80. The total job time of PARASPICE on 1 CE can be expressed as

$$T_p(1) = T_s - \Delta T_{Alg} + \Delta T_{Ovh}$$

where the second term, ΔT_{Alg} represents the performance gains from the algorithm change and the third term, ΔT_{Ovh} represents the overheads incurred by the multitasking constructs in the parallel code. The speedup (slowdown), ranging from 0.85 to 1.08, shows the effects of these two competing factors. The performance gains are primarily due to the interpretative code technique used in the Fine–grain model, whereas the overheads are mainly due to the CNCALL scheduling and the locking mechanism of the Alliant FX/80 systems. In two cases, BJT041 and MOS028, the benefit from the interpretative code technique overcomes the overheads. For the other circuits, a degrading in performance by a factor of 6% to 15% is observed.

Compared to SPICE2, the overall speedup of PARASPICE ranges from 2.41 to 3.13 for four CEs and from 2.85 to 4.53 for eight CEs. The results show that in general, the speedup improves as the circuit becomes larger. This basically reflects the fact that large circuits usually contain more parallel tasks and hence have good load balancing to offset the startup cost. It is important to note that PARASPICE is not finely tuned for the particular configuration (such as number of CEs) of the Alliant FX/80 systems. Therefore, the speedup is gained solely from the parallel algorithms employed in PARASPICE.

As mentioned before, the algorithms employed in PARASPICE are designed to exploit all possible parallelism at the problem level and their subsequent mapping to the architecture is oriented toward the adaptivity of the code to any number of processors. This strategy allows us to analyze the limitations of the algorithms and extrapolate the performance of the code for massively parallel processing. To illustrate this issue, MOS330 is used as an example in the discussion below.

The pertinent data from MOS330 are summarized as follows.

Parallel module	# of task per call	Deg. of para.	Max # of procs need	Avg. task size (μs)
SOLVE	12437	25	72	18.56
LOAD	330	330	330	486.29
TRUNC	990	990	990	73.03

The degree of parallelism is also referred to as the theoretical upper bound on speedup in the sense that it is obtained in an ideal situation, where an infinite number of processors are readily available and no synchronization overheads are incurred and all parallel tasks are of the same size. These upper bounds are the parallelism inherent in the problem — the circuit to be analyzed — and apparently are exploited fully by the parallel algorithms used in PARASPICE as discussed in Sections 3 through 5.

To preserve the parallelism exploited at the algorithm level, all parallel tasks are scheduled at the same time regardless of the system configuration. This mapping strategy is particularly suitable for massively parallel processing as exemplified by the required number of processors to achieve the maximum speedup. These numbers (72 for the parallel SOLVE module, 330 for the parallel LOAD module, and 990 for the parallel TRUNC module) reveal three regions of performance: (1) "Region 1," from 1 to 72 processors. All three modules contribute to the performance gains; (2) "Region 2," from 73 to 330 processors. Two modules, LOAD and TRUNC, contribute to the performance gains; and (3) "Region 3," from 331 to 990 processors. Only the parallel TRUNC module contributes to the speedup. After these three regions, no additional processors will be able to speed up the execution further. As a result, the performance is limited by the degree of parallelism in the parallel SOLVE module. Nevertheless, for large circuits containing thousands of transistors or more, the limiting factor will increase to a value about an order of magnitude smaller than the transistor count. Therefore, it is feasible for PARASPICE to speed up by a factor of two orders of magnitude in massively parallel processing.

Table 6.1 Overall Timing and Speedup
on Alliant FX/80

Circuit	SPICE2	PARASPICE		
		1 CE	4 CEs	8 CEs
MOS164	67.02 (1.0)	78.90 (0.85)	25.26 (2.65)	18.10 (3.70)
BJT041	78.92 (1.0)	73.40 (1.08)	25.20 (3.13)	18.65 (4.23)
MOS028	29.26 (1.0)	27.63 (1.06)	12.12 (2.41)	10.36 (2.85)
MOS165	179.09 (1.0)	196.86 (0.91)	61.32 (2.92)	41.67 (4.30)
MOS330	417.19 (1.0)	445.50 (0.94)	137.61 (3.03)	92.13 (4.53)

7. Conclusions

In this paper we have presented an approach to parallelizing the direct method for circuit simulation problems on shared–memory multiprocessors. This investigation establishes that the direct method for circuit simulation is well–suited for parallel processing and leads to the development of a collection of new algorithms pertaining to this issue.

We have demonstrated that the degree of parallelism in the LOAD and TRUNC modules is proportional to the number of transistors in the circuit and that both modules lend themselves naturally to parallelization, particularly on shared–memory multiprocessors. The results show that up to 7 times improvement can be achieved on an eight–CE Alliant FX/80. Moreover, these algorithms scale up naturally as the number of processors increases. For the SOLVE module, it has been observed that the degree of parallelism is an order of magnitude smaller than that of the LOAD module. This implies that for large circuits containing thousands of transistors, two orders of magnitude improvement in the speed of sparse solvers is feasible. The result is quite encouraging despite conventional thinking on the subject.

Based on the aforementioned parallel algorithms, a portable parallel circuit simulator called PARASPICE has been developed on the Alliant FX/80. PARASPICE is completely compatible to SPICE2 and the software retains SPICE2's robustness and general applicability. Compared to SPICE2, a speedup of 4.51 has been achieved on an eight–CE Alliant FX/80 for an 856–unknown, 330–MOSFET transistor benchmark circuit. This is about 0.65 times as fast as SPICE2 on one CPU of a Cray X–MP/48. PARASPICE is intended to be a base code for quick prototyping on a broader range of shared–memory multiprocessors.

There are a number of areas where further research can be conducted:

(1) Switching to an efficient dense solver in the Fine–grain model for the SOLVE module when the storage becomes a limiting factor;

(2) The Medium– and Coarse–grain models for the SOLVE module;

(3) A unified ANALYZE phase in the sparse solver for PARASPICE that will automatically select a suitable model for the circuit to be analyzed; and

(4) Using PARASPICE as a benchmark program for the performance evaluation of advanced architectures.

Acknowledgments

This work was supported by the Department of Energy under Grant No. DOE–DE–FG02–85ER25001.

References

[HuWi79] J.W. Huang and O. Wing, "Optimal Parallel Triangulation of a Sparse Matrix," *IEEE Trans. Circuits and Systems* (1979), Vol. 26, No. 9, pp. 726–732.

[JaNP86] G.K. Jacob, A.R. Newton and D.O. Pederson, "Direct–Method Circuit Simulation using Multiprocessors," *Proc. IEEE International Symposium on Circuits and Systems* (May 1986), Vol. 1, pp. 170–173.

[Nage75] L.W. Nagel, "SPICE2: A Computer Program to Simulate Semiconductor Circuits," Univ. of California, Berkeley, Memo No. ERL–M520, May 1975.

[Nage80] L.W. Nagel, "ADVICE for Circuit Simulation," presented at *IEEE International Symposium on Circuits and Systems* (May 1980).

[NeSa83] A.R. Newton and A.L. Sangiovanni–Vincentelli, "Relaxation–based Circuit Simulation," *IEEE Trans. on ED* (September 1983), Vol. ED–30, No. 9, pp. 1184–1207.

[SaVi87] P. Sadayappan and V. Visvanathan, "Circuit Simulation on a Multiprocessor," *Proc. CICC* (May 1987), pp. 124–128.

[Vlad82] A. Vladimirescu, "LSI Circuit Simulation on Vector Computers," Univ. of California, Berkeley, Memo No. UCB/ERL M82/75, October 1982.

[WJMM73] W.T. Weeks, A.J. Jimenez, G.W. Malhoney, D. Mehta, H. Qassemzadeh and T.R. Scott, "Algorithms for ASTAP — A Network Analysis Program," *IEEE Trans. on Circuit Theory* (November 1973), Vol. CT–20, pp. 628–634.

[Yang90a] G.C. Yang, "DSPACK: A Parallel Direct Sparse Matrix Package for Shared–Memory Multiprocessors," *1990 International Conference on Parallel Processing*, Vol.3 pp. 197–200, Aug. 1990.

[Yang90b] G.C. Yang, "PARASPICE: A Parallel Circuit Simulator for Shared–Memory Multiprocessors," *27 ACM/IEEE Design Automation Conference*, pp. 400–405, June 1990.

[YaTa85] F. Yamamoto and S. Takahashi, "Vectorized LU Decomposition Algorithms for Large–Scale Circuit Simulation," *IEEE Trans. on Computer–Aided Design* (July 1985), Vol. CAD–4, No. 3, pp. 232–239.

Multifrontal Factorization of Sparse Matrices on Shared-Memory Multiprocessors*

Kalluri Eswar   P. Sadayappan
Department of Computer and Information Science
The Ohio State University
Columbus, Ohio 43210

V. Visvanathan
Indian Institute of Science
Bangalore 560012, India

Abstract

The paper describes an approach to the parallel multifrontal factorization of unstructured sparse symmetric positive definite systems of equations. A static recursive heuristic for task-tree partitioning is used. Parallelism both within large fronts and between multiple fronts is exploited. Task granularity is maximized while maintaining good load balance and keeping synchronization costs low. Dynamic allocation of front memory is efficiently managed by maintaining multiple independent stacks. Experimental evaluation on a Cray Y-MP8/864 and upto 64 processors of a BBN Butterfly GP1000 are reported.

1 Introduction

The solution of large sparse linear systems of equations is commonly required in a number of scientific/engineering application domains such as Lattice-Gauge Theory, Computational Fluid Dynamics, Geodetic Modeling and Reservoir Simulation, Structural Mechanics and Dynamics, Electronic Device Simulation, VLSI Circuit Simulation [9, 12]. Since sparse matrix solution is often a dominant computational component in several domains, the rapid solution on parallel machines is of great interest. With highly parallel machines, the realized performance is often severely constrained by communication and synchronization overhead. This paper addresses the parallel multifrontal factorization of unstructured sparse symmetric positive-definite matrices. While implementations are only reported here for shared-memory machines (Cray Y-MP and BBN Butterfly GP1000), the method is readily applicable also for partitioned-memory machines such as the Intel iPSC.

The effective parallelization of the solution of large unstructured sparse linear systems is complicated by the structural irregularity of the dependencies between the primitive operations in the computation. Several recent efforts have been directed at the development of parallel sparse matrix solvers [1, 2, 5, 8, 10, 11]. The multifrontal approach to sparse matrix factorization has good potential for exploitation of parallelism. Duff [2] has reported on a parallel implementation of the multifrontal method on an Alliant FX/8 system. That implementation used a completely dynamic task scheduling strategy with a parameter that was used to experiment with different task granularities. We explore a different, adaptive approach to static task partitioning and scheduling in the work reported in this paper.

The paper is organized as follows. Section 2 provides some background on multifrontal sparse factorization. Section 3 elaborates on the difficulty of efficient exploitation of parallelism in the multifrontal algorithm. Section 4 develops the recursive

partitioning heuristic that partitions the multifrontal computation into tasks whose granularity is tailored to the number of processors available. Section 5 describes the parallel multifrontal algorithm after discussing the approach to inter-process synchronization and dynamic front-memory management in the algorithm. Section 6 reports on the experimental evaluation of implementations on the Cray Y-MP and the BBN Butterfly GP1000. Section 7 provides concluding remarks.

2 Sparse Matrix Factorization

Fig. 1 shows one form of the basic sparse Cholesky decomposition algorithm in the abstract. This form is a row-oriented in-place algorithm which finds a $U^T U$ decomposition of a sparse positive-definite symmetric matrix A by replacing the upper-triangular part of A by U. It involves two types of operations:

1. *Normalization* operations, involving dividing the non-zero elements of a row by the diagonal element, and,

2. *Update* operations, involving addition of a multiple of the elements of a *source* (pivot) row to the corresponding elements of a *target* row.

The *elimination tree* [3, 7] of a symmetric sparse matrix provides a succinct characterization of the row-level dependencies to be satisfied during Cholesky decomposition. Fig. 2(a) shows the structure of a sparse matrix and Fig. 2(b) shows its elimination tree (referred to henceforth as e-tree). The e-tree of an N x N matrix has N nodes, with node i being the parent of node j iff the first occurring nonzero in the lower-triangular column j of the matrix is in row i. A row k updates row l only if l is an ancestor of k in the e-tree. Conversely, the only rows that may update a row l are nodes m that are in the subtree of the e-tree rooted at l.

The approach proposed here uses a multifrontal factorization [4, 10] method. In this method, illustrated in Fig. 2, a *front* is associated with each node of the e-tree during the factorization process. When a node is ready for elimination, the update contributions to other matrix elements due to that node are formed and maintained in the dense front associated with that node. These update contributions form an upper-triangular *update matrix* with one less row than the front. All leaf nodes in the e-tree are initially ready and their update matrices can be formed (in any order) after appropriate scaling of the first row. For an intermediate node in the e-tree the update contributions from each of its child nodes are first added into its front before scaling and forming its update matrix.

*This work was partially supported by a grant from Cray Research, Inc.

The multifrontal factorization is not associated with a strict sequencing of e-tree nodes during processing. Rather, the e-tree provides only a partial order which should be satisfied by the order chosen: a node can be processed only after all its child nodes have been processed. But not all total orders satisfying this partial order are equally efficient. Any approach to implementing a multifrontal factorization has to deal with the problem of managing the memory required for the fronts. Since each row of the matrix can appear in several fronts, static allocation of space for all of the fronts is prohibitively expensive in terms of memory. Therefore, memory for the fronts has to be dynamically allocated when needed.

A postorder traversal of the e-tree allows the use of a stack of fronts. When a node is encountered in the postorder, space for its front can be allocated by pushing it onto the stack. The update matrices of its children will be right below the parent node's front on the stack. After all of them are merged into the parent front, its update matrix is formed and "slided" down the stack after "flushing" out all the child update matrices.

A different approach is to use a preorder traversal to allocate the space for the fronts, but a postorder for the processing of the fronts. In this case, a parent front will be just below a child front. The child can simply be popped off the stack after its update matrix has been merged with the parent. This removes the need for "sliding" the parent's front down the stack and thus makes the front memory management simpler. After all its child nodes have merged their update matrices into this front, its update matrix can be formed, merged into its parent's front, and this node's front then popped off the stack.

We use the preorder approach in the design of our algorithm, a sequential version of which we present in pseudocode form in Fig. 3.

The next section addresses the exploitation of parallelism during multifrontal factorization.

3 Exploiting Parallelism in Multifrontal Factorization

Parallelism in multifrontal factorization is available at two levels: between independent fronts, and within a front's computation. Fronts of nodes in the e-tree which do not have an ancestor-descendant relationship cannot affect each other and so can be processed independently. Within a front's computation, the formation of each row and indeed, of each element, of the update matrix, is completely independent of the others.

The two levels of parallelism do not, however, nest simply. A machine which can support fine-grained parallelism, the granularity desired being at the level of the update of one element, can exploit both levels of parallelism effectively. But commercial parallel machines do not yet have that capability.

Exploitation of only one or the other of the two levels of parallelism will not provide satisfactory results. For example, on a matrix whose structure is got by applying a 5-point stencil on a 31x31 grid, exploiting only inter-frontal parallelism can achieve an idealized speedup of only about 5.8, based on the total operation count and the operation count on the "critical" path in the e-tree. On the other hand, exploiting only intra-frontal parallelism on the same matrix would again be insufficient. On 64 processors, assuming uniform distribution of work, there would

be about 3000 update operations per processor and there would be over 900 synchronization points between the processors (one for each node in the e-tree). A processor would therefore perform only about 3 operations on the average between synchronizations.

A parallel algorithm should therefore be designed to judiciously exploit both levels of parallelism. Some observations about the work available at different levels of the e-tree are helpful. In general, the front sizes of nodes nearer the root are larger than those at lower levels in the e-tree, which means that they have more intra-frontal parallelism. However, the number of independent nodes is small, which means that there is not much inter-frontal parallelism near the root. On the other hand, the number of independent nodes near the leaves of the e-tree is much larger but these nodes have few operations per front. There is therefore more inter-frontal parallelism here and less intra-frontal parallelism. A good approach would therefore be for the nodes nearer the root to be worked on by larger groups of processors cooperatively and for nodes at lower levels to be worked on by smaller groups of processors (or a single processor). In order to minimize scheduling and synchronization overheads, entire independent subtrees at lower levels of the e-tree should be assigned to a single processor, if possible.

To achieve such a task assignment, that also provides a time-balanced distribution of work among the processors, a recursive partitioning algorithm is used, as explained in the next section. This algorithm performs a static partitioning of the work among the processors. Our approach thus differs considerably from that investigated by Duff [2], where a completely dynamic strategy for task assignment was used.

4 A Recursive Approach to Task Partitioning

Consider the elimination tree of Fig. 4(a). Assume that there are two processors, P_1 and P_2 available. The obvious way to start the computation is with one of the processors, say P_1, working on node 1 and P_2 working on node 6. After these are done, the only nodes which can be worked on are 2 and 7. To reduce synchronization costs, P_1 should continue working on nodes 2, 3, 4 and 5, and P_2 similarly on nodes 7, 8, 9 and 10. When nodes 5 and 10 have been processed, node 11 is the only node that can be processed. P_1 and P_2 must, therefore, cooperatively process its front. Nodes 12, 13, 14 and 15 have to be similarly processed in sequence.

Fig. 4(b) shows a binary tree in which the leaf nodes represent the processors P_1 and P_2, and the root node represents the two processors when working cooperatively. The nodes in the e-tree of Fig. 4(a) can be considered to be mapped onto the nodes of this binary tree.

This example illustrates that, in general, we should attempt to find as many independent subtrees as possible near the leaf level and assign them to different processors. Only when this cannot be done any further, because of a node like node 11 above, should processors start working cooperatively on nodes, in groups of two. Again, as many nodes as possible should be assigned to each pair. When this cannot be done any further, a bigger group of four processors should be formed by amalgamating two groups of two processors, and assigned nodes for this group to work on together. This process should go on and ultimately a single group containing all the processors will be assigned nodes near the root.

The preceding paragraph neglects to mention that assignment of nodes to groups of processors should be done keeping load balance in mind. In order to achieve this, the assignment algorithm traverses the tree top-down rather than bottom-up. It begins by assigning nodes near the root of the tree to the single group of all the processors. To achieve the goal of assigning as few nodes as possible to this group, the algorithm traverses the tree in a breadth-first fashion, trying to find the earliest point where two independent sets of subtrees of nodes having nearly equal total work occur. At this point, the algorithm divides the group of processors into two equal groups and assigns a subtree set to each. The algorithm is then recursively applied to each of the smaller groups. The process stops when we come down to the single processor level, where no further splitting is necessary and all remaining nodes are assigned to individual processors.

We now give a more formal description of the recursive partitioning algorithm. We assume that the number of processors p is a power of two. This simplifies the description but is not a restriction, for the algorithm can very easily be generalized for arbitrary p, as discussed in section 7.

The basic idea behind the approach is illustrated by Fig. 5. Fig. 5(a) shows an unbalanced non-binary elimination tree of an unstructured sparse matrix. The work distribution for multifrontal factorization of a sparse matrix may be represented using a weighted elimination tree, where each node is associated with a weight that equals the total number of elemental arithmetic operations required to process (factor) that node, i.e. merge child fronts, scale the row and form the update matrix.

Fig. 5(b) shows a balanced binary *composite logical processor tree* (CLPT) for $p = 8$. Each leaf node of the CLPT corresponds to one of the physical processors that the computation is to be mapped on. Each intermediate node of the CLPT represents a group of processors, called a *composite logical processor* (CLP), corresponding to the union of the processor groups represented by the children of that node. For uniformity, we consider leaf nodes also to be CLPs. In the description that follows, each node in the CLPT is assigned a unique number, shown within the circle in the example of Fig. 5(b).

The objective of the partitioning heuristic developed below is to map the nodes of the unstructured e-tree onto the balanced binary CLPT in such a way that:

1. For any pair of nodes i and j of the e-tree with i being an ancestor of j, $map(i)$ is either identical to $map(j)$ or is an ancestor of $map(j)$ in the CLPT.

2. Any pair of nodes of the CLPT with a common parent is balanced (to within a specified tolerance) with respect to the cumulative weight of the e-tree nodes mapped to the subtrees rooted at those CLPs.

3. E-tree nodes are mapped onto the lowest feasible level in the CLPT as possible, subject to the above two constraints.

With such a mapping, e-tree nodes assigned to leaf nodes of the CLPT are factored by the corresponding processor. E-tree nodes mapped onto non-leaf nodes of the CLPT are *cooperatively* factored by the corresponding group of processors, with the responsibilities of updating the rows of the frontal matrix being distributed among the processors.

Fig. 6 outlines the recursive e-tree partitioning algorithm. The algorithm recursively attempts to partition (a part of) the e-tree into two parts with roughly equal total arithmetic. Starting at the root, the weighted e-tree is traversed downwards, assigning traversed nodes to the root of the CLPT, until two groups of remaining subtrees with approximately equal total cumulative weight can be identified. These two groups of subtrees of the e-tree are then to be assigned in a similar recursive fashion, among the nodes of two subtrees of the root of the CLPT.

The choice of the value for the tolerance *tol* affects the partitioning and hence the synchronization costs and load balance. Based on empirical measurements, a value of 0.1 was used for *tol* in the experimental evaluation reported in section 6.

The next section develops the parallel multifrontal algorithm which makes use of the recursive partitioning algorithm described above.

5 A Parallel Multifrontal Algorithm

The recursive partitioning algorithm assigns a CLP to each node of the e-tree. The parallel computation can then be carried out as follows. The leaf nodes of the CLPT (the individual processors) can first work on the respective nodes assigned to them. When a pair of processors forming a CLP at the next level complete their work at this level, they can proceed to work cooperatively on the nodes assigned to that CLP and this process can continue until all the work is finished. This process is made efficient by grouping e-tree nodes into *clusters* as explained below.

Consider the set S of nodes of the e-tree assigned to a particular CLP. S can be partitioned into sets $\{S_1, S_2, \ldots, S_k\}$ such that $\forall i \neq j$, if $a \in S_i$ and $b \in S_j$, then a and b are independent nodes.

Each of the sets S_i is called a *cluster*, and the nodes contained in it form a connected subgraph (which is a tree) of the e-tree, assigned to the same CLP. Fig. 7 illustrates the cluster concept by an example. The e-tree depicted has the CLP number assigned to each node written in its circle. The two clusters assigned to CLP 2 are encircled.

Identifying clusters is important because the nodes of a cluster form a mini e-tree which can be processed in preorder in quite the same way that the complete e-tree is processed in the sequential algorithm.

When a leaf CLP (a single processor) processes a cluster, it uses a stack in the same way the sequential preorder algorithm does. When a group of processors forming a CLP processes a cluster, each processor performs the same preorder traversal of the cluster. Multiple independent stacks are used to manage the frontal memory. A row of a front gets placed on the stack of the processor that performs updates to it. This assignment of rows to processors does not change within all the fronts of a cluster. This means that synchronization is not needed before merging a row of an update matrix into the parent's front: the corresponding row in the parent's front will be "owned" by the processor performing the merge, and it will be on its own stack.

We define the concept of child clusters as follows. A cluster C_j is a *child* cluster of a cluster C_i if $parent(root(C_j)) \in C_i$.

When processing a cluster, update contributions from the roots of child clusters have to be merged into appropriate nodes of the cluster. The update matrices of such interface nodes, which we call pseudo-leaves, are placed in separate heap storage, and are merged in the same way that update matrices are merged on the stack. When the root node of a cluster is reached, its update matrix is written out into this heap storage for use by the parent cluster.

Based on the above ideas, we present a parallel multifrontal algorithm in pseudocode form in Fig. 8.

On p processors, this algorithm generates exactly p tasks, one for each processor.

6 Performance Evaluation

The parallel algorithm described in the previous section has been implemented to perform the Cholesky factorization of a symmetric matrix having a given sparsity structure. It has been implemented on a Cray Y-MP8/864 and a BBN GP1000 Butterfly. We only report the performance of the factorizing phase of the algorithm, since the symbolic preprocessing steps need be performed only once for matrices with a fixed sparsity structure (several matrices with identical sparsity structure are often factorized in engineering applications, such as circuit simulation [12]) and then used repeatedly for different values for the matrix elements.

6.1 Cray Y-MP8/864

The Cray Y-MP8/864 (hereafter referred to as the Cray Y-MP) is a supercomputer with eight vector processors. Parallelism in an application can be exploited in two ways on the Cray Y-MP: microtasking and macrotasking. Microtasking is well suited for applications with vectorizable innermost loops and parallelizable loops at outer levels. Here, the different iterations of the outer loop can be run on different processors with each processor executing the inner loop using vector instructions. Macrotasking, on the other hand, exploits parallelism only at the subroutine level. Different subroutine invocations which can execute concurrently are run on different processors.

Microtasking the frontal merge and update operations was first attempted. However, the average granularity of a task was considerably below the minimum work threshold above which it is effective. The high overhead of microtasking relative to the average task execution time thus made it ineffective for the test matrix (derived from a 31x31 grid with a 5-point stencil). While it might be more feasible to use simple microtasking for much larger matrices, the algorithm that was implemented should consistently be more efficient than one using microtasking to exploit only intra-front parallelism.

Using the Cray Y-MP multitasking system's library of routines, which allows task creation and various forms of synchronization between them (events, barriers, etc.), a parallelized, vectorized sparse Cholesky factorizer was programmed in Cray Fortran. The performance of the program on the test matrix 31x31 is given in Table 1.

Although the Cray Y-MP has eight processors, it was found nearly impossible, in a multi-user mode, to obtain all of them for the duration of parallel program execution. We have, therefore, been unable to get reliable results for $p = 8$.

6.2 BBN Butterfly

The BBN Butterfly is a parallel machine consisting of multiple microprocessors. Each processor has a memory associated with it and a collection of such processor-memory pairs is interconnected using a multistage network. Although memory is physically distributed, programs still view a single global (virtual) address space encompassing all the memories, using high-order interleaving. The main implication of this architecture is the difference in memory access latency associated with a reference to a location in the local memory module vis-a-vis a reference to a location on a non-local memory module.

As an initial step, we viewed the Butterfly as a uniform-memory-access machine and simply ported the Cray Y-MP version of the factorizer to the Butterfly. The only changes made were to change the declarations of certain shared arrays to implement them as scattered arrays, whose elements are distributed across the memories of the processors, instead of using the default high-order interleaved storage allocation. The Uniform System library was used for task creation, events were simulated using busy-waiting on shared variables, and barriers were implemented using atomic instructions. The performance of the resulting program on the test matrix 31x31 is shown in Table 2.

As can be seen, the speedups attained were quite low, and not nearly as good, for $p = 2, 4$, as that on the Cray Y-MP. We can see rapid saturation, and actual inversion of speedup in going from 32 processors to 64 processors. The obvious conjecture about the cause of poor speedup was the increased memory latency and contention costs on the Butterfly, as compared to the Cray Y-MP. The need to verify this conjecture, coupled with a desire to improve the performance of the algorithm by estimating, as completely as possible, the various overheads being suffered by it, motivated the methodology of parallel program analysis presented next.

6.2.1 A Methodology for Parallel Program Analysis

Let the running time of an algorithm A on a particular input I on a single processor of a shared memory multiprocessor be T_1. Let the same algorithm, when run on p processors, take time T_p. If $\frac{T_1}{T_p} < p$, less than linear speedup has been got. To model the overheads which cause this, we consider the processor-time product $\Gamma_p = T_p \times p$. This measures the total amount of processor-time resource used by the parallel computation. Letting $\Gamma_{Useful} = T_1 \times 1 = \Gamma_1$ denote the amount of processor-time resource spent on "useful" work, we can immediately blame the remaining amount of the processor-time resource used on overheads. Thus we may write $\Gamma_p = \Gamma_{Useful} + \Gamma_{Overhead}$

The overhead has to be broken down into various components, and on the Butterfly we may list the following. (We do not consider scheduling overheads for two reasons: simple experiments place the cost of scheduling a task on the Butterfly to be around 25 microseconds, and our algorithm requires the generation of only one task for each processor.)

Γ_{Code_mod} Processor-time overhead due to modification in code executed in the parallel and sequential computations.

$\Gamma_{Latency}$ Processor-time overhead due to difference in local and non-local memory latencies.

$\Gamma_{Contention}$ Processor-time overhead due to delays in memory accesses caused by contention with other processors for the same memory module.

Γ_{Synch} Processor-time overhead in synchronizing between tasks.

$\Gamma_{Imbalance}$ Processor-time overhead due to idling of processors caused by load imbalance.

We now present methods by which each of these quantities can be estimated. The essential idea is to modify the original program in different ways producing different programs whose execution will not incur one or more of the above overheads. By making a sufficient number of such suitable modifications and measuring the processor-time resource used in each, estimates for each of the above overhead components can be got by solving the appropriate equations.

Since Γ_p consists of six components, we need at least six measurements. Two measurements, Γ_1 and Γ_p, are got directly from T_1 and T_p, respectively. We make four more measurements, $\Gamma_{nosynch}$, Γ_{active}, Γ_{local_ns} and Γ_{nocomp}.

$\Gamma_{nosynch}$ is the total processor-time used by the program when run with all synchronization (waiting for source rows and barrier synchronizations) removed. Of course, the execution will produce an incorrect factorization of the matrix, but it will perform exactly the same number of arithmetic and indexing operations as the correctly synchronized parallel program. The latency overheads for non-local memory accesses will be precisely the same, and all "code modification" overheads will be identical to that incurred by the correct parallel execution. The memory contention will not be exactly the same, but can be expected to be similar. This experiment removes the imbalance and synchronization overheads.

Γ_{active} is similar to $\Gamma_{nosynch}$ except that p separate experiments are performed with only one processor being active each time. The sum of the times spent by each processor, when active alone, gives this measurement. In addition to removing imbalance and synchronization overheads, this experiment also removes contention overheads.

Γ_{local_ns} is the total processor-time used by the program when all globally shared variables are replaced by local variables, and all synchronizations are removed. This experiment lets us remove the latency overhead in addition to those components removed by Γ_{active}.

Γ_{nocomp} is the total processor-time used by the program when all steps pertaining to the actual computation are removed. This leaves only the synchronization overheads.

The descriptions of the six measurements and the relevant overhead factors are summarized in Table 3, where a $\sqrt{}$ in a particular column and row indicates the presence of the component corresponding to the column in the measurement corresponding to the row. The asterisk on the contention overhead in $\Gamma_{nosynch}$ signifies that it is not exactly the same as that in Γ_p.

The values of the individual components can be solved for by inspection of the above table. We have

$\Gamma_{Useful} = \Gamma_1$
$\Gamma_{Code\_mod} = \Gamma_{local\_ns} - \Gamma_1$
$\Gamma_{Latency} = \Gamma_{active} - \Gamma_{local\_ns}$
$\Gamma_{Contention} = \Gamma_{nosynch} - \Gamma_{active}$
$\Gamma_{Synch} = \Gamma_{nocomp}$
$\Gamma_{Imbalance} = \Gamma_p - \Gamma_{nosynch} - \Gamma_{nocomp}$

The above methodology can be very effectively used to improve the performance of a parallel program because it can point out the primary causes of poor performance. In the next section we describe its use on the factorizer program ported from the Cray Y-MP onto the BBN Butterfly.

6.2.2 Analysis of Initial Version of Factorizer

As stated earlier, we had conjectured that the increased memory latency and contention costs on the Butterfly were the cause of the poor performance. Rather than perform a complete analysis, as described in the previous section, it was quite sufficient to measure $\Gamma_{nosynch}$ for the program. It can be seen that

$$\Gamma_{nosynch} = \Gamma_{Useful} + \Gamma_{Code\_mod} + \Gamma_{Latency} + \Gamma_{Contention}$$

Table 4 shows that $\Gamma_{nosynch}$ is considerably larger than Γ_{Useful}. $\Gamma_{nosynch}$ and Γ_{Useful} are measured in processor-seconds.

We did not expect that Γ_{Code_mod} would change very much when going from the Cray to the Butterfly. Therefore, the latency and contention overheads had to be the primary factors leading to poor performance.

In the next section, we describe how the parallel program was modified to reduce the latency and contention overheads. We then give a complete analysis of the performance of the modified program, which verifies that latency and contention costs were indeed significant overheads.

6.2.3 Tuned Cholesky Factorizer for the Butterfly

The factorizer program has two kinds of globally shared data structures, whose access by the processors may require non-local memory operations: read-only symbolic information and read-write matrix values. The read-only data structures can simply be replicated across all the processors' local memories.

The data structure containing the matrix values is written into when a processor has normalized a row, generating its final values. The same row is read by other processors when they need it as a source row for updating target rows that they perform updates to. To reduce the latency and contention associated with accessing such source rows, we modified the program so that each processor using a source row first made a local copy of it and then used the local copy to update the target rows it was responsible for. This software caching made a considerable difference as borne out by the performance of the revised program on the same test matrix 31x31, which we summarize in Table 5.

These results compare very favorably with those reported by Duff [2] on a similar matrix.

6.2.4 Analysis of Tuned Cholesky Factorizer

The performance of the modified version was completely analyzed using the methodology presented earlier. The results are summarized in Table 6 where the overheads are given as percentages of Γ_p.

7 Discussion

An adaptive recursive partitioning approach to the distribution of the computation for unstructured factorization of large unstructured sparse matrices has been considered. While over 50% efficiency was achievable for 32 processors, performance gains can be seen to diminish quite rapidly on further increasing the number of processors. The analysis of section 6.2.4 shows that, for a large number of processors, the primary factor limiting performance is load imbalance. A semi-dynamic approach to scheduling of tasks holds promise in alleviating this problem and is being investigated. Task granularities are still determined by using the recursive partitioning strategy. The management of front memory however becomes more complicated when dynamic assignment of fronts to processors is done.

The generalization of the recursive partitioning approach presented here for an arbitrary (non-power-of-two) number of processors is straightforward. When the number of processors in a CLP is not even, it can be split into two groups as evenly as possible and, to achieve load balance, the work measure used in the algorithm can be weighted in the ratio of the number of processors in each group.

The parallel factorization approach presented here can also be readily applied to private-memory machines. There is a natural mapping of nodes of the CLPT to groups of processors in recursively partitionable processor networks like hypercubes or meshes. Analogous to synchronization overheads in shared-memory machines, communication overheads are reduced by this approach. Furthermore, communications are localized to nearby processors as much as possible.

References

[1] C. Ashcraft, S.C.Eisenstat, and J.W.H. Liu "A Fan-in Algorithm for Distributed Sparse Numerical Factorization," *SIAM Journal on Statistical and Scientific Computing*, vol. 11, no. 3, pp. 593-599, 1990.

[2] I.S. Duff, "Multiprocessing a Sparse Matrix Code on the Alliant FX/8," *Journal of Computational and Applied Mathematics*, Vol. 27, pp. 229-239, 1988.

[3] I.S. Duff, A.M. Erisman and J.K. Reid, *Direct Methods for Sparse Matrices*, Oxford University Press, London, 1986.

[4] I. S. Duff and J. K. Reid, "The Multifrontal Solution of Indefinite Sparse Symmetric Linear Equations," *ACM Transactions on Mathematical Software*, Vol. 9, pp. 302-325, 1983.

[5] J.A. George, M. Heath, J.W.H. Liu and E. Ng, "Sparse Cholesky Factorization on a Local-Memory Multiprocessor," *SIAM Journal of Scientific and Statistical Computing*, vol. 9, 327-340, 1988.

[6] J.A. George, J.W.H. Liu and E. Ng, "Communication Reduction in Parallel Sparse Cholesky Factorization on a Hypercube," *Hypercube Multiprocessors 1987*, pp. 576-586, M.T. Heath Ed., SIAM, 1987.

[7] A. George and J.W.H. Liu, *Computer Solution of Large Sparse Positive Definite Systems*, Prentice Hall, NJ, 1981.

[8] J. Gilbert and R. Schrieber, "Highly Parallel Sparse Cholesky Factorization," *SIAM Symposium on Sparse Matrices*, Glenden Beach, Oregon, May 1989.

[9] D.J. Kuck and A.H. Sameh, "A Supercomputing Performance Evaluation Plan" *Proceedings of the International Conference on Supercomputing*, Lecture Notes in Computer Science, #297, pp. 1-17, Athens, Greece, June 1987.

[10] R. Lucas, T. Blank and J. Tiemann, "A Parallel Solution Method for Large Sparse Systems of Equations," *IEEE Transactions on Computer-Aided Design*, Vol. CAD-6, No. 6, pp. 981-991, November 1987.

[11] E. Rothberg and A. Gupta, "Techniques for Improving the Performance of Sparse Matrix Factorization on Multiprocessor Workstations," *Proceedings of Supercomputing 90*, pp. 232-243, New York, November 1990.

[12] P. Sadayappan and V. Visvanathan, "Circuit Simulation on Shared-Memory Multiprocessors," *IEEE Transactions on Computers*, vol. C-37, no. 12, pp. 1634-1642, December 1988.

Algorithm SPARSE-CHOLESKY

```
for k := 1, n
    A_kk := √A_kk
    for j > k such that A_kj ≠ 0
        A_kj := A_kj / A_kk
    endfor
    for i > k such that A_ki ≠ 0
        for j ≥ i such that A_kj ≠ 0
            A_ij := A_ij − A_ki * A_kj
        endfor
    endfor
endfor
end SPARSE-CHOLESKY
```

Figure 1: Abstract Sparse Cholesky Decomposition Algorithm

Algorithm PREORDER-MULTIFRONTAL

```
/* Let n be the size of the matrix. For j = 1, ..., n,
   let preorder(j) give the index of the j^th node in a
   preorder traversal of the e-tree. */
for j := 1, n
    i := preorder(j)
    load initial front for node i onto stack
    if i is a leaf node then
        repeat
            t := node on top of stack
            form update matrix for node t
            store normalized row
            merge update matrix into parent's front
            delete front for node t from stack
        until t is not the last child of its parent or t is the root
    endif
endfor
end PREORDER-MULTIFRONTAL
```

Figure 3: Preorder Multifrontal Factorization Algorithm

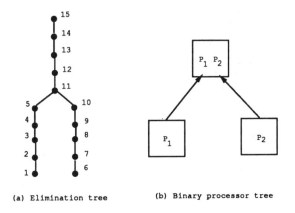

(a) Elimination tree   (b) Binary processor tree

Figure 4: Illustration of Partitioning Idea

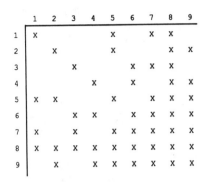

(a) A Sparse Symmetric Matrix

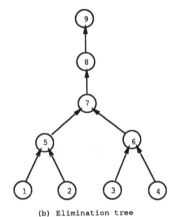

(b) Elimination tree

	1	2	3	4	5	6	7	8	9
1	1	0	0	0	1	0	2	1	0
2		1	0	0	2	0	0	3	1
3			4	0	0	4	1	1	0
4				1	0	2	0	2	1
5					1	0	1	4	1
6						4	2	1	1
7							1	1	1
8								1	1
9									1

(c) Symmetric matrix with structure of (a)

	1	5	7	8
1	1	1	2	1
5		0	0	0
7			0	0
8				0

	5	7	8
5	-1	-2	-1
7		-4	-2
8			-1

(d) Initial front at node 1   (e) Update matrix sent to node 5 by node 1

Figure 2: Illustration of Multifrontal Factorization

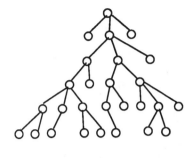

(a) Unstructured Elimination Tree

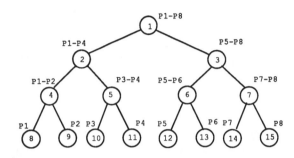

(b) Composite Logical Processor Tree

Figure 5: Mapping an e-tree onto a CLPT

Algorithm RECURSIVE-PARTITION
/* partition matrix for factorization by $p = 2^{logp}$ processors */
form the elimination tree for the matrix
$L := \{$root node of elimination tree$\}$
RECBISECT$(L, 1)$
end RECURSIVE-PARTITION

Procedure RECBISECT(L, clp)
/* L is a list of nodes and clp is the CLP to use */
if $clp \geq p$ then /* if at lowest level of CLPT */
  assign clp to all nodes in subtrees rooted at nodes in L
else
  try to partition L into two sets $L1$ and $L2$, so that:
  $| work(L1) - work(L2) | < tol \times [work(L1) + work(L2)]$
  while (L is not empty) **and** (L is not partitionable)
    $n :=$ node in L with maximum work in its subtree
    $L := L - \{n\}$
    assign node n the CLP clp
    $L := L \bigcup \{$children of n in the elimination tree$\}$
    try to partition L into equi-work subsets $L1$ and $L2$
  endwhile
  if L is non-empty **then**
    RECBISECT$(L1, 2 \times clp)$
    RECBISECT$(L2, 2 \times clp + 1)$
  endif
endif
end RECBISECT

Figure 6: Recursive Partitioning Algorithm

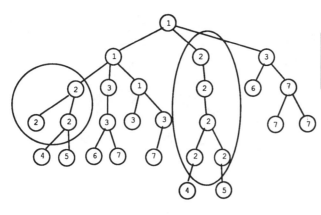

Figure 7: Illustration of Cluster Concept

Algorithm PARALLEL-MULTIFRONTAL
    for $proc := 1, p$ **in parallel**
        PARTASK($proc$)
    endfor
end PARALLEL-MULTIFRONTAL

Procedure PARTASK($proc$)
  for $l := 0, logp$
    $clp :=$ CLP containing processor $proc$ at level l in the CLPT
    synchronize with other processors in CLP clp
    for each cluster c assigned to CLP clp
      let $\{i_1, i_2, \ldots, i_k\}$ be the nodes in cluster c in preorder
      for $i := i_1, \ldots, i_k$
        if i is a non-leaf node **then**
          load initial front for node i onto stack
        else
          if i is a pseudo-leaf **then**
            merge rows from heap into front on stack
          else
            load initial front for node i onto stack
            process node i
          endif
        $t := i$
        while t is the last child of its parent **and**
             t is not the root of the cluster
          $t :=$ node on top of stack
          process node t
        endwhile
      endif
      endfor
      store update matrix for root of cluster into the heap
    endfor
  endfor
end PARTASK

Figure 8: Parallel Multifrontal Factorization Algorithm

Table 1: Performance on the Cray Y-MP

Number of processors (p)	Time (in milliseconds)	Speedup
1	37.8	-
2	19.3	1.96
4	10.8	3.50

Table 2: Performance of the Initial Version on the Butterfly

Number of processors (p)	Time (in seconds)	Speedup
1	8.575	-
2	5.358	1.60
4	2.912	2.94
8	1.663	5.16
16	1.018	8.42
32	0.695	12.3
64	1.111	7.72

Table 3: Overheads included in each Measurement

	U	CM	L	C	S	I
Γ_1	√					
Γ_p	√	√	√	√	√	√
$\Gamma_{nosynch}$	√	√	√	√*		
Γ_{active}	√	√	√			
Γ_{local_ns}	√	√				
Γ_{nocomp}						√

U = Useful; CM = Code_mod; L = Latency;
C = Contention; S = Synch; I = Imbalance.

Table 4: Analysis of Initial Version of Factorizer

Number of processors (p)	$\Gamma_{nosynch}$	$\frac{\Gamma_{nosynch} - \Gamma_{Useful}}{\Gamma_{Useful}} \times 100$
2	10.49	22.3
4	11.49	34.0
8	12.34	43.9
16	13.61	58.7
32	17.23	101.0
64	31.79	271.0

Table 5: Performance of Tuned Factorizer

Number of processors (p)	Time (in seconds)	Speedup
1	8.537	-
2	4.423	1.930
4	2.244	3.804
8	1.231	6.935
16	0.718	11.89
32	0.454	18.80
64	0.386	22.12

Table 6: Analysis of Tuned Factorizer (Overhead Percentages)

	Number of processors (p)					
	2	4	8	16	32	64
Code_mod (%)	0.28	0.58	0.16	0.46	2.13	3.46
Latency (%)	1.33	2.58	5.03	7.45	8.84	8.69
Contention (%)	0.27	0.41	0.67	1.48	1.78	1.06
Synch (%)	0.69	0.78	1.08	2.04	4.43	10.0
Imbalance (%)	0.91	0.52	6.35	14.3	24.1	42.2

AN EFFICIENT ARNOLDI METHOD IMPLEMENTED ON PARALLEL COMPUTERS

Sun Kyung Kim and A.T.Chronopoulos
Department of Computer Science
University of Minnesota
Minneapolis, Minnesota 55455.

Abstract . Main memory accesses for shared memory systems or global communications in message passing systems decrease the computation speed. In this paper, a variant of the Arnoldi algorithm is presented so that only one global communication per one iteration is required in a message passing system and one global memory sweep in a shared-memory machine per each iteration is required. We also introduce s-step Arnoldi method whose one iteration corresponds to s iterations of the standard Arnoldi algorithm. The s-step method has improved data locality, minimized global communication and superior parallel properties to the standard one[2,4]. These algorithms are implemented on a 64-node NCUBE/seven hypercube and a CRAY-2, and performance results are presented.

1. Introduction

Many important scientific and engineering problems require the computation of a small number of eigenvalues of nonsymmetric large sparse matrices. The most commonly used algorithm for solving such an eigenproblem is the Arnoldi algorithm. The Arnoldi algorithm has three basic types of operations: matrix-vector products, inner products and the vector updates.

In a distributed memory parallel machine with a high communication latency, the algorithm designer must structure the algorithm so that large amounts of computation are performed between communication steps[1]. Thus grouping together for execution the inner products of each iteration in the Arnoldi method may lead to a speed up on this type of computers. On shared memory systems with a memory hierarchy such as the CRAY-2 the data locality of the computations is very important in achieving high execution speed. The data locality of a computation is good if data be kept for "a long time" in fast registers or local memories and many operations can be performed on them[5]. Thus restructuring the three types of operations in the Arnoldi method may lead to a speed up on shared memory systems with a memory hierarchy.

In this paper we also introduce the s-step Arnoldi method. In the s-step method, s consecutive steps of the standard method are performed simultaneously. This means, for example, that the inner products (needed for s steps of the standard method) can be performed simultaneously and the vector updates are replaced by linear combinations. In the s-step Arnoldi method, the vector updates are decreased slightly compared to the standard one ,also their parallel properties and data locality are improved and the s-step Arnoldi method has only one global communication for one s-step iteration. A disadvantage of s-step Arnoldi method is that one more matrix-vector multiplication is required for one s-step iteration compared to s iterations of the standard method.

2. The restructured Arnoldi method

The method of Arnoldi can be successfully used for computing extreme eigenvalues of large nonsymmetric matrices[7]. The Arnoldi algorithm is based on the Arnoldi recursion for reducing to upper Hessenberg matrices of a real nonsymmetric matrix A. The standard Arnoldi algorithm is as follows:

Algorithm 2.1 The Arnoldi Algorithm
Choose q_1 with $||q_1|| = 1$

For $j = 1$ until Convergence **Do**
1. Compute and store Aq_j
2. Compute $h_{i,j} = (Aq_j, q_i)$, $i = 1,...,j$
3. $r_j = Aq_j - \sum_{i=1}^{j} h_{i,j} q_i$
4. Compute (r_j, r_j)
5. $h_{j+1,j} = \sqrt{(r_j, r_j)}$
6. $q_{j+1} = r_j / h_{j+1,j}$
EndFor

This algorithm generates a set of vectors Q_j and each new Arnoldi vector generated is orthogonal to all preceding Arnoldi vectors. The vector operations(on vectors of size N) for j-th iteration are $(3j+2)N + 1Mv$, and storage requirements are $(j+1)N + 1Ms$, where Mv and Ms stand for matrix vector product and matrix storage of A respectively.

The eigenvalues of the upper Hessenberg matrices H_j are called Ritz values of A in Q_j. The Ritz value λ and Ritz vector $Q_j y (=z)$ obtained from an eigenvector y of a given H_j are approximations to eigenelements of A. The residual norms of the Ritz pair λ, z can be computed by using the formula

$$|| (A-\lambda_i I)z_i || = h_{j+1,j}|e_j^T y_i|, \quad \text{for } i=1,...,j \quad (2.1)$$

,where $e_j^T =(0,0,...,0,1)$ is a j-dimensional vector and y_i is the i-th eigenvector of H_j. Also the calculation of y_i can be performed in parallel with $h_{j+1,j}$. (2.1) is used as a stopping criterion. In practice, the Arnoldi method is used iteratively together with various acceleration techniques[7,8].

In the standard Arnoldi algorithm iteration, the inner products cannot be performed in parallel. An algorithm based on restructuring the standard Arnoldi algorithm to decrease the global communication cost and to get better performance in distributed-memory message passing systems are introduced here.

In algorithm 2.1, step 2(or step 5) must be completed before the rest of the computations in the same step start. This forces double access of vectors q,r,Aq from the main memory at each iteration.

Algorithm 2.2 The restructured Arnoldi Algorithm
Choose r_0 with $r_0 \neq 0$

For $j = 0$ until Convergence **Do**
1. Compute and store Ar_j
2. Compute (r_j, r_j), (Ar_j, r_j), (Ar_j, q_i) $i = 1,...,j$
3. $h_{j+1,j} = \sqrt{(r_j, r_j)}$
   $h_{j+1,j+1} = (Ar_j, r_j) / (r_j, r_j)$
   $h_{i,j+1} = (Ar_j, q_i) / h_{j+1,j}$, $i = 1,...,j$
4. $q_{j+1} = r_j / h_{j+1,j}$
5. $r_{j+1} = Ar_j / h_{j+1,j} - \sum_{i=1}^{j+1} h_{i,j} q_i$

EndFor

Algorithm 2.2 is a variant of algorithm 2.1 and the orthonormal vectors q_j are generated in the same way as the standard Arnoldi method. Computationally the difference between Algorithm 2.1 and 2.2 is the computation of h, r_j. We need one more vector operation to compute r_j in Algorithm 2.2. Algorithm 2.2 is better for parallel processing because the $j+1$(at j-th iteration) inner products required to advance each iteration can be executed simultaneously. Also, one memory sweep through the data is required to complete each iteration allowing better management of slower memories.

3. The s-step Arnoldi algorithm

One way to obtain an s-step Arnoldi algorithm is to use the s linearly independent vectors $\{v_k^1, Av_k^1,..., A^{s-1}v_k^1\}$ in building the Arnoldi vector sequence. Let us denote by $\bar{V}_k$ the matrix of columns $\{v_k^1, v_k^2,..., v_k^s\}$. $\bar{V}_k$ is spanned by $\{v_k^1, Av_k^1,..., A^{s-1}v_k^1\}$, so that $\bar{V}_k$ is made orthogonal to all preceding subspaces $\bar{V}_{k-1}, \bar{V}_{k-2},..., \bar{V}_1$.

Each matrix $\bar{V}_k$ can be decomposed into $\bar{Q}_k * \bar{R}_k$, where $\bar{Q}_k$ is an orthonormal basis of $\bar{V}_k$ and $\bar{R}_k$ is an $s \times s$ upper triangular matrix.

Remark 3.1 Let $\bar{V}_{i_1}$ be orthogonal to $\bar{V}_{i_2}$ for $i_1 \neq i_2$. Then $V_k = \{\bar{V}_1, \bar{V}_2,..., \bar{V}_k\}$ can be decomposed into $\ddot{Q}_k * R_k$, where $\ddot{Q}_k = \{\bar{Q}_1, \bar{Q}_2,..., \bar{Q}_k\}$ and $R_k = \text{diag}(\bar{R}_1, \bar{R}_2,..., \bar{R}_k)$.

Lemma 3.1 Let $\ddot{H}_j$ be an upper Hessenberg matrix and $\bar{H}_k = R_k^{-1}\ddot{H}_j R_k$ for $j=sk$. Then $\bar{H}_k$ is similar to the matrix $\ddot{H}_j$ and $\bar{H}_k$ is a block upper Hessenberg matrix whose blocks $H_{i,j}$, for $1 \leq i,j \leq k$, are $s \times s$ matrices and the subdiagonal block $H_{i+1,i}$ has one nonzero element only at location $(1,s)$.

Proof. Similar to the proof of Lemma 6.1 in [4]. We will demonstrate this for the special case $s=3$, $k=3$. The general case is shown similarly but the description is more complicated.

$R_3^{-1}\ddot{H}_9 R_3$

We will use the following s-dimensional column notations for the blocks $H_{l,k}$, for $1 \leq l \leq k$ of the block upper Hessenberg matrix $\bar{H}_k$.

$$H_{l,k} = [\, \mathbf{h}_{l,k}^i \,], \quad \text{for } i=1,...,s.$$
$$\text{where } \mathbf{h}_{l,k}^i = [h_{l,k}^{1,i},..., h_{l,k}^{s,i}]^T$$

The following theorem shows that the matrix $\bar{H}_m$ has the same eigenvalues as the matrix A if all iterations of the s-step Arnoldi were completed.

Theorem 3.1 Let A be a $N \times N$ nonsymmetric matrix and $V_m = \{\bar{V}_1, \bar{V}_2,..., \bar{V}_m\}$ for $N=sm$. Let

$$V_m^{-1}AV_m = \bar{H}_m \tag{3.1}$$

, then $\bar{H}_m = R_m^{-1}H_N R_m$ where H_N is the Arnoldi upper Hessenberg matrix.

Proof. By multiplying both sides of the equation (3.1) by V_m we obtain:

$$AV_m = V_m\bar{H}_m \tag{3.2}$$

From Remark 3.1 we have:

$$A(\ddot{Q}_m R_m) = (\ddot{Q}_m R_m)\bar{H}_m$$

By multiplying both sides of the equation by R_m^{-1} we obtain:

$$A\ddot{Q}_m = \ddot{Q}_m R_m \bar{H}_m R_m^{-1}$$

By multiplying both sides of the equation by $\ddot{Q}_m^T$ we obtain:

$$\ddot{Q}_m^T A\ddot{Q}_m = R_m \bar{H}_m R_m^{-1}$$

From Remark 3.1 $\ddot{Q}_m$ is an orthogonal matrix and from Lemma 3.1 $R_m\bar{H}_m R_m^{-1}$ is a upper Hessenberg matrix. Therefore, by the implicit Q theorem[3], $R_m\bar{H}_m R_m^{-1}$ is the same as the Arnoldi upper Hessenberg matrix H_N. Also $\ddot{Q}_m$ is the same sequence of vectors as the standard Arnoldi vectors Q_N if the initial vector v_1 is the same. $\square$

Corollary 3.1 The block upper Hessenberg matrix $\bar{H}_k$, for $k=1,...,m-1$ (of the s-step Arnoldi) has the same eigenvalues as the standard Arnoldi matrix H_j, for $j=sk$.

Proof. By Lemma 3.1, the matrices $\bar{H}_k$ and H_j for $j=sk$ are similar. $\square$

If $k<m$ then by equating column blocks in equation (3.2) we obtain the following equation:

$$AV_k = V_k\bar{H}_k + u_k e_{sk}^T \quad \text{for } k=1,...,m-1 \tag{3.3}$$

where u_k is the residual vector. From equation (3.3) we derive the following block equations:

$$A\bar{V}_k = \sum_{l=1}^{k}\bar{V}_l H_{l,k} + u_k e_{sk}^T \tag{3.4}$$

The residual vector u_k is orthogonal to V_k. If we choose $v_{k+1}^1 = u_k$ (i.e. normalization is not applied) then the nonzero entry of $H_{k+1,k}$ is 1. Let $\mathbf{t}_{l,k}^j = [t_{l,k}^{1,j},..., t_{l,k}^{s,j}]^T$, for $1 \leq l \leq k$, denote the parameters in defining v_{k+1}^j. We now give the defining equations of the s-step Arnoldi method in the form of an algorithm.

Algorithm 3.1 s-step Arnoldi Algorithm
$\bar{V}_1 = [v_1^1, Av_1^1,..., A^{s-1}v_1^1]$
For $k=1$ **until Convergence Do**
    Select $[\, \mathbf{h}_{l,k}^i \,]$, for $1 \leq l \leq k$ and $1 \leq i \leq s$, to orthogonalize $\bar{V}_k$ against $\bar{V}_{k-1},..., \bar{V}_1$ in equation (3.4).
    This also gives:

$$v_{k+1}^1 = Av_k^s - \sum_{l=1}^{k}\bar{V}_l H_{l,k} \tag{3.5}$$

    Select $[\mathbf{t}_{l,k}^j]$, $2 \leq j \leq s$ to orthogonalize $\{Av_{k+1}^1,..., A^{s-1}v_{k+1}^1\}$ against $\bar{V}_k,..., \bar{V}_1$ which gives

$$v_{k+1}^j = A^{j-1}v_{k+1}^1 - \sum_{l=1}^{k}\bar{V}_l \mathbf{t}_{l,k}^j \quad \text{for } j=2,...,s \tag{3.6}$$

EndFor

Next, we demonstrate how to determine the parameters $h_{l,k}{}^i$, $t_{l,k}{}^j$ in the Algorithm 3.1. Equation (3.4) multiplied by $\overline{V}_l{}^T$, for $1 \leq l \leq k$ from the left yields

$$\overline{V}_l{}^T A \overline{V}_k = \overline{V}_l{}^T \overline{V}_l H_{l,k} \qquad (3.7)$$

Equation (3.6) multiplied by $\overline{V}_l{}^T$ from the left yields

$$0 = \overline{V}_l{}^T A^{j-1} v_{k+1}{}^1 - \overline{V}_l{}^T \overline{V}_l t_{l,k}{}^j \quad \text{for } j=2,...s \qquad (3.8)$$

Equations (3.7), (3.8) determine [$h_{l,k}{}^i$], $1 \leq i \leq s$ and [$t_{l,k}{}^j$], $2 \leq j \leq s$ as solutions of linear systems of size s. We will introduce some notations in order to describe these linear systems.

Remark 3.2 Let $W_k = \overline{V}_k{}^T \overline{V}_k = \{ (v_k{}^i, v_k{}^j) \}$, $1 \leq i,j \leq s$, then W_k is symmetric and it is nonsingular if and only if $v_k{}^1,...,v_k{}^s$ are linearly independent.

Remark 3.3 From equations (3.7), (3.8) and Remark 3.2 it follows that the following linear systems must be solved to determine [$h_{l,k}{}^i$], $1 \leq i \leq s$ and [$t_{l,k}{}^j$], $2 \leq j \leq s$:

$$W_l h_{l,k}{}^i = c_{l,k}{}^i, \quad \text{where } c_{l,k}{}^i = [(v_l{}^1, Av_k{}^i), \cdots, (v_l{}^s, Av_k{}^i)]^T,$$

$$W_l t_{l,k}{}^j = b_{l,k}{}^j \quad \text{where } b_{l,k}{}^j = [(v_l{}^1, A^{j-1}v_{k+1}{}^1), \cdots, (v_l{}^s, A^{j-1}v_{k+1}{}^1)]^T$$

The following Corollary reduces the matrix W_k to scalars (computed in preceding iterations) and the following inner products

$$(A^i v_k{}^j, v_l{}^j) \quad \text{for } i=1,...,s \; j=1,...,s \; l=1,...k-1 \quad \text{and} \qquad (3.9)$$

$$(A^i v_k{}^1, A^j v_k{}^1) \quad \text{for } i=0,...,s-1 \; j=i,...,s \qquad (3.10)$$

Corollary 3.2 The computation of matrix $W_k = (v_k{}^i, v_k{}^j)$, $1 \leq i,j \leq s$ can be reduced to the inner products in equations (3.9), (3.10) and previously computed scalars.

Proof. We use the block orthogonality of V_k and equations (3.5) and (3.6). The matrix of inner products W_k can be formed from the inner products in (3.9), (3.10) and the s-dimensional vectors $b_{l,k-1}{}^j$ as follows:

$$(v_k{}^i, v_k{}^j) = (A^{i-1}v_k{}^1, A^{j-1}v_k{}^1)$$

$$- \sum_{l=1}^{k-1}\sum_{r=1}^{s} t_{l,k-1}{}^{r,i-1}(v_l{}^r, A^{j-1}v_k{}^1), \text{ for } 2 \leq i, j$$

☐

Also, the vectors $c_{l,k}{}^i$, $b_{l,k}{}^j$ can be reduced to computing the inner products in equations (3.9), (3.10) and previous scalar work using a proof similar to the corollary 4.2.

We now reformulate the s-step Arnoldi algorithm taking into account the theory developed above.

Algorithm 3.2 s-step Arnoldi Algorithm
  Select $v_1{}^1$
  Compute $\overline{V}_1 = [v_1{}^1, Av_1{}^1,..., A^{s-1}v_1{}^1]$
  Compute inner products
For $k=1$ until Convergence **Do**
  1. Call Scalar1
  2. Compute $v_{k+1}{}^1 = Av_k{}^s - \sum_{l=1}^{k} \overline{V}_l h_{l,k}{}^s$
  3. Compute $Av_{k+1}{}^1, A^2 v_{k+1}{}^1,..., A^s v_{k+1}{}^1$

  4. Compute the inner products in (3.9), (3.10)
  5. Call Scalar2
  6. Compute $v_{k+1}{}^j = A^{j-1}v_{k+1}{}^1 - \sum_{l=1}^{k} \overline{V}_l [t_{l,k}{}^j]$ for $j=2,...,s$
EndFor
Scalar1 : Decompose W_k and Solves $W_l \overline{h}_{l,k}{}^i = c_{l,k}{}^i$ for $i=1,...,s$
Scalar2 : Solves $W_l t_{l,k}{}^j = b_{l,k}{}^j$, for $j=2,...,s$

From equation (3.3) it follows that in the s-step Arnoldi method, the Ritz values of A in V_k are the eigenvalues λ_k of $\overline{H}_k$, and the Ritz vectors are vectors $V_k x_k (=z_k)$, where the eigenvectors x_k of $\overline{H}_k$ associated with the λ_k. The residual norms of the Ritz value λ and Ritz vector z can be computed by using the formula $|| (A - \lambda_i I)z_i || = ||v_{k+1}|| |\bar{s}_{ki}|$, for $i=1,...,sk$ where $\bar{s}_{ki}$ is the last element of the i-th eigenvector of $\overline{H}_k$. This can be used as a stopping criterion.

We next compare the computational work and storage of the s-step Arnoldi method to the standard Arnoldi method. We only present the vector operations on vectors of dimension N and neglect the operations on vectors of dimension s.

operation	standard Arnoldi	s-step Arnoldi
Inner products	$(k-1)s^2 + \frac{s}{2}(s+1)$	$(k-1)s^2 + \frac{s}{2}(s+1)$
Vector updates	$(2k-1)s^2$	$2(k-1)s^2$
Matrix*Vector	s	$s+1$
Vector-storage	$ks+1$	$ks+1$

Table 3.1 : Vector Ops for s iterations from $s(k-1)+1$-th of the standard method and k-th iteration of s-step method

4. Numerical experiments

Large, sparse problems arise frequently in the numerical integration of partial differential equations(PDEs). Thus we borrow our model problem from this area. The test problems were derived from the following partial differential equation using the five point centered difference scheme on a uniform n x n grid with $h=1/(n+1)$.
Problem :

$$-(au_x)_x - (bu_y)_y + (cu)_x + (du)_y + gu = f$$

on the unit square, where

$$a(x,y)=e^{-xy}, \quad b(x,y)=e^{xy}, \quad c(x,y)=\beta(x+y)$$

$$d(x,y)=\gamma(x+y) \quad \text{and} \quad g(x,y)=1./(1+x+y)$$

subject to the Dirichlet boundary conditions u=0 on the boundary. Then we obtain a linear system of equation $Ax=f$ of order $N=n^2$. The right-hand side f was chosen so that the solution was known to be $e^{xy} \sin(\pi x) \sin(\pi y)$. The parameters β and γ are useful for changing the degree of symmetry of the resulting coefficient matrix of the linear systems. In this paper we set $\gamma=50.$, $\beta=1$ for a nonsymmetric matrix A.

The experiments were conducted on the NCUBE/7 with 64 processors and on the CRAY-2 multiprocessor system at the University of Minnesota. The NCUBE/7 is a particular example of a distributed-memory message passing parallel computer with the hypercube interconnection network. This NCUBE processor has a local memory of size 128KBytes and performance .3Mflops (in 64 bits). The communication delay between the processors of the NCUBE is significantly higher than the computation delay[6]. Thus the algorithms must be organized so that the number of communication points is as low as possible. The CRAY-2 is an

T_j	Code*	standard	modified	2-step
10x10	0.10204000E+02	0.9575713E+01	0.9575713E+01	0.9575713E+01
20x20	0.10204000E+02	0.10199149E+02	0.10199149E+02	0.10199149E+02
30x30	0.10203999E+02	0.10204783E+02	0.10204783E+02	0.10204783E+02
40x40	0.10203866E+02	0.10204008E+02	0.10204008E+02	0.10204008E+02
T_j	3-step	4-step	5-step	6-step
10x10	-		0.9575713E+01	-
20x20		0.10199149E+02	0.10199149E+02	
30x30	0.10204783E+02	-	0.10204783E+02	0.10204782E+02
40x40	-	0.10204008E+02	0.10204008E+02	

Table 4.1 : Largest Eigenvalues using the Arnoldi Methods on CRAY-2

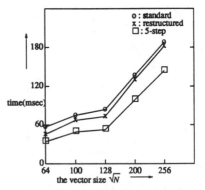

Figure 4.1 : CRAY-2 performance(msec) using 4 processors for the Arnoldi methods

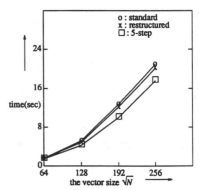

Figure 4.2 : NCUBE/7 performance(sec) using 64 nodes for the Arnoldi methods

example of a shared-memory four processor computer with memory hierarchy. All processors have equal access to a very large central memory (512 Mwords in size) and each processor has eight vector registers (64 words long) and a local memory (16K words in size). Each processor has a theoretical peak computational rate is 459 Mflops. Multitasking was used for parallelization.

In the accuracy test, Table 4.1 shows that matrices generated by the standard and restructured and s-step methods have the same largest eigenvalues. We test with problem size N=4096. We also compare the largest eigenvalues by the standard and restructured and s-step with those by SAAD's program($code^*$) using reorthogonalization and deflated iterative method[8] in tables 4.1. The stopping criterion is $\epsilon = 10^{-6}$ in SAAD's program. In the standard and restructured and s-step methods we find the largest eigenvalues after a reduced matrix of each size is generated, so these methods require minimal storage and time.

We reduced the different size matrices A for the model problem to 20x20 upper Hessenberg matrices using the standard and restructured and 5-step Arnoldi algorithms. Figure 4.1 shows the time of these methods for different size test problems on the CRAY-2 with 4 processors. Memory reference time and calls to the Multitasking Library are decreased by making possible grain size large and by decreasing synchronization points in the restructured and s-step methods, so those methods have better performance than the standard one on CRAY-2.

Figure 4.2 shows the total time of the standard and restructured and 5-step methods for different size of test problems with 64 nodes on NCUBE/7. The speedup for s-step Arnoldi method over the standard one comes from minimized global communication, and fewer vector operations even though s-step algorithm has one more matrix-vector multiplication.

5. Conclusion

The Arnoldi algorithm was restructured in this paper. The restructured algorithm decreases the global communication bottleneck of the standard Arnoldi algorithm by restructuring computations in such a way as to increase the number of inner products that are accumulated during one iteration. The restructured algorithm has better data locality by decreasing the memory contention bottleneck.

We have also introduced an s-step Arnoldi method and proved that s-step methods generate block upper Hessenberg matrices which are similar to reduction matrices generated by the standard Arnoldi method. The resulting algorithms has better data locality and parallel properties than the standard one. In s-step method, the inner products needed for s steps of the standard method can be performed simultaneously and the vector updates are replaced by linear combination. The s-step Arnoldi method requires less computational work than the standard one. For large $s > 5$ loss of accuracy for eigenvalues has been observed.

REFERENCES

[1] Cevdet Aykanat, Fusun Ozguner, Fikret Ercal and Ponnuswamy Sadayappan, "Iterative Algorithms for Solution of Large Sparse Systems of Linear Equations on Hypercubes", *IEEE Transactions on computers, Vol. 37, No.12, December (1988) 1554-1568*

[2] A.T.Chronopoulos and C.W.Gear, "s-step iterative methods for symmetric linear systems", *Journal of Computational and Applied Mathematics* 25 (1989) 153-168

[3] G. H. Golub, C. F. Van Loan, "MATRIX computations", pp. 219-225, *1989 by The Johns Hopkins university Press*

[4] Sun Kyung Kim and A.T.Chronopoulos, "A Class of Lanczos Algorithms Implemented on Parallel Computers", *to appear in Parallel Computing 1991*

[5] Gerard Meurant, "Multitasking the conjugate gradient method on the CRAY X-MP/48", *Parallel Computing 5 (1987) 267-280*

[6] Sanjay Ranka, Youngju Won, and Sartaj Sahni, "Programming the NCUBE Hypercube", *Tech.Rep. CSci No 88-13, Univ. of Minnesota, Mpls, MN, 1988*

[7] Youcef Saad, "Variation on the Arnoldi's method for computing Eigenelements of Large unsymmetric matrices". *Linear Algebra and its applications 34:269-295 (1980)*

[8] Youcef Saad, "Partial eigensolutions of Large nonsymmetric matrices", *Research Report YALUE/DCS/RR-397, June 1985*

[9] Paul E. Saylor, "Leapfrog Variants of Iterative Methods for Linear Algebraic Equations", *Journal of Computational and Applied Mathematics* 24 (1988) 169-193

Parallel Implementation of Gauss-Seidel Type Algorithms for Power Flow Analysis on a SEQUENT Parallel Computer *

G. Huang, W. Ongsakul

Dept. of Electrical Engineering
Texas A&M University
College Station, TX 77843

Abstract: In this paper we propose a parallelization procedure of the G-S algorithms for the power flow analysis. The corresponding G-S algorithm is implemented on a SEQUENT parallel computer for some IEEE test systems. The issues such as synchronization versus asynchronization, how to balance the computing load for each processor are discussed.

1. Introduction

Power flow program is used to find the steady state solution of an electric power transmission system, and is one of the most frequently used computations. They are used to assure efficient operating planning, and safer operation control [1] [2]. For more sophisticated computation such as optimal power flow [3] [4] [5] and transient stability programs [2], power flow constraints are always part of the formulation. Thus, enormous efforts have been devoted to research and development of faster power flow methods for the last 20 years [6] [7].

So far, few researchers are focusing on the subject of parallelization of G-J and G-S algorithms for power flow analysis. By sharp contrast, there are active researches in local relaxation methods and variations of the G-J and G-S algorithms to solve discretized linear partial differential equations in parallel. One such an example is the red-black partitioning scheme to parallelize the Gauss-Seidel algorithm [8]. Motivated by the red-black scheme, we propose a parallelization procedure of the G-S algorithms for the power flow analysis. The parallelization is formulated as a basic coloring problem with the requirement that no two directly connected neighboring buses have the same color. After the coloring, the processors are carrying out the G-S algorithm concurrently for each color to achieve parallel computation.

If we synchronize the parallel implementation from color to color, we can guarantee the newly updated data will be used in the next stage of computation. We name such an implementation as the color-by-color synchronized implementation. However, the synchronization involves coordination among processors and the more the synchronization steps the more the synchronization overhead and the waiting time. To reduce the waiting time, a modified implementation named system-by-system synchronized implementation is introduced. In the modified scheme, we still run the algorithm concurrently for each color, the only difference is that we do not wait until all the processors in the same color have finished their updates to continue the process. We proceed to the next color computation as soon as any of the processors finishes its update. We only wait to make sure that all the buses have finished their updates before starting the next iteration. This is the reason that it is called the system-by-system synchronized implementation. In such a way, we can save some waiting time at the expense of not guaranteeing using the newly updated data from the previous color. We can further relax the synchronization by relegating the requirement that each bus should have at least one update before we continue the process. We name this implementation as the totally asynchronous computation. In this paper, we shall concentrate on the first two implementations and more research is needed for the third implementation.

The rest of the paper will be organized as follows. In the second section, we will give a conceptual description of the

*This work is partially supported by Texas Advanced Research Program No. 4659, and NSF ECS-8900499.

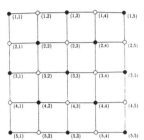

Figure 1: A uniform grid system

parallelization of G-J and G-S algorithms. In the third section, we will give a coloring scheme and have it tested on some IEEE standard systems. In the fourth section, we will describe a few variations of the implementations and have them tested on a SEQUENT machine. In the fifth section, we will analyze the data and compare them with the ideal speedup ratio. In the sixth section, we shall conclude our paper.

2. Parallelization of G-J and G-S Algorithms - A Conceptual Description

2.1 Parallelization of G-J Algorithms

Suppose the sequential G-J algorithms for the power flow analysis can be described by the update equation:
$$x_i^{k+1} = f_i(x_1^k, x_2^k, \ldots, x_n^k).$$
Detailed flow charts and equations for the rectangular G-J and G-S algorithms can be found in [2]. Some other alternatives such as polar form computation are investigated by the authors [9]. Suppose there are $n+1$ buses, where bus zero is chosen as the slack bus, and we have an idealized parallel computer architecture which has the same communication topology as the power transmission network topology from bus one to bus n. Now, each bus corresponds to a processor which can perform multiplication, addition and can communicate the computed data with the neighboring (directly-connected) processors.

Note that for the G-J algorithm, the updating equations rely only on the old data, thus the computing process can be carried out independently at each processor as long as the processor has access to the old data. One iteration of the G-J algorithm can be described as follows:

For $i = 1$ to n do in parallel,
$$x_i^{k+1} \longleftarrow f_i(x_1^k, x_2^k, \ldots, x_n^k).$$
end for.

Note in $f_i(x_1^k, x_2^k, \ldots, x_n^k)$ the required data are only from the neighboring buses, which is a small number independent of problem size n [9]. Thus, the required communication links are limited by the number of links which are directly connected to the node. And, the communication can be done in a synchronous unit. This implies that when the problem size increases the communication overhead will not increase. This property is favorable for large scale implementations. From now on, we shall define the connectivity degree of a node as the number of links which are directly connected to that node.

2.2 Parallelization of Gauss-Seidel Algorithm - A Simplified Case

Parallelization of G-S algorithm is a more complicated issue. To illustrate the idea, consider a uniform grid power system with the slack bus lying outside the grid. First we color the nodes as black and white nodes as illustrated in Fig. 1. Note that the updating equation depends on only the neighboring variables. Now as soon as the neighboring buses (differently colored buses) of a node have finished their own updates, the node uses the newly updated values to perform its own update. One iteration of the G-S algorithm can be described as follows:

For i = black do in parallel,
   $x_i^{k+1} \longleftarrow f_i(x)$ { *where x consists of the newest updates*
end for.   *from the neighboring buses of bus i*}
For i = white do in parallel,
   $x_i^{k+1} \longleftarrow f_i(x)$ { *where x consists of the newest updates*
end for.   *from the neighboring buses of bus i*}

We define the speedup ratio as the ratio between the serial G-S computing time divided by the parallel G-S computing time. That is, Speedup ratio = $\frac{T_s}{T_p}$, where T_s is the serial G-S computing time and T_p is the parallel G-S computing time. Note,

T_s = per serial iteration time × iteration
     complexity (i.e. number of iterations to converge),
T_p = per parallel iteration time × iteration complexity.

Assume for each serial G-S iteration, we start from black colored buses and then finish with the white colored buses. For this case, the serial G-S algorithm and the parallel algorithm will have the same iteration complexity since they generate the same set of data. And, the speedup ratio will be $\frac{n}{2}$ because it takes n unit time to finish one serial G-S iteration, and it takes two unit (black and white) time to finish one parallel G-S iteration.

One should observe the following:

- We save half of the processors if we use the same processors to compute a white substep and a black substep. The architecture can be organized in such a way that each processor corresponds to a pair of neighboring white and black nodes. Each processor has access to the neighboring processors to obtain the neighboring bus data. Whenever the processor finishes its update of the white bus using the black bus data, it can proceed to update the black bus using the white bus data. This architecture has $\frac{n}{2}$ processors, and the number of the communication links of each processor remain the same since we share the links.

- The G-S algorithm saves one half of the processors of the G-J algorithm, but both algorithms converge in almost the same computing time, this is because the G-S algorithm takes two unit time per iteration, but it saves half of the iteration complexity [9].

2.3 Parallelization of G-S algorithms - A General Case

For real power systems, we do not have such a uniform network topology as just described. However, we can easily extend the idea. Note that the above black-white coloring scheme decomposes the system into two separate groups of buses in which the update equation depends on only the bus data of the other group. Thus, the parallelization of Gauss-Seidel algorithm for a more general network can be posed as: How to color the buses in such a way that no two directly connected neighboring buses have the same color? If this can be done, one iteration of the Gauss-Seidel algorithm can be carried out in parallel as follows:

Do the following sequentially while varying the colors,
For i = the same color do in parallel,
   $x_i^{k+1} \longleftarrow f_i(x)$ { *where x consists of the newest updates*
end for;   *from the neighboring buses of bus i*}
end do.

One should observe the following:

- We use the same speedup ratio defined in section 2.2, but the G-S algorithm here will go through more than two colors. Also, the serial G-S algorithm is carried out from color-to-color to simplify the speedup analysis.

- To maximize the parallelism, it is better to have a minimum number of colors so that each color has a maximal amount of buses for parallel computation. To balance the computation load, each color should have almost the same number of buses to avoid idling processors.

- Assume we need C colors to partition the system. Then the computer architecture can be organized as follows: One set of C neighboring buses with different colors corresponds to a processor. Each processor requires communication links to obtain all the neighboring bus data directly connected to these C buses for new updates. The required communication links are upper bounded by the maximum degree of the nodes in the system. Thus, the processor complexity will be $\frac{n}{C}$. If n is not divisible by C, then the processor complexity will be $\lceil \frac{n}{C} \rceil$ where $\lceil k \rceil$ means the smallest integer which is greater than k.

- Using the same argument as in 2.2, we can conclude that speedup ratio will be $\lceil \frac{n}{C} \rceil$.

- For a planar topology, it is known that we need no more than four colors to color the graph as required [10]. For the power systems we find that three colors seem to be enough even though the power systems are not planar graphs. This again will be demonstrated in the table 1 of the next section.

- By experience, the three colored G-S algorithm saves two-third of the processor of the G-J algorithm, but it takes about 50 % longer to converge. This is because three unit time is needed for per G-S iteration, but the iteration complexity remains to be half of the G-J algorithm.

3. A Simple Chromatic Coloring Scheme

3.1 The Basic Coloring Problem

From section 2.3, the parallelization of G-S algorithms can be reduced to the following Basic Coloring Problem.
The Basic Coloring Problem:
Color the nodes of a connected graph in such a way that the following constraints are satisfied:
1) No two adjacent nodes should have the same color.
2) The number of colors should be minimized.
3) The number of nodes in each color should have almost the same quantity.

There are numerous coloring algorithms [11] [12] to satisfy the constraints (1) and (2) but none considers the constraint (3) except some efforts have been initiated to solve a similar problem with constraint (3) for optimal power flow analysis [4] [5] by the first author. It is known that this is an NP complete problem [12] even when only constraints (1) and (2) are considered. Hence only a heuristic algorithm is proposed here.
The Algorithm to Solve the Basic Problem
We color the nodes in the order of node connectivity degrees. The idea is that the higher the node connectivity degree, the more color restriction the node has to observe. While coloring, the algorithm is also keeping a color-by-color record of the number of buses (designated as the bus quantity) which has been assigned to the color. The least used color in the existing color set always has the first priority to be chosen. This way we can assure a more uniform distribution of the colors to satisfy the constraints (2) and (3).
The algorithm is shown below:

Initialize the set of used colors := ϕ,
initialize the set of existing colors := ϕ, and
initialize the set of the bus quantities associated with the colors as [0,0, ...,0];
Order nodes according to decreasing connectivity degrees;
Get the first node in the list,
While some nodes are not yet colored
   reinitialize the used colors of the neighboring nodes := ϕ;
   get the set of existing colors := [ ];

Table 1: Summary of Colored Bus Numbers

System	Color	Bus Number
IEEE 14 Bus	Red	1, 4, 6, 10, 14
	Blue	3, 5, 7, 11, 12
	Green	2, 8, 9, 13, (6)
IEEE 118 Bus	Red	1, 7, 9, 11, 15, 16, 18, 20, 23, 27, 30 31, 36, 37, 41, 45, 48, 53, 55, 57, 58 60, 63, 66, 69, 72, 74, 80, 83, 86, 88 92, 95, 101, 103, 106, 108, 111, 114, 118
	Blue	2, 3, 4, 6, 8, 13, 14, 17, 21, 25, 29 32, 33, 34, 35, 40, 44, 46, 49, 52, 56 61, 65, 67, 70, 73, 77, 79, 81, 85, 90 93, 96, 100, 105, 109, 112, 115, 116, 117
	Green	5, 10, 12, 19, 22, 24, 26, 28, 38, 39 42, 43, 47, 50, 51, 54*, 59*, 62, 64 68, 71, 75, 76, 78, 82, 84, 87, 89, 91 94, 97, 98, 99, 102, 104, 107, 110, 113

Table 2: Coloring Effect on Synchronized Processor Implementations of IEEE 14 Bus System

No. of CPUs	Colored		Non-Colored	
	Iterations	CPU time	Iterations	CPU time
1	132	4.17	132	4.03
5	132	1.73	190	2.12

Table 3: System-by-System Synchronized Implementations

System	No. of CPUs	Colored		Non-Colored	
		Itera-tions	CPU time	Itera-tions	CPU time
IEEE 14	1	132	4.17	132	4.03
	2	132	2.18	155	2.87
	5	153	1.32	177	1.45
IEEE 30	1	332	19.46	333	19.05
	2	332	9.98	339	10.48
	5	355	4.73	384	5.17
IEEE 57	1	420	46.30	422	45.25
	5	422	11.85	615	14.80
IEEE 118	1	1166	396.15	1163	386.10
	2	1166	200.22	1231	214.88
	4	1166	109.58	1295	119.22
	5	1177	87.08	1372	108.08

get the set of the bus quantities associated with the existing colors as:= [the bus quantities in the colors];
Get the first edge at the current node;
While there are not yet processed edges
    set of used colors of the neighboring nodes := set of
    used colors of the neighboring nodes +
    used color at node at opposite sides of the edge;
    get the next connected edge;
End While;
Assign to the current node a color not marked in the set of used colors but still in the set of existing colors and which has the smallest bus quantity;
If all the existing colors have been used, use a new color, and add the new color to the set of existing colors;
Add one to the associated bus quantity in the color;
Get the next node from the list.
End While.

The algorithm has been carried out on the standard the IEEE systems [13]. Part of the results are shown in table 1.

Sometimes, as in the IEEE 118 bus system, we discover that we need four colors to decompose the system to satisfy the constraints. But if we allow a few neighboring nodes to have the same color, then we can color the system by three colors. By this way, the parallelism is increased at the expense that a few nodes become unable to use the newly computed updates from the neighboring nodes. We shall call the modification as the dishonest coloring scheme and the associated G-S algorithm as the dishonest G-S algorithm. The dishonest coloring scheme restricts the available colors to three colors only. During the dishonest coloring process, if all the three colors are used, we will pick up the least used color to color the node even though the color has been used in some neighboring nodes. The scheme is applied to IEEE 118 bus system and the result is given in table 1. In the table, buses 54 and 59 repeat some of their neighbors' color.

4. Parallel Implementation on a Sequent Computer

Since we do not have the idealized computer as proposed in section 2, we have to adapt the G-S algorithm to our SE-QUENT parallel machine, which is a shared memory multi-instruction multi-data access (SM MIMD) machine with ten processors. One of the processors needs to be assigned as the master processor to perform the coordinating tasks; thus, at most only nine processors are available. In the following, we shall discuss the variations on the implementation; all the CPU data will be presented in units of seconds.

4.1 Synchronized Processor Implementation and the Coloring Effect

In this implementation, we synchronize all the processors (in the same color) in such a way that none of the processor will proceed until all the processors have finished their current updating tasks.

We also carry out the same synchronized computation without following the color assignment to investigate the effect of coloring. Instead, we use the synchronized processors on the system by following the order of the bus number. The complexity data of these two variations for the IEEE 14 bus system are listed in table 2 for comparison. Apparently the non-colored one is not as efficient as the colored one.

4.2 System-by-System Synchronized Implementation and the Coloring Effect

It takes some waiting time to synchronize among processors. If we can reduce the synchronization substeps, the speed can be increased. On the other hand, before we compute the mismatch values to check the convergence criterion, we need to make sure all the buses have finished their updates. We call such an implementation as the system-by-system synchronized implementation. We also looked into the effect of the coloring in this implementation. The experimental data are listed in the table 3 to demonstrate the effect of coloring.

Note that for the IEEE 14 bus system, the maximum parallelism is five, hence we use five processors to implement the algorithm. Due to the unequal number of nodes in each color, one of the color will have one idle processor. On the other hand, we can also implement on two processors to finish one color in three cycles of looping. Both implementations are tabulated for observations. Similarly, we use the appropriate processor numbers for the rest of the IEEE systems. The one processor data are also tabulated for speedup ratio information. It is obvious that the colored system-by-system implementation does have a faster computing time than the non-colored implementation.

4.3 Full Processors Usage and Idling Processor Usage

In many cases, it is difficult and sometimes impossible to have uniform parallelism for each color and thus

Table 4: Full Processor Usage Versus Idling Processor Usage of IEEE 14 Bus System

No. of CPUs	Full Processor Usage		Idle Processors	
	Iterations	CPU time	Iterations	CPU time
5	137	1.18	153	1.32

some processors are idling during computing. When this happens, it is natural to think that maybe we should fill the idling processors up with some computing load to speedup the convergence. We name the practice as the full processors usage implementation. To investigate its effect, we implemented the idea on the colored IEEE 14 bus system using system-by-system synchronization. As shown in table 1, we fill in bus

Table 5: Color-by-Color Synchronized Versus System-by-System Synchronized Complexity of IEEE 14 Bus System

No. of CPUs	Color-by-Color		System-by-System	
	Iterations	CPU time	Iterations	CPU time
2	132	2.80	132	2.18
5	132	1.73	153	1.32

Table 6: Time per Iteration and Waiting Time of Synchronized Processor Implementation of IEEE 14 Bus System

No. of CPUs	Iterations	CPU time	time per iteration	waiting time
1	132	4.17	0.0316	-
2	132	2.80	0.0212	0.0054
3	132	2.30	0.0174	0.0069
4	132	2.02	0.0153	0.0074
5	132	1.73	0.0131	0.0068
6	134	1.58	0.0118	0.0065
7	164	1.67	0.0102	0.0057
8	154	1.47	0.0095	0.0055
9	163	1.57	0.0096	0.0061

6 in the green level to have full processor usage. It is clear from the data in table 4, full processor usage does speedup the convergence.

5. Summary of Observations and Insights

We summarize our observations and obtained insights from the experiments as follows:

- Synchronization is an important issue for parallel implementations as demonstrated by the apparently different convergence behaviors shown in the tables 2 and 3.

- Color-by-color synchronization scheme is closest to the conceptual algorithm. The scheme guarantees the full use of the newly updated values from the previous color for the next color computation. As expected, it demonstrates better iteration complexity as shown in table 2. But, each iteration takes longer time since synchronization involves coordination and thus induces the waiting time among processors.

- System-by-System synchronization scheme is intended to save the synchronization overhead. Table 5 does indicate the saving, but on the other hand one can see its iteration complexity is deteriorating. This is understandable since the previous color nodes may not have finished all their updates, hence some old data are used for the current updates.

- When more processors are involved, waiting time almost remains constant for the same system as illustrated by the table 6.

- It can be observed from the table 6 that as the number of processors increases, the computing time may not decrease. This is because that the processor number does not match the parallelism of the system and thus many processors are doing G-J type updatation which deteriorates the iteration complexity. This concludes that throwing processors at a problem without careful consideration of its parallelism usually will not speedup the computation.

Table 7: Analytical and Experimental Speedup Comparison

System IEEE	Actual# of CPUs	Ideal# of CPUs	Speedup	
			Analytical	Experimental
14	5	5	5	3.16
30	5	10	5	4.12
57	5	19	4.75	3.91
118	5	40	5	4.55

- When the system dimension increases, the iteration complexity increases. But if we have enough processors with appropriate synchronization overhead, the per-iteration computing time will remain almost constant if the ideal speedup can be achieved as analyzed in section 3. For this reason, we compare the ideal speedup with the experimental speedup. The experimental speedup is defined as the ratio between one processor CPU time with the parallel CPU time. Since we do not have enough processors, we scale the ideal speedup ratio proportionally and name it as the analytical speedup ratio. Thus, the closer the experimental speedup ratio is to the analytical speedup ratio, the closer the ideal speedup ratio will be approached if we have enough processors. This argument uses the earlier observation from the table 6 that when the processors increase, the synchronized overhead will remain about the same for the same system. Table 7 indicates that the bigger the system dimension, the better the speedup performance and that the G-S algorithm indeed has potential for large scale parallel implementation even though the synchronization cost reduces the speedup ratio.

6. Conclusion

Parallelization procedures and the corresponding parallel implementation for the G-S algorithm of the power flow analysis are investigated and tested on a SEQUENT parallel computer. Some synchronization issues and the effects are demonstrated by numerous runs in the IEEE test systems. The approach performs satisfactorily and the potential of the algorithm for large scale implementations is supported by the experiments.

References

[1] A. H. El-Abiad and G. W. Stagg, "Automatic Evaluation of Power System Performance-Effects of Line and Transformer Outages," *AIEE Transactions*, Vol. PAS-81, February 1963.

[2] Glenn W. Stagg and Ahmed H. El-Abiad, *Computer Methods in Power System Analysis*, McGraw-Hill, 1980.

[3] H. W. Dommel and W. F. Tinney, "Optimal Power Flow Solutions," *IEEE Transactions*, Vol. PAS-87, October 1968.

[4] J. Zaborszky, G. Huang, and K. W. Lu, "A Textured Model for Computationally Effective Voltage and Thermal Management," *IEEE Trans. on Power Apparatus and Systems*, July, 1985.

[5] G. Huang, K. W. Lu, and J. Zaborszky, "A Textured Model/Algorithm for Computationally Efficient Dispatch and Control on the Power System," in *IEEE CDC*, Greece, December 1986.

[6] B. Stott and O. Alsac, "Fast Decoupled Load Flow," *IEEE Transactions on Power Systems*, Vol. PAS-93, No.3, May/June 1974.

[7] R. Bacher and W. F. Tinney, "Faster Local Power Flow Solution: The Zero Mismatch Approach," *IEEE Transactions on Power Systems*, Vol. 4, No. 4, November 1989.

[8] C.-C. Jay Kuo, Bernard C. Levy and Bruce R. Musicus, "A Local Relaxation Method for Solving Elliptic PDEs on Mesh-Connected Arrays," *SIAM J. Sci. Stat. Comput.*, Vol.8, No.4, July 1987.

[9] G. Huang and W. Ongsakul, "A New Relaxation Algorithm for Power Flow Analysis," *IEEE International Symposium on Circuits & Systems Conference*, New Orleans, LA, May 1990.

[10] C. Berge, *Graphs and Hypergraphs*, North Holland, Amsterdam, 1973.

[11] R. D. Dutton, R. C. Brigham, " A New Graph Coloring Algorithm," *The Computer Journal*, Vol. 24, No. 1, 1981.

[12] M. H. Williams and K. T. Miline "The Performance of Algorithms for Coloring Planar Graphs," *The Computer Journal*, Vol. 27, 1984.

[13] *AEP Test System Data*, December, 1961.

Parallelization of the EM Algorithm for 3D PET Image Reconstruction: Performance Estimation and Analysis*

C. M. Chen and S.-Y. Lee
School of Electrical Engineering
Cornell University
Ithaca, NY 14853

Abstract

The EM algorithm is one of the most suitable techniques for PET image reconstruction. However, two problems, long reconstruction time and large memory requirement, have prevented the EM algorithm from being widely used. In this paper, we present highly efficient parallelization schemes on multiprocessor systems. Subject to the constraint of minimizing memory requirement, two task and data partitioning schemes, namely, *partition-in-box* and *semihybrid-partition*, are proposed for well balanced computational load distribution. Data sharing strategies have been suggested using message passing and shared memory models. For efficient communication, we have designed a new integration algorithm, termed *Reverse Exchange*, for the hypercube topology and new integration and broadcasting algorithms for an *n-D mesh* topology. When the communication link setup time is negligible, the upper bound of the integration and broadcasting time for both classes of topologies are proportional to the size of data to be transmitted only. High parallelization efficiencies have been achieved on a message passing system, Intel iPSC/2, and a shared memory system, BBN Butterfly TC2000. Also, in an attempt to estimate the performance without actual implementation, the efficiency prediction formulas have been derived which accurately predict the performance.

1 Introduction

Computerized Tomography (CT) is a technique to reconstruct the image of internal structure of an object from projections without destroying or disturbing it. Various types of CTs have been developed for different purposes. Positron Emission CT (PET), for instance, is an imaging technique to visualize the spatial distribution of radionuclides by measuring the event counts of positron-electron annihilation inside the human body.

Like in most other types of CTs, PET images can be reconstructed by either analytic algorithms or iterative methods. Both of analytic and iterative approaches suffer from the two problems, namely, a long computation time required for the reconstruction algorithms and a very large memory inherently needed for the enormous amount of projection data. These two problems will become even more serious due to the ever-increasing demand for faster CT image reconstruction and higher resolution 3D PET images.

In our effort to cope with these difficulties, we have attempted to design a dedicated parallel architecture with the best performance/cost to provide sufficient computing power and storage for 3D PET image reconstruction. In this paper, we are primarily interested in the parallelization of the Expectation Maximization (EM) algorithm, especially the EM algorithm proposed by Shepp and Vardi [1]. Since the EM algorithm requires extremely large memory (refer to Section 2), the parallel EM algorithm and the dedicated parallel architecture to be designed should be subject to the constraint of minimizing the required memory space.

In early works, a VLSI architecture and a multiprocessor system were proposed by Jones et al. [2] and Llacer et al. [3], respectively. Both attempted to use multiple PEs in one form or another to accomplish fast PET image reconstruction with the EM algorithm. However, linear data integration in [2] and unbalanced load distribution in [3] tend to degrade the efficiency. None of these two has utilized the potential parallelism involved in the EM algorithm thoroughly. And memory space minimization has not been explicitly discussed in both systems either.

Recently, transputers have been employed for parallel implementation of the EM algorithm. Rosenberger [4] proposed a pipelined linear structure for reconstruction of the PET images using transputers. One problem with this approach is that the probability that a photon pair emitted from a location is detected by a certain detector pair is approximated to be space invariant. Another potential problem is that communication and computation can not be completely overlapped in a transputer as expected in their performance analysis. That is, it is not clear how the number of PEs employed may influence the overall performance.

Another possible way to provide computing power and storage is to perform image reconstruction on a supercomputer. Although it has been shown by Kaufman [5] that an 128×128 2D PET image can be reconstructed in several seconds on a Cray computer, the high cost/performance ratio is not preferable for most implementations and the capability to reconstruct a 3D PET image still needs to be investigated.

In this study, we have carefully considered four potential data parallelisms involved in the EM algorithm. Subject to the memory-minimizing constraint, we propose two task and data partitioning schemes, namely, *partition-in-box* and *semihybrid-partition* by efficiently exploiting two of these four data parallelisms. It has been shown in our implementations that these two schemes as well as the partition-by-box and modified partition-by-box schemes which were proposed in [6] provide very good computational load distribution. Data sharing strategies for the partition-by-box, modified partition-by-box schemes and semihybrid-partition schemes have been suggested based on three parallel processing models, namely, *message passing model*, *shared memory model* and *hybrid model*. For the message passing model, we have proposed new integration and broadcasting algorithms for the hypercube and *n-D mesh* topologies. When the communication link setup time is negligible, the upper bound of the integration and broadcasting time is only proportional to the size of data to be transmitted.

In an attempt to predict the achievable performance and analyze the factors degrading the performance without actual implementations, which is important for designing a new dedicated architecture, a new analytic approach based on the *load imbalance* and *overhead* is taken. With this approach, we have derived the *efficiency prediction formulas* closely estimating the performance which have been verified on a message passing system, Intel iPSC/2 and a shared memory system, BBN Butterfly TC2000. The influence of performance degradation factors can be analyzed by using these formulas.

In the following, the PET system model and the EM algorithm are first introduced in Section 2. The discussions of potential parallelism, and task and data partitioning schemes are provided in Section 3. The data sharing strategies for the

*This work was supported by grant number R01 CA51324 from the National Cancer Institute, NIH.

proposed task and partitioning schemes in different processing models are described in Section 4. The performance modeling and the *efficiency prediction formulas* are presented in Section 5. The implementation results, verification of the efficiency prediction and discussions are given in Section 6. Conclusion of this study is then provided in Section 7.

2   The EM algorithm

The EM algorithm for CT image reconstruction was originally proposed by Shepp and Vardi [1] based on the maximum likelihood estimation. Due to its theoretic properties, e.g., a higher S/N ratio for reconstructed images and a guarantee to converge to an estimate image with the maximum likelihood, the EM algorithm has received more attention recently. Varieties of EM algorithms have been proposed [5,7,8,9,10,11]. In this paper, we will adopt the EM algorithm originally proposed by Shepp and Vardi [1] since most of other versions have the similar algorithmic structure as the Shepp's EM algorithm. The parallelization techniques developed for the Shepp's EM algorithm in this study can be either directly applied or applied with minor modification to the other versions.

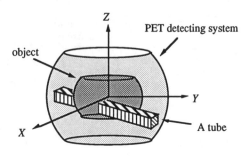

Figure 1: System model for the 3D PET system

The 3D system geometry employed in this study is a truncated sphere with the symmetrical upper and lower parts as shown in Fig. 1. The truncated sphere can be considered as a stack of coaxial rings of detectors where the radius of each ring varies to form a spherical surface. The object space can be decomposed into a number of small cubical *boxes* (voxels). Note that only those boxes inside the disk in each layer are to be reconstructed. Each positron in the radionuclide injected through blood vessels is recombined with an electron to generate a pair of photons in the opposite directions on a straight path. A pair of detectors defines a parallelopiped-like space called *tube* as illustrated in Fig. 1 and Fig. 2. The annihilation events detected by each detector pair are accumulated in the counter associated with the corresponding tube, of which final count forms the projection data.

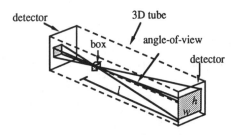

Figure 2: Definition of angle-of-view

What we would like to do is to reconstruct the internal image of the object (more precisely, the spatial distribution of radionuclide throughout the object modelled by boxes) from the projection data of all tubes. The correction equation for the EM reconstruction algorithm proposed by Shepp and Vardi can be written as follows:

$$\lambda^{new}(b) = \lambda^{old}(b) \sum_{t=1}^{T} \frac{n(t)p(b,t)}{\sum_{b'=1}^{B} \lambda^{old}(b')p(b',t)} \qquad (1)$$

where

$\lambda(b)$: the number of photon pairs emitted from box b (the image to be reconstructed),

$n(t)$: the number of photon pairs detected by tube t (projection data),

$p(b,t)$: the probability that a photon pair emitted from box b is detected by tube t,

T : the total number of tubes,

B : the total number of boxes.

Starting from an initial solution (guess) of the object image, λ, we update λ according to Eq. (1) in each iteration. The $p(b,t)$ can be represented by the angle-of-view of box b associated with tube t, defined as hw/l^2 as shown in Fig. 2. To avoid computing $p(b,t)$ on the fly which requires enormous computation, the $p(b,t)$ is precomputed for a given system geometry. The storage problem arises from the tremendous size of $p(b,t)$, $\mathcal{O}(\mu \times T \times B)$, which is much more than that of the image, $\mathcal{O}(B)$, and projection data, $\mathcal{O}(T)$, where μ is the percentage of nonzero $p(b,t)$.

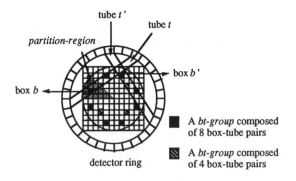

Figure 3: An illustration of the *partition-region* and the *bt-groups* in a 2D PET

To minimize the storage, only the nonzero $p(b,t)$ associated with those boxes in the *partition-region* are stored. The *partition-region* is the region enclosed by $0 \leq x \leq N_{xy}/2$, $0 \leq y \leq x$, and $0 \leq z \leq N_z/2$, where N_{xy} is the size of image in the x and y directions and N_z is that in the z direction. The *partition-region* consists of about 1/16 of the entire 3D object image. There exist other 7 or 15 box-tube pairs which have the same probability $p(b,t)$ as each box-tube (b,t) associated with a box in the *partition-region*. This technique minimizing the memory requirement will be referred to as *Minimizing Memory By Symmetry (MMBS)* in the following. The same idea as the *MMBS* technique has been employed by Kaufman [5] for 2D PET image reconstruction. For simplicity, the *partition-region* is illustrated for a 2D PET in Fig. 3. As an example, (b',t') with the same probability as (b,t) can be derived by rotating the box b and the tube t by 270 degrees counterclockwise with respect to the origin.

3 Task and Data Partitioning

One major goal in parallelization of sequential algorithms on multiprocessor systems is to achieve the largest *speedup* or highest *efficiency* such that the processing time is minimized. The *speedup* is defined as (the sequential processing time)/(the parallel processing time) and the *efficiency* as (the speedup)/(the number of PEs employed). In this study, we utilize four types of data parallelisms. Based on two of them, two task and data partitioning schemes are proposed. Since the EM algorithm requires an extremely large memory space, the task and data partitioning schemes are designed subject to the constraint of minimizing memory space.

Eq. (1) can be decomposed into the following steps, where n, $\tilde{n}$, ε, δ, and λ are the (row) vector forms of $n(t)$, $\tilde{n}(t)$, $\varepsilon(t)$, $\delta(b)$, and $\lambda(b)$, respectively, and P the matrix form of $p(b,t)$:

(1) $\tilde{n} = \lambda^{old} P$

(2) $\varepsilon(t) = n(t)/\tilde{n}(t)$, for all t,

(3) $\delta = P\varepsilon^T$,

(4) $\lambda^{new}(b) = \delta(b)\lambda^{old}(b)$, for all b.

In each iteration, first, the estimated projection data, $\tilde{n}(t)$, is computed for each tube based on the image, λ^{old}, obtained in the previous iteration (step(1)). Then, the updating factor, $\varepsilon(t)$, for each tube is determined by the ratio of the measured projection data to the estimated one (step(2)). The correction factor for a box, $\delta(b)$, is derived by a weighted (by P) sum of the updating factors of tubes passing through the box (step(3)). Finally, a corrected image is obtained for next iteration (step (4)).

Note that P is a sparse matrix and only non-zero elements are stored in practice. Since the two major computations in each iteration of the EM algorithm are large vector-sparse matrix multiplications in the steps (1) and (3), they are rich in data parallelism, i.e., many data can be processed in parallel by multiple PEs. In the task partitioning, the idea is to partition the scalar multiplications in Eq. (1) such that each PE performs about the same number of scalar multiplications, i.e., well balanced load distribution. Note that each scalar multiplication can be associated with a unique box-tube pair. There are basically four types of data parallelism involved in Eq. (1):

DP1: The all tubes passing through a box, i.e., the all box-tube pairs associated with a box, may be processed by a PE. The box-tube pairs associated with different boxes may be processed by different PEs simultaneously.

DP2: The all box-tube pairs associated with a box may be partitioned into mutually exclusive sets, i.e., no two sets contain the same box-tube pair. Different sets may be processed by different PEs at the same time.

DP3: The all boxes passed through by a tube, i.e., the all box-tube pairs associated with a tube, may be processed by a PE. The box-tube pairs associated with different tubes may be processed by different PEs simultaneously.

DP4: The all box-tube pairs associated with a tube may be divided into mutually exclusive sets. Different sets may be processed by different PEs at the same time.

It should be reminded that an important constraint in this study is to minimize the memory requirement. As suggested in Section 2, the *MMBS* technique may reduce the memory requirement for probability matrix P, which consumes most of the memory, by a factor of between 8 and 16. However, the *MMBS* technique can not be applied if the data parallelism

DP3 and DP4 are exploited. It is because there does not exist an exclusive data partition for the matrix P if the *MMBS* technique is employed. It implies that some sort of pointer scheme would be required to indicate where the required $p(b,t)$ is, which would need at least as many pointers as non-zero elements in P.

By exploiting the data parallelisms DP1 and DP3, task and data partitioning schemes, called *partition-by-box* and *partition-by-tube*, were proposed in [6]. A modified version of the *partition-by-box* scheme, called the *modified partition-by-box* scheme, was also introduced in [6] to improve load balance. The partition-by-tube scheme is expected to have the similar performance as the partition-by-box scheme. Likewise, utilizing the data parallelism DP4 is expected to yield the similar performance as that by utilizing the data parallelism DP2. And utilizing the data parallelisms DP3 and DP4 requires much more memory than utilizing the others. Therefore, we propose another two task and data partitioning schemes based on the data parallelisms DP1 and DP2, namely, *partition-in-box* and *semihybrid-partition*.

3.1 Partition-in-box

In contrast to the partition-by-box scheme, the unit for task and data partitioning is the *bt-group* based on the data parallelism DP2. For the steps (1) and (3), the *bt-groups* of each box in the partition-region are partitioned into N exclusive sets, each assigned to a PE, such that the numbers of *bt-groups* for all PEs differ at most by 1. For the task in the step (2), the n is evenly partitioned into N sets and the divisions are accordingly partitioned into N parts, each of which is computed by a PE. Likewise, for the task in the step (4), the δ is also evenly partitioned into N banks and the multiplications for all boxes are partitioned into N parts, each of which is computed by a PE. As illustrated in Fig. 4 in which 4 PEs are employed, each row in the matrix P is divided into N segments and each segment is assigned to a PE.

Compared to the partition-by-box scheme, the partition-in-box scheme is expected to have less overhead for assuring the data coherence in the step (1) since each $\tilde{n}(t)$ is possible to be updated by more than one PE at any time for the partition-by-box scheme. On the other hand, for the step (3), the partition-by-box scheme is expected to be superior to the partition-in-box scheme in that the former does not have coherence problem for the δ at all, but the latter does have.

A general problem associated with the partition-in-box scheme is that there are too many shared data. Each PE needs to interact with all other PEs at every step. As more PEs are employed, this interaction will introduce more overhead degrading the system performance. For the partition-by-box scheme, each PE needs to share data with all other PEs only in two steps (steps (2) and (3)).

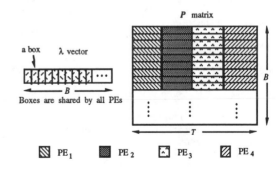

Figure 4: An illustration for the partition-in-box scheme

3.2 Semihybrid-partition

From the analysis in the previous section (Section 3.1), an interesting task and data partitioning scheme would be to first divide the boxes in the partition-region into N_1 groups as in the partition-by-box scheme. Then, in each group, each box is further partitioned into N_2 sets by using the partition-in-box scheme for the step (1) and all the boxes in each *bt-group* is partitioned into N_2 subgroups for the steps (3) and (4). Each set in a box is assigned to a PE and each subgroup is taken care of by a PE. Note that $N = N_1 \times N_2$ and $N_2 \leq 8$. This scheme is called *semihybrid-partition*.

The advantage of the semihybrid-partition is that it has very little coherence problem in each group in the step (1) due to using the partition-in-box scheme, and there is no coherence problem for the semihybrid-partition scheme in the step (3) due to using the partition-by-box scheme. Among groups in the step (1), the semihybrid-partition scheme still has the coherence problem as the partition-by-box scheme has. Moreover, more data are shared by PEs in the semihybrid-partition scheme. However, the semihybrid-partition scheme introduces less interaction among PEs since, for example, there are at most N_1 PEs attempting to update the same $\tilde{n}(t)$ simultaneously in contrast to N PEs in the partition-by-box scheme.

4 Data Sharing Strategy

One of the major performance degrading factors is the inevitable overhead due to sharing the data by PEs. The data sharing strategy mainly concerns about the questions of when and how to share the data during the data-sharing phases. We can consider three parallel processing models. They are *message passing model*, *shared memory model* and *hybrid model*. The hybrid model is a hybrid form of the message passing and shared memory models.

The message passing model requires all the shared data duplicated locally, called *local shared data*. And, all the computations are performed using the local data and local shared data. On the other hand, for the shared memory model, there is only one copy of shared data. PEs may read or update the shared data from the only copy of the shared data, called *global shared data*. Note that both models can be implemented on any type of system, i.e., a message passing system or a shared memory system.

The multiprocessor system implementing the message passing model considered in this study has a CPU and an enough number of I/O processors for each PE. The CPU and I/O processors can work independently. The shared memory in the multiprocessor system implementing the shared memory model is distributed shared memory. The shared memory access time varies with the location of shared memory. For the message passing model, obviously, it requires a means to integrate the results of all local shared data and broadcast them back to each PE. For the shared memory model, the shared data coherence needs to be ensured by the control of accesses to the global shared data.

In the previous study [6], we found that the integration and broadcasting for the partition-by-box and modified partition-by-box schemes degraded the system performance quickly as the number of PEs employed increased. The reason is that the time for the integration and broadcasting algorithms employed based on the pseudo-binary tree embedded in the hypercube [12] and a complete binary tree is proportional to $\log N$ for a given problem size. For the shared memory model, we found that the data access pattern is an important factor in minimizing the overhead in resolving the shared data coherence problem. Since the data access pattern inherent to the particular task and data partitioning employed in [6] resulted in a great amount of contention for the critical section, the performance was worse than using the message passing model.

In this study, we propose new integration and broadcasting algorithms for hypercube and *n-D mesh* topologies. When the communication link setup time is negligible, the upper bound of the integration and broadcasting time is proportional to the size of data to be transmitted. In addition, since the partition-in-box scheme has less coherence problem than the partition-by-box scheme in the step (1), we attempt to implement the semihybrid-partition scheme in the shared memory model for the step (1) and in the message passing model for the other steps. But, the matrix P is shared by all PEs in each group for minimizing memory requirement by using the shared memory model in the step (3). It is expected that this setup can reduce the size of data to be transmitted to half, which can save the communication time at the expense of very little coherence problem. And, the memory requirement for local $\tilde{n}$ is reduced. Besides, it can provide a performance comparison between the performance degradation due to the message passing and shared memory access. It is believed that these results may better the design of the data sharing strategies for the other possible schemes, e.g., the partition-in-box scheme using the shared memory model or the message passing model.

4.1 Data Sharing Strategy for Partition-by-box

The shared data for the partition-by-box scheme are $\tilde{n}$ and ε as described in [6]. Each PE has a portion of the matrix P, the vectors n, δ and λ. To minimize the integration and broadcasting time in a hypercube topology, we employ the Exchange algorithm for broadcasting, which can be found in [13]. For the integration, we propose a new integration algorithm, called *Reverse Exchange* integration algorithm, by reversing the order of communication in the Exchange algorithm. By applying Reverse Exchange and Exchange algorithms, for the shared data size of M, we have reduced each of integration and broadcasting times to $\mathcal{O}_{dx}((1 - 1/N)M) + \mathcal{O}_{ls}(\log N)$ for hypercube topology, where dx denotes data transmission and ls link setup. In contrast to this, the complexity would be $\mathcal{O}_{dx}(M(\log(N + 1) - 1)) + \mathcal{O}_{ls}(\log(N + 1) - 1)$ for using the complete binary tree and $\mathcal{O}_{dx}(M\log N) + \mathcal{O}_{ls}(\log N)$ for using the pseudo binary tree embedded in the hypercube topology. When M is very large as in our case, this reduction is very significant.

Reverse Exchange Integration Algorithm

```
ValidData = the whole local shared data (size of M)
Bit      = 1
Nid      = Node ID
while (Bit < N) {
    NidPair = Nid ⊕ Bit  (⊕ denotes exclusive OR)
    if (Nid < NidPair) {
        SendData = second half of ValidData
        RecvData = first half of ValidData
    }
    else {
        SendData = first half of ValidData
        RecvData = second half of ValidData
    }
    send SendData to NidPair
    receive RecvData from NidPair
    integrate local RecvData with received RecvData
    ValidData = RecvData
    Bit = Bit × 2
}
```

After $\log N$ steps of integration, each PE has $1/N$ integrated local shared data. Then each PE can perform $1/N$ divisions in the step (2) and produce $(1/N)$ of ε. Between the steps (2) and (3), the Exchange broadcasting algorithm is employed to make each PE have a complete copy of ε for the step (3).

In addition to the hypercube topology, we can also use a linear array topology with connectivity of 2 (or a ring topology), and an *n-D* mesh. An integration and a broadcasting algorithm, each of which requires the time proportional to $\mathcal{O}_{dx}((1 - 1/N)M) + \mathcal{O}_{ls}(N)$, have been developed for the linear array topology. When the setup time for linking every two PEs is small, N is relatively small and the M is large, we expect that the linear topology provides the similar performance as that by the hypercube topology for integration and broadcasting. The linear integration algorithm is provided for reference below:

Linear Integration Algorithm

```
ValidData = the whole local shared data (size of M)
partition ValidData into N parts evenly
Nid       = Node ID
i         = (Nid − 1) mod N
j         = (Nid − 2) mod N
while (i ≠ N) {
    send ith part to successor
    receive jth part from predecessor
    integrate  local jth part with received jth part
    i = j
    j = (j − 1) mod N
}
```

After N steps of integration, like in the hypercube topology, each PE has $1/N$ integrated local shared data. Similarly, each PE performs $1/N$ divisions in the step (2) and produce $(1/N)$ of ε. Then, the linear broadcasting algorithm which is almost the same as the linear integration algorithm can be used to distribute the $1/N$ results in each PE to all other PEs in N steps.

When N is small enough, the linear integration and broadcasting algorithms can be applied to any topology with a Hamiltonian circuit embedded, e.g., 2D mesh. For an *n-D mesh*, if N is too large, a reasonable compromise would be to apply the linear integration and broadcasting algorithms to every dimension consecutively. For a $P \times Q$ 2D mesh, for example, the time required for integration would be $\mathcal{O}_{dx}((1-1/PQ)M)+\mathcal{O}_{ls}(P+Q)$. In general, the integration and broadcasting times for an *n-D mesh* are $\mathcal{O}_{dx}((1-1/N)M)+\mathcal{O}_{ls}(nk)$, assuming k is the size in each dimension.

More efficient topologies would be those in which each PE can access data in all other PE's directly like in a completely connected topology. Such a topology would require a high cost for hardware realization. However, it may be simulated (embedded) on a shared memory system, to be referred to as *embedded completely connected topology*. The advantage of the (embedded) competely connected topology is that the overhead due to synchronization among PEs for each of integration and broadcasting steps can be eliminated which are required for the integration and broadcasting algorithms discussed above.

4.2 Data Sharing Strategy for Semihybrid-partition

The shared data are P, $\tilde{n}$, λ, ε and δ. In the step (1), the shared memory model is employed within a group. Since there is no coherence problem in each box for the partition-in-box scheme used within a group, the coherence of $\tilde{n}$ can be maintained by inserting synchronization point after execution of each box. The number of synchronization points may be further reduced by assigning as many box-tube pairs associated with a tube to the same PE as possible.

After each group finishes the step (1), the local $\tilde{n}$ in each group are integrated by using the message passing model. The ε generated in the step (2) will be broadcast to each group. In step (3), to minimize the memory requirement, the matrix P in each group are shared by all the PEs in that group by using the shared memory model.

5 Performance Analysis and Estimation

Performance modeling is very important for the design of parallel architectures and algorithms. The ability of accurately analyzing the factors degrading the performance and predicting the achievable performance enables us to improve the performance and to avoid a wrong design of new architectures. Various performance modeling techniques have been proposed, which range from analytic queueing models to the trace driven simulation. A comprehensive survey can be found in [14]. In this study, by taking advantage of the static feature of the EM algorithm, we

have developed an efficient analytic method to predict and analyze the performance. The performance is evaluated in terms of *efficiency*, which has been defined in Section 3, and the memory space requirement.

The following three schemes described in Sections 3 and 4 are considered in performance analysis. For convenience, the short notations for these schemes are used:

PbbMp : the partition-by-box scheme using the message passing model,

MPbbMp : the modified partition-by-box scheme using the message passing model,

PshMp : the semihybrid-partition scheme using the message passing model in the step (3).

5.1 Memory Space Requirement

The memory space requirement for these three schemes is provided in Table 1. The memory space requirement for the *MPbbMp* scheme is the same as that for the *PbbMp* scheme. For a large PET system, it is clear that the memory space requirement is dominated by $\mathcal{O}(\mu \times T \times B)$. However, if N is large enough compared to $\mu \times B$, the space requirement for the vectors $\tilde{n}$ and ε should not be neglected.

Table 1: Memory space requirement for the *PbbMp* (*MPbbMp*) and *PshMp* schemes in terms of orders ($\mathcal{O}$).

	P	n	δ	λ	$\tilde{n}$	ε
PbbMp	$\mu \times T \times B$	T	B	B	NT	NT
PshMp	$\mu \times T \times B$	T	B	B	$N_1 T$	$N_1 T$

5.2 Efficiency Analysis and Estimation

The *efficiency prediction formulas* (*EP Formulas*) are characterized by *computational load imbalance, shared memory access overhead imbalance* (*SMA overhead imbalance*) and some system parameters. The advantages of using these quantities are:

- For a given PET system, a task and data partitioning scheme and a data sharing strategy, the computational load imbalance and the SMA overhead imbalance can be calculated.

- For a given parallel computer system, the system parameters can be measured provided that the sequential processing time for each step is known or can be estimated in some way.

Therefore, for a static problem, the achievable efficiency may be predicted without actual implementation. Moreover, the major factors degrading the system performance can be analyzed so that the system performance may be improved easier. In design of a new architecture, it may also provide some idea of what values of these system parameters should be achieved in order to attain a desirable system performance, e.g., the ratio of the communication bandwidth to the computation speed. In deriving the *EP Formulas* for these schemes using message passing model, two modes, termed the *min mode* and the *max mode*, are considered for different system conditions. The *max mode* is defined to be the case that one of the following conditions is satisfied:

(1) All the PEs are synchronized at the very beginning of the communication.

(2) The sender may send data at any time.

(3) The first two conditions are not satisfied. The sender may send data only when the receiver is not sending data, but, the computational load difference between every send-receive pair is larger than the load for sending.

On the other hand, the *min mode* refers to the case that the sender may send data only when the receiver is not sending data and the computational load difference between every send-receive pair is smaller than the load for sending. The *EP formulas* for both modes are provided for the message passing model. For the hybrid model, the *EP formulas* for the *max mode* is provided since the condition (1) of the *max mode* is satisfied in most cases. If not, the *EP formulas* for *min mode* may be easily derived.

Limited by the allowed length of this paper, the *EP Formulas* for the *PbbMp*, *MPbbMp* and *PshMp* are presented without derivation. Note that the *EP Formula* of the *MPbbMp* scheme is the same as that of the *PbbMp* scheme, except that the computational load imbalance is considered to be zero for the former. In the following *EP Formulas*,

terms —
cluster : a set of PEs working in the shared memory model is defined as a *cluster*; in the message passing model, the size of a cluster is considered as 1,

symbols [superscripts] —
(k) : the kth step in an iteration; if it is omitted, it denotes a quantity for an entire iteration,

$(k.5)$: between steps (k) and $(k+1)$,

symbols [quantities] —
$r_{c_0}^{(k)}$: $t_s^{(k)}/t_s$, where t_s is sequential processing time; in the EM algorithm, $r_{c_0}^{(1)} \approx r_{c_0}^{(2)} \approx 0.5$,

$\rho_{c_{max}}^{(k)}$: $(t_{c_{max}}^{(k)} - t_{c_0}^{(k)})/t_{c_0}^{(k)}$, where $t_{c_{max}}^{(k)}$ is the maximum computation time among PEs and $t_{c_0}^{(k)}$ is $t_s^{(k)}/N$,

$r_{\theta_0}^{(k.5)}$: $t_{\theta_0}^{(k.5)}/t_{c_0}$, where $t_{\theta_0}^{(k.5)}$ is communication time between the kth step and the $(k+1)$th step for each PE; $r_{\theta_0} = r_{\theta_0}^{(1.5)} + r_{\theta_0}^{(2.5)}$,

$\rho_{c_{min}}^{(k)}$: $(t_{c_{min}}^{(k)} - t_{c_0}^{(k)})/t_{c_0}^{(k)}$, where $t_{c_{min}}^{(k)}$ is the minimum computation time among PEs; $\rho_{c_{min}}^{(k)} < 0$,

$r_{\varphi_0}^{(k)}$: $t_{\varphi_0}^{(k)}/t_{c_0}$, where t_{φ_0} is the average SMA overhead time for the kth step among clusters,

$\rho_{\varphi_{cl0}}^{(k)}$: $(t_{\varphi_{clmax0}}^{(k)} - t_{\varphi_0}^{(k)})/t_{\varphi_0}^{(k)}$, where $t_{\varphi_{clmax0}}^{(k)}$ is the maximum SMA overhead among clusters,

$r_{\sigma_{cl0}}^{(k)}$: $t_{\sigma_{cl0}}^{(k)}/t_{c_0}$, where $t_{\sigma_{cl0}}^{(k)}$ is the synchronization overhead time for the kth step among clusters; it is assumed that the synchronization overhead is the same for all clusters,

$r_{\theta_{cl0}}^{(k.5)}$: $t_{\theta_{cl0}}^{(k.5)}/t_{c_0}$, where $t_{\theta_{cl0}}^{(k.5)}$ is communication time between the kth step and the $(k+1)$th step for each cluster; $r_{\theta_{cl0}} = r_{\theta_{cl0}}^{(1.5)} + r_{\theta_{cl0}}^{(2.5)}$.

EP Formula 5.1 *For the* **PbbMp** *scheme,*

using the **max mode***:*

$$\eta = \frac{1}{1 + r_{c_0}^{(1)} \cdot \rho_{c_{max}}^{(1)} + r_{c_0}^{(3)} \cdot \rho_{c_{max}}^{(3)} + r_{\theta_0}} \approx \frac{1}{1 + \rho_{c_{max}} + r_{\theta_0}}.$$

using the **min mode***:*

$$\eta = \frac{1}{1 + r_{c_0}^{(1)} \cdot \rho_{c_{min}}^{(1)} + r_{c_0}^{(3)} \cdot \rho_{c_{min}}^{(3)} + r_{\theta_0}} \approx \frac{1}{1 + \rho_{c_{min}} + r_{\theta_0}}.$$

EP Formula 5.2 *For the* **PshMp** *scheme,*

using **max mode***:*

$$\eta = \frac{1}{1 + (1 + \rho_{\varphi_{cl0}}^{(1)}) \cdot r_{\varphi_0}^{(1)} + (1 + \rho_{\varphi_{cl0}}^{(3)}) \cdot r_{\varphi_0}^{(3)} + r_{\sigma_{cl0}}^{(1)} + r_{\theta_{cl0}}}.$$

For simplicity, the overhead due to I/O, which may be denoted r_{IO}, is not included in these *EP Formulas*. However, in practice, it is not necessarily negligible.

6 Implementation, Verification and Discussions

The three schemes described in Section 5 have been implemented on a hypercube multiprocessor system, Intel iPSC/2, and a shared memory system, BBN Butterfly TC2000. The iPSC/2 is a message passing system composed of one host processor and 32 PEs, and the TC2000 is a shared memory system consisting 40 PEs in the systems available to us. On the iPSC/2, we have implemented the *PbbMp* scheme using the hypercube and linear array topologies and the *MPbbMp* scheme using the hypercube topology. On the TC2000, we have implemented the *MPbbMp* and the *PshMp* schemes using the *embedded completely connected topology*. Recall that the *PshMp* scheme employs the shared memory model in the step (1).

Experiments have been designed to calculate the required system parameters for predicting the efficiency, e.g., link setup time. By using the *EP formulas*, the major factors degrading the performance are discussed.

6.1 Implementation on Intel iPSC/2

The iPSC/2 employs circuit switching communication with a bidirectional link between adjacent PEs. In every send-receive pair, only one PE is allowed to send at a time. The total amount of data for integration and broadcasting in each iteration is $T(N-1)/N$ for both of the hypercube and linear topologies. It is found that compared to the total communication time, the link setup time is negligible even in the case of such a small tested PET system. Therefore, the link setup time is neglected in the following discussions.

The efficiencies are predicted by the *EP Formula* 5.1 for the three implementations. For the *PbbMp* scheme using the hypercube topology, the *max mode* has been employed for integration since the computational load difference between most of send-receive pair is greater than the load for sending a given amount of data.

For the broadcasting, since it is assumed that all PEs are synchronized after integration and the load distribution in the step (2) is well balanced, all PEs are considered to be synchronized before broadcasting. Therefore, the *max mode* is also employed for the broadcasting. In Table 2, the predicted (η_{pr}) and measured (η_m) efficiencies are provided for different numbers of PEs employed along with the ρ_{c_0}, r_{θ_0}, and r_{IO_0}.

Table 2: Predicted (η_{pr}) and measured (η_m) efficiencies for *PbbMp* scheme using hypercube topology.

N	ρ_{c_0}	r_{θ_0}	r_{IO_0}	η_{pr}	η_m
2	7.55×10^{-4}	2.19×10^{-3}	1.74×10^{-5}	99.70%	99.79%
4	4.30×10^{-3}	6.58×10^{-3}	5.00×10^{-5}	98.92%	98.49%
8	7.08×10^{-3}	1.53×10^{-2}	1.95×10^{-4}	97.79%	97.66%
16	1.31×10^{-2}	3.29×10^{-2}	1.37×10^{-3}	95.50%	95.07%
32	3.24×10^{-2}	6.80×10^{-2}	2.88×10^{-3}	90.64%	89.51%

For the *PbbMp* scheme using the linear array topology, the *max mode* is employed for both of the integration and broadcasting. When $N = 2$, the implementation of the *PbbMp* scheme using the hypercube topology is the same as using the linear array topology. The predicted (η_{pr}) and measured (η_m) efficiencies are provided in Table 3.

In contrast to the two implementations of the $PbbMp$ scheme, the $MPbbMp$ using the hypercube topology is predicted by the $min\ mode$. It is because the load difference between every send-receive pair is less than the load for sending given data. The load imbalance, ρ_{c_0}, is ignored since it is too small. Similarly, the predicted (η_{pr}) and measured (η_m) efficiencies are provided in Table 4 together with r_{θ_0} and r_{IO_0}.

Table 3: η_{pr} and η_m for $PbbMp$ scheme using linear array topology.

N	ρ_{c_0}	r_{θ_0}	r_{IO_0}	η_{pr}	η_m
2	7.55×10^{-4}	2.19×10^{-3}	1.74×10^{-5}	99.70%	99.79%
4	4.30×10^{-3}	7.30×10^{-3}	4.30×10^{-5}	98.85%	98.70%
8	7.08×10^{-3}	1.71×10^{-2}	1.80×10^{-4}	97.62%	97.41%
16	1.31×10^{-2}	3.66×10^{-2}	7.65×10^{-4}	95.20%	94.70%
32	3.24×10^{-2}	7.57×10^{-2}	3.15×10^{-3}	89.99%	88.74%

Table 4: η_{pr} and η_m for $MPbbMp$ scheme using hypercube topology.

N	r_{θ_0}	r_{IO_0}	η_{pr}	η_m
2	2.44×10^{-3}	7.5×10^{-5}	99.75%	99.8%
4	7.30×10^{-3}	1.00×10^{-4}	99.26%	99.08%
8	1.71×10^{-2}	6.90×10^{-4}	98.25%	98.03%
16	3.66×10^{-2}	1.63×10^{-3}	96.30%	95.93%
32	7.57×10^{-2}	2.57×10^{-3}	92.74%	91.98%

Apparently, high efficiencies have been achieved by all of these three implementations. And, the $EP\ Formula$ 5.1 has successfully predicted the efficiency with the maximum error less than 2%. As expected, higher efficiencies have been attained by using the $MPbbMp$ scheme than by the $PbbMp$ scheme. How each factor degrades the system performance is clear from Tables 2, 3 and 4.

- For all these three implementations, the major factor degrading the system performance is the communication overhead, r_{θ_0}. For the $PbbMp$ scheme, the load imbalance is another important factor.

- The communication overhead, r_{θ_0}, is not completely dependent on the topology. As a matter of fact, it quite depends on when the data are sent and how data transfer is carried out at the hardware level. It is expected that for the same task and data partitioning scheme and data sharing strategy ($PbbMp$ or $MPbbMp$ scheme), if the data can be sent whenever needed, the similar performance can be achieved by both of the hypercube and linear array topologies since the link setup time is negligible.

- The $MPbbMp$ scheme using the hypercube topology and the $PbbMp$ scheme using the linear array topology have the similar communication overhead, r_{θ_0}, and I/O overhead, r_{IO_0}. However, since the load imbalance for the $MPbbMp$ scheme is negligible, the $MPbbMp$ scheme using the hypercube topology attains higher efficiency than the $PbbMp$ scheme using the linear array topology.

- The reason why $MPbbMp$ scheme is better than the $PbbMp$ scheme using the same topology is because the effect of load imbalance is larger than that of extra waiting for sending $T(N-1)/N$ data.

6.2 Implementation on BBN Butterfly TC2000

The BBN Butterfly TC2000 is a shared memory system with the Butterfly interconnection network. Basically, it is a distributed

shared memory system in the sense that the shared memory is composed of the memories in different PEs. Each PE can access the shared memory allocated in its own local memory, called *local shared memory*, or the shared memory allocated to the memory of the other PEs, called *remote shared memory*. It requires less time for a PE to access the local shared memory than the remote shared memory. In all the schemes implemented on this system, the shared data are evenly distributed in the local shared memories of all PEs.

Since all PEs can access any remote shared memory at any time provided that the coherence is preserved, the $MPbbMp$ scheme using the embedded completely connected topology has been implemented on TC2000. The $max\ mode$ of the $EP\ Formula$ 5.1 is employed since the data can be sent at any time. The send and receive mechanisms have been simulated by using the block transfer instruction (Do_bt) provided by TC2000 system. Table 5 contains the predicted (η_{pr}) and measured (η_m) efficiencies.

Table 5: η_{pr} and η_m for $MPbbMp$ scheme using the embedded completely connected topology.

N	r_{θ_0}	r_{IO_0}	η_{pr}	η_m
2	2.80×10^{-3}	2.20×10^{-4}	99.70%	98.74%
4	8.50×10^{-3}	5.90×10^{-4}	99.10%	98.02%
8	1.99×10^{-2}	1.30×10^{-3}	97.92%	96.62%
16	4.67×10^{-2}	3.20×10^{-3}	95.52%	93.55%
32	1.05×10^{-1}	1.39×10^{-2}	89.37%	87.50%

Two reasons have made this implementation achieve lower efficiencies than those in the iPSC/2. The first is the larger communication overhead, r_{θ_0}. The other is that the initialization and output ratio, r_{IO_0}, becomes more influential in TC2000 than in the iPSC/2. But, the r_{θ_0} is still the major factor which degrades the system performance.

For all these implementations using the message passing model, it is obvious that the best way to improve the efficiency is to reduce the size of data to be integrated and broadcast. As discussed before, one of the possible approaches is the $PshMp$ scheme. Another advantage of the $PshMp$ scheme is that it requires only $\mathcal{O}(N_1 T)$ of $\bar{n}$ and ε in contrast with $\mathcal{O}(NT)$ for those implemented in the message passing model. However, the $PshMp$ introduces a new overhead due to the longer remote share memory access time.

The integration and broadcasting are also implemented by using the embedded completely connected topology. The integration and broadcasting times are estimated from the time used in the previous implementation, i.e., the $MPbbMp$ scheme using the embedded completely connected topology. This calculation is based on the assumption that the maximum block transfer time is proportional to the size of data for a given number of PEs. The predicted (η_{pr}) and measured (η_m) efficiencies are provided in Table 6 for the implementation with two PEs in each cluster. In addition, the r_{θ_0}, r_{IO_0}, $r_{\varphi_0}^{(1)}$, $\rho_{\varphi_{cld}}^{(1)}$, $r_{\varphi_d}^{(3)}$, $r_{\varphi_i}^{(3)}$, and $\rho_{\varphi_{cld}}^{(3)}$ are also included in this table. Note that the sub-subscript d denotes the SMA overhead for the operations in the double precision format and i for the integer format. There is no SMA overhead imbalance for the operations in the integer format.

Table 6: η_{pr} and η_m for $PshMp$ scheme using the hybrid model.

N	r_{θ_0}	r_{IO_0}	$r_{\varphi_0}^1$	$\rho_{\varphi_{cld}}^1$	$r_{\varphi_d}^3$	$\rho_{\varphi_{cld}}^3$	$r_{\varphi_i}^3$	η_{pr}	η_m
2	NA	0.00027	0.051	0.073	0.057	0.063	0.01	88.78%	88.54%
4	0.0028	0.00057	0.051	0.085	0.057	0.088	0.01	88.40%	87.27%
8	0.0085	0.0013	0.051	0.09	0.057	0.11	0.01	87.78%	87.24%
16	0.022	0.0032	0.051	0.099	0.057	0.12	0.01	86.54%	85.96%
32	0.051	0.012	0.051	0.11	0.057	0.12	0.01	83.82%	82.14%

Although the size of data to be integrated and broadcast has been reduced to half, the performance is substantially degraded because of using the shared memory model. At most one third of degradation is caused by the integration and broadcasting. Several observations may be made from Table 6:

- The performance is degraded more in the step (3) than in the step (1) since the total number of the remote shared memory accesses is larger in the step (3).

- The total number of operations accessing the shared data from the remote shared memory is proportional to the number of PEs.

- The worse performance is mainly due to the integration and broadcasting and partly due to increased SMA overhead imbalance as the number of PEs is increased.

- This implementation shows worse performance than the others implemented in the message passing model. But, if the difference between the remote shared memory access time and local shared memory access time is reduced to some extent, e.g., 60 percents reduced, it might have better performance than the *MPbbMp* scheme using the embedded completely connected topology.

The last observation has actually disapproved the suggestion made by [15] in which the author suggested that the nonshared memory model can achieve better performance than the shared memory model on a shared memory system. We believe that which model is better depends on the task and data partitioning schemes, data sharing strategies and the system characteristics such as the difference between the remote shared memory access time and local shared memory access time.

7 Conclusions

In this study, we have proposed two task and data partitioning schemes for parallelization of the EM algorithm proposed by Shepp and Vardi for 3D PET image reconstruction. Subject to the constraint of minimizing memoroy requirement, the two schemes, *partition-in-box* and *semihybrid-partition*, provide much better load distribution than most of the previous works. Data sharing strategies for the *partition-by-box*, which was proposed in [6], and the semihybrid-partition schemes have also been suggested. For the hypercube topology, we have proposed a new integration algorithm, namely, *Reverse Exchange* integration algorithm which can integrate data of size M in $\mathcal{O}_{dx}((1 - 1/N)M)$ time with $\mathcal{O}(\log(N))$ link setup. For a n-D *mesh*, we have developed new integration and broadcasting algorithms which require $\mathcal{O}_{dx}((1-1/N)M)+\mathcal{O}_{ls}(nk)$ time. If the time for link setup is negligible, the upper bound of the integration and broadcasting time for both classes of topologies are only proportional to the size of data. In contrast to this, the complexity would be $\mathcal{O}_{dx}(M(\log(N + 1) - 1)) + \mathcal{O}_{ls}(\log(N + 1) - 1)$ for using the complete binary tree and $\mathcal{O}_{dx}(M\log N) + \mathcal{O}_{ls}(\log N)$ for using the pseudo binary tree embedded in the hypercube topology. When M is very large as in our case, this reduction is very significant.

In an attempt to predict the achievable efficiency by these proposed schemes without actually implementing on a system, a new performance modelling approach has been employed. Efficiency prediction formulas, termed *EP Formulas*, have been derived characterized by the computational load imbalance, the overhead imbalance and some system parameters. The *PbbMp*, *MPbbMp* and *PshMp* schemes have been implemented on Intel iPSC/2 and BBN Butterfly TC2000. The implementation results show that very high efficiencies can be achieved, especially using the message passing model. Using message passing model on iPSC/2, the modified partition-by-box scheme yields a better performance than the partition-by-box scheme. The hypercube topology seems to be better than the linear array topology. However, it is not necessarily always true since it depends on the restrictions for setting up the communication link and the overhead for sending data as discussed in Section 6.

Although it seems that the *PshMp* achieves worse performance because of remote shared memory access, if the remote shared memory time can be reduced to some extent, the *PshMp* might work better. These implementation results have also verified that the *EP Formulas* derived can predict the efficiencies accurately. The major factors which degrade the performance can easily be identified from these formulas.

References

[1] L. A. Shepp and Y. Vardi, "Maximum likelihood reconstruction for emission tomography," *IEEE Trans. Med. Imaging*, vol. MI-1, pp. 113–122, oct. 1982.

[2] W. F. Jones, L. G. Byars, and M. E. Casey, "Positron emission tomographic images and expection maximization : A VLSI architecture for multiple iterations per second," *IEEE Trans. Nucl. Sci.*, vol. NS-35, pp. 620 – 624, Feb. 1988.

[3] J. Llacer and J. D. Meng, "Matrix-based image reconstruction methods for tomography," *IEEE Trans. Nucl. Sci.*, vol. NS-32, pp. 855–864, Feb. 1985.

[4] F. U. Rosenberger, D. G. Politte, G. C. Johns, and C. E. Molnar, "An efficient parallel implementation of the EM algorithm for PET image reconstruction utilizing transputers," in *1990 Nuclear Science Symposium Conference Record*, Oct. 1990.

[5] L. Kaufman, "Implementing and accelerating the EM algorithm for positron emission tomography," *IEEE Trans. Med. Imaging*, vol. MI-6, pp. 37–50, Mar. 1987.

[6] C. M. Chen, S.-Y. Lee, and Z. H. Cho, "Parallelization of the EM algorithm for PET image reconstruction," in *1990 Nuclear Science Symposium Conference Record*, Oct. 1990.

[7] R. M. Lewitt and G. Muehllehner, "Accelerated iterative reconstruction for positron emission tomography based on the EM algorithm for maximum likelihood estimation," *IEEE Trans. Med. Imaging*, vol. MI-5, pp. 16–22, Mar. 1986.

[8] E. Tanaka, N. Nohara, T. Tomitani, M. Yamamoto, and H. Murayama, "Stationary positron emission tomography and its image reconstruction," *IEEE Trans. Med. Imaging*, vol. MI-5, pp. 199–206, Dec. 1986.

[9] H. Hart and Z. Liang, "Bayesian image processing in two dimensions," *IEEE Trans. Med. Imaging*, vol. MI-6, pp. 199–206, Sept. 1987.

[10] E. Levitan and G. T. Herman, "A maximum *a posteriori* probability expectation maximization algorithm for image reconstruction in emission tomography," *IEEE Trans. Med. Imaging*, vol. MI-6, pp. 185–191, Sept. 1987.

[11] P. R. Phillips, "Bayesian statics, factor analysis, and PET images–part I: Mathematical background," *IEEE Trans. Med. Imaging*, vol. 8, pp. 125–132, June 1989.

[12] S.-Y. Lee and J. K. Aggarwal, "Exploitation of image parallelism via the hypercube," in *Second Conference on Hypercube Multiprocessors*, (Knoxvile, Tn), Sept. 1986.

[13] S.-Y. Lee, H. D. Chang, K.-G. Lee, and B. Y. Ku, "Parallel power system transient stability analysis on hypercube multiprocessors," in *Proc. Int. Conf. PICA*, May 1989.

[14] P. Heidelberger and S. S. Lavenberg, "Computer performance evaluation," *IEEE Trans. Comput.*, vol. c-33, pp. 1195–1220, Dec. 1984.

[15] C. Lin and L. Snyder, "A comparison of programming model for shared memory multiprocessors," in *International Conference on Parallel Processing*, vol. II, pp. 163–170, Aug. 1990.

On the complexity of parallel image component labeling

VIPIN CHAUDHARY and J. K. AGGARWAL
Computer and Vision Research Center
The University of Texas at Austin

Abstract

We present a new parallel algorithm for labeling connected components in a binary image. We assume an eight connectivity for object pixels and a four connectivity for background pixels. The representation of the connected components used in the algorithm makes it easy to calculate certain properties of regions, i.e., area, perimeter, etc. The algorithm is implemented on a shared memory computer. The computational complexity of this parallel algorithm is $O(\lceil \log n \rceil)$ for an image of size n, and is the best achieved yet. The memory requirement is also linear in the number of object pixels. Unlike most previous papers we differentiate between the computational and communicational complexity of the algorithm. The implementations of the algorithm on several distributed memory architectures and their computational and communicational complexities are discussed. These complexities improve upon previous results. The algorithm is easily extended for gray level images without affecting the complexities. Results of the implementation on the Sequent Balance multiprocessor are presented.

1 Introduction

Labeling the objects is a fundamental task in computer vision, since it forms a bridge between low-level image processing and high-level symbolic processing [1]. One of the first known algorithms [2] performed two passes over the image. In the first pass, labels were generated for object pixels and stored in an equivalence table. During the second pass, these labels were replaced by the smallest equivalent labels. The memory requirement and computational complexity for evaluating the equivalent label were very high for images with complex regions. Lumia et al. [3] reduced the size of the equivalence table by reinitializing the table for each scan line. This algorithm also required two passes. Haralick [4] proposed an algorithm that did not require an equivalence table but, instead, required the repeated propagation of labels in the forward and backward directions. This technique improved on the memory requirement but required a large number of passes through the image for complex objects. Mandler and Oberlander [20] presented a one pass algorithm for generating border line chain codes for each component and also provide a list of adjacent regions for each region. The drawback with this algorithm is the computationally expensive corner detection. Samet and Tamminen [5] gave an algorithm applicable to hierarchical data structures, such as quadtrees, octrees, and in general bintrees. The authors used a very small equivalence table and the processing cost was linear in the size of the image. Several other techniques to accomplish related problems with varying degrees of accuracy and computation time involved have been presented [6, 7].

Most computer vision tasks require an enormous amount of computation, demanding high performance computers for practical real time applications. Parallelism appears to be the only economical way to achieve this level of computation. However, it is particularly difficult to label connected components in parallel. The process of labeling is global but depends critically on the local pixels, and the processing is irregular. Hirschberg [8] gave one of the earlier formulations of a parallel approach to

connected component labeling. Given a graph representation with n nodes, this technique required $n^2 / \log n$ processors with shared memory in an $O(\log^2 n)$ algorithm. Nassimi and Sahni [9] presented an algorithm on an n x n mesh connected computer which labeled the components in an n x n image in $O(n)$ time. Tucker [10] presented a parallel divide-and-conquer algorithm on a simple tree connected SIMD computer which required $O(n^{1/2})$ time for an image of size n. Cypher et al. suggested a EREW PRAM algorithm with a complexity of $O(\log n)$ [11] and hypercube and shuffle-exchange algorithms with a complexity of $O(\log^2 n)$ for an image of size n [12, 13]. These algorithms apply only to binary images since the concept of boundary would otherwise be imprecise. They also assume four connectivity for pixels. The number of processors used for the EREW PRAM algorithm are $3n + 2n^{1/2}$ for an images of size n, thus, putting a very low bound on the processor efficiency. Their biggest drawback is the relatively large amount of memory it needs to hold the intermediate results. Sunwoo et al. [14] also implemented a parallel algorithm on a hypercube whose complexity was linear in the size of the image. Agarwal and Nekludova [19] presented an algorithm on an EREW PRAM with a complexity of $O(\log n)$. This algorithm is described for hex connected images. Little et al. [15] gave parallel algorithms on the Connection machine which took $O(\log n)$ router cycles for an image of size n. Maresca and Li [16] presented an algorithm which labeled components in an n x n image in $O(n)$ time on a polymorphic torus. Manohar and Ramapriyan [17] gave another algorithm on a mesh connected computer with a complexity of $O(n \log n^2)$ for an n x n image.

We present a new parallel algorithm for connected component labeling with worst case complexities that are best achieved yet that depend only on the number of objects (or more precisely, the number of object pixels) in the image. Most previous algorithms have the disadvantage that their complexity is a function of both the object and the background pixels. The memory requirement of our algorithm is also linear in the number of objects. The algorithm is then implemented on the Sequent Balance shared memory multicomputer. The computational complexity of our new parallel algorithm on a shared memory computer is $O(\lceil \log n \rceil)$ for an image of size n. Unlike other parallel algorithms for connected component labeling, which are very specific to the architecture, the algorithm described in this paper applies to a very general class of architectures.

This paper is organized as follows. We first introduce the concept of connected component labeling and the notations and definitions used in our description. This is followed by a description of our data structure and the parallel algorithm. Next we discuss the implementation of this algorithm on shared memory architectures and present a complexity evaluation. The implementation of the algorithm on various distributed memory computers and their computational and communicational complexity are discussed next. Then we present the results of implementing the algorithm on the Sequent Balance computer and finally conclude with a summary. The Appendix describes the evaluations of certain communicational complexities in detail.

---

[1]This research was supported in part by IBM.

2 Connected component labeling

This section describes the algorithm for connected component labeling. We use certain assumptions about the input image for our algorithm. First, it is a binary image with object pixels represented as runs [1]. This representation is very economical and often used for image compression. We further assume an eight connectivity for the objects and a four connectivity for the background.

2.1 The notation

Henceforth, we denote object pixels by pixels, unless they are ambiguous. A *conexon* (also *path* in [1]) of length n from pixel P to pixel Q is a sequence of pixel points $P = P_0, P_1, ..., P_n = Q$ such that P_i is a neighbor (an eight connectivity) of P_{i-1}, $1 \leq i \leq n$. We say that pixel P is *connected* to pixel Q if a *conexon* exists from P to Q. This is a $conexon^k$ if it lies within rows $1...k$. A maximal set of *connected* pixels within rows $1...k$ (at least one of which is in row k) forms the $object^k$. The maximal set of *connected* pixels in row k forms the rep^k. The set of rep^k belonging to the same $object^k$ forms the $part^k$. A $touch^k$ is a rep^k that *connects* two or more rep^{k-1}, at least two of which are not in the same $part^{k-1}$.

For the sake of simplicity, we assume the input to be in the run length representation [1] from left to right and top to bottom. This does not affect the generality of the algorithm, since it is trivial to convert image data (a two dimensional array) to the run length representation. The format of the input to the algorithm is as follows:

— If a row has at least one *rep* in it, then it is a sequence of integers [**r**, **n**, l_1, r_1,...,l_n,r_n], where **r** is the row number, **n** is the number of *reps* in this row, and l_i, r_i for $1 \leq i \leq$ **n** are the left and right column numbers of the ith *reps* in row **r**, respectively.

— If there are no *reps* in the row, then the sequence for that row is null.

The output of the algorithm is a linked list of the various connected components in the input image. Each connected component is, in turn, a linked list of the *reps* it comprises.

2.2 The data structure

There are two basic structures in this algorithm, namely *reps* and *objects*. Linked lists are used to represent both these concepts. A third linked list *component* is used to represent the output — the list of connected components in the image. The *reps* in a row are assumed to be in order from left to right, i.e., the leftmost *rep* occurs first and the rightmost *rep* occurs last. Each rep^k is represented as a structure consisting of the following fields: *left* — the starting column number of rep^k; *right* — the column number where rep^k ends; *row* — the row number of this rep^k; *repnext* — a pointer to the *neighboring* rep^k; *object* — a pointer to the $object^k$ *connected* to this rep^k.

Each $object^k$ is also a linked list of structures having the following fields: *leftmost* — a pointer to the leftmost rep^k associated with this $object^k$; *end* — a pointer to the end of the linked list of *reps* from previous rows which are *connected* to this $object^k$; *start* — a pointer to the start of the linked list of *reps* in previous rows which are *connected* to this $object^k$; *label* — a positive integer assigned to this $object^k$; *objnext* — a pointer to the next $object^k$ in the linked list; *touch* — a boolean set to true if a rep^k is found to be *connected* to $object^{k-1}$.

Component is a linked list of structures with the following fields: *compo* — a pointer to the *rep* belonging to this component; *compnext* — a pointer to the next *component* in the linked list.

2.3 The algorithm

The algorithm is conceptually very simple. Every *rep* is associated with an *object*. If two *reps* are *connected*, they both point to the same *object*, i.e., the *object* with the smaller *label*. All the *reps* pointing to the same *object* form a *component*. We now describe the major procedures of the algorithm. For the detailed algorithm (complete pseudo-code), the reader is referred to [22].

To describe the algorithm we assume for simplicity that there are as many processors available as there are *rep*. The algorithm has two main parts: initialization of the data structure, and the evaluation of connected components. During initialization each processor initializes the fields of each *rep*, each of which is also an *object* as well as a *component*. Labels are assigned randomly.

The next part involves checking for connectedness of two reps. The *reps* are merged in a binary tree fashion. If rep^{k-1} A is *connected* to rep^k B, then the following condition holds:

$$(A \rightarrow left \leq B \rightarrow right + 1) \wedge (A \rightarrow right \geq B \rightarrow left - 1).$$

To combine two components corresponding to rep^{k-1} A and rep^k B, we only need to make the following changes:

B $\rightarrow object \rightarrow leftmost \rightarrow object \rightarrow end \rightarrow repnext = $ A $\rightarrow object \rightarrow leftmost \rightarrow object \rightarrow start.$

A $\rightarrow object \rightarrow leftmost \rightarrow object \rightarrow start = $ B $\rightarrow object \rightarrow leftmost \rightarrow object \rightarrow start.$

A $\rightarrow object \rightarrow leftmost = $ B $\rightarrow object \rightarrow leftmost.$

Thus, combining two components requires a constant amount of computation. Thus, if it were executed as a sequential algorithm its complexity would be linear in terms of the number of object pixels [22].

3 The implementation on shared memory architectures

Having described the algorithm for labeling connected components, we now present a strategy to implement the algorithm on a shared memory computer.

The image is partitioned horizontally into a set of subimages, preferably into as many as the number of processors. Each of these subimages has its connected components labeled concurrently by multiple processors. These subimages are then merged at several levels, as are their labels.

The subimages are numbered as consecutive integers from 0 onwards. Subimages with consecutive numbers are adjacent. Note that the binary representation of the subimage numbers differs at the least significant digit. At the first level, adjacent subimages (0 and 1, 2 and 3, ...) are merged to form new subimages. The subimage arrived at by merging two subimages inherits the lower image number. For cases with an odd number of subimages, the last subimage is merged with an empty subimage(or, passed on as is to the next level). Thus, the adjacent images at the second level differ at the second most significant digit in their binary representation. This procedure of merging is continued until all the subimages are merged into a single image. In the above discussion, we assume that a subimage is comprised of one or more rows of *reps*. However, this technique can easily be extended to include cases where the subimages are comprised of a part of the row.

Consider two adjacent subimages S_1 and S_2. Without a loss of generality, we further assume that S_1 extends from rows i to k-1 and that S_2 extends from rows **k** to **j**. The merger of these subimages involves only rows **k-1** and **k**. If rep^{k-1} is *connected* to rep^k, then the *objects* corresponding to these *reps* are assigned the same label — the smaller one. Also, if a $touch^k$ is found, then the labels of the $part^{k-1}$ and rep^k are given the same label — the one associated with $part^{k-1}$.

The conditions for connectedness of two *reps* as well as the procedure for combining two components is the same as described in the algorithm previously.

3.1 Complexity

Consider an image with n object pixels labeled by p processors. On an average, each subimage has n/p object pixels. Thus, labeling components for all the subimages takes $O(n/p)$ time. The maximum number of components that need to be merged at each level is $O(n/p)$. Since merging two components requires a constant amount of computation, the components can be merged in $O(n/p)$ time at each level. The number of levels of merging is $\lceil \log p \rceil$. Hence, the computational complexity of the entire labeling process is $O(n/p(1 + \lceil \log p \rceil))$. If we have n processors, then the complexity is $O(\lceil \log n \rceil)$.

4 The implementation on distributed memory architectures

This section discusses the parallel implementation of the connected component labeling algorithm on various distributed memory architectures. Unlike the shared memory computers, the distributed memory computers have a communication overhead. We present the computational and communicational complexity of implementing the algorithm on various distributed memory architectures. Most previous algorithms for distributed memory architectures did not consider the communicational complexity. As we will show, communicational complexity dominates.

For the complexity evaluation, we assume the size of the image to be n pixels and the number of processors to be p. The unit of communication is a pixel and it takes unit time for unit distance. We shall assume that the distance between two processors directly connected to each other is unity. All the complexity evaluations are worst case evaluations. Note that the complexities hold for both SIMD and MIMD distributed memory architectures.

4.1 A binary tree connected computer

We are given a binary tree connected computer with $p = 2q - 1$ processors, thus, we have q leaf processors. We first split the image equally such that each leaf processor has a subimage of size n/q. For now we assume that each leaf processor has its own subimage associated with it. We describe the distribution of the subimages later.

The algorithm is implemented as follows. During the first phase, each of the leaf processors computes the connected components of its associated subimage. These connected components are then merged in a binary tree fashion, as in shared memory architectures. The difference in the two is that in the case of a tree connected binary computer, the entire subimage is communicated, along with the connected component information. Finally, the connected components of the entire image are in the root processor.

The connected component labeling process in the leaf processors takes $O(n/q)$ time. After this, there are $\log q$ merging steps, each of which takes a constant amount of computation. Thus, the total computation time is $O(n/q + \log q) = O(2n/(p+1) + \log(p+1)/2)$. If we have a binary tree connected

computer with $2n - 1$ processors, then the computational complexity of the algorithm is $O(\log n)$.

The merging process occurs in phases in a binary tree fashion. The total communicational time required is $O(n/q + 2n/q + \ldots + n/2) = O(n(q-1)/q) = O(n(p-1)/(p+1))$. Thus, if we have a binary tree connected computer with $2n - 1$ processors, then the communicational complexity of the algorithm is $O(n)$.

The distribution of the subimages is the inverse process of merging performed at different levels of the binary tree connected computer. Thus, the distribution process has the same communicational complexity as merging, i.e., $O(n)$. Hence, the communicational complexity of the algorithm is $O(n)$.

4.2 A mesh connected computer

The interconnections between the processors in the mesh connected computer can either be unidirectional or bidirectional. For example, the AT&T Pixel Machine has unidirectional interconnection links, i.e., at any instance all the links can transmit data in only one direction. The computational complexities and the communicational complexities for both these types of mesh connected computers are dealt with separately. For the sake of simplicity, we consider a mesh connected computer with $p = q^2$ processors. We first split the image equally such that each processor has a subimage of size n/q^2. The distribution of the subimages is done as an inverse process of merging the subimage connected components.

The algorithm for both unidirectional and bidirectional mesh connected computers is implemented as follows. During the first phase, each of the processors computes the connected components of its associated subimage. These connected components are then merged in a binary tree fashion as in the binary tree connected computer. The exact merging scheme can be split into two operations: row and column merging. We first merge the subimage connected components of all the rows. The resulting merged subimages of row merging are then merged as a column merge. Note that each of these mergings is similar to the mergings of the binary tree connected computer. The difference is that the distances between subsequent merges do not remain equal.

4.2.1 Bidirectional mesh

For a bidirectional mesh connected computer it can be easily shown by induction that the number of mergings required is q, i.e., $q/2$ for row merging and $q/2$ for column merging. Fig. 1 gives an example of a merging process with 4 processors. The arrows indicate the direction of data movement and the superscript on the arrow indicates the relative amount of data transfer.

Computational complexity. The connected component labeling process in the processors takes $O(n/p)$ time. After this process there are q merging steps, each of which takes a constant amount of computation. Thus, the total computation time is $O(n/p + p^{1/2})$. If we have a mesh connected computer with n processors, then the computational complexity of the algorithm is $O(n^{1/2})$.

Communicational complexity. We denote the communication time for row merging by B_r and the communication time for column merging by B_c:

$$
\begin{aligned}
B_r &= O(n/p[1 + 2 + (4 + 4) + (8 + 8 + 8 + 8) + \ldots + q/2 \; terms]) \\
&= O(n/3 + 2n/3p) \\
B_c &= O(n/q[1 + 2 + (4 + 4) + (8 + 8 + 8 + 8) + \ldots + q/2 \; terms]) \\
&= O(np^{1/2}/3 + 2n/3p^{1/2})
\end{aligned}
$$

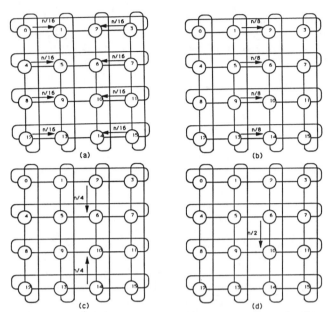

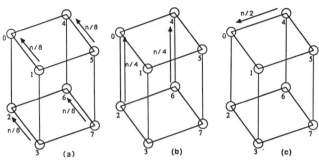

Figure 2: *Phases (a) to (c) illustrate the merging process for a hypercube of degree 3 using the pseudo binary tree. The arrows indicate the direction of data movement and the superscript on the arrow indicates the communication cost.*

Figure 1: *Phases (a) to (d) illustrate the merging process for a Bidirectional mesh with 16 processors. The arrows indicate the direction of data movement and the superscript on the arrow indicates the communicational cost.*

Thus, the total communicational complexity B is given as follows:

$$B = B_r + B_c = O(1/3[np^{1/2} + n + 2n/p^{1/2} + 2n/p])$$

If we have a mesh connected computer with n processors, then the communicational complexity of the algorithm is $O(n^{3/2})$. For unidirectional mesh connected computer with n processors, it can be easily shown that the complexities are the same.

4.3 Hypercubes
A hypercube of degree d has 2^d nodes and each node has exactly d neighbors. The distance between any two nodes is less than or equal to d. We first discuss the embedding of binary trees in hypercubes. Wu [21] presents three results which we use in our implementation of the algorithm on the hypercube.

Proposition 1 *A complete binary tree of height $d > 2$ cannot be embedded in a hypercube of degree $\leq d$ such that adjacency is preserved. In other words, a complete binary tree cannot be embedded in a hypercube with a dilation cost of 1 and an expansion cost of less than 2.*

Proposition 2 *A complete binary tree of height $d > 0$ can be embedded in a hypercube of degree $d + 1$ in such a way that the adjacencies of nodes of the binary tree are preserved.*

Proposition 3 *A complete binary tree of height $d > 0$ can be embedded in a hypercube of degree d with cost = 2; i.e., neighbors in the binary tree are mapped into nodes of, at most, distance 2 away in the hypercube.*

A complete binary tree of height d has $2^d - 1$ nodes. The smallest hypercube large enough to house a binary tree of height d is of degree d. The algorithm is implemented on the hypercube as follows. For the sake of simplicity, we consider a hypercube with $p = 2^d$ processors. We first split the image equally such that each processor has a subimage of size n/p. During the first phase, each of the processors computes the connected components of its associated subimage. These connected components

are then merged in a binary tree fashion, as in the binary tree connected computer. The distribution of the subimages is an inverse process of merging the subimage connected components.

By proposition 1, it is clear that we cannot embed a complete binary tree (to be used in our merging process) in a hypercube with a dilation cost of 1 and an expansion cost of less than 2. Proposvitions 2 and 3 give us two solutions to embed the complete binary tree in a hypercube. In the first, we can use a hypercube with twice the number of processors. In the second, the neighboring nodes in the binary tree will have a distance of 2 between them. Using twice the number of processors is undesirable, since it reduces the efficiency of processors. Being unable to preserve the adjacency of the tree nodes also increases our communicational cost. These two drawbacks are alleviated using a pseudo binary tree.

A pseudo binary tree is a binary tree structure which can easily be embedded into the hypercube topology such that a node in the hypercube may represent more than one node in the corresponding pseudo binary tree [18]. A pseudo binary tree is an efficient topology for distributing and merging subimages. The modified singlecast scheme [18] in which the controller (one of the processing elements) distributes a set of images is used to distribute the subimages. The merging process is exactly the inverse. Fig. 2 shows the phases of merging in the hypercube.

Computational complexity. The connected component labeling process in the processors takes $O(n/p)$ time. After this process there are $\log p$ merging steps, each of which takes a constant amount of computation. Thus, the total computation time is $O(n/p + \log p)$. If we have a mesh connected computer with n processors, then the computational complexity of the algorithm is $O(\log n)$.

Communicational complexity. The merging process occurs in phases in a binary tree fashion. The total communicational time required is $O(n/p + 2n/p + \ldots + n/2) = O(n(p-1)/p)$. Thus, if we have a hypercube with n processors, the communicational complexity of the algorithm is $O(n)$.

The distribution of the subimages is the inverse process of merging performed at different levels of the pseudo binary tree. Thus, the distribution process has the same communicational complexity as merging, i.e., $O(n)$. Hence, the communicational complexity of the algorithm is $O(n)$.

5 Implementation results
The algorithm described in this paper was tested on several images with all sorts of complex regions. The images contain

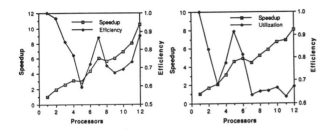

Figure 3: *Speedup and Efficiency on a 256 X 256 binary image with 172 components and on a 512 X 512 binary text image with 793 components.*

convex, concave, simply connected, and multiply connected regions with holes. For a 256 X 256 image with 172 components, the algorithm takes 351 ms on a uniprocessor. Fig. 3 gives the speedups obtained for two images. Efficiency (which is the ratio of the speedup to the number of processors used) for the two images is also shown. We obtain a maximum speedup of 10.53 using 12 processors for the 256 X 256 binary image, and the efficiency varies around 0.75 on the Sequent Balance multiprocessor. The maximum speedup obtained for the binary text image is 8.07 using 12 processors.

6 Conclusion

We present a parallel algorithm for labeling connected components whose memory requirement and computational and communicational complexities are the best achieved yet. Since no assumptions have been made as to the type of input, the algorithm works for all types of binary images. The representation of the connected components as a linked list of *reps* makes it easy to calculate certain features of regions, i.e., area, perimeter, etc. The algorithm is easily extended to gray level images by including another field indicating the gray level in the *rep*. Also, the connectivity check for two *reps* would have to compare the gray values. The algorithm implemented on a shared memory computer has a computational complexity of $O(\lceil \log n \rceil)$ for an image of size n. We discuss the results of implementing the algorithm on the Sequent Balance multiprocessor. The speedup graphs show an almost linear speedup. The computational and communicational complexities of the algorithm implemented on various distributed memory architectures, i.e., a binary tree connected computer, a unidirectional and bidirectional mesh connected computer, and a hypercube computer are computed. It is trivial to see that the algorithm can be implemented on the polymorphic torus architecture with complexities no worse than those of the mesh connected computers. These complexities are the best yet achieved. The communicational complexities are greater than the computational complexities for all the distributed memory architectures. This explains the decrease in the efficiency of processors with the increase in the number of processors [14]. The theoretical formulation of these complexities gives a better idea as to which architecture will have a better efficiency. In other words, we have an analytic expression for the communication overhead.

References

[1] A. Rosenfeld and A. C. Kak, *Digital Picture Processing.* New York: Academic Press, 1982.

[2] A. Rosenfeld and J. L. Pfaltz, "Sequential operations in digital signal processing," *J. ACM*, vol. 13, no. 4, pp. 471-494, 1966.

[3] R. Lumia, L. Shapiro, and O. Zuniga, "A new connected components algorithm for virtual memory computers," *Comput. Vision Graphics Image Process.*, vol. 22, no. 2, pp. 287-300, 1983.

[4] R. M. Haralick, "Some neighborhood operations," In M. Onoe, K. Preston Jr., and A. Rosenfeld, eds. *Some neighborhood operations in real time/parallel computing image analysis.* New York: Plenum Press, 1981.

[5] H. Samet and M. Tamminen, "An improved approach to connected component labeling of images," in *Proc. IEEE Conf. on Computer Vision Pattern Recognition*, 1986, pp. 312-318.

[6] S. W. Zucker, "Region growing: Childhood and adolescence," *Computer Graphics Image Process.*, vol. 5, no. 3, Sep. 1976.

[7] V. Chaudhary and J. K. Aggarwal, "Parallelism in computer vision - A review," In V. Kumar, P. S. Gopalakrishnan, and L. Kanal, eds. *Parallel Algorithms for Machine Intelligence and Vision*, pp. 270-309. New York: Springer-Verlag, 1990.

[8] D. S. Hirschberg, A. K. Chandra, and D. V. Sarwate, "Computing connected components on parallel computers," *Communications ACM*, vol. 22, no. 8, pp. 461-464, 1979.

[9] D. Nassimi and S. Sahni, "Finding connected components and connected ones on a mesh connected parallel computer," *SIAM Journal Comput.*, vol. 9, no. 4, pp. 744-757, 1980.

[10] L. W. Tucker, "Labeling connected components on a massively parallel tree machine," in *Proc. IEEE Conf. on Computer Vision Pattern Recognition*, 1986, pp. 124-129.

[11] R. Cypher, J. L. C. Sanz, and L. Snyder, "EREW PRAM and mesh connected computer algorithms for image component labeling," in *Proc. IEEE Workshop on Computer Arch. for Pattern Anal. Mach. Intell.*, 1987, pp. 122-128.

[12] R. Cypher, J. L. C. Sanz, and L. Snyder, "Hypercube and shuffle-exchange algorithms for image component labeling," in *Proc. IEEE Workshop on Computer Arch. for Pattern Anal. Mach. Intell.*, 1987, pp. 5-9.

[13] R. Cypher, J. L. C. Sanz, and L. Snyder, "Practical algorithms for image component labeling on SIMD mesh connected computers," in *Proc. Int. Conf. on Parallel Processing*, 1987, pp. 772-779.

[14] M. H. Sunwoo, B. S. Baroody, and J. K. Aggarwal, "A parallel algorithm for region labeling," in *Proc. IEEE Workshop on Computer Arch. for Pattern Anal. Mach. Intell.*, 1987, pp. 27-34.

[15] J. J. Little, G. Blelloch, and T. Class, "Parallel algorithms for computer vision on the connection machine," in *Proc. IEEE Int. Conf. on Computer Vision*, 1987, pp. 587-591.

[16] M. Maresca, H. Li, and M. Lavin, "Connected component labeling on polymorphic torus architecture," in *Proc. IEEE Conf. Computer Vision Pattern Recognition*, 1988, pp. 951-956.

[17] M. Manohar and H. K. Ramapriyan, "Connected component labeling of binary images on a mesh connected massively parallel processor," *Computer Vision Graphics Image Process.*, vol. 45, pp. 133-149, 1989.

[18] S. Y. Lee and J. K. Aggarwal, "Exploitation of image parallelism via the hypercube," in *Proc. of Second Conf. on Hypercube Multiprocessors*, 1986.

[19] A. Agarwal and L. Nekludova, "A parallel O(log N) algorithm for finding connected components in planar images," in *Proc. Int. Conf. Parallel Processing*, 1987, pp. 783-786.

[20] E. Mandler and M. F. Oberlander, "One-pass encoding of connected components in multi-valued images," in *Proc. IEEE Conf. on Computer Vision Pattern Recognition*, 1990, pp. 64-69.

[21] A. Y. Wu, "Embedding of tree networks into hypercubes," *Journal of Parallel and Distributed Computing*, vol. 2, pp. 238-249, 1985.

[22] V. Chaudhary and J. K. Aggarwal, "A fast parallel implementation of image component labeling," submitted to *IEEE Trans. Parallel Dist. Sys.*. Also TR-90-6-65, Computer and Vision Research Center, The University of Texas at Austin.

Detecting Repeated Patterns on Mesh Computers*

Russ Miller
Dept. of Computer Science
State University of New York
Buffalo, NY 14260
miller@cs.buffalo.edu

Steven L. Tanimoto
Dept. of Comp. Sci. and Eng., FR-35
University of Washington
Seattle, WA 98195
tanimoto@cs.washington.edu

Abstract

In certain kinds of data analysis, it is useful to automatically detect the largest repeated patterns in the data. This can be useful in describing the data, compressing it, and possibly in discovering primitives of unknown languages. We give an efficient parallel algorithm to find the largest repeated subarrays of an $n \times n$ array of data on an $n \times n$ mesh-connected computer. We also discuss related problems, such as finding the most commonly occurring patterns of a given size, and finding the largest repeated patterns whose instances are permitted to have undergone symmetry transformations.

1 Introduction

The main problem considered in this paper is that of labeling the largest replicated square subarrays of a given square array. The problem is motivated by the desire to automatically discover certain structure in bitmaps and other digital images. The detection of repeated substructures in images may lead to the discovery of language primitives, icons, or units of description that may then be used to describe or encode the image or related images. While the problem of structure discovery can be posed in many different ways, we use a simple formulation of the problem and present an efficient solution for it on a mesh-connected array of processors.

Karp, Miller, and Rosenberg [Karp et al 1972] considered several problems of repeated substructure detection using serial computers. These include the "depth-k matches problem" and "the maximum matches problem" for strings, trees, and arrays. In the case of two-dimensional arrays, they presented an algorithm that finds the greatest integer, call it k, for which there is a $k \times k$ repeated subarray in the input. Given an $n \times n$ array, their algorithm requires $O(n^2 \log k)$ time on a serial computer. In this paper, we present a parallel algorithm that runs in $\Theta(n \log k)$ time when the array is mapped one element per processor in a natural fashion onto an $n \times n$ array of processors.

2 Definitions

In this section, we define the mesh-connected computer. We also define an equivalence relation that we find useful in describing the algorithm.

2.1 The 2-Dimensional Mesh-Connected Computer

The *(2-dimensional) mesh-connected computer (mesh) of size* n^2 is a machine with n^2 simple *processing elements (PEs)* arranged in a square lattice. To simplify exposition, we assume that $n^2 = 4^c$, for some integer c. For all $i, j \in [0, \ldots, n-1]$, let $P_{i,j}$ represent the PE in row i and column j. Processor $P_{i,j}$ is connected via bidirectional unit-time communication links to its four *neighbors*, $P_{i-1,j}$, $P_{i+1,j}$, $P_{i,j-1}$, and $P_{i,j+1}$, assuming they exist. Each PE has a fixed number of registers (words), each of size $O(\log n)$, and can perform standard arithmetic and Boolean operations on the contents of these registers in unit time. Each PE can also send or receive a word of data to or from each of its neighbors in unit time. Each PE contains its row and column indices, as well as a unique identification register, the contents of which is initialized to the PE's row-major index, shuffled row-major index, snake-like index, or proximity order index, as shown in Figure 1. (If necessary, these values can be generated in $\Theta(n)$ time.)

The communication diameter of a mesh of size n^2 is $\Theta(n)$, as can be seen by examining the distance between PEs in opposite corners of the mesh. This means that if a PE in one corner of the mesh needs data from a PE in another corner of the mesh at some time during an algorithm, then a lower bound on the running time of the algorithm is $\Omega(n)$. Since it is possible for a largest repeated subsquare to be a 1×1 subsquare which exists only in processors $P_{0,0}$ and $P_{n-1,n-1}$, the largest repeated square pattern problem has an $\Omega(n)$ time lower bound.

2.2 The Equivalence Relation

In this section, we define an equivalence relation (*d-equivalent*) and the corresponding set of equivalence classes (*d-equivalence classes*), which we find useful in describing our algorithm. Define cells (i,j) and (i',j') to be *d-equivalent*, and to be in the same *d-equivalence class*, if and only if the $d \times d$ pattern with northwest cell (i,j) and the $d \times d$ pattern with northwest cell (i',j') are identical and completely contained within the array. Notice that for a given d, $1 \le d \le n$, every cell (i,j), $i, j \in [0, n-1]$, of the $n \times n$ array is in at most one d-equivalence class. Further, these d-equivalence classes can be enumerated with the labels $1 \ldots n^2$. Therefore, the problem we solve in this paper is that of finding the largest integer, call it k, such that there exists at least one k-equivalence class which contains at least 2 items.

*Research supported in part by the National Science Foundation under NSF Grants DCR-8608640, IRI-8800514, and IRI-8605889.

3 A $\Theta(n \log k)$ Time Solution to the 2-Dimensional Largest Pattern Matching Problem

In this section, we give an efficient parallel algorithm to solve the 2-dimensional largest repeated pattern problem. Specifically, we present a $\Theta(n \log k)$ time algorithm for a mesh of size n^2 to identify the largest integer, call it k, such that there are repetitions of $k \times k$ regions. Furthermore, the algorithm that we describe also identifies the equivalence classes of such $k \times k$ regions. That is, at the end of the algorithm, every processor will know whether or not it is the northwest corner of a $k \times k$ region whose pattern exists elsewhere in the original data array, and if the answer is affirmative, the PE will know the unique label corresponding to all duplicates of its pattern. (See Figure 2.)

The algorithm we present consists of two phases. The first phase works by merging equivalence class labels in a bottom-up fashion. At iteration i, 2^{i-1}-equivalence class labels, corresponding to squares of size $2^{i-1} \times 2^{i-1}$, are used to generate 2^i-equivalence class labels, which correspond to squares of size $2^i \times 2^i$. If there is at least one pattern of size $2^i \times 2^i$ that is repeated in the array, then the algorithm continues, while if there are no such patterns, then the second phase is invoked. The second phase consists of a binary search to find k in the range $[2^{i-1}, 2^i)$.

The algorithm makes use of a lemma in [Karp et al 1982] which permits gradual refinement of an equivalence relation on array positions. We restate their lemma in an Appendix to this paper.

Theorem 3.1 *Given an $n \times n$ array of data, distributed one item per processor in a natural fashion on a mesh of size n^2, in $\Theta(n \log k)$ time the largest integer, call it k, can be determined, such that there exists repetitions of $k \times k$ patterns. In addition, every processor that contains the northwest element of a repeated $k \times k$ pattern will know the label of its equivalence class.*

Proof. Let s be the size of the alphabet of possible array elements. Therefore, there are s^{n^2} different ways that the $n \times n$ array could be filled. However, for a fixed d, there are at most n^2 d-equivalence classes present in any $n \times n$ data array, since there are only n^2 positions in such an array. The algorithm is divided into two search phases. The first phase is a bottom-up merge phase which will identify an integer $a = 2^c$, c an integer, such that k is in the range $[a, 2a)$. That is, the first phase will terminate when it finds that there are repeated $a \times a$ patterns, but no repeated $2a \times 2a$ patterns. The second phase consists of a binary search for k in the range $[a, 2a)$.

At the beginning of iteration i of the bottom-up merge phase, we assume that every processor $P_{j,k}$ knows the label of its 2^{i-1}-equivalence class. Every processor creates a *master record* and a *request record* containing its position in the mesh as well as the label of its 2^{i-1}-equivalence class. A $\Theta(n)$ time mesh *random access read (RAR)* is performed so that every processor $P_{j,k}$ finds the label of the 2^{i-1}-equivalence class of its $2^{i-1} \times 2^{i-1}$ eastern neighboring region, which is rooted at processor $P_{j,k+2^{i-1}}$. Similarly, a RAR is used for each processor to find the label of the 2^{i-1}-equivalence class of its southern and southeastern $2^{i-1} \times 2^{i-1}$ neighboring regions. The concatenation of its 2^{i-1}-equivalence class label with these three labels

forms the initial label of the equivalence class for iteration i. A $\Theta(n)$ time mesh sort is used to group the equivalence classes together by these initial labels, and a $\Theta(n)$ time parallel prefix operation is used to count and relabel the equivalence classes to obtain the 2^i-equivalence class labels. (Notice that relabeling the equivalence classes at each iteration avoids the exponential explosion in the labels of the equivalence classes.) Next, a local comparison followed by a $\Theta(n)$ time semigroup operation is used to determine whether or not there exists a 2^i-equivalence class with two or more entries. If the answer is affirmative, then all processors discard their 2^{i-1}-equivalence class label, a RAR is performed so that all processors learn their 2^i-equivalence class label, and we proceed to iteration $i + 1$. If the answer is negative, we proceed to phase two of the algorithm in search of $k \in [2^{i-1}, 2^i)$.

When the first search phase terminates, it will be known that k lies in $[a, 2a)$, for $a = 2^c$, c an integer. The second search phase consists of a binary search that examines squares with dimension in the range $(a, 2a)$. That is, the squares examined in the second phase have dimensions that are not powers of two. Therefore, given the a-equivalence class of a cell (i, j), if the binary search is concerned with squares of dimension $b \in (a, 2a)$, the eastern, southern, and southeastern regions of concern to cell (i, j) are contained within cell (i, j)'s region. See Figure 3. The process of determining whether or not some b-equivalence class contains at least 2 items, as well as labeling the b-equivalence classes, is done as before.

The operations of sorting, random access read, broadcasting, and parallel prefix can all be performed in $\Theta(n)$ time on a mesh of size n^2 (c.f., [Miller and Stout 1989] and [Nassimi and Sahni 1981]). Therefore, since each iteration of the binary search takes $\Theta(n)$ time, the running time of the algorithm is as claimed. □

4 Practical Considerations

Based on the communication diameter of the mesh, the algorithm we have given is potentially a factor of $\Theta(\log n)$ from optimal. However, unlike many theoretically efficient high-level mesh algorithms, the algorithm we have given is nonrecursive and based on standard mesh operations. Therefore, one would expect the algorithm to be efficient on commercially available SIMD meshes, such as those produced by Thinking Machines and MasPar.

In fact, if one is to implement this algorithm on a real machine, certain optimizations can be made. During iteration i of the bottom-up label merging phase, processor $P_{j,k}$ needs to obtain the 2^{i-1}-equivalence class labels from processors $P_{j,k+2^i-1}$, $P_{j+2^i-1,k}$, and $P_{j+2^i-1,k+2^i-1}$. Instead of performing a RAR to satisfy each request, or even using one modified RAR (each processor generates one master record and three request records) to satisfy all three requests, these requests can be satisfied by a series of two rotations. In fact, complete rotations are not even required. First, perform a restricted row rotation (i.e., left shift) which shifts label information $2^i - 1$ processors to the left. This will satisfy the request for information from the east neighbor ($P_{j,k+2^i-1}$), and will also serve to route information from the southeast neighbor ($P_{j+2^i-1,k+2^i-1}$) to the south neighbor ($P_{j+2^i-1,k}$). Next, perform a restricted column rotation (i.e., upward shift) which shifts label information $2^i - 1$ processors upward. This serves to route the required label information from the south and southeast neighbor to the destination. A similar modification can be made to the second phase of the algorithm.

Another practical modification concerns sorting. Instead of using an asymptotically optimal $\Theta(n)$ time sort with relatively high constants, one may opt for a very efficient and practical $\Theta(n \log n)$ time sort. (The reader is referred to [Scherson and Sen 1989], and the references contained therein, for descriptions and analyses of "shear sort" algorithms. For the mesh, these algorithms rely on being able to efficiently sort rows and columns of data.)

Finally, the semigroup operation and broadcast, which are used to determine whether or not at least one equivalence class has more than one entry and to decide the next dimension to consider, can be replaced by a global-OR operation from the processors to the controller, followed by a broadcast from the controller back to the processors. The global-OR and controller broadcast operations are standard features available on most fine-grained parallel machines.

5   Related Problems

5.1   The Largest Repeated Rectangle with Given Aspect Ratio

Our algorithm need *not* be limited to finding *square* patterns that are repeated. A simple modification to our algorithm can be made to solve the problem of finding a largest repeated rectangular pattern where the rectangle has any given aspect ratio (e.g., 3 to 5). The equivalence relation $E_{3,5}$ can be built up from $E_{1,1}$, and then from $E_{3,5}$, we can compute $E_{6,10}, E_{12,20}$, ..., and eventually find, through binary search, the largest value of c such that $E_{3c,5c}$ has an equivalence class with more than one member. The running time of the algorithm is $\Theta(n \log c)$.

5.2   The Widest of Longest Repeated Rectangle Problem

By enlarging the rectangles for repeated patterns in one dimension only, repeated subarrays of size $k \times 1$ are found for k as large as possible. This information can be used to find the largest l such that there exists a $k \times l$ repeated subarray. The running time of this algorithm is $\Theta(n \log k + n \log l)$.

5.3   Repetitions with Symmetry Transformations

A variant of our algorithm can be used to find largest repeated patterns in an array, where an instance of a pattern may be related to another instance by a symmetry transformation. Sample symmetry transformations include rotation by $\pi/2$, reflection, and negation (on an element-by-element basis). Note that the group generated by rotation and reflection has 8 members, while the group generated by negation has 2 members.

Given a group of symmetry transformations with γ members, we begin the modified algorithm by computing, for the input array A, the variant arrays $t_0(A), \ldots, t_{\gamma-1}(A)$. These results are stored so that processor $P_{i,j}$ holds the $(i,j)^{th}$ element of each of the variant arrays. The equivalence relation is now considered to be defined on an extended set of positions

$$\{(u,v) : u = c \cdot n + i, v = c \cdot n + j, 0 \le c < \gamma, 0 \le i, j < n\}.$$

Each processor $P_{i,j}$ keeps track of which equivalence classes the positions (u,v) are in for $u = c \cdot n + i, v = c \cdot n + j, 0 \le c < \gamma$. Notice that each processor requires $\Theta(\gamma)$ words of storage.

Assuming that each transformation in the symmetry group can be applied to the input data in $\Theta(n)$ time, then it is not difficult to see that the time required by the modified algorithm is γ times that of the original. Finding the largest k such that there is a repeated subarray in an $n \times n$ array of data when symmetry transformations are permitted thus requires $\Theta(\gamma n \log k)$ time.

6   Possible Extensions

The following problems are some that we believe would be interesting, but for which we do not yet know of particularly efficient mesh algorithms.

1. Given a number r of elements, find a subarray containing r elements that occurs most frequently in the input array. For example, if $r = 24$ then we seek a maximally repeated rectangular pattern of dimensions either 1×24, 2×12, 3×8, 4×6, 6×4, 8×3, 12×2, or 24×1.

2. In an $n \times n$ array, find a largest (in number of elements) repeated subpattern of any shape, not necessarily rectangular.

7   References

1. Karp, Richard M., Miller, Ray E., and Rosenberg, Arnold. 1972. Rapid identification of repeated patterns in strings, trees and arrays. *Proceedings of the Fourth A. C. M. Symposium on the Theory of Computing*, pp.125-136.

2. Miller, Russ, and Stout, Quentin F. 1989. Mesh computer algorithms for computational geometry, *IEEE Transactions on Computers,* Vol. 38, No. 3, pp. 321-340.

3. Nassimi, David, and Sahni, Sartaj. 1981. Data broadcasting in SIMD computers, *IEEE Transactions on Computers*, Vol. 30, No. 2, pp. 101-107.

4. Scherson, Isaac D., and Sen, Sandeep. 1989. Parallel sorting in two-dimensional VLSI models of computation. *IEEE Transactions on Computers*, Vol. 38, No. 2, pp. 238-249.

Appendix

Here we restate Lemma 2 of [Karp et al 1972], which is used in the proof of Theorem 1.

Consider the set $S = \{(0,0), (0,1), \ldots, (i,j), \ldots (n-1, n-1)\}$ of positions for the input array X. The equivalence relation $E_{a,b}$ on S is defined as follows. We say $(i_1, j_1)E_{a,b}(i_2, j_2)$ iff $X[i_1, j_1] = X[j_1, j_2]$, $X[i_1, j_1 + 1] = X[j_1, j_2 + 1], \ldots, X[i_1 + a - 1, j_1 + b - 1] = X[j_1 + a - 1, j_2 + b - 1]$. Thus two positions are $E_{a,b}$ equivalent exactly when the $a \times b$ subarrays located there match.

The following lemma tells us that for $c \le a$ and $d \le b$, we can compute $E_{a+c,b+d}$ directly from $E_{a,b}$.

Lemma:

$$(i_1, j_1)E_{a+c,b+d}(i_2, j_2) \text{ iff}$$
$$(i_1, j_1)E_{a,b}(i_2, j_2) \text{ and}$$
$$(i_1, j_1 + d)E_{a,b}(i_2, j_2 + d) \text{ and}$$
$$(i_1 + c, j_1)E_{a,b}(i_2 + c, j_2) \text{ and}$$
$$(i_1 + c, j_1 + d)E_{a,b}(i_2 + c, j_2 + d)$$

Proof: Given two equivalent arrays, any subarray of one is equivalent to the subarray of the same size and relative position in the other. Thus, it is easy to see that equivalence of two $(a + c) \times (b + d)$ arrays implies the four equivalences of the $a \times b$ arrays cited in the lemma.

In the other direction, given the four equivalences of the subarrays of size $a \times b$, the entire $(a + c) \times (b + d)$ arrays must also be equivalent This is clear because the four subarrays cover the whole array due to the assumption that $c \le a$ and $d \le b$. $\square$

0	1	2	3
4	5	6	7
8	9	10	11
12	13	14	15

Row-Major

0	1	4	5
2	3	6	7
8	9	12	13
10	11	14	15

Shuffled Row-Major

0	1	2	3
7	6	5	4
8	9	10	11
15	14	13	12

Snake-Like

0	1	14	15
3	2	13	12
4	7	8	11
5	6	9	10

Proximity Order

Figure 1: Indexing schemes for the processors of a mesh.

A	A	B	B	Y	X	X	X
A	A	B	B	Y	X	X	X
C	C	D	D	Y	X	X	X
A	B	B	D	T	C	C	D
A	B	B	Y	A	A	B	B
Y	X	X	X	A	A	B	B
Y	X	X	X	C	C	D	D
Y	X	X	X	C	C	D	F

Input Array

1	0	0	0	4	3	0	0
0	0	0	0	0	0	0	0
2	0	0	0	0	0	0	0
0	0	0	0	0	2	0	0
0	0	0	0	1	0	0	0
4	3	0	0	0	0	0	0
0	0	0	0	0	0	0	0
0	0	0	0	0	0	0	0

$k = 3$

Output: Equivalence Class Labels and k

Figure 2: Example of equivalence class labels.

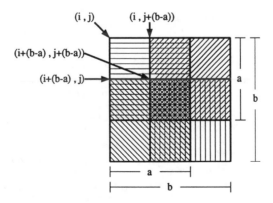

Figure 3: Creating $b \times b$ squares from $a \times a$ squares.

Parallel Processing of Incremental Ray Tracing on a Multiprocessor Workstation

S. HORIGUCHI   **A. KATAHIRA**   **T. NAKADA**

Department of Information Science
Tohoku University
Sendai 980, Japan

IBM Research
Tokyo Research Lab.
Tokyo 102, Japan

Abstract

This paper presents a novel parallel processing method for image synthesis using incremental ray tracing on a real multiprocessor workstation. The most efficient technique for image synthesis is ray tracing, which was proposed by Whitted in 1980. Ray tracing algorithms are very simple and can generate realistic images. However, they are very time-consuming, since intersection calculations between objects and rays increase exponentially as the complexity of scenes increases. Fast image synthesis for animation is one of the most important topics in computer graphics. As the area of computer applications has broadened, the complexity of images to be synthesized has greatly increased. Parallel processing of computer graphics is one way of achieving fast image synthesis. Up to now, parallel processing for image synthesis has been investigated on special-purpose systems developed for computer graphics. This paper describes a parallel processing technique for incremental ray tracing, which recalculates only the rays changed by moving objects in successive scenes of continuous image synthesis. The performance of parallel ray tracing was evaluated on the multiprocessor workstation TOP-1. Strategies for allocating pixels to processes under a multi-process operating system on the multiprocessor workstation are discussed.

Keywords: Multiprocessor Workstation, Computer Graphics, Synthesizing Images, Incremental Ray Tracing

1. Introduction

Computer Graphics[1,2] is very frequently used to generate realistic depictons of three-dimensional objects in various fields. As the area of computer graphics broadens, the need for more complex images and realistic animation is greatly increasing. Whitted[3] proposed ray tracing as a method of synthesizing images in computer graphics. Ray tracing in computer graphics uses a very simple algorithm and can produce very realistic images that include reflection and transmission phenomena on objects. However, it has the disadvantage that calculating intersections between objects and rays requires a huge computation time proportional to the complexity of the scenes. As a result, the execution of image synthesis using ray tracing is limited on conventional computer. A fast processing technique for synthesizing animated images is therefore strongly desired.

As a result of recent developments in VLSI technology, it is now possible to construct multiprocessor systems, which are expected to lead to the next generation of computer systems. Various types of multiprocessor system are under intensive investigation in both industrial and university environments.[4,5] This has created the need for research on parallel algorithms that can exploit the potentially high degree of parallelism in multiprocessor systems.

Fast and efficient parallel algorithms for computer graphics have been implemented on special systems. H. Nishimura *et al.*[6] proposed a parallel image processing system, LINKS-1, for synthesizing each pixel of an image in parallel. They calculated a subset of rays allocating to each processor with an object description of the whole space. LINKS-1 is a tightly coupled multiprocessor with common memory. As the number of processors increases, the interconnection between processors and common memory in LINKS-1 become a bottleneck in parallel processing.

S. Goldwasser[7] proposed a multiprocessor system for generalized object display. The multiprocessor architecture is designed to facilitate the real-time display and manipulation of a single three-dimensional object on a raster-scan video display. J. Woodwark[8] proposed a parallel processing method for generating images on a screen so that each processor has a corresonding region on the screen and merely creates the distributed portions of the images. However, it is impossible to allocate all object description in the object space to individual processors when the number of processors increases beyond a certain point. H. Kobayashi *et al.*[9] proposed a method for parallel processing of objects in image synthesis using ray tracing. In this system, a subspace of an object space is allocated to each processor, which calculates the local intensity on an object in this subspace. The propagation of a ray is realized by interprocessor communication. The performance of this type of parallel processing has only been simulated on a conventional computer.

Up to now, parallel processing for image synthesis has not been sufficiently investigated on general-purpose multiprocessor systems. We investigate parallel incremental ray tracing on a multiprocessor workstation to generate continous realistic animated images.

In the next section, we briefly describe the system architecture and parallel operating system of a multiprocessor workstation with ten processing elements. Parallel incremental ray tracing is discussed in Section 3. In Section 4, the speed-up achieved by using parallel processing for synthesis of images by incremental ray tracing is discussed with reference to the multi-process model in the operating system of a multiprocessor workstation. We also discuss the cache performance of the multiprocessor workstation, which has two different types of snoop protocol: an update protocol and an invalidate protocol. Section 5 presents our conclusions.

2. A Multiprocessor Workstation

Multiprocessor workstations are a very promising means of developing parallel algorithms and of satisfying the continuously increasing need for personal computing power. The multiprocessor workstation TOP-1 (Tokyo Research Parallel Processor-1)[10,11] has ten processing elements, each of which consists of an Intel 80386 and a Weitek 1167 floating-point coprocessor, a 128KB snoop cache, and a system bus interface. The workstation is a shared-memory-type multiprocessor attached to the shared bus, with up to 128MB capacity. All ten processors are completely identical, but one processor is dedicated to hard disk management via the local bus extension, to realize a high-performance, large-capacity hard disk system with 1.2GB capacity.

The unique architectural features are (1) two-way interleaved dual 64-bit buses that are supported by two snoop cache controllers per processor card, (2) a communication and interruption mechanism for notifying the system of asynchronous events, (3) a mechanism that allows two different snoop coherency protocols to coexist in the system and to be changed by software for each memory operation, and (4) an efficient arbitration mechanism that allows prioritized quasi-round-robin service with distributed control. The cache controller contains a set of counters for collecting statistical data such as the cache-hit ratio for each type of memory access, the snoop performance, and the bus performance.

Several commercial multiprocessor systems have been announced and new parallel and distributed operating systems have been implemented on them. Mach[12], an operating system kernel being developed at Carnegie-Mellon University, is intended to support distributed and parallel computing. It separates a typical *process* abstraction into a *task* and *thread*, whereas a UNIX process is effectively a single thread running within a task. A tightly coupled multiprocessor effectively executes multiple threads with a single task by using a multi-thread operating system. The operating system on the multiprocessor workstation TOP-1 is a multi-process model based on shared memory on UNIX System V.

3. Incremental Ray Tracing

One of the biggest problems in computer animations using ray tracing is the enormous number of computations needed to synthesize realistic images. To reduce the computation time, K. Hirota *et al.*[13] proposed an incremental ray tracing method that re-computes only the changed parts of the image for dynamic sequences of images. The key idea behind incremental ray tracing is to localize the influence of changed objects by using subdivided object spaces and to minimize the cost of acess to the data structure. We introduce the concept of a *locus cell* into incremental ray tracing, and extend the method to parallel processing on a multiprocessor.

Figure 1 shows a binary shade tree in conventional ray tracing. Since all the rays have to be calculated when objects in the space are changed, an enormous computation time is necessary to generate sequences of images. Most parts of continuous animated images are not necessary for re-calculating of rays. Figure 2 shows the concept of incremental ray tracing in two-dimensional object space, for ease of understanding. The solid lines correspond to the rays to be calculated by a moving object in object space. The dotted lines show the rays that can use the previous calculation result from the last image. Since the

orbits of moving objects are given in the whole space in order to synthesize images, we can estimate the object subspace that will influence the rays in the next image. We define the subspace through which the objects are moving as a *locus cell*. Locus cells include all the rays that need to be re-calculated by moving objects in the next image. We can determine which rays need to be reserved by using the locus cells in the improved incremental ray tracing method. The overhead of determining the rays to be re-calculated is not very large in relation to the number of calculations of intersections between rays and objects. This is why we introduce improved incremental ray tracing instead of conventional ray tracing for synthesis of animated images.

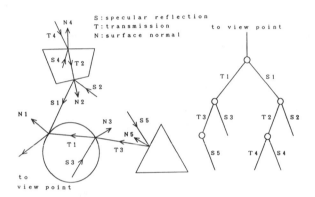

Fig.1. Binary shade tree of the ray tracing method

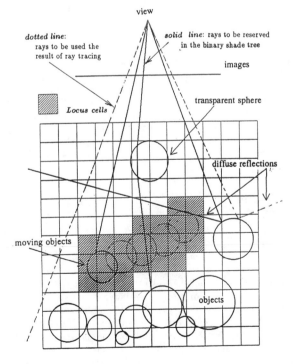

Fig.2 *Locus cells* and reserved rays in incremental ray tracing.

4. Performance of Parallel Incremental Ray Tracing

4.1 Pixel Allocation Method for Processes

Most pixels of the screen are not necessary for re-calculating of the binary shade tree of a ray. In parallel incremental ray tracing, the method of allocating pixels to processes is the main factor determing the performance of a parallel system. To achieve highly effective parallel ray tracing, a pixel allocation strategy is indispensable. It is necessary to distribute the load of calculation on processes uniformly. We implemented three pixel allocation methods on the multiprocessor workstation. Figure 3 shows these three static methods: dot allocation, sequential allocation, and block allocation. In the figure, the integers from 1 to 16 correspond to the identical number of processes. In Figure 3, the three strategies are used to assign an 8 x 8 pixel-image to 16 processes.

We implemented several test images for estimating the performance of parallel incremental ray tracing on the multiprocessor workstation TOP-1. In this paper, we used two test images. A test image is named MVSTS, in which a group of four spheres is moving over a background consisting of 121 spheres. Another test image is named ORBIT3, in which three groups, each consisting of four spheres, are moving around a large diffuse sphere. We evaluate the performance by using two test images: MVTST and ORBIT3.

Figure 4 shows the speed-up ratio for the execution time of one process as a function of the number of processes. Figure 4 (a) shows the speed-up ratio of the first frame of the continuous test image MVTST. The speed-up ratio is almost the same in the three allocation methods, since all the pixels of the image in the first frame have to be calculated by parallel incremental ray tracing. Figure 4 (b) shows the average speed-up ratio of the next ten frames after the second frame. In the block allocation method, the speed-up ratio is very small and is not proportional to the number of processes. Since most of the rays need not be recalculated by incremental ray tracing, the calculation load on each process is not uniform. Both sequential and dot allocations give a linear speed-up proportional to the number of processes. Figure 5 shows the speed-up ratio of the test image ORBIT3. In this case, almost the same results are achieved. It is clear that the execution time is dominated by the strategy of pixel allocation, and that both dot allocation and sequential are superior to block allocation in terms of balancing the load of parallel incremental ray tracing.

4.2 Memory Capacity

To realize parallel incremental ray tracing, a large memory capacity is necessary for defining the object space and reserving the binary shade tree. We discuss the memory capacities required by three types of object definition strategy in a parallel computation environment. In the first, the object definition is shared by all processes. The object difinition is located in the shared memory to be accessed by all processes. In the second, the object is defined on all proesses independently. In the third, a common part of the object is shared by all processes. Figure 6 shows the total memory capacity required by parallel incremental ray tracing for 160 x 160 pixels of the test image ORBIT3. The memory capacity required to reserve the binary shade tree of rays is constant and occupies most of the total memory capacity of parallel incremental ray tracing, as shown in the figure. The

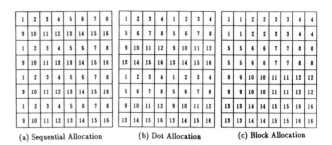

(a) Sequential Allocation   (b) Dot Allocation   (c) Block Allocation

Fig.3. Three static methods of allocating pixels to processes
    (a) Sequential Allocation
    (b) Dot Allocation
    (c) Block Allocation

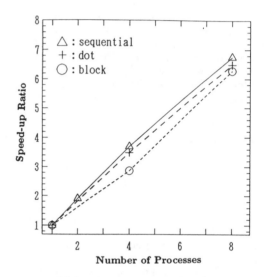

(a) Speed-up ratio of the first frame

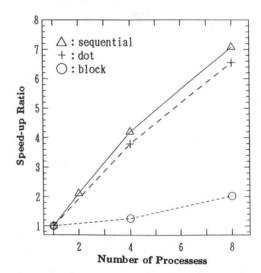

(b) Average speed-up ratio of images after the second frame

Fig.4. Speed-up ratio of pixel allocation methods as a function of the number of processes in the test image MVTST

total memory capacity depends on the strategies for defining of object space. Since most of the memory capacity is occupied by the binary shade tree, the total capacity is not so different in the three strategies of object definition. The execution times of the three strategies are almost the same for both test images.

4.3 Cache Performance

A number of snoop cache protocols have been proposed, but none is considered to be suitable in every situation. The TOP-1 cache system has two major types of operation for maintaining consistency: an *update* protocol and an *invalidate* protocol. The cache system has a unified size of 128 KB. The cache mode is specified by *a cache mode register* in the cache bus controller. Since this register is I/O mapped, each CPU can change the content of the cache mode register by an I/O instruction. The cache block is controlled by five states: *invalid, clean-private, clean-shared, dirty-private*, and *dirty-shared*. The state transition diagram of the TOP-1 protocol is shown in Fig. 7. On the shared bus, there is a signal line called Cache Hit (CH), whcih indicates a snoop hit. When a processor causes a read miss, the cache controller takes the shared bus and puts the read command onto the shared bus. If another cache has a copy, it activates the CH. The reading cache controller can determine whether the cache block is shared or not from the CH. When a processor writes to the shared block, the cache controller gets the shared bus and supplies the address and data. When the snoop write comes in, the cache controller checks the corresponding tag of the cache block. In the write-update mode, if the snoop hits in

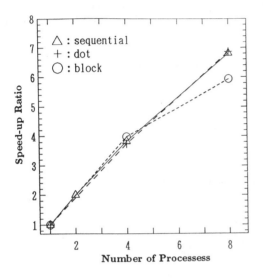

(a) Speed-up ratio of the first frame

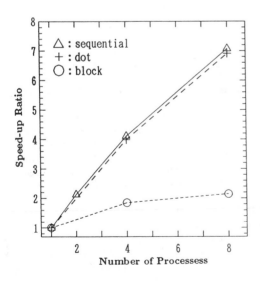

(b) Average speed-up ratio of images

Fig.5.

Speed-up ratio of pixel allocation methods as a function of the number of processes in test image ORBIT3

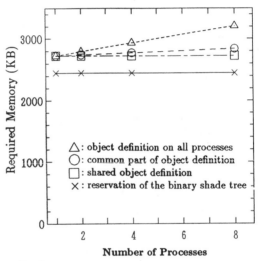

Fig.6. Total memory capacities required by object definition and the binary shade tree to be reserved in the test image OR-BIT3

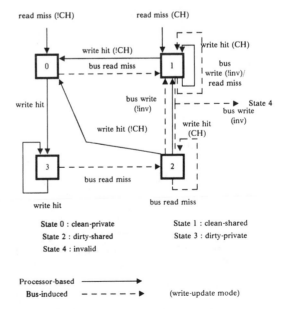

Fig.7. Semantic structure of cache protocols of the multiprocessor workstation

the cache, the cache controller updates the cache block with the latest value and activates the CH. In the write-invalidate mode, if the snoop hits in the cache, the cache controller invalidates the cache block and does not activate the CH. The cache block state of the CPU, which does the shared block write, is changed to *private* if the CH is not activated, or to *shared* if the CH is activated, as shown in Fig. 8.

The update protocol updates the cache line in other caches if the line is shared and the write operation is performed on the line in the cache. It is thus similar to Xerox's Dragon protocol. Update protocols generally yield a good performance in situations with a small number for shared blocks and high contentions for shared variables such as semaphores. The invalidate protocol invalidates all shared lines in other caches when a write occurs, as in Berkeley's SPUR. This is suited to situations with a large number of shared blocks and low contentions for them. Invalidate protocols are also preferred when a process is migrated from one processor to another, and as a result, blocks belonging to that process are shared between two caches, even if the blocks include only local data.

Fig. 8 shows the speed-up ratio and cache hit ratio as functions of the number of processes in the test image ORBIT3. In this case, the update cache protocol is adopted. The speed-up ratio is almost linear as a function of the number of processes. The instruction cache hit ratio is more than 99% and the data hit ratio is greater than 96.7% for 100 x 100 pixels of the test image ORBIT3. As a result of the high cache hit ratio, effective parallel performances are obtained on the multiprocessor workstation. The difference between the cache hit ratio for the update protocol and that for the invalidate protocol is very small in parallel incremental ray tracing on the multiprocessor.

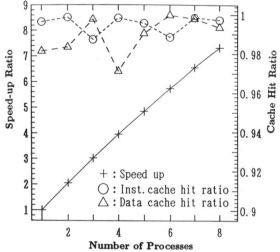

Fig.8. Speed-up ratio and cache hit ratio as a function of the number of processes in the test image ORBIT3.

5. Conclusions

There is currently a strong need in various fields for fast and efficient parallel processing for image-synthesis on multiprocessors. We actually implemented a parallel incremental ray tracing method on the TOP-1 multiprocessor workstation. Experimental evaluation proved that the performance is affected by the methods of allocating pixel to processes. It is clear that parallel incremental ray tracing realizes fast processing for synthesizing continuous images. A high cache hit ratio is obtained in the execution of parallel incremental ray tracing on a multiprocessor workstation.

Acknowledgment The authors would like to thank to Dr. N. Suzuki and Dr. S. Simizu, Tokyo Research Laboratory, IBM Japan, Ltd., for support in this research. This research is partly supported by a Grant of Aid, No. 02805042 from Ministry of Culture and Science of Japan.

References

[1] J. D. Foley and A. V. Dam:"Fundamentals of Interactive Computer Graphics," *Addison-Wesly Pub. Co., Inc.* (1982).

[2] A. Fujimoto *et al.*:"ARTS: Accelerated Ray-Tracing System," IEEE CG & Applications, Vol. 6, No. 4, pp. 16-26 (1986).

[3] T. Whitted:"An Improved Illumination Model for Shaded Display," CACM, Vol. 23, pp.343-394 (1980).

[4] R. W. Hockney and C. R. Jesshope:"Parallel Computer 2," *Adam Hilger* (1988).

[5] J. Beetem *et al.*:"The GF11 Supercomputer," *Proc. Annual Int'l Symp. on Computer Architecture*, pp.108-115 (1985).

[6] H. Nishimura *et al.*: "LINKS-1:A Parallel Pipelined Multimicrocomputer System for Image Creation," *Proc. Annual Int'l Symp. on Computer Architecture*, pp.387-394 (1983).

[7] S. M. Goldwasser: "A Generalized Object Display Processor Architecture," *Proc. Annual Int'l Symp. on Computer Architecture*, pp.38-45 (1984).

[8] J. R. Woodwark: "A Multiprocessor Architecture for Viewing Solid Models," *Display* pp.97-103 (April 1984).

[9] H. Kobayashi *et al.*: "Effective Parallel Processing for Synthesizing Continuous Images," *Proc. of CG Int'l* pp.343-352 (1989).

[10] S. Shimizu, N. Ohba, A. Moriwaki, T. Nakada, and M. Ohara: "TOP-1: A High-Performance Multiprocessor Workstation" (in Japanese), *Proc. of JSPP Information Processing Society of Japan*, pp.155-162 (Feb. 1989).

[11] S. Horiguchi and T. Nakada: "Experimental Performance Evaluation of Parallel Fast Fourier Transform on a Multiprocessor Workstation," *Proc. of Int'l on Parallel Processing* No.3, pp.97-101 (August 1990).

[12] A. Tevanian and R. F. Rashid: "MACH: A Basis for Future Unix Development," *CMU-CS-87-139* (Jun. 1897).

[13] K. Hirota and K. Murakami: "Incremental Ray Tracing (in Japanese)," *Graphics and CAD, JSIP*, vol.27, No.5, pp.1-8 (July 1987).

Computer Graphics Rendering on a Shared Memory Multiprocessor

Scott Whitman *
P. Sadayappan

Department of Computer & Information Science
The Ohio State University
2036 Neil Avenue Mall
Columbus, Ohio 43210

Abstract

This paper encompasses an analysis of parallel methods used to accelerate 3-D computer graphics rendering. Several methods are presented for comparison purposes and their performance analyzed on a 96 processor BBN Butterfly GP1000 shared memory computer. Load imbalance, network contention, communication costs, and code modification overheads are quantified for each of the implemented approaches. A memory referencing scheme is implemented which takes advantage of the local memory resident on each processor board to allow access to the graphics database. Based on performance analysis, a technique known as the task adaptive decomposition strategy represents the best parallel graphics algorithm which attains high performance with minimum setup cost.

Keywords: Scanline, BBN GP1000, parallel processing, three-dimensional, load balancing

1 Introduction

One of the goals of software developers involved in the field of computer graphics rendering has been to derive new methods that can be used to solve the computer image generation problem in a minimal amount of time. Recent techniques used in solutions to this problem involve the application of parallel processing by having a number of processing elements (PEs) work on the problem simultaneously. Simulations done by [12], [13], [8], and [6] indicate the potential for high performance in parallel graphics rendering. Specialized implementations by [5], [16], and [14] achieved reasonable performance although these approaches were not thoroughly analyzed nor did they involve any comparison to other designs. The work presented here concentrates on several software algorithmic approaches to solve this problem on commercial multiprocessors. We have chosen to concentrate on image space rendering methods since these are typically more efficient than either object space methods or ray tracing. For those who are interested in the features that either of these latter two methods has to offer, see [7] and [1] regarding object space parallel algorithms and Badouel's[2] paper for a description of a parallel ray tracing method.

In this research, we focused on the tiling (hidden surface removal, illumination, and anti-aliasing) portion of the graphics display algorithm although parallel approaches were used to speed up both the front end (reading in, transforming of data) as well as the back end (writing out the image) portions of the program.

*Current address: Lawrence Livermore National Laboratory, P.O. Box 808, L-416, Livermore, California 94550

2 Performance Evaluation

The rendering method used here incorporates stochastic sampling for anti-aliasing[4] using the Blinn[3] shading model on four separate test images. In order to allow a comparison to other renderers, several test images were used from the SPD database[10]. The tree from this database was generated with approximately 106,000 polygons while the mountain was generated with 131,000 polygons. We also wanted to test the algorithms on some real world type data so we used a stegosaurus image which was designed at The Ohio State University for an animation as well as a Chrysler Laser automobile which was designed from a CAD/CAM program. The stegosaurus contained approximately 10,000 polygons while the automobile contained approximately 46,000 polygons[1]. These images are illustrated in figure 1.

A number of factors cause less than ideal speedup in practice on a parallel architecture. In order to quantitatively understand the effect of these factors on achieved performance, a number of supplementary measurements and evaluations were made. The factors to be considered are: scheduling, memory latency, communication utilizing block transfers, memory access contention, overhead due to adaptation of the algorithm for parallel execution, and synchronization. Due to space limitations, the details on how these factors were calculated could not be included in this document. See [18] for more information.

3 Approaches to Parallel Image Space Decomposition

In this section, a number of parallel decomposition schemes are described. In all of the schemes, tasks are assigned to regions on the screen. The schemes differ in the approach to determining the size and number of these regions as well as the memory referencing strategy. Graphical coherence, as described in [15], allows one to use previously derived values in order to speed up a sequential computation. The independent parallel generation of various parameter values used in a graphics display algorithm implies redundant recomputation and the loss of graphical coherence. The tradeoff between parallelism and coherence is an important issue which is investigated herein. A scanline z-buffer algorithm is used as the hidden surface removal method within a single task since it is general purpose and sufficiently fast.

[1]The stegosaurus data was created by John Donkin of ACCAD (the Ohio State University Advanced Computing Center for the Arts and Design) for the Fernbank Museum in Atlanta. The car was created by Chrysler and obtained from Evans & Sutherland Computer Corporation.

Figure 1: Stegosaurus, Laser, Tree and Mountain

Table 1: Scanline algorithm degradation factors, Laser image

Latency	Code Modification	Contention	Load Imbalance
4.6%	7.5%	17.9%	6.3%

3.1 Data Non-Adaptive Partitioning

The data non-adaptive partitioning scheme relies on sub-dividing the image space regardless of the screen location of the input data. This method employs a dynamic scheduling mechanism whereby tasks are scheduled onto processors as each processor is available for work. The method of referencing remote data in the first two algorithms is denoted the Uniformly Distributed or UD scheme. The basic idea is that shared data is scattered throughout the memory modules in the system and reference remotely in a uniform manner.

3.1.1 Scanline Decomposition

A scanline decomposition is probably the most natural parallel partitioning scheme and it was first suggested by Hu and Foley in their paper describing a hardware parallel rendering machine[11]. A scanline is a single line of pixels (picture elements) on the screen in a frame buffer display. The basic idea involves partitioning an image such that each scanline is a task in itself (#tasks = #scanlines). Hu and Foley's research showed that dynamic assignment of single scanlines to processors results in better performance than interleaving groups of successive scanlines statically.

The percentages for each of the major overhead factors as measured with the Laser test image on 96 processors of the BBN GP1000 are given in table 1. Scheduling overhead is so small in comparison to the other factors that it was not included in this table. The Laser image was chosen as a representative example for this data although similar results could be derived for each of the other images (see [18] for more detailed results).

Lack of vertical scanline coherence is the primary contributor to the overhead in adapting this algorithm for parallel processing. This is manifest as the total degradation due to code modification. Network contention is a major contributor to performance degradation and increases as a function of the number of processors. If we can maximize coherence without sacrificing load imbalance and still achieve parallelism, we can reduce some of these factors

Table 2: Degradation factors for rectangular region decomposition, UD Scheme

Latency	Code Modification	Contention	Load Imbalance
2.6%	8.3%	15.6%	7.3%

significantly. This is achieved in the rectangular region scheme as shown below.

3.1.2 Rectangular Region Decomposition

The rectangular region decomposition algorithm is a generalization of the scanline decomposition algorithm. Instead of using single scanlines as wide as the screen for each task, a small group of contiguous scanlines is designated as a single task. Each task is the same size and constitutes a small area of the screen. The number of these areas can be increased (and their size therefore decreased) resulting in a larger number of tasks per processor in an effort to achieve good load balancing. One advantage of the rectangular region algorithm over the scanline approach is that a fixed granularity ratio is used in the former scheme. This allows better load balancing especially if the program is run on larger processor configurations; thus its scalability is superior to the parallel scanline approach. A comparison of the major degradation factors is given in table 2.

Although this approach achieves a reduction in some of the overhead factors when compared to the parallel scanline algorithm, it still has significant contention problems. A method of alleviating this problem is to minimize the use of the interconnection network. A description of how this can be done is given below.

3.1.3 Rectangular Region (LC Scheme)

This algorithm is implemented in exactly the same way as the last one, with the only exception being the remote memory referencing strategy. A brief description of this strategy, denoted the Locally Cached or LC scheme, follows. Instead of referencing globally shared data remotely, the data is cached into the local memory module prior to referencing, thus allowing data to be accessible locally. Although others have used an elaborate software caching mechanism for graphics rendering ([9] and [2]), we rely on the fact that the exact data needed for a given task can be copied directly to each processor prior to the computation of a given region. If a data element is relevant to more than one processor, it is copied to all processors which would reference it, incurring a space penalty. Although extra memory is required, a tradeoff of space versus time is necessary to achieve faster memory referencing than the previous Uniformly Distributed (UD) scheme. The cost of non-local memory access is minimized by block transferring data from its global storage location to the local memory of the processor(s) that need it.

The LC scheme is more than just a "block copy then local reference" scheme. It consists of a complicated set of operations which involve constructing data structures for block copying of data, using the hardware block transfer function built into the GP1000. The data is set up so that those polygons which are relevant to a particular region on the screen are put into a contiguous array for that region. This allows later block copying and reduces the amount of space needed for the polygons. In an image space algorithm, the location of the polygons in screen space is known prior

Table 3: Degradation factors for rectangular region decomposition, LC Scheme

Communication	Code Modification	Contention	Load Imbalance
0.04%	6.4%	8.7%	6.6%

Table 4: Degradation factors for top-down decomposition, LC scheme

Communication	Code Modification	Contention	Load Imbalance
0.03%	3.3%	22.7%	6.7%

to tiling, thereby facilitating this type of approach. The block transfer copying mechanism uses the interconnection network for a very short amount of time, thus reducing overall communication and contention. The results in the graph below bear this out in comparison to the previous remote referencing scheme. Table 3 indicates the overheads for the Laser image for this scheme. Based on the advantages of the LC Scheme, we have decided to implement the remaining partitioning methods using this memory referencing strategy.

3.2 Data Adaptive Partitioning

In a data adaptive algorithm, load balancing is achieved by constructing tasks which are estimated to take nearly the same amount of time. By using image space partitioning in a parallel graphics rendering program, tasks can be determined based on the location of data within the image. Although there are many different decomposition methods that fall under the data adaptive category, we have chosen one algorithm as a representative example for implementation. Whelan[17] and Roble[14] used similar data adaptive schemes in their work but found that the overhead required in utilizing a sophisticated heuristic was too great. The solution presented below is based on a simple heuristic which does not cost much in terms of computation time. The heuristic in this algorithm is based on the assumption that the number of polygons in a region is linearly related to the time it takes to tile that region. Using this simple heuristic, good load balancing can be achieved with minimal overhead.

The basic idea involves creating a dense rectangular mesh in which polygons are placed according to their screen space location (similar to the rectangular region decomposition). In this case, however, clusters of the mesh regions are formed and the number of polygons in each cluster are counted to note their screen space location. This creates a tree of clustered regions.

After the tree is created, it is traversed in top-down fashion and the area with the most polygons at a given point is then split into its two components. The splitting process is stopped when the desired number of tasks has been reached. The count of the number of polygons in each small area is used so it is not necessary to sequentially go through the entire list of potential polygons to determine which polygons are relevant to each area at this time. The limiting factor in the splitting is the leaf level, which is why a fairly dense mesh is created at the beginning. When the tree has been traversed, each of the regions is now available for rendering in parallel. The graph in table 4 shows the overhead comparison.

The code modification overhead is reduced in this scheme since

fewer regions are used in comparison to the rectangular region methods and coherence is maintained within these regions. The load balancing is fairly good for all processor configurations which indicates that the heuristic that is used is adequate. The reason for the increase in network contention is primarily due to the method of communication required to facilitate data transfer in this algorithm. The LC scheme requires communication from each of these small regions which form the larger clusters in order to obtain the data necessary for rendering a particular task area. There may be many small regions which are part of a single large cluster, so communication required to obtain data from each one of these regions. Although the total amount of data is not large which is evident by the communication factor shown in the graph, the number of messages is significantly higher than in the rectangular region algorithm resulting in a greater chance of collisions in the network.

3.3 Task Adaptive Partitioning

One of the problems with the algorithms discussed thus far is that they rely on a good choice for the granularity ratio. Instead of using empirical tests to determine this ratio, we can just have each processor start off with one task (thus R=1) and dynamically partition tasks during runtime to even out the load. This reduces the overhead in the front end portion of the program. In the front end, polygons are placed into regions according to their screen space location since there are less regions to start off with than in the other algorithms. After some time, a given processor will have finished its initial area and have no work left. The dynamic work partitioning then proceeds as follows:

1. When a processor needs work (call this processor P_s), it searches among the other processors for the one which contains the most amount of work left to do (call this processor P_{max}).

2. The P_s processor then sets a lock preventing any other processor from splitting P_{max}.

3. P_s partitions P_{max}'s work into two segments; the first segment goes to P_{max} and the second segment goes to P_s. In the case here, the first segment is the top portion of the rectangular area that P_{max} was already working on (thus maintaining coherence) while the second segment is the bottom portion.

4. P_s then copies the data necessary for it to work on the second segment without interrupting P_{max}.

5. P_s unsets the lock and starts doing its work.

In order to facilitate this approach, it was necessary to come up with a method for determining the amount of work a given processor has left to do. Since all of the initial areas are the same size, the number of scanlines left to render in a particular area is used as an indication of how much work there is left on a given processor. The results indicating the primary overhead contributors is shown in table 5.

Synchronization is an additional overhead in this algorithm, but it was not a significant factor in performance degradation. The communication cost in this algorithm is somewhat larger than the other LC schemes, due to the dynamic partitioning of this particular dataset. The code modification here is the smallest of all the

Table 5: Degradation factors for task adaptive algorithm

Comm.	Code Mod.	Contention	Ld. Imbal.	Synch.
4.0%	1.6%	9.8%	10.2%	2.2%

Table 6: Absolute speedup of algorithms on 96 processors on BBN Butterfly

Algorithm	Stegosaurus	Laser	Tree	Mountain
Scanline	57.8	60.8	70.6	56.5
Rect. - UD	53.1	65.1	68.4	71.9
Rect. - LC	68.4	75.3	77.4	83.4
Top-Down	58.4	68.4	74.6	79.0
Task Adaptive	69.7	72.6	68.1	82.2
Task Adapt. Time	9.89 sec	18.01 sec	24.35 sec	39.34 sec

algorithms since the number of areas generated is initially equal to P. In addition, coherence is maintained in the upper portion of a split area reducing the parallel execution overhead. Network contention was about the same as the rectangular region LC approach. Toward the end of the computation when dynamic load balancing is taking place, there is a flurry of communication and this causes network contention to increase at this point. The burst of communication is due to the dynamic splitting of small tasks at the end of the computation.

4 Discussion

Based on the data shown in these graphs, it can be seen that the task adaptive algorithm utilizing the locally cached (LC) memory referencing scheme is clearly superior for all of the images. This algorithm requires less regions at the beginning of the program than any of the other algorithms. The overhead time for the front end portion of this algorithm is small for this algorithm in comparison to the others. The rectangular region algorithm, which is slightly faster for some images in the tiling section only, requires significantly more setup prior to tiling, degrading overall performance. One might have thought that the additional time required to set up the LC scheme would outweigh its benefit in total algorithmic performance but this turned out not to be the case. Table 6 includes speedup of the algorithms on 96 processors for the different test images. Note that this is absolute speedup, that is, it is based on the "best" sequential time so that all the algorithms can be compared equally to each other. Also, below the values for each algorithm is the parallel time for the task adaptive algorithm on 96 processors to give the user an idea of the rendering time involved.

Since the BBN Butterfly is a non-uniform memory access (NUMA) machine, local memory referencing proceeds approximately an order of magnitude faster than remote referencing. With several of the algorithms, the overhead due to contention begins increasing rapidly and non-linearly with an increase in the number of processors, especially beyond 48 processors. The benefits of using the the Locally Cached memory referencing scheme become very significant after this point since this scheme provides sufficiently reduced communication and network contention overheads compared to a global referencing strategy.

Acknowledgments

This research has been supported in part by the Department of Computer and Information Science and the Ohio Supercomputer Center. Rick Parent contributed to numerous discussions and evaluations of this research effort. D. Jayasimha helped with a critique of the majority of this work. Finally, thanks are due to BBN Advanced Computers, Inc., John Price, and in particular Ed Forbes, whose countless hours of support allowed us to run tests on the BBN ACI corporate headquarters' GP1000.

References

[1] Abram, Greg *Parallel Image Generation with Anti-Aliasing and Texturing*, Ph.D. dissertation, University of North Carolina at Chapel Hill, 1986.

[2] Badouel, Didier, Bouatouch, Kadi, and Priol, Thierry "Ray Tracing on Distributed Memory Parallel Computers: Strategies for Distributing Computations and Data." Course Notes for Course 28, Siggraph (1990) pp. 185 - 198.

[3] Blinn, James F. "Models of Light Reflection for Computer Synthesized Pictures." *Computer Graphics, Proceedings of Siggraph 11*(1977) pp. 192-198.

[4] Cook, Robert L. "Stochastic Sampling in Computer Graphics." *ACM Transactions on Graphics* (January 1986).

[5] Crow, Franklin C. , Demos, Gary, Hardy, Jim, McLaughlin, John, and Sims, Karl "3D Image Synthesis on the Connection Machine." *Proceedings of the International Conference on Parallel Processing for Computer Vision and Display* (January 1988), Leeds, UK.

[6] Fiume, Eugene, Fournier, Alain, and Rudolph, Larry "A Parallel Scan Conversion Algorithm with Anti-Aliasing for a General Purpose Ultracomputer." *Computer Graphics, Proceedings of Siggraph 17*, 3 (July 1983) pp. 141-149.

[7] Franklin, Wm. Randolph and Kankanhalli, Mohan S. "Parallel Object-Space Hidden Surface Removal." *Computer Graphics, Proceedings of Siggraph 24*, 4 (August 1990) pp. 87-94.

[8] Ghosal, Dipak and Patnaik, L. M. "Parallel Polygon Scan Conversion Algorithms: Performance Evaluation of a Shared Bus Architecture." *Computers & Graphics 10*, 1 (1986) pp. 7-25.

[9] Green, S. and Paddon, D. "Exploiting Coherence for Multiprocessor Ray Tracing." *IEEE Computer Graphics and Applications* (November 1989) pp. 12-26.

[10] Haines, Eric "A Proposal for Standard Graphics Environments." *IEEE Computer Graphics & Applications 7*, 11 (November 1987) pp. 3-5.

[11] Hu, Mei-Cheng and Foley, James D. "Parallel Processing Approaches to Hidden-Surface Removal in Image Space." *Computers & Graphics 9*, 3 (1985) pp. 303-317.

[12] Kaplan, Michael and Greenberg, Donald P. "Parallel Processing Techniques for Hidden Surface Removal." *Computer Graphics, Proceedings of Siggraph* (July 1979) pp. 300-407.

[13] Parke, Frederic I. "Simulation and Expected Performance Analysis of Multiple Processor Z-Buffer Systems." *Computer Graphics, Proceedings of Siggraph 14*, 3 (July 1980) pp. 48-56.

[14] Roble, Doug R. A Load Balanced Parallel Scanline Z-Buffer Algorithm for the iPSC Hypercube. In *Proceedings of Pixim '88*, Paris, France, October 1988.

[15] Sutherland, Ivan E. , Sproull, Robert F. , and Schumacker, Robert A. "A Characterization of Ten Hidden-Surface Removal Algorithms." *Computing Surveys 6*, 1 (March 1974).

[16] Theoharis, Theoharis A. Exploiting Parallelism in the Graphics Pipeline. Tech. Rept. Technical Monograph PRG-54, Oxford University Computing Laboratory, June, 1986.

[17] Whelan, Daniel S. *Animac: A Multiprocessor Architecture for Real-Time Computer Animation*, Ph.D. dissertation, California Institute of Technology, 1985.

[18] Whitman, Scott *Utilizing Scalable Shared Memory Multiprocessors for Computer Graphics Rendering*, Ph.D. dissertation, The Ohio State University, Columbus, Ohio, 1991.

Parallel Discrete Event Simulation
Using Space-Time Memory

Kaushik Ghosh and Richard M. Fujimoto *

College of Computing
Georgia Institute of Technology
Atlanta, GA 30332

April 11, 1991

Abstract

An abstraction called space-time memory is discussed that allows parallel discrete event simulation programs using the Time Warp mechanism to be written using shared memory constructs. A few salient points concerning the implementation and use of space-time memory in parallel simulation are discussed. It is argued that this abstraction is useful from a programming standpoint for certain applications, and can yield good performance. Initial performance measurements of a prototype implementation of the abstraction on a shared-memory multiprocessor are described, and compared with a conventional, message-based implementation of Time Warp.

1   Introduction

Sequencing problems in parallel computations arise due to data dependence relationships that must be satisfied for the computation to be correct. While *conservative* synchronization mechanisms rely on blocking to *avoid* violations of dependence constraints, *optimistic* methods rely on detecting synchronization errors at runtime, and recovering using a *rollback* mechanism. Here, we are concerned with extensions to the optimistic Time Warp mechanism [Jef85].

Synchronization plays an especially important role when executing discrete event simulation programs on a parallel computer because these simulations are often irregular, and exhibit highly data dependent behavior.

In discrete event simulations, the computation consists of a number of separate *event computations*, where each event has a timestamp to denote the occurrence of some change in the state of the system being simulated. The parallel computation should yield the same results as if the events were processed sequentially in non-decreasing timestamp order.[1] Several successes have been reported in using the Time Warp mechanism to parallelize discrete event simulation programs in a variety of applications [Fuj90].

Most existing parallel discrete event simulation mechanisms assume a process-oriented view where the simulation is assumed to consist of a collection of logical processes that communicate *exclusively* by exchanging timestamped event messages [Fuj90]. This paradigm forbids the use of shared memory to hold state variables, for reasons that will be discussed later. In this paper, we investigate extensions to the Time Warp mechanism that allow the use of shared state. We argue that for certain applications, shared memory abstractions are more natural from a programmability standpoint than message-only communication.

The space-time memory abstraction supports the use of shared state for parallel discrete event simulation programs. Unlike conventional memory that is viewed as a linear array of values that are accessed using a single, *spatial* coordinate, space-time memory is a two-dimensional structure that is addressed using both a spatial and a temporal (i.e., a simulated time) coordinate.

There has been some work based on computation models utilizing shared-memory. Jones was perhaps the first to propose such an approach [JCRB89, Jon86]. Our work differs from his in that he utilizes a conservative simulation protocol, while we use an optimistic one. Use of a conservative protocol avoids the need for multiple versions of state variables, but at the expense of lost concurrency, and the necessity of relying on application specific information to determine which simulator events can be executed concurrently.

More closely related to our work is the space-time simulation method proposed by Chandy and Sherman [CS89]. As in our work, they view the simulator state as a two-dimensional space-time graph. However, they then partition this graph into regions, and assign each region to a process that is responsible for computing the values of state variables in that region. The computation proceeds until a fixed point is achieved. Reiher et. al. use a similar "temporal" decomposition for load management purposes [RBJ91]. The mechanism that we use does not rely on processes, though processes can be (and have been) added where desired. The underlying simulation mechanism used here is event-oriented rather than process-oriented.

Finally, the space-time memory abstraction discussed here was originally proposed in [Fuj89b]. There, space-time memory is used in the context

*This work was supported by Innovative Science and Technology contract number DASG60-90-C-0147 provided by the Strategic Defense Initiative Office and managed through the Strategic Defense Command Advanced Technology Directorate Processing Division, and by NSF grant number CCR-8902362.

[1]Here, we ignore the possibility of distinct events containing the same timestamp.

of a parallel computer architecture that utilizes rollback for synchronization. Here, we discuss implementation of the abstraction on a conventional general purpose shared-memory multiprocessor, and report the performance of our implementation.

2 Rationale Behind Space-Time Memory

Discrete event simulation programs utilize program variables to model the state of the system, and timestamped events that model changes in system state. It is also convenient to use logical processes to model certain components of the simulation that persist from one event to another.

For many simulations, it is convenient and natural to utilize state variables that can be accessed by distinct logical processes. For example, consider a combat simulation [WJ89]. Assume that there are two armies, each consisting of some number of combat units, fighting on some terrain. In such simulations, the battlefield is usually partitioned into an array of grid cells in order to capture the notion of physical proximity between combat units; combat units that are near to each other in the actual system correspond to logical processes that "reside" in neighboring grid cells in the simulation. A common operation performed by a combat unit is to examine the number and strength of units in neighboring grid cells, and then perform some appropriate action, e.g., attack or retreat. This suggests that the grid data structure that indicates where the combat units reside is shared among the various logical processes (combat units).

In a sequential simulator, this is trivially implemented by declaring the grid data structure to be a global variable. However, consider a parallel simulator; one cannot simply map the global variables into the shared memory of the multiprocessor. This is because combat unit C_1 at simulated time T_1 expects to see the state of the system *as it existed at time* T_1. However, another combat unit C_2 at simulated time T_2 expects to see the system as it existed at time T_2. How can the state variables simultaneously accommodate both of these processes?

We accommodate both processes through the abstraction of space-time memory (STM). While conventional memory is viewed as a one-dimensional array of values addressed with a *spatial* coordinate (e.g., a word address), space-time memory is two-dimensional, and addressed with both a *spatial* and a *temporal* (simulated time) coordinate. With space-time memory, an event at time T will ask to read the value of a state variable *as it existed at simulated time* T. A process at simulated time T is afforded a global view of the system as it existed at simulated time T, exactly the same as in the sequential simulator.

Of course, one can achieve the same effect in a conventional, message-based implementation of Time Warp by defining a logical process to imple-

ment each sector of the grid data structure, and scheduling "read" and "write" events each time an access to a shared state variable is required. This, however, may lead to poor performance because the overhead associated with scheduling an event and waiting for the corresponding reply is high relative to the simple memory reference that is required in the sequential simulator.

Suppose the memory contents of the logical process (henceforth termed LP) D_1 require updates by LPs $S_1, S_2, \cdots S_n$. The memory contents of the appropriate state vector of D_1 would then requires updates by all of the messages sent by the S_is. Further, suppose that the state variables of D_1 have to be read by the S_is before an update. In such a scenario, the S_is would have to send messages to D_1 requesting the contents of D_1's state vector, D_1 would then reply with messages to each of the S_is with the contents of its state vector, the S_is would then send back messages with updated values, and D_1 would have to update its state vector after processing each of these messages. This type of communication, often called *pull processing* [WJ89], produces many messages.

Another method of communication without the use of space-time memory would be to replicate shared data at the several LPs that may need such data. This would require no message passing for reading shared data. However, modification of shared data would require updates at all LPs that hold a copy of the data. In general, such *push processing* [WJ89], reduces the number of messages required for communication, and alleviates serial bottlenecks at the LP whose memory contents are being read, but it substantially complicates the coding of the application because of the need to keep replicated data up to date.

The use of shared objects affords a clean abstraction whereby data is available without the need for explicit message passing. (An implementation on a message-based machine like a hypercube, however, would need to pass messages to afford this abstraction.) Our protocol for reads and writes gives priority to events with lower timestamps, which tends to reduce the number of rollbacks. The copying of data from the previous version (required in a write) appears in pull/push processing as copying message contents from the state vector of one LP to another. Thus, like pull processing, space-time memory allows simpler programming than push processing, but it can be expected to achieve better performance than pull processing.

It should be noted that the STM abstraction is not restricted to shared-memory multiprocessors. The abstraction can be implemented on message-based multicomputers in much the same way as distributed shared memory [LH89].

3 The Abstraction and Its Implementation

3.1 Abstraction

The basic unit of the shared STM system is an *object*. To the programmer, an object is simply a collection of state variables. In actuality, however, each object contains successive *versions* of the state variables mapped to that object.

There are three operations that can be performed on space-time memory:

obj = MakeObj(size); creates an object of size **size** bytes and returns a pointer to that object. In our current implementation, each version of an object is of the same size.

ver = ReadObj(obj); returns a pointer to that version of the object **obj** which has the highest timestamp less than that of the event that invoked the ReadObj primitive. If no such version exists, an error status is returned.

ver = WriteObj(obj); creates a new version of the object **obj**. This is actually a read-modify-write operation. A copy of that version of the object **obj** that has the highest timestamp less than that of the writing event is returned; the timestamp on the newly created version is set to that of the event performing the write. Also, versions with timestamps greater than the timestamp of the newly created version are invalidated.

Any event can write into or read from any object in the system. However, an event cannot invoke the read/write primitive on the same object more than once. (This makes sense, since the read/write operation returns a handle to a 'page of memory', as it were, and this handle can then be used much as the file descriptor returned by a file-open for reading or writing is used, without needing to 'open the file' every time a file read/write is done.)

3.2 Implementation

An implementation of the STM abstraction has been added to an existing Time Warp discrete event simulation testbed [Fuj89a] on a GP1000 BBN Butterfly shared-memory multiprocessor.

Every event contains a list of pointers to versions that it has read. The versions that an event has written are similarly maintained. Each version includes a list of pointers to events that have read it. The lists are necessary to implement rollbacks.

The data structure for an object includes a variable that records the timestamp of the lowest timestamped event that is now writing the version. This variable will henceforth be called EarliestW. EarliestW is initialized to infinity when an object is created.

Every object has associated with it two locks: the earliest writer's time stamp (EWTS) lock must be obtained by an event before updating EarliestW; the write (WR) lock is used to maintain the validity of the data structure of the linked list of versions during searches, insertions and deletions. Separating these functions is expected to improve performance.

Versions are stored as an ordered linked list, where versions are sorted by the timestamp at which each was created.

Since a lower timestamped event cannot depend on a higher timestamped event, when a writer with timestamp T is writing an object, any reader/writer with timestamp less than T can be allowed to enter the object, provided it can obtain the proper locks.

3.2.1 Reads

The principal task of the read operation is to return a pointer to the most recent (in timestamp order) version of the object that is at least as old as the timestamp of the reader.

When an object is read, the reading event (let the timestamp on this event be T_r) attempts to obtain the EWTS lock on the object. After the EWTS lock has been obtained, if T_r is greater than EarliestW of the object, the event is made to wait in the queue of waiting readers/writers associated with this object (which is kept in increasing timestamp order), and the EWTS lock is released; else (T_r is less than the EarliestW) the event tries to obtain the WR lock of the object.

On obtaining the WR lock, the EWTS lock is released, and the list of versions of the object is traversed to locate the highest timestamped version whose timestamp is less than T_r. If such a version does not exist (a version is being read before it is written), an error status is returned. Otherwise, the WR lock is released, and a pointer to this version is returned.

3.2.2 Writes

Like the read operation, the write operation also returns a pointer to the most recent version of the object being written. Also, the write operation must invalidate newer versions of the object, and roll back events that accessed these versions.

Upon a write to an object, if there are free versions available on the processor to which the writing event belongs, the EWTS lock of the object is obtained by the writing event (timestamped T_w, say); else, fossil collection is initiated to reclaim storage from invalid versions and versions with version numbers less than the global virtual time of the simulation.

If T_w is greater than EarliestW, the event is made to wait in the queue of waiting readers/writers for the object, and the EWTS lock is released; otherwise, the event writes its own timestamp on EarliestW, and tries to obtain the WR lock on this object.

After obtaining the WR lock, the EWTS lock is released, and the list of versions is traversed to locate the highest timestamped version (V, say) whose timestamp is less than T_w and data from V is copied into the newly created version. The list of readers of V is examined to roll back events that read V and had timestamps greater than T_w.

The newly created version is inserted in its proper position in the version list of the object. Next, all versions of the object with timestamp greater than T_w are invalidated and events that accessed these versions are rolled back.

Finally, the WR lock is released, and a pointer to the newly created version is returned to the writing event.

3.2.3 Postprocessing of Events

As long as a writing event holds a pointer to some version of an object, no event with higher timestamp is allowed to enter the object; otherwise, race conditions could arise in reads and writes. Hence, whenever a event completes, we do some postprocessing to allow continuation of events that were waiting for this writing event to complete.

Therefore, after the processing of an event is over, the EWTS lock is obtained for each object that was written by the event, and the objects are examined to find out if this event's timestamp (say, T_e) is equal to the value stored in EarliestW of the object. If so, and if there are events waiting in the queue of readers/writers for the object, EarliestW is set to infinity, the event with the lowest timestamp among those waiting (say, E_l) is removed from the queue, and the EWTS lock is released. E_l will now proceed to obtain the EWTS lock, examine the EarliestW of the object and continue with the read/write protocol.

If T_e is equal to the EarliestW of the object, but there are no waiting readers/writers, EarliestW is updated to infinity, and the EWTS lock is released.

3.2.4 Rollbacks

Rolling back an event E causes events scheduled by E to be canceled. Also, if E is associated with a logical process, events of that process with timestamp greater than E must be rolled back. Finally, the reads/writes done by E must also be considered.

If E had read any version, the pointer to E in the readers list of each version it read is set to NULL.

If E had written any version, or created any object (the treatment is identical), those versions are invalidated after first obtaining the WR lock on the corresponding object. (To facilitate this, there is a pointer from each version to the corresponding object.) All versions of the object with timestamp larger than E's must be invalidated. Events that accessed these versions must also be rolled back.

3.3 Other comments on the protocol

Two locks are used because they serve different purposes, and operations on the data structures protected by the one can proceed in parallel with those on the other. This reduced lock granularity is considered useful since invalidation of versions may involve remote memory references and can therefore be time consuming.

Another notable aspect of the protocol is that it gives priority to reads/writes with lower timestamps. This tends to reduce the number of rollbacks.

4 Correctness of protocol

In this section we give the pseudo-code of the read/write protocol, and argue that the protocol is deadlock free and returns the correct versions in reads and writes.

4.1 The pseudo-code

The data structures used in the pseudo-code are :
array SemEWTS[NumberOfObjects] of binary semaphores;
array SemWrite[NumberOfObjects] of binary semaphores;
These semaphores are initialised to 1.

Reading an object::

```
T_r = timestamp on the reading event;
P(SemEWTS[i]);
if (T_r > EarliestW of object i) {
    insert the event in proper place in
        the ordered queue of events
        waiting to read or write
        the object;
    V(SemEWTS[i]);
}
else {
    P(SemWrite[i];
    V(SemEWTS[i]);
    traverse the list of versions of
        object i to find the valid
        version with highest
        timestamp < T_r;
    V(SemWrite[i]);
    return(pointer to the version
        found);
}
```

Writing an object:

```
T_w = timestamp on the writing event;
P(SemEWTS[i]);
if (T_w > EarliestW of object i) {
    insert the event in proper place in
        the ordered queue of events
        waiting to read or write
        the object;
    V(SemEWTS[i]);
}
else {
    EarliestW of object i = T_w;
    P(SemWrite[i]);
    V(SemEWTS[i]);
    create a new version;
    timestamp of new version = T_w;
    traverse the list of versions of
        object i to find the valid
        version (say, V) with highest
        timestamp < T_w;
    copy data from V into the newly
        created version;
    roll back events that read from V and
        had timestamps > T_w;
```

```
    mark the versions of object i with
      timestamps > T_w as invalid and
      roll back the events that read
      or wrote these versions;
  V(SemWrite[i]);
  return(pointer to the newly created
      version);
}
```

Post-processing of an event:

```
T_c = timestamp on the event;
for (each object i that the event wrote {
  P(SemEWTS[i]);
  if (T_c == EarliestW of object i)
    if (the queue of events waiting
        to read or write object i
        is non-empty)
      remove the event with lowest
        timestamp from this queue;
    else /*the queue is empty*/
      EarliestW of object i = ∞;
  V(SemEWTS[i]);
}
```

4.2 Correctness

Since rollbacks affect only events and versions with timestamp greater than that on a straggler, it is easy to see that the protocol cannot have livelocks due to circular-rollbacks.

Here, we shall argue that the protocol is deadlock free and returns the proper versions in reads and writes.

4.2.1 The protocol is deadlock free

If there can be a deadlock in the protocol as described above, let it involve events E_1, E_2, $\cdots E_d$. Let E_i be waiting on a semaphore held by E_{i+1} for $1 \leq i < d$ and E_d be waiting for the semaphore held by E_1.

Consider the event E_l with the lowest timestamp among $E_1 \cdots E_d$. At the instant before E_l performed its first read/write, there were no deadlocks in the system. So, E_l would get past the P(SemEWTS ...) in the pseudo-code for reads/writes after a finite delay, and none of the other events among $E_1 \cdots E_d$ could make it wait in the queue of waiting events of the object E_l is reading/writing. (We assume here that the P() operation is fair.) In the same way, E_l would also get past the P(SemWrite ...) after a finite interval. So, there will be no deadlock after the first read/write processed by E_l. Continuing similar arguments as above, we find that there will be no deadlock after the second, the third $\cdots$ the last read/write processed by E_l. Thus, there can be no deadlock due to the read/write protocol.

4.2.2 A committed event gets data from the proper version(s) in reads/writes

If a committed 'reading' event has a timestamp T, and the read returns a pointer to a version with timestamp T_r, then there should finally (i.e., when the simulation ends) be no version with timestamp less than T but greater than T_r. Similarly, if a committed 'writing' event has a timestamp T, and the write copies data from a version with timestamp T_w, then there should be no version with timestamp less than T but greater than T_w when the simulation ends. Such semantics are consistent with, and in fact required by, a completely sequential execution of the simulation, with events being processed on a non-decreasing timestamp order.

The above requirements are met by the protocol. A read returns a pointer to the immediately lower timestamped version available. If eventually this is not the correct version to be returned–as can happen if there is a write later (in real time) by an event with timestamp between that on the version returned and that on the reading event–the reading event will not be committed, since the writing event would cause reading events with greater timestamps than itself to be rolled back. Similarly, a write rolls back events that wrote versions with higher timestamps. Hence, a writing event that copied data from an incorrect version will not be committed.

5 The Sharks World Benchmark

In this section we discuss the use of STM in a well known benchmark—the Sharks World [CCU90, BL90, PRB90, NR90]. The Sharks World benchmark was designed as a simulation that captures the essence of certain problems of practical interest, e.g., military applications. This benchmark simulates a toroidal ocean, containing sharks and fish. Sharks move in straight lines and eat fish, but do not attack each other. Sharks have a fixed attack radii; any fish that lie within the attack radius of a shark are instantaneously eaten. We have made the simplifying assumption that the fish remain stationary.

For the purpose of parallel simulation, the ocean is divided into a number of sectors, and each sector is modeled as a LP in the simulation. This was done to facilitate comparison of the space-time memory approach with a conventional, message-based implementation of Time Warp. In our current implementation, the only events explicitly modeled are sector crossings by sharks. The information maintained by each sector is the number of creatures of each type that are contained within it, and the starting and ending coordinates of the sector. The information maintained about each fish is its coordinates, its oceanwide unique creature-identity, the identity of the shark that killed it, and the time at which the fish was eaten (initialized to infinity). The information kept for each shark is its coordinates, speed, direction of motion, and its oceanwide unique creature-identity.

become available. The result of these effects is the performance advantage enjoyed by the space-time memory version becomes negligible when 16 processor are used.

The contention problem described above is reduced if the number of sectors (objects) is increased, because locking is performed on a per object basis. Figures 2 and 3 show the execution times of the two versions as the number of sectors is increased to 256 and 512, respectively. It is seen that the STM-based version outperforms the message-based version when 16 processors are used for these larger numbers of sectors.

The efficiency figures (percentage of processed events that are eventually committed) for these experiments are shown in figures 4, 5, and 6 for 128, 256, and 512 sectors, respectively. In general, the message based implementations have lower ef-

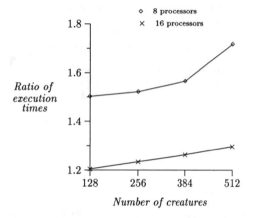

Figure 2: Comparison of the execution times of message-based and STM-based sharks world with 256 sectors

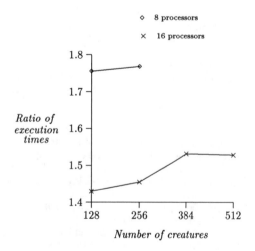

Figure 3: Comparison of the execution times of message-based and STM-based sharks world with 512 sectors

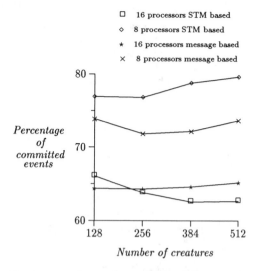

Figure 4: Comparison of the efficiencies of message-based and STM-based sharks world with 128 sectors

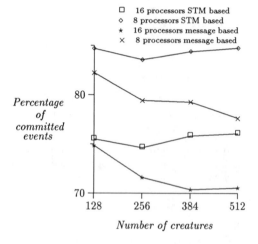

Figure 5: Comparison of the efficiencies of message-based and STM-based sharks world with 256 sectors

ficiencies because of the additional message traffic required. A straggler event will tend to roll back more events than the version using space-time memory. Any rollback arising due to out-of-sequence read/writes in the STM-based version also shows up in the message based version. However, the 'depth of rollback' would, in general, be less in STM than in message based sharks world.

Figures 3 and 6 do not show a complete set of points because the message-based version ran out of memory when 384 and 512 creatures were used on 512 sectors.

5.1 Sharks World Using Space-Time Memory

An object is associated with each sector and contains information relevant to that sector. Arrival of a shark at a sector is modeled as a write into that sector-object with timestamp equal to that of the arrival. All fish in the sector within the attacking range of the shark (considering its straight line trajectory through the sector) have their time of death and killing-shark-identity updated if the time of killing by the newly arriving shark is less than the existing time of death of the fish. Further, the time that the shark enters the 'border region' of the sector is calculated. The sector objects that are visible to the shark from the border region it enters are written into, and the fish-data for fish in the shark's attack range in those sectors are updated in a way similar to that discussed above. Finally, a sector-leaving event is scheduled for the shark. This basic scenario is repeated for the length of the simulation.

5.2 Sharks World Using Messages

The message-based implementation of the Sharks World uses pull processing. However, unlike [WJ89], our pull processing is event driven and not time driven.

In the absence of shared objects, the arrival of a shark at a sector is modeled as a message from the LP of the source sector to that of the destination sector. Associated with each event is a state vector representing the state variables of the LP to which the event belongs. The state vector includes a data structure similar to that used in each version of the space-time memory. When a shark-arrival event is processed, the destination sector (LP) updates its state vector accordingly. Attacking of fish within the sector is done exactly as described in the previous section, except that the state vector is updated rather than the version.

However, attacking fish in neighboring sectors must be implemented differently. A message requesting creature-information is sent to each neighboring sector N_i visible from the border region of the sector the shark is in (call this sector S_1). When the LP representing each N_i receives the message, it replies with a message having the state vector of N_i as its contents. Upon receiving such a message S_1 calculates times of killing of the fish in N_i within the attack range of the shark in its border region. S_1 then sends a message to N_i with these killing times (possibly updated, if fish are in range) of the fish. Upon receiving such a message, N_i compares the time of death of fish in its state vector with the corresponding time sent by S_1, and updates the state vector for those fish whose time of death in the state vector are greater than that of the message sent by S_1.

6   Performance of the Two Sharks World Simulations

Both the space-time and message-based simula-

tions were executed with the same set of input parameters for each run. All experiments were performed on a GP1000 BBN Butterfly multiprocessor. The message-based version is optimized to execute efficiently on a shared memory multiprocessor, and uses direct cancellation [Fuj89a].

The ocean is a square toroid, 65536 units on each side. Initial shark and fish coordinate locations were chosen using a uniform distribution, as were shark velocities (speeds were in the range [50,200]). The shark attack range was set to 20 units.

Each simulation run lasted 50000 time units, by which time the simulation was expected to reach a steady state.

Using a large number of sectors reduces the granularity of the simulation, but also increases the total number of events–since sector crossings by sharks are events. We used 128, 256 and 512 sectors in the experiments described here.

The ratio *time for message-based execution/time for STM-based execution* as the number of creatures are varied is shown in figure 1. Here, 128 sectors are used. The performance of STM was substantially better in runs using eight processors, and nearly the same as the message-based version for sixteen processors. Although time did not permit us to report speedup figures relative to a sequential simulation, others have shown that message-based Time Warp achieves good speedups for this problem [PRB90].

Increasing the number of processors increases contention for the shared data structures used by the space-time memory version. Since we use spin locks in our current implementation, this means

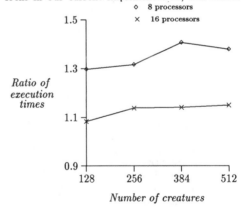

Figure 1: Comparison of the execution times of message-based and STM-based sharks world with 128 sectors

that processors spend more time waiting on locks, resulting in somewhat reduced performance. One solution to this problem is to switch execution to another process if a process finds itself in a long queue waiting to access a certain object. The message-based version does not suffer from this contention problem because it sends messages to request and distribute sector information rather than block, thereby switching execution to another process rather than waiting for the information to

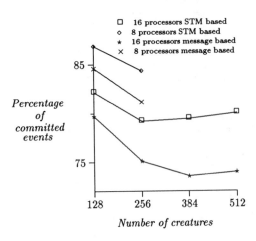

Figure 6: Comparison of the efficiencies of message-based and STM-based sharks world with 512 sectors

7   Conclusions and Future Work

The space-time memory abstraction allows parallel discrete event simulation programs to be written using shared-memory constructs. We have argued that this simplifies the coding of discrete event simulations in certain application domains (compared to 'push processing'), and often yields better performance than Time Warp programs using only message passing for communications. Initial performance measurements of a prototype implementation of space-time memory on a shared-memory multiprocessor are encouraging.

Much of our current work is focused on utilizing space time memory to automatically parallelize sequential discrete event simulation programs. This abstraction allows one to more easily parallelize simulation programs that utilize global state variables, and does not require that the simulation use processes. Other work in examining hardware implementation of the space-time abstraction as well as porting the existing software system to the C-threads package is being pursued. A more involved application–a combat simulation–is being programmed on the Space-Time Memory testbed.

Acknowledgment

We thank the anonymous referees whose comments improved the paper.

References

[BL90]   Rajive. L. Bagrodia and Wen-Toh Liao. Parallel simulation of the sharks world problem. *Proceedings of the 1990 Winter Simulation Conference*, pages 191–198, December 1990.

[CCU90] Darrel Conklin, John Cleary, and Brian Unger. The sharks world (a study in distributed simulation design). *Proceedings of the SCS Multiconference on Distributed Simulation*, 22(2):157–160, January 1990.

[CS89]   K. M. Chandy and R. Sherman. Space, time, and simulation. *Proceedings of the SCS Multiconference on Distributed Simulation*, 21(2):53–57, March 1989.

[Fuj89a] R. M. Fujimoto. Time Warp on a shared memory multiprocessor. *Transactions of the Society for Computer Simulation*, 6(3):211–239, July 1989.

[Fuj89b] R. M. Fujimoto. The virtual time machine. *International Symposium on Parallel Algorithms and Architectures*, pages 199–208, June 1989.

[Fuj90] R. M. Fujimoto. Parallel discrete event simulation. *Communications of the ACM*, 33(10):30–53, October 1990.

[JCRB89] D. W. Jones, C-C. Chou, D. Renk, and S. C. Bruell. Experience with concurrent simulation. *1989 Winter Simulation Conference Proceedings*, pages 756–764, December 1989.

[Jef85] D. R. Jefferson. Virtual time. *ACM Transactions on Programming Languages and Systems*, 7(3):404–425, July 1985.

[Jon86] D. W. Jones. Concurrent simulation: An alternative to distributed simulation. *1986 Winter Simulation Conference Proceedings*, pages 417–423, December 1986.

[LH89]   K. Li and P. Hudak. Memory coherence in shared virtual memory systems. *ACM Transactions on Computer Systems*, 7(4):321–359, November 1989.

[NR90]   David M. Nicol and Scott E. Riffe. A 'conservative' approach to parallelizing the sharks world simulation. *Proceedings of the 1990 Winter Simulation Conference*, December 1990.

[PRB90] Matthew T. Presley, Peter L. Reiher, and Steven Bellenot. A time warp implementation of sharks world. *Proceedings of the 1990 Winter Simulation Conference*, pages 199–203, December 1990.

[RBJ91] Peter Reiher, Steven Bellenot, and David Jefferson. Temporal decomposition of simulations under the time warp operating system. *Proceedings of the 1991 Workshop on Parallel and Distributed Simulation*, pages 47–54, January 1991.

[WJ89]   F. Wieland and D. R. Jefferson. Case studies in serial and parallel simulation. *Proceedings of the 1989 International Conference on Parallel Processing, Vol. 3*, pages 255–258, August 1989.

Partitioning and Mapping Nested Loops on Multiprocessor Systems*

Jang-Ping Sheu and Tsu-Huei Tai

Department of Electrical Engineering, National Central
University, Chung-Li 32054, Taiwan, R.O.C.
sheujp@ncu.dnet.ncu.edu.tw

Abstract

In this paper, a method for parallel executing nested loops with constant loop-carried dependencies on message-passing multiprocessor systems to reduce the communication overhead is presented. First, we partition the nested loop into blocks which result in little communication without concern for the topology of machines. Then, the partitioned blocks generated by the partitioning method can be mapped onto multiprocessor systems according to the specific properties of various machines. We propose a heuristic mapping algorithm for the hypercube machines.

I. Introduction

In many numerical programs, nested loops are the most time-consuming parts and usually offer the most amount of parallelism. In the past few years, several researchers studied the problem of transforming nested loops into parallel forms and mapping them onto special purpose architectures. Chen [2], Lee and Kedem [9], Liu, Ho, and Sheu [10], and Moldovan and Fortes [11] synthesized systolic arrays based on the hyperplane method. In message-passing multiprocessor systems, the communication overhead is still one order of magnitude higher than the corresponding computation [1]. Thus, an efficient parallel execution of algorithms or programs requires low ratio of communication overhead when computation is performed. Because of this requirement, various techniques have been developed for partitioning and mapping algorithms onto multiprocessor systems to reduce the communication overhead [3,5,6,12,14].

Some approaches, such as the *greatest common divisors* method, the *minimum distance* method [12], Shang and Fortes' method [14], and D'Hollander's method [3], partition the iterations of nested loops into independent blocks so that there are no dependence relations between computations that belong to different blocks. In other words, these methods separate the iterations into several blocks so that there is no data communication or synchronization between them. For many important nested loop algorithms, such as matrix multiplication, discrete fourier transform, convolution, transitive closure, and so forth, these index sets cannot be partitioned into independent blocks. Therefore, these algorithms will execute sequentially by their methods. The *grouping* method [6] partitions the algorithms into blocks with limited communication and gains more parallelism. For algorithms which cannot be partitioned into independent blocks, the grouping method will get better performance than the above partitioning methods.

In this paper, we concentrate on partitioning and mapping nested loops with constant loop-carried dependencies for execution on message passing multiprocessor systems. In the partitioning phase, we divide the nested loop into blocks which reduce the inter-block communication, without regard to the machine topology. First, the execution ordering of the iterations is defined by a given time function which is based on Lamport's hyperplane method [7]. Then, the iterations are partitioned into blocks so that the execution ordering is not disturbed and the amount of inter-block communication is minimized. In the second, or mapping phase, we map the partitioned blocks to the topology of the target machine. These blocks are mapped onto a fixed size multiprocessor system in such a manner that the blocks which have to exchange data frequently are allocated to the same

*This work was supported by the National Science Concil of the Republic of China under grant NSC 80-0408-E-008-09.

processor or neighboring processors.

II. Basic Concept and Definitions

Throughout this paper, the set of real numbers and the set of integers are denoted by $\mathbf{R}$ and $\mathbf{Z}$, respectively. The set of non-negative real numbers and the set of non-negative integers are denoted by $\mathbf{R}^+$ and $\mathbf{Z}^+$, respectively. The symbols $\mathbf{Z}^n$ and $\mathbf{R}^n$ represent the n-th cartesian powers of all integers and real numbers, respectively. Finally, we use boldface characters to represent the vertices in an n-dimensional graph; when the vertices are used in an equation, they represent the coordinates of these vertices. In this paper, we consider an n-nested loop of the form:

for $I_1 = l_1$ to u_1 by k_1
  for $I_2 = l_2$ to u_2 by k_2
    $\vdots$
    for $I_n = l_n$ to u_n by k_n
      $Statement_1$;
      $Statement_2$;
      $\vdots$
      $Statement_m$;
    end
  end
end.

Where l_j and u_j are integer-valued linear expressions possibly involving $I_1, I_2, \ldots, I_{j-1}$ for $1 < j \leq n$. Without loss of generality, we assume that $l_j \leq u_j$ and $k_j = 1$ for all $1 \leq j \leq n$. We make additional assumptions about the statements. These statements contain no I/O instructions, no transfer of control to any statement outside the loop, and no subroutine or function calls which can modify data. These assumptions are first used by Lamport [7]. The index set J^n = { $(I_1, I_2, \ldots, I_n)$ | $l_j \leq I_j \leq u_j$, for $j = 1, \ldots, n$ } is the set of loop indexes (or the iteration space).

Definition 1: Dependence vector

In an n-nested loop, suppose variable A is generated at iteration $\bar{i} = (i_1, i_2, \ldots, i_n)$ and used at iteration $\bar{j} = (j_1, j_2, \ldots, j_n)$, then the *dependence vector* $\bar{d}$ of variable A is a vector $(d_1, d_2, \ldots, d_n)^t \in \mathbf{Z}^n$ where $d_k = j_k - i_k$ for $1 \leq k \leq n$. □

Note that, the hyperplane method only be used on nested loops with constant loop-carried dependencies [6,11]. In other words, every index point of the index set has the same set of dependence vectors.

Example 1: Consider a 2-nested loop $L1$.

for $i = 0$ to 3
  for $j = 0$ to 3
$S_1 : A[i+1, j+1] := A[i+1, j] + B[i,j];$ (L1)
$S_2 : B[i+1, j] := A[i,j] * 2 + C;$
  end
end.

The index set $J^2 = \{(i,j) | 0 \leq i, j \leq 3\}$. The dependence vectors of variable A are $\bar{d}_1 = (0,1)^t$ and $\bar{d}_2 = (1,1)^t$. The dependence vector of variable B is $\bar{d}_3 = (1,0)^t$. The set of dependence vectors $D = \{\bar{d}_1 \ \bar{d}_2 \ \bar{d}_3\}$. In the index set of a nested loop, two iterations can be executed concurrently if and only if they are independent of each

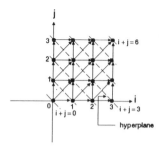
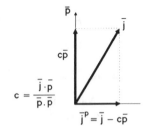
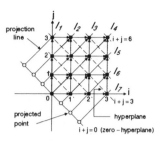
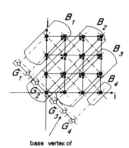

Fig. 1 The computational
structure and hyperplanes
of loop(L1).

Fig. 2 Projection of a
vector $\bar{j}$ with the
projection vector $\bar{p}$.

(a) Projected points,
projection line,
and hyperplanes.

(b) Groups and the
corresponding blocks of
loop (L1).

Fig. 3 The projected structure of
example 1 with $\Pi = (1,1)$.

other. From this point of view, Lamport proposed the hyperplane
method [7]. The index set is traversed by many hyperplanes defined
by a linear function Π if $\Pi \, \bar{d}_i > 0$ for any $\bar{d}_i \in D$. Because there is no
dependence relation between the points lying on the same hyperplane,
these points can be executed simultaneously. Given Π, the execution
time of every index point can be determined. For loop ($L1$), let the
time function $\Pi = (1, 1)$. The hyperplanes $i + j = constant$ are shown
in Fig. 1. The hyperplane method is composed of two parts, the time
transformation and space transformation [11]. The time transformation is used to determine the execution time of each index point and
the space transformation assigns the index points onto processing elements. The hyperplane method is used widely in the synthesis of
loops in systolic arrays [2,9,10,11]. Based on this method, the execution ordering of index points can be scheduled easily. Thus, the time
transformation of the hyperplane method is used in our approach.
However, the space transformation used for systolic arrays is not suitable for message-passing multiprocessor systems; we must do this in
another way. We propose the following partitioning and mapping algorithms for multiprocessor systems: First, we partition the nested
loops into larger blocks which result in little inter-block communication, without regard to the topology of the target machine. Then,
these blocks are mapped onto message-passing multiprocessor systems
according to the specific properties of various machines. Some terms
are defined in the following before the description of our algorithm.

We can represent an n-nested loop L as a directed graph Q in an
n-dimensional space. Each vertex in Q represents an iteration (also
called the index point) and has a coordinate $(i_1, i_2, \ldots, i_n)$ in the space
if the corresponding iteration has a loop index $(i_1, i_2, \ldots, i_n)$. There
is an arc from vertex $\mathbf{v}_i$ to vertex $\mathbf{v}_j$, if the iteration corresponding to $\mathbf{v}_j$ depends on the iteration corresponding to $\mathbf{v}_i$. Such a graph
is called the *computational structure* [6] of the nested loop L. The
computational structure of loop ($L1$) is shown in Fig. 1.

Definition 2: Computational structure [6]

A *computational structure*, Q, of a nested loop L is a two-tuple, Q
$= (V, D)$, where $V = \{\bar{i} | \bar{i} = (i_1, i_2, \ldots, i_n) \in J^n\}$ is the set of vertices
in Q, and D is the set of dependence vectors.

□

In a computational structure $Q = (V, D)$, a vertex $\mathbf{v}_j \in V$ is *de-
pendent* on another vertex $\mathbf{v}_i \in V$ along a vector $\bar{d}$, if $\mathbf{v}_j - \mathbf{v}_i = \bar{d}$. On
the other hand, if two vertices $\mathbf{v}_i$, $\mathbf{v}_j \in V$ and there does not exist a
path between $\mathbf{v}_i$ and $\mathbf{v}_j$, then $\mathbf{v}_i$ and $\mathbf{v}_j$ are *independent*, i.e., $\mathbf{v}_i$ is
not reachable from the vertex $\mathbf{v}_j$ and vice versa.

Definition 3: Projection vector

Let $\bar{p} = (p_1, \ldots, p_n)$ denote a *projection vector*. We define the
projection of the vector $\bar{j} = (j_1, \ldots, j_n)$ with respect to the projection
vector $\bar{p}$ as $\bar{j}^p = (j_1^p, \ldots, j_n^p)$, where $\bar{j}^p = \bar{j} - \frac{\bar{j} \cdot \bar{p}}{\bar{p} \cdot \bar{p}} \bar{p}$, i.e., $\bar{j}^p$ is the
projection of vector $\bar{j}$ onto a plane that is perpendicular to $\bar{p}$.

□

Fig. 2 is the graphical representation of the projection method.
For a computational structure $Q = (V, D)$ of a nested loop L, the
projection of a vertex $\mathbf{v}_i \in V$ and a dependence vector $\bar{d}_i \in D$ with
respect to the vector $\bar{p}$ are called the *projected point* and the *projected
dependence vector*, respectively.

Definition 4: Projection line

The *projection line* which corresponds to a projected point $\mathbf{v}$ along
the direction $\bar{p}$ is the line with the parametric equation, $\bar{j} = \mathbf{v} + t\bar{p}$,
$t \in \mathbf{R}$.

□

From the geometric view, $\bar{p}$ is the normal vector of the hyper-
planes such that $\bar{p}\bar{x} = constant$, where $\bar{x} \in \mathbf{R}^n$. The hyperplane $\bar{p}\bar{x}$
$= 0$ is called the *zero-hyperplane*. If we project a vector $\bar{j}$ with the
projection vector $\bar{p}$ by the manner defined in Definition 3 then it is
similar to projecting $\bar{j}$ onto the zero-hyperplane.

Definition 5: Projected structure

The *projected structure* $Q^p = (V^p, D^p)$ of an n-dimensional computational structure $Q = (V, D)$ with the projection vector $\bar{p}$ is a
directed graph such that,

(1) The vertex set V^p is the set formed by the projected points.

(2) D^p is the set of projected dependence vectors.

(3) There is an arc from $\mathbf{v}_i^p$ to $\mathbf{v}_j^p$, for $\mathbf{v}_i^p$, $\mathbf{v}_j^p \in V^p$, iff there exists
    a projected dependence vector $\bar{d}_k^p$, such that $\mathbf{v}_j^p - \mathbf{v}_i^p = \bar{d}_k^p$.

□

Suppose the dimension of a computational structure Q is n. Because Q is projected onto the zero-hyperplane whose dimension is
$n - 1$, the projected structure Q^p is an $n - 1$ dimensional structure.
From the definitions of the projection line and projected point, every
projected point defines a unique projection line with the projection
vector $\bar{p}$. Assume that there are two vertices $\mathbf{v}_i^p$ and $\mathbf{v}_j^p$ of a projected
structure. Vertex $\mathbf{v}_i^p$ is said to be dependent on $\mathbf{v}_j^p$, if there exists
a projected dependence vector $\bar{d}_k^p \in D^p$ such that $\mathbf{v}_i^p = \mathbf{v}_j^p + \bar{d}_k^p$.
This means that each index point on the corresponding projection
line of $\mathbf{v}_i^p$ depends on a index point which lies on the corresponding
projection line of $\mathbf{v}_j^p$ except the boundary index points of the index
set J^n.

Consider the Example 1, we use $\Pi = (1, 1)$ as the projection vec-
tor, i.e., $\bar{p} = \Pi$. Let $\bar{j}^p = (j_1^p, j_2^p)$ be the projected point corresponding
to the index point $\bar{j} = (j_1, j_2)$, and $\bar{j}^p = \bar{j} - \frac{\bar{j} \cdot \Pi}{\Pi \cdot \Pi} \Pi = \bar{j} - \frac{i_1 + i_2}{2} \Pi$,
where $0 \leq i, j \leq 3$. We get 7 projected points, $V^p = \{(-3/2, 3/2), (-1,
1), (-1/2, 1/2), (0, 0), (1/2, -1/2), (1, -1), (3/2, -3/2)\}$. Seven pro-
jection lines are defined with the projected points along the direction
$\Pi = (1, 1)$. The projected points and projection lines are depicted
in Fig. 3(a) where the projected points except $(0, 0)$ are illustrated
by open circles. From Fig. 3(a), we know that the index points lying
on the same projection line do not belong to the same hyperplane.
Furthermore, the index points of neighboring lines do not lie on the
same hyperplane, either. For instance, the index points belonging to
the line l_4 lying on the hyperplanes: $i + j = 0$, $i + j = 2$, $i + j = 4$
, and $i + j = 6$; and the index points belonging to l_3 lying on the
hyperplanes: $i + j = 1$, $i + j = 3$, and $i + j = 5$. Therefore, the
index points located on these two lines can be grouped into a block
and assigned to the same processor. Fig. 3(b) depicts the partitioning of loop $L1$. There are 4 groups and each consists of two projected

points except the boundary group G_4. Let block B_i be the set of index points which are projected to the projected points belonging to group G_i. If we assign each block to one processor, the number of data dependencies between index points is 33, and only 12 of them require inter-processor (inter-block) communication.

III. Partitioning of Index Set in Nested Loops

The partitioning algorithm is composed of two phases: the projection phase and the grouping phase. First, in the projection phase, the vertex set of Q (or the index set of L) is projected onto the zero-hyperplane, $\Pi \bar{x} = 0$. Every projected point represents some iterations whose corresponding vertices in Q are projected to this point. In other words, these iterations lie on the corresponding projection line of the projected point. Because this projection line is perpendicular to the hyperplane, these computations of the index points lying on the projection line will not be executed at the same time. Therefore, the iterations located on any one projection line can be assigned to the same processor without increasing the total number of execution steps.

Next, we group the projected points into a group whose corresponding projection lines do not interfere with each other. If there are more projected points within a group, then there are fewer data exchanges among the corresponding partitioned blocks. We want to group as many projected points into one group as we can. As a consequence, a good grouping scheme will make the size of groups and the corresponding blocks as large as possible.

Definition 6: Grouping and auxiliary grouping vectors

The *grouping vector* is a vector used for grouping the projected points into groups. All the projected points are grouped along the grouping vector. The *auxiliary grouping vectors* are a set of vectors which determine the base vertex of every group in the projected structure.

$\square$

Suppose D^p is the set of projected dependence vectors that corresponds to the dependence vectors D of a nested loop L. The matrix formed by the projected dependence vectors is denoted by $mat(D^p)$; the rank of a matrix A is represented by $rank(A)$. From the definition of a vector space, n linearly independent vectors form a basis of an n-dimensional vector space. That is, an n-dimensional vector space can be generated using exactly n linearly independent vectors. Since the dimension of the projected structure is $n-1$, there are at most $n-1$ vectors that can be used as grouping and auxiliary grouping vectors in the grouping phase. If we use more than $n-1$ vectors, then some vectors can be expressed as a linear combination of the other $n-1$ linearly independent vectors. Thus, it will cause a conflict when we determine the base vertices of groups. Now, we will explain how to choose the grouping and auxiliary grouping vectors.

When two index points are inter-dependent, there are data transfers occurring between them. If these index points belong to the same partitioned block, then the number of inter-block data exchanges will be reduced. Since the dependence relation between two projected points represents the dependence relations of the index points that are projected to these projected points, we will gather the projected points which have dependence relations among them. We want to choose the grouping and auxiliary grouping vectors from D^p such that the communication among partitioned blocks is minimized. Let r_i be the smallest positive integer such that $r_i \bar{d}_i^p \in \mathbf{Z}^n$. Then, r_i projected points can be grouped along the direction of $\bar{d}_i^p$ as a group and all the index points which are projected into this group are executed at different execution steps. This means that these index points can be mapped to the same processor. Suppose $\bar{d}_l^p$ is the projected dependence vector whose r_l is the largest, i.e., $r_l = \max_{\bar{d}^p \in D^p} \{r_i\}$. If there is more than one projected dependence vector whose r_i is equal to the largest value, then we choose one from these vectors arbitrarily. The vector $\bar{d}_l^p$ is called the grouping vector. The size of each group is $r = r_l$.

Definition 7:

The *partitioning* $G_\Pi(Q)$ of a computational structure $Q = (V, D)$ with the projection vector (time function) $\Pi = (a_1, \ldots, a_n)$ is the partitioning of all vertices of Q into disjoint blocks, $B_0, \ldots, B_{\alpha-1}$ where

α is the total number of partitioned blocks in a nested loop. Suppose the disjoint groups of the projected structure $Q^p = (V^p, D^p)$ are denoted by $G_0, \ldots, G_{\alpha-1}$ which correspond to $B_0, \ldots, B_{\alpha-1}$, respectively. Then,

(1) each group G_i, for $0 \leq i \leq \alpha - 1$, contains r projected points except the boundary groups, i.e., G_4 of Fig. 3(b);

(2) for each group G_i, $0 \leq i \leq \alpha - 1$, there exists an ordering along a direction $\bar{d}^p \in D^p$ for all vertices in G_i, i.e., $(\mathbf{v}_0^p, \ldots, \mathbf{v}_{r-1}^p)$, such that $\mathbf{v}_{j+1}^p - \mathbf{v}_j^p = \bar{d}^p$ for $0 \leq j \leq r - 2$;

(3) and each block $B_i = \bigcup_{\mathbf{v}_k^p \in G_i} \{\bar{j} \in J^n | \bar{j} = \mathbf{v}_k^p + t\Pi, t \in \mathbf{R}\}$ for $0 \leq i \leq \alpha - 1$.

$\square$

The selection of the auxiliary grouping vectors is discussed in the following. From the definition of rank, the rank of a matrix represents the maximum number of linearly independent columns (or rows) of this matrix [8]. Since $rank(mat(D^p)) = \beta$, it is possible to choose β linearly independent vectors from D^p. We select $\beta - 1$ vectors from $D^p - \{\bar{d}_l^p\}$ as the auxiliary grouping vectors such that these $\beta - 1$ vectors and the grouping vector are linearly independent. Since all the data exchanges is caused by the data dependence relations, we can reduce the amount of communications among the groups by choosing the auxiliary grouping vectors from $D^p - \{\bar{d}_l^p\}$.

Definition 8: Forward and backward neighboring groups

Let the set of auxiliary grouping vectors be denoted by Ψ. Suppose $\mathbf{u}_0^p$, $\mathbf{v}_0^p$, and $\mathbf{w}_0^p$ are the base vertices of groups G_i, G_j, and G_k. If $\mathbf{u}_0^p = \mathbf{v}_0^p - r\bar{d}_l^p$ and $\mathbf{u}_0^p = \mathbf{w}_0^p + r\bar{d}_l^p$, then G_j and G_k are the *forward and backward neighboring groups* of G_i along the grouping vector $\bar{d}_l^p$, respectively. If $\mathbf{u}_0^p = \mathbf{v}_0^p - \bar{d}_j^p$ and $\mathbf{u}_0^p = \mathbf{w}_0^p + \bar{d}_j^p$, where $\bar{d}_j^p \in \Psi$, then G_j and G_k are the forward and backward neighboring groups of G_i along the auxiliary grouping vector $\bar{d}_j^p$, respectively.

$\square$

After the grouping and auxiliary grouping vectors have been determined, we will group the projected points into groups. The idea is similar to the *region growing* method [4] in image processing that is used to find all groups of the projected structure. We group some projected points into a group and from this determine the other groups along the directions of the grouping and auxiliary grouping vectors. First, the projected points are grouped along the direction of the grouping vector. The projected structure is split into many parallel lines along the direction of $\bar{d}_l^p$. We select one line from these parallel lines arbitrarily, and choose the projected point $\mathbf{v}_0^p$ lying on the selected line to be the base vertex of the first group G_0. Then, we group r projected points along the direction of the grouping vector from $\mathbf{v}_0^p$ and generate the first group G_0. The following step is to determine the base vertices of the other groups and the members of them. We use the first group as the initial seed group. From the seed group, we get the backward and forward neighboring groups along the directions of the grouping and each of the auxiliary grouping vectors. After this, the neighboring groups found above are used as the seed groups and from these groups find all neighboring groups which have not been determined in the previous step. Using the groups found in the last step as the seed groups, we determine the other groups in the way described above repeatedly until there exists no neighboring group of the seed groups.

If there are some projected points that have not been gathered into groups, we select an ungrouped line, and group r projected points on the line into a group. Using the group as the initial seed group, we perform the grouping procedure described above until all the projected points have been gathered into groups. After all groups of the projected structure have been determined, the corresponding partitioned block of $G_i = \{\mathbf{v}_0^p, \ldots, \mathbf{v}_{r-1}^p\}$ is the set of $B_i = \bigcup_{\mathbf{v}_k^p \in G_i} \{\bar{j} \in J^n | \bar{j} = \mathbf{v}_k + t\Pi, t \in \mathbf{R}\}$.

If we group the projected points by the manner described above, then a group only depends on another group along the direction of the grouping vector and each of the auxiliary grouping vectors. In addition, a group depends on at most two groups along the direction of vectors $\in D^p - (\Psi \cup \{\bar{d}_l^p\})$. The formal partitioning algorithm is described as follows:

Algorithm 1: (Partitioning a nested loop)

Input: A *computational structure* $Q = (V, D)$ of an n-nested loop

L and a time transformation function $\Pi = (a_1, \ldots, a_n)$ found by the hyperplane method as the projection vector.

Output: A set of partitioned blocks $G_\Pi(Q) = \{B_0, \ldots, B_{\alpha-1}\}$ of the computational structure Q.

Projection Phase: /* Phase 1 */

Project the vertex set V, and the dependence vectors D, onto the zero-hyperplane with respect to Π. After projection, we get two sets: the set of projected points, V^p, and the set of projected dependence vectors, D^p. These two sets constitute the projected structure $Q^p = (V^p, D^p)$ of Q.

Grouping Phase: /* Phase 2 */

/* In Step 1 and Step 2, we select vectors as grouping and auxiliary grouping vectors, where $\beta = rank(mat(D^p))$. */

Step 1: Let r_i be the smallest positive integer such that $r_i\bar{d}_i^p \in \mathbf{Z^n}$, for $\bar{d}_i^p \in D^p$. Then, the size of every group is $r = \max_{\bar{d}_i^p \in D^p}\{r_i\}$. Assume that $\bar{d}_l^p \in D^p$ is the vector whose $r_l = r$. Then we choose $\bar{d}_l^p$ as the grouping vector. If there are many vectors $\in D^p$ whose r_i values are equal to r, then we choose one arbitrarily.

Step 2: Choose the other $\beta - 1$ projected dependence vectors as the auxiliary grouping vectors from $D^p - \{\bar{d}_l^p\}$ such that these $\beta - 1$ vectors and $\bar{d}_l^p$ ar linearly independent. Let the set of the auxiliary grouping vectors be denoted by $\Psi = \{\bar{d}_l^p, \ldots, \bar{d}_{\beta-1}^p\}$.

/* After Step 2, the projected structure can be split into many parallel lines along the direction of $\bar{d}_l^p$. */

Step 3: Select a line arbitrarily; choose a projected point which is lying on this line as the base vertex of a group. Then, starting from the base vertex, group r projected points along $\bar{d}_l^p$ into a group. Let the group be the initial seed group.

Step 4: From the seed groups, find all the backward and forward neighboring groups. Then, the groups found above are used as the seed groups. Repeat this step until there exist no neighboring group of the seed groups.

Step 5: After Step 4, if there are some lines whose projected points are not grouped, go to Step 3.

Step 6: For every group of the projected structure $G_i = \{\mathbf{v}_0^p, \ldots, \mathbf{v}_{r-1}^p\}$, the corresponding partitioned block $B_i \in G_\Pi(Q)$ is the set $\bigcup_{\mathbf{v}_k^p \in G_i}\{\vec{j} \in J^n | \vec{j} = \mathbf{v}_k^p + t\Pi, t \in \mathbf{R}\}$. □

Example 2: Consider the matrix multiplication algorithm:

```
for i = 0 to 3
  for j = 0 to 3
    for k = 0 to 3
      C[i,j] := C[i,j] + A[i,k] * B[k,j];      (L2)
    end
  end
end.
```

This program can be rewritten into the following equivalent form.

```
for i = 0 to 3
  for j = 0 to 3
    for k = 0 to 3
      A^(i,j,k)[i,k] := A^(i,j-1,k)[i,k];
      B^(i,j,k)[k,j] := B^(i-1,j,k)[k,j];
      C^(i,j,k)[i,j] := C^(i,j,k-1)[i,j];          (L3)
      C^(i,j,k)[i,j] := C^(i,j,k)[i,j] + A^(i,j,k)[i,k] *
                        B^(i,j,k)[k,j];
    end
  end
end.
```

$\bar{d}_A^P = (-1/3, 2/3, -1/3)$: Grouping vector
$\bar{d}_B^P = (2/3, -1/3, -1/3)$
$\bar{d}_C^P = (-1/3, -1/3, 2/3)$: Auxiliarly grouping vector

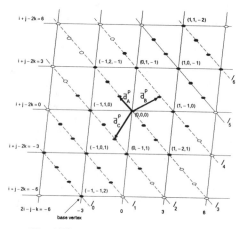

Fig. 4 The projected structure of example 2

The dependence matrix is $\begin{bmatrix} 0 & 1 & 0 \\ 1 & 0 & 0 \\ 0 & 0 & 1 \\ \bar{d}_A & \bar{d}_B & \bar{d}_C \end{bmatrix}$. Let the time function $\Pi = (1, 1, 1)$, then the zero-hyperplane is $i + j + k = 0$.

Projection Phase:

Q is projected onto the zero-hyperplane. The projected structure $Q^p = (V^p, D^p)$ is depicted in Fig. 4. There are 37 projected points shown by solid circles; the points shown by the open circles are the projected points generated by the projection when the upper (or lower) bounds of the loop indices are larger. The projected dependence vectors are $D^p = \{\bar{d}_A^p = (-1/3, 2/3, -1/3)^t, \bar{d}_B^p = (2/3, -1/3, -1/3)^t, \bar{d}_C^p = (-1/3, -1/3, 2/3)^t\}$.

Grouping Phase:

/* Since $rank(\begin{bmatrix} -1/3 & 2/3 & -1/3 \\ 2/3 & -1/3 & -1/3 \\ -1/3 & -1/3 & 2/3 \end{bmatrix}) = 2$, we select two projected dependence vectors such that one is the grouping vector and the other one is the auxiliary grouping vector. */

Step 1: The size of a block is $r = \max_{\bar{d}_i^p \in D^p}\{r_i\} = 3$. Then, there are 3 projected points in one group except the boundary groups. We select the grouping vector arbitrarily. Let $\bar{d}_l^p = \bar{d}_A^p = (-1/3, 2/3, -1/3)^t$.

Step 2: The auxiliary grouping vector is $\bar{d}_C^p = (-1/3, -1/3, 2/3)^t$.

/* After Step 2, the projected structure is split by $\bar{d}_A^p$ into 7 parallel lines, $l_0, \ldots, l_6$. */

Step 3: Select the projected point $(-1, -1, 2)$, $(-4/3, -1/3, 5/3)$, $(-5/3, 1/3, 4/3)$ from l_0 to form the group G_1 and let $(-1, -1, 2)$ be the base vertex $\mathbf{v}_0^p$ as shown in Fig. 4. The vertex $\mathbf{v}_2^p$ in this group is $\mathbf{v}_2^p = (-1, -1, 2) + 2\bar{d}_A^p = (-5/3, 1/3, 4/3)$.

Step 4: Determine the base vertices of other groups along $\bar{d}_C^p = (-1/3, -1/3, 2/3)$. The resulting groups are depicted in Fig. 5, every group is shown by a dashed boxes.

Step 5: In this example, all points can be grouped, and terminate the step.

/* Consider the group G_{10} as shown in Fig. 5; there are two groups G_7 and G_{11} depending on G_{10} along the direction $\bar{d}_A^p = (-1/3, 2/3, -1/3)$ and $\bar{d}_C^p = (-1/3, -1/3, 2/3)$, respectively. In addition, it also sends data to G_{12} and G_{13} along $\bar{d}_B^p = (2/3, -1/3, -1/3)$ which is not a grouping or auxiliary grouping vector. Hence, there are $2 \times 3 - 2 = 4$ groups that depend on the group G_{10}. */

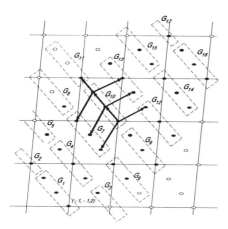

Fig. 5 Grouping the projected points of Fig. 4.

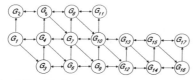

Fig. 6 Graphical representation of the partitioned groups and their communication links.

Step 6: The procedure partitions the nested loop into 17 partitioned groups. We use a node to represent a partitioned group, and if G_i will send data to G_j then there is a edge from G_i to G_j. The resulting graph is illustrated in Fig. 6.

□

IV. Mapping the Partitioned Blocks of Nested Loops onto Hypercubes

When an algorithm is executed on a message-passing multiprocessor system, the overhead due to inter-processor communication and synchronization, and idle processors due to contention for shared hardware resources can lead to poor overall performance [13]. Therefore, we have to map the parallel component of the algorithm on the processors in such a way that minimizes the time required to perform necessary inter-processor communication and the amount of processor idle time. In Section III, a nested loop is partitioned into blocks without concern for the machine topology. The next step is to map the partitioned blocks of a nested loop onto a specific target machine. We choose the hypercube topology as the target machine because of the recent interest in this configuration. Hypercubes are loosely coupled parallel processors based on the binary n-cube interconnection network. An n-dimensional hypercube computer consists of $N = 2^n$ identical processors, each provided with its own local memory, and directly connected to n neighbors .

In general, the problem size is much larger than the machine size. Without loss of generality, we assume that the number of partitioned blocks of a nested loop is larger than the number of processors in the target machine. Therefore, we first divide these blocks into clusters which are suitable for mapping onto the target machine. After this processing, the clusters are assigned to processors by some mapping algorithm.

First, we divide the total partitioned blocks into two clusters, of equal size, and divide every resulting cluster into two subclusters until there are N clusters generated. Then these N clusters can be mapped to an n-cube. Since the number of clusters increases by a factor of two after each dividing, we divide the partitioned blocks n times to generate N clusters. Our mapping approach is based on the concept illustrated above. Moreover, the partitioned blocks are first modeled by the Task Interaction Graph (TIG) model [13]. In the TIG model, the vertices represent blocks of the partitioning, and the edges of the TIG represent communication requirements between blocks.

Assume that there is a partitioning of a nested loop and it will be mapped onto an n-dimensional hypercube. The mapping algorithm proceeds in two phases:

(1) Cluster formation: The TIG is partitioned into as many clusters as the number of processors. We start with the entire partitioning as a single cluster, and successively divide each cluster into two equal size of blocks n times.

(2) Cluster allocation: The clusters generated in the first phase are numbered using the gray code scheme. After this numbering, a cluster is allocated to the processor whose binary representation is identical to the binary number of the cluster.

In general, we divides the partitioned blocks of a nested loop into clusters such that the neighboring blocks will be allocated to the same cluster. From the discussion of our partitioning method, most of the communication happens at neighboring blocks. Hence, the amount of data transmission of the clusters is reduced.

V. Conclusions

In this paper, we consider the problem of executing nested loops in parallel on message-passing multiprocessor systems. We partition the nested loop into blocks which result in little inter-block communication and then map these blocks efficiently onto the multiprocessor systems.

References

[1] W. C. Athas and C. L. Seitz, "Multicomputers: Message-passing concurrent computers," *IEEE Computers*, pp. 9-24, Aug. 1988.

[2] M. C. Chen, "The generation of a class of multipliers: Synthesizing highly parallel algorithms in VLSI," *IEEE Transactions on Computers*, Vol. 37, No. 3, pp. 329-338, Mar. 1988.

[3] E. H. D'Hollander, "Partitioning and labeling of index sets in do loops with constant dependence," in *Proceedings of 1989 International Conference on Parallel Processing*, Vol. II, pp. 139-144, 1989.

[4] R. C. Gonzalez and P. Wintz, *Digital Image Processing*, 2nd Ed., Addison-Wesley, Reading, Mass., 1987.

[5] R. Gupta, "Synchronization and communication costs of loop partitioning on shared-memory multiprocessor systems," in proceedings of 1989 *International Conference on Parallel Processing*, Vol. II, pp. 23-30, Aug. 1989.

[6] C. T. King and L. M. Ni, "Grouping in nested loops for parallel execution on multicomputers," *IEEE Transactions on Parallel and Distributed Systems*, Vol. 1, No. 4, pp. 486-499, Oct. 1990.

[7] L. Lamport, "The parallel execution of Do loops," *Communications of the ACM*, Vol. 17, No. 2, pp. 83-93, Feb. 1974.

[8] S. Lang, *Linear Algebra*, Addison-Wesley, Reading, Mass., 1986.

[9] P.-Z. Lee and Z. M. Kedem, "Synthesizing linear array algorithms from nested for loop algorithms," *IEEE Transactions on Computers*, Vol. C-37, No. 12, pp. 1578-1598, Dec. 1988.

[10] L. S. Liu, C. W. Ho, and J. P. Sheu, "On the parallelism of nested for-loops using index shift method," in proceedings of 1990 *International Conference on Parallel Processing*, Pennsylvania, Vol. II, pp. 119-123, Aug. 1990.

[11] D. I. Moldovan and J. A. B. Fortes, "Partitioning and mapping algorithms into fixed size systolic arrays," *IEEE Transactions on Computers*, Vol. C-35, No. 1, pp. 1-12, Jan. 1986.

[12] J.-K. Peir and R. Cytron, "Minimum distance: A method for partitioning recurrences for multiprocessors," *IEEE Transactions on Computers*, Vol. 38, No. 8, pp. 1203-1211, Aug. 1989.

[13] P. Sadayappan and F. Ercal, "Nearest-neighbor mapping of finite element graphs onto processor meshes," *IEEE Transactions on Computers*, Vol. C-36, No. 12, pp. 1408-1424, Dec. 1987.

[14] W. Shang and J. A. B. Fortes, " Independent partitioning of algorithms with uniform dependencies," in *Proceedings of 1988 International Conference on Parallel Processing*, pp. 26-33, 1988.

Finding Optimal Quorum Assignments for Distributed Databases

Donald B. Johnson* Larry Raab†

Dartmouth College‡

Abstract

Replication has been studied as a method of increasing the availability of a data item in a distributed database subject to component failures and consequent partitioning. The potential for partitioning requires that a protocol be employed which guarantees that any access to a data item is aware of the most recent update to that data item. By minimizing the number of access requests denied due to this constraint, we maximize availability. In the event that all access requests are reads, placing one copy of the data item at each site clearly leads to maximum availability. The other extreme, all access requests are write requests or are treated as such, has been studied extensively in the literature. In this paper we investigate the performance of systems with both read and write requests. We describe a distributed, online algorithm for determining the optimal parameters, or optimal *quorum assignments*, for a widely studied protocol, the quorum consensus protocol[7]. In addition, we demonstrate via simulation both the value of this algorithm and the effect of various read-write ratios on availability. This simulation, on 101 sites and up to 5050 links (fully-connected), demonstrates that the techniques described here can greatly increase data availability, and that the best quorum assignments are frequently realized at the extreme values of the quorum parameters.

Keywords: availability, discrete event simulation, distributed database, quorum assignment, reliability, replication.

1 Introduction

Replication has been proposed as a means of improving the availability of data in a distributed database. Although numerous copies of a data item maybe present in a computer network with replication, we wish to have them operate as if there existed only one copy of each data item. We, therefore, require that any operation, read or write, on a data item access the value most recently written to that item, independent of which physical copies have the most current value. In the event that all access requests are read requests, the best approach is clearly to put one copy of the data item at every site. The situation in which all requests are write requests, or, equivalently, when no distinction is made between reads and writes, has been studied extensively in the literature[3, 4, 6, 9, 11, 13]. In [12], we show that the benefits of replication under these circumstances is limited. In this paper, we examine the situation in which there occur both read and write accesses and they are not necessarily treated in the same manner.

*e-mail address:djohnson@dartmouth.edu

†e-mail address:raab@dartmouth.edu

‡Department of Mathematics and Computer Science. Hanover, N.H. 03755

Our examination can be viewed in two ways. We can think of ourselves as investigating how availability changes in response to changes in the read-write ratio, or how, given a particular read-write ratio, reads and writes are treated in order to maximize availability. We answer the former question in section 4 and the latter in section 3. We also examine, in section 4.5, the extent to which previous research exclusively on write requests applies to databases with a significant number of read requests.

The protocol most frequently studied is the quorum consensus protocol[7]. Unlike its predecessor, the majority consensus protocol[15], the quorum consensus protocol distinguishes between read and write transactions. Thus read throughput, or the percentage of read requests granted, can be increased by relaxing the criteria for allowing a read to proceed. Unfortunately, any relaxation of the read constraints must be matched by a corresponding tightening of the write constraints. We define more precisely these constraints, or *quorums*, and the relationship between them in section 2. Throughout the paper, availability is measured as the probability that an item can be accessed from an arbitrary site.

In [1], Ahamad and Ammar investigate optimal quorum assignments under the assumption that if two sites are operational then they can communicate. Although this eliminates the possibility of the partitioning which necessitates consistency control, it does provide some interesting analytic results which we show also hold in networks with fallible links. They prove, for instance, that the minimum or maximum of the availability function (see step 3 of Figure 1) occurs at the extreme quorum values. Figures 2-7 exhibit this same property. Ahamad and Ammar also show, as we discuss in section 4.5, that requiring a majority of votes for both read and write accesses is optimal for a wide range of network parameters. Due to the exponential computations involved in purely analytic studies, they are unable to investigate networks with more than nine copies. For networks as large as those studied in the present paper, we have shown that greater availability can be achieved with significantly more copies[11]. In [5], the quorum assignment problem is addressed in conjunction with determining the optimal vote assignment. Again assuming a non-partitionable network, the authors find the optimal quorum assignment by exhaustively searching an exponential set of candidate coteries. Numerical results are given in [5] for networks with up to seven sites.

2 Quorum Consensus Protocol

In the quorum consensus protocol[7], each copy of a replicated object is assigned a number called its *vote assignment*. Every read access must acquire a minimum number of the total votes called the *read quorum*, q_r; every write access must acquire a minimum number of the total votes called the *write quorum*, q_w. In order to ensure consistency in a

1   Assume that the following are known:
    α = fraction of accesses which are read requests,
    r_i = fraction of read accesses submitted to site s_i,
    w_i = fraction of write accesses submitted to site s_i,
    f_i = probability density function for each site s_i.

2   Let $r(v) = \sum_{i=1}^{n} r_i * f_i(v)$,   and
    $w(v) = \sum_{i=1}^{n} w_i * f_i(v)$.

3   Let
$$A(\alpha, q_r) = \alpha \sum_{k=q_r}^{T} r(k) + (1 - \alpha) \sum_{k=T-q_r+1}^{T} w(k)$$

4   Find q_r for which $A(\alpha, q_r)$ is maximized, and
    assign $q_w = T - q_r + 1$.

Figure 1: **Optimal Quorum Assignment Algorithm**

network with a total of T votes, the read and write quorums must satisfy the following conditions:

1. $q_r + q_w > T$, and

2. $q_w > \frac{T}{2}$.

The first condition guarantees that each read is aware of the most recent write. The second condition not only ensures that each write is aware of the most recent write, but also prohibits simultaneous writes.

The performance of the quorum consensus protocol is therefore dependent upon the values chosen for q_r and q_w. Condition 2 implies that $\frac{T}{2} < q_w \leq T$. In addition, conditions 1 and 2 taken together allow the assumption that $0 < q_r \leq \frac{T}{2}$, since requiring $q_r > \frac{T}{2}$ would be unnecessarily restrictive. We consider q_r to be our primary variable, and we assign $q_w = T - q_r + 1$, by condition 1.

The quorum consensus algorithm works as follows: when an access request is submitted to a site, that site collects the votes from every site in its current component. If the request is for read, then the read is granted if the number of votes collected is at least q_r. Likewise, if the request is for write, then the write granted if the number of votes collected is at least q_w. If, in either case, the required number of votes is not collected, the access is denied.

3   Finding Optimal Quorum Assignments

The algorithm in Figure 1 describes how to determine the optimal quorum assignments. In the presentation of the algorithm, $S = \{s_1, s_2, s_3, \ldots, s_n\}$ denotes the set of all sites s_i in the network, and T denotes the total number of votes in the system.

3.1   Description of the Algorithm

The optimal quorum assignment algorithm assumes that four parameters are known. The first three of these, α, r_i, and w_i, are likely to be explicit in the model or can be directly measured by the system. The fourth parameter, $f_i(v)$, is the probability that site s_i is in a component containing exactly v votes. The derivation of this probability density function is discussed in section 3.2.

In Step 2 we form two new probability density functions from the function $f_i(v)$. The first function, $r(v)$, is the probability that an arbitrary read request will be submitted to a site within a component containing v votes. Likewise, $w(v)$

is the probability that an arbitrary write request will be submitted to a site within a component containing v votes.

In Step 3 we use the cumulative distribution functions $R(q_r) = \sum_{q_r}^{T} r(k)$ and $W(q_r) = \sum_{T-q_r+1}^{T} w(k)$ to calculate availability given α and q_r. Thus $R(q_r)$ is the probability that an arbitrary read request will be granted, and $W(q_w)$ is the probability that an arbitrary write request will be granted.

The final step requires that, given α from step 1, we find the q_r which maximizes $A(\alpha, q_r)$. Since q_r can only assume integer values between 1 and $\lfloor \frac{T}{2} \rfloor$, one could naively, yet in polynomial time in the number of sites, conduct an exhaustive search for the optimal q_r. However, a number of characteristics of the function $A(\alpha, q_r)$ can be used to significantly reduce the computation time. As will be shown in section 4.3, $A(\alpha, q_r)$ is frequently maximized when $q_r = 1$ or $q_r = \lfloor \frac{T}{2} \rfloor$. This fact can be used in a numeric technique such as the so-called golden section search in one dimension (described, for example, in [14]). Other techniques can be applied to the continuous approximations of A, where $f_i(v) = \frac{dF_i(v)}{dv}$ such that $F_i(v_1) - F_i(v_2)$ is the probability that the component containing site s_i has between v_1 and v_2 votes. The Brent's Method (again, see [14]) makes use of the derivative of A, which we know from $r(v)$ and $w(v)$.

3.2   Finding Component Sizes

Step 1 of the algorithm in Figure 1 assumes that we know the probability density function $f_i(v)$. The function $f_i(v)$, the probability that site s_i is in a component containing v votes, can easily be found for some symmetric networks including ring, fully-connected, and single bus networks.

A ring of n sites with a copy at each site and one vote per site has density function

$$f_i(v) = \begin{cases} vp^v r^{v-1}(1-r) + p^v r^v & \text{if } v = n = T \\ vp^v r^{v-1}\Big((1-p) + p(1-r)^2\Big) & \text{if } v = T - 1 \\ vp^v r^{v-1}(1-pr)^2 & \text{if } 0 < v < T - 1 \\ (1-p) & \text{if } v = 0 \end{cases}$$

where p and r are the reliability of the sites and links, respectively.

The density function for a fully-connected network of n nodes involves another function, $Rel(m, r)$, which is the probability that all m sites of a fully-connected network can communicate assuming that the sites never fail and the links have reliability r. Gilbert in [8] presented the following recursive formula for $Rel(m, r)$,

$$Rel(m, r) = 1 - \sum_{i=1}^{m-1} \binom{m-1}{i-1} (1-r)^{i(m-i)} Rel(i, r)$$

Using Rel, we can express the probability density function as

$$f_i(v) = \binom{n-1}{v-1} p^v \Big((1-p) + p(1-r)^v\Big)^{n-v} Rel(v, r)$$

The density function of the single bus network depends upon the design of the network. If the architecture is such that no site can function when the bus is inoperative, then $f_i(v) = \binom{n-1}{v-1} r p^v (1-p)^{n-v}$, where r is the reliability of the bus. If, on the other hand, the bus failure does not necessitate site failure, then

$$f_i(v) = \begin{cases} p & \text{if } v=1 \\ \binom{n-1}{v-1} r p^v (1-p)^{n-v} & \text{otherwise.} \end{cases}$$

Unfortunately, it is very unlikely that f_i can be calculated efficiently in general graphs. In [11], we prove that calculating the expected size of the component containing a site i is #P-complete. Since this expectation is equivalent to the mean of the distribution f_i when the votes are uniform, and since the mean of this distribution can be calculated in polynomial time in the number of sites, finding f_i must be #P-complete.

Although #P-complete in general, it is not difficult to approximate f_i based upon past performance of the database system. This method may even be preferable to exact calculation since it requires very little computation time, is able to recognize changes in the system not anticipated by the model, and can accommodate dynamic protocols. (see [10]). Periodically, each site s_i queries every site with which it can communicate, recording the total number of votes possessed by all the sites in its component. If past history is indicative of future performance, then the values acquired in this manner approach $f_i(v)$. [1] Rather than performing broadcasts solely to acquire this vote total, site i can record the totals received while performing other functions required by the consistency control algorithm such as acquiring permission for data access.

4 Examples and Enhancements

In this section we demonstrate the use of the algorithm as presented in section 3.1 by finding the optimal quorum assignments for five read-write ratios on each of seven topologies. Figures 2-7 show the effect of the quorum assignments on availability. For every site i, we approximate the density function f_i using the on-line technique described in section 3.2.

Although we find that optimal quorum assignments significantly improve performance, we also discover that in some cases achieving the optimal availability eliminates nearly all writes. We present and demonstrate an enhancement to our algorithm that finds optimal quorum assignments given a minimum write throughput constraint.

4.1 System Model

The system we consider is composed of sites and bidirectional links subject to failure and subsequent recovery. Processors are fail-stop. Since message passing is the only inter-node communication mechanism, processor and link failures can partition the network into components. All events are modeled to occur instantaneously.

The results cited below are for a single data object with one copy at every site. We employ a uniform vote assignment of one vote per copy, since the data access distribution and component reliabilities are all uniform and the topologies are roughly symmetric. Although we investigate other

---

[1] Since non-operational sites cannot record access requests, density functions approximated in this manner yield availability A', the probability that an access request submitted to an arbitrary *operational* site will succeed, rather than A, the probability that an access request submitted to an arbitrary site will succeed. But q_r maximizes $A(\alpha, q_r)$ if and only if q_r maximizes $A'(\alpha, q_r)$.

scenarios in [11], our purpose here is not to be exhaustive but to illustrate and enhance the algorithm for determining optimal quorum assignments.

We examine a ring with 101 sites and this ring with added links. In this paper we denote by Topology i a ring with $i = 0, 1, 2, 4, 16, 256, 4949$ additional links, or chords. The exact placement of the chords can be found in [11].

4.2 Simulator

The events, site and link failures and recoveries and access requests are generated by a steady-state discrete event simulator. A detailed explanation of the necessity of simulation and of the parameters used by our simulator is contained in [11]. We list these parameters and their values below:

- The submission of data access requests by each site is modeled as a Poisson process with mean $\mu_t = 1$.

- The ratio, ρ, of the mean time-to-next-access to the mean time-to-next-failure is $\frac{1}{128}$.

- Site and link failures and recoveries are modeled as Poisson processes. The mean time-to-next-failure of each component, μ_f, is the same for both sites and links. Likewise, the mean time to recovery, μ_r, is the same for both.

- Each component is 96% reliable. Thus $\frac{\mu_f}{\mu_f + \mu_r} = .96$.

- In order to overcome the initial state, we do not monitor system state or performance until an initial $100,000$ accesses have passed.

- The simulation is run for $1,000,000$ accesses beyond the initial $100,000$.

- We assume that both read and write requests are submitted uniformly at random to every site in the network. Thus $r(v) = w(v)$. We regard a down site as a member of a component of size zero.

All simulations were run on a DEC Station 5000. Simulations were batched to a sufficient degree to achieve a 95% confidence interval with an interval half-size of at most $\pm 0.5\%$.

4.3 Optimal Quorums

Figures 2-7 show the various availability curves for seven different topologies, each with α, the percentage of access requests which are reads, equal to 0, .25, .50, and .75. The availability curves produced by topology 4949 (fully-connected) are not shown since they are nearly identical to those produced by topology 256.

The most striking observation from these figures is that all curves for a given topology converge at $q_r = \lfloor \frac{T}{2} \rfloor$. This occurs since q_r and q_w are nearly equal and therefore no distinction is made between read and write requests. We also see that for a given α the availabilities at $q_r = 1$ is independent of the topology, the read succeeds whenever the site to which the request was submitted is operational. Since the probability that a site is operational is 96%, the availability at $q_r = 1$ is .96α.

The graphs also show that all the curves, with only the exception of topology 16 at $\alpha = .75$, have maximum value at an endpoint of the curve. If the maximum occurs when

$q_r = \lfloor \frac{T}{2} \rfloor$, then this is clearly the best quorum assignment. On the other hand, a maximum that occurs at $q_r = 1$ may be unsuitable since $q_w = T$ and writes will succeed only when every copy is accessible. Remedying this shortcoming is the subject of the next section. We return to the case $q_r = \lfloor \frac{T}{2} \rfloor$ in section 4.5.

4.4 Write Constraint

If the optimal quorum assignment is unacceptable due to low write availability, then let us consider only those assignments which yield write availability of at least A_w. Such assignments require read quorums q_r such that $A(0, q_r) \geq A_w$. We can now maximize $A(\alpha, q_r)$ given this new constraint on q_r.

We demonstrate this method in Figure 4. Notice that the bottom curve on each graph is $A(0, q_r)$, and therefore we can use this curve to find the range of q_r for which $A(0, q_r) \geq A_w$. Suppose that $\alpha = 75\%$. Then the optimal availability is 72% and is achieved when $q_r = 1$. But at this point $q_w = T$ and therefore a write request will succeed only when all copies are accessible. Since this is very unlikely in a system of 101 copies, we can require $A_w \geq 20\%$, from which we find that q_r must be greater than 27 votes. Since the availability at $\alpha = 75\%$ decreases monotonically as q_r increases, $q_r = 28$ is the quorum assignment that optimizes availability when we require $A_w \geq 20\%$. The availability at this point is 50%.

4.5 Effects of Read-Write Ratio

As mentioned in the introduction, previous work has concentrated on the effects of replication without distinguishing between read requests and write requests. This approach is inherent in the majority consensus protocol[15], the coterie-based protocols[6], the primary copy protocol[2], and the quorum consensus protocol with $q_r = q_w$ [7].

From Figures 2-7, we see that one-half of the curves have maximum at $q_r = \lfloor \frac{T}{2} \rfloor$. The situations in which this is true include low read rates and highly-connected topologies, supporting the conclusion of [1] as mentioned in section 1. In these cases, the results of previous research apply directly to the case where there are both read and write accesses. On the other hand, the remaining curves indicate that we cannot ignore the read-write ratios in determining availability. In fact, this ratio can have a profound effect on the optimal quorum assignment and on consequent availability. Frequently, in fact, the assignment $q_r = \lfloor \frac{T}{2} \rfloor$, $q_w = \lfloor \frac{T}{2} \rfloor + 1$ yields the lowest availability.

5 Conclusion

The results of this paper demonstrate both the critical influence of the quorum assignment on the availability of replicated data and the ability of our algorithm to determine the optimal quorum assignment. Although we have shown that a seemingly necessary calculation is #P-complete, we have described and used a method for approximating this value on-line. In addition to being feasible, this on-line method has the advantage of changing the quorum parameters over time in response to changes in the network topology, component reliabilities, or access request distribution. This property allows our algorithm to be employed by a dynamic quorum reassignment protocol, thereby adjusting quorum assignments to exploit temporal characteristics of these parameters[10].

We have demonstrated via simulation the effectiveness of our algorithm and have shown that optimal quorums frequently occur either when both read requests and write requests require a majority of the votes or when read requests require only one vote and write requests require all votes. We have then described an enhancement for our algorithm that modifies quorum assignments of the latter type, which, while maximizing availability, incur an intolerable reduction in write throughput. Quorum assignments found in this way yield the maximum availability that can be achieved while guaranteeing some minimum write throughput.

References

[1] Mustaque Ahamad and Mostafa H. Ammar. Performance characterization of quorum consensus algorithms for replicated data. In *Proceedings of the 6th Symposium on Reliability in Distributed Software and Database Systems*, pages 161–168. IEEE, 1987.

[2] P. A. Alsberg and J. D. Day. A principle for resilient sharing of distributed resources. In *Proceedings of the 2nd Annual Conference on Software Engineering*, pages 627–644, October 1976.

[3] Daniel Barbara and Hector Garcia-Molina. The reliability of voting mechanisms. *IEEE Transactions on Computers*, C–36(10):1197–1208, 1987.

[4] Daniel Barbara, Hector Garcia-Molina, and Annemarie Spauster. Increasing availability under mutual exclusion constraints with dynamic vote reassignment. *ACM Transactions on Computer Systems*, 7(4):394–426, 1989.

[5] Shun Yan Cheung, Mustaque Ahamad, and Mostafa H. Ammar. Optimizing vote and quorum assignments for reading and writing replicated data. *IEEE Transactions on Knowledge and Data Engineering*, 1(3):387–397, September 1989.

[6] Hector Garcia-Molina and Daniel Barbara. How to assign votes in a distributed system. *Journal of the ACM*, 32(4):841–860, October 1985.

[7] D. K. Gifford. Weighted voting for replicated data. In *Proceedings 7th ACM SIGOPS Symposium on Operating Systems Principles*, pages 150–159, Pacific Grove, CA, December 1979.

[8] E. N. Gilbert. Random graphs. *Annals of Mathematical Statistics*, 30:1141–1144, 1959.

[9] Sushil Jajodia and David Mutchler. Dynamic voting algorithms for maintaining the consistency of a replicated database. *ACM Transactions on Database Systems*, 15(2):230–280, June 1990.

[10] Donald B. Johnson and Larry Raab. Finding optimal quorum assignments. Technical Report PCS-TR90-158, Dartmouth College, November 1990.

[11] Donald B. Johnson and Larry Raab. Effects of replication on data availability. *International Journal of Computer Simulation*, 1991. to appear.

[12] Donald B. Johnson and Larry Raab. A tight upper bound on the benefits of replication and consistency control protocols. In *Proceedings of the 10th Symposium on Principles of Database Systems*. ACM, May 1991. to appear.

[13] Jehan-François Pâris and Darrell D. E. Long. Efficient dynamic voting algorithms. In *Proceedings of the 4th International Conference on Data Engineering*, pages 268–275. IEEE, February 1988.

[14] William H. Press, Brian P. Flannery, Saul A. Teukolsky, and William T. Vetterling. *Numerical Recipes*. Cambridge University Press, 1986.

[15] R. Thomas. A majority consensus approach to concurrency control. *ACM Transactions on Database Systems*, 4(2):180–209, June 1979.

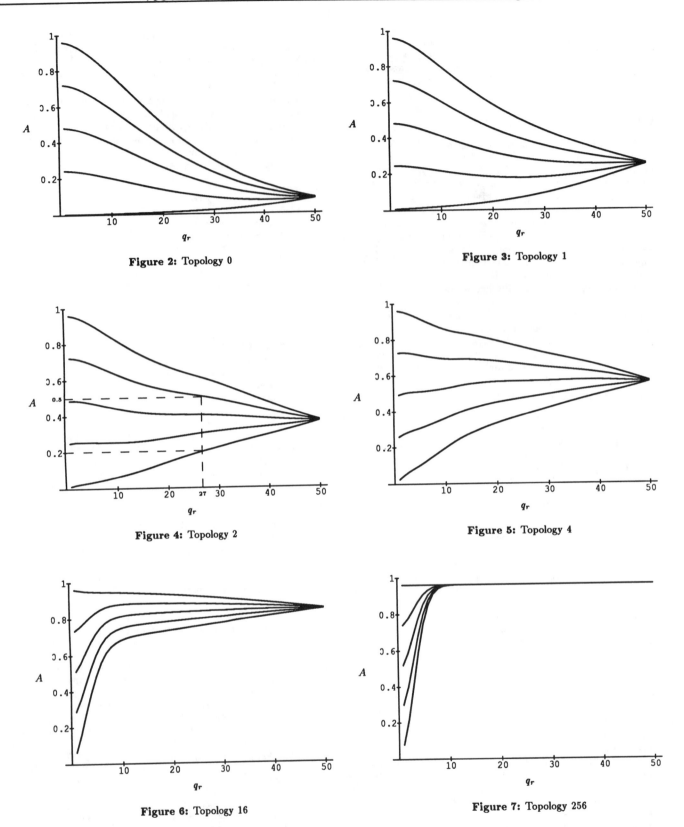

Figure 2: Topology 0

Figure 3: Topology 1

Figure 4: Topology 2

Figure 5: Topology 4

Figure 6: Topology 16

Figure 7: Topology 256

The curves of each figure represent, from bottom to top, $\alpha = 0$, .25, .50, .75, and 1.
The parameter α is the fraction of access requests which are read requests.

Multi-associativity: A Framework for Solving Multiple Non-uniform Problem Instances Simultaneously on SIMD Arrays

Martin C. Herbordt* and Charles C. Weems*

Abstract: A fundamental problem in using SIMD array processors is supporting the irregular, non-uniform, communication among processing elements (PEs) many analysis and simulation applications require. Although general communication networks allow simple implementations of these algorithms, the efficiency is often less than optimal as they do not take full advantage of the proximity inherent in most of these computations. In this paper we propose a machine independent computational framework we call *multi-associativity* that extends array-based associative processing techniques to arbitrary aggregates of PEs. One advantage is that it allows the direct application of many existing SIMD and associative array algorithms to multiple non-uniform aggregates simultaneously. Another is that multi-associativity *extends* the associative paradigm by allowing operations on and among aggregates themselves, operations not defined when the object in question is an entire array. Two consequences are support for divide-and-conquer (and a resulting further reduction in the complexity of some algorithms), and communication among aggregates. We present an efficient mapping of multi-associativity onto a SIMD processor together with sample algorithms.

1. Introduction

SIMD arrays have been found useful when applied to problems involving modeling and analysis of real-world phenomena: specific applications have included testing circuits, forcasting the weather, computer graphics, image processing, and image understanding. The primary reason is that viable solutions involve mapping subproblems onto geometrically corresponding processing elements (PEs), and that communication between PEs is usually only necessary between neighboring elements. Applications exist, however, where a proposed solution maps well onto a PE array geometrically, but where the communication patterns are irregular and non-uniform. In one scenario, different parts of the array (or PE aggregates) are mapped to objects with different shapes, resulting in different communication patterns arising within each aggregate. Tasks with this characteristic are common in image understanding, such as when segmenting an image or extracting feature characteristics.

One methodology that has been applied to solving problems on multiple non-uniform data sets is associative processing [4, 13]. For small numbers of data sets this is very efficient; however, we are often restricted to operating on one data set (or problem instance) at a time, and some important applications involve computing many thousands of features [2]. We have also had some success with what could best be described as ad hoc methods: the solutions are often efficient [14], but are not the result of a uniform technique. Willebeek-LeMair and Reeves [15] have embedded binary trees in meshes to implement broadcast and reduction primitives on non-uniform, contiguous, regions, and applied those operations with great success to parallel image segmentation. In this paper we generalize these methods.

We propose an additional level of parallelism we call *multi-associativity* as a framework for performing associative computation on mulitple data sets simultaneously, thereby taking maximal advantage of the proximity inherent in many space mapped computations. The strategy is to simulate efficiently *within aggregates of PEs simultaneously* the associative operations typically supported in hardware at the array level. These include broadcast from controller to array and feedback from array to controller (e.g. count and global-OR of responders). Once we have efficient implementations of these primitives, we can apply them to existing array algorithms, or build them up into other complex functions. For this strategy to succeed on a real SIMD array (with only a single thread of control), we must construct multi-associative primitives that have a low branching factor with respect to the shapes of the aggregates.

We begin with an introduction to associative processing and a definition of the extension to multiple aggregates. We then discuss the mapping of multi-associativity onto the Content Addressable Array Parallel Processor (CAAPP) [13], the low-level processor of the Image Understanding Architecture (IUA) being developed at the University of Massachusetts and Hughes Research Labs. Finally, numerous sample algorithms are presented.

2. Multi-associative Processing

Associative processing is an effective way for performing operations on arbitrary aggregates of PEs holding related data. The prototypical associative operation is for the controller to broadcast a query to the array, and to receive a response from the matching elements either in the form of Some/None (global OR) or a Count. But associative processing, as opposed to the familiar associative memory operations, also enables the conditional generation of symbolic tags based on the values of data, and the use of those tags to constrain further processing. Only subsets of the data are involved in any particular operation, but all pixels and features with a given set of properties are processed in parallel; the fundamental operations are [4]:

*Department of Computer and Information Science, University of Massachusetts at Amherst, Amherst, Massachusetts 01003, NetAd: herbordt@cs.umass.edu. This work was supported in part by DARPA under contracts DACA76-86-C-0015, and DACA76-89-C-0016, monitored by the U.S. Army ETL; by the AFOSR, under contract F49620-86-C-0041; and by a Coordinated Experimental Research grant from the NSF (DCR 8500332).

1. Global Broadcast—Local Compare—Activity Control

2. Some/None Response

3. Count Responders

4. Select a Single Responder

For efficiency, the fundamental associative operations must be supported in hardware. The CAAPP, for example, has specialized hardware for Some/None, Response Count, and Select Single Responder operations: the execution times of these operations are 0.1, 1.6, and 2.4 micro-seconds, respectively [12]. Global broadcast, local compare, and activity control are all standard SIMD array operations.

Associative processing is much more efficient than serial processing, but still yields an obvious bottleneck. In some models of computer vision computation [2], it is necessary to compute attributes of thousands of tokens; a simple associative system would typically perform these computations in parallel, but one token at a time. We would prefer to operate on all PE aggregates simultaneously, but this involves an additional level of parallelism; how is this possible with only one thread of control?

As a solution we propose that a set of primitives— corresponding to the fundamental associative operations presented above—can be constructed to run efficiently on multiple aggregates of PEs simultaneously. We present the implementation of those primitives in the next section; here we describe in more detail the multi-associative model, independent of specific architectures (except that we assume an SIMD array of N PEs). To facilitate the definition, we use set notation.

We begin by partitioning the array into $k \leq N$ aggregates $S_i \in \{S_1, \ldots, S_k\}$ of PEs as defined by the mapping of the data sets in the application. Next, we define associative capabilities analogous to those presented above, but that can be executed *simultaneously within each aggregate* S_i. In order for these multi-associative primitives to be meaningful, we must replace the role of the controller with an analog available in each PE aggregate: we propose that a suitable replacement is simply any specified PE or subset of PEs within that aggregate. For example, "global broadcast" from controller to array is replaced with: "$\forall i$, multicast by a selected PE (or subset of PEs $S_i^{MCast}) \subseteq S_i$, to a selected subset of receiving PEs, $S_i^{Rec} \subseteq S_i$." In the same way, "count responders" is replaced with: "$\forall i$, send count of subset of selected PEs, $S_i^{Sel} \subseteq S_i$, to the selected PE (or subset $S_i^{Rec}) \subseteq S_i$." Whenever $\left|S_i^{MCast}\right| > 1$, the signal multicast to the subset S_i^{Rec} is the OR of the values multicast by the individual elements of S_i^{MCast}.

We also define operations on the aggregates themselves: Split/Merge Aggregate and Data Transfer Between Aggregates. Split allows, for all aggregates in parallel, the S_i to be split into any number of new aggregates, depending on some predicate. Merge allows any number of aggregates to be combined into a single aggregate. We now have *six* capabilities defining multi-associative processing:

1. Multicast by a Subset/Local Compare/Activity Control

2. Some/None Response to a Subset

3. Count Responders to a Subset

4. Select a Single Responder

5. Split/Merge Aggregates

6. Data Transfer Between Aggregates

There are several benefits to this approach. The first four capabilities allow us to map all algorithms defined over associative processor arrays to associative aggregates of PEs in parallel. The Split/Merge operations add new capabilities to the associative model: there are three basic advantages.

1. Split and Merge enable the use of divide-and-conquer strategies. It is now possible, for several important algorithms, to iteratively partition the S_i into subsets, solve the subproblems, and then merge. Just as in the sequential case, we can thereby reduce the computational complexity from linear to logarithmic.

2. We can save different partial results in individual PEs, as required by parallel prefix and some reductions.

3. We can take advantage of implicit ordering resulting from the shape of an aggregate, for example, when we wish to extract corner points in order around a border.

We emphasize the major characteristic of the multi-associative model: there is still only one controller. Therefore, only algorithms with a branching factor $\ll$ than the number of associative sets will run efficiently. However, the primitive operations and most of the applications considered so far lend themselves to algorithms that meet this criterion.

3. Multi-Associativity on the CAAPP

3.1 CAAPP Architecture

The CAAPP is an associative array of one-bit processing elements (PEs). The controller, PE instruction set, and memory organization are similar to that of many SIMD processors. Hardware support is provided for Some/None and Responder-Count operations. Communication among PEs takes place in two ways: using the nearest neighbor mesh, and via a reconfigurable mesh called the coterie network. The first method is similar to that used by the the MPP [1] (and many others). In the second method, PEs transmit information by writing to a specified register connected to the coterie network. PEs then read a register which will have been set to the OR of these signals, within a local group as defined by the network configuration. In order to distiguish broadcast by PEs from the broadcast by the controller, we refer to this operation as "coterie multicast."

Each PE in the CAAPP controls four switches (simplified here from eight), enabling the creation of electrically isolated groups of PEs that share a local associative Some/None feedback circuit. These switches control access in the different directions (north, south, east, west). The isolated groups of processors, called *coteries*, have access only to the multicast signals of PEs within their own circuit. The coterie network switches are set by loading the corresponding bits of the mesh control register in each PE. Because each PE views the mesh control register as local storage, coterie configurations can be loaded from memory, or can be based on data dependent calculations. Coteries can be any contiguous set of PEs; the coterie network can also be set so that columns and rows are isolated. The row and column "buses" can be arbitrarily segmented and thus can emulate the Mesh With Reconfigurable Buses [10] and the Polymorphic-Torus [9].

3.2 Basic CAAPP Operations

Load PE Id. Always available to each PE is its ID, defined as its address in row and column coordinates.

Select. PEs all contain an *activity* register, the value of which determines whether that PE will execute the instruction currently being broadcast by the controller. A PE with an activity register set to one is said to be "Selected."

Coterie Multicast. PEs multicast on the coterie network by loading the X register with the bit to be output, whence it propagates at electrical speeds for some distance across the network. The X register value is retransmitted every machine cycle by all PEs having already received it; the signal thus resembles a wavefront, moving outward until it reaches every PE on the circuit. The signals are input by reading the X register. If multiple PEs in an electrically connected region (coterie) write to their X registers, the resulting signal is the OR of those values. Coterie Multicast requires 8 cycles per bit on a 256×256 array.

Open/Close Switches. Each PE controls its coterie switches by writing a 0 or a 1 to the corresponding bit of the mesh control register. The entire register is read from (or written to) in a single cycle.

Route and Combine. By Route, we refer to an operation where all PEs can send data to any destination PE. If multiple PEs send packets to the same destination, then those packets can be combined according to some arithmetic or boolean operator. For example, in SumCombine, the result in a receiving PE is the sum of the contents of all the packets sent there. On the CAAPP, Route and Combine are implemented in software [5] and finish in time $O(d)$ where d is the maximum distance any packet must travel.

3.3 Multi-Associative Primitives on the CAAPP

In order to use the hardware support effectively, associative sets must be restricted to contiguous sets of PEs. Although the resulting capabilities are not as general as for the full multi-associative model, they are sufficient for the sample algorithms presented in the next section.

1. Multicast, Local Compare, Activity Control. Local compare and activity control are basic operations, multicast is implemented through the coterie network.

2. Some/None Response. Direct implementation of coterie multicast.

3. Count Responders. CountResponders has been implemented in two ways: using SumCombine, and using a two dimensional reduction technique. The latter (described in detail below) is usually the algorithm of choice, as it requires only $O(\log d)$ as opposed to $O(d)$ operations, where d is the maximum dimension of an aggregate. If d is small, however—a result that can be obtained quickly by the controller, then we use the simpler SumCombine.

4. Select a Single Responder. The method is to apply the standard associative SelectMax algorithm [3] to aggregates of PEs. If the PE ID is used as the key, then a unique PE in each aggregate will be selected. At no extra cost, all the PEs in the coterie "listen in" on the process, and so know the maximum value (and therefore the ID of the leader) at completion. The algorithm is bit-serial; for k bit integers the algorithm starts with the high order bit $(k-1)$. For each of k iterations i, PEs multicast bit $k-i$: if any are ones, then all PEs with a zero turn themselves off as they have been eliminated. If there are no ones, or if they are all ones, then no PEs are eliminated on that step. By the time the low order bit is reached, only the PEs with the maximum value remain. An analogous algorithm selects the PEs

with the minimum value. SelectMax and SelectMin run in $3k$ steps, where k is the number of bits in the word.

5. Create, Split/Merge Aggregates. Associative aggregates are created, split, and merged by opening and closing the coterie network switches to form electrically isolated regions (coteries).

6. Data Transfer Between Aggregates. Neighboring aggregates communicate by closing the coterie switches between them and using coterie multicast.

4. Multi-Associative Algorithms

To avoid redundancy, the routines are often presented as applied to one data set; in all cases, however, they can be applied to any number of data sets simultaneously.

4.1 Basic Operations

Create Connected Component. Each PE fetches the label of its four nearest neighbors via the mesh network and tests them against its own value. Switches are closed in the direction of the PEs whose labels are equal—or similar according to some measure—and opened otherwise.

Separate Border. To form coteries made up of border PEs (of existing coteries), each PE sets a flag if any of its switches are open. PEs then open connections in the directions of PEs whose flags are cleared.

Separate Lines. In order to restrict coteries to horizontal (vertical) lines, leave open the N and S (E and W) switches.

Elect Leader, SelectMin/Max. The extraction of information about data sets is facilitated if that information is collected in a small number of PEs, say one per aggregate, whence it can be rapidly transmitted to the host or intermediate level processor(s). ElectLeader selects a unique PE within each aggregate by running the multi-associative primitive Select Single Responder. SelectMin and SelectMax also apply the same underlying procedure, but where the input parameter is a user specified key.

Collect Sparse Region Info, Get Sorted List. If a small set S of arbitrarily placed PEs in each aggregate contain information to be collected, then an efficient method is to iterate the use of Select Single Responder and Multicast. First run ElectLeader to select an accumulator. Then iterate the following for the number of points in S. Execute Select Single Responder to select a point from S. The selected PE multicasts its information to the accumulator, and removes itself from S. The accumulator reads and processes that information. In Get Sorted List we instead run SelectMin or SelectMax on a key; the information collected will then be ordered. The complexity of both procedures is roughly equivalent to the number of PEs in S, times the complexity of SelectMax.

Parallel Prefix on a Line. This procedure works for horizontal and vertical lines, and can be extended to work on some curves [6]. No assumptions are made as to the lengths of the lines or the starting points. First we define parallel prefix: Given an ordered set $\{x_1, \ldots, x_n\}$ of n elements, one per processor, and a binary associative operator $*$, compute the n S_i's: $S_i = x_1 * x_2 * \ldots * x_i$, leaving the ith prefix sum in the ith processor. The implementation of parallel prefix takes particular advantage of the coterie network: it requires only $\log n$ communication steps, rather than the $2 \log n$ required for a tree-connected parallel processor.

To execute parallel prefix, there must an ordering to the PEs, in this case determined by the distance from the west end-point. PEs calculate their offsets as follows: each PE checks its W coterie switch to determine whether it is on the west end of the line. If so, it multicasts its column position. All PEs on the line read the data, and subtract it from their own column position to obtain an offset.

In the first iteration, PEs whose rightmost offset bit is a 0 open their E switches, the rest open W. The "0" PEs multicast their data, the "1" PEs receive it and "sum." In the next iteration, PEs whose rightmost *two* offset bits are $< 01_2$ do not participate. PEs with an offset ending in 01_2 open W, PEs with 11_2 open E. The "01" PEs multicast, the others receive and combine. In the next iteration, PEs whose rightmost *three* offset bits are $< 011_2$ do not participate. All PEs with an offset ending in 011_2 open W, PEs with 111_2 open E. The "011" PEs multicast, the rest receive and combine. This continues for $\log N$ iterations.

Reduction on a Line. Sometimes results must be collected, but without the need to save the intermediate values. Although simpler than parallel prefix, reduction still requires a logarithmic number of steps. Therefore the simplest method is to run ParallelPrefix and ignore the intermediate values.

4.2 Sample Algorithms

FindExtremum. To find the extrema in a region, run SelectMin (or SelectMax) on the column or row address. FindExtremum requires $3 \log n$ steps.

GetSmallestCircumscribingRectangle. Follows immediately from FindExtremum.

LabelConnectedComponents. ElectLeader is run to select a single PE in each component. Because of the way that the switches are set, PEs in each coterie will receive the leader ID at no extra cost. This necessarily unique value is the label of the component.

CreateBorder-CornerList. This procedure is a multi-associative implementation of the algorithm presented in [13]. The corners (or other information contained in border PEs) are accumulated in the leader PEs such as to retain their ordering: reconstructing the shapes of underlying regions is thus greatly simplified. See [6] for details.

GetAdjacentRegionLabels. Border PEs fetch the labels of the adjacent regions. A modified version of CollectSparseRegionInfo is then run on the boundary PEs; the modification is that PEs, whose data matches that just sent, remove themselves from the set immediately rather than sending the same label again. Since most of the cells will have redundant information, GetAdjacentRegionLabels only requires a number of iterations equal to the number of border regions, not the number of border PEs.

MergeRegions. At the beginning of this operation, leader PEs in pairs of regions have already determined the label of the neighboring region with which each is to merge (if any). The leaders multicast that label to their coteries. The PEs on the border between the regions close the coterie switches in the direction of that region.

Count(Selected)PEs. We now sketch a second version of CountPEs (the first used Combine). A detailed presentation (including CountSelectedPEs) can be found in [6]. The algorithm runs in three phases. The first begins with SeparateLines to form horizontal strips. The left and right end-points of each strip are identified; the right end-points are selected to be strip accumulators. Each left end-point multicasts its column ID, the corresponding right end-point reads it and obtains the number of PEs in the line. At the beginning of the second phase, only border PEs will have relevant information. We take advantage of that fact to perform a log time reduction on coteries consisting of contiguous accumulator PEs. At the end of the second phase, very few PEs will still be carrying information. The final phase consists of running ElectLeader to select an overall region accumulator and CollectSparseRegionInfo to combine the subtotals.

The first phase requires one communication step, the second $\log d$ communication steps, where d is the maximum dimension of the largest set. However, the third phase is not so easily bounded: the number of accumulators remaining from phase two is roughly equal to the number of local minima (in terms of column ID) on the region boundary. Although this number is usually small, it is possible to construct regions where it is not. We therefore use global feedback to bound the algorithm. If the the third phase has not completed after some number of iterations of CollectSparseRegionInfo (say ten), then the remaining region counts are obtained one at a time using the global operations SelectSingleResponder (to select a remaining region) and CountResponders (to obtain its count), a process that takes only a few micro-seconds per region.

GetMean. The next two algorithms are multi-associative implementations of known associative algorithms [4]. GetMean is similar to SelectMax in that both are bit serial over a k-bit label, and both run from high to low order ($k-1$ to 0). Assume we are trying to find the mean of a label L over the PEs in a region. We sum the L's by successively counting the PEs with the ith bit of L set, and scaling that count by 2^i. We start at the high order so that scaling can be accomplished with one shift per iteration. Select PEs with bit $k-1$ of L set. Run CountSelectedPEs to get the count. Add the count to the accumulator (zero to start). Shift the accumulator left 1 bit. Repeat this process for bits $k-2\ldots0$ of L, but without shifting after the final iteration. GetMean requires a number of iterations equal to $\log(Max(L))$; each iteration contains one add, one shift, and one CountSelectedPEs operation.

GetMedian. The method used is analogous to binary search: we find the range of possible values and successively halve the interval on each iteration. Run ElectLeader to select an accumulator. SelectMin and SelectMax are run to find the lower and upper bounds (L and H), and Count-PEs to obtain C, the number of PEs in the region. Let the initial guess be $G = \frac{L+H}{2}$. Select and count the PEs with a label greater than G; depending on whether the count is higher of lower than $C/2$, the the new guess G is either $\frac{H+G}{2}$ or $\frac{L+G}{2}$. H (or L) is updated to the former value of G. The algorithm converges after $\log(H-L)$ iterations.

Histogram within Regions. Run ElectLeader to select an accumulator. For each bin, multicast the value of that bin. PEs are selected according to whether their value matches that of the bin. Run CountSelectedPEs to get the bin count. The algorithm requires a number of iterations equal to the number of bins, and each iteration contains one CountSelectedPEs operation.

ParallelPrefix. This ParallelPrefix routine applies to rectangular aggregates; this technique is also extendable to semi-convex regions, that is, regions convex in the horizontal or vertical dimension. For regions of arbitrary shape, the multi-associative implementation of parallel prefix is more complex—for example, by using the tree embedding technique of [15]—and the applications fewer. Start by separating the region into horizontal strips with SeparateLines. Execute ParallelPrefixLine on these strips. Next, form a coterie of the right end-points of the lines with SeparateLine. Execute ParallelPrefixLine on these enpoints, then move these partial results down one PE. Multicast the values back down the rows. Each PE reads the multicast and combines that value with its own to obtain the final result. ParallelPrefix requires $2 \log d + 3$ arithmetic and communications operations.

Reduction. There are two methods: (1) run ParallelPrefix and ignore the intermediate results, and (2) use Combine. As with CountPEs, the choice depends on the shapes and sizes of the regions.

ConvexHull. The convex hull of a set of points $s_i \in S$ is defined as the smallest convex set contained in S. Intuitively, the convex hull in a plane can be found by conceptually wrapping S with a rubber band and eliminating the interior points [11]. One leading method is the Jarvis March [8], which is analogous to wrapping the points in a package, one hull edge at a time. The algorithm begins with the selection of a point p known to be on the hull, e.g. the point with the greatest X value. For each point s_i, the slope of the segment $\overline{ps_i}$ is calculated. The next point known to be on the hull is selected by finding the segment $\overline{ps_i}$ making the smallest angle with respect to the positive X-axis. This process is repeated for all h points on the hull; the serial algorithm thus has a complexity of $O(hN)$ where N is the total number of points and h the number of points on the hull. In the multi-associative version, we assume that the points in S are mapped to PEs according to their row and column coordinates. A point p on the hull is found using ElectLeader, and its ID is simultaneously distributed to the rest of S. The s_i calculate the angle formed by $\overline{ps_i}$ with the column axis. Calling SelectMin with that angle as the key locates the next point on the hull. The procedure is then repeated until the PE forming the smallest angle is p, in other words until the loop has closed. The complexity is $O(h)$, the number of points on the hull, if SelectMin is counted as a constant time operation.

5. Conclusions and Future Work

Two criteria for evaluating the success of a computational framework are whether the user can simply and efficiently: (1) create the necessary algorithms for the chosen application, and (2) map the framework onto target architectures.

1. Creating Algorithms. Multi-associativity has many desirable properties: it enables the application of associative algorithms to multiple non-uniform aggregates of PEs simultaneously. In addition, it supports operations on the aggregates themselves enabling the use of divide-and-conquer, and communication among aggregates. We have demonstrated these uses with many algorithms, and we believe that many more are possible.

2. Mapping onto Hardware. We have presented a mapping of multi-associativity onto the CAAPP. Evaluating the performance is a difficult issue: the theoretical complexities have yielded promising results, especially when combined with the fact that problem instances are being solved for multiple data sets simultaneously. Another way to gauge the performance is to compare the execution time ratio of the associative to the multi-associative primitives with the number of data sets being processed simultaneously. Since the former is on the order of from 1 to 100, and the latter often in the thousands, we believe that good results are likely. Of course the only way to be certain of that performance advantage is to test the multi-associative algorithms on large numbers of applications.

Multi-associativity can also be mapped onto existing mesh-connected SIMD array processors. The existence of an efficient mapping of multi-associativity onto meshes with reconfigurable buses seems likely, although perhaps more complex than that on the coterie network. On meshes with no local broadcast, the core primitives of reduction and broadcast can be implemented using the tree embedding technique of [15]. Alternatively, broadcast can be implemented with a connected components algorithm, and reduction by the general routing technique in [7]. In either case, the lower bound complexity is restricted by the fact that the communication proceeds at one processor per move.

Acknowledgments

We would like to thank Mike Scudder for his review of an earlier version of this paper, and Jim Burrill, Deepak Rana, and Mike Rudenko for their useful comments.

References

[1] K.E. Batcher (1982): "Bit-Serial Parallel Processing Systems," *IEEE Trans. Comps.* C-31 (5).

[2] J. Brolio, B.A. Draper, J.R. Beveridge, A.R. Hanson (1989): "ISR: A Database for Symbolic Processing of Computer Vision," *IEEE Computer,* December.

[3] A.D. Falkoff (1962): "Algorithms for Parallel Search Memories," *J. ACM,* 9 (4), pp. 488-511.

[4] C.C. Foster (1976): *Content Addressable Parallel Processors,* Van Nostrand Reinhold Co. New York.

[5] M.C. Herbordt, C.C. Weems, J.C. Corbett (1990): "Message Passing Algorithms for a SIMD Torus with Coteries," *Proc. 2nd ACM Symp. Par. Algs. and Archs.*

[6] M.C. Herbordt, C.C. Weems, M.J. Scudder (1991): "A Computational Framework and Algorithms for Low-Level Support of Intermediate-Level Vision Processing," *COINS TR 91-26, U. Mass.*

[7] M.C. Herbordt, J.C. Corbett, J. Spalding, C.C. Weems (1991): "Practical Algorithms for Online Routing on SIMD Meshes," to be submitted for publication.

[8] R. Jarvis (1973): "On the Identification of the Convex Hull of a Finite Set of Points in the Plane," *Inf. Proc. Let.* 2.

[9] H. Li, M. Maresca (1989): "The Polymorphic-Torus Architecture for Computer Vision," *IEEE PAMI,* 11 (3).

[10] R. Miller, V.K. Prasanna Kumar, D. Reisis, Q.F. Stout (1988): "Meshes With Reconfigurable Buses," *Proc. MIT Conf. on Advanced Research in VLSI.*

[11] F.P. Preparata, M.I. Shamos (1985): *Computational Geometry: An Introduction,* Springer-Verlag, New York.

[12] D. Rana, C.C. Weems (1990): "A Feedback Concentrator for the Image Understanding Architecture," *Proc. Int. Conf. Application Specific Array Processors.*

[13] C.C. Weems, S.P. Levitan, A.R. Hanson, E.M. Riseman, J.G. Nash, D.B. Shu (1989): "The Image Understanding Architecture," *Int. J. Computer Vision,* 2 (3).

[14] C.C. Weems, E.M. Riseman, A.R. Hanson, A. Rosenfeld (1991): "An Image Understanding Benchmark for Parallel Computers," *JPDC,* 9.

[15] M. Willebeek-LeMair, A.P. Reeves (1990): "Solving Nonuniform Problems on SIMD Computers: Case Study on Region Growing," *JPDC,* 8.

Efficient Parallel Maze Routing Algorithms on a Hypercube Multicomputer

Cevdet Aykanat and **Tahsin M. Kurç**

Faculty of Engineering & Science, Department of CIS

Bilkent University, Ankara, Turkey

Abstract   *Lee's maze routing algorithm is parallelized and implemented on an Intel iPSC/2 hypercube multicomputer. A new parallel front wave expansion scheme is proposed. The proposed scheme performs two front wave expansions concurrently, one starting from the source cell and the other one starting from the target cell. The proposed scheme increases the processor utilization and decreases the total number of interprocessor communications. An efficient parallel sweeping scheme which avoids interprocessor communication is proposed. Path recovery phase is highly sequential by nature. A scheme, which pipelines the parallel sweeping operations by the path recovery phase, is also proposed.*

1 Introduction

Lee's maze router[1] is a well known algorithm for global wire routing in VLSI. In global routing for gate arrays, a global grid represents the wiring surface when one layer is used for net interconnections. Net terminals are located within the cells. The vertical and horizontal grid lines between neighbor cells represent the channels for wire routing. The overall objective is to realize all the net interconnections using shortest routes. Lee's maze router[1] is then applied to find the shortest interconnection route between individual net terminals in a pre-determined order. As the interconnection paths between net terminals are constructed, some of the cells will be declared as *blocked*. In this work, for the sake of simplicity, a cell is declared as *blocked* when it is used in a single wire path. Only two pin nets are considered in this paper. Path for a net can go from a cell to a *free* neighbor cell by crossing either a common vertical or horizontal channel. Hence, moves from a cell are restricted to its four adjacent cells (cells to its *North*, *East*, *South*, and *West*). Lee's maze routing algorithm consists of three phases, namely, *front wave expansion*, *path recovery*, and *sweeping* [2]. The Lee's maze routing algorithm is given below.

A queue, called expansion queue, initially contains only the source cell. A queue, called sweep queue, is initially empty. A two dimensional $N \times N$ Status array holds the status for the cells of an $N \times N$ grid. All the free cells are initially unlabeled.

Step 1. *Remove a cell c from the expansion queue. Examine the four adjacent cells of the cell c using the current information in the Status array. Discard the blocked and already labeled adjacent cells. Update the status of the unlabeled free adjacent cells as labeled in the Status array and add those cells to the expansion queue. If all the adjacent cells of the cell c are either blocked or already labeled, then add the cell c into the sweep queue. Repeat this step until target cell is reached.*

Step 2. *Follow the labels starting from the target cell until the source cell is reached. Label the visited cells as blocked.*

Step 3. *Remove a cell c from sweep queue or expansion queue. Follow the labels starting from c until a blocked or an unlabeled cell is reached. Unlabel the visited cells in the status array. Repeat this step until both sweep queue and expansion queue become empty.*

Front wave expansion phase (step 1) is a *breadth-first* search strategy starting from the source cell. During *front wave ex-*pansion phase, the labeling operation of a *free* and *unlabeled* adjacent cell is performed such that, the label points to the cell *c* being expanded. The *front wave expansion* phase terminates successfully when the target cell is labeled. The *front wave expansion* phase may also terminate when a remove operation from an empty expansion queue is attempted. Such a termination condition indicates the non-existence of a wire-path from the source to the target. The Lee's maze routing algorithm is guaranteed to find the shortest wire path between the source and the target. The *front wave expansion* phase is followed by the *path recovery* phase (step 2) and *sweeping* phase (step 3).

Since the grid size may be quite large for real VLSI problems, Lee's router algorithm is time consuming and it requires large amount of memory to hold the status of the grid cells. Hence, Lee's algorithm is a good candidate for parallelization on a distributed memory multiprocessor. In this paper, the parallelization of Lee's maze routing algorithm on a commercially available multicomputer implementing the hypercube topology is addressed.

2 Parallel Front Wave Expansion

The nature of communication required in *front wave expansion* phase corresponds to a two dimensional mesh. That is, each processor needs to communicate only to its *north*, *east*, *south*, and *west* neighbors. It is well known that, a $2^{\lfloor d/2 \rfloor} \times 2^{\lceil d/2 \rceil}$ processor mesh can be embedded into a d-dimensional hypercube [3]. The trade-off between volume of interprocessor communication and processor utilization is resolved by following the scheme proposed in [2]. The $N \times N$ routing grid is first covered by $h \times w$ square (or rectangle) subblocks starting from the top left corner and proceeding left to right, top to bottom. Scattered mapping is then applied over the coarse grid consisting of contiguous $h \times w$ grid subblocks. Hence, adjacent grid subblocks are assigned to different but neighbor processors of the mesh. The simplifications presented in [2] are also maintained in this work.

The first parallel front wave expansion algorithm implemented in this work is very similar to the parallel algorithm given in [2]. This scheme is referred to as *Expansion Starting from Source Only (Sonly)* scheme here. In spite of the given partitioning scheme, the *Sonly* scheme may result in low processor utilization especially for large h and w values. This is due to the *expansion* of a single *front wave* beginning from the source cell. Note that, finding a routing path from source to target is equivalent to finding a path from target to source. Hence, two *front waves*, one beginning from the source (*source front wave*) and the other one beginning from the target (*target front wave*), can be *expanded* concurrently. This scheme has a potential to increase the processor utilization. The parallel algorithm for the *Expansion Starting from Source and Target* (S+T) scheme is given below :

Each processor stores and maintains a local status array, a local expansion queue, a sweep queue, and four send and four receive queues for communicating with its four neighbor processors. Initially, all local queues are empty and all local cycle counts are initialized to 1. The host processor broadcasts the coordinates of the source cell and the target cell to all processors. The processor which owns the source/target cell location adds the local coordinates of the source/target cell together with a source/target front wave tag to its local expansion queue. Then each processor executes the following algorithm.

Step 1. *The cells in the local expansion queue may belong either to the current source front wave or to the current target front wave. Each processor examines these cells accordingly for expansion in four directions. The local adjacent cells of the cells being expanded are examined for adding to the local expansion queue for later expansion. The adjacent cells that are detected to belong to grid partitions assigned to neighbor processors are added to the corresponding send queues for later communication together with the tag of the cells being expanded.*

Step 2. *Each processor transmits the information in its four send queues to their destination processors.*

Step 3. *The adjacent cells in the four receive queues may belong either to the current source front wave or to the current target front wave. Each processor examines these cells accordingly for adding them to its local expansion queue for later expansion.*

Step 4. *Each processor, after incrementing its local cycle count by 1, checks whether it has received a message from the host. It proceeds to step 1 if the message has not been received yet or if the message has been received with an upper bound value greater than or equal to the current value of the local cycle count. It terminates only if the local cycle count value is greater than the upper bound value received, and signal the host about its termination.*

During the expansion process at step 1, the routing status of four adjacent cells of a cell being expanded are examined. If the current routing status of an adjacent cell is *unlabeled*, the adjacent cell is labeled with the reverse expansion direction and *tagged* with the *tag* of the cell being expanded in the local *status* array. However, if the current routing status of an adjacent cell is *blocked* or *labeled* with the same *tag* of the cell being expanded, then the adjacent cell is discarded. Otherwise, if the adjacent cell is already *labeled* with a different *tag* compared to the *tag* of the cell being expanded, it shows the *collision* of two different *front waves*. At step 3, the local cells stored in the *receive* buffers are examined in a similar way. The processor which detects the *collision* at step 1 or 3, signals the host about the collision. It also includes the current value of the local cycle count and the local coordinates of the pair of adjacent cells in the message.

The given parallel algorithm does not guarantee that all processors will be executing the same front wave expansion cycle at any instant of time. Some of the processors may be found to be *leading* some others by a number of expansion cycles. A *leading* processor may be the first processor which detects a *collision*. Hence, if the host processor terminates the *front wave expansion* as soon as it receives a *collision* message, the path to be recovered may not be the shortest path from source to target. *Lagging* processors have potential to detect collisions on earlier cycles. Hence, all *lagging* processors should be allowed to perform *expansion* until the cycle count of the *leading* processor which has detected the *collision* the first time. This is achieved by the scheme given at step 4 of the algorithm. The details of this scheme are described in [4].

In the proposed parallel algorithm, the number of *expansion* cycles to be performed in the *front wave expansion* phase is reduced by a factor of two compared to the original parallel algorithm. Hence, the total number of local communications is reduced by a factor of two, since the number of local communications per *expansion* cycle is fixed to four. The proposed algorithm is expected to reduce the total number of expanded cells. Hence, the proposed algorithm is also expected to reduce the volume of communication. However, the amount of computation required for the expansion of an individual cell is

increased in the proposed scheme. The proposed parallel algorithm will increase the processor utilization compared to the original algorithm.

Blocking *send* and *receive* messages are issued at step 2 and step 3, respectively, of the given (non-overlapped) parallel algorithm. Each processor may stay idle waiting for data from its four neighbors after sending data to them. This idle time can be reduced by rearranging the first three steps of the non-overlapped node algorithm as follows :

Step 1. *Issue four non-blocking receives to receive data into receive queues. Issue four non-blocking sends for front wave cells (at depth q) to transmit send queues to corresponding processors.*

Step 2. *Expand the front wave cells (of depth q) in local expansion queue and insert their adjacent cells (of depth q+1) either into the local expansion queue or into the corresponding send queues accordingly.*

Step 3. *Synchronize on the non-blocking receive messages issued at step 1. Then, expand the front wave cells (of depth q) in the receive queues and add their adjacent cells either into the local expansion queue or into the corresponding send queues accordingly.*

In this overlapped scheme, the transmission of data in the send queues constructed in the previous expansion cycle is initiated (step 1) before the local expansion computations (steps 2 and 3) in the current cycle. The front wave cells in the receive queues are expanded in place (step 3) instead of being added into the local expansion queue for later expansion. Hence, the set-up time and the transit time for the four send operations at step 1 are overlapped with the computations at step 2 and 3.

The *blocking receive* messages at step 3 of the non-overlapped scheme and the *synchronization* on *non-blocking* receive messages at step 3 of the overlapped scheme constitute a local synchronization between neighbor processors of the mesh. That is, processors do not proceed to the next *front wave expansion* cycle before receiving messages from all four neighbors. Due to this local synchronization, both of the parallel algorithms are guaranteed to find the shortest path between the source and the target whenever a path from source to target exists. However, these parallel algorithms will not terminate if no path exists between source and target. The schemes to provide global termination detection for such cases are discussed in [2, 4].

3 Parallel Path Recovery and Sweeping

It has been experimentally observed that $S+T$ scheme outperforms the *Sonly* scheme. Therefore, the path recovery and sweeping algorithms are derived assuming that $S+T$ scheme is used for the front wave expansion phase. At the end of the *front wave expansion phase*, the host program broadcasts the two cells C_1 and C_2 at which the collision of two front waves occurred. The processor(s) which own(s) the cells C_1 and/or C_2 starts the path recovery phase. The path recovery phase is terminated after source and target cells are reached.

During the expansion of a local boundary cell C_b at step 1 (step 2) of the non-overlapped (overlapped) scheme, the status information about the non-local adjacent cell(s) are maintained by the neighbor processor(s). The situation may be such that the cell C_b can only be expanded into non-local adjacent cell(s) which may be blocked or labeled with the same tag of the cell C_b. The decision for adding such local boundary cells into the local sweep queue requires extra communications during the front wave expansion phase. In fact, only these cells will introduce interprocessor communication during the sweeping phase. If C_b can be expanded into an at least one local adjacent cell, then there is no need to add C_b into the sweep queue even if a non-local adjacent cell of cell C_b is found to be free by a neighbour processor. Because, the parent cells on the paths from C_b back to the target or source cell will be unlabeled during the local sweeping phase starting from a grandchild cell (of

C_b) in the local expansion or sweep queue. Hence, in order to avoid interprocessor communication during the sweeping phase and the extra communication during the front wave expansion phase, local boundary cells which can only be expanded to non-local adjacent cells at step 1 (step 2) of the non-overlapped (overlapped) front wave expansion algorithm are added into the sweep queue. The first parallel *path recovery* and *sweeping* algorithm is given in the following subsection.

3.1 Non-pipelined Scheme

In this scheme, the sweeping phase starts after the completion of path recovery phase. The host and node programs for the non-pipelined scheme are given below :

Host Program :

Step 1. *Start path recovery phase by sending C_1 and C_2 to the processor(s) which own(s) these cells.*
Step 2. *Wait for source and target reached signal from nodes.*
Step 3. *Broadcast start sweep signal to nodes.*
Step 4. *Wait for sweep terminated signals from all nodes.*
Step 5. *Terminate the distributed sweep phase.*

Node Program :

Step 1. *Wait for a path recovery cell C.*
Step 2. *Follow the labels starting from cell C until source, or target, or a non-local boundary cell is reached. Label visited cells as blocked. Inform the host if source or target cell is reached. If a non-local boundary cell is reached, send this cell to the neighbor processor which owns it.*
Step 3. *Repeat steps 1 and 2 until a start sweep phase signal is received from host.*
Step 4. *Remove a cell c from the local sweep or expansion queue. Follow the labels starting from c until a blocked, or an unlabeled, or a non-local boundary cell is reached. Unlabel visited cells.*
Step 5. *Repeat step 4 until both sweep and expansion queues become empty. Then, inform the host about the termination of local sweeping phase.*

Since the path recovery phase is highly sequential by nature most of the processors wait idle during the path recovery phase. In order to reduce this idle time, an algorithm is proposed in the next subsection. In this scheme path recovery and sweep phases are pipelined in a way that processors which do not perform path recovery, can initiate the sweeping of some of the cells in its expansion and sweep queues.

3.2 Pipelined Scheme

Assume that the path from source to target is found in p cycles during the front wave expansion phase using Sonly scheme. Also assume that the cell Q is reached after q cycles of path recovery phase as is illustrated in Fig.(1). Take any cell c with depth d_C (from source) in the local expansion queue or the sweep queue constructed during the front wave expansion phase. In the worst case, the expansion path from c to source joins the *shortest path* from target to source at a cell X as is illustrated in Fig.(1). Then, the depth of the path from X to source, d_X, is $d_X = d_C - d_{CX} = d_Q - d_{XQ}$, where d_{CX} and d_{XQ} are the depths of the paths from C and Q to X respectively, and $d_Q = p - q$ is the depth of Q to source. Hence, $d_{CX} = (d_C - p + q) + d_{XQ}$. However, $d_{CX} + d_{XQ} \geq M_{CQ}$, where M_{CQ} is the manhattan distance between C and Q. Hence, $d_{XQ} \geq \frac{1}{2}(M_{CQ} + p - q - d_C)$, and $d_{CX} \geq \frac{1}{2}(M_{CQ} + d_C - p + q)$. Thus,

$$r = \frac{1}{2}(M_{CQ} + d_C - p + q) \qquad (1)$$

sweeping cycles can be performed for a cell c in the expansion or sweep queue at that instance of path recovery phase. If

($S+T$) scheme is used in front wave expansion phase, then there are two front waves. Hence, the path recovery is performed in two directions, one towards source and other towards target. Therefore, there are two Q cells, Q_s and Q_t, reached after q_s and q_t cycles of path recovery phase towards source and target, respectively. For each cell, r sweeping cycles is calculated inserting either $q = q_s$ or $q = q_t$ in Eq.(1). In the *pipelined scheme*, local sweep and expansion queues of each processor should have depth information. This depth information shows the distance of a cell from source or target according to its tag. This depth information is required to calculate the number of sweeping cycles that can be performed for a cell. The host program for the pipelined scheme is the same as the host program given for the non-pipelined scheme. The node program for the pipelined scheme is given below.

Node Program :

Step 1. *If a cell C is received for path recovery, perform path recovery until source or target or boundary is reached. If a boundary cell C_b is reached during path recovery then send C_b to corresponding neighbor processor. If C_b is tagged from target, set $Q_t = C_b$, otherwise set $Q_s = C_b$. Broadcast Q_s, q_s or Q_t, q_t. If source or target is reached during path recovery inform host.*
Step 2. *If not performing path recovery then remove a cell c from sweep queue or local expansion queue.*
   (a) Calculate the r value for the cell c.
   (b) If $r \leq 0$ then put the cell back into the sweep queue.
   (c) If $r > 0$ then perform sweeping until r becomes 0 or a blocked cell or an unlabeled cell or boundary is reached.
   (d) If $r = 0$ is reached, add the new cell into sweep queue with the new depth information.
Step 3. *If new Q_s, q_s or Q_t, q_t are received replace old ones.*
Step 4. *If end of path recovery signal is received from host then go to step 5 else go to step 1.*
Step 5. *Perform the non-pipelined sweep algorithm.*

4 Experimental Results

All schemes for parallel maze routing has been coded in C language and run on an iPSC/2 hypercube multicomputer. Timing results are obtained and displayed graphically in Figures 2-6. The meanings of the abbreviations in the figures are; **S+T** : $S+T$ scheme, **S** : *Sonly* scheme, **NO** : non-overlapped scheme, **O** : overlapped scheme, **NONPIP** : non-pipelined scheme, **PIP** : pipelined scheme. Figures 2-6 are constructed by using the averages of the timing results measured on 4 randomly generated different grids with 40-45% randomly generated blockages and randomly generated nets.

Fig.(2) displays the execution times of various *parallel front wave expansion* algorithms as a function of h=w values on an 8 processor hypercube for a 1024x1024 grid. Execution times of all algorithms decrease with increasing w at the beginning due to the decrease in the volume of communication. However, execution times begin to increase after a turn over value of w due to the increase in the processor idle time with increasing w. Speed-up curves in Fig.(3) are obtained by running various parallel *front wave expansion* algorithms using the optimal h, w value for 1024x1024 grids. Fig.(3) illustrates that speed-up increases with increasing grid size. As is seen in Fig.(2) and Fig.(3), *overlapped* scheme gives better performance compared to the *non-overlapped* scheme. Fig.(2) and Fig.(3) also illustrate that $S+T$ scheme outperforms the *Sonly* scheme. A maximum speed-up of 4.75 is obtained on an 8 processor hypercube. Fig.(4) illustrates the efficiency curves for the optimal parallel *front wave expansion* algorithm (*overlapped* $S+T$ scheme) using the optimal h, w value. As is seen in Fig.(4), the efficiency of the *overlapped* $S+T$ algorithm remains almost constant when both the number of processors and grid size are doubled. Hence, it can be concluded that, the overlapped $S+T$ scheme scales on the hypercube architecture.

Fig.(5) displays the execution times of the two parallel *path recovery + sweep* algorithms as a function of h=w values on an 8 processor hypercube for a 1024x1024 grid. Execution times of both algorithms decrease with increasing w at the beginning due to following two reasons. First, in the path recovery phase, the number of interprocessor communications decreases with increasing w values. Second, in the sweeping phase, the size of the local expansion queues decrease with increasing w values due to the decrease in the number of boundary cells. However, execution times begin to increase after a turn over value of w due to the deterioration in the load balance during the sweep phase. As is also seen in Fig.(5), the *pipelined* scheme degrades the performance slightly due to the computational overhead involved in the calculation of r values. The percent processor idle time in the *non-pipelined* scheme will increase with increasing number of processors in the hypercube. However, the computational overhead introduced by the pipelined scheme is fixed for each cell in the local queues. Hence, the *pipelined* scheme is expected to increase the performance on hypercubes with larger number of processors. Speed-up curves in Fig.(6) are obtained by running the parallel (*pipelined* and *non-pipelined*) *path recovery + sweeping* algorithms using the optimal h, w value. As is seen in this figure, almost linear speed-up is obtained for large grid sizes.

References

[1] C. Y. Lee, "An algorithm for path connections and its applications," *IRE Trans. Electronic Computers*, Vol. EC-10, Sept. 1961.pp. 346-365.

[2] Youngju Won and Sartaj Sahni, "Maze Routing On a Hypercube Multiprocessor Computer," *Proceedings of Intrl. Conf. on Parallel Processing*, St.Charles, August 1987, pp. 630-637.

[3] Y. Saad and M. Schultz, "Topological properties of hypercubes," *Research Report, YALEU/DCS/RR-389, Computer Science Dept., Yale University*, Jun. 1985.

[4] T. M. Kurç, C. Aykanat, and F. Erçal, "Parallelization of Lee's Routing Algorithm on a Hypercube Multicomputer", in the Proceedings of *The Second European Distributed Memory Computing Conference*, Munich, Germany, April 22-24, 1991.

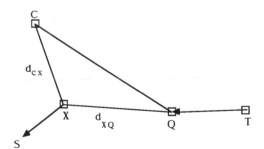

Fig.1 Calculation of r value

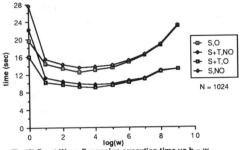

Fig.(2) Front Wave Expansion execution time vs h = w

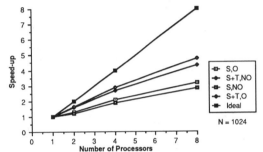

Fig.(3) Speed-up curves for Front Wave Expansion

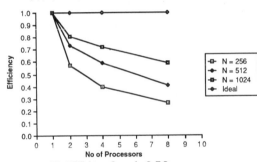

Fig.4 Efficiency Curve for S+T,O

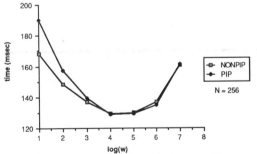

Fig.(5) Path Recovery + Sweep execution time vs h = w

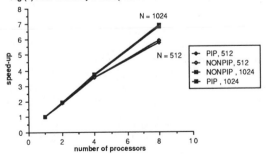

Fig.(6) Speed-up curves for Path Recovery + Sweep

On the Performance of a Deadlock-free Routing Algorithm for Boolean n-Cube Interconnection Networks with Finite Buffers[1]

Ming-yun Horng and Leonard Kleinrock
Computer Science Department
University of California, Los Angeles
Los Angeles, CA 90024-1596
(213) 825-3643

Abstract

This paper presents a mathematical model for evaluating the performance of a deadlock-free routing algorithm for Boolean n-cube interconnection networks with finite buffers. With this deadlock-free routing algorithm, all messages entering the network will be delivered correctly to their destinations without discards or deadlocks. We solve for the throughput of the network, the average message delay and the probability of acceptance of an input message. We determine the effect of the buffer size on performance and show that only a few buffers in each node are essential to yield good performance. We also show that the throughput of the network does not degrade even when the network is saturated.

Keywords: Boolean n-Cube Networks, Hypercubes, Interconnection Networks, Multiprocessor systems, Routing Algorithm, Performance Analysis.

1 Introduction

A major problem in designing a multiprocessor system is to construct a fast and efficient interconnection network among the processors. Many interconnection networks have been proposed [7], [18]. There is no single network that is considered the best for all applications. Recently, the Boolean n-cube interconnection network (also known by such names as binary n-cube, cosmic cube and hypercube) has been drawing much attention due to its powerful topological properties and its application to high speed switching networks. Several research and commercial systems have been built based on this type of network [17], [10], [8].

One important requirement is that the interconnection network should be deadlock-free. Deadlocks may occur as node buffers become full. A node with full input buffers also blocks its neighbors' output channels and increases the chance of the neighbors becoming blocked. When the movement of messages comes to a halt, the system crashes. Several deadlock prevention techniques based on the concepts of removing or avoiding cycles of channel dependency have been developed [6], [14].

However, with current VLSI technology, the channel speed in an interconnection network is approaching one gigabit per second [2]. At this high speed, the amount of time that the routing algorithm can afford to spend in making routing decisions is severely constrained. The routing algorithm must be very simple.

In this paper, by exploiting the homogeneity of the network and applying the concepts of timestamps [4] and backtracking [8], we present a simple deadlock-free routing algorithm for finite-buffered Boolean n-cube networks. With this algorithm, most messages are transmitted over their shortest paths from source to destination nodes. In a few situations, when a node does not have enough free buffers to accept all the incoming transit messages, the node sends some of its buffered messages on less direct paths to avoid congestion. Moreover, every message carries a globally unique timestamp. The oldest message will make nonstop progress to its destination without being backtracked. As soon as the oldest message has been delivered, another message becomes the oldest. Thus, every message accepted into the network is guaranteed to arrive at its destination without loss.

Boolean n-cube networks have been analyzed in the literature [5], [16], [3], [1]. However, mathematical models that consider queueing effects are rare. Moreover, in a large scale interconnection network, the storage capacity of each node is limited. Thus, a model which assumes there are an infinite number of buffers in each node cannot reflect the actual behavior of the system. Abraham and Padmanabhan [1] developed a model considering finite buffers, however, in which messages can be lost during transmission if the buffers at an output channel are full.

This paper presents a mathematical model for evaluating the performance of an algorithm with backtracking for lossless finite-buffered Boolean n-cube interconnection networks. We solve for the throughput of the network, the average message delay and the probability of acceptance of an input message. We determine the effect of the buffer size on the performance. We show that only a few buffers in each node are essential to yield good performance. We also show that the throughput of the network does not degrade even when the network is saturated. The match between the model and the simulation results is extremely good.

[1]This work was supported by the Defense Advanced Research Projects Agency under Contract MDA 903-87-C0663, Parallel Systems Laboratory.

2 Preliminaries

2.1 Operation of the Network

Topological properties of the Boolean n-cube network are discussed in [15], [17]. A Boolean n-cube network consists of 2^n nodes, each addressed by an n-bit binary number from 0 to $2^n - 1$. Nodes are interconnected in such a way that there is a link between two nodes if and only if their addresses differ in exactly one bit position. Every node has exactly n neighbors. A Boolean 4-cube network is shown in Figure 1.

Each node handles messages for several local processors. Communication among nodes is achieved by multi-hop message passing. The header of a message can be computed as the exclusive-OR of the message's source and destination addresses. This information indicates the dimensions the message must traverse before reaching its destination. A one-bit in the header corresponds to a valid channel for transmission. Whenever a message is sent along a valid channel, the corresponding one-bit is changed to zero and the message is one hop closer to its destination. When the header contains only zeroes, the message has arrived at its destination.

Also, a message can be sent along a non-valid channel while the corresponding bit is changed from zero to one. As a result, the message is sent one hop farther away from its destination. The message then needs an extra hop to move itself back along this dimension later before reaching the destination. We say a message is forwarded if it is sent along a valid channel. A message is said to be backtracked if it is sent along a non-valid channel.

Since the propagation delay between two neighboring nodes is very small, it is possible to operate the system synchronously. We assume time is divided into cycles with a duration which corresponds to the transmission time of a message to its neighbor. We assume that all messages are of the same fixed size. We further assume that a node is capable of sending messages along its n channels simultaneously. Thus, a node can receive up to n messages from its neighbors in a cycle. The communication error rate in a well protected interconnection network is assumed to be extremely small and negligible.

2.2 Deadlock-free Routing Algorithm

At the beginning of a cycle, each node of the network makes its two-phase routing decision as follows. In the first phase, every node randomly selects one of the one-bits from the header of every message the node currently has. If more than one message is selected with the same bit position, the one having the highest priority (to be defined later) is successfully assigned to the channel. Obviously, the random assignment of messages to channels in the first phase is not an optimal approach in terms of the number of messages transmitted in a cycle. However, in [9] we have shown that the mean delay of the random approach is fairly close to the lower bound. Moreover, the routing decision of this random approach can be made simultaneously for all messages in a node by a parallel

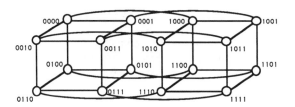

Figure 1: A Boolean 4-cube network

circuit and is therefore a simple and rapid calculation.

Let M be the buffer size of each node, where $M \geq n$. Suppose a node currently has i messages and j of them are successfully assigned to channels in the first phase. Then, $M - i + j$ buffers will be freed for any messages coming in from neighbors if there is no second phase. However, we wish to make available at least n buffers in case each of the node's n neighbors choose to send it a message in the cycle. So, in order to provide n free buffers when $M - i + j < n$, the node forces out $n - M + i - j$ messages with lower priority in the second phase. Since j channels have been assigned in the first phase, messages chosen in the second phase must be assigned only to the other $n - j$ channels. Note that $n - M + i - j \leq n - j$ in any case. If any message is assigned to a channel in the second phase with the corresponding bit in its header being one, then this message will be forwarded in the direction toward the destination. Otherwise, the message will be backtracked. A similar technique called "referral" is used in the Connection Machine [8]. Messages which fail in assignment in this cycle are kept in buffers for assignment in the next cycle. After making this two-phase routing decision, the node is ready to send messages to its neighbors.

Whenever two neighboring nodes are ready, they begin sending each other a message (if there is one) along the channel they share. We assume the channel is bidirectional. We further assume that, among these received messages, the ones destined for the node are delivered to local processors immediately. Other messages, known as *transit messages*, are saved in buffers for further transmission. After a node finishes exchanging messages with all of its neighbors, the node admits some input messages from its local processors and then continues to the next cycle.

The message priority is defined as follows. Each message is globally timestamped. The timestamp contains the time when the message was created and the source node address. Messages are queued and selected for routing in an order based on their timestamps, where older messages have higher priority over younger ones. If two messages are created at the same time, then the message with lower source address has higher priority over that with a higher source address. Since at least one of the stored messages is successfully assigned in the first phase, the message with highest priority always makes progress in every cycle. Whenever the oldest message has been delivered, another message becomes the oldest and proceeds without blocking. The network is deadlock-free.

3 Performance Analysis

In this section we develop an approximation to determine the throughput of the network, the average message delay, and the probability of acceptance of an input message.

3.1 Assumptions of the Model

We assume that a message's destination is uniformly distributed over the network, and that a local processor does not input messages to the node if the messages are for some other processors of the same node. Thus, we have

$$d_i = \text{Prob}[\textit{An input message has i 1-bits in its header}]$$

$$= \frac{\binom{n}{i}}{2^n - 1}, \quad \text{for } i = 1, 2, \dots, n$$

The expected number of hops a message must travel in the network is easily calculated as $\bar{d} = n\,2^{n-1}/(2^n - 1)$, which approaches $n/2$ when n is large. However, since messages can be backtracked during transmission, the number of hops actually traversed by a message can be larger than what is needed.

The arrivals of input messages from local processors to each node are assumed to be based on a geometric distribution with a generation rate of λ messages per cycle. We let g_i be the probability that a node's local processors generate i messages in a cycle. That is,

$$g_i = (1 - \alpha)\,\alpha^i, \quad \text{where} \quad 0 < \alpha = \frac{\lambda}{1 + \lambda} < 1.$$

We note that in the case of destination nodes being uniformly distributed over the network, the throughput of a node must be less than 2 messages per cycle. However, some of these messages might be rejected by the node if the node does not have enough buffers. We define P_A as the probability of acceptance of an input message. Clearly, the communication load of the network is determined by λ and the set of probabilities d_i.

3.2 Imbedded Markov Chain Analysis

The Boolean n-cube network is assumed to be decomposed into a set of statistically identical nodes. (clearly, one should not use this assumption in the case of unbalanced traffic.) Considering an arbitrary node separately, we have a bulk-arrival and bulk-service system as shown in Figure 2. Let M be the buffer size of the node. We assume there exists an equilibrium state for the node.

Let the sequence of random variables $X_0, X_1, X_2, \dots$ form an imbedded Markov chain, where X_m is the number of messages the node has at the beginning of cycle m. An $(M + 1) \times (M + 1)$ matrix P, which represents the one-cycle transition probability matrix of the node, is defined as

$$P = [p_{i,j}]\,,$$

where

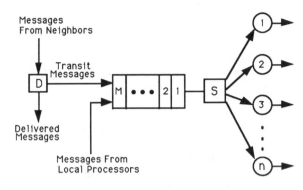

Figure 2: Structure of a Node

$$p_{i,j} = \lim_{m \to \infty} \text{Prob}\left[\, X_m = j \mid X_{m-1} = i \,\right]$$

We further define the steady state probability vector Π as

$$\Pi = [\,\pi_0,\, \pi_1,\, \pi_2,\, \dots,\, \pi_M\,],$$

where

$$\pi_i = \lim_{m \to \infty} \text{Prob}\left[\, X_m = i \,\right] \text{ for } i = 0,1,2,\dots,M$$

If the Markov chain is irreducible, aperiodic and recurrent nonnull, then Π can be uniquely determined through the following set of linear equations [11]:

$$\begin{aligned} \Pi &= \Pi P\,, \\ \sum_{i=0}^{M} \pi_i &= 1\,. \end{aligned} \tag{1}$$

3.2.1 Calculating the Matrix P

The key problem of this model is to calculate the transition probabilities $p_{i,j}$. Based on the description of the routing algorithm, it follows that the node repeatedly makes a routing decision, exchanges messages with its neighbors, and then admits some input messages from its local processors. However, without loss of generality, in the model we assume in each cycle all of the transit message are received by the node after all of its selected messages have been transmitted. See Figure 3.

We let $Y_0, Y_1, Y_2, \dots$ be a sequence of random variables, where Y_m is the number of messages the node has in cycle m after the node has transmitted all of its currently selected messages, but before it has received any transit messages. We also let $Z_0, Z_1, Z_2, \dots$ be a sequence of random variables, where Z_m is the number of messages the node has in cycle m after the node has received all of its transit messages, but before it has accepted any input messages.

Let us first determine the number of messages transmitted by the node in a cycle. Since one-bits in the header are assumed to be uniformly distributed and channel selection is made randomly, it follows that every channel of the node has an equal probability of being assigned in the first phase. We let $f_{i,j}$ be the probability that j messages are successfully assigned to channels in the first phase,

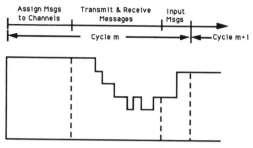

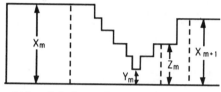

(a) Snapshot of queue fluctuations (real system)

(b) Snapshot of queue fluctuations (model)

Figure 3: Snapshot of queue fluctuations in cycle m

given that the node currently has i messages. It can be shown [9] that

$$
f_{i,j} = \begin{cases} \binom{n}{j} \sum_{k=0}^{j-1} (-1)^k \binom{j}{k} \left(\frac{i-k}{n}\right)^i & \text{if } \begin{cases} \text{if } i \geq 1 \text{ and} \\ 1 \leq j \leq \min(i,n) \end{cases} \\ 1 & \text{if } i = j = 0 \\ 0 & \text{otherwise.} \end{cases}
$$
(2)

Given that j messages are successfully assigned in the first phase, the second phase must choose $n - M + i - j$ messages if $i - j > M - n$. If we further let $f'_{i,j}$ be the probability that j messages are successfully assigned to channels at the end of both phases, given that the node currently has i messages, we have

$$
f'_{i,j} = \begin{cases} f_{i,j} & \text{if } i - j < M - n \\ \sum_{k=1}^{n-M+i} f_{i,k} & \text{if } i - j = M - n \\ 0 & \text{otherwise.} \end{cases}
$$
(3)

We now proceed to determine the number of transit messages received in a cycle. Recall that a transit message is a message received from a neighbor which needs further transmission. Assuming that the traffic load is evenly distributed and that any incoming channel is independent of any other incoming channel, and letting P_t be the probability of receiving a transit message from a channel and t_i be the probability of receiving i transit messages in a cycle, we have

$$
t_i = \binom{n}{i} P_t^i (1 - P_t)^{n-i}, \text{ for i=0,1,2,...,n.}
$$
(4)

We solve for P_t later. We further define the following three $(M+1) \times (M+1)$ matrices:

$$
D = [d_{i,j}],
$$

where

$$
d_{i,j} = \lim_{m \to \infty} \text{Prob}\,[\,Y_m = j \,|\, X_m = i\,].
$$

$$
R = [r_{j,k}],
$$

where

$$
r_{j,k} = \lim_{m \to \infty} \text{Prob}\,[\,Z_m = k \,|\, Y_m = j\,].
$$

$$
A = [a_{k,l}],
$$

where

$$
a_{k,l} = \lim_{m \to \infty} \text{Prob}\,[\,X_{m+1} = l \,|\, Z_m = k\,].
$$

We find the following equations:

$$
d_{i,j} = \begin{cases} f'_{i,i-j} & \text{if } 0 \leq j \leq i \leq M \\ 0 & \text{otherwise} \end{cases}
$$
(5)

and

$$
r_{j,k} = \begin{cases} t_{k-j} & \text{if } 0 \leq j \leq k \leq M \text{ and } k-j \leq n \\ 0 & \text{otherwise} \end{cases}
$$
(6)

and

$$
a_{k,l} = \begin{cases} (1-\alpha)\,\alpha^{l-k} & \text{if } 0 \leq k \leq l \leq M-1 \\ \alpha^{M-k} & \text{if } 0 \leq k \leq M \text{ and } l = M \\ 0 & \text{otherwise.} \end{cases}
$$
(7)

Finally, we have the one-cycle transition probability matrix P:

$$
P = DRA.
$$
(8)

Since $p_{i,i} \neq 0$ for $i = 0, 1, 2, ..., M$, the Markov chain is irreducible. It is not difficult to show that the Markov chain is also aperiodic and recurrent nonnull. Thus, the probability distribution π_i for $i = 0, 1, 2, ..., M$ can be determined from the set of linear equations described in Eq.1. However, the remaining problem is to determine the value for P_t.

3.2.2 Determining the Value for P_t

The value for P_t must satisfy the condition that in equilibrium the network message input flow equals the message output flow. We let u be the channel utilization, which is equal to the probability that a channel transmits a message in a cycle. Since a message can be assigned to a channel either in the first phase or in the second phase, we let u_1 be the probability that a channel transmits a message which is assigned in the first phase, and u_2 be the probability that a channel transmits a message which is assigned in the second phase. Clearly, $u = u_1 + u_2$. It is also clear that, given the probability distribution of π'_is, we have

$$
u = \sum_{i=1}^{M} \pi_i \sum_{j=1}^{\min(i,n)} f'_{i,j} \frac{j}{n},
$$
(9)

and

$$u_1 = \sum_{i=1}^{M} \pi_i \sum_{j=1}^{min(i,n)} f_{i,j} \frac{j}{n}. \qquad (10)$$

We further assume that every transmitted message has the same probability p of being assigned in the first phase and has the same probability q of being assigned in the second phase. That is,

$$p = \text{Prob} \begin{bmatrix} A \text{ message is assigned in the 1st phase,} \\ \text{given that it is transmitted in the cycle.} \end{bmatrix}$$
$$= u_1/u$$

and

$$q = \text{Prob} \begin{bmatrix} A \text{ message is assigned in the 2nd phase,} \\ \text{given that it is transmitted in the cycle.} \end{bmatrix}$$
$$= u_2/u,$$

where $p + q = 1$. We note that if a message is assigned in the second phase, it must be assigned to a free channel which was not assigned in the first phase. Thus, if a message with i one-bits in its header is assigned to a channel in the second phase, it will be forwarded with probability $(i-1)/(n-1)$ and be backtracked with probability $(n-i)/(n-1)$. The state transition diagram for this case is given in Figure 4. The system is in state i if, at the beginning of a cycle, the message currently has i one-bits in its header. Let h_i be the expected number of steps to move from state i to state 0. We have the following recursive relations:

$$h_i = \begin{cases} 0 & \text{if } i = 0 \\ \left(p + \frac{i-1}{n-1}q\right)h_{i-1} + \left(\frac{n-i}{n-1}q\right)h_{i+1} + 1 & \text{if } i = 2,..,n-1 \\ h_{n-1} + 1 & \text{if } i = n. \end{cases}$$

It can be shown that

$$h_i = \sum_{j=n+1-i}^{n} \delta_j, \text{ for } i = 1, 2, ..., n \qquad (11)$$

where

$$\delta_i = \begin{cases} 1 & \text{if } i = 1 \\ \prod_{j=1}^{i-1} \frac{\frac{j}{n-1}q}{1-\frac{j}{n-1}q} + \frac{n-1}{q}\sum_{m=1}^{i-1}\frac{1}{m}\prod_{j=m}^{i-1}\frac{\frac{j}{n-1}q}{1-\frac{j}{n-1}q} & \text{for } i = 2, ..., n. \end{cases}$$

Let $\overline{h}$ be the expected number of hops actually traversed by a message. That is,

$$\overline{h} = \sum_{i=1}^{n} d_i h_i. \qquad (12)$$

We further let P_ϵ be the probability that a message from a neighbor is delivered to a local processor in the node. On average a message is expected to exit from the network after moving $\overline{h}$ hops. Thus, we have

$$P_\epsilon = 1/\overline{h}. \qquad (13)$$

The message output rate of the network is given by

$$NnuP_\epsilon = Nnu/\overline{h}, \qquad (14)$$

where $N = 2^n$ is the number of nodes in the network.

Moreover, given the arrival process is geometric, it can be shown [9] that

$$P_A = 1 - \pi_M. \qquad (15)$$

Thus, the message input rate to the network is given by

$$N\lambda P_A = N\lambda(1 - \pi_M). \qquad (16)$$

In equilibrium, the input rate equals the output rate. From Eqs. 16 and 14, we have that

$$\lambda(1 - \pi_M) = nu/\overline{h}. \qquad (17)$$

This equation must be satisfied by the given P_t and the calculated probability distribution π_i. In all the examples we have studied, we have found that the input rate is a decreasing function of P_t while the output rate is an increasing function; moreover, the input rate is larger than the output rate when P_t approaches 0 and that the input rate is smaller than the output rate when P_t approaches 1. Thus, in these examples, P_t exists and is unique. See Figure 5 and [9].

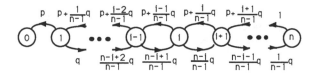

Figure 4: Transition diagram of finding i one-bits in the header

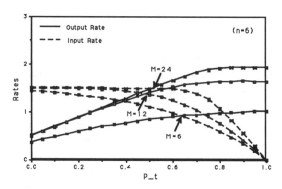

Figure 5: Input rate and output rate vs a given P_t

3.3 Delay and Throughput

The throughput of the network, γ, is defined as the total number of new messages the network accepts in a cycle. Thus,

$$\gamma = N\lambda(1 - \pi_M) \qquad (18)$$

The mean queue length in each node is given by

$$\overline{q} = \sum_{i}^{M} i\,\pi_i. \qquad (19)$$

Applying Little's result [13], the average message delay is given by

$$T = \frac{N\overline{q}}{\gamma} \qquad (20)$$

3.4 Effects of Backtracking

We now calculate the effects of backtracking. Let us define

$$P_f = \text{Prob}\,[\,a\ channel\ forwards\ a\ message\ in\ a\ cycle\,]$$
$$P_r = \text{Prob}\,[\,a\ channel\ backtracks\ a\ message\ in\ a\ cycle\,],$$

where $P_f + P_r = u$. The net progress made by a channel in a cycle is equal to $P_f - P_r$. We realize that whenever a message is backtracked, it is one hop farther away from its destination node. The message then needs an extra step of forwarding to compensate for this loss. We further let

$C = $ Number of channels in the network,
$K = $ Number of cycles in a long time period, and
$S = $ Throughput of the network in this period.

For the whole network, the expected number of one-bits (in headers) which are changed to zeroes in K cycles is easily calculated as CKP_f. The expected number of zero-bits which are changed to ones is CKP_r. As a result, the expected number of one-bits "decreased" due to the network's transmission is $CK(P_f - P_r)$, which must be equal to the number of one-bits in the headers of input messages from local processors in this period when the network is in a steady state. Thus,

$$CK(P_f - P_r) = S\overline{d} \qquad (21)$$

Moreover, the expected total number of hops made by all messages (forward or backward) in K cycles is

$$CK(P_f + P_r) = S\overline{h} \qquad (22)$$

Thus, we immediately have the following equation:

$$\frac{\overline{h}}{\overline{d}} = \frac{P_f + P_r}{P_f - P_r}. \qquad (23)$$

Solving for P_r and P_r, we have

$$P_r = \frac{u}{2}(1 - \frac{\overline{d}}{\overline{h}}), \qquad (24)$$

$$P_f = \frac{u}{2}(1 + \frac{\overline{d}}{\overline{h}}). \qquad (25)$$

The net progress of a channel in a cycle is then given by

$$P_f - P_r = u\frac{\overline{d}}{\overline{h}}. \qquad (26)$$

4 Model Verification and Discussion

The accuracy of the mathematical model is verified by comparing it with simulations. In this section, we present several results of these simulations for a Boolean 6-cube network with various numbers of buffers in each node. Other results are reported in [9]. We note that the match between the simulated results and the model is extremely good.

Figure 6 shows the utilization, the probability of forwarding, and the net progress of a channel when $M = 10$. The net progress is also compared with that in the infinite buffered network where the net progress equals the channel utilization. We note that the effect of backtracking is not serious even if the buffer size is relatively small. Figure 7 presents the probability of acceptance of an input message for various buffer sizes. Again, we see that small buffers are enough to accept most of the input messages. Figure 8 shows the average message delay. We realize that when the generation rate is low, a node with a large buffer simply behaves as a node with an infinite buffer space. Thus, an increase in buffer size does not affect the delay. However, when the generation rate is high, more buffers admit more input messages. The queue grows as the buffer size increases. Thus, the average message delay also increases. When the buffer size in each node is very small (eg. 6 in this case), the average message delay is also larger because messages are likely to be backtracked. It is important to note in Figure 9 that even if the new message generation rate is much larger than that can be accepted by the network, the throughput of the network does not degrade.

5 Optimization Issues

In most queueing systems, two performance measures, response time and throughput, compete with each other. Typically, by raising the throughput of the system, which is desirable, the mean response time is also raised, which is not desirable. Moreover, we wish to consider the blocking of newly generated messages from local processors. We combine these three performance measures into a single measure, power, which is given as follows [12].

$$Power = \frac{Throughput}{Mean\ Response\ Time}(1 - B),$$

where B is the blocking probability. For our system we have

$$Power = \frac{\gamma}{T}P_A,$$

A system is said to be operating at an optimal point if the power is maximized. In Figure 10, we show the power as a function of λ for various buffer sizes; the peak of each curve identifies the optimal generation rate for each buffer size. This result is able to serve as a guide to network flow control. In Figure 11, we observe that an increase in buffer size increases the power. However, we note that assigning a large number of buffers to a node cannot significantly improve the performance. If we wish to consider the cost of the buffer, we can further divide the power by the buffer size; using this modified measure, we show in Figure 12 that small buffers yield good performance.

6 Conclusion

We have presented a mathematical model for evaluating the performance of a deadlock-free routing algorithm for Boolean n-cube networks with finite buffers. This algorithm is simple enough to be implemented in hardware. We have shown that only a few buffers in each node are essential to yield good performance. We have also shown that the throughput of the network does not degrade even when the network is saturated.

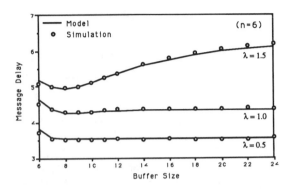

Figure 8: Average message delay vs buffer size

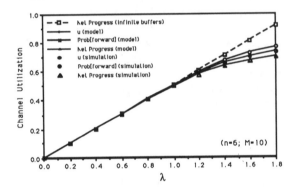

Figure 6: Channel utilization and prob. of forwarding

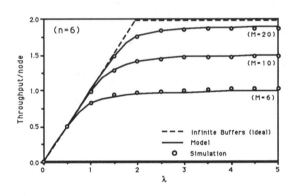

Figure 9: Throughput vs new message generation rate

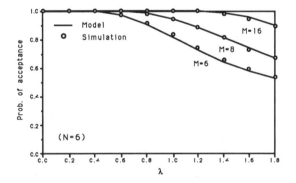

Figure 7: Prob. of acceptance of an input message

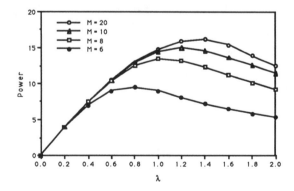

Figure 10: Power vs new message generation rate

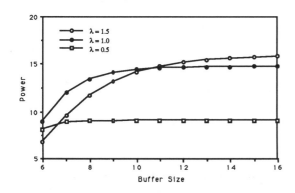

Figure 11: Power vs buffer size

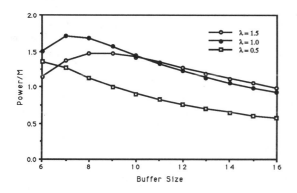

Figure 12: Power/M vs buffer size

References

[1] S. Abraham and K. Padmanabham, "Performance of Direct Binary n-Cube Networks for Multiprocessors," *IEEE Trans. Comput.*, vol. C-38, pp. 1000-1011, July 1989.

[2] W. C. Athas and C. L. Seitz, "Multicomputers: Message-Passing Concurrent Computers," *IEEE Comput.*, pp. 9-24, Aug. 1988.

[3] L. N. Bhuyan and D. P. Agrawal, "Generalized Hypercube and Hyperbus Structures for a Computer Network," *IEEE Trans. Comput.*, vol. C-33, pp. 323-333, Apr. 1984.

[4] J. Balzewicz, J. Brzezinski and G. Gambosi, "Time-Stamp Approach to Store-and-Forward Deadlock Prevention," *IEEE Trans. Commun.*, vol. COM-35, pp. 490-495, May 1987.

[5] W. J. Dally, "Performance Analysis of k-ary n-cube Interconnection Networks," *IEEE Trans. Comput.*, vol.C-39, pp. 775-785, June 1990.

[6] W. J. Dally and C. L. Seitz, "Deadlock-Free Message Routing in Multiprocessor Interconnection Networks," *IEEE Trans. Comput.*, vol. C-36, pp. 547-553, May 1987.

[7] T. Feng, "A Survey of Interconnection Networks," *IEEE Trans. Comput.*, vol. C-30, pp. 12-27, Dec. 1981.

[8] W. D. Hillis, *The Connection Machine*, MIT Press, 1985.

[9] M.-Y. Horng, *Performance Analysis of the Boolean n-Cube Interconnection Network for Multiprocessors*, Ph.D. Dissertation, Comput. Sci. Dep., Univ. California, Los Angeles, 1991.

[10] Intel Scientific Computers, *iPSC User's Guide*, No. 175455-001, Santa Clara, Aug. 1985.

[11] L. Kleinrock, *Queueing Systems, Vol. I: Theory*, John Wiley and Sons, New York, 1975.

[12] L. Kleinrock, "Power and Deterministic Rules of Thumb for Probabilistic Problems in Computer Communications," *Int. Conf. on Commun.*, pp. 43.1.1-43.1.10, June 1979.

[13] J. D .C. Little, "A Proof of the Queueing Formula $L = \lambda W$", *Oper. Res.*, vol. 9, pp. 383-387, May 1961.

[14] P. M. Merlin and P. J. Schweitzer, "Deadlock Avoidance : Store-and-Forward Deadlock," *IEEE Trans. Commun.*, vol. COM-28, pp. 345-354, Mar. 1980.

[15] Y. Saad and M. H. Schultz, "Topological Properties of Hypercubes," *IEEE Trans. Comput.*, vol. C-37, pp. 867-872, July 1988.

[16] Y. Saad and M. H. Schultz, "Data Communication in Hypercubes," *J. Parallel Distrib. Comput.*, vol. 6, pp. 115-135, 1989.

[17] C. L. Seitz, "The Cosmic Cube,", *Commun. ACM*, vol. 28, pp. 22-33, Jan.1985.

[18] H. J. Siegel, *Interconnection Networks for Large-Scale Parallel Processing*, Lexington Books, 1985.

A Comparison of SIMD Hypercube Routing Strategies

Melanie Fulgham
Robert Cypher
Jorge Sanz

IBM Research Division
Almaden Research Center
650 Harry Road
San Jose, CA 95120

Abstract

This paper presents simulation results for several oblivious SIMD hypercube routing algorithms. Simulations were performed over a wide range of machine sizes, queue sizes and communication patterns. In addition to quantifying the performance of published routing algorithms, we present modifications of known algorithms which have certain performance advantages over the original algorithms. For example, we introduce new techniques for generating randomness which attain better performance than those presented by Valiant and Brebner over a wide range of parameters. While most published simulation results are for networks of 1K processors or less, we study networks with up to 64K processors. Our results show that some important parallel communication phenomena do not manifest themselves until the network is 4K or larger.

Keywords: hypercube, SIMD computer, message passing, routing, simulation.

1 Introduction

Communication in parallel computers is one of the most important problems in parallel computing today. For this reason it is essential to understand which parallel routing techniques are efficient and practical. Although an extremely wide range of routing algorithms have been proposed [2,7,8,9,13,16,19], it is very difficult to compare them and to select the best routing algorithm for a given machine. There are two primary obstacles to the comparison of routing algorithms. First, most routing algorithms exhibit such complicated phenomena that their performance has not been characterized analytically. Although upper bounds have been proven for a few routing algorithms, these bounds are usually asymptotic and thus do not apply to fixed machine sizes. Second, different routing algorithms often assume different architectural features, so it is difficult to determine if an improvement in performance is a result of an improved algorithm or a more powerful architecture. We have attempted to overcome both of these obstacles by performing simulations of different routing algorithms on a single architectural model. The use of simulation allows a good characterization of the performance of algorithms which are too complicated to be bounded analytically, and the single architectural model provides a common platform for comparing algorithms that were originally proposed assuming different architectures.

All of our simulations are for a distributed memory parallel computer with a hypercube interconnection. We chose the hypercube topology because it has a small diameter and because it supports many algorithms efficiently [3,6,13]. In addition, the hypercube is a commercially successful architecture as demonstrated by the Connection Machine [5], the NCUBE [4,14], and the Intel iPSC [1,15]. Routing algorithms can be classified as being either *oblivious* or *adaptive*. In oblivious routing algorithms each message follows a path which is a function of its source-destination pair (plus possibly some locally generated random bits). All other routing algorithms are adaptive. Adaptive algorithms have the advantage of being able to change the route of a packet when it encounters congestion. However, the hardware needed for adaptive routing is more complex. In this paper we focus on oblivious algorithms in our simulation studies. Adaptive algorithms will be the subject of a future study. Some of the routing algorithms which we simulate use a source of randomness to make routing decisions. These algorithms are called *randomized* algorithms, and the remaining algorithms are called *deterministic* algorithms.

Several other simulation studies of the hypercube have been published [8,9,10,19]. The first two discuss adaptive routing schemes where messages are continually injected into the network. The third study guided the development of the Caltech Cosmic Cube. The last study simulates batch routing using an oblivious randomized routing scheme discussed later in this paper. In *batch routing*, an initial set of packets is routed until all of the packets arrive at their destinations. Another set of packets cannot be injected in the network until the initial set has arrived. Batch routing is also used in the Connection Machine as well as in the model presented in this paper. Batch routing is appropriate for SIMD machines since the next instruction cannot be issued until all the nodes have finished the previous instruction.

Within the class of oblivious hypercube routing algorithms, there many algorithms which have been proposed and many parameters which can be varied. Our simulations were performed over a wide range of machine sizes, queue sizes and communication patterns. In addition to quantifying the performance of published routing algorithms, we present modifications of known algorithms which have certain performance advantages over the original algorithms. For example, we introduce new techniques for generating randomness which attain better performance than those presented by Valiant and Brebner over a wide range of parameters. While most published simulation results are for networks of 1K processors or less, we study networks with up to 64K processors. Our results show that some important parallel communication phenomena do not manifest themselves until the network is 4K or larger.

The rest of this paper is organized as follows. The simulation model is defined in Section 2. Section 3 describes the simulation results and conclusions are presented in Section 4.

2 The Model

The model studied in this paper is a packet switched SIMD (single instruction stream, multiple data stream) computer with a hypercube interconnection topology. The hypercube (or n-cube) has $N = 2^n$ nodes each labeled with a binary number from 0 to $N - 1$. Two nodes are connected if and only if their binary representation differs in one bit. Each node contains a processor and its local memory. In addition, each node has $\log N$ buffers of size 1 for input and $\log N$ queues for output. Output queues are of some fixed size which is specified by the run-time input parameters of the simulation. Each output queue is connected to a neighbor's input buffer by a bidirectional channel. Since the model is SIMD each processor simultaneously services the same queue. The input buffers are serviced serially from low to high dimension. After all of the input buffers have been processed, all of the output buffers send the packets at the head of their queues in parallel. A packet is blocked and must remain in its output buffer if the input buffer at the end of its channel is full. All buffers are FIFO. This helps to keep the switches simple and efficient. Routing information for each packet is contained exclusively in the packet header. Although the above model is small and simple, it was chosen for its ability to support many routing algorithms.

Time is measured in discrete units to reflect the SIMD nature of the model. Costs are apportioned as follows: one time unit to serve an input buffer and either accept the packet (that is, deliver it to the associated processor) or place it in the tail of the appropriate output queue, and one time unit to remove the packet at the head of an output queue and send it to its neighbor.

2.1 Packet Routes

A packet route is either *direct* or *indirect*. In direct routing a packet proceeds directly to its final destination according to some specified routing algorithm. With indirect routing, the packet uses the specified routing algorithm to first go from its source to some intermediate destination and then to go from the intermediate destination to its final destination. When indirect routing is used, the two routing stages are pipelined. That is, once the packet reaches its intermediate destination, it immediately proceeds towards its final destination.

Indirect routing appears unnatural as it can cause packets to traverse longer paths than those required by direct routing. However, indirect routing has been proven useful in avoiding congestion [19]. The worst case behavior of deterministic oblivious permutation routing is $\Omega(\sqrt{N}/d)$, where d is the degree of the network [7]. However, Valiant and Brebner [19] have shown that randomness and indirect routing can be used to obtain good average case performance regardless of the permutation being routed. Specifically, they have shown that indirect routing in which the intermediate destinations are chosen with uniform probability from the N nodes and a simple routing algorithm yield an $O(\log N)$ expected time regardless of the permutation being routed and regardless of whether or not the two phases are pipelined. Thus, the intermediate destination will often be selected randomly in accordance with Valiant and Brebner's scheme. The idea of using random destinations is to avoid the congestion which could be caused by certain permutations.

2.2 Intermediate Destinations

Intermediate destinations are created in several ways. Each processor has a source of randomness and creates an intermediate destination for its packet by modifying its own ID according to some randomization scheme. In the first scheme, each processor selects, at random, a set of dimensions of a specified size. The processor then creates the intermediate destination by taking its ID and complementing the bits corresponding to the dimensions just selected. This technique is called *bit flipping*. In the second scheme, each processor selects, at random, a set of dimensions of a specified size. The processor then creates the intermediate destination by flipping a fair coin for each of the selected dimensions. If the coin shows heads, the bit in the processor ID corresponding to the current dimension is changed. This technique is called *coin flipping*. Valiant and Brebner used coin flipping on all $\log N$ dimensions to create their intermediate destinations [19].

2.3 Communication Patterns

Three types of communication patterns are simulated: *transpose*, *complement*, and *random traffic*. Each of these patterns was selected to examine a particular aspect of routing behaviour. The transpose sends each message from a processor with binary address $(a_{n-1} \ldots a_0)$ to the processor with binary address $(a_{m-1} \ldots a_0, a_{n-1} \ldots a_m)$ where $m = \lfloor n/2 \rfloor$. The transpose was chosen because when some common routing algorithms are used, it forces $\sqrt{N}$ packets to pass through a single node. Thus the transpose gives a good indication of how a routing algorithm handles communication patterns that tend to cause congestion. The complement send each message from a processor with binary address $(a_{n-1} \ldots a_0)$ to the processor with binary address $(\overline{a_{n-1}} \ldots \overline{a_0})$ where $\overline{0} = 1$ and $\overline{1} = 0$. Performing the complement does not tend to cause serious congestion, but it does force all the packets to travel the diameter of the network. Finally, random traffic consists of sending each packet to a destination selected independently and uniformly out of the set of N processors. Random traffic is a good model for pointer-based communication in which the pointer values have no relation to the hypercube structure of the machine. It should be noted that these patterns of communication are selected as representatives of traffic types. If any specific permutation is known at compile time (such as a transpose or a complement), efficient routes can be precomputed in advance using algorithms in [20] or [17].

2.4 Packet Header Algorithms

There are several packet header algorithms that determine how the header information is encoded. The header can be viewed as an array of $2 \log N$ elements where each element is either a dimension number or the special symbol, ∞. The dimensions occurring in the array are the ones that need to be changed in order for the packet to arrive at the desired (intermediate or final) destination, given a particular starting point. The symbol, ∞, is a place holder and signifies that no change in a dimension is needed. The order of the dimensions depends upon the algorithm used to create the packet header. Given a header array, the operation Pack, as defined in [19], moves all the ∞ symbols to the end of the array without changing the order of the dimensions that occur in the array. Each packet is created locally. Consequently the randomization and Packing are done locally.

Let H be a header array, s be a source, and d a destination where $(s_{n-1} \ldots s_0)$ and $(d_{n-1} \ldots d_0)$ are the binary representations of s and d. All of the packet header algorithms examine each bit position in the source and destination. For each bit position i, if s_i differs from d_i, i is placed in the header, otherwise ∞ is placed in the header. The following packet header algorithms are used in the simulator.

LOWHIGH examine the dimensions from least to most significant, then Pack the resulting array

RAND process as in LOWHIGH, then randomize the order of the dimension changes

SHIFTRAND examine the dimensions from least to most significant, then cyclically shift the array right by a number k chosen uniformly from $\{0 \ldots \log N - 1\}$, and finally Pack the resulting array

ZERO examine the dimensions from least to most significant, first for dimensions that are changing from a zero to a one bit, i.e. $s_i = 0$ and $d_i = 1$, $\log N > i \geq 0$, then again examine the dimensions from least to most significant, this time for dimensions changing from a one to a zero bit, i.e. $s_i = 1$ and $d_i = 0$, $\log N > i \geq 0$, Pack the resulting array

ZERORAND process as in ZERO, then randomize the order within each of the two groups of dimension changes, i.e. the dimension changes for bits changing from a zero to one and those changing from a one to zero

The header algorithms LOWHIGH, RAND, and SHIFTRAND are similar to CUBES, BASIC, and CUBESS in [19]. It is easy to visualize the packet routes specified in the ZERO and ZERORAND headers by hanging a hypercube by the node whose ID is all ones. This induces a leveling of the cube by Hamming distance from the top node, i.e. all nodes at the same level have the same number of ones in their ID. All packet routes travel in two directions, first up towards the top node and then down toward the bottom node whose ID is all zeroes. Given a packet with source s and destination d, the packet will change directions from up to down at the node whose ID is the logical or of s and d. This leveling is suggested in [2] in order to support efficient barrier synchronization on the hypercube.

Headers containing both an intermediate and final destination are formed from the combination of two headers created by the header algorithms described above. The header for the intermediate destination is placed in the first $\log N$ spaces of the header. It contains the dimension changes necessary for routing the packet from the starting location to the intermediate destination. The second $\log N$ spaces are filled with the header dimensions needed to move the packet from the intermediate to the final destination. Since packet routing is pipelined, the two parts of the header are combined. This is performed by concatenating all the dimension changes for the intermediate destination to those of the final destination. The result is then Packed. Headers with this format allow a switch to treat packets proceeding to an intermediate destination identically to those traveling to a final destination. This helps keep the switch logic simple.

3 Simulation Results

Statistics were measured for many quantities such as packet delays, packet time in system, queue population, and switch population. The one of immediate interest is the delivery time of the last packet to arrive at its destination. The slowest time is of importance since in an SIMD machine all processors wait for each other to complete before proceeding to the next step. When the simulations involve randomization, point estimates for expected values along with 95% confidence intervals were obtained. The graphs presented plot the number of processors vs. the estimated expected completion time. Most runs produced confidence intervals with sizes less than 2% of the point estimate. Consequently, confidence intervals have been omitted from the graphs for clarity. At time zero, each processor begins with one initialized packet. Routing is batched, so no additional packets are injected into the network. In order to compare different routing schemes fairly, all of the simulations were first performed with large queues. This allowed all of the techniques to run to completion. Specifically, queue sizes of 100 are used unless specified otherwise. Then smaller queues were studied for those techniques which did not show large amounts of congestion. In most cases, queues of size 2 or 4 were found to be sufficient.

The uniform random number generator used is a prime-modulus, multiplicative congruential generator proposed by Learmonth and Lewis [11] in 1974. This generator has passed many statistical tests and is considered one of the best available for simulation work [12]. This generator is currently implemented in the following packages: GPSS, SAS, IMSL Library and LL-RANDOMII.

Simulations were run on various models of the IBM 3090/400 series under the VM/CMS environment.

3.1 Random Traffic

We will begin by examining the results for random traffic using *direct* routing. The type of message route specified by the packet header significantly affected performance. Packets created with the header algorithms LOWHIGH, RAND, and SHIFTRAND, all performed equivalently and the best.

In contrast, the algorithms ZERO and ZERORAND were much slower due to congestion near the top node (node $N - 1$). In fact, the expected amount of congestion at the top node can be calculated analytically. Consider a randomly selected source node s and its destination d. The message starting at s will pass through the top node iff the bitwise OR of s and d is all 1's. For each bit position i where $0 \leq i < \log N$, there is a 3/4 chance that s_i OR $d_i = 1$. Thus there is a probability of $(3/4)^{\log N}$ that the message will pass through the top node and the expected number of messages passing through the top node is $N(3/4)^{\log N} = N^{1+\log_2(3/4)} \approx N^{0.585}$. Although both ZERO and ZERORAND have the same expected amount of congestion at the top node, ZERORAND did perform significantly better than ZERO. This difference is because ZERO makes all of the packets which pass through the top node also pass through the node $N/2 - 1$, while ZERORAND lets them pass through all of the nodes with a single zero in their ID.

We will now examine the results for random traffic using indirect routing. First note that indirect routing through a random destination is essentially the same as two rounds of direct routing to a random destination. The only difference is that with indirect routing the two rounds are pipelined. In general, indirect routing took less than twice as long as a single round of direct routing. For a network of 16K nodes using the LOW-HIGH, RAND, or SHIFTRAND header algorithms, the time added by indirect routing was three fourths the time required for direct routing. See Figure 1 for details.

3.2 Complement

Direct routing to the complement is collision-free when using either of the LOWHIGH or SHIFTRAND header algorithms (a proof of this fact is given in Appendix A). Since no collisions occur in either of these schemes, an output queue size of one is sufficient. Using ZERO and ZERORAND for the complement is very inefficient since all packets travel through the top node (node $N - 1$). When routing with 16K processors to a random intermediate destination and then to the complement, the algorithm ZERORAND was twice as fast as ZERO. However, ZERORAND was almost four times slower than direct routing with LOWHIGH or SHIFTRAND. Routing directly to the complement with RAND performed about the same as indirect routing using an intermediate destination created by flipping a coin on each of the dimensions using LOWHIGH (or SHIFTRAND). Since this intermediate destination is on a minimal route to the complement destination, the two schemes are almost equivalent. The only difference is that dimensions not changed by the intermediate destination are processed in a low to high order (or in a cyclically shifted low-to-high order), whereas in RAND the order of all the dimensions is random. Routing directly to the complement with RAND is equivalent to routing first to an intermediate destination and then to the complement using RAND. With 16K processors, using the Valiant and Brebner technique, the LOWHIGH, SHIFTRAND, RAND header algorithms performed almost one and a third times slower than direct routing using either of the LOWHIGH or SHIFTRAND header algorithms. Results are displayed in Figure 2.

3.3 Transpose

We examined several indirect routing schemes for the transpose. These schemes involve an intermediate destination and transpose final destination. Most of the variations change the amount and type of randomness used to generate the intermediate destination. The first set of variations involve flipping a coin on each of the dimensions in order to form an intermediate destination from the source ID. Results are shown in Figure 3.

As mentioned before, when intermediate destinations are created using coin flips with the LOWHIGH header algorithm, routing was proven to have $O(\log N)$ expected completion time [19]. Experimental results show that the maximum expected queue size grows proportional to $\log N$ [19]. Our simulation results show that RAND and SHIFTRAND perform equivalently to LOWHIGH for networks with up to 64K nodes. Even when the output queue sizes are decreased to size 2, LOWHIGH, RAND, and SHIFTRAND still route equivalently and no slower than runs with a very large queue size. As in the random traffic, ZERORAND is about twice as good as ZERO. However ZERORAND takes more than double the time of LOWHIGH, RAND, and SHIFTRAND. Both ZERO and ZERORAND required larger queue sizes to complete in a reasonable amount of time. Queues of size 4 were unacceptable, while queues of size 20 worked well for networks up to 16K. An exact lower bound on the output queue size was not determined. When using a random intermediate destination with the LOWHIGH, SHIFT-

RAND, and RAND header algorithms, the expected completion time for routing the transpose is the same as for a random permutation. This is expected since the Valiant and Brebner technique makes the routing of all permutations seem alike. However when the packets are pipelined, permutations where the intermediate destinations are on a minimal or near minimal path to the final destination (e.g. the complement) are routed faster than random permutations. This occurs because the packet is always/usually making progress towards its final destination.

In an effort to improve performance, the packet route was shortened. This is done by creating intermediate destinations from source IDs with less than $\log N$ coin flips. The dimensions on which to perform the coin flips are chosen at random, but no dimension is selected more than once. Results showed that the optimal number of dimensions to flip coins for depended on the network size. The larger the hypercube the more coin flips necessary. No coin flips, i.e. direct routing is the fastest for networks of 128 processors or less. Three bits are optimal for 256 and 512 nodes. For 1024 processors, five bits are superior. Six bits are fastest for 2048 nodes. Seven bits are optimal for both 4096 and 8192 nodes. For networks of 16K nodes, eight bits are best while nine bits are optimal for 32K nodes. For 64K nodes, ten bits are the best. Some of the results are graphed in Figure 4. An even faster way to introduce randomness is to create intermediate destinations by bit flipping. In bit flipping, any dimension that is chosen is automatically changed. Again the dimensions are chosen at random without any dimension being selected more than once. This method, which randomizes a fixed number of bits, forces the intermediate destination to be a fixed distance from the source. Results from intermediate destinations created by bit flipping half the dimensions were superior to those that created intermediate destinations from coin flips on all the dimensions (see Figure 5).

This can be explained by the following. On average the same number of bits are changed in the intermediate destinations for both schemes; however, a coin flipped intermediate destination can be up to twice as far from its source as the randomized intermediate destination. Since the routing is pipelined, the packets with intermediate destinations far from the source are slower than the other packets to start moving towards their final destination. These slower packets are the last to arrive and effect the completion time of the routing. The optimal number of dimensions to change again depends on the number of processors. No change, i.e. direct routing is the fastest if the number of processors is not larger than 128. Two bits are best for 256 and 512 processors. Three is superior for 1024 and 2048 processors. Between 4K and 32K processors, four bits results in the fastest routing. Five bits are the most efficient for 64K. See Figures 6 and 7 for details on the larger networks. Almost none of the header algorithms are able to efficiently route packets directly to the transpose (see Figures 8 and 9).

The only header scheme that efficiently routes directly with a large or small output queue size (size 4) for networks up to 32K is RAND. Output queues of size 3, show a small performance degradation starting at networks of 8K processors. The header algorithm SHIFTRAND routed effectively, but only for the networks up to 2K. At 16K nodes, direct routing with SHIFTRAND was clearly worse than routing with an intermediate destination created by coin flipping on all the dimensions. Randomizing the order of the dimension changes, as in RAND, is extremely effective in alleviating the hot spots in the network. Some of the details are in Figure 10.

This scheme even works better than using either an intermediate destination with the optimal number of changes using the LOWHIGH header algorithm or an intermediate destination with one or two randomized bits using the RAND header algorithm. This is because RAND's direct routes are the shortest possible and do not have the congestion of the deterministic direct routes. However, routing directly using the RAND header algorithm does not perform well on all permutations. For example, Valiant showed that there exists a permutation which would require at least $N^{1/4}$ time on average [18, pp. 360–361].

4   Recommendations

Choosing the fastest routing scheme requires knowledge of the communication pattern in advance. Since this information might not be available, we would like to select a single routing scheme that works well regardless of the communication pattern. Following are results and observations from our study that help guide this decision.

We simulated several communication patterns to produce various types of congestion. Different algorithms performed the best on different communication patterns. Routing efficiency of an algorithm also depended on network size.

Direct routing schemes are the most efficient when little or no congestion is present in the network. Random destinations or permutations are performed the fastest by direct routing using LOWHIGH, SHIFTRAND, or RAND header algorithms. Queues of size 4 are sufficient. The complement is routed the most efficiently using direct routing with the LOWHIGH or SHIFTRAND header algorithms. Routing is actually collision free in this case. Consequently queues of size 1 are sufficient. For direct routing of the transpose, and perhaps other permutations that have the potential for serious congestion, the RAND header algorithm works the best.

Permutations that cause congestion in the network can also be routed by using a random intermediate destination. This random destination can be created by modifying the source ID of the packet in several ways. Two of the methods are flipping the bits on chosen dimensions of the ID, or coin flipping the bits on selected dimensions. The outcome of the coin flip determines whether or not the dimension is changed. In general flipping bits is more effective than coin flipping on bits. When performing the transpose, flipping half the bits is superior to coin flipping on all the bits. The optimal number of bits to flip depends on the network size. See the Section on the transpose for specific results. If the intermediate destination is created by coin flipping on all the dimensions, then the header algorithms LOWHIGH, SHIFTRAND, RAND produce equivalent results. Queues of size 4 are sufficient for networks up to 64K with this scheme.

Direct routing on a leveled hypercube is very slow since the top node is a bottleneck. When routing the complement, all packets pass through the top node. With networks with up to 16K nodes, if the packets are routed first to a random intermediate destination, ZERORAND takes up to three times as long as LOWHIGH to route to random destinations, the transpose, or the complement. This performance may be acceptable if barrier synchronization is needed. The leveled techniques also require a larger queue size.

Acknowledgements

Many thanks to G. Shedler for his invaluable discussions and comments, Y. Birk, J. Bruck, L. Gravano, and C.T. Ho for their comments, and to M. Flickner for expert assistance in setting up and using the computing environment.

References

[1] R. Arlauskas, "iPSC/2 System: A Second Generation Hypercube", *The 3rd Conf. on Hypercube Concurrent Computers and Applications*, vol. 1, pp. 38-50, 1988.

[2] Y. Birk, P.B. Gibbons, J.L.C. Sanz, and D. Soroker, "A Simple Mechanism for Efficient Barrier Synchronization in MIMD Machines", IBM RJ 7078, Oct. 1989.

[3] R. Cypher and J.L.C. Sanz, "Massively Parallel Computing: Theory, Algorithms, Applications and Technology", Springer-Verlag, to appear.

[4] J. Hayes, T. Mudges, Q. Stout, et al, "A Microprocessor-based Hypercube Supercomputer", *IEEE Micro*, pp. 6-17, Oct. 1986.

[5] D. Hillis, "The Connection Machine", MIT Press, 1985.

[6] C.T. Ho, "The Hypercube- A Reconfigurable Mesh", *Reconfigurable SIMD Parallel Processors*, Prentice Hall, to be published.

[7] C. Kaklamanis, D. Krizanc, and A. Tsantilas, "Tight Bounds for Oblivious Routing in the Hypercube", *Proc. of the 1990 ACM Symp. of Parallel Alg. and Arch.*.

[8] S. Konstantinidou, "Adaptive, Minimal Routing in Hypercubes", *6th MIT Conf. on Adv. Research in VLSI*, pp. 139-153, 1990.

[9] S. Konstantinidou and L. Snyder, "The Chaos Router: A Practical Application of Randomization in Network Routing", *Proc. of the 1990 ACM Symposium of Parallel Algorithms and Architectures*.

[10] C.R. Lang, "The Extension of Object-oriented Languages to a Homogeneous, Concurrent Architecture", Tech. Report 5014:TR:82, Com. Sci. Dept., Cal. Instit. of Tech., 1982.

[11] G.P. Learmonth and P.A.W. Lewis, "Statistical Tests of Some Widely Used and Recently Proposed Uniform Random Number Generators", *Proc. of the 7th Conf. on Comp. Sci. and Stats. Interface*, 1974.

[12] P.A.W. Lewis and E.J. Orav, "Uniform Pseudo-Random Variable Generation", *Simulation Methodology for Statisticians, Operations Analysts, and Engineers*, vol. 1, Wadsworth & Brooks/Cole, pp. 65-99.

[13] D. Nassimi and S. Sahni, "Data Broadcasting in SIMD Computers", *IEEE Trans. on Computers*, vol. C-30, no. 2, pp. 101-107, Feb. 1981.

[14] "NCUBE, Product Report", Ncube Corporation, Headquarters, Beaverton, OR, 1986.

[15] S.F. Nugent, "The iPSC/2 Direct-Connect Communications Technology", *The 3rd Conf. on Hypercube Concurrent Computers and Applications*, vol. 1, pp. 51-60, 1988.

[16] A.G. Ranade, S.N. Bhatt, and S.L. Johnsson, "The Fluent Abstract Machine", *Advanced Research in VLSI, Proc. of the 5th MIT Conf.*, pp. 71-93, 1988.

[17] J.T. Schwartz, "Ultracomputers", *ACM Trans. on Prog. Lang. and Sys.*, vol. 2, pp.484-521, Oct. 1980.

[18] L. Valiant, "A Scheme for Fast Parallel Communication", *SIAM J. on Comput.*, vol. 11, no. 2, pp. 350-361, May 1982.

[19] L. Valiant and G.J. Brebner, "Universal Schemes for Parallel Communication", *Proc. 13th ACM Symp. on Theory of Comput.*, pp. 263-277, 1981.

[20] A. Waksman, "A Permutation Network", *Journal of the ACM*, vol. 15, no. 1, pp. 159-163, Jan. 1968.

A   Appendix

This appendix explains why LOWHIGH and SHIFTRAND achieve collision free routing when performing the complement. In LOWHIGH, each switch corrects the same dimension at the same time, thereby avoiding collisions. For the SHIFTRAND header algorithm, the idea of the proof follows. The proof is by induction on the number of routing steps. A routing step includes servicing all the input queues and routing packets at the head of all the output queues. By model definition, all switches are initialized with one packet in input queue zero. (Actually any input queue will work).

Base Case:

All output queues are empty. Consequently all packets proceed to the appropriate output queue, thereby emptying all the input queues. Since the input queues are empty, all the packets can then be sent from the output queues to the neighboring input queue.

Inductive Step:

Assume the inductive hypothesis holds for some $k > 1$ routing steps. For the sake of contradiction suppose that two packets collide in the $k + 1$ step; that is at routing step $k + 1$, the two packets want to leave on the same dimension. Since all packets must change all dimensions for the complement, this implies that both the headers were shifted by the same amount. Consequently the packets must also have been in the same node in step k. Therefore both packets should also have collided in step k. But this contradicts the induction hypothesis. $\square$

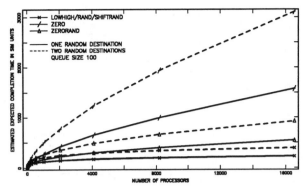

Figure 1: Routings of random destinations using various header algorithms.

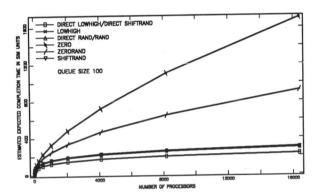

Figure 2: Complement for various header algorithms using coin flipping on all dimensions.

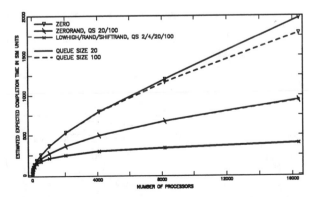

Figure 3: Transpose for various header algorithms using coin flipping on all dimensions.

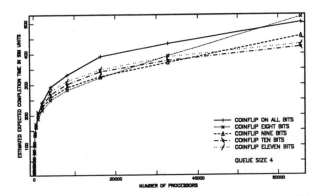

Figure 4: Transpose using LOWHIGH routing with various amounts of coin flipping.

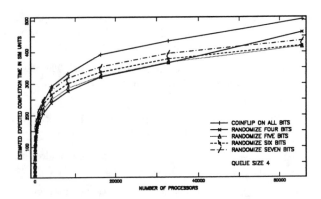

Figure 7: Transpose using LOWHIGH routing with various degrees of randomization from four to seven dimensions and for coin flipping on all dimensions.

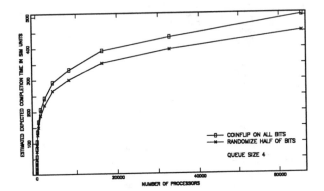

Figure 5: Transpose using LOWHIGH routing comparing randomization by coin flipping to randomization by selecting dimensions.

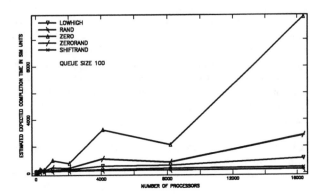

Figure 8: Direct routing of the transpose.

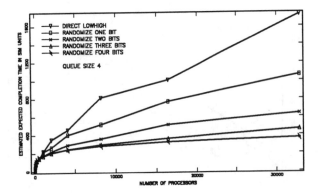

Figure 6: Transpose using LOWHIGH routing with various degrees of randomization from zero to four dimensions.

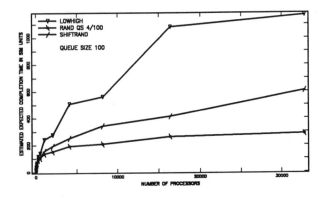

Figure 9: Direct routing of the transpose.

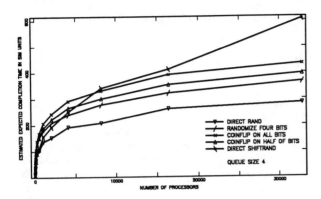

Figure 10: Comparison of various efficient routing techniques for the transpose.

Fault Tolerant Based Embeddings of Quadtrees into Hypercubes *

Narayanan Krishnakumar, Vinayak Hegde and S.Sitharama Iyengar

Computer Science Department
Louisiana State University
Baton Rouge, LA 70803

ABSTRACT

The recent advances in parallel computer architecture have also increased their complexities and thus the growing need for reliability and fault tolerance. A new class of reconfigurable mapping strategies used to recover from faults as well as maintain the quality of the mapping is introduced. The characteristic feature of our technique is that we provide an initial embedding which facilitates a reconfigured system with a constant dilation embedding. Specifically, we focus on the reconfigurable embedding of a complete quadtree into a faulty hypercube environment. The initial 2-dilation embedding and the final 3-dilation embedding after fault recovery are both optimum. It is possible to extend our methodology for a variety of computational graph structures onto various interconnection networks such as star graphs, and generalized boolean n-cubes.

Key Words: fault tolerance, reconfiguration, hypercubes, quadtrees, graph embeddings, dilation.

1.0 Introduction

In recent years we have witnessed a tremendous surge in the availability of very fast and inexpensive hardware. These have been made possible partly by novel architectural features such as pipelining, vector processing etc., and partly by using novel interconnections between processors and memories such as Hypercubes, Orthogonal Tree Networks and others. Parallel architectures based on binary hypercube topology have gained widespread acceptance in parallel computing The hypercube offers a rich interconnection topology with high communication bandwidth, low diameter and a recursive structure naturally suited to divide-and-conquer applications.

Two most important considerations in selecting data structures and algorithms for *interconnection networks* are *processor utilization* and *communication cost*. The communication needs of the computations are modeled by a graph, called the *guest graph*. This graph depicts the required interaction between the data elements of the computation. The interconnection topology of the ensemble architecture is depicted by a graph, called the *host graph*. The guest graph is embedded in the host graph for execution. The quality of an embedding is often measured by two parameters: (a) expansion and (b) dilation. The expansion is a measure of processor utilization where as the dilation affects the communication cost. We consider a binary hypercube as the host graph and a complete quadtree, the guest graph. It is known that embedding arbitrary graphs into hypercubes is NP-complete. Embedding trees have been studied by many researchers [1, 2] Throughout our discussion, the terms *mapping* and *embedding* mean the same, and are used interchangeably.

* This project is presently funded by Board of Regents Grant LEQSF-RD-a-04 and Office of Naval Research Grant N00014-91-J-1306.

Quadtrees have received a lot of attention in recent years as an efficient data structure for a variety of image processing applications. In the quadtree representation of an image the root represents an entire square field, and each node has either four children, each representing a quadrant of its parent, or is a leaf representing a quadrant. The reader is referred to [3, 4] Jones and Iyengar [4] in which efficient ways of storing quadtrees and related works are discussed. We use quadtree as an underlying data structure for illustrating our technique because of its wide scope for application, though methods described need not be restricted to one set of data structures.

The approaches taken to provide fault tolerance in interconnection networks fall into the following three categories : *architecture-based*, *algorithm-based* and *reconfiguration-based*. The motivation to develop schemes based on the first approach is that the computational graph structure is fixed and hosts may be built in a such a way that, in case of faults, it can maintain the topology of the computational graph. Spare nodes and edges may be added to the basic host architecture. The resultant architecture will facilitate fault recovery by providing subgraphs isomorphic to the original graph [5]. On the other hand, in algorithm-based techniques, the underlying graph structure is made robust using redundant paths without altering the architecture of the host [6]. Reconfiguration schemes provide intelligent mapping of the computational graph into the interconnection network so that reconfiguration becomes easy [7]. The most the reconfiguration algorithms enable the system to recover from a single failure in one-step by locating a free node in the system. The time taken by the system to reconfigure itself after the detection of faults is the primary concern of such algorithms. Due to lack of space, reader is referred to [8] for a review of related works.

The previous techniques do not address the problem of maintaining the quality of initial mapping, after reconfiguration. In this paper, we investigate methodologies to develop reconfigurable mapping strategies that maintain certain important qualities of the initial mapping. As an example, we give reconfigurable mapping algorithms for embedding a complete quadtree into the hypercube in a faulty environment. The characteristic feature of our technique is that the dilation in the reconfigured system is a constant. The remainder of the paper is organized as follows. We introduce preliminaries and basic notations used in our technique in section 2.0. In section 3.0 we describe the problem statement and discuss the related issues. A detailed description of the basic embedding (initial) algorithm with examples is given in section 4.0. The reconfiguration algorithm is explained in section 5.0. In section 6.0 we summarize the results and discuss the need for research in this area.

2.0 Notations and Preliminaries

We use the terms *nodes* and *links* for the hypercube and *vertices* and *edges* for the quadtree.

2.1 Terminologies for a Complete Quadtree

We denote a Complete Quadtree of height h, as $CQT(h)$. The root of $CQT(h)$ is at level 0, and the leaves are at level h. The number of vertices at level i is 4^i. We now introduce some terminologies pertaining to $CQT(h)$.

* The number of leaves, l is given by, $l = 4^h$.

* The number of vertices at level $h-1$ (i.e., the parents of leaves) denoted by, p, is given as: $p = 4^{h-1}$.

* The total number of vertices in $CQT(h)$ denoted by, m, is given as: $m = \dfrac{4^{h+1} - 1}{3}$.

We perform a breadth-first-search on $CQT(h)$ and assign to each vertex, its BFS-Number, called $BFSN$. For example, the $BFSN$ of root is 1, its children are 2,3,4 and 5, and so on. We refer every vertex by its $BFSN$.

2.2 Terminologies for a Binary Hypercube

We denote a hypercube with n nodes, as $H(n)$. We represent every node in $H(n)$ by a unique binary number (k-bit address) $b_k, b_{k-1} \cdots, b_1$, where k is the dimension of $H(n)$. Note that a node is adjacent to k nodes in $H(n)$. Also two nodes in $H(n)$ are adjacent when their addresses differ in exactly one of k bits. $H(8)$ is called a *cube*. The addresses of the nodes in the *cube* are given in rectangular boxes as shown in figure 3.0(b),(c).

2.3 Embedding Terminologies

The embedding function f maps each node in the guest graph $G = (V_G, E_G)$ into a *unique* node in the host graph $H = (V_H, E_H)$. V_G and V_H denote the node sets of the guest graph and the host graph respectively, and E_G and E_H the edge sets. Let v_1 and v_2 be nodes in G. Since the embedding is a 1-1 mapping of V_G onto V_H, let $f(v_1)$ and $f(v_2)$ be the nodes in H that are images of v_1 and v_2 respectively. In the following we define distance, dilation, and expansion.

* Let the distance, $d(v_i, v_j)$ between v_i and v_j be defined as the number of edges in the shortest path connecting v_i and v_j in either G or H.

* The dilation d_f is defined as
$$d_f = \max \frac{hd(f(v_1), f(v_2))}{d(v_1, v_2)}$$

* The expansion e_f is defined as
$$e_f = \frac{|V_H|}{|V_G|}$$

Our embedding algorithm builds an array, $B[1..m]$, where $B[i]$ contains the address of a unique node in $H(n)$ for the vertex of $CQT(h)$ whose $BFSN$ is i. We now state a simple lemma that is useful in following the algorithms.

Lemma 2.3.1: The height of $CQT(h)$ and the dimension, k of $H(n)$ are related as: $k = 2h + 1$.

Proof: We prove by induction on h.
Basis: The claim is true for $h = 1$. We need a *cube* ($k = 3$) to embed $CQT(1)$.
Hypothesis: We let the claim true for $CQT(h-1)$. Assume that we need a hypercube of dimension k'. Then we have $k' = 2(h-1) + 1 ....(i)$
Step: Now consider embedding $CQT(h)$ into a hypercube of dimension k. $CQT(h)$ is obtained by taking 4 copies of $CQT(h-1)$ and having a new root that is connected to all the 4 roots of $CQT(h-1)$'s. Hence the number of nodes in the hypercube to embed $CQT(h)$ is 4 times that is used to embed $CQT(h-1)$. That is, $k = k' + 2 ....(ii)$
From (i) and (ii) we get $k = 2h + 1$, and hence the proof. ∎

We now state a corollary which follows from lemma 2.3.1.

Corollary 2.3.2: The number of *cubes*, p, is equal to *one-fourth* of l (the leaves).

The addresses of the eight nodes in a *cube* have identical bits in their $k-3$ most significant position. We call the binary number formed by taking these $k-3$ bits, the *address of the cube*. All *cubes* that have 2 m.s.bs identical are said to be in a *quadrant*, and these 2 m.s.bs identify that *quadrant*. Further, every *quadrant* is recursively divided into four *sub-quadrants* by making use of the next 2 m.s.bs until no such division is possible. We now define a *similar* cube and a *square*.

* Two *cubes* are said to be *similar*, if the bits in their addresses are identical, except two successive bits starting at the same odd bit position. Note that these two differing bits will identify a *quadrant* or a *sub-quadrant*. For example, if addresses of *cube1* and *cube2* are 01 01 00 00 01 and 01 11 00 00 01 then they are *similar*.

* In any *quadrant* or *sub-quadrant* we get four *similar* cubes. In each of these *cubes*, all the eight nodes are identified using three least significant bits (l.s.bs). The nodes having three identical l.s.bs in these four *cubes* are said be *similar*. A *square* is formed among *similar* nodes.

2.4 Fault Model

Following are the underlying assumptions made regarding the faults, during the execution of our reconfiguration algorithms:

* The model tolerates node failures up to a maximum of available *free nodes* after the initial embedding. Links are assumed to be fault-free.

* Task migration is allowed.

* Any node that detects a fault may initiate the reconfiguration algorithm. We assume that reliable fault diagnosis mechanisms are available.

* Reconfiguration is carried out after the faults are detected.

* Once a *free node* is assigned (or used) in the recovery process, it can not be re-assigned for recovering any other faults in future.

* We define a class of faults, called *homogeneous faults*, that occur very frequently in computations based on tree structures as follows : Faults occur only at the leaves of the tree and further they are restricted to follow an even distribution at each level of the tree. In this paper, reconfiguration is provided for homogeneous faults.

The motivation for considering homogeneous faults in tree models is that the the leaves are computationally intensive and there is equal probability of faults at any leaf. Further, when the number of *free nodes* is less than the total number of leaves, it is natural to distribute *free nodes* evenly at each level of the tree.

3.0 Problem Statement and Motivation

Conventional embedding schemes make assignments by optimizing criteria such as dilation. Reconfigurable embedding algorithms allow the system to recover from faults by giving new assignments and still optimizing the dilation.

Our objective is to develop reconfigurable embedding algorithms for complete quadtrees into hypercubes that maintain constant dilation. We know that a complete quadtree can be embedded into its nearest hypercube with dilation 2 [1]. It is clear that the smallest hypercube (expansion < 2) has more nodes than the number of vertices in the complete quadtree. We refer to the nodes that do not participate in the embedding as *free nodes* and those that have failed as *faulty nodes*. Faults are tolerated by assigning *free nodes* to *faulty nodes*. An arbitrary assignment of a *free node* for a *faulty node* in a hypercube may result in a maximum dilation of *O(log n)*. A reconfigurable embedding technique should tolerate faults up to the total number of *free nodes*, a good technique should maintain a constant dilation.

The smallest constant of dilation provided by an embedding *which makes use of all the free nodes* is 3. This can be verified by considering the smallest complete quadtree (of height 1) and its nearest hypercube (of dimension 3), a *cube*.

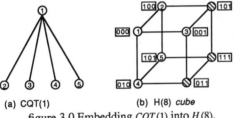

(a) CQT(1)   (b) H(8) *cube*

figure 3.0 Embedding CQT(1) into H(8).

A simple 2-dilation embedding of CQT(1) into H(8) is given in figure 3.0(b). The numbered circles of the *cube* are used in the embedding. The shaded circles are the *free nodes*. Note that one of the *free nodes* is at a distance of 3 from the node to which the root of the tree is assigned.

This motivates us to investigate the existence of reconfigurable embedding algorithms for complete quadtrees of any size (into hypercube) that maintains a dilation of 3 after the recovery from homogeneous faults.

4.0 Fault-free Embedding Algorithms

We give an initial embedding of the CQT(h) into H(n) with a dilation of 2. Section 4.1 describes the *Algorithm_2D*. The algorithm is explained in detail using the embedding of a CQT(2) into H(32) as an illustrative example. An informal discussion on the embedding technique, the actual pseudo-code for the algorithm, and its correctness of proof are given.

4.1 Algorithm_2D

This section is primarily concerned on a 2-dilation initial (pre-fault) embedding of a complete quadtree into its nearest hypercube. Ho and Johnsson [1] have shown that a CQT(h) can be embedded into H(n) with dilation 2. However, specific algorithms to do the embedding are not found in the literature. We give one such algorithm here. The embedding of CQT(1) into H(8) is given in figure 3.0(b). In figure 4.1 we show the embedding of CQT(2) into H(32).

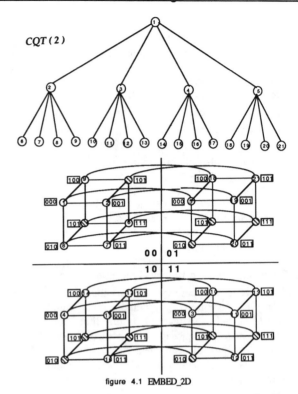

CQT(2)

figure 4.1 EMBED_2D

The hypercube, H(32) is divided into four quadrants each having a *cube*. The quadrant numbers are **00, 01, 11, 10** and are marked in their appropriate quadrants. The circles represent the processors of H(32). For the sake of clarity, the interconnections among the nodes in quadrants **00** & **01**, and **00** & **10** are shown. In the actual hypercube, there are similar interconnections among the nodes in quadrants **10** & **11**, and **01** & **11**. The bits placed in rectangular boxes near the nodes are nothing but the last 3 l.s.b's of the addresses of the nodes in the hypercube. Again, the circles with numbers in them (numbers correspond to a vertex in the quadtree) are used in the embedding process, while the shaded circles do not participate in the initial embedding. Both the embedding algorithms work in a top-down fashion on the quadtree. That is, the root vertex is assigned to a node in the hypercube first, then its children, and so on. Also, the root of the CQT(h) is always placed in the quadrant, **00**. In addition, the root (or the subroot), and its leftmost child will be placed in the same quadrant (or subquadrant), and the remaining three children are placed in each of the other three quadrants (or subquadrants). The procedure *Find_child_index* computes the BFSNs of a vertex, given its BFSN and level. We now proceed to the informal discussion on the *Algorithm Embed_2D*.

As a first step, the root (1) is assigned to the node in the quadrant, **00** at address 00000. The identifier *index* refers to the BFSN of the current root (or subroot), and *level* to level at which the current root (or subroot) is found in the quadtree. Note that the root of the quadtree is at level 0. Next, the procedure Embed_children_2D is called to assign the subroots (2),(3),(4) and (5). As we explained in the previous section, the subroot (2) is placed in the quadrant **00**, (where its parent is placed) at address 00001, while (3),(4) and (5) are placed at quadrants **11,10** and **01** at addresses 11000,10000 and 01000 respectively. The assignment of these subroots is done by

Embed_in_a_5CUBE. Finally, the children of these subroots are assigned to the nodes in the same quadrant as these subroots are assigned. This is done by Embed_in_a_3CUBE, which is the terminating step for the recursive algorithm Embed_2D. Note that in any *cube*, there will be at most two internal vertices of the $CQT(h)$. We call them r_1 and r_2. Without loss of generality, we let r_1 be the vertex at level $h-1$. Our embedding ensures that there is a *free node* adjacent to the node which is assigned to r_1.

Now we give a formal algorithm for the initial 2-dilation embedding.

```
Algorithm Embed_2D ;

/*     Input:                                              */
/*         1. A complete quadtree of height h  (CQT(h))    */
/*         2. A hypercube with n nodes (H(n) is the optimal) */
/*     Output:                                             */
/*         An array, B[1..m] of binary numbers of length k */
/*     where k = log(n) and m the number of vertices in CQT(h). */

/*     B[i] contains the address of a unique node in H(n) for  */
/*     the vertex of CQT(h) whose BFSN is i.              */

begin
      Set B[1] to be b_k b_{k-1} b_{k-2} b_{k-3} ...b_5 b_4 b_3 b_2 b_1,

            where b_i = 0 for i = k,k-1,k-2,k-3,...,1.
      Embed_children_2D(1 , 0 , k );
end.
```

```
Procedure Embed_children_2D (index, level, k) ;
begin
      c = B[index];
      Find_child_index (children, index, level) ;
      if ( k = 3 ) then
      begin
            Embed_in_a_3CUBE (children, k) ;
      end
      else if ( k = 5 ) then
            begin
                  Embed_in_a_5CUBE (children, k) ;
            end
      else
            begin
                  Embed_in_NCUBE (children, k) ;
            end
            end;
end;
```

```
Procedure Embed_in_NCUBE (children, index, level) ;
begin
      if (c_{k-1} = 1) then
      begin
            c_{k-1} = 0 ; c_{k-3} = 1; B[children[1]] = c ;
            Embed_children_2D(children[1] , (level + 1), (k - 2));
            c_{k-1} = 1 ; B[children[2]] = c ;
            Embed_children_2D(children[2] , (level + 1), (k - 2));
            c_{k-3} = 0 ; c_k = 1; B[children[3]] = c ;
            Embed_children_2D(children[3] , (level + 1), (k - 2));
            c_{k-1} = 0 ; B[children[4]] = c ;
            Embed_children_2D(children[4] , (level + 1), (k - 2));
      end
      else begin
                  c_{k-3} = 1 ; B[children[1]] = c ;
                  Embed_children_2D(children[1] , (level + 1), (k - 2));
                  c_{k-3} = 0 ; c_{k-1} = 1; B[children[2]] = c ;
                  Embed_children_2D(children[2] , (level + 1), (k - 2));
                  c_k = 1 ; B[children[3]] = c ;
                  Embed_children_2D(children[3] , (level + 1), (k - 2));
                  c_{k-1} = 0 ; B[children[4]] = c ;
                  Embed_children_2D(children[4] , (level + 1), (k - 2));
            end
end
```

```
Procedure Embed_in_3CUBE(children, k) ;
begin
      if (c_{k-2} = 0) then
      begin
            c_{k-2} = 1 ; B[children[1]] = c ;
            c_{k-1} = 1 ; B[children[2]] = c ;
            c_{k-2} = 0 ; B[children[3]] = c ;
            c_k = 1 ; c_{k-1} = 0 ; B[children[4]] = c ;
      end
      else begin
                  c_{k-1} = 1 ; B[children[1]] = c ;
                  c_{k-2} = 0 ; B[children[2]] = c ;
                  c_k = 1 ; c_{k-1} = 0 ; B[children[3]] = c ;
                  c_{k-2} = 1 ; B[children[4]] = c ;
            end
end;
```

```
Procedure Embed_in_5CUBE(children, k) ;
begin
      if (c_{k-1} = 1) then
      begin
            c_{k-4} = 1 ; B[children[1]] = c ;
            Embed_children_2D(children[1] , (level + 1), (k - 2));
            c_{k-1} = 1 ; B[children[2]] = c ;
            Embed_children_2D(children[2] , (level + 1), (k - 2));
            c_{k-4} = 0 ; c_k = 1; B[children[3]] = c ;
            Embed_children_2D(children[3] , (level + 1), (k - 2));
            c_{k-1} = 0 ; B[children[4]] = c ;
            Embed_children_2D(children[4] , (level + 1), (k - 2));
      end
      else begin
                  c_{k-4} = 1 ; B[children[1]] = c ;
                  Embed_children_2D(children[1] , (level + 1), (k - 2));
                  c_{k-4} = 0 ; c_{k-1} = 1; B[children[2]] = c ;
                  Embed_children_2D(children[2] , (level + 1), (k - 2));
                  c_k = 1 ; B[children[3]] = c ;
                  Embed_children_2D(children[3] , (level + 1), (k - 2));
                  c_{k-1} = 0 ; B[children[4]] = c ;
                  Embed_children_2D(children[4] , (level + 1), (k - 2));
            end
end;
```

```
Procedure Find_child_index (children, index, level) ;
begin
      FIRST0 = Vertex_index(level) ;
      FIRST1 = Vertex_index(level +1) ;
      for i = 1 to 4 do
            children[i] = FIRST1 + ((index - FIRST0) * 4) + (i - 1);
end;
```

```
Procedure Vertex_index (level) ;
begin
      SUM = 0 ;
      for i = 0 to (level - 1) do
      begin
            SUM = SUM + 4^i ;
      end;
      SUM = SUM + 1 ;
      return (SUM) ;
end;
```

Theorem 4.1.1: The algorithm, *Embed_2D* embeds a complete quadtree of height h into its nearest hypercube with at most two dilation.

Proof: The recursive procedure, *Embed_children_2D* is called with the three parameters: (i) BFSN of the root (or subroot), (ii) the level of the root (or subroot), and (iii) the number of bits to work with (which is initially all the k bits). Notice in the algorithm that this recursive procedure is called with different values of these parameters at different times. Since the assignment of addresses to the vertices depends on these parameters, the embedding is one-to-one. When the number of bits to work with $(=k)$ becomes 3, the recursive call terminates. Since k is an odd number (from lemma 2.3.1) and it is reduced by 2 in every recursive step, eventually the algorithm terminates.■

5.0 Reconfiguration Algorithm

In this section, we present algorithms to reconfigure the embedding after the faults are detected. The informal discussion on the reconfiguration algorithm will be based on the example embeddings we have chosen in sections 4.1. Since the reconfiguration algorithm is quite involved, we establish the concept of reconfiguration in this section and the actual reconfiguration algorithm and its pseudo-code are given in [8].

We now discuss the reconfiguration algorithm. Note that after the initial embedding, each set of consecutive four leaves (that is leaves with common parent) are placed in a *cube* in the hypercube. It is also seen that the number of *cubes* is equal to one-fourth of the number of leaves (from Corollary 2.3.2). In each of these *cubes*, there will be *free nodes* (shaded circles in figure 4.1). We now state a lemma which can be easily verified by observation.

Lemma 5.0.1: In every quadrant (or subquadrant) the subquadrant (or sub-subquadrant) **00** will have 2 *free nodes*, while **01, 11,** and **10** will have 3.

Following are the important steps involved in reconfiguration.

(1) *Find_free_node_distribution*: This refers to the availability of free nodes in every quadrant. We collect the number of *free nodes* in each of the *cubes* and form a sequence, called *FN*. *FN* is an array of size $(m-l)$. The indices of *FN* are the BFSN of the interior vertices of the quad tree. *FN*[1] contains the total number of *free nodes* in the hypercube and *FN*[2],*FN*[3],*FN*[4],*FN*[5] have the number of *free nodes* in quadrants **00,11,10** and **01** respectively. This counting process continues until a *cube* is encountered. For example, *FN*s for figure 4.1 are 11,2,3,3,3. Note that our initial embedding provides one of the valid homogeneous free node distribution.

(2) *Find_fault_node_distribution*: From the given homogeneous fault sequence obtain a fault distribution, called *FT*. This is again similar to obtaining *FN*, except that the distribution of faults may be different from the free node distribution. For example, for one valid homogeneous fault distribution, *FT*s is 11,3,3,2,3.

(3) *reconfiguration step*: The idea here is to transform the fault distribution to the free node distribution. These two distributions are stored in arrays, *FN and FT* , and their indices are the BFSN of the interior vertices in quadtree. Hence, we can represent these distributions to be arranged in the form a quadtree of height $h-1$. We use figure 5.0 to discuss the reconfiguration strategy.

Following are the steps needed to transform FT into FN:

1. The distribution lists *FN and FT* are traversed in a breadth-first fashion, and at each level the contents of these lists with the same index, are compared.

2. If the content of *FN* for *any cube* is greater or equal to that of *FT*

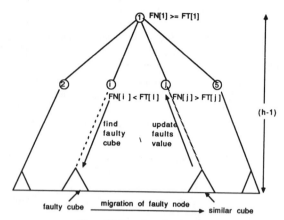

figure 5.0 Reconfiguration Steps

then the algorithm reassigns the free nodes in the same cube for the faulty processes. Note that the resulting dilation is relaxed to 3 to make use of all the available free nodes.

3. Let i be the index (or vertex in the tree). Consider the case when $FN[i] < FT[i]$. Since the total number of faults allowed is always less than that of free nodes, there must exist an index j such that i and j have a common parent and $FN[j] > FT[j]$. This situation implies that one of the *cubes* in the quadrant corresponding to the the subtree rooted at i has more faults than the free nodes. We now traverse down the subtree rooted at i and locate this *faulty cube*.

4. Locate the *similar cube* in the subquadrant where the subtree rooted at j is placed. Note that there is at least an *extra free node* in this *cube*.

5. Transfer the *extra* fault in the *faulty cube* to the *similar cube* located in the previous step. If the faulty cube is located in non-adjacent sub-quadrants, it is necessary to find a path between the faulty cube and the free cube such that intermediate cubes on the path are similar cubes. This is done using a procedure *Find_migration_path* which creates local faults in the similar cubes and migrates fault to the cube with the extra free node.

6. Update the faults distribution in both the subtrees rooted at i and j.

7. Repeat the above steps for all the levels of the quadtree.

The psuedo-code of the complete reconfiguration algorithm is given in [8] and the partial code in appendix.

Theorem 5.0.1: The reconfiguration algorithm maintains a constant dilation of 3.

Proof: We claim that the migration of *extra* fault to an *extra free node* does not result in more dilation. It can be easily seen from the fact that the migration takes place among *similar cubes*. From the definition of *similar cubes*, it can be shown that the parents of the *faulty leaf* and the *extra free node* are adjacent in the hypercube. Hence, the claim.■

6.0 Conclusion

We have introduced reconfigurable embedding techniques which tolerate faults as well as maintain the quality of embedding after the fault recovery. As an example, reconfigurable embedding algorithms for complete quadtrees on to faulty hypercubes have been presented. The initial 2-dilation embedding and the final 3-dilation embedding after fault recovery are both optimum. Results are presented for a class of faults called *homogeneous faults*. We are currently developing techniques to take care of different classes of faults. It is possible to extend our methodology for a variety of computational graph structures onto various interconnection networks such as star graphs, and generalized boolean n-cubes.

References

[1]   C.T. Ho, S.L. Johnsson, "Dilation d Embedding of a Hyper-Pyramid into A Hypercube," *Proceedings of the Supercomputing '89*, pp. 294-303, November 1989.

[2]   S.N. Bhatt, F. Chung, T. Leighton, A. Rosenberg, "Optimal Simulations of Tree Machines," *Proceedings of FOCS*, pp. 274-282, 1986 .

[3]   H. Samet, "The Quadtree and Related Hierarchical Data Structures," *ACM Computing Surveys*, vol. 16, pp. 187-260, 1984.

[4]   L. Jones, S.S. Iyengar, "Space and Time Efficient Virtual Quadtrees," *IEEE Transactions on Pattern Analysis and Machine Intelligence*, vol. 6, pp. 244-247, 1984.

[5]   S. Dutt, J.P. Hayes, "Design and Reconfiguration Strategies for Near-Optimal k-Fault-Tolerant Tree Architectures," *IEEE-Conference on Fault Tolerance*, pp. 328-333, 1988.

[6]   A. Despain, D. Patterson, "X-tree: A Structured Multiprocessor Computer Architecture," *5th Symposium on Computer Architecture*, pp. 144-151, April 1978.

[7]   S.K. Chen, C.T. Liang, W.T. Tsai, "An Efficient Multi-Dimensional Grids Reconfiguration Algorithm on Hypercubes," *International Conference on Parallel Processing*, pp. 368-373, 1988.

[8]   N. Krishnakumar, V. G. Hegde, S. S. Iyengar, "Reconfigurable Embedding Schemes : An illustration of fault tolerant embedding of quad trees into nearest hypercube," *Technical Report RRL-TR-91-001*, pp. 1-30, 1991 .

APPENDIX

```
Algorithm Homogeneous ;

/*    Input:                                                      */
/*       1. An array, B [1..m] of binary numbers of length k      */
/*    where k = log (n) and m the number of vertices in CQT(h).   */
/*    B [i] contains the address of a unique node in H(n) for     */
/*    the vertex of CQT(h) whose BFSN is i (output of Embed_2D)    */
/*       2. A sequence of homogeneous faults, FS [1..p],          */
/*    where p is the number of vertices of CQT(h) at level h-1.   */
/*    Output:                                                     */
/*       An array, B [1..m] with new addresses of hypercube nodes. */


begin
    Find free_node_distribution ;
                   /* compute the free node distribution from B   */
                   /* and m and store it in an array, FN          */
    Find_fault_node_distribution ;
                   /* compute fault node distribution from the    */
                   /* fault sequence, FS and store it in FT.      */
    Homogeneous_root (1, 0) ;
                   /* a recursive procedure to verify the faults  */
                   /* at its children for homogeneity. In this call*/
                   /* the index and level of the root of CQT(h)   */
                   /* are 1 and 0 respectively.                   */
end.
```

```
Procedure Homogeneous_root (index, level) ;
begin
   if (level < h-1) then
      begin
         Find_child_index (children, index, level) ;
                      /* From the index and the current level of the  */
                      /* root (or the subroot), find out the indices */
                      /* of its children and store them in the array*/
                      /* called, 'children'.                    */
         for i = 1 to 2 do
         begin
            free_root_index[i] = 0 ;
            free_cube_index[i] = 0 ;
            fault_root_index[i] = 0 ;
            fault_cube_index[i] = 0 ;
         end ;
         i = 1 ;
         s = 1 ;
         for j = 1 to 4 do
         begin
            difference = FT [children [j]] - FN [children [j]] ;
            if (difference = 2) then
            begin
               fault_number = difference ;
               fault_root_index[1] = children[j] ;
               fault_root_index[2] = children[j] ;
            end
            else if (difference = 1) then
               begin
                  fault_number = difference ;
                  fault_root_index[s] = children[j] ;
                  s = s + 1 ;
               end
            else if (difference < 0) then
               begin
                  free_root_index[i] = children[j] ;
                  i = i + 1 ;
               end ;
         end ;
         if (fault_number <= 0) then
         begin
            for j = 1 to 4 do
            begin
               Homogeneous_root (children [j], level+1) ;
            end ;
         end
         else begin
            if (fault_number = 2) then Fault_cubes_2
                            (fault_cube_index, fault_root_index, level) ;

         else begin
            Fault_cubes_1 (fault_cube_index[1], fault_root_cube[1], level) ;

            if (fault_root_index[2] <> 0) then Fault_cubes_1
                            (fault_cube_index[2], fault_root_cube[2], level) ;
            end ;

            Find_migration_path (fault_root_index, faulty_cube_index,
                            free_root_index, free_cube_index, level) ;



            for j = 1 to 4 do
            begin
               Homogeneous_root (children [j], level+1) ;
            end ;
         end ;
      end ;
end;
```

DILATION-6 EMBEDDINGS OF 3-DIMENSIONAL
GRIDS INTO OPTIMAL HYPERCUBES [†]

Hongfei Liu and Shou-Hsuan Stephen Huang[*]

Department of Computer Science, University of Houston, Houston, TX 77204

Abstract – Many parallel computers based on the topology of the boolean hypercube are now commercially available. A distinctive feature of the hypercube is its *universality* – programs written for simpler architectures, such as trees, grids, can be transported onto the hypercube with minimal communication overhead. These simulations are typically obtained by embedding the simpler architecture into the hypercube. This paper investigates the following embedding problem: given a three-dimensional grid, and the smallest hypercube with at least as many nodes as the grid points, how can we assign grid points to hypercube nodes in a one-to-one fashion so as to keep grid-neighbors as close to each other as possible in the hypercube. We give a simple strategy which ensures that grid-neighbors are always mapped to hypercube nodes that are within a distance of six.

Introduction

A number of parallel computer architectures, where many processing elements (PE's) are connected via an interconnection network, have been proposed or built in response to the ever-growing need for speeding up computationally intensive tasks. Processor utilization and communication time are two important considerations in selecting data structures and algorithms for the underlying architecture. Communication is one of the most expensive resources in such an architecture, and its efficient utilization is imperative.

The communication needs of the computations can be modeled by a graph called **problem graph**, it discloses the interaction between the data elements of the computation. Similarly, the topology of the host computer can be captured by a graph, the **system graph**. Each vertex of the system graph represents a processor with local memory and edge a communications link between processors. The problem graph is embedded into the system graph for execution.

Formally, an **embedding** of a problem graph $G_p = \langle V_p, E_p \rangle$ into a system graph $G_s = \langle V_s, E_s \rangle$ is a function $\phi : V_p \to V_s$. If the function ϕ is a one-to-one function then the embedding is said to be *one-to-one*. The practical considerations associated with this type of embedding bring forth the following two notions, the **dilation** of ϕ, denoted by $\delta(\phi)$, is defined as $\delta(\phi) = \max\{dist(\phi(u), \phi(v)) | (u, v) \in E_p\}$, where $dist(a, b)$ denotes the shortest path length between the nodes a and b in G_s. The **expansion** of ϕ, denoted by $\epsilon(\phi)$, is defined as $\epsilon(\phi) = |V_s|/|V_p|$, where $|V|$ denotes the cardinality of the set V. The parameter ϕ may be omitted in the context.

One of the most important parallel computers is the hypercube which has many interesting topological properties (for example, trees and grids can be efficiently simulated). A **hypercube** of dimension n is conventionally thought of as an undirected graph of 2^n nodes labeled 0 to $2^n - 1$ in binary, two nodes are connected by an edge **iff** their labelings differ in exactly one bit position. It is therefore of great importance to study the embeddings into hypercubes. Practically, we would like to have embedding with dilation δ, and expansion ϵ, as small as possible. An **optimal** hypercube for a grid is defined to be smallest the hypercube with at least as many nodes as the grid points.

A number of researchers have studied the embeddings into hypercubes. Livingston and Stout [8] gave a survey of various embeddings into hypercubes. Several papers in [9] also have certain discussion on the problem. Using an accurate characterization of communication overhead, Lee and Aggarwal [6] presented an embedding strategy and applied to the hypercubes. Ho and Johnson [5] and Chan and Chin [2] obtained embeddings of most 2-dimensional grids with both δ and ϵ no greater than 2. They also gave the lower bound on the dimensions of hypercubes which can embed multi-dimensional grids using unit dilation ($\delta = 1$). Ho and Johnson [5] also addressed some issues about the embeddings of multi-even-dimensional grids. Scott and Brandenburg [10] presented a nice *square property* of the hypercube and reached the same lower bound. Chan [1] gave a very creative way of embeddings of all 2-dimensional grids into their optimal hypercubes with $\delta \leq 2$. A dilation-7 embedding of 3-dimensional grids into optimal hypercubes is given in [3] and a dilation-$(4d + 1)$ embedding of d-dimensional grids is given in [4].

This paper will focus on the embeddings of all 3-dimensional grids into their optimal hypercubes, since the 3-dimensional grids have much more practical meaning than higher dimensional grids. Our embedding has dilation at most 6, which is an improvement over the result in [3] and the scheme is simpler. Section 2 contains a discussion of relevant results that will be used later about 2-dimensional grid embedding. Section 3 describes the 3-dimensional grid embedding scheme and its correctness. Section 4 draws the conclusion and poses some open problems.

Previous results on 2-dimensional grids

For the sake of completeness, from [4] we here reproduce and take note of some of its properties the very special *partitioning matrix* $A(\alpha, \beta)$, or simply A, an integer matrix comprised of 1's and 2's having $\bar{\alpha} = 2^{\lfloor \log \alpha \rfloor}$ rows and β columns. As it turns out that, in regards to the partitioning of the nodes of the two-dimensional grid $G = \alpha \times \beta$ into $\bar{\alpha}$ groups, which we call "chains", $a_{i,j}$ (the element in the ith row and j^{th} column of matrix A) essentially indicates how many nodes from column j of grid G will belong to chain i. Matrix A has as its first column the vector

$$\begin{pmatrix} a_{1,1} \\ a_{2,1} \\ a_{3,1} \\ \vdots \\ a_{\bar{\alpha},1} \end{pmatrix} = \begin{pmatrix} \lceil \alpha/\bar{\alpha} \rceil \\ \lfloor \alpha/\bar{\alpha} \rfloor \\ \lfloor 2\alpha/\bar{\alpha} \rfloor - \lfloor \alpha/\bar{\alpha} \rfloor \\ \vdots \\ \lfloor (\bar{\alpha} - 1)\alpha/\bar{\alpha} \rfloor - \lfloor (\bar{\alpha} - 2)\alpha/\bar{\alpha} \rfloor \end{pmatrix}$$

Equivalently, for all $1 \leq i \leq \bar{\alpha}$, $a_{i,1} = \lfloor (i-1)\alpha/\bar{\alpha} \rfloor - \lfloor (i-2)\alpha/\bar{\alpha} \rfloor$. For example, the first column vector for the grid 5×5 is

$$\begin{pmatrix} 2 & 1 & 1 & 1 \end{pmatrix}^{\tau}$$

The entire matrix is based on a cyclic shift of the first column, i.e., for all $1 \leq i < \bar{\alpha}$ and $1 \leq j < \beta$, $a_{i+1,j+1} = a_{i,j}$ and $a_{1,j+1} = a_{\bar{\alpha},j}$. In general, we have for all $1 \leq i \leq \bar{\alpha}$ and $1 \leq j \leq \beta$, $a_{i,j} = a_{(i-j) \bmod \bar{\alpha}+1,1}$:

[†] This research was supported in part by Texas Advanced Research Program Grant ARP-1080.

[*] E-Mail Address: s_huang@cs.uh.edu.

$$A = \begin{pmatrix} a_{1,1} & a_{1,2} & \cdots & a_{1,\beta} \\ a_{2,1} & a_{2,2} & \cdots & a_{2,\beta} \\ \vdots & \vdots & \ddots & \vdots \\ a_{\bar{\alpha},1} & a_{\bar{\alpha},2} & \cdots & a_{\bar{\alpha},\beta} \end{pmatrix} = \begin{pmatrix} a_{1,1} & a_{\bar{\alpha},1} & a_{\bar{\alpha}-1,1} & \cdots \\ a_{2,1} & a_{1,1} & a_{\bar{\alpha},1} & \cdots \\ \vdots & \vdots & \vdots & \\ a_{\bar{\alpha},1} & a_{\bar{\alpha}-1,1} & a_{\bar{\alpha}-2,1} & \cdots \end{pmatrix}$$

For example, the partitioning matrix for the grid 5×5 is the following 4×5 matrix:

$$A(5,5) = \begin{pmatrix} 2 & 1 & 1 & 1 & 2 \\ 1 & 2 & 1 & 1 & 1 \\ 1 & 1 & 2 & 1 & 1 \\ 1 & 1 & 1 & 2 & 1 \end{pmatrix}$$

For $p \le q$, let $Rowsum(i; p, q) = \sum_{j=p}^{q} a_{i,j}$ and let $Colsum(j; p, q) = \sum_{i=p}^{q} a_{i,j}$, for $p > q$, $Rowsum(i; p, q) = 0$ and $Colsum(j; p, q) = 0$. Then for $1 \le i, i' \le \bar{\alpha}$, $1 \le j, j' \le \beta$, $1 \le k \le \beta$ and $1 \le q + 1, q' + 1, q + k, q' + k \le \beta$, matrix A has the following properties:

(P1) $a_{i,j} \in \{1, 2\}$,

(P2) $\sum_{i=1}^{\alpha} \sum_{j=1}^{\beta} a_{i,j} = \alpha\beta$,

(P3) $|Colsum(j; 1, i) - Colsum(j'; 1, i)| \le 1$,

(P4) $Rowsum(i; 1, k) \in \{\lfloor k\alpha/\bar{\alpha} \rfloor, \lceil k\alpha/\bar{\alpha} \rceil\}$,

(P5) $|Rowsum(i; q+1, q+k) - Rowsum(i'; q'+1, q'+k)| \le 1$.

Having contructed $A(\alpha, \beta)$, we can define $chain(N)$, the chain number for node $N = (x, y)$, as

$$chain(N) = z \quad iff \quad Colsum(y; 1, z-1) < x \le Colsum(y; 1, z).$$

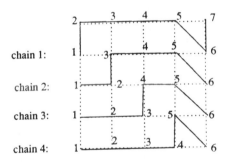

Fig. 1(a). Four chains with rank values for grid 5x5.

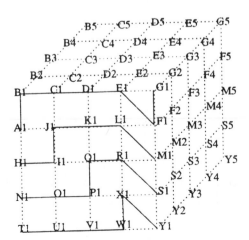

Fig. 1(b). A grid of 5x5x5.

Clearly for any node $S = (x, y)$ if $T \in \{(x, y'), (x', y)\}$, where $|x - x'| \le 1$ and y' arbitrary, using (P3) we have the following fact:

(F1) $|chain(S) - chain(T)| \le 1$.

Each chain is formed by connecting all the nodes with the same chain number from column 1 to column y sequentially, if there are two nodes with the same chain number in the same column then the bottom one goes before the top one. The jth node in each chain is said to have **rank** number j. Formally the rank number for node $N = (x, y)$ can be defined as

$$rank(N) = 1 + Rowsum(chain(N); 1, y-1) + \chi(x, y),$$

$$where \quad \chi(x, y) = \begin{cases} 1, & if\ chain(x+1, y) = chain(x, y); \\ 0, & otherwise. \end{cases}$$

Fig. 1 shows the four chains and the $rank$ values given to the nodes in the 5×5 grid.

Let $S = (x, y)$, $H = (x, y+1)$, $V = (x+1, y)$, i.e., S, H are horizontal neighbors, and S, V are vertical neighbors. It is easy to obtain the following facts:

(F2) $|rank(S) - rank(H)| \le 2$ if $chain(S) = chain(H)$,

(F3) $|rank(S) - rank(H)| \le 3$ if $chain(S) \ne chain(H)$,

(F4) $|rank(S) - rank(V)| \le 1$.

The New Embedding Strategy

We shall use the most common Gray code, the binary reflected ones, which are recursively defined as follows: Γ_n is a bijection from $\{0, 1, ..., 2^n - 1\}$ onto $\{0, 1\}^n$, given by $\Gamma_1(0) = 0$, $\Gamma_1(1) = 1$, and

$$\Gamma_n(x) = \begin{cases} 0\Gamma_{n-1}(x), & 0 \le x \le 2^{n-1} - 1; \\ 1\Gamma_{n-1}(2^n - 1 - x), & 2^{n-1} \le x \le 2^n - 1 \end{cases}$$

for $n \ge 2$.

For example, $\Gamma_2(0) = 00$ and $\Gamma_3(4) = 110$. One nice property of the Gray code used in our context is that $\Gamma_n(x)$ and $\Gamma_n(x+1)$ differ in exactly one bit position for $0 \le x < 2^n - 1$. Notice also that $\Gamma_n(0)$ and $\Gamma_n(2^n - 1)$ differ in one bit position. The subscript n may be omitted in our context.

The following fact is quoted in [3]:

(F5)   The binary-reflected Gray code $\Gamma(p)$ and $\Gamma(p+q)$ will differ by at most $\lceil \log \frac{3q}{2} \rceil$ bits.

Secondly for any integers a, b, c, and d, some inequalities which will be used in the next section are:

(F6) $\log \lceil \frac{a}{b} \rceil \le \lceil \log \frac{a}{b} \rceil$,

(F7) $\lceil \frac{ab}{d} \rceil \le c \lceil \frac{ab}{cd} \rceil$, and

(F8) $\lfloor \frac{a}{c} \rfloor \le \lceil \frac{a+b}{c} \rceil - \lceil \frac{b}{c} \rceil \le \lceil \frac{a}{c} \rceil$.

Now given a three-dimensional grid $\alpha \times \beta \times \gamma$, clearly, its optimal hypercube is of dimension $n_o = \lceil \log(\alpha\beta\gamma) \rceil$. The objective here is to assign a unique binary string (called the label) of length n_o to each node in the grid such that the Hamming distance between labels of any two grid neighbors is as small as possible. In this section, we give a simple embedding strategy which embeds any 3-dimensional grid into its optimal hypercube with dilation, $\delta \le 6$. The strategy is divided into 3 steps which determines the 3 segments of the labels accordingly, the first segments disclose the first partition of grid points into groups, the third segments indicate another partition of the grid points into groups, and the middle segments record the integer values assigned to the grid points. We divide the algorithm into three steps.

Step 1. (Determine the first $\lfloor \log \alpha \rfloor$ bits of each node's label.) Consider the three-dimensional grid $\alpha \times \beta \times \gamma$ as γ pages of $\alpha \times \beta$ grid. Construct the partitioning matrix $A(\alpha, \beta)$, and use it to partition each page, P_j ($1 \le j \le \gamma$), into $\bar{\alpha} = 2^{\lfloor \log \alpha \rfloor}$ chains, namely $chainA[i, j]$ ($1 \le i \le \bar{\alpha}$). Naturally we get $\bar{\alpha}$ sets of chains, S_i, where each set S_i contains γ identical chains, i.e, $S_i = \{chainA[i, j] | 1 \le j \le \gamma\}$. The kth node T in each chain, $chainA[i, j]$, is said to have **rank$_1$** of k, denoted as $rank_1(T) = k$ (the subscript 1 indicates Step 1). By stretching each set, S_i, of chains (flattening them on a plane) we obtain $\bar{\alpha}$ two-dimensional grids, each of these new grids is called a **layer**. Layer i, $i = 1, 2, \cdots, \bar{\alpha}$, is a $\gamma \times l_i$ grid, where $l_i \in \{\lfloor \frac{\alpha\beta}{\bar{\alpha}} \rfloor, \lceil \frac{\alpha\beta}{\bar{\alpha}} \rceil\}$ (from (P4)) and $\sum_{i=1}^{\bar{\alpha}} l_i = \alpha\beta$ (from (P2)). Each node which belong to layer i is assigned $\Gamma(i - 1)$ as the first $\lfloor \log \alpha \rfloor$ bits of its label.

Step 2. (Determine the last $\lfloor \log \gamma \rfloor$ bits of each node's label.) Let $\bar{l} = \lfloor \frac{\alpha\beta}{\bar{\alpha}} \rfloor$, $e = \alpha\beta - \bar{\alpha}\bar{l}$, generate the sequence p_i as follows: $p_i = \bar{l} + \lceil \frac{ie}{\bar{\alpha}} \rceil - \lceil \frac{(i-1)e}{\bar{\alpha}} \rceil$, where $i = 1, 2, \cdots, \bar{\alpha}$. It is easy to show that $p_i \in \{\lfloor \frac{\alpha\beta}{\bar{\alpha}} \rfloor, \lceil \frac{\alpha\beta}{\bar{\alpha}} \rceil\}$ and $\sum_{i=1}^{\bar{\alpha}} p_i = \alpha\beta$. This means that the p_i's are a permutation of the l_i's. By concatenating the $\bar{\alpha}$ layers obtained in Step 1 in an order so that the l_i's appear in the same sequence as the p_i's. For the above mentioned example, the sequence of l_i's happens to be the same as that of p_i's. Hence we get a single 2-dimensional grid, $H = \gamma \times \alpha\beta$ which contains exactly the grid points in the orginal grid. Construct the partitioning matrix $B(\gamma, \alpha\beta)$, partition H into $\bar{\gamma} = 2^{\lfloor \log \gamma \rfloor}$ longer chains, called **threads**, denoted by B_k, $k = 1, 2, \cdots, \bar{\gamma}$. Fig. 2 depicts the scenario for the partitioning of the concatenated grid 5×25 into 4 threads. Each node in thread B_k is assigned $\Gamma(k - 1)$ as the last $\lfloor \log \gamma \rfloor$ bits of its label.

Step 3. (Determine the middle $\omega = \lceil \log \frac{\alpha\beta\gamma}{\bar{\alpha}\bar{\gamma}} \rceil$ bits of each node's label.) Each thread, B_k, obtained in Step 2, can be viewed as $\bar{\alpha}$ sub-threads $B_k[i]$ concatenated one after another, see Fig. 2 for an example, each sub-thread comes from some layer. The jth ($j = 1, 2, \cdots$) node T on each $B_k[i]$ is said to have **rank$_3$** of j, denoted as $rank_3(T) = j$ (the subscript 3 here indicates Step 3). We are going to mark each node in each thread a number t, $0 \le t < m$, where $m = 2^\omega$, such that the mark difference (in mod m) between

any two neighbors (in the original 3-dimensional grid) is at most 10. Let $size(B_k[i])$ be the number of nodes in the sub-thread $B_k[i]$, we call $f_{k,i} = size(B_k[i]) - m$ the **overflow** of sub-thread $B_k[i]$ (with respect to m). where $1 \le k \le \bar{\gamma}$ and $1 \le i \le \bar{\alpha}$. Hence we obtain a $\bar{\gamma} \times \bar{\alpha}$ overflow matrix $F = (f_{k,i})$. The overflow matrix for our $5x5x5$ example is

$$F = \begin{pmatrix} 1 & 0 & -1 & 0 \\ 1 & -1 & 0 & -1 \\ 1 & -1 & 0 & -1 \\ 0 & 0 & -1 & 0 \end{pmatrix}$$

Generally speaking, the idea of our marking scehme is that each sub-thread with negative overflow is used to absorb the extra nodes (more than m) from the previous (left) sub-thread with positive overflow and the sub-thread with zreo overflow to propogate the extras to the next (right) sub-thread. Specifically, the marking scheme is as follows: the jth ($j = 1, 2, \cdots$) node of the sub-thread $B_k[i]$ is marked with the number, $(c_{k,i} + j - 1) \mod m$, where

$$c_{k,i} = \begin{cases} \max\{c_{k,i-1} + f_{k,i-1}, 0\}, & \text{if } 2 \le i \le \bar{\alpha} \\ 0. & i = 1 \end{cases}$$

denotes the number of *carry-over* nodes for the sub-thread $B_k[i]$ from the left sub-threads. Hence the matrix $C = (c_{k,i})$ is called the carry-over matrix for those sub-threads. For our example, we have

$$C = \begin{pmatrix} 0 & 1 & 1 & 0 \\ 0 & 1 & 0 & 0 \\ 0 & 1 & 0 & 0 \\ 0 & 0 & 0 & 0 \end{pmatrix}$$

and the mark of each node is shown in Fig. 3. Each node with mark t is assigned $\Gamma(t)$ as the middle ω bits of its label.

Let $B_k[p, q] = B_k[p] \cdot B_k[p + 1] \cdots B_k[q]$, where $\cdot$ denotes the concatenation, clearly, $B_k[i, i] = B_k[i]$ and $B_k = B_k[1, \bar{\alpha}]$. Also let $overflow(B_k[p, q])$ denote the summation of the overflows over the 'segment' $B_k[p, q]$, i.e.,

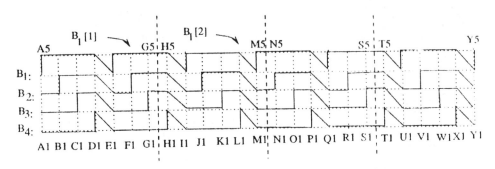

Fig. 2. Four partitioned threads for the grid 5x25.

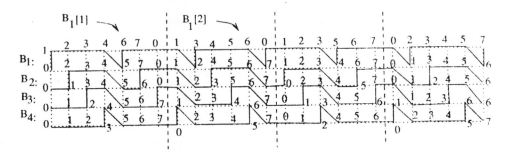

Fig. 3. Mark values on the four threads for the grid 5x25.

$$overflow(B_k[p,q]) = \begin{cases} \sum_{i=p}^{q} f_{k,i} & \text{if } 0 < p \leq q \leq \bar{\alpha} \\ 0 & \text{otherwise.} \end{cases}$$

From the above definition, we have the following result:

Lemma 1. For any $1 \leq k \leq \gamma$, $overflow(B_k) \leq 0$.

Proof. Using facts (F6) and (F7) we have

$$\begin{aligned}
overflow(B_k) &= \sum_{i=1}^{\bar{\alpha}} f_{k,i} \\
&= \sum_{i=1}^{\bar{\alpha}} size(ST(k,i)) - \bar{\alpha}m \\
&= size(T_k) - \bar{\alpha}2^{\lceil \log \frac{\alpha\beta\gamma}{\bar{\alpha}\bar{\gamma}} \rceil} \\
&\leq \lceil \frac{\alpha\beta\gamma}{\bar{\gamma}} \rceil - \bar{\alpha}2^{\log \frac{\alpha\beta\gamma}{\bar{\alpha}\bar{\gamma}}} \\
&= \lceil \frac{\alpha\beta\gamma}{\bar{\gamma}} \rceil - \bar{\alpha}\lceil \frac{\alpha\beta\gamma}{\bar{\alpha}\bar{\gamma}} \rceil \\
&\leq 0. \quad \square
\end{aligned}$$

In order to show that the above marking scheme is valid, we have to show that there is no 'shift-out' node at the very end of each thread. In other words, we have to prove the following Lemma:

Lemma 2. For any $1 \leq k \leq \bar{\gamma}$, $1 \leq h \leq \bar{\alpha}$, we have $overflow(B_k[\bar{\alpha} - h + 1, \bar{\alpha}]) \leq 0$.

Proof. Suppose there exist a k and an h such that $overflow(B_k[\bar{\alpha} - h + 1, \bar{\alpha}]) > 0$, also assume h is the minimum value for this k. Let $\bar{\alpha} = qh + r$, where $0 \leq r < h$. Then $B_k = B_k[1,r] \cdot S_1 \cdot S_2 \cdots S_q$, where $S_z = B_k[r + (z-1)h + 1, r + zh]$, $1 \leq z \leq q$. Fig. 4 gives one possible scenario. By the assumption we have $overflow(S_q) > 0$. The segment S_q spans the rightmost $\sum_{i=\bar{\alpha}-h+1}^{\bar{\alpha}} p_i = h\bar{l} + \lceil \frac{\bar{\alpha}e}{\bar{\alpha}} \rceil - \lceil \frac{(\bar{\alpha}-h)e}{\bar{\alpha}} \rceil = h\bar{l} + \lfloor \frac{he}{\bar{\alpha}} \rfloor$ columns of the partitioning matrix B for H, whereas each segment S_z, $1 \leq z < q$, spans $h\bar{l} + \lceil \frac{(r+zh)e}{\bar{\alpha}} \rceil - \lceil \frac{(r+(z-1)h)e}{\bar{\alpha}} \rceil \geq h\bar{l} + \lfloor \frac{he}{\bar{\alpha}} \rfloor$ (from fact (F8)) columns of B. By using properties (P1) and (P5) we know that $size(S_z) - size(S_q) \geq -1$ for each $1 \leq z < q$. This means that $overflow(S_z) \geq 0$. However, we have $overflow(B_k) \leq 0$ from Lemma 1. If $r = 0$ then we have already got a contradiction. Otherwise, we must have $overflow(B_k[1,r]) < 0$, notice that $B_k[1,r]$ spans the first $r\bar{l} + \lceil \frac{re}{\bar{\alpha}} \rceil$ cloumns of B. Consider the segment $S_c = B_k[\bar{\alpha} - h + 1, \bar{\alpha} - h + r]$, it spans at most (by (F8)) $r\bar{l} + \lceil \frac{re}{\bar{\alpha}} \rceil$, from properties (P1) and (P5), we know $overflow(S_c) \leq 0$. Hence we must have $overflow(B_k[\bar{\alpha} - (h-r) + 1, \bar{\alpha}]) > 0$, a contradiction to the assumption of the minimal value of h. Therefore the Lemma holds. $\square$

On the other hand, to show the marking process works nicely, we are going to prove that the maximum number of 'shift-over' nodes from sub-thread to sub-thread for each thread is bounded by a constant, namely two here.

Lemma 3. For any $1 \leq k \leq \bar{\gamma}$, and $1 \leq p \leq q \leq \bar{\alpha}$, we have $overflow(B_k[p,q]) \leq 2$.

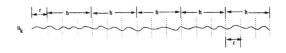

Fig. 4. An auxilary graph for Lemma 2.

Fig. 5. An auxilary graph for Lemma 3.

Proof. Assume that for some k, there exist $1 \leq p \leq q \leq \bar{\alpha}$ such that $overflow(B_k[p,q]) > 2$, also assume that $q - p$ is minimum. Let $h = q - p + 1$, and $\bar{\alpha} - q = uh + v$, where $0 \leq v < h$. Then segment $B_k[q+1, \bar{\alpha}]$ can be divided into $H_z = B_k[q + (z-1)h + 1, q + zh]$, where $1 \leq z \leq u$, and a remainder segment $B_k[\bar{\alpha} - v + 1, \bar{\alpha}]$. Fig. 5 depicts one situation. Notice that $B_k[p,q]$ spans at least $h\bar{l} + \lfloor \frac{he}{\bar{\alpha}} \rfloor$ (from (F8)) columns of the partitioning matrix B, and each segment, H_z, spans at most (by (F8)) $h\bar{l} + \lceil \frac{he}{\bar{\alpha}} \rceil$ columns of B, from the properties (P1) and (P5), we know that $overflow(H_z) \geq 0$ for $1 \leq z \leq u$. From Lemma 2 we have $overflow(B_k[p, \bar{\alpha}]) \leq 0$, if $v = 0$ then a contradiction has been found, otherwise, we must have $overflow(B_k[\bar{\alpha} - v + 1, \bar{\alpha}]) < 2$. However this segment spans $v\bar{l} + \lfloor \frac{ve}{\bar{\alpha}} \rfloor$ columns of B, and segment $B_k[q - v + 1, q]$ spans at most (by (F8)) $v\bar{l} + \lceil \frac{ve}{\bar{\alpha}} \rceil$ columns of B, again, by using properties (P1) and (P5) we have $overflow(B_k[q - v + 1, q]) \leq 0$, thus we have $overflow(B_k[p, q - v]) > 2$, a contradiction with our assumption of the minimum value of $q - p$. $\square$

Lemma 4. If S and T are two nodes on different layers and let $y = rank_1(S)$, $y' = rank_1(T)$, then $|rank_3(S) - rank_3(T)| \leq 2 + 2|y - y'|$.

Proof Using the properties (P1), (P5) and applying the *Rowsum* to the corresponding layers, we have

$$\begin{aligned}
&|rank_3(S) - rank_3(T)| \\
=&|Rowsum(chain(S); 1, y-1) + \chi(x,y) \\
&- Rowsum(chain(T); 1, y'-1) - \chi(x',y')| \\
\leq&|Rowsum(chain(S); 1, y-1) - Rowsum(chain(T); 1, y'-1)| \\
&+ |\chi(x,y) - \chi(x',y')| \\
\leq& 1 + 2|y - y'| + 1 \\
=& 2 + 2|y - y'|. \quad \square
\end{aligned}$$

For any node N in the original grid $\alpha \times \beta \times \gamma$, let $layer(N)$ denote the number of the layer to which N belongs, $thread(N)$ denote the number of the thread to which N belongs, and $mark(N)$ denote the mark value assigned to N in Step 3. For convience, let $layer\_dif(N1, N2) = |layer(N1) - layer(N2)|$, $thread\_dif(N1, N2) = |thread(N1) - thread(N2)|$, $mark\_dif(N1, N2) = |(mark(N1) - mark(N2)| \mod m$, and $rank\_dif_s(N1, N2) = |rank_s(N1) - rank_s(N2)|$, where $s = 1$ or 3. Assume the three axis of the grid are X, Y, and Z respectively as shown in Fig. 6(a).

Proposition 1. If nodes S and T are adjacent in the X dimension in the original grid, then $mark\_dif(S,T) \leq 6$.

Proof. By (F1) $layer\_dif(S,T) \leq 1$. (i) if $layer\_dif(S,T) = 0$ (cf. Fig. 6(b)) then $rank\_dif_1(S,T) = 1$, and if $thread\_dif(S,T) = 0$ then $rank\_dif_3(S,T) \leq 2$ (from (F2)), hence $mark\_dif(S,T) \leq 2$. else $thread\_dif(S,T) = 1$ (from (F1)), and $rank\_dif_3(S,T) \leq 3$ (from (F3)), hence $mark\_dif(S,T) \leq 5$ (by Lemma 3). (ii) if $layer\_dif(S,T) = 1$ (cf. Fig. 6(c)), by (F4) we have $rank\_dif_1(S,T) \leq 1$, Using Lemma 4 we have $rank\_dif_3(S,T) \leq 4$, by using Lemma 3, we have $mark\_dif(S,T) \leq 6$. $\square$

Proposition 2. If nodes S and T are adjacent in the Y dimension in the original grid, then $mark\_dif(S,T) \leq 10$.

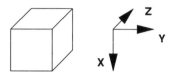

(a) Dimension assignment

(b) X-Dimensional adjacent nodes S and T belonging to the same layer.

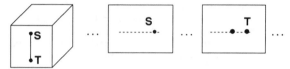

(c) X-Dimensional adjacent nodes S and T belonging to different layers.

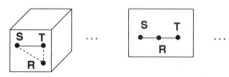

(d) Y-Dimensional adjacent nodes S and T belonging to the same layer.

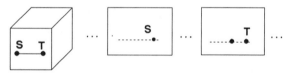

(e) Y-Dimensional adjacent nodes S and T belonging to different layers.

(f) Z-dimensional adjacent nodes S and T

Figure 6. Configurations from Step 1 to Step 2.

Proof. From (F1), $layer\_dif(S,T) \leq 1$. (i) $layer\_dif(S,T) = 0$ (cf. Fig. 6(d)), then $rank\_dif_1(S,T) \leq 2$ (from (F2)) and $thread\_dif(S,T) \leq 1$ (by (F1)). If $thread\_dif(S,T) = 0$ then $rank\_dif_3(S,T) \leq 4$ (from (P1)), hence $mark\_dif(S,T) \leq 4$. Otherwise if $thread\_dif(S,T) = 1$ then $rank\_dif_3(S,T) \leq 5$ (from (F2), (F3) and (F1)), hence $mark\_dif(S,T) \leq 7$ (from Lemma 3). (ii) $layer\_dif(S,T) = 1$ (cf. Fig. 6(e)), then $rank\_dif_1(S,T) \leq 3$ (from (F3)) and $thread\_dif(S,T) \leq 1$ (by (F1)). From Lemma 4 we know $rank\_dif_3(S,T) \leq 8$, if $thread\_dif(S,T) = 0$ then we have $mark\_dif(S,T) \leq 8$ else we have $mark\_dif(S,T) \leq 10$ by using Lemma 3. $\Box$

Proposition 3. *If nodes S and T are adjacent in the Z dimension in the original grid, then $mark\_dif(S,T) \leq 2$.*

Proof. Since we always have $layer\_dif(S,T) = 0$ (cf. Fig. 6(f)). From (F1), we know $thread\_dif(S,T) \leq 1$, and from (F4) we have $rank\_dif_3(S,T) \leq 1$ and therefore $mark\_dif(S,T) \leq 1$ if $thread\_dif(S,T) = 0$ and otherwise $mark\_dif(S,T) \leq 3$ by using Lemma 3. $\Box$

Theorem 1. *Any three-dimensional grid can be embedded into its optimal hypercube with dilation $\delta \leq 6$.*

Proof. From (F5) and the above three propositions we know that the middle ω bits for adjacent nodes, S and T, differ by at most 4 bits, plus the possible 1 bit difference for layers and 1 bit difference for threads, S and T are assigned labels with at most 6 different bits. $\Box$

Conclusions and open problems

In this paper, we have reviewed the problem of embedding grids into hypercubes. Based on the interesting partitioning matrix introduced by Chan [4], we have developed a new simple scheme for embedding 3-dimensional grids into their optimal hypercubes, the dilation is at most 6 which is the best known result to our knowledge.

Here are some open problems: (1) Can the direct mapping method in [2,5] be generalized to the 3-D or higher dimensional grid embeddings. (2) Further improvement of, such as balancing the dilations in the X, Y, Z dimensions, or employing the 'flipping' technique and/or the 'post-marking' technique, seems possible. (3) How to obtain an embedding that minimize the "radius" of a column in the grid as in problems like finding the sum or maximum of elements in a matrix (2-D or 3-D).

References

[1] Mee-Yee Chan, "Dilation-2 Embeddings of Grids into Hypercubes" *International Conference on Parallel Processing*, Vol. 3, pp. 295-298, 1988.

[2] M. Y. Chan and F. Y. L. Chin, "On Embedding Rectangular Grids in Hypercubes" *IEEE Transactions on Computers*, Vol. 37, No. 10, pp. 1285-1288, 1988.

[3] M. Y. Chan, "Embedding of 3-Dimensional Grid into Optimal Hypercubes" Technical Report UTDCS 13-88. Computer Science Department, University of Texas at Dallas, 1988.

[4] M. Y. Chan, "Embedding of d-dimensional grids into optimal hypercubes" *Proceedings of symposium on parallel algorithms and architectures*, pp. 52-57, 1989.

[5] Ching-Tien Ho and S. Lennart Johnson, "On the Embedding of Arbitrary Meshes in Boolean Cubes with Expansion Two Dilation Two" *International Conference on Parallel Processing*, Vol.3 pp. 188-191, 1987.

[6] Soo-Young Lee and J. K. Aggarwal, "A Mapping Strategy for Parallel Processing" *IEEE Transactions on Computers*, Vol. 36, No. 4, pp. 433-442, 1987.

[7] Hongfei Liu, "Embeddings of 3-dimensional grids into hypercubes", Master's thesis, Computer Science Department, University of Houston, April, 1990.

[8] Marilynn Livingston and Quentin F. Stout, "Embeddings in Hypercubes" *Mathl. Comput. Modeling*, Vol. 11, pp. 222-227, 1988.

[9] J. H. Reif (Ed), "VLSI Algorithms and Architectures" *3rd Aegean Workshop on Computing*, AWOC88, pp. 160-211, 1988.

[10] David S. Scott and Joe Brandenburg, "Minimal Mesh Embeddings in Binary Hypercubes" *IEEE Transactions on Computers*, Vol. 37, No. 10, pp. 1284-1285, 1988.

Efficient Parallel Construction and
Manipulation of Quadtrees

Frank Dehne *

School of Computer Science
Carleton University
Ottawa, Canada K1S 5B6
dehne@carleton.ca

Afonso G. Ferreira †

Laboratoire de l'Informatique du
Parallelisme
Ecole Norm. Sup. de Lyon
69364 Lyon, Cedex 07, France

Andrew Rau-Chaplin *

School of Computer Science
Carleton University
Ottawa, Canada K1S 5B6
arc@carleton.ca

Abstract

This paper studies the parallel construction and manipulation, on the hypercube multiprocessor and PRAM, of both pointer based and linear quadtrees.

While parallel algorithms for the manipulation of a variant of linear quadtrees have been previously studied in the literature, no parallel quadtree construction algorithms, for the hypercube or PRAM, have been presented. In this paper, we solve the problem of efficiently constructing pointer based and linear quadtree from images represented by either binary matrices or boundary codes.

Furthermore, previous papers considered exclusively the parallel processing of a variant of linear quadtrees, namely linear quadtrees with path encoding. In this paper, we demonstrate that, in the parallel setting, pointer based quadtrees are an attractive alternative to linear quadtrees with path encodings. We present new efficient parallel algorithms for standard quadtree operations, (such as finding the neighbors of all leaves in a quadtree, and computing the union/intersection of two quadtrees) for both the hypercube and PRAM. Our algorithms exhibit better time/processor products than the previous ones for linear quadtrees with path encoding.

Key words: parallel algorithms, image processing, quadtree, hypercube, PRAM.

1 Introduction

A *quadtree* is a well known hierarchical data structure for representing a binary image of size $\sqrt{M} \times \sqrt{M}$ ($\sqrt{M} = 2^r$ for some positive integer r). The root of the quadtree represents the entire image and has a value "black", "white", or "gray" depending on whether the entire image is black, white, or composed of both types of pixels, respectively. If the root is gray, it has four children which are roots of quadtrees recursively representing the four quadrants of the image; otherwise it has no children. For the remainder, we do not differentiate between a node of a quadtree and the portion of the image represented by that node.

There are two widely used representations of quadtrees. A *pointer based quadtree* uses the standard tree representation while a *linear quadtree* uses a linear list representation. The linear quadtree can be represented by either a preorder traversal of the nodes of a quadtree or the sorted sequence (with respect to the preorder of the tree) of the quadtree's leaves. Some linear quadtree representations of the second type require that with each leaf a code sequence representing the path from the root to that leaf is stored (*linear quadtree with path encoding*), while others store for each leaf only its size and location (*linear quadtree without path encoding*). For an overview and bibliography on quadtrees and applications we refer to the work of Samet ([16]).

Quadtrees are a very useful and widely used data structure for image processing, and quadtree algorithms for a large number of image processing tasks have been developed ([16]). Since image processing is typically data intensive, the application of parallelism to such a fundamental data structure is of both theoretical and practical interest. Recently, researchers have therefore also started to consider quadtree algorithms for parallel models of computation [2, 7, 9, 11, 12]. While some papers ([11,12]) consider parallel architectures designed (or reconfigured) particularly for quadtree manipulation, other ([9,2]) consider the general purpose architectures mesh-connected computer and PRAM, respectively. Hung and Rosenfeld ([9]) study mesh-connected computer algorithms for constructing and manipulating linear quadtrees without path encoding and obtained construction and manipulation algorithms with time complexities of $O(\sqrt{M})$ and $O(\sqrt{n})$, respectively. Table 1 lists the parameters that will be used for the remainder of this paper. PRAM algorithms for manipulating linear quadtrees with path encoding are studied by Bhaskar, Rosenfeld and Wu ([2]); the obtained results are listed in Table 3 (rightmost column). The time and space complexities listed in Table 3 for manipulating linear quadtrees with path encoding on a hypercube are obtained from [2] by using standard PRAM simulation on a hypercube, as described by Nassimi and Sahni ([13]), together with Cypher and Plaxton's deterministic hypercube sorting algorithm ([4]).

In this paper, we study two problem areas which remained unsolved in the previous literature.

In the above mentioned papers there existed, for the hypercube and PRAM, parallel quadtree manipulation algorithms, but no parallel quadtree *construction* algorithms (neither for pointer based nor for linear quadtrees) where given. Such construction algorithms, which are obviously necessary to use quadtrees on a real parallel machine, are presented in this paper. We describe algorithms for converting images represented either by a binary array or a boundary code into pointer based as well as linear quadtrees. Table 2 summarizes the obtained results.

Furthermore, all previous papers studied only the parallel processing of *linear quadtrees with path encoding*. The reason might be that a linear quadtree, being just a set of leaf nodes,

---

*   Research partially supported by the Natural Sciences and Engineering Research Council of Canada.

†   Currently on leave from the University of Sao Paulo (Brazil), project BID/USP. Research partially supported by CAPES/COFECUB (Grant 503/86-9).

seems to be easier to handle in the parallel setting, compared to maintaining and manipulating a pointer structure necessary for a pointer based quadtree. We show however that pointer based quadtrees are an efficient alternative. In fact, the parallel manipulation algorithms for pointer based quadtrees presented in this paper improve, in terms of time/space product, on the previously presented methods. In addition, they exhibit better time complexities with same number of processors, in all but degenerate cases. Table 3 summarizes the obtained results. Note that, the algorithms in [2] apply to linear quadtrees *with* path encoding. In the expected case, the height, h, of the quadtree is $O(\log N)$ ([1, 8, 10]). Hence, $N=O(N')$; i.e., the linear and pointer based quadtrees have, asymptotically, the same space requirement. In this case, we obtain improvements in the time complexity for several problems, such as computing the neighbors of all leaf nodes and the perimeter of an image [hypercube: $O(h \log N)$ vs. $O(h \log^2 N \log^2 \log N)$, PRAM: $O(h)$ vs. $O(h \log N)$] or computing the union/intersection of two quadtrees [hypercube: $O(\log N (h+\log^2 \log N))$ vs. $O(h \log^2 N \log^2 \log N)$, PRAM: $O(h+\log N)$ vs. $O(h \log N)$]. In the *worst case*, $h=O(N)$, the linear quadtree with path encoding needs to store one path requiring $O(h)$ bits, while the pointer based quadtree needs $O(h)$ pointers of $O(\log h)$ bits each; that is, $N=O(N' \log h)$. In this case, we obtain a time space trade-off between the above time complexity improvements and increased storage for pointer based quadtree algorithms. Note that, for the hypercube, the space increases by a factor smaller than the time complexity improvement, and for the PRAM both factors are equivalent.

The remainder of this paper is organized as follows. In Section 2, we discuss some preliminaries concerning the models of parallel computation and the dynamic multi-way search paradigm. In Section 3, we present efficient hypercube and PRAM algorithms for *constructing* a (pointer based or linear) quadtree from a binary image or from an image represented by its boundary code. In Section 4, we introduce efficient parallel hypercube and PRAM algorithms for manipulating pointer based quadtrees.

2 Preliminaries

Before presenting our quadtree algorithms, we introduce some notations and previous results which will be used in the remainder. We start by defining the parallel models of computation we will address henceforth.

2.1 Hypercube Multiprocessor and PRAM

A hypercube multiprocessor is a set $P_1, ..., P_p$ of p processors connected in a hypercube fashion; i.e., P_i and P_j are connected by a communication link if and only if the binary representations of i and j differ in exactly one bit. In a hypercube, there is no shared memory. The entire storage capability consists of constant size local memories, one attached to each processor ($s=O(p)$).

A CREW PRAM consists of a set $P_1, ..., P_p$ of p processors, with constant size local memories, connected to a shared memory of size s. An arbitrary number of processors can read concurrently from the same shared memory location, but concurrent write accesses are not possible.

2.2 Storing Pointer Based Quadtrees on a Hypercube Multiprocessor

While storing a pointer based quadtree on a PRAM is simple, because of its shared memory which can be used in the same way as for a standard sequential machine, we require a scheme for distributing a quadtree over the local memories of a hypercube. Consider the *level order numbering* of the nodes of a quadtree as indicated in Figure 1. For the remainder we will assume that each node with level order number i, together with the attached data and pointers to its children, is stored at processor P_i.

2.3 Multi-Way Search on a Tree

Let $T = (V, E)$ be a tree of size k, height h, and out-degree $O(1)$, and let U be a universe of possible search queries on T. A *search path* for a query $q \in U$ is a sequence $path(q)=(v_1, ..., v_h)$ of h vertices of T defined by a *successor* function f: $(V \cup \{start\}) \times U \Rightarrow V$; i.e., a function with the property that $f(start,q) \in V$ and for every vertex $v \in V$, $(v, f(v,q)) \in E$ or $(f(v,q), v) \in E$. A *search process* for a query q with search path $(v_1, ..., v_h)$ is a process divided into h time steps $t_1 < t_2 < ... < t_h$ such that at time t_i, $1 \leq i \leq h$, there exists a processor which contains (in its local memory) a description of both the query q and the node v_i. Note that, we do not assume that the search path is given in advance. We assume that it is constructed 'online' during the search by successive applications of the function f. Given a set $Q = \{q_1,...,q_m\} \subseteq U$ of m queries, $m=O(k)$, then the *multi-way search problem* consists of executing (in parallel) all m search processes induced by the m queries.[*]

The best way to visualize this process is to depict each search process as a pebble, representing the respective query and moving through the tree T. A pebble may only move along edges of T, but it it can traverse them in both directions. The multi-way search problem consists of m such pebbles moving simultaneously through the tree. Note that, each node of the tree may be 'visited', at any time, by an arbitrary number of pebbles.

On a PRAM (of size $\max\{k,m\}$) multi-way search can be easily implemented in time $O(h)$. Each query (pebble) is simply represented by one processor, navigating it through the tree. The PRAM's concurrent read capability ensures that queries visiting the same node do not interfere.

For hypercube multiprocessors, it was shown in [6] that the multi-way search problem can be solved in time $O(h \log (\max\{k,m\}))$ on a hypercube of size $\max\{k,m\}$. The algorithm presented there applies to a class of graphs called *ordered h-level graphs* (see [6] for a precise definition) which includes the class of all trees with constant degree. The global structure of this algorithm (applied to the special case of search trees) is as follows: Initially, the tree is stored as indicated in Section 2.2. The m search queries are stored in arbitrary order (with each processor storing at most one query). The m search processes for the m queries $q_1,...,q_m$ are executed simultaneously in h phases, each requiring time $O(\log (\max\{k,m\}))$. Each phase moves all queries one step ahead in their search paths. In each phase, the queries are permuted such that they are sorted with respect to the level order number of the respective node they want to visit next. Furthermore, a copy of the search tree is created and its nodes are permuted such that, at the end of each phase, each processor containing

---

[*] In subsequent sections, queries will also be referred to as *messages*.

a query q_i also stores a copy of the node the query wants to visit next. See [6] for a full description of the algorithm.

Consider the problem of changing the tree T or the set Q of queries during the execution of a multi-way search. That is, during the search (more precisely, at the end of each phase of the algorithm outlined above) leaves may be added to T, subtrees may be deleted from T, and queries may duplicate or delete themselves. This problem is referred to as the *dynamic multi-way search problem*. In [5] it has been shown that this problem can be solved, for the hypercube, such that the time complexity of each phase is still $O(\log (\max\{k,m\}))$. That is, the time complexity of the entire multi-way search procedure for the dynamic case is still $O(h \log (\max\{k,m\}))$. For the PRAM, the dynamic version also requires time $O(h \log (\max\{k,m\}))$. The problem here is that the assignment of processors to new queries and the assignment of storage space of deleted nodes to newly created ones may require a partial sum operation for each phase of the algorithm, which slows down the static solution by a factor of $O(\log (\max\{k,m\}))$.

3 Constructing Quadtrees from Images and Boundary Codes

3.1 Quadtree from Binary Image

Consider a $\sqrt{M} \times \sqrt{M}$ binary image stored on a PRAM (with M processors and $O(M)$ storage space) or a hypercube (with M processors) in row-major numbering (see Figure 2a). That is, for the hypercube, processor P_i stores the pixel with row-major number i.

The following is an outline of a general parallel algorithm, for the hypercube and PRAM, for computing a pointer based quadtree from such a binary image representation. (The implementation details will be presented afterwards.)

(1) For each pixel (in parallel) its shuffled row-major number (as indicated in Figure 2b) is computed.
(2) All pixels are sorted by shuffled row-major number.
(3) A complete 4-ary tree, with the sorted sequence of pixels as leaves, is built.
(4) From each leaf a message is sent along the path to the root of the tree. The messages move synchronously upwards from level to level. At each level, the following is executed:
   If all four messages reaching a node x come from black {white} children, then x is set to black {white} and its children are marked "to be deleted". If the messages reaching x are from children with different color, x is set to gray.
(5) All nodes marked "to be deleted" are deleted, the remaining nodes are compressed to form a consecutive sequence, and all pointers are updated.

Theorem 1 *The pointer based quadtree representation of a $\sqrt{M} \times \sqrt{M}$ binary image can be computed in time $O(\log^2 M)$ and $O(\log M)$ on a hypercube and PRAM, respectively, with $s=p=M$.*

Proof: From the definition of quadtrees it follows that the tree generated by the above algorithm is the correct quadtree. What remains to be shown is that the above steps can be implemented, on the hypercube and PRAM, within the claimed time complexity bounds. Step 1 requires only the local computation of the shuffled row-major number of the respective pixel at each processor. For a $\sqrt{M} \times \sqrt{M}$ image, this takes $O(\log M)$ local computation steps (for the PRAM and

hypercube). Step 2 requires time $O(\log M \log^2 \log M)$ and $O(\log M)$ on the hypercube [4] and PRAM [3], respectively. Step 3 can be implemented by building the tree level by level, starting with the leaves (which are given). Since it is a complete tree, at each stage the addresses of the nodes of the subsequent level can be immediately computed. This results in an $O(h)=O(\log M)$ time algorithm for the PRAM. For the hypercube, routing the nodes of the subsequent level to their respective positions involves a *concentrate* and *distribute* operation of [13]. Thus, Step 3 requires time $O(\log^2 M)$ on the hypercube. Step 4 is a multi-way search operation as outlined in Section 2.3, with traveling messages represented by query processes. Hence, it requires time $O(h \log M)=O(\log^2 M)$ and $O(\log M)$ on the hypercube and PRAM, respectively. Note that, Step 4 does not change the topology of the tree but marks only the nodes to be deleted. In Step 5, the marked nodes are deleted by compressing the sequence of the remaining (non marked) nodes. This can be accomplished, on the hypercube and PRAM, by an $O(\log M)$ time *concentrate* operation [13]. The problem of updating the pointers (address references between tree nodes) can be solved by resorting the tree to its original shape, communicating the new addresses between adjacent nodes (using, on the hypercube, the *concentrate* and *distribute* operation of [13]), and recompressing the tree. Hence, Step 5 requires time $O(\log M \log^2 \log M)$ and $O(\log M)$ on the hypercube and PRAM, respectively.   □

Linear quadtrees without path encoding can be constructed in essentially the same way by marking in Step 4 also gray nodes as "to be deleted". For linear quadtrees with path encoding, we also need to compute (between Steps 4 and 5) the path encoding for each leaf by applying one additional multi-way search procedure.

Corollary 1 *The linear quadtree representation (with or without path encoding) of a $\sqrt{M} \times \sqrt{M}$ binary image can be computed in time $O(\log^2 M)$ and $O(\log M)$ on a hypercube and PRAM, respectively, with $s=p=M$.*

3.2 Quadtree from Boundary Code

Consider an image I described by a *boundary code* of length b; i.e., a sequence $a_1, ..., a_b$ of b *boundary elements* $a_i \in \{r,l,u,d\}$ as shown in Figure 3 (see [14]). The image I consists of the entire area inside the *boundary line* defined by the boundary code. The unit size pixels of I that are adjacent to the boundary line are called *boundary pixels* (see Figure 3). For the remainder, let S_I denote a smallest (isothetic) square containing I. Note that S_I has a width of at most b.

Our parallel algorithm for computing the pointer based quadtree from the boundary code consists of two phases, each of which is outlined below.

Phase 1 computes a quadtree template representing only the boundary pixels of I. What remains to be done in Phase 2 is the creation of leaf nodes corresponding to the black and white area inside and outside the boundary line, respectively. The missing children of an internal node x, at the end of Phase 1, will be referred to as *absent children* of x. Note that, all absent children are leaves.

Phase 1:
(1) For each boundary element, its absolute address (in S_I) is computed, and the adjacent boundary pixels are created (see [14]).
(2) The shuffled row-major number of each boundary pixel with respect to S_I is computed.

(3) All boundary pixels are sorted with respect to their shuffled row-major number.

(4) A quadtree with the above sequence of boundary pixels as leaves is built. For nodes with less than four children, for each missing child a node marked "absent" is created.

In order to build the final quadtree from the template created in Phase 1, we recall the following from [14].

Lemma 1 ([14]) *After Phase 1, if an absent child of a node is black {white}, then all other absent children of a node are black {white}.*

Lemma 2 ([14]) *After Phase 1, consider a node, x, with at least one absent child. Choose an absent child R adjacent to a non-absent (black or gray) sibling Q, and a non-absent leaf q in the subtree rooted at Q which is adjacent to R. If q is white then R is white. If q is black and adjacent to the boundary line, then R is white if the boundary line is between q and R, and black if the boundary line does not separate them. If q is black and not adjacent to the boundary line, then R is black.*

The following outlines the remainder of the algorithm.

Phase 2:
(1) From each leaf, a message is sent to the root of the tree. The messages move synchronously upwards from level to level. For each node, a value *Nodetype* is determined which indicates for each side of its respective quadrant whether it is completely inside the image *I*, completely outside of *I*, or intersected by the border line (see also [14]). Note that, the *Nodetype* value for every boundary pixel (leaf of template quadtree) is given; for every internal node, given the *Nodetype* values of all its children, its *Nodetype* value can be easily determined in constant time.
For each internal node x with at least one absent child, the absent children are created and their values are determined as follows:
(a) An absent child R adjacent to a non-absent child Q is selected. The color of R is determined according to Lemma 2. However, the color of R is determined directly from the *Nodetype* value of Q rather than from the leaf q referred to in Lemma 2. All other absent children of x are assigned the same color as R (Lemma 1).
(b) The *Nodetype* values of the previously absent children are determined. Finally, the *Nodetype* value of x is computed.
(2) From each leaf, a message is sent to the root of the tree. The messages move synchronously upwards from level to level. (This is ensured by wait loops for messages starting at leaves of smaller depth.) At each level, the following is executed:
If all four messages reaching a node x come from black {white} children, then x is set to black {white} and its children are marked "to be deleted". If the messages reaching x are from children with different color, then x is set to gray.
(3) All nodes marked "to be deleted" are removed, the remaining nodes are compressed to form a consecutive sequence, and all pointers are updateed.

Theorem 2 *The pointer based quadtree representation of a binary image described by a boundary code of length b can be computed in time $O(\log b \ (h+ \log^2\log b))$ and $O(h \log b)$ on a hypercube and PRAM, respectively, with s=p=b.*

Proof: The correctness of the algorithm follows from [14]. What remains to be shown is that the individual steps listed in the above two phases can be implemented with the claimed time complexity. We start with describing a hypercube and PRAM implementation of *Phase 1*. For Step 1, the x-coordinates of the absolute addresses are computed by assigning a value 1, -1, 0, and 0 to the boundary elements r,l,u,d, respectively, and computing the partial sums of this sequence. All y-coordinates are computed analogously. For each boundary element, the creation of the boundary pixels requires only information about the directly adjacent boundary elements; otherwise, it is a local $O(1)$ time operation. Hence, for the hypercube and PRAM, Step 1 can be executed in $O(\log b)$ time. Step 2 requires $O(\log b)$ local computation steps at each processor. Step 3 requires time $O(\log b \log^2\log b)$ and $O(\log b)$ on the hypercube [4] and PRAM [3], respectively. Step 4 can be implemented by building the tree level by level, starting with the leaves (which are given). At each level, every node (initially leaves) examines its three neighbors to the right and left and determines (using the shuffled row-major numbering and current level information) with whom a common ancestor is to be created. This can be implemented on the hypercube with $O(\log b)$ time per level (and obviously in the same time on the PRAM), by using a constant number of partial sum as well as concentrate and distribute [13] operations. At the beginning of *Phase 2*, we have a quadtree template representing only the boundary pixels of the image *I*. The nodes corresponding to the black and white area inside and outside the boundary line, respectively, are now created by successive dynamic multi-way search procedures. In Step 1, a dynamic multi-way search procedure is used to add and update the absent children. [hypercube, PRAM: $O(h \log b)$]. Step 2 and Step 3 the same Step 4 and Step 5, respectively, of the algorithm in Section 3.1. Therefore, Step 2 can be implemented on a hypercube and PRAM in time $O(h \log b)$ and $O(h)$, respectively; Step 3 requires time $O(\log b \log^2\log b)$ and $O(\log b)$, respectively. □

Linear quadtrees without path encoding can be constructed in essentially the same way by marking in Step 2 of Phase 1 also gray nodes as "to be deleted". For linear quadtrees with path encoding, we also need to compute (between Steps 2 and 3 of Phase 2) the path encoding for each leaf by applying one additional multi-way search procedure.

Corollary 2 *The linear quadtree representation (with or without path encoding) of a binary image represented by a boundary code of length b can be computed in time $O(\log b \ (h + \log^2\log b))$ and $O(h \log b)$ on a hypercube and PRAM, respectively, with s=p=b.*

4 Operations on Quadtrees

4.1 Finding Neighbors in Quadtrees and Computing Region Properties

One of the main advantages of using the pointer based quadtree is that, once the quadtree has been constructed, parallel searching algorithms on quadtrees can be easily adapted from the existing sequential methods by using the dynamic multi-way search technique outlined in Section 2.3. One of the most important building blocks of quadtree applications are neighbor finding techniques. For a leaf x representing a quadrant ✕, a *neighbor* of x is a leaf y representing a quadrant that is adjacent to ✕ (with respect to the image) and has at least the same size as ✕. The *multiple neighbor finding problem* consists of finding the neighbors of all leaves of the quadtree.

Theorem 3 *Given a pointer based quadtree of size N stored on a hypercube or PRAM with s=p=N, then the multiple neighbor finding problem can be solved in time $O(h \log N)$ and $O(h)$, respectively.*

Proof: The sequential method described in [15] for finding the neighbor y of one single leaf x traverses the tree from x upwards, along path $\pi(x)$, to the lowest common ancestor of x and y; then it descends downwards to y by using the "mirror image" of the upwards path $\pi(x)$. The main problem with parallelizing this method to parallel traversals for all leaves of the tree, using multi-way search, is that a message used in multi-way search may only be of constant size and, thus, cannot store the path $\pi(x)$. Assume w.l.o.g. that the right neighbor of x is to be determined. Let α denote the right border of the quadrant associated with x, and let β denote the line defined by extending α. We observe that a query can also be routed from a leaf x to its right neighbor y (along the same path as described above) as follows: The query moves upwards from x until it reaches a node whose associated quadrant intersects β. Then, it descends downwards by selecting always the child whose associated quadrant is adjacent to α. Hence, a query process to be routed from x to its neighbor y needs to store only α and β. With this, multiple neighbor finding reduces to multi-way search and, thus, the theorem follows. □

Once the neighbors of each leaf in all four directions have been determined, the calculation of, e.g., the perimeter of the image follows immediately (see [2]).

Corollary 3 *Given a pointer based quadtree of size N stored on a hypercube or PRAM with s=p=N, then the perimeter of the associated image can be computed in time $O(h \log N)$ and $O(h)$, respectively.*

Remark. Notice that, numerous region properties of images such as the area or centroid, which are simply associative functions of the leaves (and do not need neighboring information), can be immediately calculated by partial sum operations (see [2]). This requires time $O(\log N)$ on a hypercube with $s=p=N$ or a PRAM with $s=N$ and $p=N$/log N.

4.2 Rotating Quad Trees By 90°

Given a pointer based quadtree T, the following algorithm computes the quadtree T' for the image of T rotated by 90° on a hypercube or PRAM, with $s=p=N$.

(1) For each node, the position of the rotated associated quadrant is computed.
(2) For each rotated quadrant, the shuffled row-major number (with respect to the partitioning into quadrants of the same size) is computed.
(3) The nodes are sorted by major key *level* and minor key *shuffled row-major number*.
(4) All nodes are resorted to their original position in the old tree. Each node sends its new address to its parent.
(5) All nodes are again sorted by major key *level* and minor key *shuffled row-major number*.

Theorem 4 *Given a pointer based quadtree T of size N stored on a hypercube or PRAM with s=p=N, then the quadtree T' representing the image, associated with T, rotated by 90° can*

be computed in time $O(h+\log N \ \log^2\log N)$ and $O(h+\log N)$, respectively.

Proof: The correctness of the algorithm follows from the observation that if a node v is the parent of a node w in T then the node in T' representing the rotated quadrant of v is also the parent of the node in T' representing the rotated quadrant of w. The computation of the shuffled row-major number in Step 2 requires $O(h)$ local computation steps at each processor. The remainder of the algorithm reduces to a constant number of sorting operations. Therefore, the time complexities follow. □

4.3 Constructing the Union, Intersection, and Complement of Quadtrees

The union (intersection) of two quadtrees T_A and T_B is defined as the quadtree $T_{A \cup B}$ ($T_{A \cap B}$) representing the image composed of the bitwise OR (AND) of the two original images. The complement of a quadtree T_A is defined as the quadtree $T_{\neg A}$ representing the image composed of the bitwise NOT of the original image. In this section, we study the parallel computation of the union, intersection, and complement of two pointer based quadtrees. Below, we introduce some definitions that will be used in the remainder of this section.

A tree T_{A+B} is called an *overlay* of T_A and T_B if it is the smallest 4-ary tree such that for each node v of T_A or T_B there exists a node $\delta(v)$ in T_{A+B} representing the same image area (assuming that T_{A+B} represents an image subdivision defined in standard quadtree fashion). The *combined level order numbering* of T_A and T_B is defined as follows: For each node v of T_A or T_B, the combined level order number $\eta_{A+B}(v)$ is the level order number of $\delta(v)$ in T_{A+B}. The *shuffled row major number of a node* v of T_A (or T_B) is the shuffled row major number of the associated quadrant with respect to the subdivision of the image plane into quadrants of the same size.

We assume that both quadtrees are stored by level order number as indicated in Section 2.2. As a preprocessing, we convert this storage scheme into a *combined level order numbering scheme* where every node v of T_A or T_B is stored at processor number $\eta_{A+B}(v)$. Note that, every processor stores at least one node, but at most two nodes, one of each tree. The new relative order of the nodes of one tree, say T_A, is the same as their order in the initial level order numbering of T_A. The combined level order numbering scheme can be obtained as follows: All nodes are sorted by major key level (height within their tree) and minor key shuffled row major number. For any two nodes with the same major and minor keys stored in two adjacent processors P_i and P_{i+1}, the node in P_{i+1} is moved to P_i. Finally the contents of the processors are shifted leftwards so that processors without data are avoided.

Given this storage scheme for the two quadtrees T_A and T_B, the following is an outline of a parallel algorithm for computing the quadtree $T_{A \cup B}$. Our algorithm uses dynamic multi-way search (see Section 2.3) with three different types of messages: "compare", "copy" and "update" messages.

(1) From each of the roots of T_A and T_B a wave of "compare" messages is sent towards the leaves. That is, a "compare" message is sent to each root and, each node receiving a message, duplicates it and sends one to each child (within its own tree). Messages move synchronously downwards from level to level. During this process, a new tree T is created, which will subsequently be converted into $T_{A \cup B}$.

At each level, the following is executed:

(a) Each node x receiving a "compare" message, compares itself with the respective node y (representing the same image area) of the other tree. The node y is stored at the same processor P as node x and receives a "compare" message at the same time as node x does. Unless x and y are the roots of T_A and T_B, respectively, let $parent(x)$ and $parent(y)$ denote their respective parents. Note that, $parent(x)$ and $parent(y)$ are both "gray" nodes stored at the same processor P' and, previously, received a "compare" message at the same time.

Case 1: x and y are both "gray":
A new "gray" node z for T representing the same quadrant as x and y is created and stored at processor P. Note that, $parent(x)$ and $parent(y)$ previously created a "gray" node z' for T. This node z' is made the parent of z in T.

Case 2: x or y is "black":
A new "black" node z for T representing the same quadrant as x and y is created and stored at processor P. The "gray" node z' created by $parent(x)$ and $parent(y)$ is made the parent of z in T. The two "compare" messages which reached x and y are not forwarded but deleted.

Case 3: One node, x or y, is gray and the other node is white:
A new "gray" node z for T representing the same quadrant as x and y is created and stored at processor P. The "gray" node z' created by $parent(x)$ and $parent(y)$ is made the parent of z in T. The "compare" message which reached the "white" node is deleted. The "compare" message which reached the "gray" node, is changed to a "copy" message, duplicated, and forwarded to all children.

Case 4: x and y are both "white":
A new "white" node z for T representing the same quadrant as x and y is created and stored at processor P. The "gray" node z' created by $parent(x)$ and $parent(y)$ is made the parent of z in T. The two "compare" messages which reached x and y are not forwarded but deleted.

(b) Each node x receiving a "copy" message (in the other tree there exists no node y representing the same quadrant) creates a new node z for T with the same color as x and representing the same quadrant. The node z' created by $parent(x)$ and $parent(y)$ is made the parent of z in T. A "copy" message is sent to each child, or the message is deleted if x is a leaf.

(2) From each leaf an "update" message is sent to the root of the tree. The "update" messages move synchronously upwards from level to level. (This is ensured by wait loops for messages starting at leaves of smaller depth.) At each level, the following is executed:

If all four "update" messages reaching a node x come from black {white} children, then x is set to black {white} and its children are marked "to be deleted". If the "update" messages reaching x are from children with different color, x is set to gray.

(3) All nodes marked "to be deleted" are deleted, the remaining nodes are compressed to form a consecutive sequence, and all pointers are updated.

Computing the intersection of two pointer based quadtrees is analogous. All steps of the above algorithm remain unchanged except for Cases 2, 3, and 4 where "black" and "white" should be exchanged.

Theorem 5 *Given two pointer based quadtrees with a total number of N nodes stored on a hypercube or PRAM with $s=p=N$, then the union {intersection} of these quadtrees can be computed in time $O((h+ log^2 log N) log N)$ and $O(h+log N)$, respectively, where h denotes the maximum height of the two trees.*

Proof: In order to observe the correctness of the algorithm we first study the intermediate tree T created at the end of Step 1. Consider two nodes x and y in T_A and T_B representing the same quadrant. Then, a node z in T is created in Step 1a (a "compare" message reaches x and y), and it is easy to see that through Cases 1 to 4 the right color, representing the union {intersection} of x and y, is assigned to z. Consider, on the other hand, a node x for quadrant $\times$ in, say, T_A with no node in T_B representing the same quadrant. Then, T_B has a leaf y for a quadrant y containing $\times$. Let x' be the ancestor of x representing quadrant y. If y is "black" {"white"} then no node needs to be created in T, which is guaranteed by the deletion of the "compare" messages reaching x' and y (Step 1a, Case 2). If y is "white" {"black"} then the entire subtree rooted at x' has to be copied into T. This is achieved by the "copy" messages started at x' (Step 1a, Case 3 and Step 1b).

In order to prove the claimed time complexity, we first observe that the preprocessing reduces to a sorting operation followed by a concentrate [13] and partial sum for the hypercube and PRAM, respectively. Hence, its time complexity is $O(log N \, log^2 log N)$ and $O(log N)$ on the hypercube and PRAM, respectively. The combined level order numbering scheme used to store the trees T_A, T_B, and T allows simultaneous multi-way search on all three trees, because T_A, T_B, and T are subtrees of T_{A+B}, and all nodes are stored with respect to their level order number in T_{A+B} (see Section 2.2 and 2.3). Hence, Step 1 can be implemented on a hypercube using the dynamic multi-way search procedure outlined in Section 2.3. That is, Step 1 requires time $O(h \, log N)$ on the hypercube. We observe that, during Step 1, at any time no tree node is visited by more than one message. Therefore advancing all messages from one level of the tree to the next level can be implemented, on the PRAM, in time $O(1)$. This is due to the fact that for assigning processors to messages we do not require a partial sum operation as in the general case, but we can use a fixed scheme where every processor is assigned to one node and responsible for the message visiting that node. Hence, Step 1 requires time $O(h)$ on the PRAM. Steps 2 and 3 are equivalent to Steps 4 and 5 of the algorithm in Section 3.1. Hence, from Theorem 1, their time complexity is $O((h+ log^2 log N) log N)$ and $O(h+log N)$ on the hypercube and PRAM, respectively. $\square$

References

[1] J. L. Bentley and D. F. Stanat, "Analysis of range searches in quad trees," *Information Processing Letters*, Vol. 3, No. 6, 1975, pp. 170-173.

[2] S. K. Bhaskar, A. Rosenfeld, and A. Y. Wu, "Parallel processing of regions represented by linear quadtrees," *Computer Vision, Graphics, and Image Processing*, Vol. 42, 1988, pp. 371-380.

[3] R. Cole, "Parallel merge sorting", *SIAM J. of Computing*, Vol. 17, N° 4, 1988, pp. 770-785.

[4] R. Cypher and C. G. Plaxton, "Deterministic sorting in nearly logarithmic time on a hypercube and related computers," to appear in Proc. *ACM Symposium on Theory of Computing*, 1990.

[5] F. Dehne, A. Ferreira, and A. Rau-Chaplin, "Parallel branch and bound on fine grained hypercube multiprocessors," to appear in *Parallel Computing*.

[6] F. Dehne and A. Rau-Chaplin, "Implementing data structures on a hypercube multiprocessor and applications in parallel computational geometry," *Journal of Parallel and Distributed Computing*, Vol. 8, 1990, pp. 367-375.

[7] S. Edelman and E. Shapiro, "Quadtrees in concurrent prolog," in Proc. *International Conference on Parallel Processing*, 1985, pp. 544-551.

[8] R. A. Finkel and J. L. Bentley, "Quad trees - a data for retrieval on composite keys," *Acta Informatica*, Vol. 4, No. 1, 1974, pp. 1-9.

[9] Y. Hung and A. Rosenfeld, "Parallel processing of linear quadtrees on a mesh-connected computer," *Journal of Parallel and Distributed Computing*, Vol. 7, 1989, pp. 1-27.

[10] K. J. Jacquemain, "The complexity of constructing quad-trees in arbitrary dimensions," in Proc. *7th Conference on Graphtheoretic Concepts in Computer Science (WG81)*, 1982, J. Mühlbacher (Ed.), pp. 293-301.

[11] M. Martin, D. M. Chiarulli, and S. S. Iyengar, "Parallel processing of quadtrees on a horizontally reconfigurable architecture computing system," in Proc. *International Conference on Parallel Processing*, 1986, pp. 895-902.

[12] G.-G. Mei and W. Liu, "Parallel processing for quadtree problems," in Proc. *International Conference on Parallel Processing*, 1986, pp. 452-454.

[13] D. Nassimi and S. Sahni, "Data broadcasting in SIMD computers," *IEEE Transactions on Computers*, Vol. 30, No. 2, 1981, pp. 101-106.

[14] H. Samet, "Region representation: quadtrees from boundary codes," *Communications of the ACM*, Vol. 23, No. 3, 1980, pp. 163-170.

[15] H. Samet, "Neighbor finding techniques for images represented by quadtrees," *Computer Graphics and Image Processing*, Vol. 18, No. 1, 1982, pp. 37-57.

[16] H. Samet, "The quadtree and related hierarchical data structures," *Computing Surveys*, Vol. 16, No. 2, 1984, pp. 187-260.

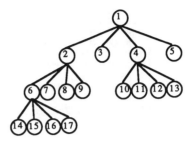

Figure 1. Level Order Numbering of the Nodes of a Quadtree

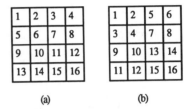

(a)   (b)

Figure 2. (a) Row-Major Numbering (b) Shuffled Row-Major Numbering.

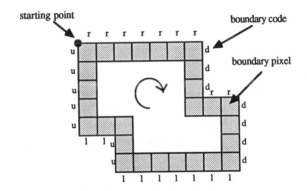

Figure 3. Boundary Code and Boundary Pixels of an Image

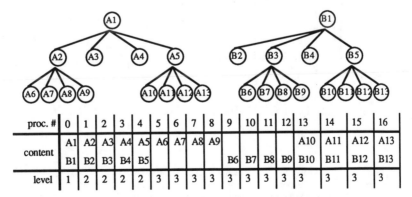

proc. #	0	1	2	3	4	5	6	7	8	9	10	11	12	13	14	15	16
content	A1	A2	A3	A4	A5	A6	A7	A8	A9					A10	A11	A12	A13
	B1	B2	B3	B4	B5					B6	B7	B8	B9	B10	B11	B12	B13
level	1	2	2	2	2	3	3	3	3	3	3	3	3	3	3	3	3

Figure 4. Combined Level Order Numbering Scheme.

M no. of pixels in the original image	t	time complexity
b length of the boundary code	s	total memory space
N size of the *explicit quadtree*	p	no. of processors
N' size of the *linear quadtree with path encoding*		
n size of the *linear quad tree without path encoding*		
h height of the quadtree		

Table 1. Overview of Parameters.

problem	pointer based	quadtree	linear	quadtree
	hypercube	PRAM	hypercube	PRAM
convert image to quad tree $(s = p = M)$	$t = O(\log^2 M)$	$t = O(\log M)$	$t = O(\log^2 M)$	$t = O(\log M)$
convert boundary code to quad tree $(s = p = b)$	$t = O(\log b \ (h+ \log^2 \log b))$	$t = O(h \ \log b)$	$t = O(\log b \ (h+ \log^2 \log b))$	$t = O(h \ \log b)$

Table 2. New Parallel Quadtree Construction Methods.

problem	pointer based	quadtree	linear	quadtree
	hypercube	PRAM	hypercube	PRAM
determine neighbors of all leaf nodes / compute perimeter	$s = p = N$ $t = O(h \ \log N)$	$s = p = N$ $t = O(h)$	$s = p = N'$ $t = O(h \ \log^2 N' \log^2 \log N')$	$s = p = N'$ $t = O(h \ \log N')$
			$s = p = O(4^h) \geq O(N)$ $t = O(h \ \log N' \ \log^2 \log N' + \log^2 N' \ \log^2 \log N') \ §$	$s = p = O(4^h) \geq O(N)$ $t = O(h + \log N') \ [2]$
comp. area / centroid	$s = p = N$ $t = O(\log N)$	$s = N,$ $p = N/\log N,$ $t = O(\log N)$	$s = p = N'$ $t = O(\log N')$	$s = p = N'$ $t = O(\log N') \ [2]$
rotate by $i*90°$	$s = p = N$ $t = O(h+ \log N \ \log^2 \log N)$	$s = p = N$ $t = O(h+ \log N)$		
compute union / intersect.	$s = p = N$ $t = O(\log N \ (h + \log^2 \log N))$	$s = p = N$ $t = O(h + \log N)$	$s = p = N'$ $t = O(h \ \log^2 N' \ \log^2 \log N') \ §$	$s = p = N'$ $t = O(h \ \log N') \ [2]$
compute complem.[†]	$s = p = N$ $t = O(\log N)$ $[t = O(1)]$	$s = p = N$ $t = O(1)$	$s = p = N'$ $t = O(h \ \log N' \ \log^2 \log N') \ §$	$s = p = N'$ $t = O(h)$ $[2]$

Table 3. Parallel Quadtree Manipulation Methods (New Results Highlighted).

---

§ Follows from [2] by standard PRAM simulation on a hypercube as described in [12], together with [4].

† This operation is trivial for pointer based quadtrees, and listed for completeness only. The hypercube time complexity assumes $O(1)$ time instruction broadcast (as, e.g., on the Connection Machine).

A Fast Parallel Algorithm to Compute Path Functions for Cographs

R. Lin
Department of Computer Science
SUNY at Geneseo
Geneseo, NY 14454

S. Olariu[†]
Department of Computer Science
Old Dominion University
Norfolk, VA 23529-0162

Abstract
The minimum path cover problem involves finding a minimum number of vertex-disjoint paths which together cover all the vertices of a graph. We show that the problem of computing a number of path functions on cographs including a minimum path cover, a hamiltonian path and a hamiltonian cycle can be solved fast in parallel. Specifically, with an n-vertex cograph G represented by its parse tree as input, our algorithm finds a minimum path cover in G in $O(\log n)$ CREW time using $O(n^2/\log n)$ processors.
Key Words: ring protocols, code optimization, list ranking, tree contraction, parallel algorithms, VLSI, scheduling, operating systems, cographs, CREW-PRAM.

1. Introduction

A computational problem with a number of practical applications is the *minimum path cover problem*. This involves finding a minimum number of vertex-disjoint paths which together cover all the vertices of a graph. Along with many other important problems in graph theory, the minimum path cover problem and many of its variants are NP-complete [8]. The path cover problem finds application to VLSI, ring protocols, code optimization, and mapping parallel programs to parallel architectures [16].

A graph G that admits a path cover of size one is referred to as *hamiltonian*. If the unique path that covers G can be extended to a cycle, G is said to possess a hamiltonian cycle. It is well-known that the problem of determining whether a graph G has a hamiltonian path or cycle is one of the most difficult problems in computational graph theory.

A class of graphs arising in a wide spectrum of practical applications [1,2,7] is the class of *cographs*, or complement-reducible graphs. The cographs are defined recursively as follows:
• a single-vertex graph is a cograph;
• if G is a cograph, then its complement $\bar{G}$ is also a cograph;
• if G and H are cographs, then their union is also a cograph.

As it turns out [7], the cographs admit a unique tree representation up to isomorphism. Specifically, we can associate with every cograph G a unique rooted tree $T(G)$ called the *cotree* of G, featuring the following properties:
(c1) every internal node, except possibly for the root, has at least two children; furthermore, the root has only one child if, and only if, the underlying graph G is disconnected.
(c2) the internal nodes are labeled by either 0 (0-nodes) or 1 (1-nodes) in such a way that the root is always a 1-node, and such that 1-nodes and 0-nodes alternate along every path in $T(G)$ starting at the root;
(c3) the leaves of $T(G)$ are precisely the vertices of G, such that vertices x and y are adjacent in G if, and only if, the lowest common ancestor of x and y in $T(G)$ is a 1-node.

We assume the PRAM model which consists of autonomous processors, each having access to a common memory [18]. At each step, every processor performs the same instruction, with a number of processors masked out. In a CRCW-PRAM several processors may simultaneously access the same memory location in reading or writing; in a CREW-PRAM simultaneous access in reading, but not in writing is allowed; in an EREW-PRAM both reading and writing are exclusive.

Recently, Adhar and Peng [2] presented a parallel algorithm to find a minimum path cover, a hamiltonian path, and a hamiltonian cycle in cographs. Their algorithm runs in $O(\log^2 n)$ CRCW time and using $O(n^2)$ processors. The purpose of this work is to show that the problem of finding a minimum path cover, a hamiltonian path, and a hamiltonian cycle for cographs can be solved efficiently in parallel. Our algorithm runs in $O(\log n)$ CREW time using $O(n^2/\log n)$ processors.

The paper is organized as follows: Section 2 describes the basic tools that are used throughout the paper, namely list ranking, Euler-tour on trees, tree contraction, along with a new result about minimum path covers in graphs which lays the theoretical basis for our subsequent algorithm; Section 3 gives a detailed description of the algorithm; Section 4 summarizes the results and proposes a number of open problems.

2. The Stepping Stones

The problem of *list ranking* is to determine in parallel the rank of every element in a given linked list, that is, the number of elements following it in the list. List ranking has turned out to be one of the fundamental techniques in parallel processing, playing a crucial role in a vast array of important parallel algorithms [1,3,5-6,13]. In particular, Cole and Vishkin[5] and Anderson and Miller[3] have showed that list ranking can be done optimally in $O(\log n)$ EREW time using $O(n/\log n)$ processors.

The well-known *Euler-tour technique* developed in [17] allows one to compute a number of tree functions by reducing them to list ranking. To make our presentation self-contained, we shall present now the details of a variant of this technique. To begin, we replace every node u in T with $d(u)+1$ copies of u, namely $u^1, u^2, ..., u^{d(u)+1}$. Next, letting $w_1, w_2, ..., w_{d(u)}$ stand for the children of u in T, with w_i ($1 \leq i \leq d(u)$) having d_i children, set for all $i=1,2,...,d(u)$:
• $\text{link}(u^i) \leftarrow w_i^1$,
• $\text{link}(w_i^{d_i+1}) \leftarrow u^{i+1}$.

Assuming that the root of T has t children, what results is a linked list starting at $root(T)^1$ and ending at $root(T)^{t+1}$ with every edge of T traversed exactly once in each direction. Therefore, the total length of the resulting linked list is $O(n)$.

It is worth noting that the linked list can be obtained in $O(\log n)$ time using roughly $d(u)/\log n$ processors associated with every node in T. Clearly, this translates into a total number of $O(\sum_{u \in T} d(u)/\log n) = O(n/\log n)$ processors.

The problem of *tree contraction* involves reducing in parallel a given tree to its root by a sequence of vertex removals. Along with list ranking, tree contraction is already recognized as one of the fundamental techniques in parallel processing [1,6,9-12,15]. It has found important applications in dynamic expression evaluation, isomorphism testing, among many others.

Abrahamson et al [1] argued that the tree contraction technique can be used to devise optimal parallel algorithms for the following problems on cographs: computing the size of the largest clique, number of cliques, number of maximal independent sets, number of cliques of the largest size, number of independent sets of largest size, and the problem of identifying a clique of maximum size. All these problems can be solved in $O(\log n)$ EREW time using $O(n/\log n)$ processors by solving a corresponding tree contraction problem.

All graphs in this paper are finite with no loops or multiple edges [4]. Let G be an arbitrary graph whose vertex-set partitions into non-empty, disjoint sets A and B with $r=|A| \leq |B|=t$, and such that every vertex of A is adjacent to all the vertices in B. Let $P_B = \{p_1, p_2, ..., p_s\}$ ($s \geq 1$) be a minimum path cover for B. For convenience, enumerate the vertices of A arbitrarily as $v_1, v_2, ..., v_r$; similarly, enumerate the vertices of B as

$$w_1, w_2, ..., w_t \tag{0}$$

by first writing down the vertices of p_1 (in the same order as they appear in p_1), followed by the vertices in p_2, and so on.

The next result establishes a property of the minimum path cover of G which will be instrumental in our path cover algorithm for cographs.

Theorem 1. G has a minimum path cover of size $\max\{1, s-r\}$.

Proof. We first argue that G cannot have a path cover of size k with $k < \max\{1, s-r\}$. For assume that such a path cover exists. Consider removing from this path cover all the vertices in A. What results is a set of paths which is clearly a path cover for B.

† This author was supported, in part, by the NSF grant CCR-8909996.

Since the removal of a vertex in A will increase the number of paths by at most one, we obtain a path cover for B of size at most $k+r$. Now the assumption that $k < \max\{1, s-r\}$ guarantees that $k+r < s$, contradicting the minimality of P_B.

To complete the proof of Theorem 1, we shall present an algorithm that actually returns a path cover of G of size $\max\{1, s-r\}$. In outline, our algorithm proceeds in the following two steps.

Step 1. The idea of this step is to use vertices in A to "stitch" together disjoint paths in P_B. If all the vertices in A are exhausted at the end of this step, then we return the resulting set of paths.

In case Step 1 ends and there are leftover vertices in A, then we proceed to the next step. For further reference we note that all the vertices $v_1, v_2, ..., v_k$ ($k=s-1$) have been used in Step 1.

Step 2. The idea of this step is to incorporate the vertices $v_{k+1}, v_{k+2}, ..., v_r$ into the path p returned by Step 1, to create a unique path that covers all the vertices of G.

Formally, Step 2 identifies the j-th edge $w_i w_{i+1}$ ($1 \leq j \leq r-s$) in B, and replace this edge by the edges $w_i v_{k+j}$ and $v_{k+j} w_{i+1}$ (i.e inserts the vertex v_{k+j} between w_i and w_{i+1}). Finally, the path p is extended by the edge $w_i v_r$.

To justify the correctness of Step 2, note that P_B contains at least $t-s$ edges. To see that this is the case, note that there are t vertices in B and so P_B must contain at least $t-1$ edges. Since B is covered by s paths, $s-1$ of these edges are missing. Therefore, P_B must contain at least $t-s$ edges, as claimed.

Further, our assumption that $r=|A| \leq |B| = t$ guarantees that B contains sufficient edges to accommodate the first $r-s$ of the leftover vertices in A.

At this moment it is important to note that vertex v_r is treated differently: the edge $w_i v_r$ is added as the last edge in the unique path p that covers all the vertices in G. The reason for this is simple: in case G admits a hamiltonian cycle (see Corollary 1.1), this cycle is obtained instantly by adding the edge $w_1 v_r$ to p. The details are spelled out by the following procedure.

Procedure Path_Cover(A, P_B)

```
0.  begin
1.      k ← min{r, s−1};
2.      p ← p₁;
3.      for i ← 1 to k do
4.          p ← append(p, vᵢ, pᵢ₊₁);
5.      if k = r then
6.          return({p, p_{k+2}, ..., p_s})
7.      j ← 1; i ← 1; {start Step 2}
8.      while j ≤ r−s do
9.          if wᵢwᵢ₊₁ is an edge then
10.             j ← j+1;
11.             remove the edge wᵢwᵢ₊₁ from p;
12.             add the edges wᵢv_{k+j} and v_{k+j}wᵢ₊₁ to p;
13.         endif
14.         i ← i+1;
15.     endwhile;
16.     add the edge v_r w_t to p;
17.     return(p)
18. end; {Path_Cover}
```

This completes the proof of Theorem 1. □

Corollary 1.1. *If $r=s-1$ then G admits a hamiltonian path. If $r>s-1$ then G admits a hamiltonian cycle.*

Proof. Trivially, in case $r=s-1$, at the end of the Step 1 all the vertices in A and all the paths in P_B have been exhausted, confirming that G has a hamiltonian path. In case $r>s-1$ then, when Step 1 ends, A contains leftover vertices; consequently, G has a path cover of size one returned in line 17 of procedure Path_Cover. Note that, by our construction, the first vertex of path p is in B, while the last vertex of the path p is in A. Now the fact that every vertex in A is adjacent to all the vertices in B guarantees that the edge $v_r w_1$ can be

added to p to obtain a hamiltonian cycle of G. □

3. The Algorithm

We shall adopt the terminology of [1]. To make our exposition self-contained we present a number of definitions. First, we assume the trees represented by an unordered array with every node in the tree featuring a parent pointer along with a doubly linked list of children. Next, it turns out that we can restrict ourselves to binary trees. To see this, note that every ordered rooted tree T can be transformed into a full binary tree BT (i.e. every internal node of BT has precisely two children) as follows [5]: if a node x has degree k in T then, in BT, we add $k-2$ identical copies of x, namely $x_1, x_2, ..., x_{k-2}$ in such a way that, with x_0 standing for x,

- the parent of x_i is x_{i-1} whenever $i \geq 1$;
- the left child of x_i is the $(i+1)$st child of x in T;
- the right child of x_i is x_{i+1} in case $i \leq k-3$, and the k-th child of x in T otherwise.

As pointed out in [1], there is no cost associated with the construction of BT, since all we need do is to reinterpret the existing pointers in T.

Given a binary tree T, a *tree contraction sequence* is a sequence of trees $T=T_1, T_2, ..., T_m$ such that T_i is obtained for T_{i-1} by one of the following basic operations:

- prune(v) - leaf v of T_{i-1} is removed;
- bypass(v) - a node v having exactly one child is removed from T_{i-1}, with the unique child of v replacing v.

Abrahamson et al [1] show that every full binary tree has an optimal contraction sequence of length $O(\log n)$, and that this sequence can be obtained in $O(\log n)$ EREW time using $O(n/\log n)$ processors.

We are now in a position to show how the tree contraction technique can be used to obtain an optimal parallel algorithm to compute a minimum path cover for cographs. For this purpose, consider an n-vertex cograph G represented by its cotree $T(G)$. We binarize $T(G)$ as described above and let $BT(G)$ stand for the resulting full binary tree. For each node x of $BT(G)$ we let $BT(x)$ stand for the subtree of $BT(G)$ rooted at x; $L(x)$ stands for the set of all the leaves in $BT(x)$, and $G(x)$ represents the subgraph of G induced by $L(x)$. Throughout this paper we shall deal with a generic node x of $BT(G)$; whenever this happens, it will be assumed that x has left and right children y and z, respectively. Our algorithm proceeds in the following five stages.

Stage 1. Perform an Euler-tour of $BT(G)$ and use the information returned to compute for every internal node x of $BT(G)$ the cardinality of $L(x)$. For every 1-node x of $BT(G)$ do the following:

if $|L(y)| \leq |L(z)|$ **then** mark y

else swap($T(y), T(z)$); mark z;

In other words, we ensure that for every 1-node, its left subtree contains at most as many leaves as the right subtree. Furthermore, every left child of a 1-node in $BT(G)$ is marked. For convenience, we continue to refer to the new tree as $BT(G)$. Call a node v of $BT(G)$ *special* if some ancestor of v has been marked in the operation described above. Nodes of $BT(G)$ which are not special are referred to as *regular*.

Next, performing an Euler-tour again mark all the special nodes in $BT(G)$, construct a linked list of all the leaves in $BT(G)$, number the leaves in preorder as $1, 2, ..., n$, and recompute the set $L(v)$ for every node v (note that the cardinality of $L(v)$ remains the same and is not recomputed).

Note: In the next stages of our algorithm it is convenient and helpful to image that left subtrees of 1-nodes in $BT(G)$ have been shrunk to their roots that, however, record all the nodes in the corresponding subtree.

Stage 2. In the second stage, we invoke the tree contraction algorithm in [1] to construct an optimal contraction sequence $T_1, T_2, ..., T_m$ for $T(G)$. This sequence will be used explicitly in the next stages. In addition, we compute a certain algebraic expression associated by $BT(G)$; specifically, we associate with every node of $BT(G)$ a non-negative integer as follows.

(1) if x is a regular leaf then $p(x) \leftarrow 1$;

(2) if x is a regular 0-node then $p(x) \leftarrow p(y) + p(z)$;

(3) if x is a regular 1-node then $p(x) \leftarrow \max\{1, p(z) - |L(y)|\}$.

(4) if x is special then $p(x) \leftarrow |L(x)|$.

The interpretation of $p(x)$ is given by the following result.

Theorem 2. For every regular node x of $BT(G)$, $p(x)$ represents the number of vertex-disjoint paths in a minimum path cover of $G(x)$.

Proof. Let x be a regular node in $BT(G)$. We proceed by induction on the height of x, that is, the number of edges in the longest path from x to a leaf of $BT(G)$. If the height of x is 0, then the conclusion follows from (1).

Assume the statement true for all regular nodes in $BT(G)$ with height less than the height of x. Let x have left and right children y and z, respectively. If x is a 0-node, then by (c3), $G(x)$ is disconnected and so the number of vertex-disjoint paths in a minimum path cover of $G(x)$ equals the number of vertex-disjoint paths in a minimum path cover of $G(y)$ plus the number of vertex-disjoint paths in a minimum path cover of $G(z)$. By the induction hypothesis, this is just $p(y) + p(z)$.

Now x is a 1-node; by the induction hypothesis, $p(z)$ is the number of vertex-disjoint paths in a minimum path cover of $G(z)$. By virtue of (c3), every vertex in $G(y)$ is adjacent to all the vertices in $G(z)$. Now let the procedure Path_Cover described in Section 2 compute a minimum path cover of $\overline{G}(x)$. By Theorem 1, this path cover has the size $\max\{1, p(z) - |L(y)|\}$, as claimed. This completes the proof of Theorem 2. □

Note that Theorem 2 guarantees that the value $p(root)$ computed at the root of $BT(G)$ gives the number of vertex-disjoint paths in a minimum path cover of G. We now argue that the value $p(root)$ can be computed efficient in parallel. More precisely we have the following result.

Theorem 3. Given an arbitrary n-vertex cograph represented by its cotree, the minimum number of vertex-disjoint paths that cover all the vertices of G can be computed in $O(\log n)$ EREW time using $O(n / \log n)$ processors.

Proof. The correctness of our approach was established in Theorem 2. To argue for the complexity, note that Stage 1 can be trivially performed in $O(\log n)$ time using $O(n / \log n)$ processors.

Thus we only need show that Stage 2 requires $O(\log n)$ time and $O(n / \log n)$ processors. Specifically, we have to show that computing $p(root)$ can be done in the time of tree contraction. To show that this is the case, at each node u of $BT(G)$ we store an expression of the form $f_u(x)$ where x is an indeterminate representing the value of the unevaluated $p(u)$. Suppose that u has two children v and w where w is a leaf (i.e. the value of $p(w)$ is a known constant c), and let $f_w(x)$ and $f_v(x)$ be the expressions stored by u and v respectively. Let u contain an operation #. After prune(w) and bypass(u) have been performed the expression stored at v is now $f_u(f_v(x)\#c)$.

The crucial observation is that for the purpose of computing $p(u)$ at every node u of $BT(G)$, the expression stored at u is of the form

$$f_u = \max\{a, x + b\} \qquad (5)$$

where a and b are constants. Initially, every node u of $BT(G)$ stores $f_u = \max\{1, 1 - |L(u)|\} = 1$. Note, in particular, that in this scheme every leaf of $BT(G)$ stores 1 as it should.

Let u, v, w be as before. To justify (5) we only need show that if prior to prune(w) and bypass(u) u stores $\max\{a, x + b\}$ and v stores $\max\{a', x + b'\}$, then after these operations, v stores $\max\{a'', x + b''\}$ for some constants a'' and b''. We distinguish between the following two cases.

Case 1. u is a 1-node.

By (3), after prune(w) and bypass(u) the expression stored at v is

$$\max\{a, \max\{1, \max\{a', x + b'\} - c\} + b\} = \max\{a'', x + b''\}$$

for constants a'' and b'' such that $a'' = \max\{\max\{a, 1 + b\}, a' - c + b\}$ and $b'' = b' + b - c$. (Here, we assumed for simplicity that w is the right child of u; the other case is similar.)

Case 2. u is a 0-node.

Now (2) implies that after prune(w) and bypass(u) the expression stored at v is

$$\max\{a, \max\{a', x + b'\} + c + b\} = \max\{a'', x + b''\}$$

with $a'' = \max\{a, a' + b + c\}$ and $b'' = b' + b + c$.

Consequently, the reconstruction of the tree after a prune and bypass operation can be performed in $O(1)$ time. Thus, computing $p(root)$ can be done in the time of tree contraction as claimed. Since no read/write conflicts occur, the computation can be performed in the EREW-PRAM model. This completes the proof of Theorem 3. □

Furthermore, by traversing the contraction sequence once more, we can, in fact, compute the value of $p(x)$ for every regular node of $BT(G)$. Since every update takes $O(1)$ time and the contraction sequence involves $O(\log n)$ trees, the entire computation can be performed in $O(\log n)$ EREW time using $O(n / \log n)$ processors.

From now on we assume that at every internal node u of $BT(G)$ the value of $p(u)$ is available. In addition, we introduce some notation. Consider an arbitrary regular node x of $BT(G)$. We enumerate the paths in a minimum path cover of $G(x)$ as 1, 2, ..., $p(x)$. Trivially, every leaf u in $L(x)$ belongs to exactly one such path (we don't know which, as yet); let this path be $path(u, x)$. Note that $path(u, x)$ stands for the identity of the path in a minimum path cover of $G(x)$, that the leaf u belongs to.

The purpose of the next stage of our algorithm is to compute $path(u, root)$ for every leaf u in $BT(G)$. For this purpose, we assume that every leaf has one processor associated with it; this processor will keep track (and record, as appropriate) the identity of the path $path(u, root)$.

Stage 3. In the third stage, using the contraction sequence obtained in Stage 2, we traverse the tree $BT(G)$ computing, a functional form at every internal node and updating path information when a prune and a bypass operation is performed.

First, we associate with every internal node x a functional form $f(x)$ defined as follows:

(6) if x is a 0-node then $f(x) \leftarrow f(z) + p(y)$;

(7) if x is a 1-node then $f(x) \leftarrow \max\{1, f(z) - |L(y)|\}$.

It is easy to see that relations (6) and (7) specify general formulas by which to update all the path numbers of leaves in $L(z)$. This update, however, is not done for a given leaf until a prune operation involving that leaf is performed. Note that when a leaf u is pruned the following situations can occur:

• u is the left child of a 1-node: in this case, clearly, all the nodes in $L(u)$ will belong to path 1 (the first path) in $L(par(u))$ ($par(u)$ is the current parent of u);

• u is the right child of a 1-node: in this case, clearly, all the nodes in $L(u)$ will update the path they belong to using (7);

• u is the left child of a 0-node: in this case, no change is needed;

• u is the right child of a 0-node: in this case all the path numbers in $L(u)$ have to be augmented by $p(v)$, where v is the current sibling of u. Note that $p(v)$ is available since it was computed at the end of Stage 2.

Next, when a node is bypassed the functional form of the bypassed node is passed on to the node at the lowest end of the bypassing edge which will update its own functional for accordingly. It is easy to confirm that when the tree contraction is done, every leaf u of $BT(G)$ knows the identity of $path(u, root)$. Furthermore, each update takes $O(1)$ time if all the processors have concurrent access to the functional form of their (dynamic) parent. To summarize, we state the following result.

Theorem 4. Given an arbitrary n-vertex cograph represented by its cotree, for every leaf u of $BT(G)$ the identity of the path in a minimum path cover of G that u belongs to, can be determined in $O(\log n)$ CREW time using $O(n)$ processors. □

For further reference, Stage 3 can also be used to compute the number of vertices in each of the paths in a minimum path cover of G. To see this, note that when every leaf u knows $path(u, root)$, we only need sort the leaves in increasing order of their path number and then using standard list ranking compute the size of every path. Clearly, this can be obtained in $O(\log n)$ CREW time using, say,

O(n) processors.

In the fourth stage of our algorithm we need the identity of $path(u,x)$ for every leaf u and every internal node x of $BT(G)$ for which $u \in L(x)$. Here, it is helpful to imagine that every internal node x stores $L(x)$ as a vector of size $|L(x)|$. For every leaf u in $L(x)$, the corresponding entry in this vector records value of $path(u,x)$. However, to compute this information efficiently we need more processors. Specifically, imagine that we assign O($n/\log n$) processors with every leaf of $BT(G)$, for a total of O($n^2/\log n$) processors. It is easy to see (refer to [1] for a discussion) that now traversing the contraction sequence obtained in Section 2, we can record for every leaf the desired information in O($\log n$) CREW time.

Stage 4. The idea of this stage is very simple. We compute for every *regular* leaf u of $BT(G)$ the position of u in $L(root)$ as we are about to explain. More formally, imagine that the leaves in $L(root)$ are enumerated in a way consistent with (0), that is, we first enumerate all the leaves that occur in the first path in a minimum path cover of G, followed by the leaves in the second and so on. As it turns out, using the tree contraction sequence computed in Section 2 and O(n) processors we can compute in O($\log n$) time the position of every regular leaf in $L(root)$. We present the details next.

As before, we associate with every internal node x of $BT(G)$ a functional form $g(x)$ as follows.

(8) if x is a 0-node then $g(x) \leftarrow g(z) + |L(y)|$;

(9) if x is a 1-node then $g(x) \leftarrow g(z) + \min\{g(z)-path(u,z),\max\{0,|L(y)|-p(z)\}\}$.

To justify relations (8) and (9), note that if x is a 0-node then the position in $L(x)$ of a leaf in $L(y)$ does not change; however, for all leaves in $L(z)$, their position in $L(x)$ is obtained from their position in $L(z)$ by adding $|L(y)|$.

Similarly, if x is a 1-node then no leaf in $L(y)$ is regular, so we only need worry about leaves in $L(z)$. It is easy to confirm that the formula that updates the position of a leaf in $L(z)$ is given by (9). The details amount to a case by case analysis and are left to the reader.

Consequently, (8) and (9) are general formulas by which we can update the position information of a leaf as we move from $L(z)$ to $L(x)$. At the same time, note that in (8) and (9) $L(y)$, $L(z)$, $p(z)$ and $path(u,z)$ are *constants* that have been computed in the previous stages. Now proceeding exactly as in Stage 3 we end up with the position of every leaf in $L(root)$, as claimed. To summarize our findings we state the following result.

Theorem 5. Given an arbitrary n-vertex cograph represented by its cotree, for every regular leaf u of $BT(G)$ its position in $L(root)$ can be determined in O($\log n$) CREW time using O(n) processors. □

Stage 5. Finally, the last stage of our algorithm attempts to find the position in $L(root)$ of all the leaves in $BT(G)$. Recall that in Stage 4 the position of every regular leaf has been determined. A similar argument shows that for every internal node x of $BT(G)$ the position of every regular leaf in $L(x)$ can be computed in O($\log n$) CREW time using all the O($n^2/\log n$) processors. In addition, as noted at the end of Stage 3, we know the exact number of leaves in each path of $L(x)$. Therefore, by a simple prefix computation we can find, in each of the $p(x)$ paths the positions that must be filled with special leaves. The crucial observation here is that these can be placed in the unfilled positions in any order. Now O(n) processors can place the special leaves in the unfilled slots in O($\log n$) time, and the problem is now completely solved. To summarize our findings we state the following result.

Theorem 6. Given an arbitrary n-vertex cograph represented by its cotree, a minimum path cover for G can be computed in O($\log n$) CREW time using O($n^2/\log n$) processors.

Corollary 6.1. Given an arbitrary n-vertex cograph represented by its cotree, the hamiltonian path (cycle) problem can be solved in O($\log n$) CREW time using O($n^2/\log n$) processors.

Proof. Follows directly from Corollary 1.1. □

4. Conclusions and open problems

We have presented a fast parallel algorithm to compute a number of path functions on cographs including a minimum path cover, a hamiltonian path and a hamiltonian cycle. Specifically, with an n-vertex cograph G represented by its parse tree as input, our algorithm finds a minimum path cover in G in O($\log n$) CREW time using O($n^2/\log n$) processors. However, a number of questions remain open. In [7] a $\Theta(n)$ sequential algorithm is proposed to solve the minimum path cover for cographs. To the best of our knowledge no such algorithm has been obtained for a parallel model of computation. A second interesting problem is to obtain the cotree representation of a given cograph as fast as possible. Typically, a cograph recognition algorithm also creates the cotree representation. However, the best such algorithm to date runs in O($\log n$) EREW time using O($n^2+mn/\log n$) processors [14], which, in view of the linear-time sequential algorithm in [7], is not optimal.

References

1. K. Abrahamson, N. Dadoun, D. G. Kirkpatrick, and T. Przytycka, A simple parallel tree contraction algorithm, *Journal of Algorithms* 10 (1989) 287-302.

2. G. S. Adhar and S. Peng, Parallel Algorithm for Path Covering, Hamiltonian Path, and Hamiltonian Cycle in Cographs, *Proc. International Conference on Parallel Processing*, 1990, pp. III-364 - III-365.

3. R. J. Anderson, G. L. Miller, Deterministic parallel list ranking, *Aegean Workshop on Computing*, 1988.

4. J. A. Bondy, U. S. R. Murty, Graph Theory with Applications, North-Holland, Amsterdam, 1976.

5. R. Cole and U. Vishkin, Approximate parallel scheduling. Part I: The basic technique with applications to optimal parallel list ranking in logarithmic time, *SIAM Journal on Computing*, 17, (1988) 128-142.

6. R. Cole and U. Vishkin, The accelerated centroid decomposition technique for optimal tree evaluation in logarithmic time, Ultracomputer Note 108, TR-242 Department of Computer Science, Courant Institute, NYU, 1986.

7. D. G. Corneil, H. Lerchs, and L. Stewart Burlingham, Complement Reducible Graphs, *Discrete Applied Mathematics*, 3, (1981), 163-174.

8. M. R. Garey and D. S. Johnson, Computers and Intractability, A Guide to the Theory of NP-completeness, Freeman, San Francisco, 1979.

9. H. Gazit, G. L. Miller, and S. H. Teng, Optimal tree contraction in the EREW model, *Princeton Workshop Book*, Plenum Press, New York, 1989.

10. A. Gibbons and W. Rytter, Optimal parallel algorithms for dynamic expression evaluation and context-free recognition, *Information and Computation*, 81 (1989), 32-45.

11. X. He, Efficient parallel algorithms for solving some tree problems, *Proc. 24-th Allerton Conf. on Communication, Control, and Computing*, 1986, 777-786.

12. X. He and Y. Yesha, Binary tree algebraic computation and Parallel algorithms for simple graphs, *Journal of Algorithms* 9 (1988) 92-113.

13. C. P. Kruskal, L. Rudolph, and M. Snir, Efficient parallel algorithms for graph problems, *Algorithmica*, 5 (1990) 43-64.

14. R. Lin and S. Olariu, A fast recognition algorithm for cographs, *Journal of Parallel and Distributed Processing*, to appear.

15. G. L. Miller and J. Reif, Parallel tree contraction and its applications, *Proc. 17th Annual Symposium on the Theory of Computing*, 1985, pp. 478-489.

16. S. Moran and Y. Wolfstall, Optimal Covering of Cacti by vertex-disjoint paths, Dept. of Computer Science, Technion, Israel, Tech. Report 501, March 1988.

17. R. E. Tarjan and U. Vishkin, An efficient parallel biconnectivity algorithm, *SIAM Journal on Computing*, 14 (1985) 862-874.

18. U. Vishkin, Synchronous parallel computation - a survey, TR. 71, Department of Computer Science, Courant Institute, NYU, 1983.

A Fault Tolerant Routing Algorithm in Star Graph Interconnection Networks

Sumit Sur

Department of Mathematics
Colorado State University
Ft. Collins, CO 80523

Pradip K. Srimani

Department of Computer Science
Colorado State University
Ft. Collins, CO 80523

Abstract

We propose an efficient fault-tolerant routing strategy for the newly developed star graph interconnection topology by using a depth first search strategy. The proposed algorithm attempts to route a message from the source to the destination along an optimal path and is guranteed to trace a path as long as the source and the destination are not disconnected. We provide exact mathematical expressions for the probabilities that the algorithm will compute the optimal path for a given number of faulty links in the network. It is shown that the algorithm routes a message along an optimal path with a very high probability.

1 Introduction

An efficient message routing scheme is one of the most important features of a distributed memory multicomputer system. When a multicomputer (set of nodes connected by an interconnection network) is used in a mission-critical application and it is conceivable that some nodes or links in the system fail, it is essential that the routing scheme must also be fault tolerant, i.e. should be able to transmit messages as long as the source and destination nodes are connected ideally along an alternate optimal path if one such exists. Any such routing algorithm in a network is intimately related to the underlying topology in the sense that it should exploit the interconnection links in order to compute an alternate path between source and destination in case of occurrence of faults. Since hypercubes have been the most popular choice for interconnection topology to design multicomputing systems due to their structural regularity and other design features, numerous studies have been made to design optimal routing schemes for them. Chen and Shin [4, 5] have proposed the most formal algorithms for both optimal and adaptive routing in hypercubes and have done exhaustive performance analysis. Their works also contain an exhaustive bibliography on related works on hypercubes. Other networks have also been studied [6, 9, 10, 11].

Recently, a new interconnection topology, called the star graphs, has been reported in the literature [1, 2, 3]. It seems to be a very attractive alternative for the hypercubes in terms of various desirable properties of an interconnection network. It has been shown that not only the star graphs can accommodate more processors and with less interconnection and less communication delay, but these star graphs are also optimally fault tolerant and strongly resilient [2] like the hypercubes. Our purpose in the present paper is to propose a depth first search routing algorithm for point to point message transmission in these star graphs that can route messages along an optimal path with a very high probability in presence of faulty links in the network. We also provide exact mathematical analysis of the performance of our algorithm.

We develop the routing scheme using depth first first search approach that is suitably guided to exploit the rich interconnection structure of star graphs. The algorithm is distributed in the sense that each node needs to know only the status of its adjoining links. The information about the path already traversed by a message is kept in a stack that is sent along the message; this stack is useful when the algorithm is forced to take a non optimal path (in order to avoid to visiting a node more than once). Our algorithm computes an optimal path with a very high probability in presence of a given number of link failures and it computes a path as long as the source and destination nodes are connected.

2 Star Graphs

A star graph S_n, of order n, is defined to be a symmetric graph $G = (V, E)$ where V is the set of $n!$ vertices, each representing a distinct permutation of n elements and E is the set of symmetric edges such that two permutations (nodes) are connected by an edge iff one can be reached from the other by interchanging its first symbol with any other symbol. For example, in S_3, the node representing permutation ABC have edges to two other permutations (nodes) BAC and CBA. Throughout our discussion we denote the nodes by permutations of English alphabets. For example, the identity permutation is denoted by $I = (ABCD...Z)$ (Z is the last symbol, not necessarily the 26th). Graph theoretic terms not defined here can be found in [7] and a detailed treatment of star graphs can be found in [1, 3]. It is easy to see that any permutation of n elements can also be specified in terms of its cycle structure with respecvt to the identity permuation. For example, $CDEBAF = (ACE)(BD)(F)$. It has been shown in [3] that the minimum distance $d(\pi)$, from a given node (permutation) π to the identity permutation is given by:

$$d(\pi) = \begin{cases} c + m & \text{when A is the first symbol} \\ c + m - 2 & \text{when A is not the first symbol} \end{cases}$$

Since star graphs are vertex symmetric [3], we can always view the distance between any two arbitrary nodes as the distance between the source node and the identity permutation by suitably renaming the symbols representing the permutations. Hence, in our subsequent discussion about a path from a source node to a destination node, the destination node is always assumed to be the identity node without any loss of generality. Next we attempt to characterize the nodes in S_n that have maximum distance $\lfloor 3(n-1)/2 \rfloor$ from I. We state two lemmas; proofs are omitted due to space limitation.

Lemma 1 *For odd n, a node π (permutation) in S_n is at maximum distance from I, iff its cycle structure satisfies the following: (i) A is the first symbol in π i.e. a 1-cycle (ii) All other cycles are 2-cycles*

Lemma 2 *For even n, any permutation π in S_n has the maximum distance D_n iff its cycle structure satisfies one of the following two: (i) A is the first symbol in π, three other symbols form a 3-cycle and the rest form 2-cycles, (ii) A is not in the first position in π, and all cycles are of length 2 (n/2 of them).*

3 Depth First Search Routing

In this section we present an efficient routing algorithm based on depth first search to trace a path from a given source node to the destination (identity) node. We assume that each node has only limited knowledge about the failure of the nodes in the entire graph; each node knows the condition (faulty or not) of only its adjacent nodes. If a node is faulty, all the links or edges adjacent on it are also assumed faulty i.e., unusable for message transmission. The algorithm has the following characteristics:

- It can send a message between two nonfaulty nodes in a star graph in presence of an arbitrary number of faulty nodes as long as the source and the destination nodes are connected.

- Due to the limited information on failure kept at each node and the depth first nature of the algorithm it is possible that the algorithm will not detect an optimal path when there is one for the given fault pattern.

- The probability with which the algorithm will find a optimal path for a given number of node failures is very high.

Before we present the algorithm we want to make the following observations:

(a) Since we are interested in routing messages from a source node u to a destination node v, we can safely assume that the destination node is always the identity node (permutation), as discussed in the previous section. And hence all cycle structures are assumed to be with respect to the identity permutation. This is done for convenience of description only; it does not affect the general applicability of the algorithm.

(b) Since a message must reach the destination via several intermediate nodes, whenever any node receives a message, it attempts to trace an optimal path to the destination node and sends the message along that path (if the node is not the destination itself).

(c) In a star graph the adjacent nodes of a given node u are determined by interchanging the first symbol in u with any other symbol in the permutation. It has been shown in [1, 3] that an optimal path from an arbitrary node u to the identity permutation can be traced by using two simple rules: if the first symbol of the node is not A, do an interchange to place it in its proper position; else interchange A with some symbol which is yet to be in its proper position. It is to be noted that the first symbol X (when not A) can also be interchanged with any other symbol Y when X and Y don't belong to the same cycle and Y is not in its proper position.

(d) In order to avoid visiting the same intermediate node more than once (except when backtracking is needed to trace a non-optimal path), a dynamic stack VN (of visited nodes) is appended to the message in which nodes are recorded in the order they are visited by the message. This stack VN is set to empty at the originating node of the message.

(e) Each node is aware of its own permutation as well as the cycle structure of the permutation.

Definition 1 *For a given node (permutation) of a star graph, the i-th DF1 symbol is defined to be the i-th symbol from the right which is not in its proper position with respect to the identity permutation.*

Definition 2 *For a given node (permutation) of a star graph, the i-th DF2 symbol is defined to be the i-th symbol from the right.*

Example: Consider the permutation $BFAEDCG$. The $DF1$ symbols are C, D, E, and so on while the first $DF2$ symbol is G, second $DF2$ symbol is C and so on (G is already in its proper position with respect to the identity permutation).

Definition 3 *For a given node (permutation), i-th optimal symbol is defined to be the i-th symbol from right that is not in its proper position and does not belong to the same cycle as A.*

Example: In the above example, where the node has the cycle structure $(ABFC)(DE)(G)$, D is the first optimal symbol, E is the second optimal symbol and there is no more optimal symbol.

Definition 4 *For a given node u, the* **available** *optimal symbol is defined to be the first optimal symbol in u in a right to left sense such that the link between u and v (obtained by interchanging this symbol and the first symbol of u) is not faulty (u and v are always adjacent in star graph). It is possible that for a given set of faulty links a node u may not have an available optimal symbol.*

Note: Available DF1 and DF2 symbols can also be similarly defined for a node u for a given set of faulty links.
Example: In the above example, assume that the links from $BFAEDCG$ to the nodes $GFAEDCB$, $CFAEDBG$, and $DFAEBC$ are faulty. Then D is the available $DF1$ symbol as well as the available $DF2$ symbol while the available optimal symbol is E. If the link to the node $EFABDCG$ is also faulty, there is no available optimal symbol for $BFAEDCG$.

The Algorithm
/* This algorithm is invoked by each node u whenever it receives a message of the type (message, VN). If u is the originating node, the stack VN is set to empty. The purpose of this algorithm is to compute the next node along the path to the destination (identity) node and send the message to this next node. "send" is the primitive that pushes the present node u to the stack VN and then sends the message on the next node along with VN. The algorithm assumes that the node u is aware if any of its adjacent links are faulty. */

```
if u is the identity node
                 then stop; (destination is reached)
if X is not A then
    if the link from u to v (obtained by placing X
      in its correct position) is not faulty
      then send to v and stop

    else if there is an vailably optimal symbol Y
    in u then send to v (obtained by interchanging
    X and Y in u) and stop endif
else if there is an available DF1 symbol Y in u
         then send to v (obtained by interchanging
         Y and X in u) and stop endif
endif;
```

/* If the algorithm has not terminated yet, all optimal paths to the destination from this node u are blocked by faulty links. This node u then attempts to send the message along a non optimal path to the destination. */

```
if there is a DF2 symbol Y in u such that the link from
   u to v (obtained by interchanging X and Y in u) is
   not faulty and v does not appear in the stack VN
   then send to v and stop
else if the stack is not empty
         then pop the topmost node w from the stack VN
```

```
          and send to w and stop
      else terminate with the error flag "source and
          destintion  are disconnected
      endif
  endif.
```

4 Performance Analysis of The Routing Algorithm

In this section we compute the probability with which our routing algorithm will trace the optimal path from the source node to the destination given that there are f faulty nodes in the star graph. Although the algorithm can compute a path as long as as the source and the destination are connected, our interest in this paper stops when the algorithm signals that an optimal path cannot be found due to presence of some faulty links. We make the following general observations:

- A star graph S_n of dimension n has $L = n!(n-1)/2$ links. If f denotes the number of faulty links in the graph, there can be in total $\binom{L}{f}$ different arrangements of the faulty links.

- We assume that all of the above fault configurations are equally likely.

- For a given source node u in the star graph, not all of the f faulty links may not affect the routing algorithms ability to compute the optimal path (the faults that are located on any possible optimal path from u to the destination will conceivably affect the performance of the algorithm). We use k to denote this relevant number of faults.

- Due to the depth first nature of the algorithm, it may not be able to compute an optimal path even when one exists in the graph for the given fault configuration.

- Without any loss of generality we assume that the source node is at a maximum distance from the destination node. From the results of section 2 we know that such nodes have different cycle structures depending on whether n is odd or even.

- For a given fault pattern and a given source node, the DFS algorithm either finds an optimal path to the destination or it is blocked after traversing part of an optimal path. The algorithm is blocked when no optimal symbol is available at any intermediate node (of course, the in such situation the algorithm begins to trace a non-optimal path or detect that source and destination are disconnected; but we are not interested in a non optimal path here). If we draw all the optimal paths and the partial paths to different blocked nodes generated by the DFS algorithm for a given source node, we get a *possibility tree* PT_u of the source node u (note that nodes in this tree may not be distinct; we view this graph as a tree for convenience of description later).

- Since optimal routing in a star graph heavily depends on the cycle structure of the source node, the probability of our algorithm's computing the optimal path will be different for n odd and even.

Next we state three theorems that enable us to evaluate the performance of the proposed depth first search algorithm. We omit the proofs and the interesting analysis that lead to the theorems for lack of space; the interested reader can find them in [12]. It is to be noted that the analysis is done separately for two cases of n odd and n even. The cycle structures of the nodes that are at a maximum distance from the identity node are different and the analysis exploit many combinatorial properties of the cycle structures of the permutations.

Theorem 1 *Consider an arbitrary set of f faulty links in a star graph S_n, n odd, and a message is to be routed by the DFS algorithm from a node u to a node v where the distance between u and v is maximum i.e., $d(u,v) = $ diameter of S_n. Then the probability of optimal routing $P(n,f)$ is given by*

$$P(n,f) = \frac{1}{\binom{L}{f}} \sum_{k=0}^{min\{f,\rho(c)\}} \phi_k(c) \binom{L-D-k}{f-k}$$

where $c = \frac{n-1}{2}$, $L = $ total number of links in $S_n = \frac{n!(n-1)}{2}$ and $D = $ diameter of S_n, n odd $= \frac{3n-3}{2}$ and $\rho(c) = c^2 + c - 1$.

Theorem 2 *Consider an arbitrary set of f faulty links in a star graph S_n, n even and a message is to be routed from a node u to a node v such that distance between u and v is maximum and all cycles in u (with respect to v) are 2-cycles. Then the probability $P(n,f)$ of optimal routing by the DFS algorithm is given by*

$$P(n,f) = \frac{1}{\binom{L}{f}} \sum_{k=0}^{min\{f,\lambda(c)\}} \psi_k(2,c-1) \binom{L-D-k}{f-k}$$

n	f=1	f=2	f=3	f=4	f=5
5	0.991667	0.983229	0.974691	0.966057	0.957331
7	0.999868	0.999735	0.999603	0.999471	0.999338
9	0.999999	0.999997	0.999996	0.999994	0.999993

Table 1: Numerical values of $P(n,f)$ with n odd

n	f=1	f=2	f=3	f=4	f=5
4	0.944444	0.887302	0.829132	0.770478	0.711861
6	0.998889	0.997776	0.996661	0.995544	0.994426
8	0.999986	0.999972	0.999957	0.999943	0.999929
10	1.000000	1.000000	1.000000	1.000000	0.999999

Table 2: Numerical values of $P(n,f)$ with n even, Case (A)

where $c = \frac{n}{2}$, $L = $ total number of links in $S_n = \frac{n!(n-1)}{2}$ and $D = $ diameter of $S_n = \frac{3n-4}{2}$, when n is even, and $\lambda(c) = c^2 - c$.

Theorem 3 *Consider an arbitrary set of faulty links in a star graph S_n, n even and a message is to be routed from a type III node with c cycles to the identity node at a maximum distance. Then the probability $P(n,f)$ of optimal routing is given by,*

$$P(n,f) = \frac{1}{\binom{L}{f}} \sum_{k=0}^{min\{f,\xi(c)\}} \delta_k(c) \binom{L-D-k}{f-k}$$

where $c = \frac{n-2}{2}$, $L = $ total number of links in $S_n = \frac{n!(n-1)}{2}$ and $D = $ diameter of $S_n = \frac{3n-4}{2}$ when n is even, and $\xi(c) = c^2 + 2c - 1$.

Numerical examples of the probabilities that the proposed algorithm will compute an optimal path are shown in tables 1,2,3 for different cases. It can be seen that the algorithm does compute the optimal path from a source node at a maximum distance from the destination node with a rather high probability in presence of faulty links.

n	f=1	f=2	f=3	f=4	f=5
4	0.888889	0.787302	0.694678	0.610474	0.534165
6	0.998889	0.997774	0.996656	0.995533	0.994407
8	0.999986	0.999972	0.999957	0.999943	0.999929
10	1.000000	1.000000	1.000000	1.000000	0.999999

Table 3: Numerical values of $P(n, f)$ with n even, Case (B)

5 Conclusion

We have proposed a depth first search based routing algorithm for star graph connected computer networks. The algorithm can route messages from a source node to the destination in presence of multiple faulty links in the network. The algorithm is guranteed to generate a path as long as the source and the destination are connected and for a given number of arbitrary faulty links the algorithm computes an optimal path with a very high probability.

References

[1] S.B. Akers and B. Krishnamurthy, "A group-theoretic model for symmetric interconnection networks," **IEEE Trans. Comput.**, Vol.38, 555-565, April 1989.

[2] S.B. Akers and B. Krishnamurthy, "The Fault tolerance of star graphs," **Proceedings of 2nd Intl. Conf on Supercomputing**, 1987.

[3] S.B. Akers, D. Harel and B. Krishnamurthy, "The star graph: an attractive alternative to n-cube," **Proceedings of Intl. Conf on Parallel Processing**, pp. 393-400, 1987.

[4] M.-S. Chen and K.G. Shin, "Depth-first search approach for fault tolerant routing in hypercube multicomputers," **IEEE Trans. Par. & Dist. Sys.**, Vol. 1, pp. 152-159, April 1990.

[5] M.-S. Chen and K.G. Shin, "Adaptive fault tolerant routing in hypercube multicomputers," **IEEE Trans. Comput.**, Vol. 39, pp. 1406-1416, December 1990.

[6] A.H. Esfahanian and S.L. Hakimi, "Fault tolerant routing in De Bruijn communication networks," **IEEE Trans. Comput.**, Vol. C-34, pp. 777-788, September 1985.

[7] F. Harary, **Graph Theory**, Addison-Wesley, Reading, MA, 1972.

[8] D.E. Knuth, **The Art of Computer Programming**, Vol. I, Addison-Wesley, 1981.

[9] D.K. pradhan, "Fault-tolerant multiprocessor link and bus network architectures," **IEEE Trans. Comput.**, Vol. C-34, pp. 33-45, 1985.

[10] D.K. Pradhan, "Dynamically restructurable fault tolerant processor network architectures," **IEEE Trans. Comput.**, Vol. C-34, pp. 434-447, May 1985.

[11] K.G. Shin and M.-S. Chen, "Performance analysis of distributed routing strategies free of ping-pong-type looping," **IEEE Trans. Comput.**, Vol. C-36, pp. 129-137, 1987.

[12] S. Sur and P.K. Srimani, "A DFS strategy to fault tolerant routing in star graphs and its performance evaluation," Tech. Report, Department of Computer Science, Colorado State University, 1991.

Embedding Binary Trees in Orthogonal Graphs *

Isaac D. Scherson
Department of Information and Computer Science
University of California - Irvine
Irvine, CA 92717

Chunyao Huang
Department of Electrical Engineering
Princeton University
Princeton, NJ 08544

Abstract

Since binary trees play an important role in computational algorithms, and orthogonal graphs characterize a large class of hardware configuration of parallel systems, the embedding of the former on the latter is of practical as well as theoretical importance. An *Embedding Procedure* is suggested. It results in an embedding with dilation cost of 1 and expansion cost of $\frac{2^m}{2^m-1}$ for a full binary tree $B(m)$ of depth m. From the embedding procedure, an isomorphic binary tree $Bg(m)$ is generated with a node labeling order similar to the traversal order of a breadth first spanning tree. When applied to the case of orthogonal graphs describing Multidimensional Access memory configurations, the embedded isomorphic tree $Bg(m)$ can be traversed with only two link modes. Each access mode in an MDA memory corresponds to a subtree of $Bg(m)$ and can be identified by a locating procedure.

1 Introduction

An important problem in the area of parallel processing is that of mapping algorithms onto parallel processing structures. A good mapping strategy can significantly reduce the communications overhead and increase the overall system performance. The complexity of mapping arbitrary data dependency graphs onto regular system graphs is complex and has been tackled by a number of investigators. In this paper we consider the simpler case of mapping regular trees onto a unified representation for a class of systems, namely orthogonal graphs [13, 12]. As trees are useful data structures in many computational algorithms, and it can be shown that arbitrary trees can be transformed into binary tree representation [8, 5], we focus on the mapping of the latter onto orthogonal graphs.

Studies of tree embeddings into hypercube structures can be found in [16, 15, 2]. Various metrics to evaluate some embedding costs have been proposed in [16, 4, 3]. We follow their terminology to measure the costs of our *Embedding Procedure* .

In the following discussion, we will briefly review basic concepts of binary trees, orthogonal graphs and embedding cost measurements. The *Embedding Procedure* is then proposed. Because an interesting shared memory system arises from the generalization of OMP [10, 7], namely Batcher's MDA memories [1, 13, 14], we exemplify the application of the *Embedding Procedure* in such a system. It turns out that MDA memories can be analyzed by spannings of binary trees, and different access patterns can be mapped into corresponding tree traversals.

2 Binary Trees and Orthogonal Graphs

Basic properties of binary trees can be found in [5, 9, 6]. Recall that a *full* binary tree $B(m)$, of *depth* m, is a binary tree with m levels and $(2^m - 1)$ nodes. The *root* is at level 1 and the nodes at level i are denoted as $(v_{i1}, v_{i2}, \ldots, v_{i2^{i-1}})$.

Details of orthogonal graphs are described in [13]. The following introduces the notation used in this paper.

Let Vm be the set of all binary vectors of length m, and $Q = \{0, 1, 2, \ldots, m - 1\}$. We define a mask vector $Zn \in Vm$ whose n least significant bits are 1, while the most significant $m - n$ bits are 0.

A left rotation operation $\bullet$ is defined such that for any $q \in Q = \{0, 1, 2, \ldots, m - 1\}$, and any $R \in Vm$, $q \bullet R$ results in a left rotation of R by q bits.

Let $\oplus$ be the boolean operation of bitwise exclusive OR, and $\wedge$ be the bitwise AND. Two vectors W and X are orthogonal mode q, denoted as $W \perp_q X$, if and only if $(W \oplus X) \wedge (q \bullet Zn) = 0m$, where $n < m; W, X, Zn, 0m \in Vm,$. $0m$ is a binary vector whose elements are all 0's.

An *orthogonal graph* $G(n, m, Q*)$, $Q* \subseteq Q$, is an undirected graph with 2^m nodes. A *link* (edge) exists between two distinct nodes W and X if and only if there exists a $q \in Q*$ such that W and X are orthogonal mode q. Note that every node in G can access $2^{m-n} - 1$ neighbors for each link mode $q \in Q*$.

3 The *Embedding Procedure*

According to [16, 4], an embedding f of a graph $Gx = (Vx, Ex)$ in another graph $Gy = (Vy, Ey)$ is a one-to-one function $f : Vx \rightarrow Vy$. The *dilation* cost of f is $max\{distance(f(A), f(B)) \mid (A, B) \in Ex\}$.

An embedding f is said to *preserve adjacency* if its dilation cost is 1. The *expansion* cost of f is $\frac{|Vy|}{|Vx|}$ (the ratio of the number of nodes in Vy to the number of nodes in Vx.)

We introduce the *Embedding Procedure* in Theorem 1 as follows.

Theorem 1 *A full binary tree $B(m)$, with depth m and $2^m - 1$ nodes can be embedded in an orthogonal graph $G(n, m, Q*)$ with a dilation cost of 1 and expansion cost of $\frac{2^m}{2^m-1}$, if and only if $m - n \geq 2, Q* \subset Q = \{0, 1, 2, \ldots, m - 1\}$, and $\#Q* = m - 1$ ($Q*$ has m-1 elements).*

Proof: Consider first the case for $m - n = 2$. We prove the *if* part by induction.
Induction Base:

Let $m = 2$. and let v_{21}, v_{22} be the two children of the root v_{11} in $B(2)$. We can use $G(0, 2, \{0\})$ to label v_{11} as 00, v_{21} as 11, and v_{22} as 10.

*This work was supported in part by the Air Force Office of Scientific Research under grant number AFOSR-90-0144.

A binary labeled tree $Bg(2)$ is created. $Bg(2) \subset G(0, 2, \{0\})$. Since $B(2)$ and $Bg(2)$ are isomorphic, $B(2)$ can be mapped into $Bg(2)$ with adjacencies preserved. The expansion of this embedding is $\frac{4}{3} = \frac{2^m}{2^m - 1}$, because $G(0, 2, \{0\})$ has 4 nodes. Theorem 1 is true for $m = 2$.

Induction Hypothesis:

Let $m = k$. $B(k)$ can be embedded into $G(n_k, k, Q_{k*})$ with a dilation cost of 1 and expansion cost of $\frac{2^m}{2^m - 1}$. $k - n_k = 2, \#Q_{k*} = k - 1$. In the embedding process, we generate a binary tree, labeled with binary numbers, $Bg(k)$. $Bg(k) \subset G(n_k, k, Q_{k*})$, and is isomorphic to $B(k)$.

Induction Steps: (the *Embedding Procedure*)

1. For $m = k + 1$, we keep all the binary node labels in $Bg(k)$, and add a 0 bit as the new MSB (Most Significant Bit) for every node in $Bg(k)$.

2. Note that in $Bg(k)$, the $(k - 2)$ LSB's (Least Significant Bit) of each node in the 2^{k-1} leaves comprise the 2^{k-2} different binary sequences of $\{0, 1, 2, \ldots, 2^{k-2} - 1\}$. Furthermore, $n_{k+1} = n_k + 1$, or $m - n_{k+1} = k + 1 - n_{k+1} = 2$ holds.

3. In this step, we generate the leaves of $Bg(k + 1)$. If a leaf of $Bg(k)$ is now relabeled as $(0, 1, b_{k-2}, b_{k-3}, \ldots, b_1, b_0)$, $b_i \in \{0, 1\}$, after applying a new link mode q_{k+1} to Q_{k+1*} to cover the first two MSB's, we can generate the left and right child of $(0, 1, b_{k-2}, b_{k-3}, \ldots, b_1, b_0)$ as $(1, 1, b_{k-2}, b_{k-3}, \ldots, b_1, b_0)$ and $(1, 0, b_{k-2}, b_{k-3}, \ldots, b_1, b_0)$ following the results of the previous step and the properties of the orthogonal graph, $G(n_{k+1}, k+1, Q_{k+1*})$.

4. These new 2^k children are the 2^k leaves of $Bg(k + 1)$, $Bg(k + 1) \subset G(n_{k+1}, k+1, Q_{k+1*})$. $B(k + 1)$ and $Bg(k + 1)$ are isomorphic, thus $B(k + 1)$ is embedded in $G(n_{k+1}, k+1, Q_{k+1*})$ with adjacencies preserved [16]. $G(n_{k+1}, k+1, Q_{k+1*})$ has 2^{k+1} nodes and $B(k + 1)$ has $2^{k+1} - 1$ nodes, the expansion of the embedding when $m = k + 1$ is $\frac{2^{k+1}}{2^{k+1} - 1} = \frac{2^m}{2^m - 1}$.

5. Since we add a new link mode q_{k+1} to Q_{k+1*} to cover the first two MSB's, $\#Q_{k+1*} = \#Q_{k*} + 1 = k + 1 = m$.

6. From the above induction steps, we know that $B(k + 1)$ can be embedded in $G(n_{k+1}, k+1, Q_{k+1*})$.

The induction steps completed, Theorem 1 is true for $m = k + 1$. In the induction process, a new link mode q_{k+1} is added when we generate the leaves of $Bg(k + 1)$. Thus we keep the elements in Q_{k+1*} disjoint to one another, and $m - 1$ is the minimum for $\#Q*$.

The *only if* part is proved as follows.

Given an orthogonal graph, $G(n, m, Q*)$, with $m - n = 2$, $Q* \subset Q = \{0, 1, 2, \ldots, m-1\}$, and $\#Q* = m - 1$, generate a full binary tree of depth m, $Bg(m)$ from the orthogonal graph.

We start by choosing the node whose binary label is $(b_{m-1}, b_{m-2}, \ldots, b_0) = (0, 0, \ldots, 0)$ as the root, v_{11}, of $Bg(m)$. Then we generate v_{21}, v_{22}, two children of the root at level 2 by using the link mode q_2, and make the binary node labels of these two children different in their LSB's. Apply q_3 and continue to generate the children of v_{21} and v_{22} at level 3 (v_{31}, v_{32} and v_{33}, v_{34}, respectively.)

Similarly, we can use q_i and generate the 2^{i-1} nodes at level i of $Bg(m)$. At level m, q_m is in fact the same as q_0. Therefore,

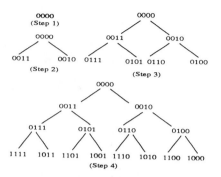

Figure 1: From $G(2, 4, \{0, 2, 3\})$ to $Bg(4)$.

$Bg(m)$, a full binary tree of depth m and $2^m - 1$ nodes, is generated from the orthogonal graph $G(n, m, Q*)$.

Figure 1 shows an example of m=4, mapping $G(2, 4, \{0, 2, 3\})$ to $Bg(4)$.

Given $G_1 = G(n_1, m, Q*)$, $G_2 = G(n_2, m, Q*)$, if $m - n_1 = 2$ and $m - n_2 > 2$, then $G_1 \subset G_2$. Thus the above proof also holds for $m - n > 2$. $\odot$

After the embedding is made, the *Embedding Procedure* in Theorem 1 generates a binary tree according to certain interesting rules described in Corollary 1.

Corollary 1 *Let $G(n, m, Q*)$ be an orthogonal graph, $m - n \geq 2$, and let $Bg(m)$ be a full binary tree generated by the Embedding Procedure in Theorem 1. The node labelling order of $Bg(m)$ is similar to the traversal order of a breadth first spanning tree with the following two constraints: (1) right children are visited before left children, and (2) children in the same depth level are visited according to the ascending order of their parents (in the previous level).* $\odot$

These properties are illustrated in Figure 2. Traversal orders could be found according to the horizontal arrow pairs in the figure. It is shown in Figure 2 that the difference between left child and right child labels at level i is $t = 2^{i-2}$.

4 Mapping Binary Trees onto MDA Memories

Batcher's Multidimensional Acess (MDA) memory system was the basis for the STARAN associative computer [1]. It is described by an orthogonal graph $G(n, 2n, \{0, 1, \ldots, 2n - 1\}$ where nodes are memory modules and edges correspond to processing element buses [14]. In such a system, 2^n processors access a $2^n \times 2^n$ array of memory modules, in either one of $2n$ access modes, by rows columns or stencils. In general, a processor labeled X accesses a memory module labeled Y, in access mode q, if and only if $X \perp_q Y$. When we map MDA memories into the binary trees by the *Embedding Procedure*, some properties of tree traversals and access modes are found. The number of link modes needed could be a measurement of tree traversal costs.

Theorem 2 *Consider an embedding obtained from the Embedding Procedure of Theorem 1, and disable all non-tree related links between any two levels i, j, $j - i \geq 2$ (we say*

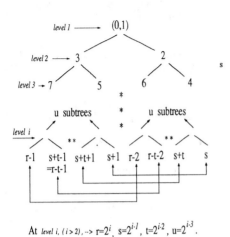

Figure 2: The orders of node labels in Bg(m).

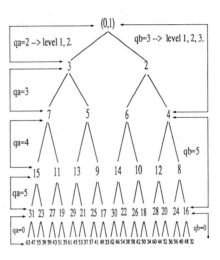

Figure 3: Tree traversals of Ga and Gb with binary connection constraints.

we enforce binary connections). At most three link modes are needed to traverse the binary tree $Bg(2n)$ embedded in $G(n, 2n, \{0, 1, 2, \ldots, 2n-1\})$.

Proof: Let q_t^i be the ith element of $Qt*$, which is a set of link modes available to traverse $Bg(2n)$. If we select $q_t^1 = n$, and $q_t^{i+1} = (q_t^i + n - 1) \bmod 2n$, then these link modes can generate the largest binary trees from $G(n, 2n, \{0, 1, 2, \ldots, 2n-1\})$. In other words, $Qt*$ becomes a minimum set of link modes to traverse $Bg(2n)$. We write this minimum set as $minQt*$.

In $Bg(2n)$, each q_t^i can generate a n-level subtree starting at depth $(2i-1)$. To simplify the computation, we can ignore the root and the overlapping level generated by q_t^{i+1} and q_t^i. Then q_t^i corresponds to a disjoint $(n-1)$-level subtree starting at depth $2i$.

If the root is not included, then the depth of $Bg(2n)$ is $2n-1$. Let $\#minQt*$ be the number of elements in a minimum set of link modes to traverse $Bg(2n)$:
$\#minQt* = \lceil \frac{2n-1}{n-1} \rceil = \lceil \frac{2(n-1)}{n-1} + \frac{1}{n-1} \rceil = 3$. ⊙

Figure 3 illustrates a tree traversal example for an MDA memory with $2n = 6$. For simplicity, we choose the last element in $minQt*$ to be 0. According to the Embedding Procedure, leaves could be generated from $qb = 0$. Note that qa is a link mode of Ga such that $m - n = 2$ (recall the relationship between Ga and the Embedding Procedure), and has different coverage of tree levels from qb.

If we don't restrict ourselves to the physical structures of binary trees, then obviously, less link modes are required as follows.

Corollary 2 If we relax the binary connection constraint in the Embedding Procedure , then the total number of link modes required to traverse $Bg(2n)$ is no more than 2.

Proof: We follow the same notation as in Theorem 2. Choose $q_t^1 = n, q_t^2 = 0$. These two link modes cover $2n$ bits of

each node's binary labels in $G(n, 2n, \{0, 1, 2, \ldots, 2n-1\})$ and $Bg(2n)$.

Since there are no binary connection constraints in the Embedding Procedure , we can traverse $Bg(2n)$ within two link modes. ⊙

Theorem 3 Each access mode in the MDA memory, denoted as $G(n, 2n, \{0, 1, 2, \ldots, 2n-1\})$, corresponds to a subtree of $Bg(2n)$ which is built according to the Embedding Procedure of Theorem 1.

Proof(1):

Assume we have 2^n PE's (Px's), and 2^{2n} memory modules (My's), x, y $\in V_{2n}$. Px has access to My in access mode q, if and only if $(q \bullet cast(x)) \perp_q (y)$ [13]. If $x \in V_n, n < m$, the function cast(x) will fill a binary vector x with $(m-n)$ leading zeros and make x an element in Vm.

According to the above equations, we can transfom y's into the same numbering system (binary or decimal) as $Bg(2n)$. Thereafter, the mapping between y's and nodes in $Bg(2n)$ can be found by applying the properties of labelling orders in Corollary 1. Thus we can locate its corresponding subtree in $Bg(2n)$ for each access mode q in the MDA memory. ⊙

We can use a different approach to prove Theorem 3 as follows.

Proof(2):

Create a 2^n x 2^n array, M(2^n, 2^n), and apply the MDA memory rule, i.e., MDA memory modules are accessed along columns such that each processor connects to 2^q memory elements every 2^q rows, for all q $\in$ [0, n-1]. For q́ $\in$ [n, 2n-1], the access pattern is the transpose of access mode q if and only if q́ = q mod n.

In each access mode, we can easily identify all the memory module numbers that can be access by every PE.

Mapping these memory module numbers into $Bg(2n)$, we can locate the corresponding subtree in $Bg(2n)$. $\odot$

Some mapping rules between access modes and subtrees could be written as follows.

Corollary 3

Consider an MDA memory $G(n, 2n, \{0, 1, 2, \ldots, 2n - 1\})$ and its associated full binary tree $Bg(2n)$. If $q = 0$, then each $PE(x)$ accesses a subtree $Bg(2n, x, 0)$. $Bg(2n, x, 0)$ consists of a node x and a full binary tree of depth n, $Bg(2n, x, 0, n)$, in $Bg(2n)$. The root of the $Bg(2n, x, 0, n)$ is $(x + 2^n)$.

Proof:

This property can be derived from the mapping rules between the 2^n PE's and the 2^n x 2^n memory array, $M(2^n, 2^n)$, in Theorem 3 and the node labelling orders of $Bg(2n)$ in Corollary 1. $\odot$

Corollary 4 *Consider*

an MDA memory $G(n, 2n, \{0, 1, 2, \ldots, 2n - 1\})$ and its associated full binary tree $Bg(2n)$. When $q = 2n - 1$, each $PE(x)$ accesses a subtree $Bg(2n, x, 2n - 1)$. If x is an even number, then $Bg(2n, x, 2n - 1)$ is composed of a node x plus a depth-n full binary tree $Bg(2n, x, 2n - 1, n)$ whose root is $(2^{n-1} + \frac{x}{2})$. If x is an odd number, then $Bg(2n, x, 2n - 1)$ is composed of the 2^n leaves of a full binary tree, $Bg(2n, x, 2n - 1, n + 1)$, of depth $n + 1$. The root of $Bg(2n, x, 2n - 1, n + 1)$ is $(2^{n-1} + \frac{x-1}{2})$.

Proof:

Similar reasoning to Corollary 3: it can be derived from the mapping rules between the 2^n PE's and the 2^n x 2^n memory array, $M(2^n, 2^n)$, in Theorem 3 and the node labelling orders of $Bg(2n)$ in Corollary 1. $\odot$

For instance, when link mode q is 5, the 8 leaves of a depth-4 full binary tree whose root is 6 are accessed in this link mode.

References

[1] K. E. Batcher, *The Multidimensional Access Memory in STARAN*, IEEE Transactions on Computers, Vol.C-26, No.2, Feb. 1977, pp. 172-177.

[2] T. Bier and K. Loe, *Embedding of Binary Trees into Hypercubes*, Journal of Parallel and Distributed Computing, 6 (1989), pp. 679-691.

[3] W. Chen, M. F. Stallmann and E. F. Gehringer, *Hypercube Embedding Heuristics: An Evaluation*, International Journal of Parallel Programming, Vol. 18, No. 6, 1989.

[4] J. W. Hong, K. Mehlhorn, and A. L. Rosenberg, *Cost Trade-offs in Graph Embeddings with Applications*, J. Assoc. Comput. Mach. 30 (1983), pp. 709-728.

[5] E. Horowitz and S. Sahni, *Fundamentals of Data Structures*, Reading, Computer Science Press, Inc., 1984.

[6] E. Horowitz and S. Sahni, *Fundamentals of Computer Algorithms*, Reading, Computer Science Press, Inc., 1985.

[7] K. Hwang, P. Tseng and D. Kim, *An Orthogonal Multiprocessor for Parallel Scientific Computations*, IEEE Transactions on Computers, Vol. 38, No. 1, Jan. 1989.

[8] A. L. Rosenberg and J. Hong, *Graphs That Are Almost Binary Trees*, SIAM J. Comput., 11, 2 (May 1982), pp. 227-242.

[9] S. Sahni, *Concepts in Discrete Mathematics*, Computer Science Press, Inc., 1981.

[10] I. D. Scherson and Y. Ma, *Analysis and Applications of the Orthogonal Access Multiprocessor*, Journal of Parallel and Distributed Computing, 7 (1989), pp. 232-255.

[11] I. D. Scherson, *Definition and Analysis of a Class of Spanning Bus Orthogonal Multiprocessing Systems*, Proceedings of the 1990 ACM Computer Science Conference, February 19-22, 1990, Washington DC, pp. 194-200.

[12] I. D. Scherson, *Orthogonal Graphs and the Analysis and Construction of a Class of Multistage Interconnection Networks*, Proceedings of the 1990 International Conference on Parallel Processing, August 1990.

[13] I. D. Scherson, *Orthogonal Graphs for the Construction of a Class of Interconnection Networks*, IEEE Transactions on Parallel and Distributed Systems, Vol. 2, No. 1, January 1991, pp. 3-19.

[14] I. D. Scherson, *Multidimensional Access Shared Memory Parallel Processing Systems*, Proceedings of the 1991 IEEE International Parallel Processing Symposium, April 1991.

[15] A. Wagner, *Embedding Arbitrary Binary Trees in a Hypercube*, Journal of Parallel and Distributed Computing, 7 (1989), pp. 503-520.

[16] A. Y. Wu, *Embedding of Tree Networks into Hypercubes*, Journal of Parallel and Distributed Computing, 2 (1985), pp. 238-249.

Performance Analysis of Layered Task Graphs*

*Hong Jiang** and Laxmi N. Bhuyan*
Department of Computer Science
Texas A&M University
College Station, Texas 77843-3112

Abstract

In this paper we evaluate the performance of task graphs when executed on a multiprocessing system. The proposed technique exploits the notions of variations of parallelism and average parallelism that are inherent in any parallel algorithm. The approach is based on decomposition that transforms an arbitrary task graph into a layered graph, analyzes each layer separately, and combines the results in a weighted manner. The technique captures the effect of variations of parallelism while significantly simplifying the otherwise complicated analysis. The technique is shown to be more effective and efficient than previous techniques.

1. Introduction

The behavior of an algorithm can be characterized by its parallelism as a function of time. In this paper, we combine the idea of average parallelism [1,2] with that of variation of parallelism [3] to achieve an improved accuracy and simplicity in performance evaluation of parallel algorithms. The variation of parallelism characterizes the change in parallelism with respect to time which is not captured by the average parallelism. Any parallel or distributed algorithm can be represented as a task graph where the nodes represent the tasks and the arcs represent the data dependencies among the tasks. Hence instead of analyzing a particular parallel algorithm directly and explicitly, the proposed technique focuses on task graphs that represent typical parallel algorithms.

Almost all parallel algorithms can be represented by task graphs that are acyclic and structured in a layer-by-layer fashion. A large number of parallel computations are themselves layered in nature. An example is the so called *Series–Parallel Task System* [5], that is formed by only series and parallel combinations of tasks. On the other hand, for parallel algorithms that cannot be represented directly in a layered fashion, a modified breadth-first search scheme can be used to convert the non-layered task graph into a layered one [6]. In this paper, we shall concentrate on acyclic layered task graphs and queueing network approach for solution.

It is generally known that the queueing network model of the parallel processing system executing a task graph with a given precedence relation does not possess product form solution [8] because of the presence of internal concurrency, i.e., one job can split into two or more jobs. Exact solution by explicit enumeration of the network states is not feasible because of the exponentially expanding state space. In [6], the authors have studied speedup bounds as well as scheduling policies for a rather restricted class of task graphs executing on a simple multiprocessing architecture (with communication and scheduling overhead ignored). Recently, a class of task graphs called Series-Parallel Task System was analyzed in [5] where a multiple-class queueing network model was proposed to predict the performance of such task graphs executing on a shared memory multiprocessor. The model can become computationally expensive as the size of the task graph increases. The time and space complexities are $O(N^2K + N^3)$ and $O(N^2K)$ respectively, with N being the number of tasks and K number of service centers [5]. In [1], average parallelism is used to characterize the task graphs and bounds on performance metrics such as speedup, etc., have been derived.

In this paper, we propose an approximate but simple model to analyze layered parallel processing task systems. By exploring the average and variation of parallelism inherent in a task graph, the proposed technique transforms the task graph into a layered task graph. The latter is then evaluated in two stages. In the first stage, each layer is analyzed locally, using the average parallelism of that layer of the graph. The analysis at this stage is based on a single-class closed queueing network with the population size equal to the average parallelism of that layer. The executions of two neighboring layers of a task graph do overlap to a certain degree. Unlike the assumption made in [6], we take into consideration the overlapping effect by deriving two very simple bounds, one for the case of zero-overlapping (hence corresponding to the one in [6]) and the other for maximum overlapping (similar to the technique proposed in [3]). These bounds are then used to extrapolate an approximate performance measure. The results of the first stage are then combined in the second stage to solve the task graph globally, using the variation of parallelism, through a weighted approach. Although the bounds can be related to [3, 6] in concept, the analysis presented here is totally different. The proposed technique is applicable to general layered task graphs in which a task may have any number of data dependencies. Moreover, our model is computationally very efficient in that it has a time and space complexity of $O(NK)$ each.

In what follows, we discuss the transformation of an arbitrary task graph into a layered one and develop the basic strategy in Section 2. The analytical techniques are developed in Section 3. In Section 4, we demonstrate the effectiveness of our analytical model by applying it to a typical class of task graphs and validating it with the simulation results. Finally, some concluding remarks are given in Section 5.

2. Transformation of A Task Graph

The parallelism of a task graph during its execution is a function of time. The parallelism at any given time t, denoted by P_t, is defined as the number of tasks that can be executed concurrently at time t. In the hypothetical machine consisting of an unbounded number of processors, the instantaneous parallelism is also same as the number of active processors at any time t. Let T_0 be the time the first task is scheduled to be executed and T_∞ be the execution time of an algorithm with unbounded number of processors. Thus the parallelism of any task graph can be expressed in terms of its instantaneous parallelism along its critical path L (i.e. from time T_0 to $T_0 + T_\infty$). A diagram, called a *parallelism profile*, is used to present such an expression graphically. As an example, the parallelism profile of a binary tree task graph is presented in Fig. 1. This diagram also gives a good indication of the work load the task graph may impose on an underlying architecture.

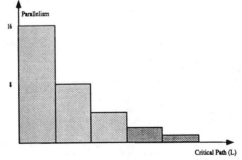

Fig. 1. Parallelism Profile of A Binary Tree Task Graph.

Based on the parallelism profile, the original task graph can be transformed into a layered graph G as follows. First, the critical path L is divided into n ($n \geq 1$) segments, $L_1, L_2, ..., L_n$, of which the lengths may be arbitrary depending on the nature of the

---

* This research was supported in part by NSF grant MIP-9002353.

** Will be with the Dept. of CSE, Univ. of Nebraska-Linocln, starting Aug. 91

task graph as explained later. The ith layer of G, G_i, contains precisely a subgraph of the original task graph such that each task in G_i must be active somewhere along L_i. Should a task be active in both L_j and L_{j+1}, it is divided into two tasks at the boundary point and are connected with an arc from G_j to G_{j+1}. Finally, each G_i is transformed into a rectangular task graph, using the idea of average parallelism [1,2]. Each such rectangular task graph in turn may consist of multiple layers (*microlayers*) of tasks. Shown in Fig. 2 is an example of layered task graph with each layer being a rectangular graph that consists of a number of *microlayers*. Inside a layer G_i of G, the number of tasks per microlayer is a constant and is equal to π_{av_i}. This is also called the average parallelism of the subgraph G_i [1, 3].

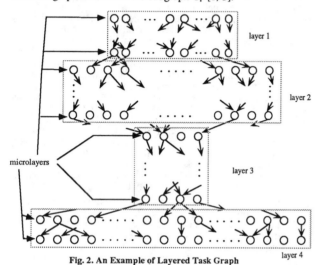

Fig. 2. An Example of Layered Task Graph

Given the method of transformation, what value of n, the number of layers in G, and the length of each L_i should we choose? The answer lies in finding the variation of parallelism of the task graph, which is defined next. In general, the more variation of parallelism a task graph has, the bigger the number n should be in order to better capture the effect of any sharp change in parallelism. However, a large n implies a larger complexity in solution. On the other hand, a small variation of parallelism suggests a longer L_i and a smaller n so that computation effort can be reduced without compromising the accuracy. The *variation*, or the distribution, of parallelism is defined in terms of the fraction of time, along the critical path L, during which the task graph exhibits a certain amount of parallelism. Thus the sum of such fractions over all possible parallelisms a task graph can exhibit is equal to unity. The variation measure is used as a weight for performance measures of the layered task graph when the layers are combined to form the final performance measure of the original task graph.

Next we discuss the transformation of a subgraph of G into a rectangular graph. The following rules govern such a transformation.

1). The width of rectangular graph is equal to the ratio of the execution time of the original graph on a uniprocessor and that of the same graph on a multiprocessor with an unbounded number of processors [1,3].

2). The average execution time of a task is same and the length of the critical path in both graphs remains same.

3). The average data dependency characteristics of the original graph is maintained in the rectangular graph. More specifically, the total number of arcs in the latter is made the same as that in the former. This ensures that the overall communication requirement is maintained.

The motivation behind this transformation is that the transformed task graph, with much simpler structure, when executed on a parallel processing system, would yield approximately the same mean performance as does the original graph [1, 3].

3. Analytical Model

After transforming a task graph into a layered task graph, the proposed technique takes two steps to solve the problem. We analyze each layer in isolation and then combine the results to form a global view of the task graph's performance. In this section we first introduce the analytical model based on average parallelism idea that carries out the first step of an analysis. Then the technique using the concept of variation of parallelism is presented to carry out the second step of the analysis.

The multiprocessing architecture under study is as shown in Fig.3. It consists of a central controller and several processors connected to a single bus. The processors have local memories and are capable of executing tasks in parallel. The controller has a scheduling unit (SU) to perform run time scheduling of the tasks. To this end, the scheduler uses the data dependencies among the tasks to find which task can be initiated for processing. When a task completes execution in a processor, the result data is sent to the scheduler which then checks if the data makes a task ready (satisfying all data dependencies) for execution on a processor. Otherwise, the task waits until the arrival of the required data.

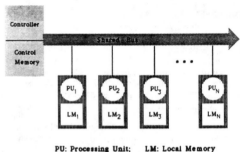

PU: Processing Unit; LM: Local Memory

Fig. 3. A Simple Parallel Processing System

If we neglect the communication time over the bus, a simple queueing model of this architecture is shown in Fig.4. All the processors constitute a processing unit (PU) which is modeled as multiple server queue and the scheduler is modeled as a single server queue. The node represented by the box in Fig.4 signifies the memory where the tasks wait for the arrival of their partner(s). Henceforth, this unit will be referred to as the delay unit. As soon as the last data (or equivalently the completion of a task that generates this data) of a waiting task arrives the task is sent to the PU for execution. We assume that the execution times of tasks are exponentially distributed random variables with the same mean. Also, the scheduling time is exponentially distributed but with a different mean. Finally, we assume that the service centers in our queueing network models have infinite buffer and serve jobs on an FCFS basis. The model neglects all I/O activities for simplicity. Extension of this simple architecture is presented in [4].

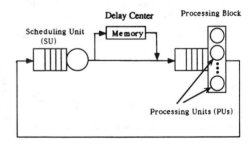

Fig. 4. A Queueing Network Model of the Architecture in Fig. 3.

3.1. Analysis of Subgraph G_i.

Here we consider one layer (G_i) of the layered task graph, shown in Fig. 2. We make the following assumptions.

1. The execution time of each task is an exponentially distributed random variable with a mean denoted by $1/\mu_2$. Also, the service time of the scheduler is exponentially distributed with mean $1/\mu_1$.

2. The same task graph is repeatedly executed so that the steady state is maintained.

3. The execution of the tasks inside G_i is mostly microlayer by microlayer. Should there be any *overlapping* between microlayers it only occurs between the two neighboring microlayers. That is, at any given time all the active tasks belong to at most two consecutive microlayers.

The last assumption is not entirely unrealistic. It was evidenced in one of the simulation studies [2] that an overwhelming majority of the simultaneously executed tasks belonged to a single pair of neighboring microlayers. With the above assumptions, the problem of analyzing the task graph reduces to that of finding the average performance of a single microlayer of tasks and the *degree of overlapping* between two neighboring microlayers. The first part is trivial, given the simple queueing network and existing solution techniques (MVA for example [7]). The second part, that of finding the degree of overlapping between two microlayers of arbitrary data dependencies, however, is nontrivial. Nonetheless, the two extremes are easy to recognize. The maximum overlapping occurs when the dependencies, if any, are exclusively *one-to-one*, as shown in Fig.5a. On the other hand, if every task on one microlayer has exactly π_{av_i} data dependencies from its neighboring microlayer, a minimum (zero) overlapping takes place, as shown in Fig.5b. The incorporation of the two extremes can facilitate deriving upper and lower bounds respectively of the performance of the rectangular task graph. We first analyze the two extremes and then extrapolate the results between the two bounds to obtain a first order approximation of the performance of the rectangular task graph.

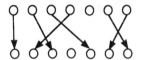

a). Maximum Overlapping   b). Minimum Overlapping

Fig. 5. Overlapping between Two Neighboring Microlevels

Maximum Overlapping:

By the above definition of maximum overlapping, the completion of any task of microlayer j at a processing unit immediately makes ready its successor task (if any) in microlayer $j+1$ for execution provided there is at least one free processing unit and no tasks of microlayer j are waiting (by assumption 3). This suggests a closed queueing network that is obtained from that of Fig.4 by removing the delay center and assuming a population equal to π_{av_i} of the task graph (i.e. the width of G_i). The center SU models run time scheduling and communication overhead entailed by each task. This queueing network can be easily solved by any multiserver mean value analysis algorithm [7]. The system response time, R_{max_i}, of this queueing network is therefore the mean response time of one microlayer of the task graph G_i with maximum overlapping. Let N_i be the number of microlayers of the rectangular graph, then a lower bound time of executing the rectangular task graph, T_{lower_i}, is simply:

$$T_{lower_i} = N_i \times R_{max_i} \qquad (1)$$

Minimum (Zero) Overlapping:

Since there is no overlapping in this case, the mean response time T_{upper_i}, an upper bound, of the task graph is:

$$T_{upper_i} = N_i \times (R_{min_i} + R_{sch}) \qquad (2)$$

where N_i is the same as in (1), R_{sch} is the time experienced by a task due to runtime scheduling and communication overhead which is assumed in the beginning of this subsection to be

exponentially distributed with mean $1/\mu_1$, and R_{min_i} is the mean execution time of one entire microlayer of tasks and is expressed as follows.

Let n be the number of processors in the processing center and $T_d(k)$ the expected total time for k departures from the processing center, given that all the n processors are busy, and $t_1, t_2, \ldots t_{\pi_{av_i}}$ be random variables representing, respectively, execution times of tasks 1 through π_{av_i} of a particular microlayer. Based on the assumption that all t_j's $(1 \le j \le \pi_{av_i})$ are mutually independent, identically exponentially distributed random variables, we have[8]:

$$R_{min_i} = \begin{cases} \overline{\max}(t_1, t_2, \ldots, t_{\pi_{av_i}}) & \pi_{av_i} \le N \\ \overline{\max}(t_1, t_2, \ldots, t_n) + T_d(\pi_{av_i} - n) & \text{otherwise} \end{cases} \qquad (3)$$

where $\bar{x}$ denotes the expected value of x and

$$\overline{\max}(t_1, \ldots, t_k) = \sum_{i=1}^{k} \begin{bmatrix} k \\ i \end{bmatrix} (-1)^{i+1} \frac{1}{i\mu} \qquad (4a)$$

and

$$T_d(k) = \int_0^\infty \tau \cdot \frac{k \, d\left[(1 - e^{-n\mu\tau})\right]}{d\tau} \, d\tau = \frac{k}{n\mu} \qquad (4b)$$

where μ is the service rate of a processor and $\bullet$ stands for convolution. R_{min_i} is the expected length of the interval starting from the moment a microlayer of tasks arrive at an empty (of tasks) processing center to the point when the center becomes empty again.

Extrapolation:

Given the lower and upper bounds, the exact overlapping between two neighboring microlayers lies somewhere between the two. As an approximation, it is reasonable to assume that the exact overlapping is uniformly distributed between the two bounds according to the mean data dependency of the layer G_i. This gives rise to the following evaluation of a layer's mean performance.

Let d_j, $j = 0, 1, \cdots, \pi_{av_i}$, be the distribution (fraction) of tasks having in-degree equal to j. Then the mean response time of one microlayer of the task graph G'_i is given as follows.

$$\bar{R}_i = \frac{\left[\pi_{av_i} - \sum_{j=1}^{\pi_{av_i}} j \cdot d_j\right] \cdot R_{max_i} + \left[\sum_{j=1}^{\pi_{av_i}} j \cdot d_j - 1\right] \cdot R_{min_i}}{\pi_{av_i} - 1} \qquad (5)$$

where R_{max_i} and R_{min_i} are the response times of one microlayer of the task graph G_i with maximum and minimum overlapping respectively. The mean response time of the rectangular subgraph (layer G_i) is therefore

$$R_i = \bar{R}_i \times N_i$$

3.2. Analysis of Layered Task Graphs

Consider a layered task graph obtained following the definitions given in the previous section. Each layer G_i in this task graph represents the average load of a portion of the original graph where it exhibits certain degree of parallelism, i.e., in the neighborhood of π_{av_i}. Also, as mentioned earlier, the entire graph maintains the mean performance of the original graph. Thus one can make the stipulation that, if the marginal effect between neighboring layers can be ignored or otherwise calculated, the performance yielded by executing each individual layer in isolation should sum up to the mean performance of the original graph. To completely ignore such marginal effect may be unrealistic in some cases, whereas to accurately capture it can also be very difficult due to the unequal width of the boundary layers. A careful observation of the layered task graph reveals, however, that this marginal effect is nothing but the overlapping effect between the two microlayers that boarder the two neighboring layers (subgraphs G_i and G_{i+1}). We can therefore take advantage of the fact that the marginal effect of any two neighboring microlayers of tasks within each layer (G_i) has been taken into account by considering the "overlapping" manifested on *each* microlayer of the layer. Thus the overlapping effects of the two boundary microlayers, with respect to their own layer, have been taken into account respectively. This suggests that the marginal effect (between two layers of the layered graph) is at least partially captured by the analysis of the corresponding subgraphs.

The above assumption enables us to analyze the layered task graph in a simple way, provided that each layer of the layered graph has been evaluated in isolation. We simply combine the results of each individual layer in a weighted approach, using the variation of parallelism of the original task graph. We consider two system parameters, namely the response time and the processing power, as our main performance measures. The response time, denoted by R, is defined as the time from the first task in the task graph being scheduled until the last task completes its execution. The processing power, denoted by U, is defined as the mean number of active (or busy) processors. Let the layered task graph have N layers, numbering from 1 through N, and let R_i and U_i be the response time and processing power respectively when layer i is being executed. Then R and U are given by

$$R = \sum_{i=1}^{N} R_i, \qquad (6)$$

$$\text{and} \qquad U = \sum_{i=1}^{L} \frac{R_i}{R} \cdot U_i. \qquad (7)$$

Here the processing power is weighted with respect to response time. This is because a particular layer will in general experience different response time when executed on different parallel processing system.

4. An Application

In this section, the analytical methods introduced in previous sections are applied to evaluate the performance of a typical class of task graphs.

4.1. Description of the problem

Consider a class of tree type of task graphs which we define by a quintuple below.

$G = < l, P_0, T_0, \alpha, \beta >$ where

l: number of layers of the tree, counting from top down, i.e. $0, 1, \ldots, l-1$,

P_0: number of tasks at layer 0,

T_0: mean execution time of tasks at layer 0,

α: rate at which the mean execution time of tasks changes from layer to layer, i.e. $T_{i+1} = \alpha T_i$, and

β: rate at which the number of tasks changes from layer to layer, i.e. $P_{i+1} = \beta P_i$. When β is restricted to be integer, it indicates the number of tasks forked by a task if $\beta > 1$. On the other hand if $1/\beta > 1$ is an integer, a task is joined to by $1/\beta$ tasks. Examples of these two cases are shown in Fig.6a and 6b respectively.

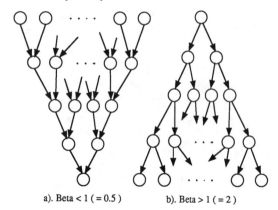

a). Beta < 1 (= 0.5)   b). Beta > 1 (= 2)

Fig. 6. Examples of Tree Type Task Graph G

We also consider task graphs obtained by combining the above two cases. This extension is called *combined* tree type $\bar{G}$. Fig.7 shows an example of $\bar{G}$.

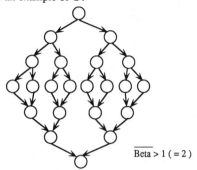

$\overline{\text{Beta}} > 1 \, (= 2)$

Fig. 7. An Example of Combined Tree Type Graph G

Our choice of tree type task graphs was motivated by two factors. First, this type of task graphs, or certain combinations of them, can represent the series-parallel task systems, mentioned in Section 1, which in turn represent a large number of parallel computations. Secondly, this class of task graphs have a large variation of parallelism and thus provides an approximate worst case scenario for evaluating the effectiveness of our analytical approach. Moreover, many computer science problems belong to divide-and-conquer type of algorithms which correspond to tree type of task graphs.

According to the principles outlined in section 2, a tree type task graph G is first transformed into a layered graph by dividing G into disjoint subgraphs G_is along the critical path. Each G_i, consisting of a few microlayers, is then transformed into a rectangular task graph, based on rules (1) through (3) of Section 2.

4.2. Results and Discussions

We experimented on G and $\bar{G}$ type task graphs with numerous combinations of values of β and α by simulating their executions on our parallel processing system. These results were used to validate the results obtained from the analytical model. In particular, we consider six cases of G where $\beta = 0.5$ and $\beta = 2$, each with $\alpha = 1/2$, $\alpha = 1$, and $\alpha = 2$ respectively, and three cases of $\bar{G}$ which are the extensions from the above G with $\bar{\beta} = 2$. In each such case we consider six subcases where the original task graph is transformed into a i-layer layered task graph, for $i = 1, 2, 3, 4, 6, 12$. This allows us to study the effectiveness of the layered approach, as well as the tradeoff between complexity and accuracy.

Due to space constraint, only a very small subset of the above experimental results that are most indicative of the worst case scenarios are presented here in the form of figures. Response time is chosen as the performance measure and isplotted against the number of processors. In all cases, analytical results are checked against simulation results. The results for graphs $G = < 12, 1, 10, 1, 2 >$, $< 12, 1, 2000, 0.5, 2 >$ are illustrated in figures 8 and 9. Extensive results are reported in [4]. The mean service time of the scheduler is assumed to be 0.3 unit in all the above experiments.

It is observed that given enough number of layers (≥ 6) the model consistently generates reasonably accurate results, with a maximum error of less than 8 % and an average error of less than 1.3 %. The effectiveness and importance of characterizing the variation of parallelism by a number of levels are clearly indicated by results. The average parallelism method [1, 2] that corresponds to the number of layers equal to 1 in the tables, does not produce acceptable results. However, it is not necessary to divide the task graph into too many layers. The reason for this is twofold. Firstly, the more layers a layered graph has, the more computational steps are needed to solve the model. Secondly, it should be realized that not all task graphs have a large variation of parallelism, hence a small number (possibly one) of layers in the layered graph suffices to give accurate results. This is evidenced by the case shown in Fig.8. In general, by observing the variation of parallelism of a task graph, one must judiciously choose the number of layers when analyzing a task graph.

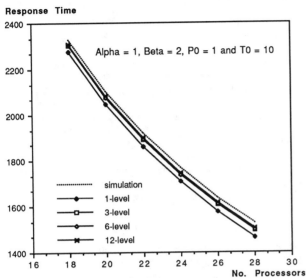

Response Time

**Fig. 8. Response of Tree Type Task Graph
Case I: Alpha = 1 and Beta = 2**

Response Time

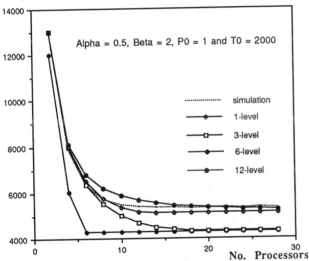

**Fig. 9. Response of Tree Type Task Graph
Case II: Alpha = 0.5 and Beta = 2**

Finally, the complexity of the proposed technique is assessed as follows. Let N be the number of tasks in the original task graph, K the number of service centers in the underlying multiprocessor, and L the number of layers in the layered graph. Then,

a). The transformation of a general task graph into a layered task graph takes $O(N)$ time and space to scan the task graph to get the parallelism profile;

b). The analysis of an individual layer L_i requires an application of the multiserver MVA algorithm to the single class closed queueing network. This takes time $T(L_i) = O(\pi_{av_i}K)$, where

π_{av_i} is the width of an individual layer (i.e. the population of the single customer class). The space required for this application is $S(L_i) = O(K)$.

c). The number of layers L varies, ranging from one to M, the maximum number of layers in the task graph which can be no more than N (i.e. the case of sequential task graph), depending on the nature of the graph.

Thus, the total time and space complexity can be determined, based on a) through c), as

$$T(N) = O(N) + \sum_{i=1}^{L} T(L_i) \le O(N) + \sum_{i=1}^{M} T(L_i) = O(NK) \qquad (8)$$

$$S(N) = O(N) + \sum_{i=1}^{L} S(L_i) \le O(N) + \sum_{i=1}^{M} S(L_i) \le O(NK) \qquad (9)$$

For comparison, the technique presented by Mak and Lundstrom entails a time and space complexities of $O(N^2K + N^3)$ and $O(N^2K)$ respectively [5].

5. Conclusion

In this paper we have proposed a simple but approximate analytical model to evaluate the performance of task graphs executed on a parallel processing system. The model is based on a decomposition approach that transforms an arbitrary task graph into a layered graph where each layer is in turn transformed into a rectangular subgraph. The first transformation process is guided by the notion of variation of parallelism, whereas the second one is based on the average parallelism idea. Each layer is analyzed in isolation through a simple product form queueing network. The results are then combined in a weighted manner, again making use of the variation of parallelism information. The analytical model produces improved results, in terms of complexity, accuracy and applicability. The accuracy depends largely on the type of task graph and the number of layers one chooses for the analysis. It is possible to find an appropriate number of layers that a layered task graph should have in order to make the analysis most cost effective.

6. References

[1] Eager, D. L., Zahorjan, J., and Lazowska, E. D., "Speedup versus Efficiency in Parallel Systems," *IEEE Transactions on Computers*, Vol.C-38, No.3, March 1989, pp. 408-423.

[2] Ghosal, D., "A Unified Approach to Performance Evaluation of Dataflow and Multiprocessing Architectures," *PhD Dissertation* The Center for Advanced Computer Studies, Univ. of Southwestern Louisiana, Summer 1988.

[3] Jiang, H., Bhuyan, L.N, and Ghosal, D., "Approximate Analysis of Multiprocessing Task Graphs", *Proceedings of International Conference on Parallel Processing*, Charles, IL, August 1990, pp. III-228-235.

[4] Jiang, H., "Performance Evaluation of Shared Memory Multiprocessing Architectures," *Ph.D Dissertation* (in preparation), Dept. of Computer Science, Texas A&M University, 1991.

[5] Mak, Victor W. and Lundstrom, Stephen F., "Predicting Performance of Parallel Computations," *IEEE Transactions on Parallel and Distributed Systems*, vol. 1, no. 3, July 1990, pp. 257-270.

[6] Polychronopoulos, Constantine D. and Banerjee, Utpal, "Processor Allocation for Horizontal and Vertical Parallelism and Related Speedup Bounds," *IEEE Transactions on Computers*, vol. C-36, no. 4, April 1987, pp. 410-420.

[7] Reiser, Martin, "Mean-Value Analysis and Convolution Method for Queue-Dependent Servers in Closed Queueing Networks," *Performance Evaluation*, 1981, pp. 7-18.

[8] Trivedi, Kishor S., *Probability and Statistics with Reliability, Queueing, and Computer Science Applications* Prentice-Hall, Inc., Englewood Cliffs, NJ, 1982.

Reconfigurable Mesh Algorithms For The Area And Perimeter Of Image Components*

Jing-Fu Jenq
University of Minnesota *and*

Sartaj Sahni
University of Florida

Abstract

Parallel reconfigurable mesh computer algorithms are developed to obtain the area and perimeter of image components. For an $N{\times}N$ image, our algorithms take $O(logN)$ time on an $N{\times}N$ RMESH.

Keywords and Phrases

reconfigurable mesh computer, parallel algorithms, image processing, area and perimeter of image components.

1 Introduction

Miller, Prasanna Kumar, Resis and Stout [4, 5, 6] have proposed a variant of a mesh connected parallel computer. This variant, called a reconfigurable mesh with buses (RMESH), employs a reconfigurable bus to connect together all processors. This is described in our companion paper [3] which is included in these proceedings. In this paper, we develop RMESH algorithms to compute the area and perimeter of the components of an image. In several applications [1] it is necessary to know the area and perimeter of each of these components (the area of a component is the number of pixels in it and the perimeter is the number of pixels on the component contour).

2 Area And Perimeter Of Connected Components

Initially, each pixel of the binary image matrix $I[0..N-1,0..N-1]$ is labeled by its component number. Specifically, each entry of I is a record with at least the two fields: *value* and *comp*. $I[i,j].value$ is a 0/1 pixel value and $I[i,j].comp$ gives the component to which this pixel belongs. If $I[i,j].value = 0$, then $I[i,j].comp = 0$. If $I[i,j].value = 1$, then $I[i,j].comp > 0$.

The area and perimeter of each component can be determined efficiently on hypercube and mesh connected computers by performing a sort. Consider the case of area determination. The pixels are first sorted by the field *comp*. Next the first and last pixel in each sequence with the same *comp* value is identified. The distance between these can be obtained by performing a data concentration [7] of the ID of the processors containing the last pixel in each sequence. While the same technique can be applied to an RMESH, more efficient algorithms result from a different technique. On an $N{\times}N$ RMESH, the area and perimeter can be determined in $O(logN)$ time while it takes $O(N)$ time to sort N^2 elements.

2.1 Area

2.1.1 CRCW PRAM Algorithm

It is instructive to first consider a CRCW PRAM version of our algorithm. We assume, for simplicity, that N is a power of 2. Our algorithm employs the divide-and-conquer approach [2]. Initially, we assume that each pixel is independent of the others; then we combine together blocks of pixels to obtain larger blocks. Two kinds of block combinations are performed. In one, we combine together two horizontally adja-

* This research was supported in part by the National Science Foundation under grants DCR-84-20935 and MIP 86-17374

cent $2^i{\times}2^i$ blocks. In the other, two vertically adjacent $2^i{\times}2^{i+1}$ blocks are combined. Notice that when two horizontally adjacent blocks of size $2^i{\times}2^i$ each are combined, we get a single block of size $2^i{\times}2^{i+1}$. Further, when two vertically adjacent blocks of size $2^i{\times}2^{i+1}$ are combined we get a block of size $2^{i+1}{\times}2^{i+1}$. Beginning with $2^0{\times}2^0$ blocks, we alternately combine pairs of horizontal adjacent blocks and pairs of vertical adjacent blocks until only one $N{\times}N$ block remains.

With each pixel $[i,j]$ we associate two additional fields: *update* and *area*. *update* is a Boolean field and *area* is an integer field which will eventually be the number of nonzero pixels in the component $I[i,j].comp$. Initially, we have:

$$I[i,j].area = I[i,j].value, \ 0 \le i,j < N$$

When two blocks are combined, the *area* fields of the boundary pixels are updated to correspond to the number of pixels in the new combined block that have the same *comp* value. Following each combination, the following is true:

If $[i,j]$ is a boundary pixel of one of the two blocks just combined, then $I[i,j].area$ is the number of pixels in the new block with *comp* value equal to $I[i,j].comp$ unless $I[i,j].comp = 0$. In this latter case $I[i,j].area = 0$.

Consider the case of horizontal combination. Assume that two $2^i{\times}2^i$ blocks are being combined and that for every boundary pixel $[i,j]$ of each block, we have:

$I[i,j].area =$ number of pixels in the block with *comp* value equal to $I[i,j].comp$ unless $I[i,j].comp = 0$.

If $[i,j]$ is a boundary pixel in block $A(B)$ and $I[i,j].comp > 0$, then its *area* value changes iff there is a pixel on the boundary of block $B(A)$ with the same *comp* value. This follows from the definition of a connected component. If no pixel on the boundary of $B(A)$ has *comp* value $I[i,j].comp$, then no pixel in this block can have this *comp* value. Let $[u,v]$ be a pixel on the boundary of $B(A)$ such that $I[i,j].comp = I[u,v].comp \ne 0$. Then the updated *area* value for pixels $[i,j]$ and $[u,v]$ is $I[i,j].area + I[u,v].area$. Pairs $[i,j]$ and $[u,v]$ of matching pixels are found by dividing the boundary of each block into four lines: 2 horizontal and 2 vertical. Call these $top(x), bottom(x), left(x), right(x), x \in \{A,B\}$. Note that the lines are not disjoint. For example, $top(A)$ and $left(A)$ share one pixel (at the top left corner). All 16 combinations of lines from A and B are used to determine matching pairs. Each combination has the form $((Y(A),Z(B)),$ $Y,Z \in \{top,bottom,left,right\}$. The code of Figures 1 and 2 describes how *area* is updated using a CRCW PRAM that has 2^{i+1} processors. For this to work correctly, it is necessary that the *area* values be read by all PEs before any PE attempts to write an *area* value. The complexity is $O(1)$. The code for the case of a vertical combination is the same. Since this combination has to be done $logN$ times starting with blocks of size $1{\times}1$ and ending with a single block of size $N{\times}N$, the complexity of the procedure to compute area for boundary pixels is $O(logN)$.

Once we have combined blocks as described above then it is the case that the area of any component n is

$I[i,j].update := $ false, $0 \le i,j < N$
for $sideA \in \{top, bottom, left, right\}$ **do**
  for $sideB \in \{top, bottom, left, right\}$ **do**
    $CombineLines(sideA, sideB)$;
Figure 1 Combine blocks A and B

---

procedure *CombineLines* $(sideA, sideB)$;
{update *area* for pixels on boundary lines *sideA* and *sideB*}
{of blocks of A and B }
Let $|sideA|$ and $|sideB|$, respectively, be the number of pixels
on boundary line *sideA* of A and boundary line *sideB* of B;
PE (c,d) examines the c'th pixel, $0 \le c < |sideA|$ of *sideA* of A
and the d'th pixel, $0 \le d < |sideB|$ of *sideB* of B.
Let these pixels, respectively, be $[i,j]$ and $[u,v]$;
if $I[i,j].comp = I[u,v].comp$
then case
  $I[i,j].update$ **and not** $I[u,v].update$:
    $I[u,v].update := true$; $I[u,v].area := I[i,j].area$;
  not $I[i,j].update$ **and** $I[u,v].update$:
    $I[i,j].update := true$; $I[i,j].area := I[u,v].area$;
  not $I[i,j].update$ **and not** $I[u,v].update$:
    $I[i,j].update := true$; $I[u,v].update := true$;
    $I[i,j].area := I[i,j].area + I[u,v].area$;
    $I[u,v].area := I[i,j].area$;
  endcase;
end;
Figure 2 Combining two boundary lines

$$\max\{I[i,j].area \mid I[i,j].comp = n\}$$

To get the condition where $I[i,j].area$ is the area of the component $I[i,j].comp$, $0 \le i,j < N$ we can run the block combination process backwards. The $N \times N$ block is decomposed into 2, each of these is then decomposed into 2, and so on until we have N^2 1×1 blocks.

2.1.2 RMESH Algorithm

The RMESH algorithm works like the CRCW PRAM algorithm. We need to provide only the details for the code of Figure 2 (i.e., procedure CombineLines). Figure 3 gives the RMESH code for the case of horizontal combination. An $N \times N$ RMESH is assumed and PE (i,j) of the RMESH represents pixel $[i,j]$, $0 \le i,j < N$. The code for a vertical combination is similar. The complexity for both is $O(1)$. So, the complete area determination algorithm takes $O(logN)$ time.

2.2 Perimeter

This can be done by preprocessing the image so that $I[i,j] = 1$ iff $[i,j]$ is a boundary pixel. This preprocessing is straightforward and requires each pixel to examine the pixels (if any) on its north, south, east, and west boundaries. Following the preprocessing, we see that the perimeter and area of a component are the same. Hence, the $O(logN)$ algorithm of the preceding section can be used.

3 References

[1] R. O. Duda and P. E. Hart, *Pattern Classification and Scene Analysis*, John Wiley and Sons, 1973.

[2] E. Horowitz and S. Sahni, *Fundamentals of Computer Algorithms*, Computer Science Press, Inc., 1978.

procedure *CombineLines*$(sideA, sideB)$;
{RMESH version }
diagonalize the *update, comp,* and *area* values of *sideB* of block B and broadcast on row buses to all PEs on the same row in block A;
the PEs of block A read their row buses and store the values read in variables *updateB, compB,* and *areaB*, respectively;
diagonalize the *update, comp,* and *area* values of *sideA* of block A and broadcast on column buses to all PEs on the same column in block A;
the PEs of block A read their column buses and store the values read in variables *updateA, compA,* and *areaA*;
{now the PE in position $[a,b]$ of block A has the information from the a'th pixel of *sideA* of A and b'th pixel of *sideB* of B}
Each PE (a,b) of block A does the following:
if *compA = compB* **then**
  case
    updateA **and not** *updateB*: *updateB := true* ;*areaB := areaA*;
    not *updateA* **and** *updateB*: *updateA := true* ;*areaA := areaB*;
    not *updateA* **and not** *updateB* : *updateA := true* ; *updateB := true* ; *areaA := areaA + areaB* ; *areaB := areaA* ;
  endcase;
{ broadcast back to *sideB*}
set up row buses in the AB combined block;
every PE (a,b) of block A for which *updateB* (a,b) is true disconnects its W switch and broadcasts *areaB*;
the diagonal PEs of block B read their buses and if a value is read, this is broadcast to the appropriate PE of *sideB* using the reverse of a diagonalize, this PE in turn updates its *areaB* value and sets its *update* value to true;
{ broadcast to *sideA* }
this is similar to that for *sideB*;
Figure 3 RMESH version of *CombineLines*

---

[3] J. Jenq and S. Sahni, "Reconfigurable mesh algorithms for the Hough transform", *Proc. International Conf. on Parl. Processing*, 1991.

[4] R. Miller, V. K. Prasanna Kumar, D. Resis and Q. Stout, "Data movement operations and applications on reconfigurable VLSI arrays", Proceedings of the 1988 International Conference on Parallel Processing, The Pennsylvania State University Press, pp 205-208.

[5] R. Miller, V. K. Prasanna Kumar, D. Resis and Q. Stout, "Meshes with reconfigurable buses", Proceedings 5th MIT Conference On Advanced Research IN VLSI, 1988, pp 163-178.

[6] R. Miller, V. K. Prasanna Kumar, D. Resis and Q. Stout, "Image computations on reconfigurable VLSI arrays", Proceedings IEEE Conference On Computer Vision And Pattern Recognition, 1988, pp 925-930.

[7] D. Nassimi and S. Sahni, "Data broadcasting in SIMD computers", IEEE Transactions on Computers, vol C-30, no. 2, Feb. 1981, pp 101-107.

A Practical Parallel Convex Hull Algorithm

Chandrasekhar Narayanaswami*
Electrical, Computer, and Systems Engineering Dept.
Rensselaer Polytechnic Institute
Troy, New York 12180, USA

Abstract

A practical parallel algorithm for the 2-D convex hull problem is described. Its efficiency due to the fast rejection of interior points by a linear preprocessing technique. Its minimal reliance on parallel sorting makes it practical. A *uniform grid* is cast over the scene and points are assigned to the grid cells. *Grid cell* dominances are then computed in parallel in an optimal fashion and points in the interior cells are eliminated. This technique may be used as a preprocessing step to improve the speed of the existing optimal parallel convex hull algorithms. The convex hull of the remaining points is then computed using a parallel version of the Graham scan.

For points chosen randomly from uniform distributions within a unit square, a unit circle, and an annulus, more than 90% of the points are eliminated. The convex hull of 200,000 points is computed in 12.2, 8.7, and 13.88 seconds, respectively, for the above mentioned point distributions, using all 15 processors of a Sequent Balance 21000 parallel computer. The algorithm shows close to linear growth with the input size for the above distributions and the preprocessing phase shows close to linear speedup.

1 Introduction

Given a set of points in the plane, the convex hull problem is to determine the smallest convex polygon containing all the points. Optimal parallel algorithms, using $O(n)$ processors that achieve the time bound of $O(\log n)$, for a CREW PRAM model of computation have been independently developed by Aggarwal et al. [1], and Goodrich [3]. Miller and Stout [4] provide optimal algorithms for various parallel architectures.

The purpose of this paper is to present a simple algorithm. Extensive experimental results are provided in [5].

2 The Algorithm

We are given a set of n points, $S = \{p_1, p_2, \cdots, p_n\}$ in the plane. We are also given a commercially available tightly

---

*Now at: IBM Austin, 11400 Burnet Road, Austin, Texas 78758

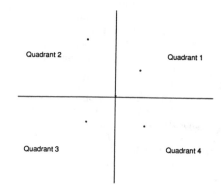

Figure 1: The Four Quadrants for a Point. The point at the origin is inside the convex hull.

coupled parallel machine containing m processors and adequate shared memory. Note that $m << n$, and so our model is equivalent to that of a CREW PRAM with a small number of processors.

The main steps in our parallel algorithm are the following:

1. Fast rejection of interior points.

2. Computation of the convex hull of the remaining points.

2.1 Rejection of Points Inside the Convex Hull

We use the simple observation that if we orient a set of axes about any point p_i and make p_i the origin of this coordinate system, and if each of the four quadrants in this coordinate system contains points from the input set, then p_i is inside the convex hull (see Figure 1). A point p_i *dominates* point p_j in the third quadrant, denoted as p_i *dom*^{3} p_j, if p_j lies in the third quadrant in a coordinate system of which p_i is the origin. Dominances in other quadrants are defined in a similar manner. Thus, points which dominate points in all four quadrants are inside the convex hull. A point which is on the convex hull has the property that it does not *dominate* any other point in at least one quadrant, i.e., there is at least one empty quadrant.

Counting the number of points dominated by each of the points (2-D dominance problem), in each of the four quadrants to determine the interior points is naive, expensive, and unnecessary.

Instead of solving the 2-D dominance problem we cast a uniform grid [2] of $G \times G$ square cells on the scene and eliminate clusters of points at once by doing the dominance computations only for the grid cells. Figure 2 shows the four quadrants for a cell $C_{i,j}$, (i increases from bottom to top and j increases from left to right in the grid), and the

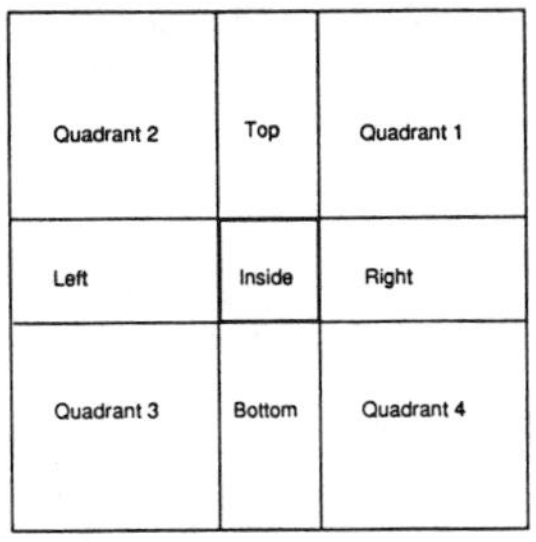

Figure 2: The Different Regions for a Cell. Figure shows the regions to the right, left, top, and bottom of a cell. The four quadrants for a cell are also shown.

regions which are to the top, bottom, right, and left of it. The dominance relations for the cells are analogous to those for the points, e.g., if cells $C_{i,j}$ and $C_{k,l}$ both have some points in them and $i > k$ and $j > l$ then $C_{i,j}$ dominates $C_{k,l}$ in the third quadrant. The dominance value of a cell $C_{i,j}$ for a particular quadrant is 1 if $C_{i,j}$ dominates a cell in that quadrant. We compute the dominance values for the cells and determine the *interior cells*. Interior cells are those cells that dominate other cells in all the four quadrants.

The main steps in computing the dominance values for the cells are as follows:

1. Determine the cells which contain points; $in(C_{i,j})$ is 1 if there is at least one point in it. To ensure that each point is stored in exactly one cell, only the bottom and left boundaries of a cell belong to the cell. A predetermined partitioning scheme is used to distribute the points among the m processors.

2. Determine whether there are points which lie in the cells to the left, right, top, and bottom of each cell. For example, $right(C_{i,j})$ is 1 if there is at least one point in all the cells in that row which are to the right of $C_{i,j}$. The recurrence equations are as follows ($\vee$ is the Boolean OR operator).

$$R(C_{i,j}) = right(C_{i,j}) = right(C_{i,j+1}) \vee in(C_{i,j+1})$$
$$L(C_{i,j}) = left(C_{i,j}) = left(C_{i,j-1}) \vee in(C_{i,j-1})$$
$$T(C_{i,j}) = top(C_{i,j}) = top(C_{i+1,j}) \vee in(C_{i+1,j})$$
$$B(C_{i,j}) = bot(C_{i,j}) = bot(C_{i-1,j}) \vee in(C_{i-1,j})$$

This evaluation is done in parallel by distributing the rows and columns of the grid among the processors. Further, each of these four relations can be determined in parallel.

3. Compute the dominance value for each interior cell $C_{i,j}$, using the following equations.

$$dom^1(C_{i,j}) = dom^1(C_{i+1,j+1}) \vee R(C_{i+1,j+1})$$
$$\vee \ T(C_{i+1,j+1}) \vee I(C_{i+1,j+1})$$
$$dom^2(C_{i,j}) = dom^2(C_{i+1,j-1}) \vee L(C_{i+1,j-1})$$
$$\vee \ T(C_{i+1,j-1}) \vee I(C_{i+1,j-1})$$
$$dom^3(C_{i,j}) = dom^3(C_{i-1,j-1}) \vee L(C_{i-1,j-1})$$
$$\vee \ B(C_{i-1,j-1}) \vee I(C_{i-1,j-1})$$
$$dom^4(C_{i,j}) = dom^4(C_{i-1,j+1}) \vee R(C_{i-1,j+1})$$
$$\vee \ B(C_{i-1,j+1}) \vee I(C_{i-1,j+1})$$

The only dependence of the dominance value of a cell with that of the dominance value of other cells occurs in a diagonal fashion. Hence this step is done in parallel by scanning the grid of cells in a diagonal fashion by interlacing the diagonals among the processors. Further, each of these four sets of dominance values can be computed in parallel.

4. Eliminate points in all the interior cells. Gather all the points in the cells that are not interior cells. This step is parallelized by using parallel array contraction.

2.2 Parallel CH Determination of Remaining Points

The points $\{p_i\}$ which remain after the point rejection step are sorted angularly around a pivot I_p, which is inside the convex hull, to form a star polygon. The angularly sorted list of points (stored in a shared array) is distributed equally among the m processors so that each processor gets a chain of the star polygon formed by the above step. Each processor then sequentially converts its simple chain into a convex chain. New chains are created from points remaining from the previous iteration and assigned to the processors. The algorithm terminates when the number of remaining points in two consecutive iterations is the same.

3 Conclusion

The uniform grid was used effectively (speedups more than 9 with 15 processors [5]) to eliminate interior points. Since more than 90% of the points are eliminated often, this phase linearizes the performance of the algorithm.

The simplicity of this technique makes it well suited for hardware implementation.

References

[1] A. Aggarwal, B. Chazelle, L. Guibas, C. ÓDúnlaing, and C. Yap. Parallel Computational Geometry. *Algorithmica*, 3:293–327, 1988.

[2] V. Akman, W. R. Franklin, M. Kankanhalli, and C. Narayanaswami. Geometric Computing and the Uniform Grid Data Structure. *Computer Aided Design*, 21(7):410–420, September 1989.

[3] M. T. Goodrich. *Efficient Parallel Techniques for Computational Geometry*. PhD thesis, Purdue University, West Lafayette, Indiana, 1987.

[4] R. Miller and Q. F. Stout. Efficient Parallel Convex Hull Algorithms. *IEEE Transactions on Computers*, 37(12):1605–1618, December 1988.

[5] C. Narayanaswami. *Parallel Processing for Geometric Applications*. PhD thesis, Rensselaer Polytechnic Institute, Troy, New York 12180, December 1990.

Ranking, Unranking, and Parallel Enumerating of Topological Orders

B.Y. Wu and C.Y. Tang

Institute of Computer Science, National Tsing Hua University

Hsinchu, Taiwan 30043, R.O.C.

E-mail: CYTANG@CS.nthu, edu.tw.bitnet

Abstract : Given a precedence forest, we develop serial algorithms for ranking and unranking, and a parallel algorithm for enumerating. The time complexities of ranking and unranking are both $O(n\log n)$, where n is the number of nodes in the forest. The time complexity of the parallel enumerating algorithm is $O(\lceil M/N \rceil \times n + n\log n)$, where M is the number of the total topological orders, and N is the number of processors. The parallel algorithm is cost optimal when $N < M/\log n$ and without any communication between processors.

Keywords : ranking, unranking, parallel algorithms, topological orders.

1. *Introduction*

For a precedence graph G (acyclic digraph), a topological order is a sequence $\alpha_1, \alpha_2, .., \alpha_n$ of all the nodes in G, such that if (α_i, α_j) is an edge in G then $i < j$. There are several topics about topological orders have been widely discussed in the literatures, such as generating a topological order [4], generating all topological orders [2,3,5], and parallel generating a topological order [1]. But there is no algorithm proposed for ranking, unranking, and parallel enumerating topological orders. For solving combinatorial problems, sometimes we may want to enumerate and evaluate all the topological orders. Furthermore, in parallel machine. efficient parallel enumerating should be considered.

In this paper, we consider the ranking, unranking, and parallel enumerating of all the topological orders on a precedence forest. After relabelling the nodes, we develop lexicographic ranking, unranking, and next algorithms. A parallel enumerating algorithm which is constructed by combining the unranking and the next algorithms is also developed. Furthermore, our unranking algorithm can be used to generate a random topological order with uniform distribution.

By using a modified AVL-tree, our unranking and ranking algorithms are with time complexity $O(n\log n)$. Our next algorithm takes only $O(n)$ time. The parallel enumerating algorithm is designed in such a way that each processor generates topological orders which ranks are in a consecutive interval. The parallel enumerating algorithm has time complexity $O(\lceil M/N \rceil \times n + n\log n)$ and without any communication between processors, where M is the number of topological orders, and N is the number of processors.

2. *Preliminaries*

In this paper, $|A|$ denotes the number of elements in set A, and the number of nodes if A is a graph. A precedence tree is a rooted directed tree. We use $T(r)$ to denote the precedence tree(or subtree) with root r. Let $R = (r_1, r_2, ..., r_k)$, in which $r_1 < r_2 < ... < r_k$, be all the roots of trees, the precedence forest, denoted by $PF(R)$, is $\bigcup_{r_i \in R} T(r_i)$. The reduced forest $PF(R_i)$ is obtained from

$PF(R)$ by deleting r_i, that is, if $s_1 < s_2 < ... < s_j$ are the sons of r_i then $R_i = (r_1, ..., r_{i-1}, s_1, s_2, ..., s_j, r_{i+1}, ..., r_k)$.

For $PF(R)$, we first label the nodes as follow : Let S be a depth first search sequence of the nodes, the node is labelled as k if it is the k-th node in S.

We use $TOF(R)$ to denote the set of all the topological orders, and $M(R)$ to denote $|TOF(R)|$.

Definition (lexicographic order, rank, unrank) : Let $A = a_1 a_2 .... a_n$ and $B = b_1 b_2 .... b_n$. A precedes B lexicographically, denoted by $A < B$, if and only if $\exists 1 \leq j \leq n$, $a_i = b_i$ for $\forall i < j$, and $a_j < b_j$. Let $A \in TOF(R)$. The rank of A is $rank(A,R) = |\{B | B < A, B \in TOF(R)\}|$. Let $\alpha < M(R)$, the unrank of α in $TOF(R)$ is $unrank(\alpha, R) = A$ such that $rank(A,R) = \alpha$.

The next lemma comes from [4].

Lemma 1 : $M(R) = n! / (\prod_{q \in PF(R)} no(q))$, where $no(q) = |T(q)|$.

For a set of data (x_1, y_1), $(x_2, y_2), ..., (x_k, y_k)$, where $x_1 < x_2 < ... < x_k$, we can use a modified AVL-tree to store them. We can efficiently do the following operations in this modified AVL-tree :

(1) Prefix_sum(x) : Given x, find $x_i = x$ for some i, and return $y_1 + y_2 + ... + y_i$.

(2) Search(α) : Given α, return x_j such $\sum_{i < j} y_i < \alpha \leq \sum_{i < j+1} y_i$.

(3) Insert a data (x,y).
(4) Delete a data (x,y).

Each of the above four operations can be done in $O(\log n)$ time, and we will not go through the details in this paper.

3. *Ranking*

Let $P = p_1 p_2 ... p_n \in TOF(R)$, where $R = (r_1, r_2, ..., r_k)$. We can obtain that : $p_1 = r_i$ for some i, and $rank(P,R) =$

$$\sum_{h<i} |\{r_h q_1 ... q_{n-1} | q_1 ... q_{n-1} \in TOF(R_h)\}| + rank(p_2...p_n, R_i)$$

$$= \sum_{h<i} M(R_h) + rank(p_2...p_n, R_i)$$

$$= \sum_{h<i} no(r_h) \times M(R)/|PF(R)| + rank(p_2...p_n, R_i).$$

Based upon the above formula, we can find $rank(P,R)$ iteratively. At the j-th iteration, we find $r_i = p_j$ in the roots of the forest, compute $\sum_{h<i} no(r_h) \times M(R)/n$, and then

update the forest. By keeping the roots in a modified AVL–tree (with $(x,y) = (r_i, no(r_i))$), we can see that all the computations can be speeded by a prefix_sum operation. Moreover, to update the forest is just to delete r_i and insert its sons into the AVL–tree.

Algorithm Ranking(P,PF(R));
insert all the roots into AVL tree;
compute M=M(R); N=|PF(R)|; rank:=0;
for j:=1 to n do
    $a:=$prefix_sum$(p_j)-no(p_j)$;rank:=rank+$M \times a/N$;
    $M:=M \times no(p_j)/N$; $N:=N-1$; delete(p_j,AVL);
    for each s (is a son of p_j) do
        insert(s,AVL);

Theorem 1 : The time complexity of Ranking is O(nlogn).

4. *Unranking*

Let unrank$(\alpha,R)=P=p_1 p_2 \ldots p_n$. We can also find $p_1,p_2,\ldots$ iteratively : Let $R=(r_1,r_2,\ldots,r_k)$. The TOF(R) can be partitioned into T_i $1 \leq i \leq k$, and $T_i=\{r_i Q | Q \in TOF(R_i)\}$. We can observe that to find i such $P \in T_i$ is equivalent to find i such $\sum_{h<i} |T_h| < \alpha \leq \sum_{h<i+1} |T_h|$. Once the $p_1=r_i$ is found, the problem is reduced to unrank$(\alpha - \sum_{h<i} |T_h|, R_i)$.

Similar as in Ranking, we can keep all the roots in an AVL–tree with $(x_i,y_i)=(r_i,no(r_i))$, and since $|T_h|=M(R_h)= no(r_h) \times M(R)/n$, the main work in each iteration is to search r_i such that

$\sum_{h<i} no(r_h) \times M(R)/n < \alpha \leq \sum_{h<i+1} no(r_h) \times M(R)/n$. And this

work can be done by a search operation in AVL–tree.

Algorithm Unranking(α,PF(R));
insert all the roots into R_AVL;N:=|PF(R)|;
compute $M=n!/(\Pi\ no(r))$ for all $r \in PF(R)$;
for i:=1 to n do
    $\beta:=\lceil \alpha \times N/M \rceil$; $p_i:=$search(β);
    $t:=$prefix_sum(p_i); $\alpha:=\alpha-(t-no(p_i)) \times M/N$;
    $M:=M \times no(p_i)/N$; $N:=N-1$; delete(p_i,AVL);
    for each son s of p_i do
        insert(s,AVL);

Theorem 2 : Unranking can be done in O(nlogn).

5. *Parallel Enumerating*

In this section, we discuss how to enumerate all topological orders of PF(R) in parallel. By Unranking algorithm, we can independently generate all the topological orders. If there are only N processors, N<M(R), we can partition the topological orders into N parts and each processor generates one part independently by Unranking algorithm. This approach takes $O(\lceil M(R)/N \rceil \times nlogn)$ time and doesn't need communication.

If the parts are divided in such a way that the ranks in

each part are consecutive and we can generate the next topological order, then, we can do it more efficiently. In fact, we develop an O(n) serial algorithm to find the next.

Let $r \in PF(R)$, $fst(r)=unrank(0,(r))$ is the first topological order of the T(r), and it can be easily showed that $fst(r)=r,r+1,r+2,\ldots,r+no(r)-1$. We also use last(r) is the last topological order of T(r). Let α and β be two nodes. $\alpha \rightarrow \beta$ denotes α is an ancestor of β. Let $P \in TOF(R)$, Next(P)=unrank(rank(P,R)+1,R). In Lemma 2, '/' means concatenation.

Lemma 2 : Let $P \in TOF(R)$ and rank(P,R)$\neq$M(R)-1, then
(1) P can be written as :
    $P=P_0/x/last(a_1)/last(a_2)/\ldots/last(a_k)$,   where
    $x<a_1>a_2>\ldots>a_k$, and not$(x \rightarrow a_1)$ and not$(a_i \rightarrow a_j)$.
(2) we can find the maximum j, such $x<a_j$ and not$(x \rightarrow a_j)$.
(3) if x has m sons, then $a_{j+1},a_{j+2},\ldots,a_{j+m}$ are the sons.
(4) Next(P)=$P_0/a_j/fst(a_k)/fst(a_{k-1})/\ldots/fst(a_{j+m+1})/fst(x)/q_1 q_2 \ldots q_{\delta-1}/fst(a_{j-1})/\ldots/fst(a_1)$, where $a_j q_1 q_2 \ldots q_{\delta-1}=fst(a_j)$ and $\delta=no(a_j)$.

We omit the proof of the above lemma and algorithm Next(P) which can be easily developed from the lemma and takes only O(n) time.

Now, we can construct a parallel algorithm to enumerate TOF(R). In this algorithm, the all topological orders are divided into N parts, and each part which contains consecutive topological orders is enumerated by one processor. Each processor uses Unranking to find the first one in its part and then uses Next to find all the others in its part. The algorithm is listed below :

Algorithm Par_enumerate;
for each processor Y_i do in parallel
    compute M(R); $\alpha:=\lceil M(R)/N \rceil \times (i-1)+1$;
    P:=Unrank(α,PF(R)); output(P);
    for j:=1 to $\lceil M(R)/N \rceil-1$ do
        P:=Next(P); output(P);

Theorem 3 : The algorithm Par_Enumerate generates all topological orders of a given precedence forest in $O(\lceil M/N \rceil \times n+nlogn)$ time and without communication between processors. The algorithm is cost optimal when N<M/logn.

Reference

1. M.C. ER, "A Parallel Computation Approach to Topological Sorting", *The Computer Journal*, Vol.26, No.4, 1983, pp.293–295.
2. D. Kalvin, Y.L. Varol, "On the Generation of All Topological Sorting", *Journal of Algorithm*, 1983(4), pp.150–162.
3. D.E. Knuth, J.L. Szwarcfiter, "A Structured Progra to Generate All Topological Sorting Arrangements", *IPL*, Vol.2, pp.153–157.
4. D.E. Knuth, *The Art of Computer Programming* Vol.3, Addison–Wesley, Reading, Mass. 1973.
5. Y.L. Varol, D. Rotem, "An Algorithm to Generate A Topological Sorting Arrangements", *The Computer Journal*, Vol.24, No.1, 1981, pp.83–84.

Fast and efficient parallel algorithms for single source lexicographic depth-first search, breadth-first search and topological-first search

Pilar de la Torre
Department of Computer Science
University of New Hampshire
Durham, New Hampshire 03824

Clyde P. Kruskal
Department of Computer Science
University of Maryland
College Park, Maryland 20742

(EXTENDED ABSTRACT)

The problems

Optimal sequential solutions are known for the following fundamental graph problems: Given a directed graph G and a distinguished source vertex s, compute a depth-first search numbering, a breadth-first search numbering, and a topological sort numbering of G starting at s. Their solutions play a central role in the development of numerous algorithms for graphs. Despite considerable research efforts the parallel complexity of these problems remains open.

This paper addresses *lexicographic* versions of these problems, each of which has the following input: an n-vertex directed graph G, an ordering on the edges incident to each vertex, and a distinguished vertex s. The ordering on the edges incident to a particular vertex is the relative order in which the incident edges are to be examined by the standard sequential depth-first, breadth-first, and topological-first search algorithms. The *lexicographic depth-first search (lex-dfs)* problem is to compute the numbering of G produced by the sequential depth-first search algorithm starting at s. The *lexicographic breadth-first search (lex-bfs)* problem is to compute the numbering of G produced by the sequential breadth-first search algorithm starting at s. Knuth's topological sorting algorithm [10] can be viewed as the numbering produced by the lexicographic version of a search algorithm that we called *topological-first search* [4]. The *lexicographic topological-first search (lex-tfs)* problem is to compute the numbering of G produced by Knuth's topological sorting algorithm starting at s [10].

The solution to each of these problems is the output of a greedy sequential algorithm. This is a feature known to be shared by many **P**-complete problems [3], which are currently believed unlikely to belong to the class **NC**. (A problem is in **NC** if there is an algorithm that solves instances of size n in $\log^{O(1)} n$ time using $n^{O(1)}$ processors on a PRAM.)

Previous work

Although the lexicographic depth-first search problem for general graphs is already **P**-complete for planar graphs [2], the lexicographic depth-first search problem for DAGs was placed

in **NC** independently in [4] and [8]. The algorithm in [4] computes the lexicographic depth-first search numberings from all sources in $O(\log^2 n)$ time using $O(n^3 / \log^2 n)$ on the EREW PRAM. Within the same resource bounds and using a different approach, the algorithm in [8] solves the single source problem for the CREW PRAM.

The question as to whether the lex-breadth-first search problem is in **NC** was posed as an open problem in [9] and settled independently in [4] and [8]. The algorithm in [4] computes the lex-breadth-first search numberings from all sources in $O(\log^2 n)$ time using $O(n^3 / \log n)$ processors on the EREW PRAM. On the other hand, [8] gives an algorithm that computes the lex-breadth-first search numbering from a single source in $O(\log^2 n)$ time using $O(n^3 / \log n)$ processors on the CREW PRAM. The lex-topological-first search problem has been recently placed in **NC** by an algorithm that computes the numberings from all sources in $O(\log^2 n)$ time using $O(n^3 / \log n)$ processors on the EREW PRAM [4].

In contrast, the (nonlex) depth-first search for DAGs, (nonlex) breadth-first search for directed graphs, and (nonlex) topological sort can be solved using $M(n)$ processors, where $M(n)$ is the number of EREW PRAM processors required to multiply two $n \times n$ integer matrices in $O(\log n)$ time, currently $M(n) = O(n^{2.376})$. The best previously known algorithm for (nonlex) dfs problem for DAGs and reducible flow graphs runs in $O(\log^3 n)$ time on the EREW PRAM [11]. The best algorithms for the single source (nonlex) bfs problem [7] and for computing a (nonlex) topological sort numbering for a DAG take $O(\log^2 n)$ time [6].

Main results

In the extended version of this article [5], we present parallel algorithms for the following problems: lexicographic depth-first search for reducible flow graphs[1] (RFGs), lexicographic breadth-first search for directed graphs, and lexicographic topological-first search for directed acyclic graphs (DAGs). Our algorithms attain the same complexity as the best known parallel solution for the simpler problem of *directed reachability*.

The research of this author was supported in part by the National Science Foundation under Grants DCR-8600378 and CCR-9010445.

[1]The class of reducible flow graphs, which includes the class of directed acyclic graphs, arises naturally in connection with code optimization and data flow analysis [1].

Each of these algorithms runs in $O(\log^2 n)$ time using $M(n)$ processors on the EREW PRAM where, as indicated above, $M(n) = O(n^{2.376})$ currently. This is a significant improvement upon the best previously known parallel solutions, which required $\Omega(n^3/\log^2 n)$ processors to run in $\Omega(\log^2 n)$ time on a EREW PRAM.

The central building block of our solutions is an algorithm that, in $O(\log^2 n)$ time and using $n^2/\log n$ EREW PRAM processors, reduces the lexicographic depth-first search problem for directed acyclic graphs to the transitive closure problem. Our algorithm is inspired by Ramachandran's algorithm [11] for building a (non-lex) depth-first search of a DAG from source s which runs in $O(\log^3 n)$ time using $M(n)$ processors on the EREW PRAM. As indicated in [5], attempting to follow Ramachandran's approach to compute the lex-depth-first search tree would lead to failure. In contrast, the divide-and-conquer strategy of our algorithm is based on computing a special path which we call the *lex-splitting path*: This path, which is contained in lex-depth-first search tree of the input graph, is used by our algorithm to efficiently decompose the problem into smaller problems that are then solved recursively. Furthermore, observing that it is not necessary to recompute the reachability matrix from scratch at each recursive call enables us to reduce the $O(\log^3 n)$ running time of the (nonlex) depth-first search solution for DAGs [11] to $O(\log^2 n)$ without increasing the number of processors.

Our reduction from lex-depth-first search for DAGs to transitive closure enable us to perform the following transformations in $O(\log^2 n)$ time using $n^2/\log n$ processors of an EREW PRAM: (1) A reduction from lex-depth-first search for reducible flow graphs to transitive closure and single-source shortest paths; (2) A reduction from the lex-breadth-first search forest problem for directed graphs to transitive closure and single-source shortest paths; (3) A reduction from the problem of computing Knuth's topological sorting to transitive closure and single-source longest paths.

Each of these reductions yields the desired $O(\log^2 n)$ time and $M(n)$ processors algorithm for the corresponding problem. Table 1 summarizes our results.

Remarks

The complexity of our algorithms can be expressed in terms of the complexity of three central problems for directed graphs: transitive closure, the single-source shortest paths problem, and the single-source longest paths problem for DAGs. Improved

algorithms for these problems would automatically improve our algorithms for the lex-first problems.

Another interesting aspect of our algorithms is that they close a previously existing "gap" between the best known upper bounds for the parallel complexities of the lexicographic and nonlexicographic versions of each problem. Whether the lexicographic versions are significantly "harder" than their non-lexicographic counterparts remains an attractive open problem of parallel computation.

References

[1] A. A. Aho and J. D. Ullman. *Principles of Compiler Design*. Addison–Wesley, 1977.

[2] R. Anderson and E. W. Mayr. Parallelism and the maximal path problem. *IPL*, 24:121–126, 1987.

[3] R. J. Anderson. *The complexity of parallel algorithms*. PhD thesis, Stanford University, 1984.

[4] P. de la Torre and C. P. Kruskal. Fast parallel algorithms for all sources lexicographic search and path-finding problems. *Manuscript (submitted)*, 1988. (Technical Report UMIACS-TR-89-69, University of Maryland.).

[5] P. de la Torre and C. P. Kruskal. Fast and efficient parallel algorithms for single source lexicographic depth-first search, breadth-first search, and topological-first search. *Manuscript (submitted)*, 1989. (Technical Report TR-91-07, University of New Hampshire.).

[6] E. Dekel, D. Nassimi, and S. Shani. Parallel matrix and graph algorithms. *SIAM J. Comput.*, 10(4):657–675, 1981.

[7] H. Gazit and G. L. Miller. An improved parallel algorithm that computes the BFS numbering of directed graphs. *IPL*, 28:61–65, 1988.

[8] R. Greenlaw. Fast parallel algorithms for greedy breadth and depth first search. *Manuscript (submitted)*, 1988.

[9] M. Hoover and W. L. Ruzzo. A compendium of problems complete for P. *Draft*, November 1985.

[10] D. E. Knuth. *The art of computer programming, 1: Fundamental algorithms*. Addison-Wesley, 1968.

[11] V. Ramachandran. Fast and processor-efficient parallel algorithms for reducible flow graphs. Technical Report ACT-103, Univ. of Illinois at Urbana-Champaign, 1988.

	previous nonlex		previous lex		lex from this paper	
	processors	time	processors	time	processors	time
dfs DAGs	$M(n)$	$\log^3 n$	$n^3/\log^2 n$	$\log^2 n$	$M(n)$	$\log^2 n$
dfs RFGs	$M(n)$	$\log^3 n$	—	—	$M(n)$	$\log^2 n$
bfs	$M(n)$	$\log^2 n$	$n^3/\log n$	$\log^2 n$	$M(n)$	$\log^2 n$
tfs	$n^3/\log^2 n$	$\log^2 n$	$n^3/\log n$	$\log^2 n$	$M(n)$	$\log^2 n$

Table 1

A PARALLEL PERCEPTRON LEARNING ALGORITHM[*]

Tzung-Pei Hong

Institute of Computer Science and Information Engineering
National Chiao-Tung University
Hsin-Chu, 30050, Taiwan, R.O.C.
E-Mail: 7617523@TWNCTU01.BITNET

Shian-Shyong Tseng

Institute of Computer and Information Science
National Chiao-Tung University
Hsin-Chu, 30050, Taiwan, R.O.C.
E-Mail: SSTSENG@TWNCTU01.BITNET

ABSTRACT

In this paper, we will propose a parallel perceptron learning algorithm based upon the single-channel broadcast communication model. Large speed-up can be expected in our algorithm, since training instances are processed in parallel instead of one by one.

1. Introduction

Connectionist (neural network) models[3][7] are drawing interesting as useful tools for A.I. Among them, "perceptron model"[6][7] is the simplest and quite suitable for implementing classification systems, and a connectionist expert system can sometimes be formed by applying several perceptrons[3]. In the past, the weight vector of a perceptron was learned from the called "perceptron learning algorithm" [2][6][7] by considering the training instances one by one, so its efficiency is limited. Due to the dramatic increase in computing power and the concomitant decrease in computing cost over the last decade, learning from training instances by applying parallel processing techniques becomes a feasible way for conquering the problem of low speed in learning[5].

2. Perceptron

Basically, a perceptron is a network consisting of a layer of input cells and a output cell, but not intermediate cells as shown in Figure 1.

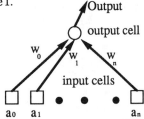

Figure 1. Perceptron

In Figure 1, a single linear discriminant function is applied:

$$g(A) = A \cdot W = a_0 w_0 + w_1 a_1 + \ldots + w_n a_n = \sum_{j=0}^{n} w_j \cdot a_j$$

where $A = (a_0, a_1, \ldots, a_n)$ is an input vector, and

$W = (w_0, w_1, \ldots, w_n)$ is a weight vector.

If $g(A) > 0$, the output cell responses +1; the input pattern A belongs to this class. If $g(A) < 0$, the output cell responses -1; the input pattern A does not belong to this class.

If there exists a single linear discriminant function which can model all input patterns, then the set of input patterns are called linearly separable; otherwise they are called nonlinearly separable[2][6].

---

[*] This research was supported by the National Science Council of the Republic of China under NSC79-0408-E009-17.

The learning of the perceptron model is to get the weight vector W from a set of n training instances automatically; every instance is represented by an input vector A (a_0, a_1, ..., a_m) and its correct classification or label C (+ or -). A famous weight-learning algorithm, called "the perceptron learning algorithm"[2][3][7], is described below:

Perceptron learning algorithm:
INPUT: A set of n linearly separable training instances.
OUTPUT: A weight vector consistent with all the training set.
STEP 1: Initially set W to be the zero vector.
STEP 2: Let W to be the current perceptron weight vector. Randomly pick out a training instance I with the attribute vector A and the corresponding classification C. If W can correctly classify I, do nothing; else, form a new weight vector $W' = W + A * C$.
STEP 3: Set $W = W'$ and go to Step 2 until W can classify all the training instances.

Because the training instances are processed one by one, the performance of the algorithm is largely limited. Below, to cope with this problem, we will propose an algorithm based upon a single-channel broadcast communication model to process the training instances in parallel.

3. Single-channel broadcast communication model

A single-channel broadcast communication model[1][4] shown in Figure 2 basically has a number of processing elements sharing one communication channel through which each processing element can communicate with each other.

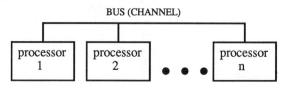

Figure 2. A single-channel broadcast communication model

All of the processing elements may try to broadcast by trying to get hold of the common bus. Of course, this means that there will be a conflict owing to the simultaneous request of more than one processing element. There are some kinds of mechanisms[1] to resolve this conflict by allowing only one processing element to broadcast. When this processing element succeeds in grabbing the common bus, it broadcasts and all other processing elements listen.

4. Parallel perceptron learning algorithm

During the process of correcting the weight vector in the original perceptron learning algorithm, the training examples will be randomly picked out and tried until a chosen one is not consistent with the current weight vector. When it is inconsistent, we change the weight vector and repeat the above procedure. Therefore, some time for the correction of weight vector is wasted, which may be saved by parallel processing approach, if the picked instance is consistent with the current weight vector.

Parallel perceptron learning algorithm:

INPUT: A set of n linearly separable training instances.

OUTPUT: A weight vector consistent with all the training set.

STEP 1: Initially set W to the zero vector and put every instance in an individual processing element.

STEP 2: Let W be the current weight vector to be broadcast. For every processing element P_i, check whether W can correctly classify the training instance in P_i. If not, the processing element P_i calculates a new weight vector $W_i' = W + A_i * C_i$ and competes for the right of broadcasting.

STEP 3: The processing element which gets hold of the right of broadcasting puts its W' in the common bus and set W equal to W'.

STEP 4: Go to Step 2 until a null message is found.

5. Experimental analysis

In this section, an experiment implemented by Turbo–Pascal at the IBM PC/AT is conducted to simulate the parallel perceptron learning algorithm and get the speed–up. First, a weight vector and a set of training instances (not including labels) were randomly generated using a uniform pseudo random number generator. According to the weight vector, all the training instances were given the correct classification (label). Therefore, each set of training instances generated by this way is linearly separable.

By experimenting 500 times respectively for various n with 10 attributes, the average speed–up is shown in Figure 3. From Figure 3 we find that the average speed–up S_{ave} increases when n increases, and further that S_{ave} can be approximated by the equation $S = 1.48 * N^{0.91}/\ln N$ with only a small and tolerable error.

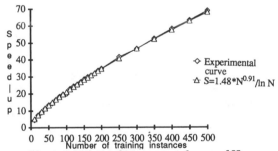

Figure 3. Relation between speed–up and N

6. Nonlinearly separable training instances

In the above discussion, we assume the training instances are linearly separable. For managing nonlinearly separable training instances, an improvement of the perceptron learning algorithm, called "pocket learning algorithm"[2][3] has been proposed. In the pocket learning algorithm, instead of finding a weight vector of classifying all the training instances correctly, a weight vector is found if it can correctly classify the maximum number of consecutive training instances. Therefore, it must keep so far the best weight vector in the pocket besides the current weight vector. When running time is up and there is no weight vector which can classify all the training instances, the best weight vector is output. Similarly, we can devise a parallel pocket learning algorithm based upon the broadcast communication model with a special hardware to find the number of the processing elements competing for the right of broadcasting in every cycle. The algorithm is described below:

Parallel pocket learning algorithm:

INPUT: A set of n training instances.

OUTPUT: A weight vector consistent with the maximum number of training instances.

STEP 1: Initially set W and the pocket weight vector P to be the zero vector and put every instance and P in an individual processing element.

STEP 2: Let W be the current weight vector to be broadcast. For every processing element Q_i, check whether W can correctly classify the training instance in Q_i. If not, the processing element Q_i calculates a new weight vector $W_i' = W + A_i * C_i$ and competes for the right of broadcasting.

STEP 3: The special hardware counts the number of the processing elements competing for the right of broadcasting and compares it with the stored count. If the new count < stored count, it represents the current weight vector can classify more training instances than before, set stored count = new count and signal every processing element to set the pocket weight vector P to be W.

STEP 4: The processing element which gets hold of the right of broadcasting puts its W' in the common bus and set W equal to W'.

STEP 5: Go to Step 2 until a null message is found or running time is up.

7. Conclusion

We have proposed a parallel perceptron learning algorithm based upon the single–channel broadcast communication model. Experimental result shows the average speed–up can get to $O(n^{0.91}/\log n)$. Also, a generalized parallel learning algorithm has been proposed which can solve both separable and non–separable training instances.

At last, we will point out that, in the broadcast communication model, each processing element actually executes the same work as the work of a conventional perceptron. Therefore, how to design a new architecture by combining the broadcast communication model and perceptron together to increase the degree of parallelism further is a very interesting problem.

Acknowledgement

We would like to thank Mr. Chain–Wu Lee for his assistance in obtaining the data for part of the figures.

REFERENCES

[1] R. Dechter and L. Kleinrock, "Broadcast communication and distributed algorithm," **IEEE Transaction on Computers**, Vol. C–35, No. 3, March 1986.

[2] S. I. Gallant, "Optimal linear discriminants," **Proceeding of the 8th International Conference on Pattern Recognition**, 1986.

[3] S. I. Gallant, "Connectionist expert system," **Communication of the ACM**, February 1988.

[4] A. G. Greenberg, "On the time complexity of broadcast communication schemes," **Proceeding of the 14th Annual ACM Symposium on Theory of Computing**, 1982.

[5] T. P. Hong and S. S. Tseng, "A parallel concept learning algorithm based upon version space strategy," **Proceeding of the International Phoenix Conference on Computers and Communications**, 1990.

[6] N. J. Nilsson, **Learning Machines**, McGraw–Hill, New York, 1965.

[7] D. E. Rumelhart and J. L. Mcclelland, **Parallel Distributed Processing: Explorations in the Microstructures of Cognition**, Vol. 1, MIT Press, Cambrige, Mass., 1986.

A Multidestination Routing Scheme for Hypercube Multiprocessors

Zhonggang Li and Jie Wu

Department of Computer Engineering, Florida Atlantic University, Boca Raton, FL 33431

Abstract

We proposed here a uniform nonredundant algorithm for three types of distributed communications in hypercube multiprocessors: (1) one to one destination (unicast); (2) one to many destinations (multicast); (3) one to all destinations (broadcast). The routing paths can be dynamically selected depending on some specific routing criteria. The proposed method is designed in such a way that it can be easily extended to incorporate fault tolerance features.

I. Introduction

An n-dimensional hypercube (n-cube) system consists of $N=2^n$ processors that are addressed distinctly by n-bit binary numbers, $a_{n-1}a_{n-2}....a_0$, from 0 to 2^n. Two processors are directly connected via a link if and only if their binary addresses differ in exactly one bit position [1].

There are in general three types of communications of interest in this enviroment: *one-to-one*, *one-to-many*, and *one-to-all* [2]. *Time steps* and *traffic* are the main criteria which measure the performance of communication at the system level. Time steps measures the total number of links the message traverses to reach each of the destinations. Traffic measures the total number of links the message traverses to reach all destinations.

In this paper we provide a general purpose (or uniform) nonredundant communication scheme for hypercubes. *Traffic and time steps* can be guaranteed to be optimal in the cases of unicast, broadcast and multicast when the destinations are within a single subcube. It is conjectured that finding an optimal solution for destinations of a group of subcubes is of NP complexity. Based on this scheme, a fully fault tolerant routing algorithm has been derived[3].

II. Notation and Preliminaries

In the following, node a has address $a_{n-1}a_{n-2}...a_0$. The ith rightmost bit a_i is the bit on coordinate i. Let Ω be a boolean space {0,1} and n the dimension of the hypercube. The node set N is defined as {a| $a=a_{n-1}a_{n-2}...a_0, \forall a_i \in \Omega$, $i \in [0,n-1]$}. Q_n is used to represent an n-cube and Σ is the ternary set {0,1,*}.

Definition 1: A *package* P is a set of nodes and has the address $P_{n-1}P_{n-2}...P_0$. More specifically, P={a| $a=a_{n-1}a_{n-2}...a_0$, where $a_i=P_i$ if $P_i \neq *$, and $a_i=1$ or 0 if $P_i=*$, $i \in [0,n-1]$}

Node a is said to belong to package P if $a \in P$. We can also say a is one node of package P.

Let D={0,1,...,n-1} be a coordinate space. A *Global Coordinate Sequence*, GCS, is a permutation of D.

Definition 2: The *Coordinate Sequence Set* Γ based on D is defined as: Γ={CS| CS=($c_0,c_1,...,c_k$), $k \leq n-1$, $c_i \in D$ for $i \in [0,k]$ and $c_i \neq c_j$ for $i \neq j$ i,j $\in [0,k]$}.

$d(P)$ is the dimension of the subcube represented by package P. $d(P)=\sum_{i=0}^{n-1} d_i(P)$, where $d_i(P)=0$, if $P_i \in \Omega$ and $d_i(p)=1$, if $P_i=*$.

Node b is the i-neighbour of node a, $b=\oplus_i a$, if $b_j=a_j$ for $j \neq i$ and $b_j=\sim a_j$ for $j=i$ (Here, $\sim$ is the logic operation NOT).

Definition 3: The *Hamming distance* between two packages P,Q with addresses $P=P_{n-1}P_{n-2}...P_0$ and $Q=Q_{n-1}Q_{n-2}...Q_0$ is defined as:

$$H(P,Q)= \sum_{i=0}^{n-1} h(P_i,Q_i), \text{ where h is defined in Figure 1(a).}$$

Definition 4: The *exclusive-or* $\oplus$ between two packages P,Q is defined as: $R=P \oplus Q$, where $R_i=P_i \oplus Q_i$ is defined in Fig1(b).

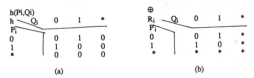

Fig1. h and $\oplus$ operations

Definition 5: $R=R_{n-1}R_{n-2}...R_0$ is the result of a *merge operation* on two packages with addresses $P=P_{n-1}P_{n-2}...P_0$ and $Q=Q_{n-1}Q_{n-2}...Q_0$, if $\exists i \in [0,n-1]$, $P_i=\sim Q_i$, and $P_j=Q_j$ for $j \neq i$, $j \in [0,n-1]$. The result of this operation is denoted as $R=P+Q$, where $R_i=*$ and $R_j=P_j$ for $j \neq i$, $j \in [0,n-1]$.

Packages in the destination set should be disjoint to guarantee nonredundant communication.

Definition 6: A package set P_Set is called a *disjoint package set* if $P \cap Q=\phi$, $\forall P,Q \in$ P_Set and $P \neq Q$.

Definition 7: Assuming CS1=(c1(0),c1(1),c1(2),...,c1(m1)) and CS2=(c2(0),c2(1),...,c2(m2)) with CS1,CS2 $\in \Gamma$, m2 $\leq$ m1, CS2 is said to be a *subordinate* of CS1 or CS2«CS1, iff $\forall i \exists j$ (c1(j)=c2(i)) and $\forall i_1,i_2 \in [0,m2]$ ($i_1<i_2$) $\exists j_1,j_2 \in [0,m1]$ ($j_1<j_2$) such that c1(j_1)=c2(i_1), c1(j_2)=c2(i_2).

III. A Uniform Communication Algorithm

For multicasting communication the destination nodes are very likely to be close to each other. Thus, the destination set, D_Set, is considered as a set of subcubes, i.e. packages. Moreover, these packages should form a disjoint set to ensure that every destination node will be reached only once. Therefore, the D_Set forms a disjoint P_Set. As a special case, a D_Set with a single node is unicasting. Also a destination set with only one package P, with P=**..*, represents broadcasting.

Once a node receives a message, it will check whether itself belongs to a package in the D_Set of the message. We refer to D_Set_In as the D_Set of the message received. Thus, if the current node address "a" belongs to a package P in D_Set_In, the communication process will pass the message to its local processor. If d(P)>0, (more than one node in P), the *splitting operation* & and the *complement splitting operation* ~& defined next are applied to split P into several packages. For the remaining packages in D_Set_In, the communication process should choose directions to send the copy of the message along optimal paths. We define *D_Set_Out[i]*, $i \in [0,n-1]$, as the D_Set in the head of the message which will be sent out along the ith link.

Definition 8: For package P, node a in P and $P_m=*$ with $m \in [0,n-1]$, $\&_m^a P$ is the result of the splitting operation on package P with reference to node a along coordinate m.

$$(\&_m^a P)_i = \begin{cases} P_i & (i \neq m, i \in [0,n-1]) \\ \\ a_i & (i=m). \end{cases}$$

$\sim\&_m^aP$ is the result of the complement splitting operation on package P with reference to node a along coordinate m.

$$(\sim\&_m^aP)_i = \begin{cases} P_i & (i\neq m,\ i\in[0,n-1]) \\ \sim a_i & (i=m). \end{cases}$$

Lemma 1: $\&_m^aP\cap\sim\&_m^aP=\phi$ and $P=\&_m^aP+(\sim\&_m^aP)$

Lemma 2: $a\in\&_m^aP$ and $\oplus_m a\in\sim\&_m^aP$

We can split P recursively along every coordinate in $\{i|$ $P_i=*,\ i\in[0,n-1]\}$. Following the order given by a GCS (how to select the GCS is discussed in [3]), CS1=$(c1(0),c1(1),...,$ $c1(m_1-1))$«GCS where $P_i=*, \forall i\in[0..m_1-1]$, P can be split into a set of disjoint packages: $\{\sim\&_{c1(0)}^aP,\ \sim\&_{c1(1)}^a(\&_{c1(0)}^aP),$ $\sim\&_{c1(2)}^a(\&_{c1(1)}^a(\&_{c1(0)}^aP)),...,\ \sim\&_{c1(m1-2)}^a(\&_{c1(m1-3)}^a(\&_{c1(m1-4)}^a..(\&_{c1(0)}^aP)..),\ \&_{c1(m1-1)}^a(\&_{c1(m1-2)}^a(\&_{c1(m1-3)}^a..(\&_{c1(0)}^aP)..)$ $\}$. In order to simplify the notation, we use $\&_{(c1(0),c1(1),...,c1(i))}^aP$ to represent $\&_{c1(i)}^a(\&_{c1(i-1)}^a(\&_{c1(i-2)}^a...(\&_{c1(0)}^aP)..)$. Since the splitting process is applied always to current node a, we omit a in $\&_m^aP$ in the later discussion, unless explicitly stated otherwise.

Lemma 3: Assuming $a\in P$, CS1«GCS, CS1=$(c1(0),c1(1),...$ $,c1(m_1-1))$, where $P_{c1(i)}=*, \forall i\in[0..m1]$, and $m1=d(P)$, we have $\&_{(c1(0),c1(1),c1(2)...,c1(m1-1))}^aP=a$.

Based on the above lemmas, the last term in the result of the recursive splitting operation on package P is always the current node a itself, and we have the following theorem.

THEOREM 1: Assuming $a\in P$, CS1«GCS, CS1=$(c1(0),c1(1),$ $...,c1(m_1-1))$ where $P_{c1(i)}=*, \forall i\in[0..m_1-1]$, and $m1=d(P)$ then

$$P=a+\sum_{i=0}^{m_1-1}\sim\&_{(c1(0),c1(1),...,c1(m_1-1))}^aP$$

When node a receives the message with D_Set_In assuming that $\exists P\in$ D_Set_In, $a\in P$, the message will be passed to the local processors first. Meanwhile, P is split recursively if $d(P)>0$. Complementary terms $\sim\&_{(c1(0),c1(1),..,c1(i))}P$ are appended to D_Set_Out[c1(i)]. Thus, the c1(i)-th neighbor $\oplus_{c1(i)}a$ will receive this message from a. By lemma 2, $\oplus_{c1(i)}a\in$ $\sim\&_{(c1(0),c1(1),..,c1(i))}P$. This means that the splitting process continues at node $\oplus_{c1(i)}a$. Therefore, this process can be further spread to other neighboring nodes, and multicasting (including broadcasting) can be achieved by this recursive process.

In broadcasting, the source node naturally belongs to the D_Set=$\{***..*\}$.

Here we propose a *labeled routing* scheme. CS(L) is a special GCS, such that CS(L)=$(L,L-1,L-2,...,0,n-1,n-2,...,$ $L+2,L+1)$. When the message is created at the source node, a label L is defined and kept as the head of the message. Then for every current node a, CS(L) is used to determine the order to split package P in P_Set_In, if $a\in P$, and to select a closest direction for package P, if $a\notin P$.

The labeled routing scheme proposed is by no means the best scheme. It is just a convenient way to represent a GCS.

Algorithm for a uniform communication scheme:

```
/*  current node address: a=a_{n-1}a_{n-2}...a_0
    D_Set of input messages: D_Set_In
    D_Set of output messages along [0..n-1] link: D_Set_Out[0..n-1] */
begin
    Receive ((D_Set_In,L),Message);
        /* GCS=CS(L)=(L,L-1,L-2,..,0,n-1,n-2,..,L+1) */
    if ∃P∈ D_Set_In, a∈P then begin
```

Transfer (Message) to the Local Processor;
```
        D_Set_In=D_Set_In-{P};
        CS1=(c1(0),c1(1),...,c1(d(P)-1))«CS(L)
            where P_{c1(i)}=*, i∈[0,d(P)-1].
        Q=P;
        for i=0 to d(P)-1 do  begin
            R=~&_{c1(i)}Q;   Q=&_{c1(i)}Q;
            Append(R, D_Set_Out[c1(i)]);
        end        /* end of recursive splitting */
    end
    for all P∈ D_Set_In do  begin
        CS2=(c2(0),c2(1),...,c2(H(P,a)-1))«CS(L)
            where h(P_{c2(i)},a_{c2(i)})=1,i∈[0,H(P,a)-1]
        Append (P, D_Set_Out[c2(0)]);
    end
    for ∀i∈ [0..n-1] D_Set_Out[i]≠φ do
        Send ((D_Set_Out[i],L),Message) along the i-th link
end.
```

For example, in a 5-cube, node 00110 receives a message from its local processor with D_Set_In=$\{1*0**,010*0\}$. Assume that for this message L=2. Then GCS=CS(L)= $(2,1,0,5,4,3)$. Thus, CS2 in node 00110 are $(2,4), (2,3)$ for $1*0**$ and $010*0$ respectively. CS1 in node 10010 is $(1,0,3)$ and CS1 in node 10000 is $(3,0)$. The routing tree for this case is presented in Fig2.

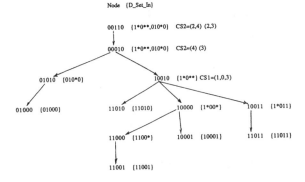

Fig2 An example of multicast

In fact, once the message enters the node which belongs to a package in the P-set, the remaining routing paths of P constitute a tree [3]. Normally, the structure of the tree is dependent on the definition of L. As a special case, the binomial tree is formed when L=n-1.

Theorem 2: For a specific package $P\in$ D-Set, the routing tree formed by the above algorithm is time and traffic optimal for any GCSs (including the labeled routing scheme).

Theorem 3: For a unicasting from node a to node b with H(a,b)=n, n node disjoint optimal paths from a to b can be obtained by changing L from 0 to n-1 in the labeled routing scheme.

References

[1] Y. Saad and M.H. Schultz, "Topological Properties of Hypercubes", *IEEE Trans. on Computers*, vol. 37, no. 7, July 1988, pp. 867-872.
[2] Y. Lan, A.H. Esfahanian and L.M. Ni, "Multicast in Hypercube Multiprocessors", *Journal of Parallel and Distributed Computing*, vol 8, Aug.1990, pp. 30-41.
[3] Z.G. Li and J. Wu, "A Fully Fault-Tolerant multidestination Routing Scheme for Hypercube Multiprocessors", *Technical Report, Dept. of Comp. Eng., Florida Atlantic Univ., Boca Raton FL, Feb 1991*.

A Recursive Mutual Exclusion Algorithm for Multiprocessor Systems with Shared Memory

Tai-Kuo Woo
Department of Computer Science
Jacksonville University
Jacksonville, FL 32211

Kenneth Block
Department of Computer and Information Science
University of Florida
Gainesville, FL 32611

Abstract

A recursive mutual exclusion algorithm for multiprocessor systems is described. Processes communicate with each other through reading and writing shared memory. A process interested in the critical section calls the recursive procedure to compete with its neighbor. As long as the process wins, the procedure continues to call itself to compete at the next level. The procedure terminates when it reaches the top level, signaling that the process is allowed to enter the critical section.

1 Introduction

Since Dijkstra introduced the mutual exclusion problem in 1965 [3], many software solutions have been proposed [4, 7, 2, 6, 8, 9]. However, software solutions didn't get much attention until the advent of shared-memory multiprocessor computers. Among the software solutions, Peterson's algorithm is simple and elegant [9]. However, the generalization for n processes presented by Peterson requires $O(n^2)$ operations. Block and Woo [1] propose a more efficient generalization of Peterson's algorithm which requires $O(n*m)$ operations, when m out of n processes are competing for the critical section. In this paper, we reduce the number of operations to $O(log\ n)$.

In Peterson's algorithm, a process will set its status to being interested in the critical section and then mark itself as the blocked process. It will wait until either the other process is not interested or it is no longer the blocked process. Both Peterson's and Block and Woo's algorithms require putting multiple filters one after each other, with each filter holding one process. A process interested in the critical section has to compete with all other interested processes at all levels. In this

paper, we present an algorithm in which a process only competes with one other process at each level. Since each filter can hold half of the interested processes, only log n filters are required. Also, a process does not have to check up on all other interested processes. As a result, the number of operations required is $O(log\ n)$.

2 The Algorithm

The algorithm proposed in this paper requires 3n shared variables. The array *in* consists of 2n elements where interested processes indicate their intention to enter the critical section. The array *block* is used to block processes when more than one process is competing. The protocol for process i wishing to enter the critical section is as follows. Note that the algorithm assumes that there are n processes 0..n-1.

```
      /* shared variables */
0     in    : array [0..2n − 1] of 0..1;
1     block : array [0..n − 1] of 0..n − 1;

      /* local variables */
2     l : 1..log₂ n;
3     a : array[1..log₂ n + 1] of 1..2n − 1;
/* protocol for process number i */

4     a₁ := i;
5     for l := 1 to log₂ n do
6         in[aₗ] := 1;
7         block[aₗ/2] := i;
8         while (block[aₗ/2] =
i) and (in[aₗ + (−1)^mod(aₗ,2)] ≠ 0) do;
9             aₗ₊₁ := aₗ/2 + n;
10    end do;
11    critical section
12    for l := log₂ n downto 1 do
13        in[aₗ] := 0;
```

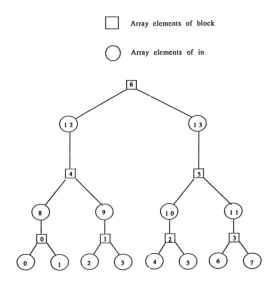

□ Array elements of block

○ Array elements of in

Figure 1: An Illustration of the Algorithm

In general, a process wishing to enter the critical section has to go through $log\ n$ levels. On each level it will compete with its neighboring process. When it starts competing, it first stores its process number in a_1 (4). Then it enters the first level (5). It marks in to indicate its intention (6) and then blocks itself (7). It remains blocked until either it finds out its competitor $in[a_l+(-1)^{mod(a_l,2)}]$ is no longer interested in the critical section or its competitor bumps it out ($block[a_l] \neq i$) (8). When it is bumped out of the while loop, it computes the position in the array in for the next level (9). The same procedure is repeated until it exits the for loop (10). When it exits the critical section, it sets the element on each level to zero (12), (13). The example below demonstrates the protocol. In Figure 1, there are eight processes 0..7. Suppose process 5 is interested in the critical section. On the first level, it marks $in[5]$ and blocks itself by setting $block[2]$. It then competes with process 4. If process 4 is not interested in the critical section, process 5 will enter the next level. If process 4 is also interested in the critical section, the process that sets $block[2]$ first gets bumped out the while loop. Suppose process 5 sets $block[2]$ first and gets bumped out the while loop. It will calculate its position in in for the second level, which is $in[10]$. At this level, it will compete with the winner of the processes, process 6, and process 7. If process 5 marks $block[5]$ first, it will enter the third level. On the third level, it will mark $in[13]$ and compete with the process that marks $in[12]$, which is one of the first four processes. If it still gets bumped out the while loop, it will enter the critical section. Since all the positions it marks in array in are stored in array a, no calculations are needed when it resets them to zero.

The protocol shown above can be converted to the recursive protocol shown below. The shared variables are the same. The local variable l is initialized to 0.

In the protocol a process interested in the critical section sets the same variables as in the protocol above. However, when it is bumped out of the while loop, the process will either advance to the next level by making a recursive call or, if it has reached the top level ($log_2 n$), enter the critical section.

$Procedure\quad Critical\_section(i, l: integer);$
$\quad l := l\ +\ 1;$
$\quad in[i] := 1;$
$\quad block[i/2] := i;$

$while\ (block[i/2] = i\ \ and\ \ in[i+(-1)^{mod(i,2)}] \neq 0)\ \ do;$
$\quad if\ \ level < log_2 n\ \ then$
$\qquad\qquad Critical\_section(i/2 + n, l)$
$\quad\quad else$
$\qquad\qquad\quad enter\ \ the\ \ critical\ section;$
$\quad in[i] := 0;\ \{\ Exit\ the\ critical\ section\ \}$

References

[1] Block, K. R., Woo, T. K.: A More Efficient Generalization of Peterson's Mutual Exclusion Algorithm. 35(1990), Information Processing Letters, pages 219-222 (1990).

[2] DeBrujin, N. G.: Additional Comments on a Problem in Concurrent Programming and Control. Communications of the ACM, Vol. 10, No. 3, pages 137-138 (1967).

[3] Dijkstra, E. W.: Cooperating Sequential Processes. Technical Report EWD-123, Technological University, Eindhoven, the Netherlands (1965).

[4] Dijkstra, E. W.: Solution of a Problem in Concurrent Programming Control. Communications of the ACM, Vol. 8, No. 9, page 569 (1965).

[5] Dijkstra, E. W.: An Assertional Proof of a Program by G. L. Peterson. EWD-779, Burroughs Corp., 1981.

[6] Eisenberg, M. A., Mcguire, M. R.: Further Comments on Dijkstra's Concurrent Programming Control Problem. Communications of the ACM, Vol. 15, No. 11, page 999 (1972).

[7] Knuth, D. E.: Additional Comments on a Problem in Concurrent Programming Control. Communications of the ACM, Vol. 9, No. 5, pages 321- 322 (1966).

[8] Lamport, L.: A New Solution of Dijkstra's Concurrent Programming Problem. Communications of the ACM, Vol. 17, No. 8, pages 453-455 (1974).

[9] Peterson, G. L.: Myths about the Mutual Exclusion Problem. Information Processing Letters, Vol. 12, No. 3, pages 115-116 (1981).

COST-OPTIMAL PARALLEL ALGORITHMS FOR CONSTRUCTING B-TREES

Biing-Feng Wang[+], Gen-Huey Chen[+], and M. S. Yu[++]
+Department of Computer Science and Information Engineering,
National Taiwan University, Taipei, Taiwan, Republic of China
++Department of Applied Mathematics,
National Chung-Hsing University, Taichung, Taiwan, Republic of China

Abstract--Two cost-optimal algorithms are presented for constructing a B-tree for a sorted list of N keys. One, based on the EREW model, uses $N/\log\log N$ processors and requires $O(\log\log N)$ time and the other, based on the CREW model, uses N processors and requires $O(1)$ time.

1. Introduction

Based on the EREW model, Moitra and Iyengar [2] have designed parallel algorithms that use N processors to construct balanced binary search trees in $O(1)$ time. In [1], also based on the EREW model, Dekel *et al.* have designed parallel algorithms that use N processors to construct balanced m-way search trees. The time complexities of Dekel *et al.*'s algorithms are $O(1)$ under an assumption that computing the k-th, $0 \le k \le \log N$, power of m takes $O(1)$ time. The main disadvantage of the m-way search tree is that it may become unbalanced when it is not static, and thus logarithmic accessing time can not be guaranteed.

Recently, Wang and Chen [3] have designed parallel algorithms for constructing 2-3 trees, which are B-trees of order 3. In this paper, as an extension of their results, we present two cost-optimal parallel algorithms for constructing B-trees of arbitrary order m. One, based on the EREW model, uses $N/\log\log N$ processors and requires $O(\log\log N)$ time and the other, based on the CREW model, uses N processors and requires $O(1)$ time.

2. Notations and Definitions

Definition 1: A *B-tree of order m*, $m \ge 3$, is a tree satisfying the following properties: (1) Every node has at most m children. (2) Every node, except for the root, has at least $\lceil m/2 \rceil$ children. (3) The root has at least 2 children. (4) All leaves are on the same level. (5) A node with w children contains w-1 keys.

For the convenience of description, we use notation $<i, j>$ to denote the j-th node from the left at the i-th level (see Figure 1). A node $<i, j>$ with w keys and $w+1$ pointers can be represented as $(P_{<i,j>,1}, K_{<i,j>,1}, P_{<i,j>,2}, K_{<i,j>,2}, \dots , P_{<i,j>,w}, K_{<i,j>,w}, P_{<i,j>,w+1})$, where $K_{<i,j>,1} < K_{<i,j>,2} < \dots < K_{<i,j>,w}$ denote the w keys, $P_{<i,j>,1}$ points to the subtree whose keys are smaller than $K_{<i,j>,1}$, $P_{<i,j>,v}$, $1 < v < w+1$, points to the subtree whose keys are between $K_{<i,j>,v-1}$ and $K_{<i,j>,v}$, and $P_{<i,j>,w+1}$ points to the subtree whose keys are greater than $K_{<i,j>,w}$.

Note that more than one B-tree of order m can store a given sorted list. For example, Figure 1 shows two different B-trees of order 3 that can store the sorted list (1, 2, 3, ..., 27). Thus, it is necessary to specify the exact one that our algorithms will construct for a given sorted list. For a given sorted list of size N, the uniquely constructed B-tree of order m has the following properties: (1) It has the minimal height $n = \lceil \log_m(N+1) \rceil$ (In this paper, the height of a tree is defined as the number of its levels). (2) Root owns $r = \lceil (N+1)/m^{n-1} - 1 \rceil$ keys. Note that $m^{n-1} \le N \le m^n-1$; hence, $1 \le r \le m-1$. (3) There exists a unique integer c, $1 < c \le n$, such that all the nodes, except the root, above the c-th level own m-1 keys and all the non-leaf nodes on and below the c-th level

own $\lceil m/2 \rceil$-1 keys. (4) Let s be the number of keys contained in the leaf node $<n, 1>$, $\lceil m/2 \rceil \le s \le m-1$. Then, each of the other leaf nodes may own s or s-1 keys, but may not own more keys than the leaf nodes on its left.

For example, the B-tree of order 3 that the proposed algorithms construct for the list (1, 2, 3, ..., 27) is shown in Figure 1(a). In this example, $n=4$, $r=1$, $c=3$, and $s=2$.

The existence and uniqueness of the constructed B-tree is shown in the next section. In the following, we define notations that are used throughout this paper.
m: The order of the constructed B-tree.
N: The number of keys in the sorted list. Assume $N>m$.
$T_{N,m}$: The constructed B-tree.
n: The height of $T_{N,m}$.
r: The number of keys that are stored in the root of $T_{N,m}$.
c: The marked level of $T_{N,m}$.
$V_{N,m}$: The number of nodes in $T_{N,m}$.
$L_{N,m}$: The number of leaf nodes in $T_{N,m}$.
s: The number of keys that are contained in the leaf node $<n, 1>$ of $T_{N,m}$.
$LL_{N,m}$: The number of leaf nodes in $T_{N,m}$ that own s keys.
$RANK_{N,m}(K_{<i,j>,v})$: The rank of the key $K_{<i,j>,v}$ in the given sorted list. It equals the number of keys (inclusive of $K_{<i,j>,v}$ itself) that precede $K_{<i,j>,v}$ in inorder traversal of $T_{N,m}$.

For example, consider Figure 1(a) again, where $m=3$, $N=27$, $n=4$, $r=1$, $c=3$, $V_{27,3}=21$, $L_{27,3}=12$, $s=2$, $LL_{27,3}=4$, $RANK_{27,3}(K_{<2,1>,1})=6$, and $RANK_{27,3}(K_{<2,1>,2})=12$.

3. Properties of the Constructed Trees

Theorem 1: For an abritrary $N > m$, $T_{N,m}$ uniquely exists. And, for the uniquely constructed B-tree $T_{N,m}$, $n = \lceil \log_m(N+1) \rceil$, $r = \lceil (N+1)/m^{n-1} - 1 \rceil$, $c = \lceil \log_{m\lceil m/2 \rceil}(((N+1)*m)/((r+1)*\lceil m/2 \rceil^n)) \rceil$, $V_{N,m}=(r+1)*((m^{c-1}-1)/(m-1)+m^{c-2}*(\lceil m/2 \rceil^{n-c+1}-1)/(\lceil m/2 \rceil-1))+1$, $L_{N,m} =(r+1)*m^{c-2}*\lceil m/2 \rceil^{n-c}$, $s = \lceil N-(r+1)*m^{c-2}*\lceil m/2 \rceil^{n-c}+1/L_{N,m} \rceil$, and $LL_{N,m}=N-(r+1)*m^{c-2}*\lceil m/2 \rceil^{n-c}+1-(s-1)*L_{N,m}$.

Theorem 2: For each node $<i, j>$ in $T_{N,m}$, (a) when $i=1$, $P_{<i,1>,v} = <2, v>$ for $1 \le v \le r+1$ and undefined for $r+1 < v \le m$; (b) when $1 < i < c$, $P_{<i,j>,v} = <i+1, (j-1)*m+v>$ for $1 \le v \le m$; (c) when $c \le i < n$, $P_{<i,j>,v} = <i+1, (j-1)*\lceil m/2 \rceil+v>$ for $1 \le v \le \lceil m/2 \rceil$, and undefined for $\lceil m/2 \rceil < v \le m$; (d) when $i=n$, $P_{<i,j>,v} =$ undefined for $1 \le v \le m$.

Theorem 3: $RANK_{N,m}(K_{<i,j>,v}) = v*(s+1)*m^{c-i-1}*\lceil m/2 \rceil^{n-c} - max\{0, v*m^{c-2}*\lceil m/2 \rceil^{n-c}-LL_{N,m}\}$ if ($i=1$ and $1 \le v \le r$), $((j-1)*m+v)*(s+1)*m^{c-i-1}*\lceil m/2 \rceil^{n-c}-max\{0, ((j-1)*m+v)*m^{c-i-1}*\lceil m/2 \rceil^{n-c}-LL_{N,m}\}$ if ($1<i<c$ and $1 \le v \le m-1$), ($(j-1)*$

$\lceil m/2 \rceil + v) * (s+1) * \lceil m/2 \rceil^{n-i-1} - max\{0, ((j-1)* \lceil m/2 \rceil + v) * \lceil m/2 \rceil^{n-i-1} - LL_{N,m}\}$ if $(c \le i < n$ and $1 \le v \le \lceil m/2 \rceil - 1)$, $(j-1)*(s+1)+v - max\{0, j-1-LL_{N,m}\}$ if $(i=n, 1 \le j \le LL_{N,m}$, and $1 \le v \le s)$ or $(i=n, LL_{N,m} < j$, and $1 \le v \le s-1)$, and undefined otherwise.

The proof of *Theorem 1, 2, and 3* can be found in [4].

4. Parallel Construction Algorithms

We assume that $T_{N,m}$ is stored in a linear array of length $V_{N,m}$. We need to specify a linear ordering for the nodes of $T_{N,m}$ such that the node with order k, $1 \le k \le V_{N,m}$, is represented by the k-th element of the linear array. A simple approach to do so is to number the nodes according to their breadth-first search order. For example, Table 1 shows the specified linear ordering for the nodes of the B-tree that was depicted in Figure 1(a). Let $ORDER_{N,m}(<i, j>)$ denote the order of node $<i, j>$ in $T_{N,m}$. We have the following lemma.

Lemma 1: Let $<i, j>$ be a node of $T_{N,m}$. Then, $ORDER_{N,m}(<i, j>) = 1$ if $i=1$, $1+(r+1)*(m^{i-2}-1)/(m-1)+j$ if $2 \le i \le c$, and $1+(r+1)*(m^{c-2}-1)/(m-1)+(r+1)*m^{c-2}*(\lceil m/2 \rceil^{i-c}-1)/(\lceil m/2 \rceil -1)+j$ if $c < i \le n$.

Denote the node with a specified order k by $NODE_{N,m}(k)$, we have the following lemma.

Lemma 2: For a given order k, $1 \le k \le V_{N,m}$, if $NODE_{N,m}(k) = <i, j>$, then (a) when $k=1$, $i = 1$ and $j = 1$; (b) when $1 < k \le 1+(r+1)*(m^{c-1}-1)/(m-1)$, $i = \lceil \log_m((k-1)*(m-1)/(r+1)+1) \rceil +1$ and $j=k - (1+(r+1)*(m^{i-2}-1)/(m-1))$; (c) when $1+(r+1)*(m^{c-1}-1)/(m-1) < k \le V_{N,m}$, $i=\lceil \log_{\lceil m/2 \rceil}((k-(1+(r+1)*(m^{c-2}-1)/(m-1)))*(\lceil m/2 \rceil -1)/((r+1)*m^{c-2})+1) \rceil +c-1$ and $j=k-(1+(r+1)*(m^{c-1}-1)/(m-1)+(r+1)*m^{c-2}*(\lceil m/2 \rceil^{i-c}-1)/(\lceil m/2 \rceil -1))$.

Since each node of $T_{N,m}$ is uniquely represented by an element of a linear array of length $V_{N,m}$, for the k-th element of the linear array, $1 \le k \le V_{N,m}$, we have to determine $NODE_{N,m}(k)$, $RANK_{N,m}(K_{<i,j>,v})$, $1 \le v \le m-1$, and $ORDER_{N,m}(P_{<i,j>,u})$, $1 \le u \le m$, where $<i, j> = NODE_{N,m}(k)$. The time complexity of the computation are dependent on how fast the set S of values $m^{n-1}, \lceil m/2 \rceil^n, m^{c-2}, \lceil m/2 \rceil^{n-c+1}, \lceil m/2 \rceil^{n-c}, m^{c-i-1}, \lceil m/2 \rceil^{n-i-1}, m^{c-i}, \lceil m/2 \rceil^{n-i}, m^{i-2}, \lceil m/2 \rceil^{i-c+1}, \lceil m/2 \rceil^{i-c}$, and m^{c-1} can be computed.

4.1 Construction Algorithm on the EREW model

Fact 1: Let $<i_k, j_k> = NODE_{N,m}(k)$ and $<i_{k+1}, j_{k+1}> = NODE_{N,m}(k+1)$. Then, $i_{k+1} - i_k = 0$ or 1.

As a result of *Fact 1*, when we have obtained the set S of values for $i = i_k$, we can obtained the set S of values for $i = i_{k+1}$ in additional constant time. The cost-optimal parallel algorithm is simply to let each processor process $\log\log N$ consecutive elements. Each processor will process the assigned elements in increasing order of their indices and therefore it will take $O(\log\log N)$ time for the first one and $O(1)$ time for each of the others.

4.2 Construction Algorithm on the CREW model

Before constructing $T_{N,m}$, N processors are used to fill in two tables $POWER_{\lceil m/2 \rceil}[1 ..n]$ and $POWER_m[1..n]$, where the contents of $POWER_{\lceil m/2 \rceil}[i]$ and $POWER_m[i]$, $1 \le i \le n$, are the values of $\lceil m/2 \rceil^i$ and m^i respectively. The table $POWER_m[1..n]$ is filled as follows. ($POWER_{\lceil m/2 \rceil}[1..n]$ can filled similarly.) Each processor P_k, $1 < k \le N$, is first to compute $u_k = \lceil \log_m k \rceil$ and $l_k = \lfloor \log_m k \rfloor$. If $u_k = l_k$, P_k fills $POWER_m[u_k]$ with the value of k. Then (since $m^{n-1} \le N \le m^n - 1$), processor P_N fills $POWER_m[n]$ with the value of $m*POWER_m[n-1]$. After the two tables have been established, $T_{N,m}$ can be constructed in $O(1)$ time by letting each processor P_k, $1 \le k \le V_{N,m}$, process the k-th element simultaneously. Now the set S of values for any i can be obtained in $O(1)$ time by looking up the tables.

References

[1] E. Dekel, S. Peng, and S. S. Iyengar, "Optimal parallel algorithms for constructing and maintaining a balanced m-way search tree," *International Journal of Parallel Programming*, vol. 15, no. 6, pp. 503-528, 1986.

[2] A. Moitra and S. S. Iyengar, "Derivation of a parallel algorithm for balanced binary trees," *IEEE Transactions on Software Engineering*, vol. SE-12, no. 3, pp. 442-449, March 1986.

[3] B. F. Wang and G. H. Chen, "Cost-optimal parallel algorithms for constructing 2-3 trees," *Journal of Parallel and Distributed computing*, vol. 11, pp. 257-261, 1991.

[4] B. F. Wang and G. H. Chen, "Efficient parallel algorithms for constructing B-trees," Tech. Rep., National Taiwan University, 1991.

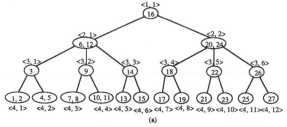
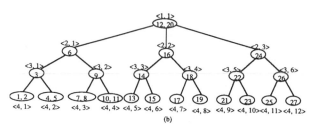

Figure 1. Two different B-trees of order 3 that store the sorted list (1, 2, 3, ..., 26, 27).

Node $<i, j>$	$<1,1>$	$<2,1>$	$<2,2>$	$<3,1>$	$<3,2>$	$<3,3>$	$<3,4>$	$<3,5>$	$<3,6>$	$<4,1>$	$<4,2>$
Order k	1	2	3	4	5	6	7	8	9	10	11
Node $<i, j>$	$<4,3>$	$<4,4>$	$<4,5>$	$<4,6>$	$<4,7>$	$<4,8>$	$<4,9>$	$<4,10>$	$<4,11>$	$<4,12>$	
Order k	12	13	14	15	16	17	18	19	20	21	

Table 1. The specified linear ordering for the nodes of the B-tree that was depicted in Figure 1(a).

An NC Algorithm
for
Recognizing Strict 2-threshold Graphs

L. Y. Tseng and W. D. Hao
Department of Applied Mathematics
National Chung-Hsing University, R.O.C.

ABSTRACT

A graph G is called a *strict 2 − threshold graph* if its edge-set can be partitioned into two threshold graphs T_1 and T_2 such that every triangle of G is also a triangle of T_1 or of T_2. The strict 2-threshold graph is a class of perfect graphs. A polynomial-time sequential algorithm for recognizing these graphs had been shown by Mahadev and Peled. Their algorithm recognizes these graph by checking eight forbidden configurations. In this paper, we show some properties of the strict 2-threshold graphs and propose a parallel algorithm on the CREW PRAM model. Our approach is significantly different from that of Mahadev and Peled's. Our algorithm is constructive in the sense that it will specify all possible standard representations of G if G is a strict 2-threshold graph. It uses $O(n^4/\log n)$ processors and $O(\log n)$ time to recognize the strict 2-threshold graphs and to specify T_1, T_2, where n is the number of vertices. Therefore, recognizing the strict 2-threshold graph is in NC.

Key words: strict 2-threshold graphs, recognition, perfect graphs, parallel algorithms

Section 1.   Introduction

The strict 2 -threshold graphs form a subclass of the 2-threshold graphs, which had been proved to be perfect graphs[4], so the strict 2-threshold graphs are perfect graphs. Intuitively, the strict 2-threshold graph is constructed by two threshold graphs in some way. Threshold graphs are also perfect graphs[3]. Some characterizations of the threshold graph can be found in [1, 3] and a good survey of the research in the field of the perfect graph can be found in [5]. One characterization for the threshold graph will be stated in Lemma 1.1. But we shall first discribe the concept of *comparability*. Let K be a subset of vertices and x be a vertex. $N_K(x)$ denotes the set of all vertices in K that are adjacent to x. We say that vertices x and y are *comparable* in K if $N_K(x) \subseteq N_K(y) \cup \{y\}$ or $N_K(y) \subseteq N_K(x) \cup \{x\}$.

Lemma 1.1 [1]   G is a threshold graph if and only if $V(G)$ can be partitioned into a clique K and a stable set S such that every two vertices in S are comparable in K.

We shall call (K, S) a representation of the threshold graph G. The problem of recognizing a 2-threshold graph was reported to be NP-complete[6]. For various results on the 2-threshold graph, see [4].

Section 2.   Definitions

A graph G is called a *split graph* if and only if $V(G)$ can be partitioned into a clique K and a stable set S[3]. (K, S) is called a representation of the split graph G.

Definition 2.1   A graph G is called a *strict 2-split graph* if and only if it is not a split graph itself and its edge set can be partitioned into two split graphs G_1 and

G_2 such that every triangle of G is also a triangle of G_1 or of G_2. Let $(K_1, S_1), (K_2, S_2)$ be the representations of G_1 and G_2 respectively, then $((K_1, S_1), (K_2, S_2))$ is a representation of G. In this representation, edges are considered to be within K_1, within K_2, between K_1 and S_1, or between K_2 and S_2.

Definition 2.2   Let G be a threshold graph and (K, S) be one of its representations. If there exists at least one vertex in K not adjacent to any vertex in S, then (K, S) is called a *standard representation*.

Lemma 2.1   Every threshold graph G with the threshold dimension 1 has a standard representation.

Definition 2.3   Let G be a strict 2-threshold graph and (T_1, T_2) be an edge partition of G. Let (K_1, S_1) and (K_2, S_2) be representations of T_1 and T_2 respectively. Then $((K_1, S_1), (K_2, S_2))$ is called a *representation* of G.

Lemma 2.2   Let $((K_1, S_1), (K_2, S_2))$ be a representation of a strict 2-threshold graph or a strict 2-split graph. Then $\mid K_1 \cap K_2 \mid \leq 1$ and $V − (K_1 \cup K_2)$ is a stable set.

Definition 2.4   Let $((K_1, S_1), (K_2, S_2))$ be a representation of a strict 2-threshold graph. This representation is said to be *standard* if there exist a vertex v_1 in K_1 which is not adjacent to S_1 and a vertex v_2 in K_2 which is not adjacent to S_2. And either $v_1 = v_2 = v_c$ or $v_1 \neq v_c, v_2 \neq v_c$ holds in case $K_1 \cap K_2 = \{v_c\}$. Note that $K_1 \cap S_2, K_2 \cap S_1, S_1 \cap S_2$ may not be empty.

Definition 2.5   A graph G is called a *candidate strict 2-split graph* if and only if the vertex set V can be partitioned into K_1, K_2 and S, where K_1, K_2 are cliques and S is a stable set, such that $K_1 \cup K_2 \cup S = V$, $\mid K_1 \cap K_2 \mid \leq 1$ and $(K_1 \cup K_2) \cap S = \phi$.

Section 3.   Some Properties

In this section, we state some properties of the candidate strict 2-split graph, the strict 2-split graph and the strict 2-threshold graph.

Lemma 3.1   Let $((K_1, S_1), (K_2, S_2))$ be a representation of a strict 2-threshold graph or a strict 2-split graph. Let $K_1 \cap K_2 = \phi$. If there exists a vertex v in $K_1 \cap S_2$, then any vertex in $N_{k_2}(v)$ must not adjacent to any vertex in $K_1 − \{v\}$.

Theorem 3.1   Let G be a strict 2-threshold graph. There exists a standard representation for G.

Lemma 3.2   Let G be a threshold graph with the threshold dimension 1. Let $adj(v) = \{w|vw$ is an edge $\}$ and $\overline{adj}(v) = adj(v) \cup \{v\}$. There exists a vertex v such that $\overline{adj}(v)$ is a clique K and $(K, V − K)$ is a standard representation of G.

Theorem 3.2   Let G be a strict 2-threshold graph and $((K_1, S_1), (K_2, S_2))$ be a standard representation of G. Then K_1, K_2 must appear in the following set of cliques.

$C = \{K|K$ is $\overline{adj}(v)$ if $\overline{adj}(v)$ is a clique or K is one of the maximal cliques in $\overline{adj}(v)$ if $adj(v)$ contains exactly two separate cliques. v is any vertex in $V(G)\}$

Let G=$((K_1, S_1), (K_2, S_2))$ be a strict 2-threshold graph . From [6], we know that G can not contain the configuration shown in Figure 1. The dotted lines indicate nonexistent edges.

Theorem 3.3   Let $G = (K_1, K_2, S)$ be a candidate strict 2-split graph. G is a strict 2-split graph with

K_1, K_2 as the cliques in its representation if and only if it does not contain the following forbidden configurations F_1, F_2 if $|K_1 \cap K_2|=0$; F_3, F_4 if $|K_1 \cap K_2|=1$. These forbidden configurations are shown in Fig. 2.

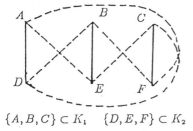

$$\{A,B,C\} \subset K_1 \quad \{D,E,F\} \subset K_2$$

Fig. 1. A forbidden configuration for a strict 2-threshold graph

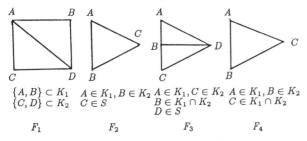

$\{A,B\} \subset K_1$ $A \in K_1, B \in K_2$ $A \in K_1, C \in K_2$ $A \in K_1, B \in K_2$
$\{C,D\} \subset K_2$ $C \in S$ $B \in K_1 \cap K_2$ $C \in K_1 \cap K_2$
                                       $D \in S$

F_1 F_2 F_3 F_4

Fig. 2. Forbidden configurations for strict 2-split graphs

Section 4. The Parallel Algorithm and the Complexity Analysis

Our parallel algorithm is on the CREW PRAM model. There are two fundamental operations in this algorithm. The first operation is to get a submatrix of the adjacency matrix. It spends us $O(\log n)$ time by using $O(n^2/\log n)$ processors. The second operation is row addition (or column addition). Every row(or column) of the adjacency matrix have at most n numbers. There is a parallel algorithm which can add n numbers in $O(\log n)$ parallel time by using $O(n/\log n)$ processors.

Our algorithm is described in the following:

INPUT : The adjacency matrix $M(n \times n)$ of G.

OUTPUT: Is G a strict 2-threshold graph? If it is, output its standard representation.

Step 1. Test if the graph itself is a stable set. If it is, output "G is a stable set and hence a threshold graph with threshold dimension 0." and stop. Otherwise, go to step 2.

Step 2. Find the set C of cliques as stated in Theorem 3.2.

Step 3. For each clique K in C, test if the graph is a threshold graph $(K, V - K)$ with threshold dimension 1. If it is, output "G is a threshold graph with threshold dimension 1." and stop. Otherwise, go to step 4.

Step 4. Construct all clique pairs (K_1, K_2) with $K_1, K_2 \in C$ and $|K_1 \cap K_2| \le 1$.

Step 5. For each clique pair (K_1, K_2) obtained in step 4, test if $(K_1, K_2, V - K_1 - K_2)$ is a candidate strict 2-split graph. Test if $V - K_1 - K_2$ is a stable set. If it is, go to Step 6. Otherwise, $(K_1, K_2, V - K_1 - K_2)$ can not be a representation for a candidate strict 2-split graph. Discard this clique pair (K_1, K_2).

Step 6. Detect if the candidate strict 2-split graph $(K_1, K_2, V - K_1 - K_2)$ obtained in step 5 has the for-

bidden configuration depicted in Figure 1. If it has the forbidden configuration, the clique pair (K_1, K_2) can not act as the clique pair in a representation for a strict 2-threshold graph. In this case, discard (K_1, K_2). Otherwise, go to step 7.

Step 7. Test if the candidate strict 2-split graph $(K_1, K_2, V - K_1 - K_2)$ is a strict 2-split graph. Test if the candidate strict 2-split graph $(K_1, K_2, V - K_1 - K_2)$ contains the forbidden configurations described in Theorem 3.3. If this candidate strict 2-split graph contains a forbidden configuration, it can never be a strict 2-split graph. Discard the clique pair (K_1, K_2). Otherwise, go to step 8.

Step 8. Partition the edges outside the clique pair (K_1, K_2) into G_1 and G_2 in order to obtain the representation $(G_1, G_2) = ((K_1, S_1), (K_2, S_2))$ for the strict 2-split graph.

Step 9. Test if this strict 2-split graph $((K_1, S_1), (K_2, S_2))$ is a strict 2-threshold graph. If it is, output "G is a strict 2-threshold graph." and stop.

The time complexity of each step is $O(\log n)$. The processor complexity of step 1 is $O(n^2/\log n)$. From step 2 to step 4, $O(n^3/\log n)$ processors are needed and from step 5 to step 9, $O(n^4/\log n)$ processors are needed. Therefore, our algorithm recognizes the strict 2-threshold graph in $O(\log n)$ parallel time by using $O(n^4/\log n)$ processors.

Section 5. Concluding Remarks

In this paper, we state some properties of the strict 2-threshold graphs. By using these properties, we propose an NC algorithm to recognize these graphs. Our algorithm is constructive. It specifies all possible standard representations for the graph once it is recognized to be strict 2-threshold. This is very useful because the structure of the graph may help us to solve other problems on this graph. Furthermore, our algorithm is time optimal.

References

[1] V. Chvátal and P.L. Hammer, Aggregation of inequalities in integer programming, Ann. Discr. Math. 1 (1977) 145-162.

[2] R. Cole, Parallel merge sort, SIAM J. Comput. 17 (1988) 770–785.

[3] M.C. Golumbic, Algorithmic Graph Theory and Perfect Graphs (Academic Press, New York, 1980).

[4] P.L. Hammer, N.V.R. Mahadev and U.N. Peled, Some properties of 2-threshold graphs, Networks 19 (1989) 17-23.

[5] W.-L. Hsu, Perfect Graphs, Technical Report TR-88-16, Institute of Information Science, Academia Sinica, R.O.C. (1988).

[6] N.V.R. Mahadev. and U.N. Peled, Strict 2-threshold graphs, Discrete Applied Mathematics 21 (1988) 113-131.

Using Separators Instead of Dynamic Programming in Approximation Algorithms for Planar Graphs

Fang Wan

Gregory E. Shannon

Department of Computer Science

Indiana University

Bloomington, Indiana 47405

April, 1991

Abstract. This paper describes a deterministic parallel algorithm for finding a near-optimal approximate maximum independent set in any planar graph. The set's cardinality is within a $\frac{k-1}{k}$ factor of optimal. Given a breadth-first search tree on a planar graph, the parallel algorithm uses $O(k^2 \log n)$ time and $64^k n$ processors on the CRCW-PRAM model and is the most efficient to date. The key technique is using separators to replace the function of dynamic programming used in the related sequential algorithm by Baker. The same technique for planar graph extends to other approximations for hard optimization problems such as minimum vertex cover.

1. Introduction. Finding approximate solutions to NP-complete or NP-hard optimization problems is an important issue in both theory and practice and has proven to be non-trivial [5]. However, Baker found an efficient sequential near optimal approximate solutions to a group of NP-complete and NP-hard problems on planar graphs, such as finding a maximum cardinality independent set. How to solve the same problems *efficiently* with similar precision in parallel has proven to be a difficult question.

For the definition of *CRCW-PRAM*, *DNC algorithm* and *linear-processor algorithm*, see [6].

Baker's sequential algorithm finds an independent set within a $\frac{k-1}{k}$ factor of optimal in $O(8^k n)$ time. In her algorithm, a planar graph is decomposed into k-outerplanar graphs whose structures are relatively simple and whose maximum independent sets can be found in $8^k n$ time by a dynamic programming (DP) technique. However, as pointed out by Chrobak and Naor [4], there is currently no efficient way in parallel to decompose a planar graph into k-outerplanar graphs without invoking a breadth-first searching (BFS) and to perform the DP on k-outerplanar graphs using less than n^2 processors. Between the two bottle necks, the DP one is the worse since there is a DNC BFS algorithm using $n^{1.5}$ processors [9].

Our DNC algorithm also finds an independent set within a $\frac{k-1}{k}$ factor of optimal. It uses $O(k^2 \log n)$ time and $64^k n$ processors on a CRCW PRAM when a BFS tree of a planar graph is given. If k is $\log \log n / 6$, then the approximate solution converges to optimal when n increases and only $O(n \log n)$ processors are necessary. In the same way that Baker's algorithm does, our algorithm implies parallel algorithms for finding approximate solutions to other hard problems such as minimum vertex covers in a planar graphs.

Our key technique is to avoid using DP. We decompose a planar graph into belt shaped connected components based on their depths in a BFS tree, with the tree given as part of the input. These *belts* can be triangulated into graphs of small diameters. We then show that a planar graph of small diameter can be decomposed into components of small boundary using separators. Most importantly, when we merge these components back into the original graph, all intermediate components also have small boundaries. Therefore we can compute maximum independent sets on these basic components and intermediate components using brute-force methods. With this technique, we are able to achieve a parallel approximate maximum independent set algorithm using only $O(n)$ processors when k is a constant. How to overcome the BFS bottle neck to achieve a linear-processor parallel approximate maximum independent set algorithm is still an open problem.

Like Baker's algorithm, many efficient sequential graph algorithms depend on DP. Although many researchers have investigated parallel algorithms for various DP-based algorithms [2, 1], those parallel algorithms are designed for specific classes of DP problems and do not apply to our problem. Our result, however, does provide a new approach for removing DP from an algorithm and replacing it with manipulations of separators. This technique is very appropriate for a wide range of problems on planar graphs.

Chrobak and Naor presented a linear-processor DNC algorithm for finding a large independent set in a planar graph [4]. However, the size of independent set from their algorithm has a maximum relative error of $\frac{1}{2}$ compared with $\frac{1}{k}$ for our and Baker's algorithm. Also, their techniques do not appear to extend to other related problems in the same way that ours do.

2. Algorithm and Analysis. A *cycle (path) separator* of a planar graph is a simple cycle (path) containing no more than two third of vertices of the graph either inside (right side) or outside (left side). A graph is a *k-radius graph* if it has a BFS tree whose maximum height is no larger than k. And the root of the BFS tree is called a *k-center* of the graph. We use $d(v_i, v_j)$ for the shortest distance between vertices v_i and v_j. In this paper we only deal with planar graphs and always assume embedded planar graphs because parallel complexities of embedding a planar graph [10] do not dominate our algorithm.

Our approximate maximum independent set algorithm, like Baker's sequential algorithm, is based on the well known Pigeon Hole Principle. We decompose vertices of a planar graph into k disjoint subsets. Any of the k vertex subsets divides the planar graph into simple shaped maximal connected components called *k-depth* graphs. We will show that finding maximum independent set in a k-depth in parallel is easier than in a k-outerplanar graph. By the Pigeon Hole Principle, one of the k subsets contains at most $1/k$ of the independent vertices of a given maximum independent set. Therefore k-depth graphs separated by this subset contain at least $(k-1)/k$ of vertices of the given maximum independent set. A union of maximum independent sets in these k-depth graphs is an independent set of the original planar graph and has a cardinality at least as large as a $(k-1)/k$ factor of the best possible.

2.1. k-depth graphs. Given a vertex, v_0, in a planar graph, G, we decompose vertices of G into subsets, $V_{i,j} = \{v \mid d(v_0, v) = j \cdot k + i\}$, where $0 \le i \le k - 1$. We use V_i to denote $\bigcup_{0 \le j} V_{i,j}$. The k-depth graphs related to V_i are maximal connected components after V_i is removed from G.

Each k-depth graph can be triangulated into a graph whose diameter is k. We can link a vertex of the smallest depth to other vertices of the same depth before triangulating the rest part.

Lemma 1 *Every k-depth graph can be triangulated into a k-radius planar graph.*

We decompose a planar graph into k-depth graphs because planar graphs of small diameter can be father decomposed into graphs of small boundary.

2.2. k-split trees. Given two graphs joining at k vertices on their boundaries, we can find the maximum independent set of their union by using a brute force if we know all maximum independent sets of the two graphs which are decided by configurations of the k vertices. A k-split tree is a binary tree which decides how a graph is decomposed into small components. Each node represents a subgraph and a separator that divides the subgraph into two child nodes; the total size of the boundary of the subgraph and its separator is less than k. Given a k-split tree of a small k, finding a maximal independent set by brute force is efficient enough.

Theorem 1 *Given a planar graph, G, of n vertices and its k-split tree of h height, a maximum independent set of G can be found in $O(k \cdot h)$ time on $2^k n$ processors.*

2.3. Find a maximum independent set in a k-depth graph. We can show that every k-radius planar graph has a $6k$-split tree. It was shown shown that every k-radius planar graph has a $O(k)$ size separator and the separator can be found in parallel [7, 8]. But how to split a k-radius planar graph as described by a $6k$-split tree was unknown.

Given a BFS tree of a connected planar graph of n vertices, we assume a preorder listing, $L_{0,n-1} = \{v_0, \cdots, v_{n-1}\}$, of the n [11]. Edges, (v_i, v_j), are called *spans* if $0 \le i \le \lfloor n/2 \rfloor$ and $\lceil n/2 \rceil \le j \le n - 1$. We can embed a k-radius planar graph, alone with its BFS tree, on the surface of a cone and only let spans lay on the bottom surface. Two spans and paths from the root of the BFS tree to the spans form a boundary of a component called a *caliper*. Every caliper has two *arms* separated by spans. A caliper that has only two spans is called a *minimal caliper*. Two calipers are neighbors if and only they share one span and only joint each other on their boundaries.

Lemma 2 *Given any k-radius planar graph, G, of n vertices and n_0 spans and its BFS tree, G can be decomposed into minimal calipers in $O(\log n_0)$ time on n processors.*

Lemma 3 *As defined above, calipers of a k-radius planar graph of n vertices have the following properties: (1) The union of two neighbor calipers is still a caliper; (2) The boundary of any caliper is no larger than $4k$; (3) Any two neighbor calipers share no more than $2k$ vertices on their boundary; (4) An arm of a minimal caliper has no more than $\frac{n}{2}$ vertices inside.*

We can split distinct arms of minimal calipers simultaneously and iterate the process until we find a $6k$-split tree. The way that a graph is split is also the way that the graph is merged back from the lowest level components of the related $6k$-split tree.

Theorem 2 *Every k-radius planar graph of n vertices has a $6k$-split tree. Given a BFS tree of a k-radius planar graph, a $6k$-split tree of the graph is of $O(\log n)$ height and can be found in $O(\log n)$ time on n processors.*

Corollary 1 *A maximum independent set of a k-radius or a k-depth planar graph of n vertices can be found in $O(k \log n)$ time with $64^k n$ processors.*

2.4. Conclusions and remarks.

Theorem 3 *Given a BFS tree of a planar graph, a maximal independent set in the graph can be found in $O(k^2 \log n)$ time using $64^k n$ processors. The cardinality of the independent set is within a $(k-1)/k$ factor of optimal.*

Trivially we can use the above scheme, with the same parallel complexities, for finding near optimal approximate minimum vertex cover in planar graphs. The cardinality of the vertex cover is within a $(k+1)/k$ factor of optimal. We believe it is the same with some other optimization problems on planar graphs, such as maximum tile salvage, partition into triangles, maximum H-matching, minimum dominating set and minimum edge dominating set [5, 3].

Reference

[1] A. Aggarwal, D. Kravets, J. Park, and S. Sen. Parallel searching in generalized monge arrays with applications. In *Proceedings of the 3rd ACM Symposium on Parallel Architectures and Algorithms* (1990) 259–267.

[2] M.J. Atallah, S.R. Koaraju, L.L. Larmore, G. Miller, and S. Teng. Constructing trees in parallel. In *Proceedings of the 2nd ACM Symposium on Parallel Architectures and Algorithms* (1989) 421–431.

[3] B. Baker. Approximation algorithms for NP-complete problems on planar graphs. In *Proceedings of the 24th IEEE Symposium on Foundations of Computer Science* (1983) 265–273.

[4] M. Chrobak and J. Naor. An efficient parallel algorithm for computing a large independent set in a planar graph. In *Proceedings of the 2nd ACM Symposium on Parallel Architectures and Algorithms* (1989) 379–387.

[5] M. Garey and D. Johnson. *Computers and Intractability: A Guide to the Theory of NP-Completeness.* W.H. Freeman and Company, New York, 1979.

[6] R. Karp and V. Ramachandran. A survey of parallel algorithms for shared-memory machines. In *Handbook of Theoretical Computer Science*, North-Holland, Amsterdam, 1990.

[7] R. Lipton and R. Tarjan. A separator theorem for planar graphs. *SIAM J. of Appl. Math.*, 16:346-358, 1979.

[8] G. Miller. Finding small simple cycle separators for 2-connected palanar graphs. *J. Comput. Syst. Sci.*, 32(3):265-279, June 1986.

[9] V. Pan and J. Reif. Fast and efficient solution of path algorithm problems. *J. Comput. Syst. Sci.*, 38(3):494-591, 1989.

[10] V. Ramachandran and J. Reif. An optimal parallel algorithm for graph planarity. In *Proceedings of the 30th Annual Symposium on Foundations of Computer Science* (1989) 282–287.

[11] R. Tarjan and U. Vishkin. An efficient parallel biconnectivity algorithm. *SIAM J. on Computing*, 14(4):862-874, 1985.

Dynamic Detection of Forest of Tree-Connected Meshes

Esther Jennings   Andrzej Lingas

Department of Computer Science, Lund University, Box 118, S-221 00 Lund, Sweden

Lenka Motyčková

Department of Computer Engineering, Brno Institute of Technology, Czechoslovakia.

Abstract
We describe a distributed algorithm which recognizes meshes connected by bridges in a network graph. A forest of tree-connected meshes is dynamically maintained by local updates of meshes and bridges as the network changes.

1. Introduction
Since the identification of special structures within a network can be used in to optimize distributed algorithms, we present a dynamic algorithm for recognition of tree of meshes. A *bridge* is defined to be a link whose removal separates the network; and a *mesh* is a maximal sub-network not containing a *bridge*.

We present an algorithm which runs in two modes. In mode 1 (initial labeling), the entire network is recognized as a tree of meshes. In mode 2 (maintenance), network changes are processed to maintain a forest of tree-connected meshes dynamically.

The labeling procedure utilizes an echo algorithm described in [3] for finding bi-connected components.

2. Network Model
Each processor is autonomous and interacts with its neighbors by message passing. The processors have unique identities. There is no central controlling process nor global storage nor global clock. Our algorithm utilizes message passing and parallel graph traversal techniques taking advantage of potentially simultaneous activies. Processors operate asynchronously. Messages sent over a link incur an arbitrary but finite delay. The order of messages is preserved; no messages are lost nor duplicated except during link failures. Messages lost during a link failure do not affect our algorithm since the link failure makes the information contained in those messages obsolete. If messages should arrive at a processor simultaneously, they are processed in an arbitrary sequential order. We assume that the network behavior is reasonable, i.e. network changes are not too frequent, which makes updates possible.

3. Forest of Tree-connected Meshes
We represent the network as a graph $G = (V, E)$ where V is the set of processors and E is the set of links. We use the labeling procedure (described below) to determine a rooted spanning tree of G. Information about the edges of the spanning tree is stored distributedly in the nodes.

Let $T = (M, B)$ be a tree of meshes, where M is the set of meshes and B is the set of bridges in G. Furthermore, $M = \{m_1, m_2, \ldots, m_j\}$, in which the elements of M are disjoint subsets of nodes in V (the vertex set of G), such that $\bigcup_q m_q = V$ and $m_q \cap m_l = \emptyset$, $1 \leq q, l \leq j$ and $q \neq l$.

The goal of our algorithm is to identify the bridges and meshes in the network. Given G, we first determine a rooted spanning tree of G. While determining the rooted spanning tree, bi-connected components and bridges are identified. Our algorithm labels all nodes and all edges of the graph G so that each node carries the identity of the mesh to which it belongs and each bridge is labeled as such to distinguish it from edges within meshes.

4. The Labeling Procedure
We use the echo algorithm [3] to traverse G and determine a rooted spanning tree. The echo algorithm has two phases. In the first phase (down-ward traversal) *DISCOVER* messages are sent, and in the second phase (echo or upward traversal), *RETURN* messages are sent.

To start, we select one of the nodes to be the root of the spanning tree. The elected root starts by sending a *DISCOVER* message to itself. Each node receiving a *DISCOVER* message for the first time propagates the message in parallel over all its active links except the incoming link for the message. The echo phase starts when a *DISCOVER* message reaches a node previously visited by a *DISCOVER* message. While echoing, *RETURN* messages are sent in the opposite direction of the *DISCOVER* messages. When all of the expected *RETURN* messages of a node have arrived at that node, the messages are merged into a new *RETURN* message to be passed further back.

Two sets of nodes are associated with each *RETURN* message. One of these is a set of *terminal nodes* (nodes that start the echoing phase), and the other is a set of *intermediate nodes* which the *RETURN* message passes through in the echoing phase. By comparing these sets of nodes, the bi-connected components are identified at cut-nodes of the network. The echoing phase terminates when the root of the spanning tree over the entire network is reached. Every node in G now has a *Mesh_Id* that identifies the mesh it belongs to. T, an embedding of G in which the meshes and bridges are identified, is now obtained. We assume that the network is connected and static when we run the labeling procedure in mode 1.

5. Dynamic Network Change Processing

In a dynamically changing network, the appropriate procedure for handling each network change event is employed. The possible basic network changes are: (1) a link connecting two nodes within the same mesh has failed, (2) a link connecting two nodes within the same mesh has recovered, (3) a link between two meshes (a bridge) has failed, and (4) a link between two meshes has recovered.

In Case (1), we determine whether the mesh decomposes into two meshes joined by a bridge. We apply the same labeling procedure locally on the mesh. When several link-failures of this type occur simultaneously while the mesh is being re-labeled (the root is FROZEN in this case to indicate that the mesh is re-configuring), requests for another local re-labeling are remembered at the root of the mesh until the current re-labeling is finished. When the current re-labeling finishes, these pending *START_DISCOVER* requests are processed by executing the labeling procedure only once more.

In Case (2), on a link-up between two nodes within a mesh, we only need to update the information at the processors incident to the link.

In Case (3), the tree of meshes decomposes into a forest of tree-connected meshes.

In Case (4), a link-up event between two meshes is considered. This causes either a new mesh to be formed, or the connection of two trees. Several events of this type can be processed in parallel. Let x and y denote the processors incident to the recovered link. We need to compute the lowest common ancestor (LCA) of x and y. We then include it, together with nodes on the upward paths from x and y to their

LCA, into the new mesh. The *LCA*, if found, is the root of the new mesh. If an *LCA* of these two nodes does not exist, this is the case where two trees are connected.

We propose an iterative method for finding the LCA. At each iteration, the message travels twice the distance of the previous iteration up the spanning tree. To avoid traveling too far unnecessarily, x and y synchronize via a *SYNC* message upon the completion of each iteration. The same procedure is invoked by both x and y. The process halts under two conditions: (i) two different tree-roots are reached, without finding the lowest common ancestor, (ii) the *LCA* is found. A *NEW_MESH* message is sent from the root of the new mesh to update the *Mesh_Id* of the new members of its mesh.

The proof of correctness is omitted here due to space limitations. For a detailed analysis of the communication and time complexity of our algorithm, readers are refered to our full length paper. To summarize, the communication complexity of the algorithm is $O(|E| + \sum_k |E_j|)$, where $|E|$ and $|E_j|$ are the number of links in the network and the affected mesh respectively. The time complexity of the algorithm is $O(D + \sum_k d_j)$ where D and d_j are the diameters of the network and the affected mesh respectively.

6. Discussion

Cidon and Gopal presented a distributed algorithm for dynamic detection of tree subgraphs in computer networks [1]. The communication complexity of their algorithm is $O(kn)$ where k is the number of link recoveries or failures, and they claim that the algorithm takes finite time to execute. Our algorithm solves the open problem of detecting a forest of tree-connected meshes in a network. Another open problem posed by Cidon and Gopal is to recognize mesh connected cliques in a dynamically changing network.

References

[1] Israel Cidon and Inder S. Gopal, *"Dynamic Detection of Subgraphs in Computer Networks"*. Algorithmica (1990), pp.277-294.

[2] To-Yat Cheung, *"Graph Traversal Techniques and the Maximum Flow Problem in Distributed Computation"*, IEEE Transactions on Software Engineering, Vol. SE-9, No. 4, July, 1983.

[3] Ernest J. H. Chang, *"Echo Algorithms: Depth Parallel Operations on General Graphs"*, IEEE Transactions on Software Engineering, Vol. SE-8, No. 4, July 1982.

A Parallel Approximation Algorithm for 0/1 Knapsack

Thomas E. Gerasch[*]

The MITRE Corporation

McLean, VA 22102-3481

Abstract

The 0/1 knapsack problem is an NP-Complete problem for which considerable interest has been generated in developing approximation algorithms. A parallelization of a well-known ε-approximation scheme for the knapsack problem, in the EREW PRAM model of computation, is presented here. The approximation scheme requires $O(n \log(n/\varepsilon) + n + \log n)$ time and $2n/\varepsilon$ processors, where n is a knapsack problem instance's size and ε is the required percentage error from optimal satisfied by the approximate solution. The time and processor requirements of the ε-approximation scheme presented here are reasonable, and the memory reference model used is the weakest such for PRAMs.

Keywords: knapsack, parallel approximation algorithm, EREW PRAM, dynamic programming.

1. Introduction

The 0/1 integer knapsack problem is a well-known, combinatorial optimization problem. An instance of the problem consists of an assignment of values for the profits and weights, p_i and w_i, respectively, and a capacity C. The problem is to maximize $\Sigma_{i=1}^{n} x_i p_i$, subject to $\Sigma_{i=1}^{n} x_i w_i \leq C$,

where $x_i \in \{0,1\}$ for $1 \leq i \leq n$. A feasible solution to a problem instance is a set of x_i values, $x_i \in \{0,1\}$, for which the weight constraint is satisfied. As with many combinatorial optimization problems, the decision form of the 0/1 knapsack problem is known to be NP-Complete [GJ79]. The knapsack problem falls into a class of problems for which there are a variety of classes of approximations [HS78], the strongest of which is the ε-approximation scheme [IK75, LE79]. An ε-approximation scheme produces an approximate solution to a problem instance whose profit value P satisfies P*-P$\leq_\varepsilon$ P*, where P* denotes the maximum profit attainable over all feasible solutions.

There have been several parallelizations of the knapsack ε-approximation scheme [GRK86,GT88,PR87] for parallel random access machines (PRAM)[FW78]. One such parallel algorithm, in the concurrent read, exclusive write (CREW) memory model PRAM [PR87] uses static interval partitioning [SS76] of the profit space to obtain an $O(n^2/p\varepsilon)$ time ε-approximation scheme using $1 \leq p \leq 2n/\varepsilon$ processors. A recursive divide-and-conquer approach using a refinement of the parallel merging used in this algorithm requires $O(\log^2 n + \log n \log(1/\varepsilon))$ time using $O(n^3/\varepsilon^2)$ processors. A further refinement of the merging procedure leads to an algorithm requiring at most $n^{2.5}/\varepsilon^{1.5}$ processors [GRK86].

The algorithm presented here is in the weakest PRAM model, using exclusive read and exclusive write (EREW) access to memory. It directly parallelizes the optimization loop of the original ε-approximation scheme [IK75]. The EREW PRAM ε-approximation scheme uses $2n/\varepsilon$ processors and takes $O(n \log(n/\varepsilon) + n + \log n)$ time.

[*]This research was partially supported under The MITRE Corporation's Sponsored Research Program.

2. Knapsack ε-Approximation Scheme

The ε-approximation scheme appropriately scales the profit values of the knapsack items to guarantee the desired maximum percentage error when translating from the scaled problem's profit values back to the original profit values. An optimal solution is found to the scaled problem using a dynamic programming technique while applying a dominance rule that limits the growth of solutions. The scaling factor is calculated so that the optimal solution for the scaled problem is dependent only on n and $1/\varepsilon$. The following is an outline of the ε-approximation scheme for the knapsack problem.

Knapsack ε-Approximation Scheme

[1] {Estimate P^*} Find P_0 such that $P_0 \leq P^* \leq 2P_0$.
[2] {Scale the knapsack items}
    $K := \varepsilon P_0/n$ {scaling factor}
    for all items **do**
        $q_i := \lfloor p_i /K \rfloor$ {scaled profit values}
    end for all
[3] Find an optimal solution to the knapsack problem instance having profit values q_i and weight values w_i for $1 \leq i \leq n$, and having capacity C. The optimal solution $\{x_i: 1 \leq i \leq n\}$ to the scaled problem instance is the feasible solution for the original problem instance.

Theorem 1. The algorithm outlined is an ε-approximation scheme for the 0/1 integer knapsack problem.

Proof: See [LE79].

3. Parallel Knapsack ε-approximation Scheme

The estimate for P^* can be obtained by first sorting the items into nonincreasing profit/weight ratios. The sorting step can be accomplished in $O(\log n)$ steps using n processors [AKS83]. If $m = \max\{k: 1 \leq k < n, \Sigma_{i=1}^{k} w_i \leq C, \Sigma_{i=1}^{k+1} w_i > C\}$,

then $P_0 = \max\{\Sigma_{i=1}^{m} p_i, \max_i\{p_i\}\}$ satisfies $P_0 \leq P^* \leq 2P_0$ [LE79].

The summations of profit and weight values are easily obtained using a summative parallel prefix operation [KRS85,BE87], which can be implemented in the EREW PRAM model using n processors in $O(\log n)$ steps. The computation of the scaled profit values is also easily parallelizable in the EREW PRAM model, requiring constant time using n processors.

The maximum profit obtainable for the scaled problem instance is bounded above by $2n/\varepsilon$ [LE79]. For each possible total scaled profit value q, $0 \leq q \leq 2n/\varepsilon$, let T(q) denote a 4-tuple (P(q), W(q), new_loc(q), last_item(q)). Each T(q) represents a feasible solution. For a scaled profit value q, $W(q) \leq C$ is the weight of the feasible solution and P(q) is its profit using the original profit values. The feasible solutions for the optimization step are represented in linked lists. The element last_item(q) is used to index the last item placed in the feasible solution whose total (scaled) profit value is q. The element new_loc(q) is used as a temporary location to index new locations in memory when a new item is to be added to a feasible solution. The new_loc and last_item array elements index locations in a pool of memory that is used to hold the items in the linked lists comprising the feasible solutions. This pool of memory is represented using two arrays item_index(0..M) and previous_item(0..M), where $M=n(2n/\varepsilon)$. Each list element consists of a knapsack item index held in the item_index component, and a pointer to the previous element in the linked list, held in the previous_item component.

The parallelization of the dynamic programming loop for the scaled knapsack problem employed in the ε-approximation scheme is a direct parallelization of the optimization loop of the original scheme [IK75]. In the original serial algorithm, as each new item is considered, feasible solutions are considered in decreasing order of total scaled profits. This means that feasible solutions are updated from lower indexed positions which contain only items indexed between 1 and i-1. This implies that the steps of the optimization loop can be performed in parallel, and, in fact, in a synchronous fashion.

Parallel Knapsack Dynamic Programming

```
[1]  for all i, 1 ≤ i ≤ 2n/ε do T(i) := ∅ end for all
        T(0) := (0,0,NULL,NULL)
[2]  for 1 ≤ i ≤ n do {serial loop over knapsack items}
[3]     for all j, 0 ≤ j ≤ 2n/ε -qᵢ, do
[4]        if T(j) ≠ ∅ and W(j)+wᵢ ≤ C then
[5]           if T(j+qᵢ) = ∅ or W(j+qᵢ) > W(j)+wᵢ then
[6]              {solution j with item i dominates solution j+qᵢ}
                 new_loc(j+qᵢ) := (i-1)(2n/ε) + (j+qᵢ)
                 item_index(new_loc(j+qᵢ)) := i
                 previous_item(new_loc(j+qᵢ)) := last_item(j)
                 last_item(j+qᵢ) := new_loc(j+qᵢ)
                 P(j+qᵢ) := P(j)+pᵢ
                 W(j+qᵢ) := W(j)+wᵢ
              endif
           endif
        end for all
     end for
```

Step 1 of the parallel dynamic programming is obviously parallelizable in the EREW PRAM model. The body of the **for all** loop in step 3 uses $2n/\varepsilon+1$ processors. Those processors whose processor numbers satisfy $0 \leq j \leq 2n/\varepsilon - q_i$ execute steps 4-6, with processor j being responsible for one of the representative sequences of steps. Processor j executes the next step only if the conditions tested prior to the step are satisfied. Step 6 presents the details of the updates of feasible solutions that are dominated.

The pool of memory used to implement the linked lists consists of two arrays, item_index and previous_item, both indexed from 0 to $M=n(2n/\varepsilon)$. Allocation from these arrays for use in the linked lists is done in blocks of size $2n/\varepsilon$. Array elements are not returned to the pool for reuse. This allocation strategy takes constant time for each iteration. (An allocation strategy that assigns locations from the memory pool only to processors executing step 6 can be implemented in $O(\log(2n/\varepsilon))$ time using a summative parallel prefix operation.)

In steps 3-6 access to the values of q_i, p_i, and w_i can be provided by generating a copy of each value for each processor; this can be done on an EREW PRAM in $\log(2n/\varepsilon)$ steps using a parallel prefix operation [BG87, KRS85]. Processors access distinct locations of the various arrays during steps 4-6, owing to the uniformity of array element indexing resulting from the uniform offset by q_i between pairs of feasible solutions in the steps. The previous discussion shows that the optimization loop takes $O(n \log(2n/\varepsilon))$ time using $2n/\varepsilon$ processors.

Locating the optimal solution to the scaled problem in step 3 of the ε-approximation scheme can be done using a parallel prefix operation to find the maximum-indexed, non-empty

solution. The traversal of the linked list to provide the indices of the items to complete step 3 of the ε-approximation scheme takes time $O(n)$ using a single processor.

Theorem 2. The parallelization of the knapsack ε-approximation scheme for the synchronous EREW PRAM model of computation takes $O(n \log(n/\varepsilon) + n + \log(n/\varepsilon))$ time using $2n/\varepsilon$ processors.

4. Conclusions

A parallelization of a well-known ε-approximation scheme for the 0/1 knapsack problem has been presented in the synchronous EREW PRAM model of computation. The algorithm requires $O(n \log(n/\varepsilon) + n + \log(n/\varepsilon))$ time and uses $2n/\varepsilon$ processors. The algorithm is in a weaker memory model for PRAMs than other previous parallelizations of the ε-approximation scheme, and its time and processor requirements are reasonable and permit implementation on currently available SIMD architectures.

References

[AKS83] Atjai, M., Komlos, J., and E. Szemeredi, 1983, "An O(NlogN) Sorting Network," *Proceedings of the 15th Annual ACM Symposium on the Theory of Computing*, pp. 1-9.

[BG87] Blelloch, G., 1987, "Scans as Primitive Parallel Operations," *Proceedings of the 1987 International Conference on Parallel Processing*, pp. 355-362.

[FW78] Fortune, S., and J Wyllie, 1978, "Parallelism in Random Access Machines," *Proceedings of the 10th Annual Symposium on the Theory of Computing*, pp. 114-118.

[GJ79] Garey, M., and D. Johnson, 1979, *Computers and Intractability: A Guide to the Theory of NP-Completeness*, W. H. Freeman, San Francisco, CA.

[GRK86] Gopalkrishnan, P., Ramakrishnan, I., and L. Kanal, 1986, "Parallel Approximate Algorithms for the 0/1 Knapsack Problem," *Proceedings of the 1986 International Conference on Parallel Processing*, pp. 444-451.

[GT88] Gerasch, T., 1988, "An SIMD Parallel ε-Approximation Scheme for 0/1 Knapsack," *Proceedings of the 2nd Symposium on the Frontiers of Massively Parallel Computation*, pp. 151-154.

[HS78] Horowitz, E., and S. Sahni, 1978, *Fundamentals of Computer Algorithms*, Computer Science Press, Rockville, MD.

[IK75] Ibarra, O., and C. Kim, 1975, "Fast Approximation Algorithms for the Knapsack and Sum of Subsets Problems," *Journal of the ACM* 22, 4 (Oct.), pp. 463-468.

[KRS85] Kruskal, C., Rudolf, L., and M. Snir, 1985, "The Power of Parallel Prefix," *Proceedings of the 1985 International Conference on Parallel Processing*, pp. 180-185.

[LE79] Lawler, E., 1979, "Fast Approximation Algorithms for Knapsack Problems," *Mathematics of Operations Research* 4, 4 (Nov.), pp. 339-356.

[PR87] Peters, J., and L. Rudolph, 1987, "Parallel Approximation Schemes for Subset Sum and Knapsack Problems," *Acta Informatica*, vol. 24, pp. 417-432.

[SS76] Sahni, S., 1976, "Algorithms for Scheduling Independent Tasks," *Journal of the ACM* 23, 1 (Jan), pp. 116-127.

EFFICIENT EXECUTION OF HOMOGENEOUS TASKS WITH UNEQUAL RUN TIMES ON THE CONNECTION MACHINE

Azer Bestavros   Thomas Cheatham *

DEPARTMENT OF COMPUTER SCIENCE
HARVARD UNIVERSITY
CAMBRIDGE, MA 02138.

Introduction

Many scientific applications [5] require the execution of a large number of identical *tasks*, each on a different set of data. Such applications can easily benefit from the power of SIMD architectures (*e.g.* the Connection Machine) by having the array of Processing Elements (PEs) execute the task in parallel on the different data sets kept on the different PEs.

It is often the case, however, that the task to be performed involves the repetitive application of the same sequence of steps, *a body*, for a number of times that depends on the input or computed data. For example, in simulating the transmission of Neutrons through a plate of Beryllium [4], a while loop is entered, calculating for each iteration, the next horizontal position of a given Neutron, and testing whether it has escaped, or has been absorbed. If so, it is deactivated. Otherwise, another iteration is executed with possibly different initial conditions due to possible collision with Beryllium atoms. To accurately determine the reflection, absorption, and pass through probabilities, the experiment has to be conducted on millions of Neutrons. Using a SIMD architecture where each Neutron is assigned to a PE becomes a necessity.

The usual technique used to encode such applications uses a coarse grain *task-level synchronization*. The idea is to execute repetitively the body of the loop in a SIMD manner [1]. When a PE is done with its task, it simply deactivates itself so it will not participate in the following iterations. This process continues until no PEs are active. For example, in the Neutron/Beryllium experiment, each Neutron is assigned to a PE. The PE keeps executing the body of the while loop until it determines that its Neutron has escaped or has been absorbed. The randomness associated with each Neutron's journey results in a large *variance* of the total number of iterations. Using task-level synchronization, the utilization of the array of PEs degrades substantially. In particular, the average execution time of a task becomes the *maximum* of the execution times of all the PEs (or tasks). The utilization of the array of PEs degrades even more when *virtual processing* is used. Virtual processing [7] allows programmers to scale-up their problems without worrying about the limited number of PEs in the available SIMD architecture.

In this paper, we propose a *body-level synchronization* scheme that would boost the utilization of the array of PEs while keeping the required overhead to a minimum. In particular, we aim at achieving an average task execution time that is the *mean* of the execution times of all the PEs. We mathematically analyze the proposed technique and show how to fine-tune its parameters to optimize its performance for a given application. Contrary to task-level synchronization, our technique becomes even more efficient when virtual processing is used. In this paper, we base our presentation and discussion on the Connection Machine architecture. Our methodology, however, can be easily applied to any other SIMD architecture.

*This work was supported by DARPA N00039-88-C-0163

Statement of the Problem

We assume that there is a set of tasks $T_1, \cdots, T_K$ to perform and that task T_j is accomplished by carrying out the following 3 steps:

1. Get some input and/or execute some initialization code.
2. Execute a body of computational steps for a number of times.
3. Execute some exit code and/or report results.

We refer to the body in step 2 as the α-cycle. We refer to the initialization and exit code in steps 1 and 3 as the ι-code and ω-code, respectively, and as the γ-code, collectively.

The basic issue addressed here is how to minimize the computational cost of carrying out the K tasks and disposing of the results, where it is assumed that K is large, relative to the number of available physical processors, P. If the number of steps, n_j, to complete task T_j is not a constant then it is probably a bad idea to simply assign the work to K virtual processors (even if there is sufficient memory to do so), because the cost would be on the order of:

$$\frac{K}{P} \max_{1 \leq j \leq K} n_j$$

where $\frac{K}{P}$ is the VP ratio [7]. What we hope to achieve is a cost closer to the average:

$$\frac{K}{P} \sum_{j=1}^{K} \frac{n_j}{K}$$

The usual technique used to encode such applications on SIMD architectures is to have each virtual processor take on the execution of one of the tasks as follows:

1. Inputs (if any) are distributed, and all PEs execute the ι-code.

2. The α-cycle is executed repetitively. When a PE is done with its task, it simply deactivates itself so as not to participate in following α-cycles. This step repeats until no PEs remain active.

3. All PEs execute the ω-code, and results are gathered.

The problem with the above procedure is that, as tasks are completed and PEs become inactive, the utilization of the PE array degrades dramatically.

Proposed Methodology

Instead of repetitively executing α-cycles until all tasks are finished, we execute α-cycles for a constant number of times m. At the end of these m cycles, tasks that finished their requested α-cycles are allowed to execute the ω-code and report their results. PEs associated with completed tasks become free and are allowed to start new tasks by getting input data and executing the ι-code. Once this is done, another round of m α-cycles is executed, and the same process repeats. We use the term β-cycle to mean the following sequence:

1. The input data (if any) is distributed to free PEs. All such PEs are labeled as busy and start by executing the ι-code.

2. The α-cycle is executed exactly m times. Only busy PEs participate. When a PE is done with its task, it simply deactivates itself so it will not participate in the remaining α-cycles.

3. Deactivated PEs execute the ω-code, and, if necessary, the results are gathered. Such PEs are declared free.

Performance Analysis

Average task execution time
Let t_α be the time it takes to execute an α-cycle and t_γ be the overhead time associated with a β-cycle. Furthermore, let m be the number of α-cycles executed in every β-cycle. The time necessary to execute one β-cycle is given by:

$$t_\beta = mt_\alpha + t_\gamma \tag{1}$$

Let n be the discrete random variable denoting the number of α-cycles necessary to execute a task Q. The probability density function of n is problem dependent. For the purpose of this paper, we assume that n follows a uniform distribution $[1, N]$, where $N = rm + s$. The total number of β-cycles necessary to terminate the task Q is $\lceil \frac{n}{m} \rceil$. Using equation 1, it follows that the total time it takes to terminate Q is:

$$t_Q = \lceil \frac{n}{m} \rceil (mt_\alpha + t_\gamma) \tag{2}$$

The expected task execution time, T_Q, is given by [2]:

$$T_Q = E(t_Q) = \frac{r+1}{2}(mt_\alpha + t_\gamma)(1 + \frac{s}{N}) \tag{3}$$

Two factors contribute to T_Q: the term $(r+1)t_\gamma$ reflects the effect of β-cycles overhead, whereas the term $(1 + \frac{s}{N})$ reflects the effect of synchronization overhead. For a constant m, and as the value of N (and consequently r) increases, the effect of t_γ, the overhead associated with β-cycles, becomes an issue. This, of course, depends on the ratio between t_γ and t_α. The smaller the value of r the lesser the effect of β-cycles overhead. On the other hand, as the value of N decreases, the synchronization overhead becomes significant. Obviously, some kind of a balance is needed.

Optimizing the Average Task Execution time
If we assume that N is much larger than m, and since s is necessarily smaller than m, we get that $\frac{s}{N} \approx 0$ and $r \approx \frac{N}{m}$. Using these approximations in equation 3, we get [2]:

$$T_Q = \frac{N+m}{2}(t_\alpha + \frac{1}{m}t_\gamma) \tag{4}$$

The value of m that minimizes the average task execution time in equation 4 can be calculated to be:

$$m = \sqrt{N\frac{t_\gamma}{t_\alpha}} \tag{5}$$

Equation 5 is valid as long as $N \gg m$. This condition, however, can be easily satisfied. For instance, if $N = 100$ and $\frac{t_\gamma}{t_\alpha} = 0.25$ (very conservative), we get $m = 5$, which, indeed, is much smaller than N. In [2] we have derived similar results for optimizing the utilization of the PE array. We refer the interested readers to that reference for further details.

Connection Machine Implementation

In order to judge the performance of our technique we have to estimate the overhead time, t_γ. The bulk of t_γ will be spent distributing data to the free PEs and collecting results from finished tasks. A straightforward technique for performing these tasks is using a parallel *rendez-vous* algorithm [3]. The idea is to keep the data for the tasks to be started in a virtual processor set (the input-VP-set) and, using a Parallel Prefix ranking procedure, each free processor is assigned a unique index. That index should be used to get the data for a new task from the input-VP-set. The time it takes to compute the rendez-vous index is 894 time units (see [7], pages 24-35.) The time it takes to send the data from the input-VP-set to the set of free PEs is 1111. Thus, distributing the data to the free PEs should take approximately 2000 time units. Gathering the results from the finished tasks can be done using the same *rendez-vous* technique, thus bringing the total overhead time to 4000. Now, let's assume an applications where the body of the computation takes $t_\alpha = 1000$ units of time (An addition or a multiplication on the CM takes approximately 100 units.) Moreover, assume that the number of iterations required per task is uniformly distributed and ranges from 1 to $N = 10000$, and that the total number of tasks to be executed is $P = 64,000$.

Substituting with the above numbers in Equation 5, we get an optimum value of $m = 200$. Now, substituting in Equation 4, we get an average execution time per task of 5,202,000. If *task-level* synchronization were used, this number would bounce to almost 10,000,000, almost double the time we achieve. The merits of our methodology become even more compelling if we look at the *total* execution time to finish all 64,000 tasks, using the available 4K PEs. Using *task-level* synchronization, it would take approximately 160,000,000 units of time, compared to only 83,232,000 using our *body-level* synchronization. All of the above gains become even more accentuated when, for an application like the Neutron/Beryllium experiment mentioned earlier, the number of tasks p is orders of magnitude larger and the number of iterations per task is unbounded.

Conclusion and Future work

Body level synchronization can be easily and efficiently implemented by compilers. In particular, handling virtual processor sets in SIMD architectures (*e.g.* the CM) can be made much more efficient if our approach rather than the usual task level synchronization approach is adopted. Our technique extends quite well to support arbitrarily *nested* computations [2]. More work remains in order to reduce the overhead associated with distributing/gathering data to/from PEs. In particular, techniques to avoid the expensive "send" communication – using "news" communication instead – should greatly reduce the overhead time, and further improve the performance. We are experimenting with a number of such alternatives (using pipelining and buffering) [1,2].

References

[1] Azer Bestavros, Thomas Cheatham, and Dan Stefanescu, "Parallel approaches for bin packing on the Connection Machine." *IEEE SPDP'90*, Dallas, Texas, December 1990.

[2] Azer Bestavros and Thomas Cheatham, "The Workers Model of Computation for the Connection Machine," *Internal Report, Department of Computer Science, Harvard University*, In progress.

[3] W. Hillis, G. Steele, "Data Parallel Algorithms," *Communications of the ACM*, December 1986.

[4] Frederick Onion, "PMCML: A Parallel Monte Carlo Modeling Language", *B.A. Thesis, Department of Computer Science*, Harvard University, May 1990.

[5] G. Marchuk, G, Mikhailov, M. Nazaraliev, M. Darbinjan, R. Kargin, and B. Elepov, *The Monte Carlo methods in atmospheric optics*, Springer-Verlag, Berlin, 1980.

[6] R. Rubinstein, *Simulation and the Monte Carlo Method*, John Wiley & Sons, Inc., 1981.

[7] "The Connection Machine Parallel Instruction Set – Ver 5.2", *Thinking Machines Corporation*, October 1989.

On a Massively Parallel ε-Relaxation Algorithm for Linear Transportation Problems[1]

Xiaoye Li and Stavros A. Zenios
Department of Decision Sciences
University of Pennsylvania, Philadelphia, PA 19104.

Abstract. We present the coordinate step algorithm (ε-relaxation) of Bertsekas and Eckstein [1] for linear transportation problems. We implement the algorithm on a Connection Machine CM-2 with 32K processing elements, and compare its performance with the parallel implementation of the network simplex given by Miller, Pekney and Thompson [3]. Computational results are reported for randomly generated test problems with 1 million arcs and 2 thousand nodes.

1 Introduction. The linear transportation problem [**LTP**] is the following:

$$Min \sum_{(i,j)\in\mathcal{E}} c_{ij}x_{ij} \qquad (1)$$

Subject to
$$\sum_{(i,j)\in\mathcal{E}} x_{ij} = s_i, \ \forall i \in V_O \qquad (2)$$

$$\sum_{(i,j)\in\mathcal{E}} x_{ij} = d_j, \ \forall j \in V_D \qquad (3)$$

$$0 \le x_{ij} \le u_{ij}, \ \forall (i,j) \in \mathcal{E} \qquad (4)$$

We use V_O to denote the set of origin nodes, V_D the set of destination nodes, $m_O = |V_O|$, $m_D = |V_D|$, $\mathcal{E} \subseteq \{(i,j)|i \in V_O, j \in V_D\}$ the set of arcs, x_{ij} the flow on arc $(i,j) \in \mathcal{E}$, s_i the supply of node $i \in V_O$, d_j the demand of node $j \in V_D$, u_{ij} the upper bound on the flow on arc (i,j). The surplus of a node is defined by $g_i = s_i - \sum_{(i,j)\in\mathcal{E}} x_{ij}, \forall i \in V_O$, or $g_j = \sum_{(i,j)\in\mathcal{E}} x_{ij} - d_j, \forall j \in V_D$.

[1]Research funded by NSF grant SES-91-00216 and AFOSR grant 91-0168.

2 The ε-Relaxation Algorithm. The dual problem of [LTP] is $Max\ q(p)$, subject to no constraint on p. The dual functional q is $q(p) = \sum_{(i,j)\in\mathcal{E}} q_{ij}(p_i^O - p_j^D) + \sum_{i\in V_O} s_i p_i^O + \sum_{j\in V_D} d_j p_j^D$, where each q_{ij} is the scalar function defined by

$$q_{ij}(p_i^O - p_j^D) = min_{0 \le x_{ij} \le u_{ij}}(c_{ij} + p_j^D - p_i^O)x_{ij} \qquad (5)$$

where $p = (p^O|p^D) \in R^{m_O+m_D}$ is the vector of dual prices.

The main idea of ε-relaxation algorithm is as follows: Force small single-node price increases even if they worsen the dual cost. The rationale is that if the cost deterioration is small then the algorithm can eventually approach the optimal solution. It can be shown [1] that an **exact** solution of the problem can be obtained in a finite number of iterations when the problem data are integers. We define first the notion of ε-*complementary slackness*: for any given vector p and $\varepsilon > 0$, we say that an arc (i,j) is

$$\begin{aligned}
\varepsilon - \text{inactive}, & \quad \text{if } p_i^O < p_j^D + c_{ij} - \varepsilon \\
\varepsilon^- - \text{balanced}, & \quad \text{if } p_i^O = p_j^D + c_{ij} - \varepsilon \\
\varepsilon - \text{balanced}, & \quad \text{if } p_j^D + c_{ij} - \varepsilon \le p_i^O \le p_j^D + c_{ij} + \varepsilon \\
\varepsilon^+ - \text{balanced}, & \quad \text{if } p_i^O = p_j^D + c_{ij} + \varepsilon \\
\varepsilon - \text{active}, & \quad \text{if } p_i^O > p_j^D + c_{ij} + \varepsilon
\end{aligned}$$

We say that a vector pair of primal/dual variable (x,p) satisfies ε-*complementary slackness* (ε-CS) if for each arc (i,j), $x_{ij} = 0$ if (i,j) is ε-inactive, $0 \le x_{ij} \le u_{ij}$ if (i,j) is ε-balanced, and $x_{ij} = u_{ij}$ if (i,j) is ε-active. The ε-relaxation algorithm works by maintaining a flow-price pair (x,p), that obeys ε-complementary slackness. It repeatedly selects nodes l with positive surplus g_l, and sets the corresponding price p_l to a value which is within ε of some maximizer of the dual cost with respect to p_l, with all other prices held constant. The flow vector is accordingly adjusted to maintain ε-compementary slackness. It was shown that if $\varepsilon < \frac{1}{N}$ then x is optimal [1]. For a node $i \in V_O$ the flow of an arc (i,j) will increase if the arc is ε^+-balanced and $x_{ij} < u_{ij}$ (ε^+-*unblocked*). For a node $j \in V_D$ the flow an arc (i,j) will decrease if the arc is ε^--balanced and $x_{ij} > 0$ (ε^--*unblocked*). The price of node i (j) at the end of the iteration is set to $\bar{p}_i^O$ ($\bar{p}_j^D$) + ε, where $\bar{p}_i^O$ ($\bar{p}_j^D$) is one of the maximizing points of the dual function along the i (j) th price coordinate. This price adjustment is called an *up iteration*.

The Exact Form of Up Iteration

Step 1. Let (x,p) satisfy ε-CS.

Step 2. (Scan arcs incident to node i with $g_i > 0$) Select a node j such that (i,j) is an ε^+-unblocked arc and go to Step 3. If no such arc can be found go to step 4.

Step 3. (Decrease surplus of node i by increasing x_{ij} on outgoing arcs) Let $\delta = min\{g_i, u_{ij} - x_{ij}\}$, update x_{ij} : $x_{ij} \leftarrow x_{ij} + \delta$, $g_i \leftarrow g_i - \delta$, $g_j \leftarrow g_j + \delta$. If $g_i = 0$ and $x_{ij} < u_{ij}$, stop; else go to Step 2.

Step 4. (Increase price) Set $p_i^O \leftarrow min_{\xi\in R_i^+}\xi$, where $R_i^+ = \{p_j^D + c_{ij} + \varepsilon|(i,j) \in \mathcal{E}, x_{ij} < u_{ij}\}$. If $g_i > 0$ and $R_i^+ = \emptyset$, the problem is infeasible; else go to Step 1.

3 Dense Implementation. We assume that the input graph $\mathcal{G}$ is dense. To represent an $m_O \times m_D$ transportation problem the CM-2 is configured as a 2-dimensional NEWS grid of dimensions $\lceil m_O \rceil_2 \times \lceil m_D \rceil_2$, where $\lceil n \rceil_2$ is n rounded up to the nearest power of two. Row i is associated with node i in the set of origin nodes. Column j is associated with node j in the set of destination nodes. In particular, the processor of the VP set with NEWS coordinate (i,j) stores the data for arc $(i,j) \in \mathcal{E}$. The memory of each virtual processor with NEWS address (i,j) is partitioned into several CM data fields illustrated in Figure 1. With the above layout of the memory, one iteration of the algorithm is executed as follows:

Step 1. Activate all PEs with $g_i > 0$; select ε^+-unblocked arcs/processors along axis_1, denote such processor sets as B_i; let $\omega = max\{g_i, \sum_{j\in B_i}(u_{ij} - x_{ij})\}$, update x_{ij} along axis_1: $x_{ij} \leftarrow x_{ij} + g_i\frac{u_{ij}-x_{ij}}{\omega}$; update surplus g_j for $j \in V_D$;

Step 2. Increase all p_i that satisfied $g_i > 0$ prior to Step 1, setting $p_i \leftarrow min\{p_j + c_{ij} + \varepsilon \mid (i,j) \in \mathcal{E}$ and $x_{ij} < u_{ij}\}$;

Step 3. Activate all PEs with $g_j > 0$; select ε^--unblocked arcs/processors along axis_0, denote such processor sets as B_j; let $\omega = max\{g_j, \sum_{i \in B_j} x_{ij}\}$, update x_{ij} along axis_0: $x_{ij} \leftarrow x_{ij} - g_j \frac{x_{ij}}{\omega}$; update surplus g_i for $i \in V_O$;

Step 4. Increase all p_j that satisfied $g_j > 0$ prior to Step 2, setting $p_j \leftarrow min\{p_i - c_{ij} + \varepsilon \mid (i,j) \in \mathcal{E}$ and $x_{ij} > 0\}$.

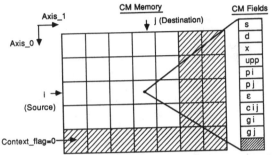

Figure 1: Memory configuration of the Connection Machine NEWS grid for the dense implementation.

C	16K (VP_ratio=64)		32K (VP_ratio=32)	
	Iteration	CM Time	Iteration	CM Time
10	100	12.75	100	6.34
100	400	52.29	433	27.58
1000	3000	289.83	2583	163.12
10000	7033	705.69	6733	423.83

Table 1: Running times of 1024 × 1024 problem. The supply/demand range is 100.

4 Computational Results.

All the test problems are generated by the Connection Machine random number generator. $\mathcal{G}$ is assumed to be fully dense. The generator provides control over several characteristics of the test problems: (1) size of the problem, $m_O = m_D$, (2) largest coefficient for arc cost, c_range, (3) maximum supply/demand at each node, sd_range. The algorithm will terminate when $max_{i \in V_O, j \in V_D}\left\{\frac{|g_i|}{s_i}, \frac{|g_i|}{d_j}\right\} < 10^{-3}$. The results reported are the average taken over five test runs. We use C/Paris to program CM-2 and a SUN4/360 as the front-end. The CM time in seconds is recorded, which accounted for more than 90% of the total execution time.

Table 1 gives the number of iterations and the running times of the algorithm for problems of size 1024 with various cost range. For a fixed size problem, when varying the cost range from small to large, we observe big difference of the running times. This shows that the complexity of the algorithm is increasing with respect to the maximum absolute value of the arc costs. To evaluate the improvement in performance that could be realized if a larger machine were available we solved one test problem with various VP ratios. From the degradation in performance we may es-

timate the solution time for solving a given problem on a fully configured 64K CM-2 when the VP ratio is half of the ratio needed on a 32K CM-2. The results are summarized in Table 2, including estimated time on the 64K machine.

CM-2	VP_Ratio	Iteration	CM Time
8K	32	2667	186.41
16K	16	2650	100.41
32K	8	2700	63.10
64K*	4		35.05

Table 2: Running times for a 512 × 512 problem. Cost range is 1000, supply/demand range is 100.

5 Concluding Remarks.

For large dense problems the ε-relaxation algorithm effectively takes advantage of the massive parallelism of the Connection Machine. The empirical results obtained so far lead us to believe that this approach is fairly robust especially when exploiting modest modifications to the original relaxation procedures. The results obtained on 32K CM-2 are competitive with those reported in [3] for which they used primal algorithm on a 14 processor BBN Buterfly computer. Our implementation on a full 64K CM-2 will outperform the primal algorithm on a small scale parallel machine by a factor of two, see Table 3.

Size	Primal simplex by Miller et al.[3]	ε-relaxation	
		32K	64K*
500	38.5	47.34	27.85
1000	168.4	126.31	74.30

Table 3: Comparisons of runing times of different algorithms. Cost range and supply/demand range are 1000.

References

[1] D. P. Bertsekas and J. Eckstein. Dual coordinate step methods for linear network flow problems. *Mathematical Programming*, 42:203–243, 1988.

[2] A. Goldberg and R. Tarjan. Solving minimum cost flow problems by successive approximation. *Proc. 19th ACM STOC.*, 7–18, 1987.

[3] D. Miller, J. Pekny, and G. Thompson. *Solution of Large Dense Transportation Problems Using A Parallel Primal Algorithm.* Report No. 546, Graduate School of Industrial Administration, Carnegie Mellon University, Pittsburgh, PA 15213, 1988.

[4] J. M. Wein and S. Zenios. On the massively solution of the assignment problem. *J. of Parallel and Distributed Computing*, 1990.

Wild Anomalies in Parallel Branch and Bound

Alan P. Sprague

Department of Computer and Information Sciences

UAB

Birmingham, AL 35294

email: sprague@cis.uab.edu

Abstract

Anomalies in the performance of the parallel branch and bound algorithm have been a continuing concern of researchers. Anomalies experienced in parallel solutions of Knapsack or other problems have been relatively mild; wild anomalies have heretofore been found only in cleverly constructed examples. We show that for instances of the Knapsack problem where profit density is held constant, strongly anomalous behavior is commonplace. This claim is backed by both experimental evidence (on the Sequent Balance) and theory.

1. Introduction

We assume familiarity with the sequential branch and bound algorithm: work proceeds as a succession of interations; in each iteration one node is expanded. In the parallel algorithm, parallelism is introduced at the iteration level: each processor takes a different node for expansion.

Pseudocode for the parallel algorithm strongly resembles the sequential algorithm. A single processor places the root on the liveset; then all processors simultaneously expand nodes in a "while" loop. This execution can be synchronous or asynchronous. Nodes which are available to be expanded are held in a liveset. The liveset may be maintained as a centralized data structure, or can be decentralized.

Certain design issues in the algorithm are architecture dependent. In a distributed memory machine the communication costs of maintaining a centralized liveset would be prohibitive, so the liveset is normally partitioned (one piece per processor), and load balancing becomes an issue. In a shared memory machine, contention for access to the liveset can make it attractive to partition the liveset.

The parallel algorithm which has been implemented on a 30–processor Sequent Balance 21000 is as follows. The algorithm is asynchronous, so one processor may complete an iteration and proceed to the next more rapidly than another processor. The liveset is broken into v pieces: where $v=1$, the liveset is then centralized and where $v>1$ it is decentralized. Hence whenever a processor needs to remove or insert a node on the liveset, it must first decide which piece of the liveset it will access. The parallel algorithm is the following:

Place the root on a liveset.
Do in parallel:
   While (not quit)
      Decide which liveset to remove a node from.
      Remove the best node from the liveset (if it has a node).
      If failure to obtain a node, abort this iteration.
      For each feasible child w of the node:
         Compute an optimistic estimate of w.
         If the optimistic estimate is better than the local incumbent:
            If w is a solution
               Make w the local incumbent.
            Else
               Decide which liveset to add w to.
               Add w to that liveset.

Each processor has an incumbent, and in addition, each liveset also has one. Processors do not directly communicate incumbent values with each other. Instead, each time a processor accesses a liveset, to remove or add a node, the incumbent of the processor is compared with the incumbent of the liveset, and the inferior of the two is replaced by the better. For further information regarding locks for liveset, termination detection, etc., see [5].

In the experimental work reported in Section 2, the number of livesets was made equal to the number of processors. This was done to limit contention for the livesets.

2. Experimental Results

This section presents execution results for three Knapsack problem instances, when executed by the asynchronous algorithm on the Sequent Balance 21000. These examples suggest that, among instances of the Knapsack Problem where profit density is constant, strongly anomalous behavior is the rule, not the exception. Some data is presented on CPU time (which however does not include the overhead involved in creating processors), but our emphasis is on the number of nodes expanded.

The first example displays the behavior one has come to expect of the parallel branch–and–bound algorithm. Table 1 shows execution times for the problem instance of Knapsack which will be called Example 1. In it the number of items is 18. In the table, each data point is obtained from 10 runs on the Sequent. When the number of processors is held fixed, the maximum execution time differs from the minimum execution time by less than 1%. As can be seen, approximately linear speedup was seen in the runs, using single–processor execution of the parallel algorithm as the reference point. The number of nodes expanded differs by far less than 1% from run to run, even as the number of processors changes.

It is perhaps disappointing that Example 1 was not generated randomly. Randomly generated examples typically do not behave like this one; the peculiar behavior of Example 2 below will be seen to be more typical. Based on the theory of [5], Example 1 was handcrafted so that in every execution, approximately the same number of nodes would be expanded. The method of handcrafting is explained in [5].

The problem instance of Example 2 has 20 items, profit density constant, and weights random integers between 1 and 10000. Results for Example 2 are displayed in Fig. 1. For each fixed number of processors, Example 2 was executed 20 times. In the figure, each vertical bar stretches from the minimum to the maximum number of nodes expanded; a short crossbar marks the mean execution time. The number of nodes expanded by a single processor is constant from run to run (which is to be expected), and the number expanded by two processors is nearly constant. However, the number of nodes expanded by more than 2 processors is wildly erratic. These results yield anomalies galore. Comparing the number of nodes expanded by a single processor with the maximum number of nodes expanded by 4 processors we see that the number of nodes increases by a factor of 30.8. Comparing the best execution with 8 processors with the worst execution with 4 processors, we see a decrease by a factor of 38 in the number of nodes expanded.

3. Theory

A theoretical explanation of the behavior exhibited in the examples is provided in [5]. In particular, the theory (1) predicts that problem instances of the family from which Examples 2 and 3 are drawn will typically exhibit strongly anomalous behavior, and (2) explains why Example 1 lacks anomalous behavior. The explanation of (1) is broken down as: (1a) For most problem instances of knapsack, with 20 items, profit density constant, and weights chosen as integers between 1 and 10000, an optimal solution fills the knapsack; (1b) An example of knapsack in which profit density is constant, and an optimal solution fills the knapsack, exhibits strongly anomalous behavior. The explanation of (2) relies on the theory developed in [1, 2].

The major experimental phenomenon that the theory does not account for is the very small standard deviation experienced in runs using 2 processors. It is likely that with only 2 processors, race conditions did not arise to any substantial amount.

References

1. T–H. Lai and S. Sahni, Anomalies in parallel branch-and-bound algorithms, Comm. ACM 27 (1984), 594–602.

2. T–H. Lai and A.P. Sprague, Performance of parallel branch–and–bound algorithms, IEEE Trans. Comput. 34 (1985), 962–964.

3. G–J. Li and B.W. Wah, Coping with anomalies in parallel branch–and–bound algorithms, IEEE Trans. Computers 35 (1986), 568–573.

4. M.J. Quinn and N. Deo, An upper bound for the speedup of parallel best–bound branch–and–bound algorithms, BIT 26 (1986), 35–43.

5. A.P. Sprague, Wild anomalies in parallel branch and bound, T.R. CIS, UAB, 1991.

Processors	1	2	4	8	16
Execution time	1512	754	390	200	104

Table 1. Execution times for Example 1, on the Sequent. In each run, between 113677 and 113694 nodes were expanded.

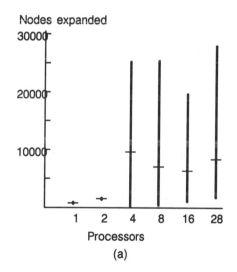

Processors

(a)

Processors	1	2	4	8	16	28
Minimum	816	1344	211	654	1101	1802
Maximum	816	1410	25178	25505	19899	28142
Mean	816	1354	9613	7144	6221	8456
Std deviation	0	23	7489	6807	5369	6124

(b)

Fig. 1. The number of nodes expanded, while solving Example 2. Each vertical bar shows the minimum, average, and maximum number of nodes expanded, based on 20 runs.

A Combined Clustering and Parallel Optimization Approach to the Traveling Salesman Problem

Bernd Freisleben and Matthias Schulte

Dept. of Computer Science, University of Darmstadt, Alexanderstr. 10, D–6100 Darmstadt, Germany

1 INTRODUCTION

The general subject of combinatorial optimization is aimed at finding the optimal configuration of elements among a large number of possible configurations. The *traveling salesman problem* (TSP) is the most well known problem of that kind. It is defined over n cities and the distances between each pair of them; the goal is to find the shortest tour (represented by a permutation of the n integers) required for a salesman to visit every city exactly once and return to the city from where he started.

In many real–life applications of the TSP it is often sufficient to arrive at solutions which are slightly worse than the optimum, but are fast to compute.

In this paper we present a parallel approach for computing approximate solutions to large TSPs. Our proposal is based on a general clustering technique to decompose a large TSP into several smaller ones and a probabilistic search heuristic called *adaptive threshold accepting* which is applied in parallel to each of the subproblems. Other approximation methods like simulated annealing [4] and threshold accepting [1] usually work on a given TSP instance as a whole; due to their inherently sequential nature they are hard to parallelize [5]. The feasibility of our approach is demonstrated by an efficient implementation on a multi–transputer system. The quality and performance of the solutions obtained in a well known 532–cities–TSP of major American cities [3] are discussed. The results indicate that it is possible to arrive at good solutions in very short time, even if the problem instances are large.

2 THE DECOMPOSITION TECHNIQUE

Our approach is based on decomposing the problem into smaller pieces which can then be assigned to several processors for parallel optimization.

The algorithm we propose uses an iterative cluster merging technique combined with an analogy of the *center of gravity* calculation rule known from mechanical physics. A cluster is defined by its center coordinates and its size, equivalent to the center of gravity and the weight, respectively. In an initializing step, one cluster is created for each city with the center coordinates assigned to the city coordinates and the size set to unary. Next, the following steps are repeated until all clusters have at least a previously defined minimal size:

1. Determine the mimimal cluster size s among all clusters built so far. If s is greater than or equal to the required minimum, terminate the iteration.

2. For every cluster i with size s find the closest neighbor j and select i with the minimal distance to j.

3. Calculate the new center coordinates using the center of gravity analogy. For Euclidian coordinates the following formula applies (the Y coordinates are calculated analogously):

$$X_{\text{new}} = \frac{\text{size}_i * X_i + \text{size}_j * X_j}{\text{size}_i + \text{size}_j}$$

4. Total the sizes of the merged clusters and assign this value to the new cluster size.

5. Combine the lists of members of the merged clusters.

This process is controlled by the parameter *minimal cluster size*, which defines a lower bound for the number of cities included in a single cluster. Figure 1 shows the number and size of the clusters obtained in the 532–cities TSP published in [3].

min_size	#clusters	MIN	MAX	AVE
10	33	10	28	16
15	20	15	45	26
20	16	21	55	31
25	12	27	60	44

Figure 1: Cluster Sizes of the 532–Cities TSP

Before the subproblems are distributed to the different processors, it is necessary to perform some preparing tasks. Since we want to optimize several subtours and then connect them to form a roundtrip, we need to know in which order they have to be traversed. This problem can be tackled by finding the Hamiltonian cycle through the center of the clusters, which is a TSP itself. *Adaptive threshold accepting*, described in detail in the next section, is used for that purpose. For now it is sufficent to know that we receive the successor and predecessor of each cluster within the roundtrip in order to be able to determine the linking cities between the subtours. This is done by finding the closest members of two consecutive clusters. Note that there is the restriction that each city can serve as a link only once, to guarantee a valid roundtrip for the original problem after the subtours are connected.

Once the starting and ending points of each subtour are defined, we are ready for optimization. This step can be done in parallel for all subproblems, because the subsets of cities for each cluster have no elements in common. After all optimized clusters have been returned, the subtours are connected to obtain an intermediate result of the original TSP. In a concluding step, this tour is submitted to all processors for independent optimization and the shortest roundtrip computed is chosen as the final result.

3 ADAPTIVE THRESHOLD ACCEPTING

A *2-change* modification technique [2] and an acceptance strategy are the essential components of our optimization algorithm. The former is needed to generate a new solution from a given tour which is then accepted or discarded by the latter. In case of an acception, the proposed solution serves as the starting point for the next cycle. This process is repeated until a termination condition (usually a significant decrease of accepted solutions) is satisfied.

Adaptive threshold accepting is an adaptive version of the acceptance strategy presented in [1] and eliminates the drawbacks of a problem instance dependent threshold schedule required in the original proposal. The algorithm compares the quality of a proposed solution with the quality of the current tour. Superior proposals are always accepted and the threshold is increased by a predefined percentage of the improvement. An inferior tour is only accepted if the decrease in quality does not exceed the threshold. In this case the threshold is lowered by the full amount of the quality change as penalty.

```
create initial tour;
th = 0;

loop:   new_tour = modify ( old_tour );
        delta = qual ( new_tour ) - qual ( old_tour );

        if ( delta < th )
            old_tour = new_tour;

        if ( delta < 0 )
            th = th - ( delta - delta / factor );
        else
            th = th - delta;

        if ( term_condition )
            terminate;
endloop.
```

Figure 2: Adaptive Threshold Accepting

Figure 2 shows a possible implementation of this mechanism, in which the parameter *factor* controls the development of the threshold values. This parameter is crucial for the tradeoff between runtime and quality of the final solution. High factors encourage the inspection of more permutations since improvements are generously rewarded. This results in a better chance to overcome temporarily inferior solutions during the search process, but this advantage has to be payed with increasing runtimes. Low factors are more restrictive and produce quick results, but they raise the probability that the process terminates in a local optimum. Considering the effect of the factor on the search process, a proper setting of this parameter allows to adjust the algorithm to suit the desired optimization goals.

4 IMPLEMENTATION AND PERFORMANCE

The different program modules have been mapped onto a master/slave organization implemented on a 72–transputer system. The master runs the clustering procedure and distributes the subproblems to each of the slaves. These execute the optimization algorithm in parallel and return the obtained subtours to the master who in turn ties them together. The intermediate result for the total tour is then propagated to all slaves for a concluding run of the optimization algorithm. The best tour is then selected by the master to determine the final result.

min_size	intermediate	t [sec]	final	t [sec]
15	94.89%	97	96.01%	576
20	94.91%	102	96.82%	603
25	96.78%	255	97.77%	742

Figure 3: Quality and Runtime of the Solutions

Figure 3 shows the results that have been obtained for the 532–cities TSP mentioned above. The percentages presented represent the quality of the solutions in comparison to the known global optimum (100%). The values indicate that the proposed clustering and parallel optimization approach allows to obtain relatively good results very fast, i.e. after the optimized subtours have been built together.

5 CONCLUSIONS

In this paper we have presented an adaptive threshold accepting algorithm for approximately solving large traveling salesman problems. We have shown that it is possible to decompose a large TSP into smaller subproblems and run the algorithm in parallel on each of them. This approach does not only maintain the serial decision sequence of our probabilistic search heuristic required to obtain high quality solutions, but also ensures that good solutions can be computed in a very short time.

6 REFERENCES

[1] G. Dueck and T. Scheuer. Threshold Accepting. Technical Report TR 88.10.011, IBM Heidelberg, 1988.

[2] S. Lin. Computer Solution of the Traveling Salesman Problem. *Bell Systems Technical Journal*, 44:2245–2269, 1965.

[3] M. Padberg and G. Rinaldi. Optimization of a 532–City Symmetric Traveling Salesman Problem by Branch and Cut. *Operation Research Letters*, 6(1), 1987.

[4] S.Kirkpatrick, C.D. Gelatt, and M.P.Vecchi. Optimization by Simulated Annealing. *Science*, 220:671–680, 1983.

[5] E.E. Witte, R.D. Chamberlain, and M.A. Franklin. Parallel Simulated Annealing Using Speculative Computation. In *Proc. of the 1990 Int. Conference on Parallel Processing*, pages 286–290, 1990.

Timing Analysis of a Parallel Algorithm for Toeplitz Matrices on a MIMD Parallel machine

I. Gohberg [†]   I. Koltracht [‡] *

A. Averbuch [†]   B. Shoham [†]

[†] School of Mathematical Sciences
Tel-Aviv University
Tel-Aviv 69978, Israel

[‡] The College of Liberal Arts and Sciences
The University of Connecticut
Storrs, Connecticut 06268, USA
E-Mail: amir@taurus.BITNET

Abstract

In this paper performance analysis of a parallel Levinson-type algorithm for Toeplitz matrices is given. A modified version of the parallel algorithm is presented to improve performance. The algorithm is implemented on a shared-memory MIMD (non-vector) machine. The derivation of the parallel algorithm is presented. The speedup limitation is investigated along with the optimal number of processors needed to a given Toeplitz matrix.

1 Introduction

The solution of linear systems of equations with a Toeplitz coefficients matrix $R = \{r_{i-j}\}_{i,j=0}^{N}$ is important in many applications in science, engineering and mathematics (see [B] and references therein).

In some applications the Toeplitz matrix is positive definite (i.e. $R = \{r_{|i-j|}\}_{i,j=0}^{N}$) and the desired solution is the vector γ_N for which

$$R\gamma_N = e_N = [0, \ldots, 0, 1]^T. \qquad (1)$$

In 1947 Levinson [L], proposed a fast algorithm for solving Eq. (1). The one advantage of this algorithm is its relatively low operations count and memory requirement ($O(N^2)$ and $O(N)$ respectively).

A significant effort was recently put into further speed-up of the solution of Toeplitz systems, in particular, by the use of modern computer architectures (see [B] for parallel architectures and [K] for vector machines).

One parallel version of the Levinson algorithm was suggested in [GKKL] as follows:

Parallel Levinson Algorithm:

1. Start with $\gamma_0 = 1$, $\gamma_0(j) = -r_j$, $j = \pm 1, \ldots, \pm N$

2. For $k = 1, \ldots, N$ compute

$$\beta_k = [1 - \gamma_{k-1}^2(-1)]^{-1}$$

$$\gamma_k(0) = \beta_k \gamma_{k-1}(-1)\gamma_{k-1}(k-1)$$

$$\gamma_k(k) = \beta_k \gamma_{k-1}(k-1)$$

$$\gamma_k(j) = \beta_k[\gamma_{k-1}(j-1) + \gamma_{k-1}(-1)\gamma_{k-1}(k-j-1)]$$
$$j = k - N, \ldots, N, \; j \neq 0, k \qquad (2)$$

3. Thus $R_N \gamma_N = e_N$

This algorithm requires $O(N)$ arithmetic operations on $O(N)$ parallel processors where the computation of $\gamma_k(j)$ components is done in parallel. The main difference between the parallel and the standard Levinson algorithms is that the reflection coefficients do not have to be computed as inner products which, in principle, could allow more efficient implementation of the parallel algorithm on a parallel and vector machines.

In this paper we analyze such an implementation on a MIMD shared bus & shared memory multiprocessor architecture. Since the

*The work by the first two authors was supported by the NSF grant #DMS 8801961

communication among the processors is done through a shared bus, (so that only one processor can use at a time), the communication overhead is very significant in our implementation.

In this paper we present a modification of the parallel Levinson algorithm that needs less communication which is related to the computation of γ_k, therefore the modified algorithm better exploits the machine and tailored to its architecture.

2 The Modified Parallel Levinson Algorithm

Following [GKKL] consider the sequence of $(N+1) \times (N+1)$ matrices,

$$R_k = \begin{bmatrix} R_k^l & | & 0 \\ --- & | & --- \\ R_k^c & | & I \end{bmatrix} \quad k = 0, \ldots, N \qquad (3)$$

where $R_k^l = \{r_{|i-j|}\}_{i,j=0}^{k}$ and $R_k^c = \{r_{|i-j|}\}_{i=k+1 \, j=0}^{N \quad k}$ and is a $(N - k) \times (k+1)$ Toeplitz matrix. Since $R_N = R$ is positive definite, it follows that R_k is invertible.

Consider also solution vectors γ_k and φ_k, for $k = 0, \ldots, N$ defined by:

$$R_k \gamma_k = e_k, \quad R_k \varphi_k = e_0, \qquad (4)$$

where $e_k = [0, \ldots, 0, 1, 0, \ldots, 0]^T$ with 1 in the $k+1 th$ position. It is shown in [GKKL] that

$$\varphi_{k-1}(k) = c_k = -\sum_{j=0}^{k-1} r_{j+1}\gamma_{k-1}(j), \qquad (5)$$

$$\gamma_k = \begin{bmatrix} 0 \\ \gamma_{k-1}(0) \\ \vdots \\ \gamma_{k-1}(N-1) \end{bmatrix} + c_k \varphi_k \triangleq \bar{\gamma}_{k-1} + c_k \varphi_k, \qquad (6)$$

$$\varphi_k = \begin{bmatrix} \varphi_{k-1}(0) \\ \varphi_{k-1}(k-1) \\ 0 \\ \varphi_{k-1}(k+1) \\ \vdots \\ \varphi_{k-1}(N) \end{bmatrix} + c_k \gamma_k \triangleq \bar{\varphi}_{k-1} + c_k \gamma_k. \qquad (7)$$

(These equations can be verified directly). Thus

$$[\gamma_k , \varphi_k] \begin{bmatrix} 1 & -c_k \\ -c_k & 1 \end{bmatrix} = [\bar{\gamma}_{k-1} , \bar{\varphi}_{k-1}] \qquad (8)$$

and hence $[\gamma_k , \varphi_k] = \dfrac{1}{1 - c_k^2}[\bar{\gamma}_{k-1} , J\bar{\varphi}_{k-1}] \begin{bmatrix} 1 & c_k \\ c_k & 1 \end{bmatrix}$, (9)

where J is a matrix with $J_{i,N-i+1} = 1$ $1 \le i \le N$, and zero elsewhere.

Modified Parallel Levinson Algorithm

1. Start with

$$\begin{bmatrix} \gamma_0(0) & \varphi_0(0) \\ \vdots & \vdots \\ \gamma_0(N) & \varphi_0(N) \end{bmatrix} = \begin{bmatrix} 1 & 1 \\ -r_1 & -r_1 \\ \vdots & \vdots \\ -r_N & -r_N \end{bmatrix} \qquad (10)$$

2. For $k = 1, \ldots, N$ let $c_k = \varphi_{k-1}(k)$ and compute

$$[\vec{\gamma}_k \quad \vec{\varphi}_k] = \frac{1}{1 - c_k^2} \begin{bmatrix} 0 & \varphi_{k-1}(0) \\ \gamma_{k-1}(0) & \varphi_{k-1}(1) \\ \vdots & \vdots \\ \gamma_{k-1}(k-2) & \varphi_{k-1}(k-1) \\ \gamma_{k-1}(k-1) & 0 \\ \gamma_{k-1}(k) & \varphi_{k-1}(k+1) \\ \vdots & \vdots \\ \gamma_{k-1}(N-1) & \varphi_{k-1}(N) \end{bmatrix} \begin{bmatrix} 1 & c_k \\ c_k & 1 \end{bmatrix} \qquad (11)$$

3. Thus $R_N \gamma_N = e_N$ and $R_N \varphi_N = e_0$ where $e_0 = [1, 0, \ldots, 0]^T$.

This algorithm requires also $O(N)$ arithmetic operations on $O(N)$ processors. The communications are, however, simplified. Except for $c_k = \varphi_{k-1}(k)$, which has to be sent to the local memories of each processor, the rest of communications is local. Fig. 1 illustrates the parallel structure of the algorithm.

A general solution of linear system of Toeplitz equations $R_N x = f$ can be solved in parallel (for derivation see [GKKL]).

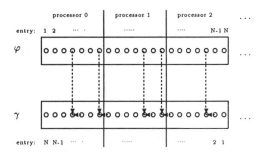

Figure 1: The implementation of the Modified Parallel Levinson algorithm

3 The Parallel Implementation

The parallel Toeplitz algorithms are implemented on a shared bus &shared memory MIMD (non-vector) multiprocessor[G] using the software tool called Virtual Machine for MultiProcessors[VMMP]. The VMMP is a software package, which provides a coherent set of services for parallel application programs.

The implementation is based on partitioning the vectors γ, φ and X into equal size slices, assign each slice to a processor, which computes the value of the components residing in its slice.

4 Numerical Results and Performance Analysis

For the purpose of parallel speedup measurements a serial program was executed on a single processor, without any system service call to VMMP.

The results given in Table 1 are for solving the Toeplitz system $R_N \gamma_N = e_N$, and in Table 2 for the solution of the linear system $R_N x = f$ where f is any vector.

We observe that the speedup increases with the problem size. This is due to the increase of the utilization of each processor.

For each fixed N we consider the following processing timing model:

$$T_{p,N} = zPN + y\frac{N^2}{P} \qquad (12)$$

where $T_{p,N}$ is the processing time for the Modified Parallel Levinson algorithm, and z is the overhead time for each of the N recursion steps as in Eq. 11 The constant y is the time needed for computing $\frac{N}{P}$ components of γ_k and φ_k (this is done N recursion times), and P is the number of processors. Neither z nor y depend on N or p.

Using Eq. (12) we can find P_{opt} for a given N:

$$\frac{\partial T}{\partial P} = zN - \frac{yN^2}{P^2} \Longrightarrow P_{opt} = \sqrt{\frac{y}{z}} N$$

For example see Table 3.

Since neither z nor y depend on N or p we can use Eq. (12) and predict the processing time if the number of processors is optimal to a given N. We observe that processing time depends on N in a non-linear manner. The non-linearity in the speedup measurements can be understood for the parallel algorithm where $\gamma_k(k - j - 1)$ and $\gamma_k(j - 1)$ has to be sent to different processor via the global memory.

Moreover, increasing the number of processors will increase the number of broadcasts of data and thus we do not get a linear speedup (recall that the processors are connected by one shared data bus).

Table 1: Processing time for solving the Toeplitz system $R_N \gamma_N = e_N$ using the modified parallel Levinson algorithm

Matrix size (N)	Performance parameters	Parallel Processing			Serial Processing
		7 procs.	2 procs.	1 proc.	
512	Proc. time (sec.)	2.400	5.033	7.866	7.767
	Speed Up	3.24	1.54		
	Efficiency	0.46	0.77		
1024	Proc. time (sec.)	7.766	19.934	31.433	31.167
	Speed Up	4.01	1.56		
	Efficiency	0.57	0.78		
8192	Proc. time (sec.)	404.066	1275.267	2044.100	2023.367
	Speed Up	5.01	1.59		
	Efficiency	0.72	0.80		
16384	Proc. time (sec.)	1593.766	4904.500	8211.006	8100.233
	Speed Up	5.08	1.65		
	Efficiency	0.73	0.83		

Table 2: Processing time for solving the Toeplitz system $Ax = f$ using the modified parallel Levinson algorithm

Matrix size (N)	Performance parameters	Parallel Processing			Serial Processing
		7 procs.	2 procs.	1 proc.	
512	Proc. time (sec.)	1.967	4.500	6.467	6.334
	Speed Up	3.22	1.41		
	Efficiency	0.46	0.71		
1024	Proc. time (sec.)	6.500	16.767	25.833	25.466
	Speed Up	3.92	1.57		
	Efficiency	0.56	0.76		
8192	Proc. time (sec.)	315.433	1068.133	1680.200	1663.300
	Speed Up	5.26	1.55		
	Efficiency	0.75	0.78		
16384	Proc. time (sec.)	1242.700	4087.266	6753.200	6610.800
	Speed Up	5.32	1.62		
	Efficiency	0.76	0.81		

Table 3: Optimal number of processors for the modified parallel Levinson algorithm.

N	256	512	4046	16384	5×10^4	10^5
P_{opt}	4	7	10	11	19	26

5 Summary

We have presented and implemented an efficient algorithm for solving Toeplitz system on a shared bus& shared memory parallel MIMD machine.

One of our main observations is that the ability to obtain a substantial speedup depends on the matrix size. For large matrix the efficiency is over 75%.

We observe that in order to get a speedup the size of the matrix should be large enough and that their is an optimal number of processor to a given matrix such that the speedup level will increase when increasing the number of processors.

It would be useful to implement this algorithm on other types of multiprocessors, especially on message passing machine, since the most consuming time part of the synchronization will be done in a constant time regardless to the number of processors.

References

[L] N. Levinson (1947), The Wiener RMS error criterion in filter design and prediction, *J. Math. Physics, 25, pp. 261-270.*

[B] J. Bunch, Stability of methods for solving Toeplitz systems of equations, *SIAM J. Sci. Stat. Comput., (6:2), pp. 349-364.*

[GKKL] I. Gohberg, T. Kailath, I. Koltracht, P. Lancaster (1987), Parallel algorithms of linear complexity for machines with recursive structure, *Lin. Alg. Appl., 88, pp.271-315.*

[K] I. Koltracht (1987), The Levinson algorithms on the CYBER 205, Cuper-C, *Newsletter, v.3:2, pp. 4-10.*

[G] E. Gabber (1987), The MMX Parallel Operating System and Its Processor, *Dept. of Computer Science, Tel-Aviv University.*

[VMMP] E. Gabber, VMMP: A Practical Tool for the Development of Portable and Efficient Programs for Multiprocessors, *IEEE Trans. on Parallel and Distributed Systems, Vol. 1, No. 3, pp. 304-317, July 1990.*

LARGE 1-D FAST FOURIER TRANSFORMS ON A SHARED MEMORY SYSTEM

YEDIDIAH SOLOWIEJCZYK & JOHN PETZINGER
ORYX CORPORATION
PARAMUS, NEW JERSEY 07652

ABSTRACT

It is demonstrated that large 1-D FFTs can be readily parallelized on a multiprocessor SIMD shared memory system by utilizing smaller transforms (<4K). Furthermore, the use of a full crossbar switch is shown to provide for peak memory bandwidth between processors and memories for algorithms such as the FFT.

KEY WORDS: Large FFT, SIMD, Shared Memory, Crossbar Switch.

INTRODUCTION

Present-day numerical analysis is requiring the use of larger FFTs, often exceeding a million points [1]. While a wide variety of FFT algorithms designed for sequential computers [2] efficiently solve transforms of modest sizes (e.g. < 4096 points), these techniques usually become bottlenecked both in computation and I/O bandwidth when applied to larger size transforms. Although parallelism has been recognized as one of the most promising avenues for increasing computational as well as I/O bandwidths in high performance computing, the adaptation of the FFT onto such architectures imposes new challenges [3]; e.g. it is well-known that the data inputs into FFT butterflies are distributed over the entire array, and serious communication bottlenecks are therefore possible.

LARGE 1-D FAST FOURIER TRANSFORMS

Large 1-D FFTs can be readily calculated from smaller transforms through the use of a "six step" algorithm [1], which goes as follows:
(i) transpose the input data set, considered to be a (M x N) matrix (complex), into a (N x M) matrix.
(ii) calculate the M point FFT on each of the N rows.
(iii) multiply the resulting (N x M) matrix by the the appropriate twiddle factors; exp(-j2ii*nc*nr/(N*M)), where nc is the column variable and nr is the row variable.
(iv) transpose the resulting (N x M) matrix into a (M x N) matrix.
(v) calculate the N point FFT on each of

the M rows and.
(vi) transpose the resulting transformed complex (N x M) matrix into a (M x N) matrix (or equivalently an M*N vector).

Although the above "six step" recipe makes it possible to develop large 1-D FFTs, severe communication bottlenecks can still occur on parallel architectures having distributed memories (e.g. hypercubes having their data distributed equally among their PEs). This paper demonstrates that the above recipe maps very efficiently on a crossbar switch based SIMD shared memory architecture (Oryx SSP) if the data is mapped as coerced odd width complex matrices. The execution of a highly parallel and computationally efficient radix-4 FFT is also discussed.

LARGE 1-D FFTs ON THE ORYX SSP

The Oryx SSP is a high performance coarse grain SIMD parallel processor designed for computationally intensive applications. It can be configured as a 1, 2, 4, or 8 PE system connected via a crossbar switch to an equal number of parallel memories (PMEMs). Each PE can communicate with any of the parallel memories on any clock cycle (Figure 1).

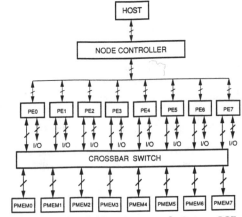

Figure 1. Architecture of Oryx SSP.

Each PE contains (i) 2 single cycle floating point multiplier/adder chips, (ii) 1 single cycle logical ALU which can generate addresses for table driven algorithms (e.g. FFT), (iii) 1 address generator ALU, and (iv) a bit reverser chip. Each PE has its own data and address resident microcode and microsequencers. This allows all PEs to run synchronously as well as providing time skewing techniques which are useful in executing multiple butterfly operations free of memory clashes. The system runs from a single 100 nsec clock (e.g 160 MFLOPS as well as 80 MWORDS/sec is provided by an 8 PE system). The 3 LSB of a 27 bit wide address field are used to steer each PE to a separate memory. Memory clashes are avoided by properly planning that no two PEs attempt to fetch or write to the same memory on the same clock cycle. A fully loaded system can accomodate up to 128 32 bit MWORDS. On this architecture, each PE can execute a complex vector-vector multiply in 6 clock cycles whereas a radix-4 butterfly is done in 17 clock cycles. Multiple FFT butterflies can be implemented in parallel (free of memory clashes) by invoking a 2 clock cycle time skew on each PE. This approach provides for a very efficient radix-4 FFT (e.g. 1K FFT runs in 582 usec on a 4 PE system). Complex matrices are mapped row-wise across parallel memories. In order to provide all PEs the ability to access contiguous complex row or column elements, matrices are always coerced to contain an odd number of complex columns. The additional complex column elements inserted in even sized matrices contains either zeros or "garbage" values which are never used in any of the matrix operations. This mapping technique provides the mechanism for transposing matrices at the full memory transfer bandwidth. Transposition of a complex matrix can be achieved by having the even PEs transfer the real components of a row into their appropriate column locations while the odd PEs are doing the same for the imaginary components. This result is clearly possible because a full crossbar switch allows all PEs to access distinct parallel memories on each clock cycle. Efficient matrix transpositions and radix-4 FFTs are responsbile for a very efficient implementation of the "six step" recipe. The twidddle table necessary in this application is generated only once thus it constitutes a small overhead in most applications. The product of the twiddle matrix with the column-wise transformed input matrix is also executed at its peak memory as well as computational bandwith (6 clock cycles/PE).

RESULTS
————

Large FFT's (up to 1 MWORDS) were run on a 4 PE Oryx SSP. Table 1 illustrates a runtime comparison between the Oryx SSP and other commercially available supercompters. As one notes from Table 1, a 4 PE Oryx SSP (80 MFLOPS) delivers twice the performance delivered by the CM-2 (7 GIGAFLOPS) and approximately one half the performance of a 1 PE Cray-YMP (333 MFLOPS). In both cases the Oryx SSP is noted to outperform the other systems on a per MFLOP rating. These results demonstrate that the Oryx SSP architecture is capable of delivering an exceptionally high computational as well as I/O bandwidth for applications requiring data sets distributed over the entire array.

SIZE	16K	64K	256K	1024K
ORYX	22	93	349	1,561
CM-2	38	161	688	2,950
CRAY-YMP	13	56	244	1,064

Table 1. Comparative Runtime (msec).

CONCLUSION
——————

A highly parallel rendition of large 1-D FFTs has been implemented on a crossbar-based SIMD multiprocessor shared memory system. A comparitive runtime study against other high end supercomputers (e.g. Cray YMP and the CM-2) demonstrates that the Oryx architecture is very efficient in solving problems requiring the use of FFTs.

REFERENCES
——————

[1] D. Bailey, "FFTs in External Hierarchical Memory ", Journal of Supercomputing, in press.
[2] L. Rabiner and B. Gold, "Theory and Application of Signal Processing", Prentice Hall, N.J., 1975.
[3] Y. Solowiejczyk and J. Petzinger, "The Radix-4 FFT On A Multiprocessor Shared Memory System", 19th Int'l Conference on Parallel Processing, pp. III-362.

An Overlapped FFT Algorithm for Hypercube Multicomputers

Cevdet Aykanat and **Argun Derviş**
Faculty of Engineering & Science, Department of CIS
Bilkent University, Ankara, Turkey

1 Introduction

The purpose of this paper is to investigate the efficient parallelization of *1D FFT* algorithm on coarse grain hypercube multicomputers. In order to achieve speedup on such architectures, the algorithm must be designed so that both computations and data can be distributed to the processors with local memories in such a way that computational tasks can be run in parallel, balancing the computational loads of the processors. In a parallel machine with high communication latency, the algorithm should be structured so that large amounts of computation are done between successive communication steps. Another important factor is the ability of the parallel system to *overlap* communication and computation. Hence, the algorithm must be structured so that the communication can be overlapped with computation. The algorithm presented here achieves efficient parallelization by considering all these points in designing an efficient parallel *1D FFT* algorithm for hypercube multicomputers.

2 Equal Load Balance FFT Algorithm

The decimation-in-time decomposition scheme is investigated for parallelization. The input is in *bit-reversed* order and the output is in *normal* order. At the kth stage, $N/2$ *simplified-butterfly* computations are performed on partially transformed p, q pairs seperated by 2^k. This process is repeated $n=\log_2 N$ times for $0 \leq k < n$. The *simplified-butterfly* computation requires a complex multiplication of the q-point with a complex coefficient and one complex addition and subtraction for transforming the p and q points respectively. Since $N/2$ *simplified-butterfly* computations are performed at each stage, the complexity of the sequential FFT algorithm is $(5N \log_2 N)t_c$. Here, t_c represents the execution time for an individual real multiplication or addition operation.

The straightforward mapping scheme is easily achieved by assigning successive $M=N/P$ FFT points to succesive processors in decimal ordering. In this mapping scheme, interprocessor communication is not required during the first $(n\text{-}d)$ stages $(0 \leq k < n - d)$ since butterfly pairs are assigned in groups to the same individual processors. However, d concurrent *exchange* communication steps are required during the last d stages $(n\text{-}d \leq k < n)$ since individual p, q points of each butterfly pair are assigned to neighbor processors. This mapping scheme achieves perfect load balance during the first $(n\text{-}d)$ stages since each processor is assigned equal number $(N/2P)$ of butterfly pairs. In order to maintain perfect load balance during the last d stages, the static mapping scheme is altered at the beginning of each stage of the last d stages. At the very beginning of each stage, each processor holding only updated values for q-points exchange local $N/2P$ q-points with the $N/2P$ p-points of its neighbor processor which holds all the p points of its butterfly pairs at that stage. In fact, processors effectively exchange the responsibility of the further *FFT* computations associated with those exchanged *FFT* points. Hence, each processor holds equal number of p and q points after the exchange operation. Thus, this scheme achieves perfect load balance since each processor performs equal number $(N/2P)$ complex multiplications. Hence, the parallel complexity of this scheme is,

$$T_{prog1} = [\frac{5N}{P} \log_2 N]t_c + [t_{su} + \frac{N}{2P}t_{tr}] \log_2 P \qquad (1)$$

Here, t_{su} represents the message startup overhead, and t_{tr} is the time taken for the transmission of a floating-point word between two neighbor processors. In the complexity model given above, two concurrent *send* communication operations (for an *exchange* operation) between a pair of processor are assumed to be overlapped completely. This *dynamic* mapping scheme is illustrated in Fig.(1) for a 16-point *FFT* on a 4-processor hypercube. The pseudo code for the node program of the proposed parallel *FFT* algorithm is given in Prog.(1). The $SEQFFTk$ given in Prog.(1) performs the in place computations corresponding to the kth stage of an N-point FFT using the table lookup method for complex coefficients ($Wfac$).

3 Overlapped FFT Algorithm

There are strong data dependecies in the *FFT* algorithm. The update of each *FFT* point requires communication in the last d-stages of the algorithm. Thus, communication and computation in the *FFT* algorithm cannot be overlapped easily. Zhu has proposed a scheme for overlapping communication with the computation of the complex coefficients in [1]. However, computation of the coefficients as they are needed is not an efficient scheme compared to the table lookup scheme. Walker has proposed a scheme in [2] for overlapping communication and computation for the *FFT* algorithm using the *basic-butterfly* scheme which requires two complex multiplication per butterfly pair. In this paper, we propose a scheme which overlaps the communication with one fifth of the computations involved in a stage of the *FFT* algorithm which uses the *simplified-butterfly* and the table lookup scheme.

The pseudo-code for the node program of the parallel *FFT* algorithm which overlaps communication and computation is given in Prog.(2). As is seen in Prog.(2), each processor classifies its computational task at each stage into two categories: those updates to be *sent* to the destination processor in the following stage and other updates to be kept as local in the following stage. Then, each processor first performs the computations associated with those points required by the destination processor in the next stage. Hence, each processor first performs $N/2P$ complex multiplications associated with its local $N/2P$ q-points. Then, each processor updates either the values of its local p-points or q-points into a send buffer (XSB array) simply by checking the $(\ell+1)^{th}$ bit value of its processor index. Here, $\ell + 1$ denotes the channel over which the exchange operation required in the next stage. Upon completion of these $N/2P$ updates, each processor issues an *non-blocking send* to initiate the transmission of the updated $N/2P$ *FFT*-point values to the destination processor. After initiating the *send* operation, each processor completes the computation associated with that stage by updating other half of its local *FFT*-points that will be kept local in the following stage. Upon completion of the second type updates each processor issues an in-place *blocking receive* to complete the already initiated exchange operation.

The proposed scheme introduces a storage overhead of size $N/2P$ per processor due to the local send buffer XSB array. The only computational overhead is the loop overhead since two *for-loops* are required instead of one. The number of floating-point computations is exactly equal to the number of computations required in Prog.(1). As is seen in Prog.(2), $N/2P$ complex additions/subtractions shown in the second inner *for-loop* of the second outer *for-loop* are overlapped with communication. Hence, communication is overlapped with one fifth of the computations involved in a stage. Thus, the parallel complexity of the proposed algorithm is

$$T_{prog2} = [\frac{5N}{P}\log_2 \frac{N}{P}]t_c + [\frac{4N}{P}\log_2 P]t_c + \quad (2)$$

$$[Max\{\frac{N}{P}t_c, (t_{su} + \frac{N}{2P}t_{tr})\}]\log_2 P$$

Hence, for sufficiently large N/P, $[N/P]t_c \geq t_{su} + [N/2P]t_{tr}$ complete overlap of communication can be achieved. This value is computed to be $N/P \geq 256$ by inserting the machine specific parameters (i.e., t_c, t_{su} and t_{tr}) for the iPSC/2 [3].

4 Experimental Results

The programs presented in this paper have been coded in C language and run on an 8-node iPSC/2 hypercube multicomputer for various $N = 2^n$ data sizes, $64 \leq N \leq 64K$. Prog.(2) gives better performance results compared to Prog.(1) for $N/P \geq 4K$. The discrepancy between the expected and observed results are due to two major reasons. First, the loop overhead introduced by Prog.(2) is assumed to be negligible in the analytic derivation. However, this loop introduces a considerable amount of overhead increasing linearly with N. The second reason is that, Eq.(2) is derived by assuming that a complete overlap of communication with computation is feasible when the total computation time per processor exceeds the total time for a preceeding *non-blocking send* operation. However, as is also indicated in [3] a complete overlap cannot be achieved due to the internal architecture of an individual iPSC/2 processor.

Fig.(2) shows the efficiency curve obtained from Prog.(2). As is seen in Fig.(2), efficiency remains over 90% when $N/P \geq 256$ *FFT* points mapped to an individual processor. Fig.(3) shows percentage improvement in the communication phase of the Prog.(2) with respect to Prog.(1). Because of the two reasons explained above, for small N no improvement is observed. Since computation increases more rapidly than communication for a fixed start-up time, the percentage of improvement begins to drop down after a maximum point is reached for each P.

References

[1] J.P. Zhu, "An Efficient FFT algorithm on Multiprocessors with Distributed Memory," in *The Fifth Distributed Memory Computing Conference*, Vol. 1, pp 358-363, Jan. 1990.

[2] D. W. Walker, "Portable Programming within a Message-Passing Model: the FFT as an Example," in *Third Conference on Hypercube Concurrent Computers and Applications*, Pasadena, CA, pp. 1438-1450, January 1988.

[3] L. Bomans, and D. Roose, "Benchmarking the iPSC/2 hypercube multiprocessor," in *Concurrency : Practice and Experience*, Vol. 1(1), pp. 3-18, September 1989.

Program 1 : Parallel N-pt FFT Algorithm

```
/* Computation over the first (n − d − 1) bits   */
n := log₂ N;   d := log₂ P;   M := N/P;   m := log₂ M;
for k :=0 to n − d − 2 do Call SEQFFTk (X, Wfac, M, k)
/* d concurrent exchange communication steps */
for k :=n − d − 1 to n − 2 do
    ℓ := k − (n − d);   dnode := mynode ⊕ 2ˡ;
    for (p :=0 to M/2-1) and (q :=M/2 to M-1) do
       temp := Wfac × X(q)
       X(q) := X(p) − temp
       X(p) := X(p) + temp
    endfor
    if ((ℓᵗʰ bit of mynode) = 1) then do
       csend from (X(p): p=0,1, ..., M/2 − 1) to dnode
       crecv into (X(q): q=0,1, ..., M/2 − 1) from dnode
    else
       csend from (X(q): q=M/2, M/2 − 1, ..., M − 1) to dnode
       crecv into (X(p): p=M/2, M/2 − 1, ..., M − 1) from dnode
    endif
endfor
/* Perform M/2 Butterfly comp. over the local (m − 1)ᵗʰ bit */
Call SEQFFTk (X, Wfac, M, m − 1)
```

Program 2 : Overlapped N-pt FFT Algorithm

```
/* Computation over the first (n − d − 1) bits   */
n := log₂ N;   d := log₂ P;   M := N/P;   m := log₂ M;
for k :=0 to n − d − 2 do  Call SEQFFTk (X, Wfac, M, k)
/* d concurrent exchange communication steps */
for k :=n − d − 1 to n − 2 do
    ℓ := k − (n − d);   dnode := mynode ⊕ 2ˡ⁺¹;
    if ((ℓ + 1)ᵗʰ bit of mynode) = 1) then do
       for (p :=0 to M/2-1) and (q :=M/2 to M-1) do
          X(q) := Wfac × X(q)
          XSB(p) := X(p) + X(q)
       endfor
       isend from (XSB(p): p=0, 1, ..., M/2 − 1) to dnode
       for (q :=M/2 to M-1) and (p :=0 to M/2-1) do
          X(q) := X(p) − X(q)
       endfor
       crecv into (X(p): p=0, 1, ..., M/2 − 1) from dnode
    else
       for (q := M/2 to M-1) and (p :=0 to M/2-1) do
          X(q) := Wfac × X(q)
          XSB(p) := X(p) − X(q)
       endfor
       isend from (XSB(p): p=0, 1, ..., M/2 − 1) to dnode
       for (p :=0 to M/2-1) and (q := M/2 to M-1) do
          X(p) := X(p) + X(q)
       endfor
       crecv into (X(q): q=M/2, M/2 − 1, ..., M − 1) from dnode
    endif
endfor
/* Perform M/2 Butterfly comp. over the local (m − 1)ᵗʰ bit */
Call SEQFFTk (X, Wfac, M, m − 1)
```

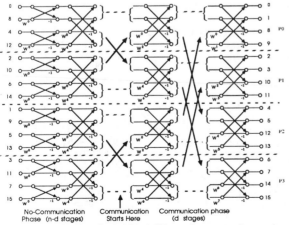

Fig.(1) Dynamic Mapping of 16-pt FFT on a 4-processor hypercube

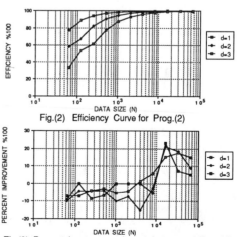

Fig.(2) Efficiency Curve for Prog.(2)

Fig.(3) Percent improvement in last d-stages of Prog.(1)

A MASSIVELY PARALLEL LINEAR SYSTEM SOLVER
FOR GENERAL AND STRUCTURAL ANALYSIS USES

Rong C. Shieh and Thomas Kraay

MRJ, Inc.
10455 White Granite Drive
Oakton, Virginia 22124

ABSTRACT - This paper describes the development, implementation, and benchmarking of an efficient, large, dense linear system solver, LSS, on massively parallel Connection Machine CM-2 model computer. It also presents the results of a comparison study on numerical solution accuracy and efficency of LSS versus MSC/NASTRAN code in structural analysis of two example problems.

THE CM-2 LINEAR SYSTEM SOLVER (LSS)

Connection Machine CM-2 systems are commercially available with 65,536 processors, each one equipped with w_p to 128K bytes of dedicated RAM. The CM-2 LSS code is capable of solving real linear algebraic equations in arbitrary precision. Single (32-bits), double (64 bits), and quadruple (128 bits) precision modes of computation were exercised to support this study. Up to 16K equations on MRJ's 16K CM-2 system and 65,000 equations for the full 64K CM-2 system are capable of being solved. The technique used in solving the linear matrix equation

$$[A] \ [X] = [B] \qquad (1)$$

for unknowns [X] was Gaussian elimination method [2] with "pseudo" full pivoting.

Effectively, as many columns as could fill half of the Connection Machine's memory were read from the data vault. The largest allowable matrix element in this portion of the matrix was chosen as the pivot. The other half of memory was used to store specific ratios derived from the current portion necessary to reduce remaining sections of the matrix not yet read into memory. Diagonal elements are stored in the next available central processing unit (CPU) as they result.

Once a set of columns is totally reduced, the ratios described above are written to the data vault. A new section is then read into core, reduced using the historical ratios previously written to the vault, and finally diagonalized itself. After the entire matrix has been reduced, dividing each constituent vectors of the [B]-matrix by the corresponding diagonal element yields the desired unknown vectors.

ACCURACY AND EFFICEINCY OF LSS VS. MSC/NASTRANCODE IN STRUCTURAL RESPONSE ANALYSIS

(1) Figure 1 shows the deformed and undeformed configurations of a three segmented, stepped, short, thin-walled tube (shell) structure subjected to a transverse linear temperature (T) in the x - direction with temperature gradient of -1°F/inch and reference temperature of T_o = 70°F. The tube structure is constrained at each of its bottom edge nodal points by a hinge support. A widely used commercial structural analysis computer code, MSC/NASTRAN [1], was first used on the Multiflow mini-super computer system to obtain distortion solution and also generate (2K x 2K) stiffness matrix A and (2K x 1) force vector [B], which are related to the displacement vector [U] = [X] by Eq. (1). These stiffness matrix and force vector data are subsequently repacked and transmitted to the CM-2 system via magnetic tapes.

Table 1 shows CM-2 LSS code-calculated, translational and rotational displacement component solution results, U_x and R_y (in the least square mean sense) of the top (free end) cross section and U_x' and R_y' at nodal point D in Fig. 1, using the single (32 bits), double and quadruple precision modes of computation. Also shown are the corresponding MSC/NASTRAN results using double precision calculational mode on the Multiflow system. The single precision results are seen to be erroneous because of involvement of an ill-conditioned, structural stiffness coefficient matrix, which is often the case in structural analysis. All of the double precision results of the LSS code calculations were found to coincide exactly with the corresponding quadruple precision results up to 12 significant digits for the mean cross-sectional displacements, and 10 significant digits for the nodal point displacements.

In comparison, the NASTRAN double precision results are exact to only two and one significant figure(s) for the mean cross sectional and nodal point displacements, respectively. The round-off error of the NASTRAN U_x' and R_y' - results in Table 1 are seen to be excessive (-3.0 and -3.7%, respectively), which will get worse as the matrix size and/or density increase. Therefore, there is a definite need to further study the round-off errors behavior of MSC/NASTRAN as a variation of matrix size (greater than 2000) and/or matrix density.

The LSS calculated results in Table 1 were obtained using only 2048 CM-2 processors. Table 2 presents the corresponding run time results. The CM-2 system at MRJ is only equipped with 32 bit Weitek coprocessors. Consequently, the 32 bit precision results were obtained using the floating point coprocessors; the 64 and 105 bit cases were run without the benefit of floating point hardware. With floating point hardware, the results would have been faster by a factor of about 4 ~ 5. The latter is reflected in the extrapolated CPU time result (75 sec) of Table 2 for the case of double precision calculation on the 64K processors with 64 bit Weitek coprocessors. The current version of LSS is specifically written for the 8K processor case on CM-2A computer.

(2) There exists a number of cases or classes of structural problems in which a dense linear system is involved in structural analysis, notably as a result of condensing or transforming the governing equations. Two statically condensed linear systems from the original 4000 and 8000 DOF to 1998 DOF ones were solved using the sparse linear system solver of MSC/NASTRN code on Mulitflow/Trace 7/200 model computer, and the corresponding CPU time results are given in Table 3 as cases M2 and M3. The LSS code on the fully equipped CM-2 system is seen to be 5 and 5.3 times as efficient as the Multiflow/200 version of MSC/NASTRAN in solving the 2K DOF dense linear systems, with matrix densities of 68 and 88%, respectively. Also, based on the Multiflow/200 version speed-up factor of 25 over the VAX/780 version of MSC/NASTRAN, the LSS code is 125 and 135 times as fast as the latter in solving the above dense linear systems.

COMPARISON OF CM-2 LSS VERSUS CRAY2 MINV/MINC CODE EFFICEINCY

Shown in Fig. 2 are the CPU time (t) results vs. matrix size of CM-2 LSS and [5] code Cray-2 MINV for matrix inversion based on using the following formulas:

$$t_L = 18.75 \ N^2 \tag{2a}$$
$$t_M = k \ N^3 \quad (k=7.23 \text{ for } N_p=1; = 4.44 \text{ for } N_p=4) \tag{2b}$$

respectively, where N is in unit of 1000 (1K) and N_p = No. of processors. The coefficient of LSS CPU time (t_L) formula, Eq. (2a), was obtained by use of t_L = 75 seconds at N = 2K on the assumption that only negligibly small additional time is needed in matrix inversion from that of linear equation using LSS. Similarly, the MINC CPU time formula for single processor case in Eq. (2b), was obtained from the CPU time data of 62.0 sec [3] at N = 2.048K by use of a cubic equation applicable to large matrix size case. (More accurately, t_M is proportional to $N^3(1 + 3/N)$.) For the four processor case of Cray-2 computer, k value (=4.44) was obtained from that of the one-processor case by a factor of 1.63, which was established from the data given in [4]. According to these formulas and (Fig 2), the CM-2 LSS code would be more efficient than the CRAY-2 MINV (or MINC) code in matrix inversion if the matrix size, N, is greater than 2.6K and 4.2K, respectively using one and four (whole) processors of Cray-2 computer.

The present paper is an excerpt of MRJ IR&D Report No. 0709-035-01 (Sept. 1990).

REFERENCES

1. Anon, MSC/NASTRAN User's Manual, Version 65C for Multiflow Trace 7/200 computer, March 9, 1989, McNeal - Schwendler Corp., Los Angles, CA.

2. Golbu, Gene H. and Van Loan, C.F., Matrix Computations, Second Edition, The Johns Hopkins University Press, Baltimore, 1989, Chapter 3.

3. Bailey, D. and Fugerson, H., "A Strassen - Newton Algorithm for High - Speed Parallelizable Matrix Inversion," Proceedings of Supercomputing 88, IEEE, 1988, pp. 419-424.

4. Jack J. Dongarra "Perfomance of Various Computers using Standard Linear Equations Software in a Fortran Environment," TM No. 22. Argonne National Laboratory, Dec. 29, 1988.

Table 1. Translational and Rotational Displacements, U_x and R_y (in a "mean" sense) of the Top Cross Section and U'_x and R'_y at x = 20", y= 0, and z = 30" (Nodal Point D in Figure 1B)

COMPUTER CODE	CM2/LSS			MULTIFLOW/NASTRAN
PRECISION (BITS)	SINGLE (32)	DOUBLE (64)	QUADRUPLE (105)**	DOUBLE (64)
U_x (10^{-3} rad)	-2.9283138	2.1161141630035144	2.1161416303311830	2.122304 (0.29%)*
R_y (10^{-4} rad)	-4.8129374	-1.777292209718076	-1.777292209559699	-1.775794 (-0.02%)
U'_x (10^{-3} in)	-0.15396219	1.9647324152788	1.9647324416799	1.905610 (-3.0%)
R'_y (10^{-2} rad)	1.2788257	1.406122284320883	1.406122284329757	1.354882 (-3.7%)

* Numbers in parentheses are the errors relative to the quad precision results of CM-2 LSS.

** 105 consists of 96 bit significand, 8 bit mantissa (sufficient for this problem), and 1 sign bit.

Table 2: CM-2 LSS CPU Time Results for Solving 2000 Linear Equilibrium Equations Governing the Thermal Distortions of the Baseline Finite Element Structural Model (Figure 1A)

Precision (Bits)	Number of Processors (n)		
	2K	64K	64K + 64 Bit Weiteks
Single (32)	20 min	~ 38 sec	N/A
Double (64)	3 hr. 19 min.	~ 6 min	~ 75 sec
Quad (105)	5 hr. 23 min.	~ 10 min	N/A

Notes: (1) All CPU time results are applicable to 100% dense linear systems.

(2) The 2K processor case CPU time results are actually obtained ones while the other CPU time results are extrapolated ones using the formula $t_{2nK} = t_{2K}/n$ ($1 \le n \le 32$), where the subscript number 2nK or 2K denotes the number of processors (in unit of K) used. Actual times would be slightly larger than the extrapolated ones shown.

Table 3: Comparison of Multiflow MSC/NASTRAN and CM-2 LSS CPU Time (sec) Results in Solving Dense, Linear System

Computer Code	F.E. Model	Mesh Size$^{(1)}$	Matrix Size	Matrix Density	CPU Time	Time Ratio
CM-2 LSS	M1	21x16	2000	100%	75	1
Multiflow	M2	43x32	1498$^{(2)}$	68%	317.09	5.0
MSC/NASTRAN	M4	70x40	1998$^{(2)}$	88%	400.29	5.3

(1) Mesh network size designation N_L x N_C in which N_L and N_C correspond to mesh network intervals in the longitudinal/ radial and circumferential directions, respectively.

(2) Condensed from 4000 DOF (for M2 case) and 8000 DOF (for M3 case) to 1998 DOF.

Figure 1   Undeformed and Deformed Structural Configurations of the Baseline Finite Element Model

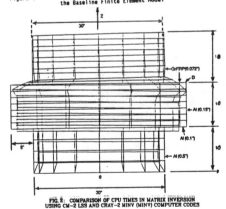

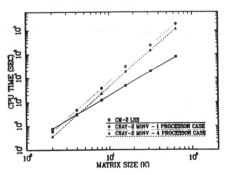

FIG. 2: COMPARISON OF CPU TIMES IN MATRIX INVERSION USING CM-2 LSS AND CRAY-2 MINV (MINC) COMPUTER CODES

PARALLEL COMPUTATION OF THE MODIFIED EXTENDED KALMAN FILTER

Mi Lu,*
Xiangzhen Qiao
Texas A & M University

Guanrong Chen
University of Houston

Abstract −− In this paper, we introduce a square-root algorithm based on the singular value decomposition (SVD) for the modified extended Kalman filter (MEKF), and develop a VLSI architecture for its implementation. Comparing with other existing square-root Kalman filtering algorithms, our new design is numerically more stable and has nicer parallel and pipelining characteristics when it is applied to the MEKF. Moreover, it achieves higher efficiency. For n-dimensional state vector estimations, the proposed architecture consists of $O(3/2n^2)$ processing elements and completes an iteration in time $O((s+8)n)$, in contrast to the time complexity of $O((s+3)n^3)$ for a sequential implementation, where $s \approx \log n$.

1. Introduction

For a discrete-time linear dynamic model in state-space description with Gaussian noise disturbance inputs, the Kalman filter [6] provides a real-time recursive algorithm for estimating the state vectors of the system using only available noisy observation data. For nonlinear dynamic systems, the extended Kalman filter (EKF) and modified extended Kalman filter (MEKF) [4] are used so that the filtering performance can be significantly improved. Some square-root filtering methods were proposed [1] to improve the numerical performance of the Kalman filter. Some efforts have also been made on developing parallel Kalman filtering architectures [3,8]. In this paper, we propose an SVD-based square-root algorithm for the implementation of MEKF, and develop a VLSI parallel processing computing system for the MEKF implementation.

2. The Modified Extended Kalman Filter (MEKF)

The modified extended Kalman filter (MEKF) was introduced in [4] to improve the performance of the extended Kalman filter (EKF). The MEKF uses an improved real-time linearization procedure in the sense that the center for each updated linear Taylor approximation is derived from an optimal Kalman filtering algorithm. The MEKF algorithm consists of two subsystems, which corresponds respectively to two state-space descriptions with different dimensions. The two models can be solved in parallel starting with the same initial estimate as shown in Figure 1. These two linear systems can be solved to yield optimal estimates by simply applying the standard Kalman filtering algorithm. In order to obtain a numerically stable and computationally efficient parallel architecture for the implementation of the MEKF, we propose to use an SVD-based square-root Kalman filtering algorithm.

3. The SVD-based Square-root Kalman Filtering Algorithm

The Kalman filter provides a real-time algorithm for estimating the n-dimensional state vector x_k of a discrete-time linear system $x_{k+1} = A_k x_k + B_k u_k + \Gamma_k \xi_k$, given a q-dimensional observation vector $v_k = C_k x_k + \eta_k$, where u_k is a $p \times 1$ control (or forcing) input, and ξ_k and η_k are two sequences of uncorrelated zero-mean Gaussian white noise with $Var(\xi_k) = Q_k$, a $p \times p$ non-negative definite matrix, $Var(\eta_k) = R_k$, a $q \times q$ positive definite matrix, and $E(\xi_k \eta_l^t) = 0$ for all k and l. In the above, A_k, B_k, Γ_k and C_k are known matrices of dimensions $n \times n$, $n \times p$, $n \times p$, and $q \times n$, respectively. The initial state x_0 is also assumed to be uncorrelated with ξ_k and η_k The standard Kalman filter estimates the state vectors of the system from a sequence of (noisy) measurements. The estimates $\hat{x}_k$ are updated by using a recursive formulas.

As has been observed, this filtering algorithm is numerically unstable in general. For this reason, several square-root computational algorithms have been proposed. For the discrete-time state-space model, the square-root covariance Kalman filter computes an optimal estimate of the state vector $\hat{x}_{k+1}$ using up-dated observation v_k. This is described as follows [8].

Time-update phase:
$$\hat{x}_{k \backslash k-1} = A_{k-1} \hat{x}_{k-1 \backslash k-1} + B_{k-1} u_{k-1},$$
$$P_{k,k-1} = A_{k-1} P_{k-1,k-1} A_{k-1}^t + \Gamma_{k-1} Q_{k-1} \Gamma_{k-1}^t. \quad (1)$$

Measurement-update phase:
$$w_k = v_k - C_k \hat{x}_{k \backslash k-1},$$
$$z_k = R_{e,k}^{-1/2} w_k,$$
$$\hat{x}_{k \backslash k} = \hat{x}_{k \backslash k-1} + R_{f,k} z_k,$$
$$\begin{bmatrix} R_k^{1/2} & C_k P_{k,k-1}^{1/2} \\ 0 & P_{k,k-1}^{1/2} \end{bmatrix} H_1 = \begin{bmatrix} R_{e,k}^{1/2} & 0 \\ R_{f,k} & P_{k,k}^{1/2} \end{bmatrix}. \quad (2)$$

Here, H_1 is an orthogonal matrix, $R_{e,k} = R_k + C_k P_{k,k-1} C_k^t$, and $R_{f,k} = P_{k,k-1} C_k^t R_{e,k}^{-t/2}$. The notation $A^{1/2}$ is used for a matrix satisfying $A = A^{1/2} A^{t/2}$ where $A^{t/2} = (A^{1/2})^t$ and $A^{-1/2} = (A^{1/2})^{-1}$ [8]. In the existing proposed methods, it is common that the matrices whose square-roots are to be taken are assumed to be positive definite. However, most of the matrices whose square-roots are to be taken in the Kalman filtering process are only non-negative but never be positive definite. Consequently, the above-mentioned approach is not applicable. Taking this issue into account, we propose to use an SVD-based square-root covariance Kalman filter algorithm for the MEKF discussed above. This consists of the following recursive computational steps:

1) $P_{0,0}^{1/2} = (Var(x_0))^{1/2}$ and $\hat{x}_{0 \backslash 0} = E(x_0)$.
2) For $k = 1, 2, ...,$ do
  i) time-update phase:
    a) $P_{k,k-1}^{1/2}$, the square-root of the matrix
$$\begin{bmatrix} A_{k-1} P_{k-1,k-1}^{1/2} & \Gamma_{k-1} Q_{k-1}^{1/2} \end{bmatrix} \begin{bmatrix} A_{k-1} P_{k-1,k-1}^{1/2} & \Gamma_{k-1} Q_{k-1}^{1/2} \end{bmatrix}^t,$$
    b) $\hat{x}_{k \backslash k-1} = A_{k-1} \hat{x}_{k-1 \backslash k-1} + B_{k-1} u_{k-1}$;
  ii) measurement-update phase:
    a) $\begin{bmatrix} R_k^{1/2} & C_k P_{k,k-1}^{1/2} \\ 0 & P_{k,k-1}^{1/2} \end{bmatrix} H_1 = \begin{bmatrix} R_{e,k}^{1/2} & 0 \\ R_{f,k} & P_{k,k}^{1/2} \end{bmatrix}$
    using an orthogonal factorization,
    b) $w_k = v_k - C_k \hat{x}_{k \backslash k-1}$,
    c) $z_k = R_{e,k}^{-1/2} w_k$,
    d) $\hat{x}_{k \backslash k} = \hat{x}_{k \backslash k-1} + R_{f,k} z_k$.

In consideration of the numerical stability, we have the following discussions.

3.1 The computation of $P_{k,k-1}^{1/2}$

Let $B = \begin{bmatrix} A_{k-1} P_{k-1,k-1}^{1/2} & \Gamma_{k-1} Q_{k-1}^{1/2} \end{bmatrix}$, which is an $n \times (n+p)$ real matrix. When $r = rank(B) = n \leq (n+p)$, it is always possible to factorize B using some orthogonal transformation, namely $B = QR$ with Q being orthogonal and R upper triangular. In the case where the rank is not full, that is: $r < n$, the QR factorization does not necessarily produce an orthogonal

---
*Supported by the National Science Foundation under grant no. MIP8809328.

basis for the linear span of the matrix B, and the most reliable technique for handling the rank deficiency is the singular value decomposition (SVD). As usual, the SVD of an $n \times m$ matrix B is defined as $B = V\Sigma U^t$, where $U \in \mathbf{R}^{m \times m}$, $V \in \mathbf{R}^{n \times n}$ are orthogonal matrices and $\Sigma = diag(\sigma_1, \cdots, \sigma_n) \in \mathbf{R}^{n \times m}$. Since $B^t = U\Sigma^t V^t$, without loss of generality we may assume that $n \leq m$. The square-root of BB^t can be computed from the

SVD of B^t. Let $B_s = V \begin{bmatrix} \sigma_1 & & & \\ & \sigma_2 & & \\ & & \ddots & \\ & & & \sigma_n \end{bmatrix}$. Then B_s is

an $n \times n$ square matrix satisfying $B_s B_s^t = BB^t$. This implies that matrix B_s is a square-root of matrix BB^t, which can be calculated from the SVD of either matrix B or B^t.

3.2 The SVD of the $m \times n$ matrix B^t ($n \leq m$)

When $m > n$, we propose to handle the rectangular SVD problem as follows. First, we compute the QR factorization of B^t: $B^t = Q \begin{bmatrix} R \\ 0 \end{bmatrix}$, where $R \in \mathbf{R}^{n \times n}$ is upper triangular. Then, we compute the square SVD of R:

$$W^t R V = \tilde{\Sigma} = diag(\sigma_1, \cdots, \sigma_n) \in R^{n \times n}. \qquad (3)$$

By defining $U = Q \begin{bmatrix} W & 0 \\ 0 & I \end{bmatrix}$, we have $U^t B^t V = \Sigma = diag(\sigma_1, \cdots, \sigma_n) \in \mathbf{R}^{m \times n}$. This process computes the SVD in $O(smn^2 + mn^2)$ time-steps.

3.3 The computation of $P_{k,k}^{1/2}$

The matrix R_k in equation (2) is always positive definite. So $R_k^c = R_k^{1/2}$, where R_k^c is the Cholesky factorization of matrix R_k. Hence, we may assume that matrix $R_k^{1/2}$ is a lower triangular matrix with positive diagonal elements. Let

$$DD = \begin{bmatrix} R_k^{1/2} & C_k P_{k,k-1}^{1/2} \\ 0 & P_{k,k-1}^{1/2} \end{bmatrix}$$, we may choose an orthogonal ma-

trix H_1 to annihilate the submatrix $C_k P_{k,k-1}^{1/2}$. Once this is done, the matrices $R_{e,k}^{1/2}$, $R_{f,k}$, and $P_{k,k}^{1/2}$ are obtained inside the annihilated matrix DD as shown in equation (2). This algorithm consists of two main parts: using the SVD to calculate $P_{k,k-1}^{1/2}$; using an orthogonal transformation to compute matrix $P_{k,k}^{1/2}$ from matrix $P_{k,k-1}^{1/2}$. The dependence relation of this algorithm is given in Figure 2,

4. Parallel computation of the MEKF

The SVD-based square-root algorithm consists of the following computations: SVDs, matrix-vector multiplications, matrix-matrix multiplications, the Cholesky factorization, orthogonal factorizations, and the back-substitution for solving triangular linear systems. All of these computations are highly parallel, and can be implemented efficiently by systolic arrays. From the dependence relation of this algorithm, we see that after the calculations of $R_{f,k}$, $P_{k,k}^{1/2}$ and $R_{e,k}^{1/2}$, the update of $\hat{x}_{k\backslash k}$ and the calculation of $P_{k,k+1}^{1/2}$ can be done concurrently. This means that all the calculations can be divided into two parts. Thus, the MEKF problem can be performed by using two standard Kalman filters parallelly, which are cooperated in the way shown in Figure 1.

The process for calculating the MEKF is illustrated in Figure 3. Array 1 (KF_1) and array 2 (KF_2) have the same structure but possess different dimensions. Every array can be divided into two parts, as shown in Figure 2, which work in parallel. In this algorithm, the time needed for a complete iteration of the sequential executions is $O((s+3)n^3)$, where n is the dimension of the state vector and $s \approx \log n$. In combining with the parallel architecture that we propose, however, only $O((s+8)n)$ time is needed in completing an iteration, using $O(3/2n^2)$ processing elements.

Reference

1. G. J. Bierman, Factorization Methods for Discrete Sequential Estimation, Academic Press, New York, 1977.

2. N. A. Carlson, "Fast triangular formulation of the square-root filter," AIAA J., Vol. 11, No. 9, 1259-1264, 1973.

3. M. J. Chen and K. Yao, "On realization of least-square estimation and Kalman filtering", Proc. of 1st Int. Workshop on Systolic Arrays, Oxford, pp. 161-170, 1986.

4. C. K. Chui, G. Chen and H. C. Chui, "Modified extended Kalman filtering and a real-time parallel algorithm for system parameter identification," IEEE Trans. on Automatic Control, Vol. 35, No. 1, pp. 100-104, Jan. 1990.

5. G. H. Golub and C. Van Loan, Matrix Computation, Johns Hopkins Press, Baltimore, 1983.

6. R. E. Kalman, "A new approach to linear filtering and prediction problems," Trans. ASME, J. Basic Eng., Vol. 82D, pp. 34-45, 1960.

7. M. Morf and T. Kailath, "Square-root algorithms for least-square estimation," IEEE Trans. on Automatic Control, Vol. AC-20, No. 4, pp. 487-497, 1975.

8. T. Y. Sung and Y. H. Hu, "Parallel VLSI implementation of Kalman filter", Proc. of IEEE/AIAA 7th Digital Avionics Syst. Conf., pp. 496-503, 1986.

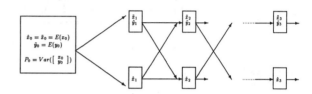

Figure 1. Parallel algorithm for the MEKF

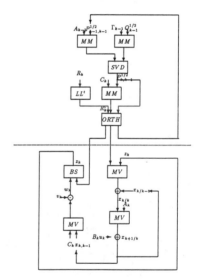

Figure 2. Dependence relations

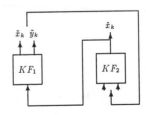

Figure 3. System architecture for the MEKF.

PARALLEL PROCESSING OF SPARSE MATRIX SOLUTION USING FINE GRAIN TASKS ON OSCAR (Optimally Scheduled Advanced Multiprocessor)

H. Kasahara, W. Premchaiswadi, M. Tamura, Y.Maekawa and S. Narita

Dept. of Electrical Engineering, Waseda University
3-4-1 Ohkubo Shinjuku-ku, Tokyo, 169, Japan

Abstract

This paper describes a parallel processing scheme for the solution of unstructured sparse systems of linear equations using fine grain tasks and the performance of the scheme on a multiprocessor system named OSCAR[1]. First of all, a special purpose compiler based on the proposed scheme generates near fine grain tasks from a loop free code, or a block of scalar assignment statements, generated by the code generation method[2][3]. Next, it statically schedules the fine grain tasks to processors by using a scheduling algorithm named CP/DT/MISF (Critical Path/ Data Transfer/ Most Immediate Successors First)[1] which can take into consideration data transfer overhead among processors. Finally, it generates an efficient parallel machine code for sparse matrix solution which minimizes the data transfer overhead and the synchronization overhead.

1   OSCAR's Architecture

OSCAR is a multiprocessor system with centralized and distributed shared memories as shown in Fig.1. Up to 16 PEs and Control & I/O processor (CIOP) are connected to three common memories (CMs) by three buses. Each PE is 32-bit custom-made RISC processor. It also has a 256-kw data memory(DM), two banks of 128-kw program memories(PMs), a 2-kw dual port memory(DPM) used as a distributed memory, a 4-kw stack memory(SM) and a DMA controller. It executes each instruction in one clock. This one clock execution allows a compiler to accurately estimate task processing time for static scheduling. Each CM is a simultaneous readable memory of which the same address can be read by three PEs at the same clock. The maximum data transfer rate attained by the buses is 60 megabytes/second.

Furthermore, OSCAR has three kinds of data transfer modes, such as one PE to one PE direct data transfer mode and one to all PEs data broadcasting mode by using the distributed shared memories and the ordinary indirect data transfer using the common memories.

2   A Parallel Processing Scheme of Sparse Matrix solution

This section describes a parallelizing compilation scheme for efficient solution of sparse matrices on OSCAR.

2.1   Direct Solution Methods

In this paper, the direct methods like the Gaussian Elimination and the Crout are applied for solving a system of linear equations Ax=b, where A is sparse but not necessarily banded or block diagonal. For the solution of the sparse matrices in the circuit simulation, the code generation method, or the symbolic generation method[3], together with the matrix reordering[2][4] has been successfully employed for many years. Generally, it decreases the solution time on a sequential machine markedly compared with an ordinary Fortran program for direct solution methods. In light of this fact, the proposed scheme processes the loop-free code generated by the code generation technique in parallel.

2.2   Parallelizing Compilation Scheme

The proposed parallelizing compilation scheme consists of the following four steps.

2.2.1   Task generation

A special purpose compiler automatically generates the loop-free code shown in Fig.2(b) for a sparse matrix in Fig.2(a). Next, the compiler decomposes the codes into tasks, each of which is a basic unit assigned to a processor element. In this paper, the statement level granularity, or near fine grain, is chosen taking into account OSCAR's processing capability and data transfer capability. In Fig.2(b), each statement is treated as a task having multiple floating point operations.

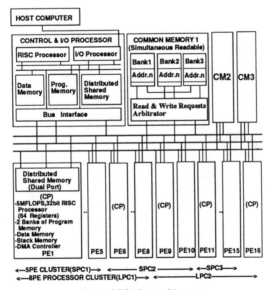

Fig.1: OSCAR's architecture

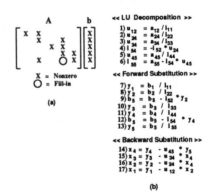

Fig.2: An example of generated fine grain tasks for sparse matrix solution

2.2.2 Task Graph Generation

Among the generated tasks, there exist data dependencies, such as, flow, output and anti-dependences [5]. The data dependences, or the precedence constraints, can be represented by a "task graph" [6] as shown in Fig.3, in which each task corresponds to a node. In Fig.3, figures inside a node circle represents task number, i, and those beside it presents a task processing time on a PE, t_i. An edge directed from node N_i to N_j represents partially ordered constraint that task T_i precedes task T_j. When we also consider a data transfer time between tasks, each edge has a variable weight. Its weight, t_{ij}, will be a data transfer time between task T_i and T_j if T_i and T_j are assigned to different PEs. It will be zero or a time to access the registers or the local data memory if the tasks are assigned to the same PE.

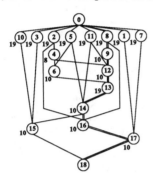

Fig.3: Task graph of Fig.2

2.2.3 Static Scheduling Algorithm

In order to process the set of tasks on a multiprocessor system efficiently, the assignment of tasks onto the processors and the execution order among the tasks assigned to the same processor must be determined optimally.

This scheduling problem, however, has been known as a "strong" NP-hard problem. In light of this fact, a variety of heuristic algorithms and a practical optimization algorithm[6] have been proposed.

In the proposed compilation scheme, a heuristic scheduling algorithm CP/DT/MISF considering data transfer, which is an improved version of CP/MISF[6], has been adopted taking into account a compilation time and quality of the generated schedule.

2.2.4 Parallel Machine Code Generation

For efficient parallel execution on an actual multiprocessor system, the optimized parallel machine code tailored to the target machine should be generated by using a scheduled result. The statically scheduled result gives us the information about tasks to be executed on each PE, the execution order of tasks on the PE, the estimated waiting time for data from the preceding tasks assigned to other PEs, the tasks to be synchronized and so on. By using the information, the compiler can generate the optimized parallel machine code which minimizes execution time including data transfer and synchronization overheads.

3 Performance Evaluation

Fig.4 shows the parallel processing time for solution of sparse matrices, such as, 200 by 200 with 1.0% non zeroes and 500 by 500 with 0.3% non zeroes on OSCAR. The solid lines represent measured processing time on OSCAR, and the dotted lines represent estimated processing time by execution simulation of the generated parallel machine code. "LM used" means that the local memories are used for passing shared data among tasks assigned to the same PE and the distributed shared memories are used for the direct data transfer and the data broadcasting among PEs. "CM used" means that the common memories are used for passing all shared data among tasks like an ordinary shared memory multiprocessor system.

Here, the following three important results should be noted.

The first result is that the use of the local memories and the distributed shared memories significantly reduces the solution time compared with the use of the common memory. It shows that the data transfer mode optimization based on the static schedule is effective to reduce the data transfer and the synchronization overhead.

The second result is that there exist no differences between the measured processing time and the simulated processing time. It means that the compiler can generate exact execution schedules. In other words, the machine code optimization by the compiler for solution of sparse linear equations is very effective.

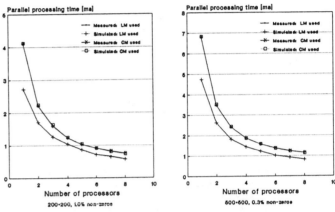

Fig.4: Parallel processing time on OSCAR

The third result is that the solution time was reduced remarkably by the use of multiple processors in every case. For example, solution time using local memory for 200 by 200 matrix was reduced from 2.71 ms for one PE to 1.04 ms for four PEs and 0.58 ms for eight PEs. Also, the solution time using local memory for 500 by 500 matrix was reduced from 4.73 ms for one PE to 1.45 ms for four PEs and 0.83 ms for eight PEs. These results mean that the proposed compilation scheme for sparse matrix solution is useful.

4 Conclusions

This paper has presented the parallelizing compilation scheme for the solution of sparse matrices and the performance of the scheme on OSCAR. From the evaluation, it has been confirmed that the proposed scheme reduces the solution time remarkably by using an arbitrary number of processors.

Reference

[1] H.Kasahara, H.Honda and S.Narita, "Parallel Processing of near fine grain tasks using static scheduling on OSCAR," on Supercomputing'90, pp.856-864, Nov. 1990.

[2] I.S.Duff, A.M.Erisman, and J.K.Reid, "Direct Method for Sparse Matrices," London, England: Oxford University Press, 1986.

[3] F.G.Gustavson, W.Liniger and R.Willoughby, "Symbolic Generation of an Optimal Crout Algorithm for Sparse Systems of Linear Equations", J. ACM, Vol.17, No.1, pp.87-109, 1970. July 1985.

[4] H.M.Markowitz, "The Elimination Form of Inverse and Its Application to Linear Programming," Management Science, Vol.3, pp.255-269, April 1957.

[5] D.A.Padua, and M.J.Wolfe, "Advanced Compiler Optimizations for Supercomputers," C.ACM, Vol.29, No.12, pp.1184-1201, Dec.1986.

[6] H.Kasahara and S.Narita, "Practical multiprocessor scheduling algorithms for efficient parallel processing," IEEE Trans. Comput., vol.C-33, pp.1023-1029, Nov. 1984.

Parallel Performance Evaluation of General Engineering Applications

Timothy J. Tautges[†]

Abstract. A new expression is derived for the overall speedup of a code made up of several parallel computational phases using different parallel solution strategies. The expression is used to derive the Speedup Improvement Equation, which quantifies the increase in overall speedup resulting from phase optimizations. Speedup data from a parallel nuclear severe accident code are used to show the interdependence of phase optimizations and their influence on overall speedup.

1 Introduction

Advances in the state of the art in parallel software algorithms and programming tools have allowed the application of parallel processing to general engineering codes. The distinguishing feature of these codes is that they consist of not one but many computationally intensive phases which each may be best suited to a different parallel solution strategy. While speedup relations exist for each of the strategies, these relations do not apply to a parallel code using several strategies. An overall speedup relation is needed, since this is the most important performance measure for these applications.

This paper derives a relation which gives the overall speedup for code as a function of computational phase speedups. Phase speedups can be evaluated using performance relations specific to the algorithms used in those phases. Expanding this relation results in the Speedup Improvement Equation, which predicts the effect on overall speedup of optimizations made to individual code phases. This relation is used to discuss the interdependence of individual phase optimizations in the severe nuclear accident code HECTR, executed in parallel on the Alliant FX/80.

2 The Speedup Improvement Equation

Consider a code made up of several expensive computational phases. In order to attain good overall speedups, a suitable parallel algorithm must be used to optimize each phase. This results in different parallel algorithms being used inside the same code. While performance relations usually exist for each of the parallel algorithms, it is difficult to extend any of the relations to treat the other parallel algorithms. Rather, an expression for the overall code performance as a function of individual phase performances can be derived, as follows.

The serial execution time of a program made up of individual computational phases is simply

$$T = T_1 + T_2 + \ldots = \sum T_i. \tag{1}$$

Each phase is parallelized, with its own characteristic speedup. If the phases are assumed to be executed in sequence, then the parallel execution time is

$$T' = \frac{T_1}{S_1'} + \frac{T_2}{S_2'} + \ldots = \sum \frac{T_i}{S_i'}. \tag{2}$$

[†]Formerly at the University of Wisconsin-Madison. Work performed under appointment to the Nuclear Engineering and Health Physics Fellowship Program administered by Oak Ridge Associated Universities for the US Department of Energy, and also made possible by a grant from the US Nuclear Regulatory Commission.

The speedup for the overall program is just the ratio of T and T':

$$S' = \left[\sum \frac{f_i}{S_i'} \right]^{-1}, \tag{3}$$

where

$$f_i = \frac{T_i}{\sum T_i} = \text{fraction of } T \text{ spent in phase i.} \tag{4}$$

Eq.(3) is a simple expression for the overall speedup of this program, as a function of phase speedup factors and execution time fractions. In fact, if two phases are defined, a parallel phase with ideal speedup and a serial phase with a speedup of unity, this equation reduces to the maximum speedup given by Amdahl's law.

Eq.(3) can be used to describe the speedup of any parallel program, with suitable definition of code phases and the input parameters f_i and S_i. It can also be used for different versions of the same parallel program. In this situation the phase speedups change between different parallel code versions, while the serial execution time fractions f_i remain constant.

Improving the overall speedup involves optimizing particular phases of code, which may or may not already be optimized. The effects of these optimizations on overall speedup need to be found, since overall performance is important for these codes. Expanding Eq.(3) in a Taylor series yields a suitable expression.

Assuming two phases, there are two independent variables in Eq.(3), the phase speedups S_1' and S_2'. Calculating the influence of phase optimizations implies two versions of code, before and after the optimization being studied. These versions are referred to with a prime and double prime, respectively. When Eq.(3) is expanded around S_1' and S_2', the result is (see [1] for details):

$$\frac{dS}{S} = \sum_{i=1}^{2} f_i' \frac{dS_i}{S_i'} + \sum_{i=1}^{2} \left[(f_i^{*\prime})^2 - f_i^{*\prime} \right] \left(\frac{dS_i}{S_i^*} \right)^2 + 2 f_1^{*\prime} f_2^{*\prime} \frac{dS_1}{S_1^*} \frac{dS_2}{S_2^*}, \tag{5}$$

where

$$\begin{aligned} S', S_i' &= \text{Overall and phase speedups in first parallel version,} \\ dS, dS_i &= \text{Overall and phase speedup improvements} \\ S_i^* &= \frac{1}{2}(S_i' + S_i''), \end{aligned}$$

$$\text{and } f_i' = \frac{T_i'}{\sum T_i'} = S' \frac{f_i}{S_i'}. \tag{6}$$

Eq.(5) is referred to as the Speedup Improvement Equation, or SIE.

The SIE shows that the influence of improvements to one phase on the overall speedup depends on two parameters, f_i' and (dS_i/S_i'). The former is the execution time fraction for phase i *in the version to be optimized*. The latter is just the fractional increase in the speedup of phase i.

Eq.(5) applied to the process of optimizing a parallel program shows some interesting trends. As the speedup of an individual phase increases, the execution time fraction of that phase decreases, and those of all other phases increase. This means that successive speedup improvements to the same phase have decreasing influence on the overall speedup because that phase's execution time fraction decreases. Also, as the execution time fraction for one phase decreases, those of the other phases increase, and so subsequent improvements to the remaining phases have a larger effect on overall speedup. These trends show the interdependence of phases within an overall parallel code. The relative importance of phases must be re-evaluated as optimization proceeds, in order to predict the benefits of further optimizations.

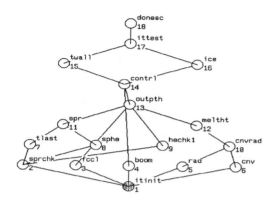

Figure 1: Task dependency graph for the HECTR program. Execution proceeds from ITINIT to DONESC, in the upward direction.

3 The HECTR Severe Accident Code

The speedup expression in Eq.(3) and the SIE were developed to analyze the parallel performance of the HECTR code [2], which simulates the conditions in the containment building during a severe nuclear accident. HECTR contains numerical models for relevant physical phenomena and nuclear plant safety systems, which are solved in a loosely coupled manner.

The HECTR code is partitioned into parallel tasks each representing an individual physics or safety system model; this method is referred to as a functional partitioning of HECTR tasks. Dependencies between the models result in the dependency graph in Figure 1. The CONTRL task, which is solved alone, accounts for up to 65% of the overall execution time. Therefore this task is optimized further using DO loop partitioning.

HECTR is typical of many other engineering codes, where the execution time is distributed across several computational areas, each amenable to different parallel processing strategies. The CONTRL task is amenable to data partitioning, while the remaining tasks lack expensive DO loops but do have functional parallelism.

4 Applications of the SIE

The partitioning of the parallel HECTR code conforms well to the assumptions made in deriving the SIE. Two distinct phases of execution can be defined, the dp and fp phases. The dp phase uses data partitioning, and is just the CONTRL task, while the remaining tasks are grouped into the fp phase, which uses functional paritioning. This definition emphasizes the use of the two partitioning strategies in the same code.

The HECTR code is parallelized in a series of discrete steps, with functional paritioning done first and then progressively smaller parts of the CONTRL task data partitioned afterwards. The implementation of data and functional partitioning result in versions 2.5 and 3.0–3.3 of parallel HECTR, respectively.

The SIE is validated by predicting the improvement in overall speedup for increasing degrees of dp phase optimization, and comparing these predictions with the actual speedups for each version of HECTR. The overall speedup improvements predicted by the SIE are all within 10% of the overall speedup increases measured with the actual parallel code, for dp and fp phase speedup improvements of up to 100%.

Table 1: Phase speedups and fp phase effectiveness for versions 2.5–3.3 of parallel HECTR.

HECTR Version	S_{dp}	S_{fp}	$(dS/S)_{fp}$ (%)	f'_{fp}
2.5	1.48	0.98	12.7	0.33
3.0	1.50	1.94	20.5	0.49
3.1	1.53	2.83	26.3	0.58
3.2	1.51	3.07	26.6	0.60
3.3	1.50	3.42	27.2	0.63

In deriving the SIE, the phase speedups are assumed to be independent. This assumption can be checked using the dp and fp phase speedups (S_{dp} and S_{fp}, resp.) for all versions of parallel HECTR, which are shown in Table 1. Increasing the optimization to the dp phase improves that phase's speedup, while the fp phase speedup remains almost constant.

While phase speedups are independent, their influence on overall speedup is not. This can be shown by calculating the effect of fp phase optimization on overall speedup, denoted by $(dS/S)_{fp}$, for varied degrees of dp phase optimization. The results are shown in the third column of Table 1. As the dp phase speedup increases, the effectiveness of fp phase optimization also increases, even though the speedup of this phase is unchanged. Since the influence of phase optimization depends on both the speedup increase *and* the execution time fraction of that phase, the latter must be changing. This is verified in the last column of Table 1, which shows that increasing the degree of dp phase optimization increases the execution time fraction of the fp phase, thus increasing the influence of any optimizations made to that phase.

5 Conclusions

The overall speedup of a parallel code is a function of the speedups of individual computational phases each weighted by their execution time fraction. This simple speedup relation is expanded to produce the Speedup Improvement Equation, which shows that the improvement to overall speedup resulting from phase optimizations depends on both the speedup improvement and the execution time fraction of the optimized phases. The SIE also shows that although phase speedups are independent, their influences on overall speedup are linked to other phases through their execution time fractions. As a code is optimized, the execution time fractions and thus the relative importances of the unoptimized phases of code increase. The SIE can also be used to determine which of several possible code optimizations will result in the largest increase to overall speedup.

References

[1] Timothy J. Tautges. *The Parallel Processing of Nuclear Power Plant Severe Accident Simulation Codes.* PhD thesis, The University of Wisconsin-Madison, November 1990.

[2] S. E. Dingman, A. L. Camp, C. C. Wong, D. B. King, and R. D. Gasser. HECTR version 1.5 user's manual. Technical Report SAND86-0101, NUREG/CR-4507, Sandia National Laboratories, April 1986.

Parallel Recognition of Two-Dimensional Images

M. Nivat†and A. Saoudi‡

(†) L.I.T.P, Université Paris VII, France

(‡) L.I.P.N, Université Paris XIII, France

Introduction

Recognition of general context-free languages plays an important role in syntax analysis of programming languages and pattern recognition. During the past decades, there has been considerable interest in efficient recognition and parsing of context-free languages. One can distinguish two well-known general recognition algorithms. The first one was proposed independently by Cocke, Kasami, and Younger (see [AU 74] and [Yo 67]) which requires grammar in Chomsky normal form. Since every context-free language can be generated by a context-free grammar in Chomsky normal form, this is not a fundamental restriction. The second algorithm was proposed by Earley [Ea 70] as an extension of Knuth's LR(k) algorithm.

Recently, there has been considerable interest in studying array grammars that generate sets of two-dimensional arrays (i.e. images). This is due to the fact that such grammars can be viewed as a syntactic model for generating digital pictures. Altough an array is a two-dimensional generalization of string, there are some problems involved in generalizing string grammars to arrays. The major problem is how to define rewriting rules for generating images. For this problem, there are several kinds of image rewriting systems (see [KSS 72], [NSD 89], [Ro 71] and [Ro73]). Another problem which is related to pattern recognition, is how to recognize a given image as a pattern generated by a given image grammar. The last problem is known as a recognition problem. More precisely, the recognition problem (for images) is to decide whether or not an image is generated by a given context-free image grammar.

This paper presents several results on the time complexity of the parallel recognition problem for Two-dimensional context-free grammars.

1. Parallel recognition of context-free grammars

The Cocke, Kasami and Younger algorithm for parsing context-free grammars in Chomsky normal form. This algorithm is beautifuly simple and can be extended to two-dimensional images. The method of Cocke-Kasami-Younger consists of constructing triangular matrix of size $n \times n$, where n is the length of the input string. A context-free grammar is a structure $G = <V, \Sigma, R, S>$, where V is the set of variables(i.e. nonterminals),Σ is the alphabet, R is a set of productions (i.e. rules), and S $(S \in V)$ is the start symbol.

If U and W are sets of nonterminal symbols, then we define the product of U and W,$U \otimes W$, by :

$U \otimes W = \{ X : \exists Y \in U \text{ and } \exists Z \in W \text{ such that } X \to YZ \text{ is a rule}\}$

Theorem 1. The recognition problem of context-free languages can be computed in $O(n^2)$ time on CREW-PRAM with $O(n)$ processors.

The following parallel algorithm solve the recognition problem for context-free grammars. It is a parallel version of Cocke-Kasami-Younger algorithm.

Algorithm 1.(PCKY)

(1) **for** $i := 1$ **to** n **pardo**

(2) $\quad T[i,1] := \{A : A \to u[i] \text{ is a rule}\}$

(3) **for** $j := 2$ **to** n **pardo**

(4) $\quad$ **for** $i := 1$ **to** $n - j + 1$ **do**

(5) $\quad$ **begin**

(6) $\quad\quad T[i,j] := \emptyset$

(7) $\quad\quad$ **for** $k := 1$ **to** $j - 1$ **do**

(8) $\quad\quad\quad T[i,j] := T[i,j] + T[i,k] \otimes T[i+k,j-k]$

(9) $\quad$ **end**

(10) **parend**

(11) **if** $S \in T[1,n]$ **then** accept **else** reject

When the PCKY algorithm halts, $T[i,k] = \{X : X \overset{*}{\Rightarrow} u[i,k]\}$. This algorithms requires $O(n^2)$ time and $O(n)$ processors.

2. Two-Dimensional Context-free image grammars

We will define the notion of context-free image grammar which generalize the notion of context-free grammars. The purpose of this generalization is to define a specification tool for two-dimensional patterns, where a pattern can be represented by two-dimensional arrays (i.e. images).

Definition 1. A context-free image grammar is a structure $G = <V, \Sigma_I, \Sigma, S, R_H, R_V>$ such that V is the set of non-terminal symbols, Σ_I is the intermediate alphabet, Σ is the set of terminal symbols, S is the axiom, R_H is the set of horizontal rules,and R_V is the set of vertical rules. One can view a context-free image grammar as a collection of two context-free grammars on strings. A context-free image grammar generate a two-dimensional images in two steps. In the first step, the horizontal rules are used to generate an intermediate string over the alphabet Σ_I. In the second step, the vertical rules are used to generate an intermediate two-dimensional image (i.e. array) over the alphabet Σ by rewriting each symbol, occuring in the intermediate string, into vertical string (i.e. column).

A context-free image grammar is said to be in Chomsky normal form if all its rules are Chomsky rules.

Lemma 1. For each context-free image grammar there exists an equivalent context-free image grammar in Chomsky normal form.

3. Sequential Recognition of Images

Context-free image grammars are more complicated than string context-free grammars because they deal with two-dimensional objects. We will show that recognition of objects generated by two-dimensional context-free image grammars is solvable in polynomial time on Random Access Machines (i.e. RAM). For more details about RAM see [AHU 74]. This class of 2D-image grammars allows to generate classes of interesting shapes such checkers of different sizes.

Let G be a context-free image grammar and I be an image over Σ. Then we denote by $I[k]$ the column k which can be viewed as string (i.e. word) .

Proposition 1. The recognition problem for context-free image sets requires $O(n^4)$ time and $O(n^2)$ space on Random Access Machine. To prove this result one can consider the following extension of Cocke-Kasami-Younger algorithm to context-free image grammars:

Algorithm 2.

$\{Vertical\ recognition\}$
(1) for $v := 1$ to n do
(2)　$CKY(I[v], V[v, 1..n, 1..n])$
$\{Horizontal\ recognition\}$
(3) for $i := 1$ to n do
(4)　$H[i, 1] := \{X\ :\ X \to B$ is a Horizontal rule and $B \in V[i, 1, n]\}$
(3)　for $j := 2$ to n do
(4)　　for $i := 1$ to $n - j + 1$ do
(5)　　$H[i, j] := \emptyset$
(6)　　　for $k := 1$ to $j - 1$ do
(7)　　　　$H[i, j] := H[i, j] + H[i, k] \otimes H[k, j]$
(8)　　end
(9) end
(10) if $S \in H[1, n]$ then accept else reject

Theorem 2. The recognition problem for context-free sets of 2D-images can be solved on RAM in $O(nt(n))$ time, where $t(n)$ is the time required for recognizing context-free languages on RAM.

4. Parallel recogntion of two-dimensional images

We now use parallel models, such PRAM and array automata, to study the parallel complexity of recognition problem for context-free sets of 2D-images .

Proposition 2. The recognition problem for context-free sets 2D-images can be solved on CREW-PRAM in $O(n^3)$ time with $O(n)$ processors.

Rytter [Ry 85] proved that the parallel time complexity of context-free languages is polylogarithmic. More precisely, he proved that the recognition problem, for a string of lenght n, can be solved on CREW-PRAM in $O(\log^2(n))$ time with $O(n^6)$ processors. This allows us to obtain the following results.

Theorem 3. The recognition problem for context-free sets of 2D-images on X-PRAM (X=CRCW,CREW, EREW) with $P(n)$ processors in $O(nt(n))$ time, where $t(n)$ is the time required for recognizing context-free grammars on Y-PRAM (Y=CREW,CRCW) with $P(n)$ processors.

In [Ko 75], Kosaraju uses array automata as a model of parallel processing and prove that the Cocke-Kasami-Younger algorithm can be used to recognize context-free languages in time $O(n)$ on two-dimensional array automata and in time $O(n^2)$ on one dimensional array automata. Using the Kosaraju's algorithm, one can prove the following result :

Theorem 4. The recognition problem of context-free image sets can be solved in time $O(n^2)$ on two-dimensional array automata and in $O(n^3)$ on one dimensional array automata.

By using three-dimensional array automata, the Kosaraju's algorithm, one can obtain a linear algorithm for solving the recognition problem of context-free sets of 2D-images.

Theorem 5. The recognition problem of context-free sets of 2D-images can be solved, by using CKY algorithm, in linear time on three-dimensional array automata.

References

[AHU 74] A. Aho, J. Hopcroft, J. Ullman, *The design and Analysis of Computer Algorithms*, Addison-Wesley (1974).

[Ea 70] J. Earley, *An efficient context-free parsing algorithm*, Comm. of the ACM 13(1970).

[KSS 72] K. Kristivasan, G. Siromoney and R. Siromoney, *Abstracts Families of Matrices and Picture Languages*, Computer Graphics and Image Processing 1 (1972)234-307.

[Ko 75] S. R. Kosaraju, *Speed of recognition of context-free languages by array automata*, SIAM J. of Comuting, vol 4 (1975)331-340.

[NSD 89] M. Nivat, A. Saoudi and R. Dare, *Parallel Generation of Finite Images*, Int. J. of Pattern Recognition and Artificial Intelligence, vol. 3 no. 3(1989).

[Ro 71] A. Rosenfeld, *Isotonic Grammars, Parallel Grammars and Picture Languages*, in Machine Intelligence, (B. Meltzer and D. Mitchie, Eds.), Univ. of Edinburgh Press 1971.

[Ro 73] A. Rosenfeld, *Array grammar normal form*, Inf. and Control 23 (1973) 173-182.

[Ry 85] W. Rytter, *The complexity of two-way pushdown automata and recursive programs*, in Combinatorial algorithms on words,(A. Apostolico and Z. Galil eds.), Springer-Verlag(1985)341-356.

[RS 89] W. Rytter and A. Saoudi, *On parallel complexity of image languages*, to appear in Information Processing Letters.

[Si 86] J. C. Simon, *Patterns and Operators : The Foundations of Data representation*, North Oxford Academic Publishers, (1986).

Parallel Incremental LR Parsing

N. Viswanathan   Y. N. Srikant

Department of Computer Science & Automation,
Indian Institute of Science, Bangalore, India.
e.mail:csa!srikant@vigyan.ernet.in

Abstract

A new parallel parsing algorithm for block structured languages, also capable of parsing incrementally, is presented. The parser is for LR grammars. A shared memory multiprocessor model is assumed. We associate processors to parse corrections independently with minimum reparsing. A new compatibility condition is used by the associated processors to terminate parsing, and prevent redoing the work of other processors. We give an efficient way of assembling the final parse tree from the individual parses. Our compatibility condition is simple and it can be computed at the parser construction time itself. Further, the compatibility condition can be tested while parsing, in constant time.

1 Introduction

Let $G = (N, \Sigma, P, S')$ be an augmented LR(1) grammar with N the set of nonterminals, Σ the set of terminals, P the set of productions, and S' the start symbol.

Let $w = x_0 y_1 x_1 y_2 x_2 \ldots y_n x_n$ be a sentence generated in L(G) and let $w' = x_0 y_1' x_1 y_2' x_2 \ldots y_n' x_n$ be a string obtained from w by substituting y_i' for y_i, where $i = 1, 2, \ldots, m$. Suppose that w has been parsed and we have its parse tree T.

Our aim is to examine the question whether we would be able to obtain the parse tree T_{new} of w' by modifying only around the corrected portions of T, using n-incremental LR(1) parsers in parallel. [GM80,Jal82,Yeh83] deal with the theory behind the sequential incremental parsing of w'. [AANL89] is an annotated bibliography on parallel parsing. Gafter [Gaf88] describes an incremental SLR(1) parser on a shared memory multiprocessor system.

2 Where and how to start?

As in sequential incremental parsing we start our incremental LR(1) parsers at 1 symbol before the corrections. We use the simple subtree operations divide, delete, replace and undoreduction described in [Jal82], to operate on the parse tree. The parsing stack of the individual parsers is restored by the subtree operation divide(T,last(x_{i-1})), executed by all the parsers P_i ($i = 1, \ldots, n$) in parallel. We assume T to be in the shared memory. The correction y_i' is then parsed. After parsing y_i', each P_i has to tackle two issues: (a) To parse in an incremental way, identify those parts of x_i, which may be partially or entirely parsed in the same way in both w' and w. (b) When y_{i+1}' is reached, P_i has to terminate.

To avoid reparsing parts of x_i we adopt Yeh's generalization of Wegman's skipping heuristic [Yeh83,Weg80]. The skipping heuristic provides a necessary and sufficient condition for directly using those subtrees of T whose frontier is in x_i. In using the heuristic we note that we need not maintain a table of INCIDENT_TREES as done by Yeh, because we use the parse tree approach and hence the subtrees of T can be directly examined one after another.

3 Parser termination

The termination condition in sequential incremental parsing is called matching reduction [GM80] and it requires that the stack configurations become identical. This condition restricts the applicability of parallel incremental parsing. If the dissimilarity in the stack is only due to the alternate derivations from the same nonterminal or a sequence of nonterminals, we characterize such a situation as compatible transition states of a nonterminal as defined below.

Definition 1: The transitions of a nonterminal B from states s_i and s_j are compatible iff any valid remaining input with its prefix derived from B and accepted from s_i can also be accepted from s_j and vice versa.

The idea is that if we can identify such compatible states of nonterminals efficiently, then we can use this information to terminate parsing even when the stacks are not identical. In the explanation below, we refer to the nonterminal A on the left hand side of any production $k : A \to X_1 \ldots X_n$, where $X_i \in N \cup \Sigma$ and $A \in N$, as the source nonterminal of the symbols X_i, $(1 \leq i \leq n)$, with respect to production k. We define pfollow of any symbol X_i in the production k to be $pfollow(X_i) = X_{i+1} \ldots X_n$, $(1 \leq i < n)$ and $pfollow(X_n) = \epsilon$. We identify three classes of compatible transition states of a nonterminal.

Right Recursive Compatibility: Suppose production k of the grammar G is right recursive in nonterminal B, that is $k : B \to X_1 \ldots B$. Let s_a and s_r be the transition states of B in the DFA constructed for the parser. Let us denote the transition of B from state s_r by B_r.

We make the claim that any valid remaining input with its prefix derived from B and accepted from state s_r can also be accepted from s_a and vice versa, whenever the source nonterminal transition state of B_r is s_a. The dissimilarity in the stack is only due to multiple derivations from the recursive nonterminal B. Therefore transition states s_a and s_r of B are compatible.

Alternative Compatibility: Suppose we have two grammar productions $k_1 : L \to X_1 \ldots B \ldots X_n$ and $k_2 : L \to Y_1 \ldots B \ldots Y_m$.

In the DFA constructed for the parser, assume that the corresponding transitions of the nonterminal B are at states s_i and s_j with the source nonterminal L's transition state as s_a.

We make the claim that any valid remaining input with

its prefix derived from B and accepted from state s_i can also be accepted from s_j and vice versa, whenever $pfollow(B)$ in productions k_1 and k_2 are identical. The dissimilarity in the stack is only due to alternate derivations from the nonterminal L. Then transition states s_i and s_j of B are said to have alternative compatibility. Two other special cases of alternative compatibility are possible. In one case a dissimilarity in the stack can occur due to alternate derivations from nonterminal L to B, either through or not through another nonterminal C. In the other case, it is due to alternate derivations to B either through L and C, or through L and D.

Modified Recursive Compatibility: Consider productions $k_1 : C \rightarrow X_1 \ldots BY_1 \ldots A \ldots X_n$ and $k_2 : A \rightarrow Z_1 \ldots BY_1 \ldots A$.

A is a right recursive nonterminal. Let the transition states of B be s_i and s_j and let the transition states of A and C be s_b and s_a respectively. If $pfollow(B)$ in production k_1 is equal to $pfollow(B) * pfollow(A)$ in productions k_2 and k_1 respectively, where $*$ is the concatenation operator, then any valid remaining input with its prefix derived from B and accepted from s_i can also be accepted from s_j if B is derived through A whose source state is s_a.

3.1 Compatibility testing

We do not identify all the possible compatible transition states of a nonterminal but only those which can be identified and used efficiently. We derive the compatible transition states information *before actual parsing*, while the states of the LR(1) DFA are constructed. The compatibility information is stored in a lookup table for each nonterminal. After parsing the insertions or deletions the compatibility checking procedure is called. The check procedure uses a table **CompTab** to identify whether the stack top state and the **preceding state** [Jal82] of the root symbol of subtree t currently examined can be compatible. The compatibility conditions are then derived from another table and tested. Whenever compatibility condition is tested true, all the subtrees before the next correction are pushed into the parsing stack and the parsing is terminated. Otherwise, the skipping heuristic is tested to avoid the reparsing of each subtree and P_i proceeds to parse the next correction.

Applying the algorithm to parallel parsing is straightforward. We consider the given program to consist of "corrections" to a trivial original program with empty modules. We associate one processor to each module and incrementally parse the "corrections" introduced in the module. Once the modules are parsed the final parse tree can be constructed in parallel.

4 Final parse tree construction

Only those processors which have tested compatibility true are used for constructing T_{new} from the individual parses. The construction involves no reparsing. The algorithm is given below. The correctness of the algorithm is guaranteed by our compatibility definition.

Algorithm for construction of T_{new}

/* Active processors have their compatibility flag set and their index range is $0 \ldots m-1$. Notations defined in [Jal82] are used */

Clear R_stack;
for $i := 0$ to $(\log\lceil m \rceil - 1)$ do
begin
for (processors with index j s.t. $j \bmod 2^{i+1} = 0$) pardo
if $(j + 2^i \leq m - 1)$ then
begin
    P_j divides T_{j+2^i} at correction a_{j+2^i};
    /* corrections $a_0 \ldots a_{m-1}$ are assumed to be those to which the processors had been initially assigned */
    $(t_{q1}, t_{q2}) \leftarrow$ divide(T_{j+2^i}, a_{j+2^i});
    push t_{q2} on R_stack;
    for (each $t = $ pop(R_stack)) do
    begin
    perform all reductions with first(yield(t));
    push t on the P_stack of P_j;
    end
end
$T_{new} := T_0$;
/* final parse tree is the tree constructed in P_0 */
end

An upper bound for the speedup possible when we parse in parallel can be obtained by using the probabilistic methods followed in [CK85,SD90].

References

[AANL89] R. op den Akker, et.al. *An Annotated Bibliography on Parallel Parsing*. Memoranda Informatica 89-67. Department of Computer Science, University of Twente, Netherlands.

[CK85] J. Cohen and S. Kolonder. *IEEE Trans. on SE*, 11(1), 1985. pp. 114-124.

[Gaf88] N.M. Gafter. In *Intl. Conf. on Parallel Processing*, IEEE, 1988. pp. 577-584.

[GM80] C. Ghezzi and D. Mandriolli. *Journal of ACM*, 27(3),1980. pp. 564-579.

[Jal82] F. Jalili. *Design of Incremental Compilers*. PhD thesis. University of Pennsylvania, 1982.

[SD90] D. Sarkar and N. Deo. pp. 677-683. *IEEE Trans. on SE*, 16(7), 1990.

[Weg80] M. Wegman. 21st *Annl. Symp. on FOCS*, 1980. pp. 320-327.

[Yeh83] D. Yeh. *Incremental Syntactic and Semantic Analysis of Modified Programs*. PhD thesis. The Norwegian Institute of Technology, 1983.

An Asynchronous Distributed Approach to Test Vector Generation Based on Circuit Partitioning on Parallel Processors

Sumit Ghosh
LEMS, Division of Engineering
Brown University
Providence, R.I. 02912

Tapan Chakraborty
ERC, AT&T Bell Labs
Princeton, NJ 08540

Abstract

A critical aspect of digital electronics is the testing of the manufactured designs for correct functionality. The testing process consists of first generating a set of test vectors, then applying them as stimuli to the manufactured designs, and finally comparing the output response with that of the desired response. A design is considered acceptable when the output response matches the desired response and rejected otherwise. The process of generation of the set of test vectors for a given digital design and a given underlying fault model is termed test vector generation. In the digital electronics industry today, test vector generation programs such as the STG [1] are based on algorithms that strictly execute serially on uniprocessor computers. As a result, for moderately complex designs, test generation is acutely time consuming. This paper presents an asynchronous distributed approach to test vector generation of combinational digital designs that is executable on parallel processors. Test generation is compute intensive and the approach presented in this paper distributes the computation load on multiple processors to achieve fast test vector generation. In this approach, the digital design under test is partitioned with the components in each partition assigned to a processor of the parallel processor system. Conceptually, the test generation algorithm is replicated for every component and a component executes its share of test generation – justification, sensitization, and conflict resolution, upon receiving appropriate messages from other components and propagates messages to subsequent components. Test vector generation is complete when consistent logical values have been assigned at each of the primary inputs of the design. This approach has been verified through an implementation on the Sequent computer system at Bell Labs. Performance measures based on a few example circuits indicate a substantial speedup of the test generation process over the current industrial test generation approaches.

2 The Process of Test Vector Generation

This section first examines the limitation associated with the conventional algorithms and then presents a distributed approach to test vector generation. The proposed approach utilizes the nine-value test generation algorithm proposed in [2] and distributes the three basic processes – sensitization, justification, and conflict resolution, of the nine-value algorithm among every component of the digital design under test. Every component executes asynchronously upon receiving appropriate messages from other components and performs its share of the overall task of test generation without requiring global knowledge of the progress of test generation. The advantages of nine-value test generation algorithm [2,3] includes its guarantee of generating a test sequence, if one exists, for every stuck-at fault in a synchronous design having a synchronizing sequence, higher degrees of freedom provided by the nine-valued value system – 0/0, 0/1, 0/X, 1/0, 1/1, 1/X, X/0, X/1, X/X, and that sensitization along any path from the fault site to a primary output port using the nine-value model simultaneously takes into account all of the possibilities of sensitizing this path as a part of multiple path sensitization. This paper is confined to combinational digital designs and therefore the two latter advantages of the nine-value algorithm [2,3] are pertinent to this research. A drawback of the nine-value algorithm is increased computation which is appropriately addressed by the increased computational power of the parallel processor system.

2.1 Analysis of the Serial Test Vector Generation

Consider the combinational digital design in Figure 1 consisting of NAND gates G1 through G8. The electrical paths labeled 1 through 4 are primary inputs while that labeled by 12 is the primary output. Assume a stuck-at logical 1 fault associated with the input pin 8 of gate G8. In the traditional nine-value algorithm, first a sensitized path is obtained from the fault site – path 8, to the primary output port – 12. The requirements for sensitization are the assignments of X/1 at each of the input ports 9, 10, and 11 and 0/1 at input port 8 of gate G8. The next step consists of justification i.e., the assertion of nine-value assignments at the input ports of gates successively in the direction of the primary input ports. For the design in Figure 1, each of the gates G4 through G7 must be justified serially for the assignments 0/1, X/1, X/1, and X/1 asserted at their output ports. Then the set of gates G1 through G3 must be justified serially corresponding to the assignments asserted at their output ports as a result of justification of the gates G4 through G7. In the process of

justification, it is possible to experience apparently conflicting assignments at the electrical path 6 following the justification of gates G5 and G6. Under these circumstances, the justification process is rolled back to either or both of G5 and G6 to explore alternate assignments at their input ports. Where the gates G5 and G6 are unable to provide non conflicting assignments at port 6, the justification process must roll back further to gate G8 to explore alternate assignments at the ports 9 through 11. If the gate G8 is unable to provide a set of assignments such that the subsequent assignments at 6 are non conflicting, the fault in Figure 1 is considered untestable. Where conflicting assignments are experienced at the output port 6 of G2 by the gates G5 and G6 and the process of justification is rolled back to gates G5 and G6, previously asserted assignments at the input port 1 and 4 by G5 and G6 respectively may need to be invalidated. Similarly, where justification is rolled back to gate G8, previously asserted assignments at input ports {11} and {7,3} of gates G8 and G7 may require invalidation. Thus, the occurrence of conflict at a gate may require invalidation of previously asserted assignments at other gates of the digital design which implies global knowledge of the progress of test generation. Assuming the absence of conflicts in the justification process, the conventional algorithm requires CPU time corresponding to the sensitization of gate G8 followed by the gates G4 through G7 and finally the gates G1 through G3. It may be observed that although each of the gates in the set G4 through G7 may in principle be justified independent of others, the traditional approach insists on their serial execution. Similarly, the justification of each of the gates in the set {G1,G2,G3} may be performed independent of others in the set as soon as non conflicting assignments are asserted at their output ports by the gates G4 through G7. Where the resolution of a conflict at the path 6 (say) involves the assertion of new assignments at each of the paths 9, 10, and 11 by gate G8, the conventional approach is limited to the serial execution of three independently executable justification subprocesses.

2.2 Parallelizing Test Vector Generation Process

In this approach, the digital design under test is partitioned with the components of each partition assigned to a processor of the parallel processor system. Each component of the digital design is represented through an entity and the task of test vector generation for the design is distributed among the entities. Every entity shares the conventional (normal) input and output ports of the original component. In addition, an entity includes special input and output ports corresponding to the conventional output and input ports respectively of the component. The test vector generation process is distributed into three subprocesses – justification, sensitization, and conflict resolution, associated with every entity and two additional processes associated with the primary inputs and outputs respectively. An entity receives messages corresponding to sensitization through its normal input ports, processes the commands encapsulated in the messages, and then propagates the sensitization command to subsequent entities in the direction of the primary output. When the sensitization command eventually reaches a primary output port, a sensitized path has been obtained in the nine-value test generation algorithm [2]. Accordingly, the assignment X/X is asserted at the special input ports of every entity whose output port is of type primary. The commands for justification are received in the form of messages through the special input ports of an entity which are then processed and the justification request is propagated to subsequent entities in the direction of the primary input ports. Associated with either of the sensitization and justification commands are the nine-valued assignment values. Corresponding to a component with a fanout exceeding unity, separate justification commands may be received from other entities in the direction of the primary output. Where the intersection of the assignment values received through separate justification commands is empty, the assignments are in conflict and the entity must initiate conflict resolution. To resolve the conflict, the entity propagates messages to the succeeding entities connected to its fanout requesting alternate assignments. Where available, the entity receiving the request propagates alternate assignments to the entity that issued the original request. When a entity receives such a request but alternate assignments are unavailable, it issues a message to the succeeding entities (towards the primary output) in the direction of the primary output requesting that alternate assignments be propagated to it. When this recursive process finally reaches the primary output and alternate assignments are yet unavailable, the conflict is unresolvable and test generation fails. In the event an alternate assignment is available, it is propagated to the entity that detected the conflict and the latter verifies whether the conflict is resolved. Following the resolution of a conflict, the justification process for the entity resumes. It is observed that corresponding to an alternate assignment, other previously asserted assignments may be invalidated.

Where non conflicting justification requests are received at every primary input port, it may apparently seem that test generation is complete. Such a conclusion may be erroneous in the asynchronous distributed approach as illustrated as follows. Consider the scenario where non conflicting assignments have been asserted at all but one primary input port (i) of a design. Assume further that a conflict is detected by a model in the digital design. The model successfully resolves the conflict and, consequently, a non conflicting assignment is asserted at the only remaining primary input port (i) of the design. In the process of resolving the conflict, a previously asserted assignment is canceled which invalidates the previous assignment at a primary input port j (j ≠ i). Given that the algorithm permits distributed asynchronous execution of the models, it is conceivable that the non conflicting assignment at input port i is received prior to the arrival of the message indicating the invalidation of the prior assignment at primary input port j. This necessitates the development of a methodology that is capable of determining with absolute certainty when non conflicting assignments are asserted at every primary input port accompanied by the total absence of any conflicting assignments in the entire digital design. In this methodology, when non conflicting assignments are asserted at a primary input port by one or more models, a special message – "token", is propagated to the models by the processor associated with the primary input port. A model propagates tokens to subsequent models through its normal output ports when tokens have been received at all of its normal input ports. The tokens associated with a model are destroyed when the model is rejustified or alternate assignments are asserted in response to conflict resolution implying that the previously asserted assignments are invalidated. Test generation is complete when tokens are received at every primary output port. Since tokens are asserted in response to non conflicting assignments at the primary input ports and as tokens are destroyed when a model is either rejustified or new assignments are asserted at its special output ports, the arrival of tokens at every primary output port implies consistent assignments at all primary inputs and the total absence of any inconsistencies in the test generation for the circuit under test.

Consider the digital design in Figure 1 where each of the components G1 through G8 are assigned to a processor of the parallel processor system. In addition, assume a processor associated with the four primary inputs 1 through 4 and another processor associated with the primary output port 12. Furthermore, assume each of the electrical paths 1 through 12 are modeled through bidirectional communication paths. The electrical path 6 representing a connection between the output of G8 and the input ports of G5 and G6 is expressed through two unique bidirectional communication paths between {G8,G5} and {G8,G6} respectively. Similarly the electrical paths 1 through 4 are expressed through two, three, two, and two unique bidirectional paths respectively. Corresponding to a stuck-at logical 1 fault associated with the electrical path 8, gate G8 is sensitized which in turn asserts the assignment X/1 at each of the input ports 9 through 11 of G8. Gate G8 propagates the assignment 0/1 to G4 and X/1 to G5 through G7 respectively along with the command "justify". Each of the processors associated with G4 through G8 simultaneously perform justification and the appropriate assignments are propagated to the gates G1 through G3 respectively. Gates G1 and G2 may be immediately justified and the assignments propagated to the processor associated with the primary inputs. Gate G2 must first ensure that the assignments propagated from G5 and G6 are consistent. In the event of inconsistency, G2 propagates messages first to G5 requesting that G5 propagate alternate assignments, if any, to G2. During justification of a gate such as G5, it is conceivable that more than one set of input assignments may be realized. The first assignment set is propagated to G2 and the primary input port 1 and the alternate assignments are stored within G5. Where none of the alternate assignments from G5 matches with the assignment from G6, G2 may propagate a message to G6 requesting that G6 propagate an alternate assignment, if any, to G2. During conflict resolution, G2 ensures that all possible combinations of assignments from G5 and G6 are considered. Where G2 eventually realizes consistent assignments

by G5 and G6, it is justified and the assignments at its input ports are propagated to the primary input ports 2 and 3. The processor associated with the primary input ports verifies that the assignments asserted at each of 1 through 4 by gates G1 through G7 are consistent. In the event of inconsistency, conflict resolution techniques discussed earlier are utilized. Corresponding to a consistent assignment at the primary input ports 1 through 4, tokens are asserted at all of the input ports of G1 through G3 and a few input ports of G4 through G7. Then, G1 through G3 assert tokens at their normal output ports with the result that tokens are available at all of the input ports of G4 through G7. Finally, a token is asserted by G3 at its primary output and test generation is complete. The nine-valued assignments at the primary input ports 1 through 4 constitute the test vector set for the stuck-at logical 1 fault at the input port 8 of G8.

3 Performance of the Distributed Approach

This section presents an analysis of the performance data collected from the test generation of s few example combinational circuits. The designs considered in this experiment include a carry look-ahead generator (TI SN54182), a 4 to 16 demultiplexer (TI SN54154), a decoder (TI SN54138), a 4-bit magnitude comparator (TI SN5485), and a slightly modified circuit borrowed from Fujiwara's book [3]. Each circuit is executed by the distributed algorithm on the Sequent parallel processor for different numbers of partitions and different faults. The number of processors of the Sequent system utilized during a particular execution equals the number of partitions plus 2. Consequently, the smallest number of processors used for execution is 3. The CPU time required for distributed execution is the time required by the special entity (executing on another processor) until a consistent solution has been obtained. In the event of an unsolvable conflict detected by a processor, th CPU time is the time used by that processor until detection of the conflict. The same circuit is also executed by the Bell Labs sequential test generation algorithm – STG, and the CPU time required is noted. While STG runs on a SUN 3/60 workstation, Sequent utilizes custom-designed processors that are observed to be slower by a factor of 4.85. The CPU times from the Sequent are appropriately scaled.

The performance data for the example circuits are presented in Tables 1 through 5. While the first row expresses the CPU times, the second row represents the speedup factor. The speedup factor is defined as the ratio of the CPU time required by STG to that required by the distributed algorithm for a particular partitioning (and the appropriate number of processors). The second column in the tables refer to the actual CPU times used by the STG program while the subsequent columns express the scaled CPU times and speedup factors for varying number of processors of the parallel processor system.

Table 1 indicates a maximum speedup factor of 9.3 with 7 processors while Tables 2 through 5 indicate maximum speedup factors of 9.5, 7.5, 6.9, and 5.6 with 7, 13, 4, and 5 processors respectively. The partitioning of the circuits in this research are done arbitrarily and study of their influence on the performance is under investigation.

References

[1] Wu-Tung Cheng and Tapan J. Chakraborty, "Gentest: An Automatic Test Generation System for Sequential Circuits," IEEE Computer, April 1989, pp. 43-49.

[2] P. Muth, "A Nine-Value Circuit Model for Test Generation," IEEE Transactions on Computers, Vol C-25, No. 6, June 1976, pp. 630-636.

[3] Hideo Fujiwara, Logic Testing and Design for Testability, The MIT Press, 1985.

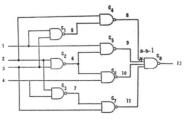

Figure 1 : Test Vector Generation of a Combinational Digital Design.

Table1: Modified Fujiwara's Circuit

N	STG	3	4	7	8	10.
CPU Times	0.5s	0.12s	0.12s	0.06s	0.06s	0.12s
Speedup Factor	1.0	4.6	4.6	9.3	9.3	4.6

Table2: SN54182: Carry Lookahead Generator

N	STG	3	5	7	11	16.
CPU Times	0.76s	0.16s	0.16s	0.08s	0.1s	0.12s
Speedup Factor	1.0	4.75	4.75	9.5	7.6	6.33

Table3: SN54154: 4 to 16 Multiplexer

N	STG	3	6	8	13	18.
CPU Times	0.9s	0.51s	0.25s	0.16s	0.12s	0.14s
Speedup Factor	1.0	1.8	3.6	5.6	7.5	6.4

Table4: SN54138: Decoder/Demultiplexer

N	STG	3	4	6	7	11.
CPU Times	0.55s	0.16s	0.08s	0.16s	0.16s	0.12s
Speedup Factor	1.0	3.4	6.9	3.4	3.4	4.6

Table5: SN5485: 4-Bit Magnitude Comparator

N	STG	3	4	5	6	15.
CPU Times	0.9s	0.41s	0.27s	0.18s	0.16s	0.20s
Speedup Factor	1.0	2.2	3.3	5.0	5.6	4.5

Granularity Analysis for Parallel 3D Coronary Arteriography

Authors[1]: Alok Sarwal, Sin Guan Tan, Fusun Ozguner, Dennis L. Parker

Abstract

Coronary arteriography is a technique used for evaluating the state of the coronary arteries[1] and to assess the need for bypass surgery and angioplasty. This technique represents an easily reproducible method for following the temporal changes in coronary morphology. Due to computation overhead, substantial time is required for accurate 3D quantification of an arterial tree, thus making it prohibitive for the clinician to utilize. A feasibility study concluded[2] that the reconstruction of the 3D image and arteriographic measurements can be made close to real time by parallel implementation. X-ray data of two and three views of a human patient and a pig arterial cast were used for this study. Load balanced mapping algorithms for different algorithms at different levels of granularity are described. Results on the speedup are presented.

1.0. Introduction

One general scheme for parallel 3D reconstruction from multiple views of coronary arteries from X-ray images is outlined below. The arterial tree is first segmented into branches (segments). This is followed by edge and center line detection of individual segments. Area of cross-section is computed along each branch by densitometric calculations for the region bounded by the geometric edges[2]. These independent segments are subsequently merged together for a complete 2D view. Corresponding segments between 2D views (2 or more) can be combined to give a 3D description of each branch[3]. This represents a coarse grain data parallel[4,5] application with potential for high speedup.

Automatic segmentation requires a global description of the geometry of the arterial tree in each view. The steps involved for such segmentation requires a 2D edge detection and representation of the edges in a suitable data-structure. Edge detection in 2D space can be done with Canny's 2D separable edge detector. One data-structure for geometric description of the arterial tree can be created using crackcodes[6]. Crackcodes provide a means of describing the outline of an object. Fourier descriptors[7] provide a technique for describing the frequency content of an object. Thus, Fourier descriptors provide another tool for describing the boundary of the arterial tree. These two algorithms are applied with fine grain parallelism.

2.0. Coarse Grain Approach

The coarse grain approach involves computing the 2D plane tree description and obtaining the 3D reconstruction by mapping individual segmented branches to the various processors. The edges and center line are detected by extraction of the branch data which is convolved with a suitable 1D kernel followed by dynamic programming. Each view is processed by distributing its branches to yield a 2D

[1] **Alok Sarwal & Sin Guan Tan:** UES Inc., 5162 Blazer Pkwy., Dublin, Ohio 43017.
Fusun Ozguner: Dept. of Elecrical Engineering, The Ohio State Univ., Columbus, Ohio 43210.
Dennis L. Parker: Dept. of Medical Informatics, Univ. of Utah, Salt Lake City, Utah 84143.

[2] Small Business Innovation Research project supported by the National Institute of Health through grants # 1R43HL42208-01 (phase 1) and # 2R44HL42208-02 (phase2).

plane tree. This process has to be repeated for up to a sequence of 30 such sets, each set having 2 or more views, to capture a complete heart cycle. The appropriate branches are extracted from the raw image based on the mapping strategy. The distribution also establishes a data-structure for subsequent recombination of the branches at the host. The 3D reconstruction algorithms are applied to the center lines from multiple views of a particular branch. This operation can also be performed on each branch independently of other branches. It is estimated that about 1800 branches have to be computed for 2D plane tree descriptions and the multiple views then combined together for about 600 branches spread out over a sequence of 30 images. Thus, there is good potential for continually keeping the data pipeline full.

2.1. Mapping

The benchmark was done for the cast of a pig arterial tree. The X-ray image has 512 by 512 pixels each with 10 bit accuracy. Segmentation is done to provide seven branches which are mapped to up to 4 processors. The raw image and the target structures are communicated to the processors where extraction of the appropriate local branches occurs. Load balancing of the computation of the branches assigned to each processor is done by a cost function that considers the time of communication of the target structure from the host and the computation on that branch. The length of the branch can vary substantially (26 to 71 pixels) as segmentation is done with some consideration to the anatomic features. Thus, an optimal mapping function is important. The overhead for communication from host to all processors is about 2.7 seconds, and is relatively high. The real benefit is obtained if geometry of larger number of branches of an image are computed simultaneously. This will allow the next image to be pipelined resulting in this communication being completely overlapped with the computation of the previous set of branches. It was also noticed that the speed up obtained is a direct function of the optimality of the mapping and can be improved if mapping larger number of branches.

3.0. Fine Grain Approach

In considering the problem of automating the segmentation process using global techniques, we have identified several key image processing routines that are primary to any such efforts. Along these lines, we have parallelized two modules i.e. crackcode and Fourier descriptor generation. Within the scheme of automated segmentation, these parallelized modules are fine grain in nature.

Crackcodes can be used to described the arterial tree after execution of a global edge detection technique. For example, a unit square can be described by the codes 'N W S E' corresponding to the compass directions (assuming the top of an image corresponds to north). In one scan of the image function, we determine all crackcodes.

Fourier descriptors is an accepted technique for contour description and pattern recognition [7]. To implement Fourier descriptors, we first construct a crackcode representation. The nth Fourier descriptor, U(n), is determined according to the equation:

$$U(n) = \frac{L}{(n2\pi)^2} \sum_{k=1}^{L} c(\Delta d) e^{-j\frac{n2\pi k}{L}} \quad (1)$$

where L is the perimeter length of the object, k is the perimeter traversal index, and $c(\Delta d)$ is a constant dependent upon the crackcode. The value of $c(\Delta d)$ will not be given for the sake of brevity.

3.1. Mapping for Crackcodes

The overall strategy adopted was to partition the image in vertical strips and assigning the set of partitions to the processors. The computation cost in generating crackcodes is due to scanning the image and searching the list of crackcodes for the proper crackcode to update. With consideration to the former, we expect to partition the image equally among the processors. However, in most cases, objects tend to be centered in their frames. Thus, there exists the potential for processors to have unequal work loads since the processors assigned to process the outer fringes of the image will end up with fewer edges and potentially fewer crackcodes. The alternate strategy is to consider the edge density per column of the image obtained from Canny's 2D edge detector and ensure a uniform distribution of edges to each processor. Thus, the number of columns assigned to each processor will vary, but the search through the respective crackcode lists will be uniformly expensive. Let $n(i)$ be the number of edges assigned to the ith processor, x be the average number of edges per processor, and N the total number of edges already assigned. The columns are assigned to the ith processor until the following condition becomes true:

$$n(i) > x \ OR \ N > x * i$$

Thus, we keep a local and global view of our assignment and provide a well balanced computation load for each processor. These mapping strategies result in no inter-processor communication. Results have been provided for both cases.

3.2. Mapping for Fourier Descriptors

We can determine all coefficients independently, as is evident from equation (1). Our mapping strategy is to assign an equal number of crackcode moves to each processor as the computation of the Fourier descriptors depend on these moves. This strategy will help ensure that the processors are equally loaded. However, this mapping is a np-compete problem and the optimal solution can only be determined by an exhaustive search. Therefore, complete crackcodes were mapped to processors on the greedy principle beginning with the crackcode with the largest number of moves. The human arterial tree (256 by 256 image) has 190 crackcodes with number of moves ranging from 4 to 1154. There is no inter-processor communication necessary for this approach and so it will provide better overall performance with larger number of crackcodes.

4.0. Results

In the coarse grain approach, segmentation was performed manually on a view of pig cast arteries to yield a set of branches. Results are presented for generating a 2D plane tree in parallel. The speed up for this case was about 2.8 for 4 processors as is shown in Fig. 1. This performance will improve for larger number of branches per processor. The time taken for computation on 4 processors is 1.2 sec. Communication overhead of about 2.7 sec. is due to loading the first image in the sequence to the processors, but this overhead will be negligible for subsequent images as they will be pipelined.

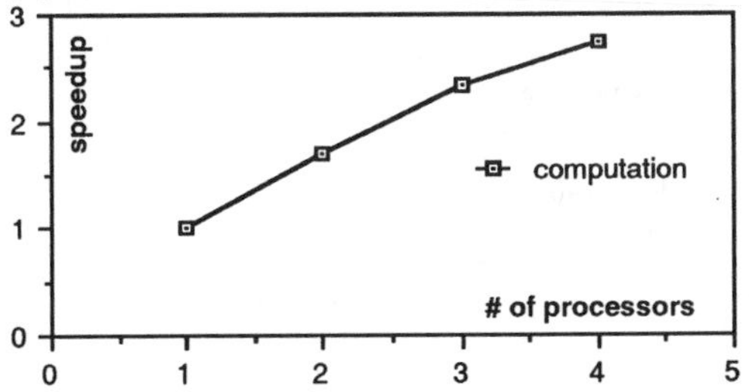

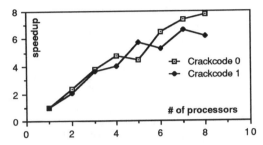

Fig. 1: Coarse Grain Benchmark: Pig Cast (512 by 512)

The speed up of the fine grain parallelized modules was almost linear up to 5 processors for crackcode and Fourier descriptors algorithms. With 8 processors, the speedup is 7.8 and 6.2 for the two crackcode mapping strategies and 5.3 for the Fourier descriptors, as shown in Fig. 2 and 3. The amount of time taken to process the crackcode routine with one processor is about 3.4 sec. The time taken to generate Fourier descriptors is 15.00 sec. It should be noted that the Fourier descriptor timings are for generating 20 coefficients.

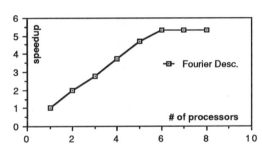

Fig. 2: Benchmarks for Crackcode: Human (256 by 256)

Fig. 3: Benchmark for 20 F.D: Human (256 by 256)

References

1. Parker, D.L., Wu, J., Pope, D.L., Bree, R.E., Caputo, G.R., Marshall, H.W., "Three dimensional measurements of coronary arteries using multiview digital angiography", Rotterdam Conference on Quantitative Coronary Arteriography, (June 1987)..

2. Parker, D.L., Pope D.L., Bree, R.E., Marshall H.: "Three dimensional reconstruction and cross-section measurements of coronary arteries using ECG correlated digital coronary arteriography", Progress in Digital Angiocardiography, Ed., P.H. Heintzen, M.D., Martinus Nijhoff, Dordrecht, The Netherlands, (1987).

3. Pope, D.L., Bree, R.E., Parker, D.L., "Cine 3D reconstruction of moving coronary arteries from DSA images", IEEE Computers In Cardiology, 277-280, (1987).

4. Bond, A.H., Fashena, D., "Parallel vision techniques on the hypercube computer", 3rd. Conf. on Hypercube Concurrent Computers and Applications, 1007-1010, (1988).

5. Lee, S.Y., Aggarwal, J.K., "Parallel 2D convolution on a mesh connected array processor", IEEE Computer Society Conf. on Computer Vision and Pattern Recognition, 305-310, (1986).

6. Tan, Sin Guan, "Image Feature extraction: Line detection and organization", M. S. Thesis The Ohio State University, 1990.

7. Fu, K. S., Persoon, E., "Shape discrimination using Fourier descriptors", IEEE Trans. on Pattern Analysis and Machine Intelligence, vol. PAMI-8, 388-397, May 1977.

TABLE OF CONTENTS - FULL PROCEEDINGS

A1

A3

A5

A8

POSTER SESSION

A11

A12

A13